AF323833

PACIFIC SYMPOSIUM ON
BIOCOMPUTING 2003

PACIFIC SYMPOSIUM ON
BIOCOMPUTING 2003

Kauai, Hawaii
3–7 January 2003

Edited by

Russ B. Altman
Stanford University, USA

A. Keith Dunker
Washington State University, USA

Lawrence Hunter
University of Colorado Health Sciences Center, USA

Tiffany A. Jung
Stanford University, USA

Teri E. Klein
Stanford University, USA

Published by

World Scientific Publishing Co. Pte. Ltd.

5 Toh Tuck Link, Singapore 596224

USA office: Suite 202, 1060 Main Street, River Edge, NJ 07661

UK office: 57 Shelton Street, Covent Garden, London WC2H 9HE

British Library Cataloguing-in-Publication Data
A catalogue record for this book is available from the British Library.

BIOCOMPUTING
Proceedings of the 2003 Pacific Symposium

Copyright © 2002 by World Scientific Publishing Co. Pte. Ltd.

ISBN 981-238-217-8

This book is printed on acid-free paper.

Printed in Singapore by World Scientific Printers (S) Pte Ltd

PACIFIC SYMPOSIUM ON BIOCOMPUTING 2003

These are the proceedings of the eighth Pacific Symposium on Biocomputing (PSB), held on the island of Kauai, Hawaii. The proceedings for this and previous PSB meetings are also available in electronic form at http://psb.stanford.edu/psb-online/ and are indexed in the Medline resource provided by the National Library of Medicine of the U.S. The online proceedings offer more than 350 manuscripts presented since 1996 at PSB. Because PSB sessions are organized in a grass roots fashion, the intellectual challenges of our field can be chronicled by an analysis of the sessions presented each year. Early conferences were dominated by protein structure, HMMs and sequence analysis. More recently, natural language processing, human genetic variation, and comparative genomics have emerged. It is particularly gratifying to note that the impact of PSB papers has been growing, and our published corpus in certain niches (such as natural language processing for biology and reconstruction of metabolic networks) has become influential in these subdisciplines.

PSB is sponsored by the International Society for Computational Biology (http://www.iscb.org/). Meeting participants benefit once again from travel grants from the ISCB, U.S. Department of Energy, National Library of Medicine/National Institutes of Health, and Applied Biosystems. In addition, GeneticXchange is a corporate affiliate.

We thank Dr. Marvin Cassman for his plenary address on the future of computational biology, and Dr. Latanya Sweeney for her plenary address on issues of database privacy and confidentiality. Tiffany Jung, the newest member of our PSB editorial team, has created the printed and online proceedings, while also administering the meeting. We would especially like to acknowledge the contributions of the session organizers who solicited papers and reviews, and ensured that the quality of the meeting remains high. The session organizers (and their associated sessions) are:

Serafim Batzoglou and Lior Pachter (Gene Regulation)

Peter Karp, Pedro R. Romero, Eric Neumann, Alexander J. Hartemink (Genome, Pathway, and Interaction Bioinformatics)

Sean D. Mooney and Patricia C. Babbitt (Informatics Approaches in Structural Genomics)

Liping Wei, Inna Dubchak, and Victor Solovyev (Genome-wide Analysis and Comparative Genomics)

Lynette Hirschman, Carol Friedman, Robin McEntire, and Cathy Wu (Linking Biomedical Language, Information and Knowledge)

Francisco M. De La Vega, Kenneth K. Kidd, and Isaac S. Kohane (Human Genome Variation: Haplotypes, Linkage Disequilibrium, and Populations)

Olivier Bodenreider, Joyce A. Mitchell, and Alexa T. McCray (Biomedical Ontologies)

We are also happy to host two discussion panels on "Graduate and Undergraduate Bioinformatics Education" (moderated by Richard Hughey) and "Genetic Diversity and DNA-based Identification" (moderated by James Sikela and Eric Juengst). The latter topic is particularly relevant in light of a special report at the end of these proceedings that describes the informatics tools that helped catalog and identify the remains of victims of the World Trade Center collapse on September 11, 2001.

As usual, the PSB organizers and session leaders relied on the assistance of those who capably reviewed the submitted manuscripts. A partial list of reviewers is provided in the following pages, and we also thank those who have been left off this list inadvertently or who wished to remain anonymous.

We encourage participants to consider submitting proposals for future PSB sessions and tutorials in order to ensure that the meeting continues to provide a forum for the early discussion and publication of new directions in biocomputing.

Aloha!

Pacific Symposium on Biocomputing Co-Chairs   *October 1, 2002*

Russ B. Altman
Department of Genetics & Stanford Medical Informatics, Stanford University

A. Keith Dunker
Department of Biochemistry and Biophysics, Washington State University

Lawrence Hunter
Department of Pharmacology, University of Colorado Health Sciences Center

Teri E. Klein
Department of Genetics & Stanford Medical Informatics, Stanford University

Thanks to reviewers...

Finally, we wish to thank the scores of paper reviewers. PSB requires that every paper in this volume be reviewed by at least three independent referees. Since there is a large volume of submitted papers, paper reviews require a great deal of work from many people. We are grateful to all of you listed below and to anyone whose name we may have accidentally omitted or who wished to remain anonymous.

Goncalo Abecasis
Aram Adourian
Laura Almasy
Russ Altman
Patsy Babbitt
Keith Ball
D. Rey Banatao
Ziv Bar-Joseph
Andrew W. Bergen
Judith A. Blake
Dario Boffelli
Relly Brandman
Michael Brudno
James A. Butler
Sharon A. Caraballo
Roland Carel
Michelle Whirl Carrillo
Joseph Chan
Sourav Chatterji
Kei-Hoi Cheung
John Chodera
Mark Craven
Eugene Davydov
Patrik D'Haeseleer
Chuong Do
Joyce Duan
Jon Dugan
A. Keith Dunker
Dani Fallin
Jane Fridlyand
Carol Friedman
Irene Gabashvili

John H. Gennari
David Gladstein
Chern-Sing Goh
Susumu Goto
Sridhar Govindarajan
Igor Grigoriev
Bing Hai
Bailin Hao
Jose Haresco
Vasileios Hatzivassiloglou
Markus Herrgard
Lynette Hirschman
Zhangzhi Hu
Conrad Huang
Hongzhan Huang
Lawrence Hunter
Trey Ideker
Tigran Ishkhanov
Stephen Johnson
Vladimir Kapitonov
Peter Karp
Tae-Sung Kim
Teri E. Klein
Evelyn S.C. Koay
Isaac Kohane
David Konerding
Cindy Krieger
Keith Laderoute
Alan Lapedes
Suin Lee

Wentian Li
Pat Lincoln
Hongfang Liu
Irene Liu
Xiaole Liu
Gabriela Loots
Yves Lussier
Inderjeet Mani
Robin A. McEntire
Pavel Morozov
Newton Morton
Lukas Mueller
Emily Mundorff
Mark A. Musen
Prakash M. Nadkarni
Yoichi Nakayama
Eric Neumann
Dahlia Nielsen
William Noble
Matej Oresic
Lior Pachter
Andrew J. Pakstis
John Park
Timothy B. Patrick
Itsik Pe'er
Scott Pegg
Tom Plasterer
James Pustejovsky
Usha Reddy
Aviv Regev
Kinkead Reiling
Rachel L. Richesson

Alan Robinson
Igor Rogozin
Pedro Romero
Alan Roter
Sarah A. Ryan
Andrey Rzhetsky
Asaf Salamov
Armindo Salvador
Mirna Samano
Aaron Sarver
Anne Sarver
Vincent Schachter
Eran Segal
Igor Seledtsov
Imran Shah
Ilham Shahmuradov
John Shon
Albert V. Smith
Barry Smith
Jacky Snoep
Victor Solovyev
Noel Southall
Terence Speed
Nati Srebro
David States
Martin Steinhauser
Robert Stevens
Chen Su
Jane Su
Xiaoping Su
David Sullivan
Baris E. Suzek
Koichi Takahashi
Tatiana Tatarinova
Kurt Thorn
Ralf Tolle
Eugene van Someren
Eberhard Voit
Mike Walker
Todd Wareham
Allison Waugh
Bruce Weir
L.F.A. Wessels
Glenn Williams
Cathy Wu
Tom Wu
Malin Young
Hongyu Zhang
Andrey Zharkikh
Yan Zhou
Pierre Zweigenbaum

CONTENTS

Preface   v

GENE REGULATION

Session Introduction   3
    S. Batzoglou and L. Pachter

Multiclass Cancer Classification Using Gene Expression
Profiling and Probabilistic Neural Networks   5
    D.P. Berrar, C.S. Downes, and W. Dubitzky

Inferring Gene Regulatory Networks from Time-Ordered Gene
Expression Data of *Bacillus subtilis* Using Differential Equations   17
    M.J.L. de Hoon, S. Imoto, K. Kobayashi, N. Ogasawara,
    and S. Miyano

Genome-Wide Analysis of Bacterial Promoter Regions   29
    E. Eskin, U. Keich, M.S. Gelfand, and P.A. Pevzner

MOPAC: Motif Finding by Preprocessing and Agglomerative
Clustering from Microarrays   41
    R. Ganesh, D.A. Siegele, and T.R. Ioerger

Improved Gene Selection for Classification of Microarrays   53
    J. Jaeger, R. Sengupta, and W.L. Ruzzo

Kernel Cox Regression Models for Linking Gene Expression
Profiles to Censored Survival Data   65
    H. Li and Y. Luan

Extracting Conserved Gene Expression Motifs from Gene
Expression Data   77
    T.M. Murali and S. Kasif

x

Decomposing Gene Expression into Cellular Processes   89
    E. Segal, A. Battle, and D. Koller

GENOME, PATHWAY, AND INTERACTION BIOINFORMATICS

Session Introduction   101
    P. Karp, P.R. Romero, E. Neumann, and A.J. Hartemink

Suitability and Utility of Computational Analysis Tools:
Characterization of Erythrocyte Parameter Variation   104
    R.E. Altenbaugh, K.J. Kauffman, and J.S. Edwards

Foundations of a Query and Simulation System for the Modeling
of Biochemical and Biological Processes   116
    M. Antoniotti, F. Park, A. Policriti, N. Ugel, and B. Mishra

Incorporating Biological Knowledge into Evaluation of Causal
Regulatory Hypotheses   128
    L. Chrisman, P. Langley, S. Bay, and A. Pohorille

Assessment of the Reliability of Protein-Protein Interactions and
Protein Function Prediction   140
    M. Deng, F. Sun, and T. Chen

Boundary Formation by Notch Signaling in Drosophila
Multicellular Systems: Experimental Observations and
Gene Network Modeling by Genomic Object Net   152
    H. Matsuno, R. Murakami, R. Yamane, N. Yamasaki,
    S. Fujita, H. Yoshimori, and S. Miyano

Influence of Network Topology and Data Collection on
Network Inference   164
    V.A. Smith, E.D. Jarvis, and A.J. Hartemink

INFORMATICS APPROACHES IN STRUCTURAL GENOMICS

Session Introduction — 176
S.D. Mooney and P.C. Babbit

Simultaneous Sequence Alignment and Tree Construction Using Hidden Markov Models — 180
R.C. Edgar and K. Sjölander

Towards Discovering Structural Signatures of Protein Folds Based on Logical Hidden Markov Models — 192
K. Kersting, T. Raiko, S. Kramer, and L. De Raedt

Automated Construction of Structural Motifs for Predicting Functional Sites on Protein Structures — 204
M.P. Liang, D.L. Brutlag, and R.B. Altman

Prediction of Boundaries Between Intrinsically Ordered and Disordered Protein Regions — 216
P. Radivojac, Z. Obradović, C.J. Brown, and A.K. Dunker

Identifying Structural Motifs in Proteins — 228
R. Singh and M. Saha

A Path Planning-Based Study of Protein Folding with a Case Study of Hairpin Formation in Protein G and L — 240
G. Song, S. Thomas, K.A. Dill, J.M. Scholtz, and N.M. Amato

Profile-Profile Alignment: A Powerful Tool for Protein Structure Prediction — 252
N. von Öhsen, I. Sommer, and R. Zimmer

Protein Threading by Linear Programming — 264
J. Xu, M. Li, G. Lin, D. Kim, and Y. Xu

GENOME-WIDE ANALYSIS AND COMPARATIVE GENOMICS

Session Introduction   276
L. Wei, I. Dubchak, and V. Solovyev

Towards Identifying Lateral Gene Transfer Events   279
L. Addario-Berry, M. Hallett, and J. Lagergren

Whole Genome Human/Mouse Phylogenetic Footprinting of
Potential Transcription Regulatory Signals   291
E. Cheremushkin and A. Kel

MAP: Searching Large Genome Databases   303
T. Kahveci and A. Singh

Towards the Development of Computational Tools for
Evaluating Phylogenetic Network Reconstruction Methods   315
L. Nakhleh, J. Sun, T. Warnow, C.R. Linder,
B.M.E. Moret, and A. Tholse

Identification of Regulatory Binding Sites Using Minimum
Spanning Trees   327
V. Olman, D. Xu, and Y. Xu

Intrasplicing — Analysis of Long Intron Sequences   339
S. Ott, Y. Tamada, H. Bannai, K. Nakai, and S. Miyano

Trajectory Clustering: A Non-Parametric Method for
Grouping Gene Expression Time Courses, with
Applications to Mammary Development   351
T.L. Phang, M.C. Neville, M. Rudolph, and L. Hunter

Algorithms for Multiple Genome Rearrangement by
Signed Reversals   363
S. Wu and X. Gu

DIGIT: A Novel Gene Finding Program by Combining
Gene-Finders   375
T. Yada, T. Takagi, Y. Totoki, Y. Sakaki, and Y. Takaeda

LINKING BIOMEDICAL LANGUAGE, INFORMATION AND KNOWLEDGE

Session Introduction   388
L. Hirschman, C. Friedman, R. McEntire, and C. Wu

Evaluation of the Vector Space Representation in Text-Based
Gene Clustering   391
*P. Glenisson, P. Antal, J. Mathys, Y. Moreau,
and B. De Moor*

Playing Biology's Name Game: Identifying Protein Names in
Scientific Text   403
D. Hanisch, J. Fluck, H.-T. Mevissen, and R. Zimmer

Mining Terminological Knowledge in Large Biomedical Corpora   415
H. Liu and C. Friedman

A Biological Named Entity Recognizer   427
M. Narayanaswamy, K. E. Ravikumar, and K. Vijay-Shanker

Linking Biomedical Language Information and Knowledge
Resources: GO and UMLS   439
*I.N. Sarkar, M.N. Cantor, R. Gelman, F. Hartel,
and Y.A. Lussier*

A Simple Algorithm for Identifying Abbreviation Definitions
in Biomedical Text   451
A.S. Schwartz and M.A. Hearst

xiv

HUMAN GENOME VARIATION: HAPLOTYPES, LINKAGE DISEQUILIBRIUM, AND POPULATIONS

Session Introduction   463
F.M. De La Vega, K.K. Kidd, and I.S. Kohane

Selection of Minimum Subsets of Single Nucleotide
Polymorphisms to Capture Haplotype Block Diversity   466
H.I. Avi-Itzhak, X. Su, and F.M. De La Vega

On the Power to Detect SNP/Phenotype Association in Candidate
Quantitative Trait Loci Genomic Regions: A Simulation Study   478
J.M. Comeron, M. Kreitman, and F.M. De La Vega

Errors and Linkage Disequilibrium Interact Multiplicatively when
Computing Sample Sizes for Genetic Case-Control
Association Studies   490
D. Gordon, M.A. Levenstien, S.J. Finch, and J. Ott

An MDL Method for Finding Haplotype Blocks and for
Estimating the Strength of Haplotype Block Boundaries   502
M. Koivisto, M. Perola, T. Varilo, W. Hennah,
J. Ekelund, M. Lukk, L. Peltonen, E. Ukkonen,
and H. Mannila

PyPop: A Software Framework for Population Genomics:
Analyzing Large-Scale Multi-Locus Genotype Data   514
A. Lancaster, M.P. Nelson, D. Meyer, G. Thomson,
and R.M. Single

Joint Bayesian Estimation of Mutation Location and Age Using
Linkage Disequilibrium   526
B. Rannala and J.P. Reeve

SNP Analysis and Presentation in the Pharmacogenetics of
Membrane Transporters Project   535
>*D. Stryke, C.C. Huang, M. Kawamoto, S.J. Johns,
E.J. Carlson, J.A. Deyoung, M.K. Leabman, I. Herskowitz,
K.M. Giacomini, and T.E. Ferrin*

Methods for Analysis and Visualization of SNP Genotype Data
for Complex Diseases   548
>*A. Tsalenko, A. Ben-Dor, N. Cox, and Z. Yakhini*

BIOMEDICAL ONTOLOGIES

Session Introduction   562
>*O. Bodenreider, J.A. Mitchell, and A.T. McCray*

Functional Discrimination of Gene Expression Patterns in
Terms of the Gene Ontology   565
>*L. Badea*

Towards a Broad-Coverage Biomedical Ontology Based on
Description Logics   577
>*U. Hahn and S. Schulz*

Evaluation of Ontology Merging Tools in Bioinformatics   589
>*P. Lambrix and A. Edberg*

Semantic Similarity Measures as Tools for Exploring the
Gene Ontology   601
>*P.W. Lord, R.D. Stevens, A. Brass, and C.A. Goble*

Linking Molecular Imaging Terminology to the Gene
Ontology (GO)   613
>*P.K. Tulipano, W. Millar, and J.J. Cimino*

xvi

A Methodology to Migrate the Gene Ontology to a Description
Logic Environment Using DAML + OIL   624
    C.J. Wroe, R. Stevens, C.A. Goble, and M. Ashburner

SPECIAL PAPER

Introduction   636

Development Under Extreme Conditions: Forensic Bioinformatics
in the Wake of the World Trade Center Disaster   638
    H.D. Cash, J.W. Hoyle, and A.J. Sutton

Session Introductions and Peer Reviewed Papers

GENE REGULATION

SERAFIM BATZOGLOU
Department of Computer Science
Stanford University
Palo Alto, CA 94305
serafim@cs.stanford.edu

LIOR PACHTER
Department of Mathematics
UC Berkeley
Berkeley, CA 94720
lpachter@math.berkeley.edu

The year 2003 marks the completion of the sequencing phase of the human genome, and will coincide with the 50th anniversary of Watson and Crick's description of the structure of DNA. Despite such remarkable and rapid progress in elucidating the structure and composition of our genome, the problem of understanding the complexity of its function is largely unsolved. The unraveling of the regulatory code is certainly one of the next great frontiers in molecular biology.

Progress in the understanding of gene regulation will have to be driven by experimental data. Unfortunately, unlike genomic sequence data, gene expression experiments are not as "clean." EST data is notoriously messy, and gene chip experimental data is non-trivial to analyze. Furthermore, there are very few experimentally confirmed transcription factor binding sites, especially in the human genome, although there is a bit of data in other organisms. The situation is similar for interaction pathways, where there are few well studied and characterized examples.

The computational challenge is formidable and at the same time progress is essential, both in order to facilitate the understanding of experimental data, and also to drive the experimental efforts themselves. There are numerous examples of such efforts in the papers represented in this track, consisting of creative and original approaches both for the analysis of data and for the generation of hypotheses for experiment. It is interesting to note that the mathematical and computational techniques used include such diverse methods as differential equations, probabilistic methods, combinatorics and statistics

The variety of organisms studied is large. Some researchers are clearly concentrating on first understanding simpler regulation systems, thus following the advice of the mathematician George Polya who suggested: "if there is a problem you can't solve, then there is an easier problem you can't solve: find it."

Thus, Eskin, Gelfand and Pevzner have wisely concentrated on analyzing bacterial promoter regions, and De Hoon et al. have examined time-ordered gene expression data from *Bacillus Subtilis*, both efforts resulting in successful analysis for which there is a combination of existing data for verification and the realistic possibility of further experiment.

A large amount of research is currently being devoted to gene expression analysis, which is not surprising since there is a lot of chip array data to analyze now, and it is provides a first-order glimpse into the structure of regulation.
The paper by Berrar et al. suggests a new probabilistic approach to clustering expression data while Murali and Kasif explore new directions for detecting conservation of expression patterns. Jaeger et al. and Ganesh et al., in two separate papers demonstrate that pre-filtering can improve clustering performance on expression data. These analyses are important and should be of immediate interest to biologists who are generating expression data.

Expression analysis is just a preliminary step towards understanding entire regulatory networks, and there is already interesting computational work in this direction. The paper by Segal, Battle and Koller presents probabilistic model of cellular processes and show how to learn the processes from gene expression data. The paper by Li and Yuan on the relationship between gene expression profiles and survival data is a refreshing application of gene expression data not commonly examined by computational biologists.

MULTICLASS CANCER CLASSIFICATION USING GENE EXPRESSION PROFILING AND PROBABILISTIC NEURAL NETWORKS

DANIEL P. BERRAR, C. STEPHEN DOWNES, WERNER DUBITZKY

School of Biomedical Sciences, University of Ulster at Coleraine,
BT52 1SA, Northern Ireland
E-mail: {dp.berrar, cs.downes, w.dubitzky}@ulster.ac.uk

Gene expression profiling by microarray technology has been successfully applied to classification and diagnostic prediction of cancers. Various machine learning and data mining methods are currently used for classifying gene expression data. However, these methods have not been developed to address the specific requirements of gene microarray analysis. First, microarray data is characterized by a high-dimensional feature space often exceeding the sample space dimensionality by a factor of 100 or more. In addition, microarray data exhibit a high degree of noise. Most of the discussed methods do not adequately address the problem of dimensionality and noise. Furthermore, although machine learning and data mining methods are based on statistics, most such techniques do not address the biologist's requirement for sound mathematical confidence measures. Finally, most machine learning and data mining classification methods fail to incorporate misclassification costs, i.e. they are indifferent to the costs associated with false positive and false negative classifications. In this paper, we present a probabilistic neural network (PNN) model that addresses all these issues. The PNN model provides sound statistical confidences for its decisions, and it is able to model asymmetrical misclassification costs. Furthermore, we demonstrate the performance of the PNN for multiclass gene expression data sets. Here, we compare the performance of the PNN with two machine learning methods, a decision tree and a neural network. To assess and evaluate the performance of the classifiers, we use a *lift*-based scoring system that allows a fair comparison of different models. The PNN clearly outperformed the other models. The results demonstrate the successful application of the PNN model for multiclass cancer classification.

1   Introduction

The diagnosis of complex genetic diseases such as cancer has traditionally been made on the basis of non-molecular criteria such as tumor tissue type, pathological features, and clinical stage. In the past several years, DNA microarray technology has attracted tremendous interest in both the scientific community and in industry. Several studies have recently reported on the application of microarray gene expression analysis for molecular classification of cancer [1,2,3]. Microarray analysis of differential gene expression has been used to distinguish between different subtypes of lung adenocarcinoma [4] and colorectal neoplasm [5], and to predict clinical outcomes in breast cancer [6,7] and lymphoma [8]. J. Khan et al. used an artificial neural network approach for the classification of microarray data, including both tissue biopsy material and cell lines [9]. Various machine learning methods have been investigated for the analysis of microarray data [10,11]. The combination of gene microarray technology and machine learning methods promises new insights into mechanisms of living systems. An application area where these techniques are expected to make major contributions is the classification of cancers according to clinical stage and biological behavior. Such classifications have an immense impact on prognosis and therapy. In our opinion, a classifier for this task should address the following issues: (1) The classifier should provide an easy-to-interpret measure of confidence for its decisions. Thereby, the final diagnosis rests with the medical expert who judges if the confidence of the classifier is high enough. In one scenario, a classification that relies on a confidence of 75% might be acceptable, whereas in another, the medical expert only accepts classifications of at least 99%. (2) The classifier should take into account asymmetrical misclassification costs for false positive and false negative classifications. For example, suppose a tissue sample is to be classified as either benign or malign. A false positive classification may result in further clinical examinations, whereas a false negative result is very likely to have severe consequences for the patient. Consequently, the classifier should ideally be very "careful" when classifying a sample to the class "benign". The misclassification costs depend on the problem at hand and have to be evaluated by the medical expert. Machine learning methods that are able to address both issues are very rare. In this paper, we present a model of a probabilistic neural network for the classification of microarray data that addresses both issues.

Many publications report on cancer classification problems where the number of classes is rather small. For example, the classification problem of J. Khan et al. comprised four cancer classes [9], and the classification problem of T. Golub et al. comprised only two classes [1]. However, multiclass distinctions are a considerably more difficult task. S. Ramaswamy et al. recently reported on the successful application of *support vector machines* (SVM) for multiclass cancer diagnosis [2].

2 Probabilistic neural networks

Probabilistic neural networks (PNNs) belong to the family of radial basis function neural networks. PNN are based on Bayes' decision strategy and Parzen's method of density estimation. In 1990, D. Specht proposed an artificial neural network that is based on these two principles [12]. This model can compute nonlinear decision boundaries that asymptotically approach the Bayes' optimal. Bayesian strategies are decision strategies that minimize the expected risk of a classification. The Bayesian decision theory is the basis of many important learning schemes such as the naïve Bayes classifier, Bayesian belief networks, and the EM algorithm. The optimum decision rule that minimizes the average costs of misclassification is called *Bayes' optimal decision rule*. It can be proven that no other classification method using the same hypothesis space and the same prior knowledge can outperform the Bayes' optimal classifier on average [13]. The following definition is adapted from T. Masters [14]:

Definition 1: *Bayes' optimal classifier*

Given a collection of random samples from n populations. The prior probability that a sample $\vec{x}$ belongs to population k is denoted as h_k. The costs associated with a misclassification of a sample belonging to population k is denoted as c_k. The conditional probability that a specific sample belongs to population k, $p(k \mid \vec{x})$, is given by the probability density function $f_k(\vec{x})$. An unknown sample $\vec{x}$ is classified into population i if

$$h_i \cdot c_i \cdot f_i(\vec{x}) > h_j \cdot c_j \cdot f_j(\vec{x})$$

for all populations $j \neq i$.

We refer to this decision criterion as *Bayes' decision criterion*. This criterion favors a class if the costs associated with its misclassification are high (c_i). Furthermore, the rule favors a class if it has a high prior probability (h_i). Finally, the rule favors a class if it has high density in the vicinity of the unknown sample ($f_i(\vec{x})$). The prior probabilities h are generally known or can be estimated. The misclassifications costs c rely on a subjective evaluation. The probability density functions, however, are unknown in real-world applications and have to be estimated. D. Specht proposed to use Parzen's method for non-parametric estimation of the density using the set of training samples. D. Parzen proved that the estimated univariate probability density converges asymptotically to the true density as the sample size of the training data increases [15]. The estimator for the density function contains a weighting function that is also known as *kernel function* or *Parzen window*. The kernel is centered at each training sample. The estimated density is the scaled sum of the kernel function for all training samples. Various kernel functions are possible [16], but the most common kernel is the Gaussian function [14]. The scaling parameter σ defines the width of the bell curves and is also referred to as *window width*, *bandwidth*, or *smoothing factor* (the latter one is most commonly used in the context of PNNs). Equation 1 shows the estimated density for the multivariate case and the Gaussian as kernel function:

$$\hat{f}_j(\vec{x}) = \frac{1}{\left(\sqrt{2\pi}\right)^{dim} \sigma^{dim} m_j} \sum_{i=1}^{m_j} exp\left(-\frac{(\vec{x} - \vec{x}_{ij})^T \cdot (\vec{x} - \vec{x}_{ij})}{2\sigma^2}\right) \tag{1}$$

where $\hat{f}_j$:   estimated density for the j-th class

$\vec{x}$:   test case

$\vec{x}_{ij}$: i-th training sample of the j-th population / class

dim :   dimensionality of $\vec{x}_{ij}$

σ :   smoothing factor

T :   transpose

m_j :   number of training cases in the j-th class

8

D. Specht proposed a four-layered feed-forward network topology that implements Bayes' decision criterion and Parzen's method for density estimation. The operation of the basic PNN is best illustrated on a simple architecture as depicted in Figure 1:

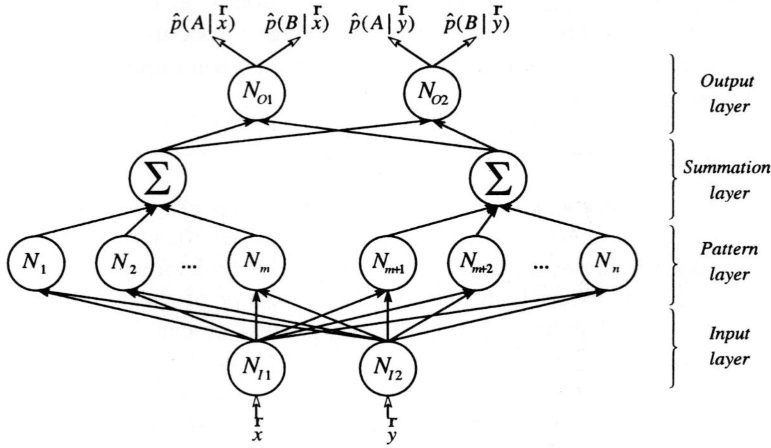

Figure 1:   Architecture of a four-layered PNN for *n* training cases of 2 classes.

The *input layer* of the PNN in Figure 1 contains two input neurons, N_{I1} and N_{I2}, for the two test cases, $\vec{x}$ and $\vec{y}$. The *pattern layer* contains one *pattern neuron* for each training case, with an exponential activation function. A pattern neuron N_i computes the squared Euclidean distance $d^2 = (\vec{x} - \vec{x}_{ij})^T \cdot (\vec{x} - \vec{x}_{ij})$ between a new input vector $\vec{x}$ and the *i*-th training vector of the *j*-th class. This distance is then transformed by the neuron's activation function (the exponential). In the PNN of Figure 1, the training set comprises cases belonging to two classes, A and B. In total, m training cases belong to class A. The associated pattern neurons are $N_1...N_m$. For example, the neuron N_3 contains the third training case of class A. Class B contains $n - m$ training cases; the associated pattern neurons are $N_{m+1}...N_n$. For example, the neuron N_{m+2} contains the second training case of class B. For each class, the *summation layer* contains a *summation neuron*. Since we have two classes in this example, the PNN has two summation neurons. The summation neuron for class A sums the output of the pattern neurons that contain the training cases of class A. The summation neuron for class B sums the output of the pattern neurons that contain the training cases of class B. The activation of the summation neuron for a class is equivalent to the estimated density function value of this class. The summation neurons feed their result to the output neurons in the *output layer*. These neurons are threshold discriminators that implement Bayes' decision criterion. The output neuron N_{O1} generates two outputs: the estimated conditional probability that the test case $\vec{x}$ belongs to class A, and the estimated conditional probability that this case belongs to class B. The output neuron N_{O2} generates the respective estimated probabilities for the test case $\overset{r}{y}$. Unlike other feed-forward neural networks, e.g., multi-layer perceptrons (MLPs), all hidden-to-output weights are equal to 1 and do not vary during processing. For the present study, we use the same smoothing factor σ for all classes. Whereas the choice of the kernel function has no major effect on the performance of the PNN, the choice of σ has a significant influence. The smaller the smoothing factor, the more influence have individual training samples. The larger the smoothing factor, the more blurring is induced. It has been shown that neither limiting case provides optimal separation of class distributions [12]. Clearly, averaging multiple nearest neighbors results in a better generalization than basing the decision on the first nearest neighbor only. On the other hand, if too many neighbors are taken into account, then the PNN generalizes weakly as well. The optimal smoothing factor can be determined through cross-validation procedures. However, the choice of the smoothing factor always implies a trade-off between the variance and the bias of a kernel-based classifier. Further techniques for adapting σ and for improving the basic model of the PNN can be found in [17,18].

3   Analysis of the leukemia data set

The leukemia data set includes expression profiles of 7,129 human DNA probes spotted on Affymetrix Hu6800 microarrays of 72 patients with either acute myeloid leukemia (*AML*) or acute lymphocytic leukemia (*ALL*) [1]. Tissue samples were collected at time of diagnosis before treatment, taken either from bone marrow (62 cases), or peripheral blood (10 cases) and reflect both childhood and adult leukemias. Furthermore, a description of cancer subtypes, treatment response, gender, and source (laboratory) was given. RNA preparation, however, was performed using different protocols. The gene expression profiles of the original data set are represented as $\log_{10}$ normalized expression values. This data set was used as a benchmark for various machine learning techniques at the First Critical Assessment of Microarray Data Analysis at the Duke University in October 2000 (CAMDA 2000). The data set was divided into a training and a validation set. Table 1 shows the number of cases in the data sets:

Table 1.   Distribution of cancer primary classes (*AML* and *ALL*) and subclasses in the training and the test set (*N/a*: no cancer subclass specified).

Primary class	ALL		AML					
Subclass	*B-cell*	*T-cell*	*M1*	*M2*	*M4*	*M5*	*N/a*	*Σ*
# of cases in training set	19	8	3	5	1	2	0	38
# of cases in validation set	19	1	1	5	3	0	5	34

The original data set of 7,129 genes contains some control genes that we excluded from further analysis. After this preprocessing, each sample consists of a row vector of 7,070 expression values. The classification of the leukemia subclasses is an even more challenging task than the classification of the primary classes (*ALL* and *AML*) in the CAMDA 2000, because the subclass distributions are very skewed in the training and the validation set. In a leave-one-out cross-validation procedure, we tested different values for the smoothing factor. We initialized σ with 0.01. The first case of the training set, $\overset{1}{x}$, was used as the hold-out case, and the remaining 37 cases were forwarded to the pattern layer. We assume equal misclassification costs for all cancer classes and classify $\overset{1}{x}$ using Bayes' decision criterion (cf. Definition 1). Then, the second case was retained as the hold-out case, and the remaining cases were moved to the pattern layer. This procedure was repeated for 100 different values for the smoothing factor, ranging from 0.01 to 1.00. After all cases had been classified in turn, we performed a sensitivity analysis. Ideally, the sensitivity for each class should be maximal. Consequently, the optimal smoothing factor maximizes the sum of all sensitivities. Based on this criterion, the PNN performed best for a smoothing factor of 0.03 on the training set. Therefore, we chose this smoothing factor to classify the cases of the validation set. Table 2 shows the resulting confusion matrix for the classification of the cancer subclasses.

Table 2. Confusion matrix for the classification of the leukemia subclasses.

		M1	*M2*	*M4*	*M5*	*B-cell*	*T-cell*	*N/a*	Σ
						Real class			
	M1	1	1	-	-	3	-	1	6
	M2	-	4	1	-	-	-	-	5
	M4	-	-	-	-	-	-	-	-
Classification	*M5*	-	-	-	-	-	-	2	2
	B-cell	-	-	2	-	16	1	2	21
	T-cell	-	-	-	-	-	-	-	-
	N/a	-	-	-	-	-	-	-	-
	Σ	1	5	3	-	19	1	5	34
	sensitivity	1.00	0.80	0.00	-	0.84	0.00	0.00	
	specificity	0.85	0.97	1.00	0.94	0.67	1.00	1.00	

The PNN is very sensitive to the class *M1* and *M2*, but fails to classify the *M4* cases correctly. This can be explained by the fact that only one case of this subclass is contained in the training set. The sensitivity for the subclass *B-cell* is relatively high (0.84). However, the PNN misclassified the *T-cell* case a *B-cell* case, although the training set contained 8 *T-cell* cases. Given this relatively large number of cases of *T-cell* cases in the training set, this result is rather disappointing. In the validation set, 5 cases are of type *AML*, but no further subclass specification is given. In the training set, this class is not present, thus the PNN is not able to predict this class. Interestingly, 3 of these 5 cases are correctly classified as cases of type *AML* (1 case is classified as *M1*, 2 cases are classified as *M5*).

4   Analysis of the NCI60 data set

The NCI60 data set comprises 1,416 gene expression profiles of 60 cell lines [19]. The data set includes nine different cancer classes: central nervous system (*CNS*, 6 cases), breast (*BR*, 8 cases), renal (*RE*, 8 cases), lung (*LC*, 9 cases), melanoma (*ME*, 8 cases), prostate (*PR*, 2 cases), ovarian (*OV*, 6 cases), colorectal (*CO*, 7 cases), and leukemia (*LE*, 6 cases). The gene expression data comprise mainly ESTs of known and unknown function given by the negative logarithm of the ratio between the red and green fluorescence of the signals. The $60 \times 1,416$ microarray matrix contains 2,033 missing values in total. Different methods for missing value imputation in the context of microarrays have been discussed. We propose the following missing value imputation method: Let $v(c_i, g)$ denote the gene expression value for case c_i and gene g. If $v(c_i, g)$ is a missing value, then replace it by the mean of all values $v(c_j, g)$ where the cancer class of c_i and c_j is the same. This method makes explicit use of the class membership of each sample and is based on the following rational: O. Troyanskaya et al. resumed that k-nearest neighbor (kNN) methods provide for the best estimation of missing values in microarrays [20]. A major problem with kNN methods is the adequate choice of the number of neighbors (k) to be taken into account. It is probable that a gene is similarly expressed in samples of the same cancer type. For missing value imputation, we therefore consider only the neighbors that belong to the same class. For some genes, our imputation method was not possible. For example, the expression values of *topoisomerase II alpha-log* are missing for both cases of class *PR*. In total, the missing value imputation was not possible with the described method for 11 genes. These genes were excluded from further analysis.

Feature selection and dimension reduction techniques are both used to remove features that do not provide significant incremental information. In the context of microarray data, such features can be redundant genes. For example, if two genes are similarly co-regulated, then they provide the same basic information, and the removal of one of these genes does, in general, not result in a loss of information for a classifier. Numerous studies have revealed that in high-dimensional microarray data, feature selection and dimension reduction methods are essential to improve the performance of a classifier (for a general discussion, see [21]). Many publications report on dimension reduction techniques such as principal component analysis (PCA) that is based on singular value decomposition [22]. To assess the performance of our model, we tested the PNN in a leave-one-out cross-validation procedure (1) on the original data set, and (2) on a reduced data set, comprising only a set of principal components. We compared the performance of the PNN with the performance of two other machine learning methods: the decision tree C5.0 [23], and a neural network: the multi-layer feedforward perceptron (MLP), trained with the backpropagation algorithm [24]. The training of the MLP was stopped when no further optimization was possible. The MLP comprised one hidden layer, containing

12

7 neurons for classifying the original data, and 4 neurons for classifying the reduced data. We applied the leave-one-out cross-validation procedure as described above to all models; i.e. each model is trained on all but one sample (*hold-out* case), and then we used the model to predict the class of the hold-out case. We iterated this procedure until each case was used as hold-out case.

4.1   Analysis of the NCI60 original data set

After data cleansing, the original data set consisted of 60 cell-line samples (9 cancer classes), and 1,405 features (expression values of genes and ESTs). We assumed equal misclassification costs for all classes. Given the relatively small number of cases per class, we chose a relatively small value for the smoothing factor. Table 3 shows the confusion matrix for $\sigma = 0.3$.

Table 3.   Confusion matrix for the NCI60 original data set.

		Real class									
		CNS	BR	RE	LC	ME	PR	OV	CO	LE	Σ
Classification	CNS	5	1	-	-	-	-	-	-	-	6
	BR	1	5	1	1	-	-	1	-	-	9
	RE	-	-	7	2	1	-	-	-	-	10
	LC	-	1	-	5	-	1	1	-	-	8
	ME	-	-	-	-	7	-	-	-	-	7
	PR	-	-	-	-	-	-	-	-	-	-
	OV	-	1	-	-	-	-	4	-	-	5
	CO	-	-	-	1	-	1	-	7	-	9
	LE	-	-	-	-	-	-	-	-	6	6
	Σ	6	8	8	9	8	2	6	7	6	60
	sensitivity	0.83	0.63	0.88	0.56	0.86	0.00	0.67	1.00	1.00	
	specificity	0.98	0.92	0.98	0.94	1.00	1.00	0.98	0.96	1.00	

The sensitivity and specificity for the classes *CO* and *LE* are very high, whereas the PNN was not able to classify the *PR* cases. This can be explained by the leave-one-out cross-validation procedure: When a *PR* case is used as the hold-out case, the training set comprises only one *PR* case. In total, the PNN misclassified 14 cases (23.3%). However, if we accept only those classifications that rely on a confidence of at least $\hat{p} = 0.8$, then the model misclassifies only 2 cases. Both C5.0 and MLP performed very weakly on the original data set (the respective confusion matrices are not shown). Their classification performance improved significantly on the reduced data set. Table 4 summarizes the performance of the three models on both the original and the reduced data set.

4.2 Analysis of the NCI60 reduced data set

We used PCA without mean centering. In our analysis, we used the first 23 principal components as hybrid variables; these variables explain more than 75% of the total variance. The sensitivities and specificities of the PNN are very similar to those that resulted from the original data set and are therefore not shown.

So far, we evaluated the performance of the PNN on the basis of its classification accuracy. However, accuracy-based evaluation metrics alone are inadequate to evaluate the performance of a classifier. A tacit assumption in the use of these accuracy measures is that the class distributions among the cases are constant and relatively balanced. The *lift* is a measure that takes different class distributions into account and is the preferred method for evaluating a classifier's performance [21].

Definition 2: *class lift* and *total lift*

Given the set of class labels, $C = \{c_1, c_2, ..., c_m\}$ and the set of cases, $S = \{x_1, x_2, ..., x_n\}$. Let $act(x_j)$ denote the actual class (label) of case x_j and $prd(x_j)$ the class (label) predicted for x_j by a classifier. Then the *class lift* for a particular class c_i, $lift(c_i)$, is measured by the prior probability, $p(act(x_j) = c_i)$, of class c_i occurring in S, and the conditional probability, $p(act(x_j) = c_i \mid prd(x_j) = c_i)$ of class $act(x_j) = c_i$ given the prediction, $prd(x_j) = c_i$, as follows:

$$lift(c_i) = \begin{cases} 0, \; \textit{if class } c_i \textit{ is not predicted} \\[2mm] \dfrac{p\left(act(x_j) = c_i \mid prd(x_j) = c_i\right)}{p\left(act(x_j) = c_i\right)} \quad \textit{otherwise} \end{cases} \qquad total\ lift = \frac{1}{m} \cdot \sum_{i=1}^{m} lift(c_i)$$

Table 4 shows the *lifts* resulting from the three models for the original and the reduced data set.

Table 4. *Lifts* for the classification of the NCI60 data set (p.c.: principal component).

Class	Maximum lift	Class lift of PNN		Class lift of C5.0		Class lift of MLP	
		All data	23 p.c.	All data	23 p.c.	All data	23 p.c.
CNS	10.00	8.33	8.33	1.67	8.33	0.00	2.00
BR	7.50	4.17	3.75	2.14	3.75	1.67	1.25
RE	7.50	5.25	5.83	1.67	3.21	0.00	1.89
LC	6.67	4.17	5.56	2.50	1.03	0.00	1.82
ME	7.50	6.56	6.56	3.75	5.63	1.07	3.75
PR	30.00	0.00	0.00	0.00	0.00	0.00	0.00
OV	10.00	8.00	8.33	0.00	5.56	0.00	1.67
CO	8.57	6.67	7.50	3.43	6.43	1.43	3.43
LE	10.00	10.00	10.00	10.00	8.57	1.00	6.67
Total lift	10.86	6.01	6.21	2.80	4.72	0.57	2.50

The *lift* can be interpreted as a *score*: the more difficult the classification of a case, the higher the potential score for the classifier. Table 4 also shows the *maximum lift*, i.e. the highest score that a classifier can obtain. Although the decision tree and the neural network performed much better on the reduced data set than on the original data set, the PNN still outperformed both models. However, it should be noted that other feature selection methods might significantly improve the performance of the decision tree and the neural network. But it is interesting that the PNN performs similarly on both the original and the reduced data set. It seems that – compared with the other models – the PNN is less sensitive to noise.

5   Discussion

We consider the ability to provide sound confidence levels and the ability to model asymmetrical misclassification costs as the two most important qualities of PNN in the context of microarray analysis. PNN have shown excellent classification performance in other applications, and perform equally or better than other types of artificial neural networks (ANNs). In contrast to other types of ANNs, e.g. MLPs, PNN are not "black boxes": The contribution of each pattern neuron to the outcome of the network is explicitly defined and accessible, and has a precise interpretation. The training of MLPs involves heuristic searches like the *steepest descent* method. These heuristics involve small modifications of the network parameters that result in a gradual improvement of system performance. Heuristic approaches are associated with long training times with no guarantee of converging to an acceptable solution within a reasonable timeframe. The training of PNN involves no heuristic searches, but consists essentially of incorporating the training cases into the pattern layer. However, finding the best smoothing factor for the training set remains an optimization problem. PNNs tolerate erroneous samples and outliers. Sparse samples are adequate for the PNN. Other types of ANN and many traditional statistical techniques are hampered by outliers. Finally, when new training data become available, PNN do not need to be reconfigured or retrained from scratch; new training data can be incrementally incorporated in the pattern layer.

A disadvantage of PNNs is the fact that all training data must be stored in the pattern layer, requiring a large amount of memory. But in general, today's standard PCs have a sufficiently large main memory capacity for an efficient implementation of PNN. In applications where large amounts of training cases are available, this argument against PNNs becomes relevant. But the problem can be circumvented by using cluster centroids as training cases, or by resorting to a parallel processor implementation.

Although the output of the PNN is probabilistic, we should keep in mind that the probabilities are estimates and conditional on the learning set. Future work will focus on an exhaustive comparison of state-of-the-art classifiers in multiclass cancer classification problems.

References

1. Golub T.R., Slonim D.K., Tamayo P., Huard C., Gaasenbeek M., Mesirov J.P., Coller H., Loh M.L., Downing J.R., Caligiuri M.A., Bloomfield C.D., Lander E.S., Molecular classification of cancer class discovery and class prediction by gene expression monitoring. *Science* **286**:531-537, (1999).
2. Ramaswamy S., Tamayo P., Rifkin R., Mukherjee S., Yeang C.H., Angelo M. Ladd C., Reich M., Latulippe E., Mesirov J.P., Poggio T., Gerald W., Loda M., Lander E.S., Golub T.R., Multiclass cancer diagnosis using tumor gene expression signatures, *Proc. Natl. Acad. Sci. USA.* **98**(26):15149-15154, (2001).
3. Tibshirani R., Hastie T., Narasimhan B., Chu G., Diagnosis of multiple cancer types by shrunken centroids of gene expression, *Proc. Natl. Acad. Sci. USA.* **99**(10):6567-6572, (2002).
4. Bhattacharjee A., Richards W.G., Staunton J., Li C., Monti S., Vasa P., Ladd C., Beheshti J., Bueno R., Gillette M., Loda M., Weber G., Mark E.J., Lander E.S., Wong W., Johnson B.E., Golub T.R., Sugarbaker D.J., Meyerson M., Classification of human lung carcinomas by mRNA expression profiling reveals distinct adenocarcinoma subclasses. *Proc. Natl. Acad. Sci. USA* **98**(24):13790-13795, (2001).
5. Selaru F.M., Xu Y., Yin J., Zou T., Liu T.C., Mori Y., Abraham J.M., Sato F., Wang S., Twigg C., Olaru A., Shustova V., Leytin A., Hytiroglou P., Shibata D., Harpaz N., Meltzer S.J., Artificial neural networks distinguish among subtypes of neoplastic colorectal lesions. *Gastroenterology* **122**:606-613, (2002).
6. West M., Blanchette C., Dressman H., Huang E., Ishida S., Spang R., Zuzan H., Olson J.A., Marks J.R., Nevins J.R., Predicting the clinical status of human breast cancer by using gene expression profiles. *Proc. Natl. Acad. Sci. USA* **98**(20):11462-11467, (2001).
7. van't Veer L.J., Dai H.Y., van de Vijver M.J., He Y.D.D., Hart A.A.M., Mao M., Peterse H.L., van der Kooy K., Marton M.J., Witteveen A.T., Schreiber G.J., Kerkhoven R.M., Roberts C., Linsley P.S., Bernards R., Friend S.H., Gene expression profiling predicts clinical outcome of breast cancer. *Nature* **415**:530-536, (2002).
8. Shipp M.A., Ross K.N., Tamayo P., Weng A.P., Kutok J.L., Aguiar R.C.T., Gaasenbeek M., Angelo M., Reich M., Pinkus G.S., Ray T.S., Koval M.A., Last K.W., Norton A., Lister T.A., Mesirov J., Neuberg D.S., Lander E.S., Aster J.C., Golub T.R., Diffuse large B-cell lymphoma outcome prediction by gene-expression profiling and supervised machine learning. *Nature Medicine* **8**:68-74, (2002).
9. Khan J., Wei J.S., Ringnér M., Saal L.H., Ladanyi M., Westermann F., Berthold F., Schwab M., Antonescu C.R., Peterson C., Meltzer P.S., Classification and

diagnostic prediction of cancers using gene expression profiling and artificial neural networks. *Nature Medicine* 7(6):673-679, (2001).

10. Lin S.M. and Johnson K.F. (eds.), *Methods of Microarray Data Analysis.* Kluwer Academic Publishers, Boston, (2002).

11. Berrar D., Dubitzky W., Granzow M. (eds.), *A Practical Approach to Microarray Data Analysis*, Kluwer Academic Publishers, Boston, Dec.2002.

12. Specht D.F., Probabilistic Neural Networks. *Neural Networks*, vol. 3, (1990) pp. 109-118.

13. Mitchell T.M., *Machine Learning.* McGraw-Hill Book Co., Singapore, (1997) pp. 174-175.

14. Masters T. *Advanced Algorithms for Neural Networks.* John Wiley & Sons, Academic Press, (1995).

15. Parzen E. On Estimation of a Probability Density Function and Mode. *Ann. Math. Stat.* **33**, (1962), pp.1065-1076.

16. Silverman B.W., *Density estimation for statistics and data analysis.* Monographs on Statistics and Applied Probability 26, Chapman & Hall, (1986).

17. Specht D.F., Enhancements to the probabilistic neural networks. Proc. of the IEEE Int. Joint Conf. on Neural Networks, Baltimore, MD., vol. 1, (1992) pp. 761-768.

18. Zaknich A., A vector quantisation reduction method for the probabilistic neural network. IEEE Proc. of the Int. Conf. on Neural Networks (ICNN), Houston/Texas, USA, (1997) pp. 1117-1120.

19. Scherf U., Ross D., Waltham M., Smith L., Lee J., Tanabe L., Kohn K., Reinhold W., Myers T., Andrews D., Scudiero D., Eisen M., Sausville E., Pommier Y., Botstein D., Brown P., Weinstein J., A gene expression database for the molecular pharmacology of cancer. *Nature Genetics* **24**(3):236-244, (2000).

20. Troyanskaya O., Botstein D., Altman R., "Missing value estimation", in: Berrar D., Dubitzky W., Granzow M. (eds.): *A Practical Approach to Microarray Data Analysis*, Kluwer Academic Publishers, Boston, Dec.2002.

21. Dudoit S. and Fridlyand, "Introduction to classification in microarray experiments", in: Berrar D., Dubitzky W., Granzow M. (eds.): *A Practical Approach to Microarray Data Analysis*, Kluwer Academic Publishers, Boston, Dec.2002.

22. Wall M.E., Rechtsteiner A., Rocha L.M.: "Singular value decomposition and principal component analysis", in: Berrar D., Dubitzky W., Granzow M. (eds.): *A Practical Approach to Microarray Data Analysis*, Kluwer Academic Publishers, Boston, Dec.2002.

23. RuleQuest Research Data Mining Tools. http://www.rulequest.com

24. Bishop C.M., *Neural Networks for Pattern Recognition*, Oxford, Oxford University Press, (1995).

INFERRING GENE REGULATORY NETWORKS FROM TIME-ORDERED GENE EXPRESSION DATA OF BACILLUS SUBTILIS USING DIFFERENTIAL EQUATIONS

MICHIEL J.L. DE HOON[1], SEIYA IMOTO[1], KAZUO KOBAYASHI[2], NAOTAKE OGASAWARA[2], SATORU MIYANO[1]

[1]*Human Genome Center, Institute of Medical Science, University of Tokyo 4-6-1 Shirokanedai, Minato-ku, Tokyo 108-8639, Japan*

[2]*Graduate School of Biological Science, Nara Institute of Science and Technology, 8916-5 Takayama, Ikoma, Nara 630-0101, Japan*

We describe a new method to infer a gene regulatory network, in terms of a linear system of differential equations, from time course gene expression data. As biologically the gene regulatory network is known to be sparse, we expect most coefficients in such a linear system of differential equations to be zero. In previously proposed methods, the number of nonzero coefficients in the system was limited based on ad hoc assumptions. Instead, we propose to infer the degree of sparseness of the gene regulatory network from the data, where we use Akaike's Information Criterion to determine which coefficients are nonzero. We apply our method to MMGE time course data of *Bacillus subtilis*.

1 Introduction

The recently developed cDNA microarray technology allows gene expression levels to be measured for the whole genome at the same time. While the amount of available gene expression data has been increasing rapidly, the required mathematical techniques to analyze such data is still in development. Particularly, deriving a gene regulatory network from gene expression data has proven to be difficult.

In time-ordered gene expression measurements, the temporal pattern of gene expression is investigated by measuring the gene expression levels at a small number of points in time. Periodically varying gene expression levels have for instance been measured during the cell cycle of the yeast *Saccharomyces cerevisiae*.[1] The gene response to a slowly changing environment has been measured during the diauxic shift of the same yeast.[2] Other experiments consider the temporal gene expression pattern due to an abrupt change in the environment of the organism. As an example, the gene expression response was measured of the cyanobacterium *Synechocystis* sp. PCC 6803 after to sudden shift in the intensity of external light.[3,4]

Several methods have been proposed to infer gene interrelations from expression data. In cluster analysis,[2,5,6] genes are grouped together based on the similarity between their gene expression profiles. Inferring Boolean or

Bayesian networks from measured gene expression data has been proposed previously,[7,8,9,10,11] as well as modeling gene expression data using an arbitrary system of differential equations.[12] To reliably infer such an arbitrary system of differential equations, however, a long series of time-ordered gene expression data would be needed, which currently is often not yet available.

Instead, we will construct a linear system of differential equations from gene expression data. This approach maintains the advantages of quantitativeness and causality inherent in differential equations, while being simple enough to be computationally tractable.

Previously, modeling biological data with linear differential equations was considered theoretically by Chen.[13] In this model, both the mRNA and the protein concentrations were described by a system of linear differential equations. Such a system can be described as

$$\frac{\mathrm{d}}{\mathrm{d}t}\underline{x}(t) = \underline{\underline{\Lambda}} \cdot \underline{x}(t), \tag{1}$$

in which the vector $\underline{x}(t)$ contains the mRNA and protein concentrations as a function of time, and the matrix $\underline{\underline{\Lambda}}$ is constant with units of $[\text{second}]^{-1}$. This equation can be considered as a generalization of the Boolean network model, in which the number of levels is infinite instead of binary.

In cDNA microarray experiments, usually only the gene expression levels are determined by measuring the corresponding mRNA concentrations, while the protein concentration is unknown. We therefore focus on a system of differential equations describing gene interactions only. A matrix element Λ_{ij} then represents the effect of gene j on gene i, $[\Lambda_{ij}]^{-1}$ being the reaction time.

To infer the coefficients in the system of differential equations from measured data, it was previously suggested[13] to discretize the system of differential equations, substitute the measured mRNA and protein concentrations, and solve the resulting linear system of equations to find the coefficients Λ_{ij} in the system of linear differential equations. The system of equations is usually underdetermined. Using the additional requirement that the gene regulatory network should be sparse, Chen showed that the model can be constructed in $O\left(m^{h+1}\right)$ time, where m is the number of genes and h is the number of nonzero coefficients allowed for each differential equation in the system.[13]

The parameter h is chosen ad hoc, which has two unexpected consequences. As each row in the matrix $\underline{\underline{\Lambda}}$ will have exactly h nonzero elements, every gene or protein in the network has h parent genes or proteins, and consequently no genes or proteins can exist at the top of a network. Secondly, every gene will inevitably be a member of a feedback loop. While feedback loops are likely to exist in gene regulatory networks, their existence should be determined from the measured data instead of created artificially.

Bayesian networks, on the other hand, do not allow the existence of loops. Bayesian networks rely on the joint probability distribution of the estimated network to be decomposable in a product of conditional probability distributions. This decomposition is possible only in the absence of loops. We further note that Bayesian networks tend to contain many parameters, and therefore need a large amount of data for a reliable estimation.

We therefore aim to find a method that allows the existence of loops in the network, but does not require their presence. Using Eq. 1, we construct a sparse matrix by limiting the number of nonzero coefficients that may appear in the system. Instead of choosing this number ad hoc, we estimate which coefficients in the interaction matrix are zero from the data by using Akaike's Information Criterion (AIC), allowing the number of gene regulatory pathways to be different for each gene.

Our method can be applied to find a network between individual genes, as well as a regulatory network between clusters of genes. As an example, we infer a gene regulatory network between clusters of genes using time course data of *Bacillus subtilis*. Clusters are created using the k-means clustering algorithm. The biological function of the clusters can be determined from the functional categories of the genes belonging to each cluster.

2   Method

We consider a regulatory network between m genes in terms of a linear system of differential equations (Eq. 1), where the vector $\underline{x}(t)$ contains the expression ratios of the m genes at time t. This system of differential equations can be solved as

$$\underline{x}(t) = \exp\left(\underline{\underline{A}}t\right) \cdot \underline{x}_0 ,\tag{2}$$

in which $\underline{x}_0$ contains the gene expression ratios at time zero. In this equation, the matrix exponential is defined in terms of a Taylor expansion as[14]

$$\exp\left(\underline{\underline{A}}\right) \equiv \sum_{i=0}^{\infty} \frac{1}{i!}\underline{\underline{A}}^i .\tag{3}$$

As Eq. 2 depends nonlinearly on $\underline{\underline{A}}$, it will be difficult to solve for $\underline{\underline{A}}$ in terms of the measured data $\underline{x}(t)$. An approximate solution can be found by replacing the differential equation (Eq. 1) by a difference equation:

$$\frac{\Delta \underline{x}}{\Delta t} = \underline{\underline{A}} \cdot \underline{x} ,\tag{4}$$

20

or

$$\underline{x}\left(t+\Delta t\right)-\underline{x}\left(t\right)=\Delta t\cdot\underline{\underline{\Lambda}}\cdot\underline{x}\left(t\right)\ , \tag{5}$$

which is of the form considered by Chen.[13] To be able to statistically determine the sparseness of matrix $\underline{\underline{\Lambda}}$, we explicitly add an error $\underline{\varepsilon}\left(t\right)$, which will invariably be present in the data:

$$\underline{x}\left(t+\Delta t\right)-\underline{x}\left(t\right)=\Delta t\cdot\underline{\underline{\Lambda}}\cdot\underline{x}\left(t\right)+\underline{\varepsilon}\left(t\right)\ . \tag{6}$$

By using this equation, we effectively describe a gene regulatory network in terms of a multidimensional linear Markov model.

We assume that the error has a normal distribution independent of time:

$$f\left(\underline{\varepsilon}\left(t\right);\sigma^2\right)=\left(\frac{1}{\sqrt{2\pi\sigma^2}}\right)^m\exp\left\{-\frac{\underline{\varepsilon}\left(t\right)^{\mathrm{T}}\cdot\underline{\varepsilon}\left(t\right)}{2\sigma^2}\right\}\ , \tag{7}$$

with a standard deviation σ equal for all genes at all times. The log-likelihood function for a series of time-ordered measurements $\underline{x}_i$ at times t_i, $i\in\{1,\ldots,n\}$ at n time points is then

$$L\left(\underline{\underline{\Lambda}},\sigma^2\right)=-\frac{nm}{2}\ln\left[2\pi\sigma^2\right]-\frac{1}{2\sigma^2}\sum_{i=1}^{n}\underline{\hat{\varepsilon}}_i^{\mathrm{T}}\cdot\underline{\hat{\varepsilon}}_i\ , \tag{8}$$

in which

$$\underline{\hat{\varepsilon}}_i=\underline{x}_i-\underline{x}_{i-1}-\left(t_i-t_{i-1}\right)\cdot\underline{\underline{\Lambda}}\cdot\underline{x}_{i-1} \tag{9}$$

is the measurement error at time t_i estimated from the measured data.

The maximum likelihood estimate of the variance σ^2 can be found by maximizing the log-likelihood function with respect to σ^2. This yields

$$\hat{\sigma}^2=\frac{1}{nm}\sum_{i=1}^{n}\underline{\hat{\varepsilon}}_i^{\mathrm{T}}\cdot\underline{\hat{\varepsilon}}_i\ . \tag{10}$$

Substituting this into the log-likelihood function (Eq. 8) yields

$$L\left(\underline{\underline{\Lambda}},\sigma^2=\hat{\sigma}^2\right)=-\frac{nm}{2}\ln\left[2\pi\hat{\sigma}^2\right]-\frac{nm}{2}. \tag{11}$$

To find the maximum likelihood estimate $\underline{\underline{\hat{\Lambda}}}$ of the matrix $\underline{\underline{\Lambda}}$, we use Eq. 9 to write the total squared error $\hat{\sigma}^2$ as

$$\hat{\sigma}^2=\frac{1}{nm}\sum_{i=1}^{n}\left[\left(\underline{x}_i^{\mathrm{T}}-\underline{x}_{i-1}^{\mathrm{T}}\right)\cdot\left(\underline{x}_i-\underline{x}_{i-1}\right)+\left(t_i-t_{i-1}\right)^2\underline{x}_{i-1}^{\mathrm{T}}\cdot\underline{\underline{\Lambda}}^{\mathrm{T}}\cdot\underline{\underline{\Lambda}}\cdot\underline{x}_{i-1}\right.$$

$$\left.-2\left(\underline{x}_i^{\mathrm{T}}-\left(t_i-t_{i-1}\right)\underline{x}_{i-1}^{\mathrm{T}}\right)\cdot\underline{\underline{\Lambda}}\cdot\underline{x}_{i-1}\right]\ , \tag{12}$$

and take the derivative with respect to $\underline{\underline{\Lambda}}$. We find a linear equation in $\underline{\underline{\Lambda}}$:

$$\hat{\underline{\underline{\Lambda}}} = \underline{\underline{B}} \cdot \underline{\underline{A}}^{-1}, \tag{13}$$

in which the matrices $\underline{\underline{A}}$ and $\underline{\underline{B}}$ are defined as

$$\underline{\underline{A}} \equiv \sum_{i=1}^{n} \left[(t_i - t_{i-1})^2 \cdot \underline{x}_{i-1} \cdot \underline{x}_{i-1}^{\mathrm{T}} \right] ; \tag{14}$$

$$\underline{\underline{B}} \equiv \sum_{i=1}^{n} \left[(t_i - t_{i-1}) \cdot (\underline{x}_i - \underline{x}_{i-1}) \cdot \underline{x}_{i-1}^{\mathrm{T}} \right] . \tag{15}$$

In the absence of errors, the estimated matrix $\hat{\underline{\underline{\Lambda}}}$ is equal to the true matrix $\underline{\underline{\Lambda}}$. We know from biology that the gene regulatory network and therefore $\underline{\underline{\Lambda}}$ is sparse. However, all of the elements in the estimated matrix $\hat{\underline{\underline{\Lambda}}}$ may be nonzero due to the presence of noise, even if the corresponding elements in the true matrix $\underline{\underline{\Lambda}}$ are zero. We may decide to set a matrix element equal to zero if the resulting increase in the total squared error, as given by Eq. 12, is small.

Formally, we would use Akaike's Information Criterion[15,16]

$$\mathrm{AIC} = -2 \cdot \begin{bmatrix} \text{log-likelihood of the} \\ \text{estimated model} \end{bmatrix} + 2 \cdot \begin{bmatrix} \text{number of estimated} \\ \text{parameters} \end{bmatrix} \tag{16}$$

to decide which matrix elements should be set equal to zero. The AIC avoids overfitting of a model to data by comparing the total error in the estimated model to the number of parameters that was used in the model. The model with the lowest AIC is considered to be optimal. The AIC is based on information theory and is widely used for statistical model identification, especially for time series model fitting.[17]

We use a mask $\underline{\underline{M}}$ to set matrix elements of $\hat{\underline{\underline{\Lambda}}}$ equal to zero:

$$\hat{\underline{\underline{\Lambda}}}' = \underline{\underline{M}} \circ \hat{\underline{\underline{\Lambda}}}, \tag{17}$$

where $\circ$ denotes the Hadamard (element-wise) product,[14] and the mask $\underline{\underline{M}}$ is a matrix whose elements are either one or zero. The corresponding total squared error $\hat{\sigma}^2$ can be found by replacing $\hat{\underline{\underline{\Lambda}}}$ by $\hat{\underline{\underline{\Lambda}}}'$ in Eq. 12. The total squared error, given the mask $\underline{\underline{M}}$, can be minimized by solving the set of equations

$$\text{if } M_{ij} = 1: \quad \left[\hat{\underline{\underline{\Lambda}}}' \cdot \underline{\underline{A}} \right]_{ij} = B_{ij};$$

$$\text{if } M_{ij} = 0: \quad \hat{\Lambda}'_{ij} = 0; \tag{18}$$

22

yielding the maximum likelihood estimate $\hat{\underline{\underline{\Lambda}}}'$. In this equation, $\underline{\underline{A}}$ and $\underline{\underline{B}}$ are determined from Eqs. 14 and 15 using the measured gene expression levels $\underline{x}_i$.

We then calculate the AIC corresponding to $\underline{\underline{M}}$ by substituting the estimated log-likelihood function from Eq. 11 into Eq. 16:

$$\text{AIC} = nm \ln \left[2\pi\hat{\sigma}^2\right] + nm + 2 \cdot \left(1 + \left[\begin{array}{c} \text{sum of the mask} \\ \text{elements } M_{ij} \end{array}\right]\right), \qquad (19)$$

the estimated parameters being $\hat{\sigma}^2$ and the elements of the matrix $\hat{\underline{\underline{\Lambda}}}$ that we allow to be nonzero. From this equation, we see that while the squared error decreases, the AIC may increase as the number of nonzero elements increases. A gene regulatory network may now be inferred from gene expression data by finding the mask $\underline{\underline{M}}$ that yields the lowest value for the AIC.

For any but the most trivial cases, the number of possible masks $\underline{\underline{M}}$ is extremely large, making an exhaustive search to find the optimal mask infeasible. Instead, we propose a greedy search. Initially, we choose a mask at random, with an equal probability of zero or one for each mask element. We attempt to reduce the AIC by changing each of the mask elements M_{ij}. This process is continued until we find a final mask, for which no further reduction in the AIC can be achieved. We repeat this algorithm many times starting from different (random) initial masks, and choose the final mask $\underline{\underline{M}}$ that has the smallest corresponding AIC. If this optimal mask is found in several tens of trials, we assume that no better masks exist.

3   Results

We will demonstrate our technique of finding a gene regulatory network using gene expression data that were recently measured in an MMGE gene expression experiment of *Bacillus subtilis*.[18] MMGE is a synthetic minimal medium containing glucose and glutamine as carbon and nitrogen sources. In this medium, the expression of genes required for biosynthesis of small molecules, such as amino acids, is induced. The expression levels of 4320 ORFs were measured at eight time points at one hour intervals in this experiment, making two measurements at each time point.

3.1   Data preprocessing

To reduce the effect of measurement noise present in the data, the expression levels of each gene were compared to the measured background level. Genes with an average gene expression level lower than the average background level in either the red or the green channel were removed from the analysis.

Global normalization was then applied to the 3823 remaining genes, and the base-2 logarithms of the gene expression ratios were calculated. Since we are only interested in genes with appreciably changing expression levels during the experiment, we applied a statistical test to the measured log-ratios to determine if they are significantly different from zero. Usually, the t-test would be performed at every time point to determine which log-ratios are significantly different from zero. However, a t-test would be unreliable in this experiment, as there are only two measurements at each time point. We therefore devised a statistical test incorporating the measurements at all eight time points.

Under the null hypothesis, we assume that a gene was not affected by the experimental manipulation. The measured log-ratios at different time points are then equivalent. We further assume that the log-ratios have a normal distribution with zero mean. The standard deviation is then estimated from all $8 \times 2 = 16$ measurements:

$$\hat{\sigma}_{j|\mathrm{H_0}} = \sqrt{\frac{1}{2n} \sum_{i=1}^{n} \sum_{k=1,2} (x_{ji}\,[k])^2}, \tag{20}$$

in which $x_{ji}\,[k]$ denotes the data value of measurement k at time point i for gene j. At each time point, we calculate the average log-ratio as

$$\bar{x}_{ji} = \frac{1}{2} \sum_{k=1,2} x_{ji}\,[k]. \tag{21}$$

Under the null hypothesis, $\bar{x}_{j\cdot}$ (the average of two gene expression log-ratios at a time point) is a random variable with a normal distribution with zero mean and an estimated standard deviation $\hat{\sigma}_{j|\mathrm{H_0}}/\sqrt{2}$. The joint probability for $\bar{x}_{j\cdot}$ to be larger in absolute value than the measured values $\bar{x}_{ji}$ is then

$$\begin{aligned}
P = \prod_{i=1}^{n} P_i &= \prod_{i=1}^{n} p\left(|\bar{x}_{j\cdot}| > |\bar{x}_{ji}|\right) \\
&= \prod_{i=1}^{n} \left[1 - \mathrm{erf}\left(\frac{|\bar{x}_{ji}|}{\hat{\sigma}_{j|\mathrm{H_0}}/\sqrt{2}}\right)\right],
\end{aligned} \tag{22}$$

in which erf is the error function. For a single factor P_i in this product, we would normally choose a significance level α, and reject the null hypothesis if $P_i < \alpha$. Accordingly, we adopt the criterion that $P < \alpha^n$ for rejection of the null hypothesis. This allows us to determine whether the expression levels of a gene change significantly during the experiment by making use of all the available data for that gene.

24

We chose a significance level $\alpha = 0.00025$ such that the expected number of false positives ($0.00025 \times 3823 = 1$) is acceptable. By applying this criterion to the 3823 genes, we found that 684 genes were significantly affected.

3.2 Clustering

The 684 genes were subsequently clustered into five groups using k-means clustering. The Euclidean distance was used to measure the distance between genes, while the centroid of a cluster was defined by the median over all genes in the cluster. The number of clusters was chosen such that a significant overlap was avoided. The k-means algorithm was repeated 1,000,000 times starting from different random initial clusterings. The optimal solution was found 81 times. The full clustering result is available at http://bonsai.ims.u-tokyo.ac.jp/~mdehoon/publications/Subtilis/clusters.html.

In order to determine the biological function of the clusters that were created, we considered the functional category in the SubtiList database[19,20] for all genes in each cluster. Table 1 lists the main functional categories for the five clusters that were formed.

Figure 1 shows the log-ratio of the gene expression as a function of time for each cluster. While the expression levels of clusters I, II, and V change considerably during the time course, clusters II and III have fairly constant expression levels. Cluster IV in particular can be considered as a catchall cluster, to which genes are assigned that do not fit well in the other clusters.

3.3 Network construction

From the measured log-ratios of those twelve genes, we constructed the matrices $\underline{\underline{A}}$ and $\underline{\underline{B}}$ and calculated the matrix $\underline{\underline{\hat{\Lambda}}}$. The process of calculating a mask $\underline{\underline{M}}$, starting from a random initial mask, was repeated 1000 times. The optimal solution was found 55 times. It is therefore unlikely that there are other masks with a lower AIC. Note that the total number of possible masks is $2^{25} = 33,554,432$.

The network that was found is shown in Figure 2. The number of parents of a cluster in the network varies between zero and five. Clusters III and IV appear as the top of the network, while clusters I, II, and V are connected in a loop. Note that this network can neither be generated by the previously proposed method,[13] nor by a Bayesian network model.

The two strongest interactions in the network are the positive and negative effect of cluster IV on cluster V and cluster II respectively. This suggests that the opposite behavior of the gene expression levels of cluster II and V are caused by cluster IV, instead of a direct interaction between clusters II and V.

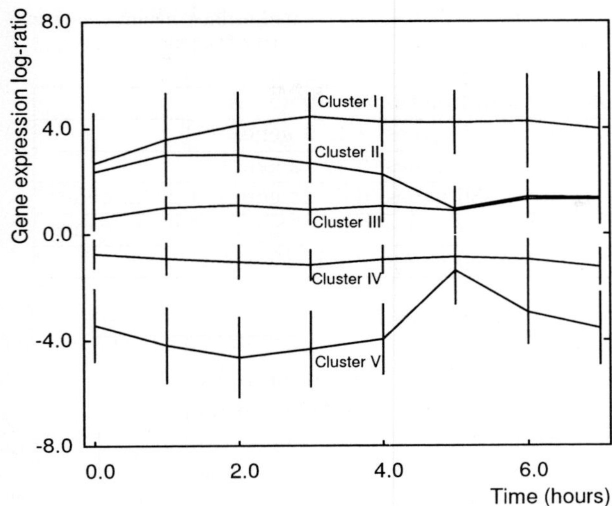

Figure 1: The log-ratio of the gene expression as a function of time for each cluster, as determined from the measured gene expression data.

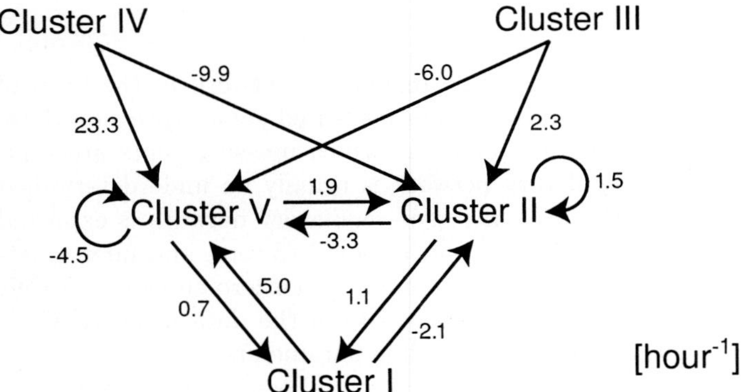

Figure 2: The network between the five gene clusters, as determined from the MMGE time-course data. The values show how strongly one gene cluster affects another gene cluster, as given by the corresponding elements in the interaction matrix $\hat{\underline{\underline{\Lambda}}}'$. In effect, this matrix represents how rapidly gene expression levels respond to each other. As an example, a change in the gene expression level of Cluster I would cause the expression level of Cluster V to change considerably within $1/(5.0 \text{ hour}^{-1}) = 12$ minutes, if the expression levels of Clusters II, III, and IV are unchanged.

Table 1: The main functional categories for the five clusters created using k-means clustering. The functional categories refer to the SubtiList database at Institut Pasteur.

Cluster	Number of genes	Main functional categories
I	42	2.2: 11 genes; 1.1: 9 genes
II	62	1.2: 15 genes; 2.2: 12 genes
III	187	5.1: 30 genes; 6.0: 23 genes; 1.2: 22 genes
IV	343	5.1: 40 genes; 5.2: 39 genes; 1.2: 33 genes
V	50	1.2: 15 genes; 2.1.1: 15 genes

Functional categories

1.1:	Cell wall.
1.2:	Transport/binding proteins and lipoproteins.
2.1.1:	Metabolism of carbohydrates and related molecules — Specific pathways.
2.2:	Metabolism of amino acids and related molecules.
5.1:	Similar to unknown proteins from *Bacillus subtilis*.
5.2:	Similar to unknown proteins from other organisms.
6.0:	No similarity.

4　Discussion

We have shown a method to infer a gene regulatory network in the form of a linear system of differential equations from measured gene expression data. Due to the limited number of time points at which measurements are typically made, finding a gene regulatory network is usually an underdetermined problem. Since biologically the resulting gene regulatory network is expected to be sparse, we set some of the matrix entries equal to zero, and infer a network using only the nonzero entries. The number of nonzero entries, and thus the sparseness of the network, was determined from the data using Akaike's Information Criterion without using any ad hoc parameters.

Describing a gene network in terms of differential equations has three advantages. First, the set of differential equations describes causal relations between genes: a coefficient Λ_{ij} of the coefficient matrix determines the effect of gene j on gene i. Second, it describes gene interactions in an explicitly numerical form. Third, because of the large amount of information present in a system of differential equations, other network forms can easily be derived from it. In addition, we can link the inferred network to other analysis or visualization tools, such as *Genomic Object Net*[22].

In previously described methods, either loops cannot be found (such as in Bayesian network models) or the method artificially generates loops in the network. While the method proposed here allows loops to be present in the network, their existence is not required. Loops are found only if warranted by the data. When inferring a regulatory network between gene clusters using time-course data of *Bacillus subtilis* in an MMGE medium, we found that some of the clusters were part of a loop, while others were not.

If the number of genes m is equal to or larger than the number of experiments n, the matrix $\underline{\underline{A}}$ in Eq. 18 is singular. The problem is then underdetermined, and an interaction matrix $\hat{\underline{\underline{\Lambda}}}$ can be found with zero total error $\hat{\sigma}^2$ and an AIC of $-\infty$. This breakdown of our proposed method can be avoided by applying it to a sufficiently small number of genes or gene clusters, or by limiting the number of parents in the network.

References

1. P.T. Spellman, G. Sherlock, M.Q. Zhang, V.R. Iyer, K. Anders, M.B. Eisen, P.O. Brown, D. Botstein, and B. Futcher, "Comprehensive identification of cell cycle-regulated genes of the yeast Saccharomyces cerevisiae by microarray hybridization" *Mol. Biol. Cell* **9** (1998) 3273–3297.
2. J.L. DeRisi, V.R. Iyer, and P.O. Brown, "Exploring the metabolic and genetic control of gene expression on a genomic scale" *Science* **278** (1997) 680–686.
3. Y. Hihara, A. Kamei, M. Kanehisa, A. Kaplan, and M. Ikeuchi, "DNA microarray analysis of cyanobacterial gene expression during acclimation to high light" *The Plant Cell* **13** (2001) 793–806.
4. M.J.L. de Hoon, S. Imoto, and S. Miyano, "Statistical analysis of a small set of time-ordered gene expression data using linear splines" *Bioinformatics*, in press.
5. M.B. Eisen, P.T. Spellman, P.O. Brown, and D. Botstein, "Cluster analysis and display of genome-wide expression patterns" *Proc. Natl. Acad. Sci. USA* **95** (1998) 14863–14868.
6. P. Tamayo, D. Slonim, J. Mesirov, Q. Zhu, S. Kitareewan, E. Dmitrovsky, E.S. Lander, and T.R. Golub, "Interpreting patterns of gene expression with self-organizing maps: Methods and application to hematopoietic differentiation" *Proc. Natl. Acad. Sci. USA* **96** (1999) 2907–2912.
7. S. Liang, S. Fuhrman, and R. Somogyi, "REVEAL, a general reverse engineering algorithm for inference of genetic network architectures" *Proc. Pac. Symp. on Biocomputing* **3** (1998) 18–29.
8. T. Akutsu, S. Miyano, and S. Kuhara, "Inferring qualitative relations in genetic networks and metabolic pathways" *Bioinformatics* **16** (2000)

727–734.

9. N. Friedman, M. Linial, I. Nachman, and D. Pe'er, "Using Bayesian networks to analyze expression data" *J. Comp. Biol.* **7** (2000) 601–620.

10. S. Imoto, T. Goto, and S. Miyano, "Estimation of genetic networks and functional structures between genes by using Bayesian networks and nonparametric regression" *Proc. Pac. Symp. on Biocomputing* **7** (2002) 175–186.

11. S. Imoto, S.-Y. Kim, T. Goto, S. Aburatani, K. Tashiro, S. Kuhara, and S. Miyano, "Bayesian network and nonparametric heteroscedastic regression for nonlinear modeling of genetic network" *Proc. IEEE Computer Society Bioinformatics Conference* (2002) 219–227.

12. E. Sakamoto and H. Iba, "Evolutionary inference of a biological network as differential equations by genetic programming" *Genome Informatics* **12** (2001) 276–277.

13. T. Chen, H.L. He, and G.M. Church, "Modeling gene expression with differential equations" *Proc. Pac. Symp. on Biocomputing* **4** (1999) 29–40.

14. R.A. Horn and C.R. Johnson, *Matrix Analysis.* Cambridge University Press, Cambridge, UK (1999).

15. H. Akaike, "Information theory and an extension of the maximum likelihood principle" Research Memorandum No. 46, Institute of Statistical Mathematics, Tokyo (1971). In B.N. Petrov and F. Csaki (editors), *2nd Int. Symp. on Inf. Theory.* Akadémiai Kiadó, Budapest (1973) 267–281.

16. H. Akaike, "A new look at the statistical model identification" *IEEE Trans. Automat. Contr.* **AC-19** (1974) 716–723.

17. M.B. Priestley, *Spectral Analysis and Time Series.* Academic Press, London (1994).

18. Microbial Advanced Database Organization (Micado). http://www-mig.versailles.inra.fr/bdsi/Micado/.

19. I. Moszer, P. Glaser, and A. Danchin, "SubtiList: a relational database for the Bacillus subtilis genome" *Microbiology* **141** (1995) 261–268.

20. I. Moszer, "The complete genome of Bacillus subtilis: From sequence annotation to data management and analysis" *FEBS Letters* **430** (1998) 28–36

21. T.W. Anderson and J.D. Finn, *The New Statistical Analysis of Data.* Springer Verlag, New York (1996).

22. H. Matsuno, A. Doi, Y. Hirata, and S. Miyano, "XML documentation of biopathways and their simulation in Genomic Object Net" *Genome Informatics* **12** (2001) 54–62. *Genomic Object Net* is available at http://www.GenomicObject.net.

GENOME-WIDE ANALYSIS OF BACTERIAL PROMOTER REGIONS

ELEAZAR ESKIN[1], URI KEICH[2], MIKHAIL S. GELFAND[3], PAVEL A. PEVZNER[2]

[1] *Department of Computer Science, Columbia University, New York, NY 10027,*
`eeskin@cs.columbia.edu`

[2] *Department of Computer Science and Engineering, University of California at San Diego, La Jolla, CA 92093-0114,* `{keich,ppevzner}@cs.ucsd.edu`

[3] *IntegratedGenomics-Moscow, P.O.Box 348, Moscow, 117333, Russia,*
`gelfand@integratedgenomics.ru`

Identifying prokaryotic promoter sequences is notoriously difficult and for most sequenced bacterial genomes the promoter sequences are still unknown. Since experimental analysis trails behind sequencing, genome-wide computational promoter discovery is often the only realistic way to discover these sequences in newly sequenced bacterial genomes. However, genome-wide samples for promoter discovery may be very large and corrupted complicating promoter discovery. We discuss three aspects of genome-wide promoter discovery: sample generation, signal finding algorithms, and scoring signals. We applied our new MITRA algorithm to analyze samples of divergent and convergent genes in 20 bacterial genomes and found strong putative dyad signals in 17 out of the 20 genomes. Moreover, in 12 out of 20 genomes the found signals are identical or similar to the known regulatory patterns (Pribnow-Gilbert boxes and CRP binding sites). Since many of putative signals correspond to previously known elements of bacterial transcriptional regulation, the remaining discovered signals are good candidates for unknown regulatory elements.

1 Introduction

A fundamental challenge of molecular biology is understanding the regulation of gene expression, in particular, on the level of transcription. In prokaryotes, genes encoding transcription factors may constitute up to 10% of the genome, as in *Pseudomonas aeruginosa* [1]. Thus an important component of genomic analysis is automated identification of transcription signals such as promoters. Since the experimental analysis trails behind the sequencing of new genomes, we are interested in discovery of regulatory sequences from complete genomes. This paper describes a method for discovering putative regulatory sites by fully automated genome-wide sequence analysis.

Discovering putative regulatory sites from complete genomes is a very difficult problem [2]. These difficulties are threefold and include (i) difficulties in sample generation, (ii) algorithmic difficulties in scaling to large and corrupted

">

samples, and (iii) statistical difficulties in assessing the significance of patterns that are discovered.

Regulatory signals can be modeled either as patterns or as profiles. This paper focuses on pattern approaches but the described algorithms are applicable to both patterns and profiles. Our algorithm will return a set of patterns which hopefully correspond to actual transcription signals in the genome. We use the term *monad* pattern to refer to a contiguous l-mer that occurs in the sample with up to d mismatches. Allowing for mismatches takes into account the fact that binding sites exactly matching the pattern are rare and often physiologically undesirable. We use the term $(l, d) - k$ pattern to denote a monad of length l that occurs in the sample k times (with up to d mismatches). Since bacterial regulatory signals are often build from two monad parts that occur with a fixed (or almost fixed) distance from each other, we model these binding sites by *dyad* patterns. A dyad pattern $(l_1 - (s_1, s_2) - l_2, d) - k$ are two monad patterns of length l_1 and l_2 respectively, that occur k times in the sample with a minimum separation distance of s_1 and a maximum separation distance of s_2. If $s_1 = s_2$ (a dyad with a fixed distance between the monads) we use the notation $(l_1 - s - l_2, d) - k$ patterns. This paper focuses on dyad signals that are common elements of bacterial transcriptional regulation.

There are many approaches to discovering monad signals [3,4,5,6,7,8] and dyad signals [9,10,11,12]. Recently there has been an emergence of powerful sample-driven approaches to monad pattern discovery [13,14,15] that are efficient enough to handle large genomic samples. For dyad signals, the sample-driven approaches include the algorithms presented in Marsan and Sagot 2000 [16] and MITRA [15]. For a given l_1, l_2, s_1, s_2, and k, these methods can find all $(l_1 - (s_1, s_2) - l_2, d) - k$ dyad patterns in the sample and are efficient enough to apply to samples of the total size exceeding $100,000$ nucleotides.

Our sample generation approach relies on comparative analysis of intergenic regions between divergently and convergently transcribed genes. We first take advantage of the relative location of genes in order to determine the regions where the binding sites most likely occur. Although regulatory elements are located upstream of genes, most upstream regions of bacterial genes do not contain promoters. Genes in bacterial genomes often form operons and only the intergenic region upstream of the first gene in the operon contain regulatory elements. Since the operon structure in bacterial genomes is rarely known, it is not clear how to automatically generate samples of regulatory regions. Our sample generation approach is based on the observation that intergenic regions between consecutive genes transcribed in divergent directions are guaranteed to be upstream regions of operons. By similar reasoning, intergenic regions between two convergently transcribed genes usually do not contain binding sites. We use the intergenic regions between genes that are transcribed in convergent directions as our background sample. Since both divergent and convergent intergenic regions are selected from the same genome, the convergently transcribed intergenic regions allow us to estimate a background distribution for upstream regions with regulatory elements (the intergenic region between divergently transcribed genes).

A key ingredient for discovery of putative regulatory signals is a method to assess the statistical significance of a potential signal. The problem is nontrivial since our samples are large, biased, and contain low complexity regions. There exists a number of approaches to assessing the statistical significance of patterns. They include the shuffling approach [17], building statistical models to estimate the probability of a pattern [18] or profile [3,4], and using a background sample to assess the significance of a pattern depending on whether or not it is over represented in the sample [19]. Our ability to discover signals depends on the reliability of the method for determining the statistical significance of observed patterns in such large samples. For example consider the experimentally confirmed promoter represented by the Pribnow-Gilbert dyad $TTGACA-17-TATAAT$ in the $B.\ subtilis$ genome. This dyad (with 2 allowed mismatches) occurs 143 times in our sample. However, it appears anywhere from 34 to 64 times at other separation distances from 3 to 23 nucleotides. Most of them are likely to be simply random events instead of having any biological meaning. It indicates a need for a new scoring approach that combines the traditional statistical analysis of monad patterns with the analysis of spacing and positional parameters.

Our scoring approach estimates the significance of patterns in the target sample against the patterns in the background sample. For each pattern, we compute the strength of the signal which measures the difference between the number of occurrences of the pattern and the expected number of occurrences based on the background distribution (strength score). However, even with scoring method that takes into account a background sample, it is still difficult to determine which patterns correspond to biologically meaningful signals. Our key idea is to incorporate two types of additional information to help make this determination. Firstly, we contrast every dyad pattern with a fixed separation distance against a dyad pattern with a "random" spacer (dyad score). Secondly, since some regulatory elements are positional (i.e., tend to occur at the same relative position) we also analyze the relative position of the signal to the start of the gene (positional score). Although we do not have reliable information about the transcription start position of the gene, we can still obtain a rough estimate of the relative positions of the signal using the translation start instead. Each type of information on its own is generally not sufficient to make a determination on whether or not a signal is an actual binding site. In fact, some actual binding sites are not positional signals and some actual dyad binding sites have looser restrictions on their separation distance than other dyad binding sites. However, the combination of the three types of information helps us to decide whether or not a signal is a putative binding site.

We apply our new MITRA algorithm to analyze samples of divergent and convergent genes in 20 bacterial genomes and find particularly strong putative dyad signals in 9 out of the 20 genomes and signals that correspond to known binding sites in 12 of the 20 genomes. Details of the MITRA algorithm are presented in [15]. Detailed information about all of the signals reported in this paper is available at: http://www.cs.columbia.edu/compbio/mitra/.

2 Sample Generation

The main difficulty in the analysis of complete bacterial genomes is scarcity, or even lack, of the experimental data about location of regulatory regions and the operon structure. Thus, for any given gene, it is difficult to decide, whether it is the first gene in an operon (and thus the transcription factor binding sites are upstream of this gene), or it is preceded by other genes. Therefore, simply taking upstream regions for every gene in a bacterial genome would lead to extremely corrupted samples and failure of the motif finding algorithms.

A better approach to sample generation was first proposed by Washio et al., 1998 [20] and later explored by Sagot and colleagues in [2] and [21]. It is based on the observation that the divergently transcribed genes are guaranteed to be the most upstream genes of the respective operons. Thus, the target sample we used to search for the regulatory signals consists of genomic fragments between divergently transcribed genes. Similarly, a region between convergently transcribed genes cannot be an upstream region for any gene, and such regions formed a background sample (i.e., sample without binding sites).

The sequences of 20 complete bacterial genomes (Table 1) were downloaded from the ERGO database [22]. The choice of the genomes was dictated by (i) availability of experimental information for some of these genomes and (ii) availability of several genomes from one taxonomic group. We used the limited set of experimentally confirmed promoters to verify that our predictions agree with the available data. We used genomes from the same taxonomic group (Gram-positive bacteria from the Bacillus/Clostridium group, mycoplasmas, chlamidiae, proteobacteria from the α, β and γ divisions, ϵ-proteobactera) to check whether our promoter predictions for these genomes produce similar putative patterns.

To create our samples, we extract the last 310 bases of the intergenic region and the reverse complement of the first 310 bases. We remove the 10 bases closest to the gene to delete the strongly conserved Shine-Dalgarno signal that would dominate our results. In the cases of alternatives caused by overlapping genes the shortest intergenic fragments were selected. In cases where the intergenic region is longer than 620 nucleotides, a portion of the intergenic region is left out of the target sample. We perform the same procedure for creating the background sample that model the regions without the regulatory elements.

3 Finding Statistically Significant Signals

We use the MITRA algorithm to detect all "statistically significant" $(l_1 - (s_1, s_2) - l_2, d) - k$ patterns. MITRA is fully described in Eskin and Pevzner, 2002[15] and is easily adapted to use the scoring method described below. Although MITRA was used for these experiments, any algorithm (such as Marsan and Sagot, 2000[16]) that can recover all $(l_1 - (s_1, s_2) - l_2, d) - k$ patterns, properly modified to incorporate the scoring functions described below, would produce equivalent results.

We incorporate three types of information into assessing the significance of a signal: signal strength score, dyad separation score and positional score.

We use the background sample obtained from intergenic sequences between genes transcribed in convergent directions to estimate background distribution. We first describe our scoring method for monad patterns and then extend it to dyads patterns. For a pattern P we define $p_P = \frac{n_P}{n_B}$ as the number of l-mers in the background sample that are within d mismatches of the pattern, n_P divided by the total number of l-mers in the background sample, n_B. We smooth our estimates for p_P using Dirichlet priors[23] and adjust our estimates of p_P to take into account the differences in nucleotide composition between convergent and divergently transcribed intergenic regions.

Let n_T be the size of the target sample. Given that the pattern P occurs (with mismatches) o_P times in the target sample, we define the score of the pattern as $s_P = \frac{o_P - n_T p_P}{\sqrt{n_T p_P (1 - p_P)}}$. For a single pattern, the score can be interpreted as the number of standard deviations from the mean if we assume a binomial distribution. The pattern score is simple and efficient enough to incorporate into MITRA for the exhaustive search to discover *all* top scoring patterns. Instead of returning all patterns that occur k times, we instead specify a minimal score threshold t. For a pattern P and minimum score threshold t, the minimum number of occurrences for the pattern k_P to make into the ranked list would be $k_P = n_T p_P + t \sqrt{n_T p_P (1 - p_P)}$.

We score the dyad patterns D composed of monad patterns P_1 and P_2 in a similar way. For each dyad pattern D, we estimate p_D from estimates of the probabilities of the patterns P_1 and P_2. Since the mismatches of an instance of the dyad can be spread to both monads, we need to estimate a probability for each monad occurring with a certain number of mismatches. As above, we compute the counts for each occurrence of the pattern with i mismatches over the background sample and divide by the size of the sample. We use p_P^i to denote the probability of a pattern P occurring with i mismatches. We then estimate the probability for the dyad pattern D using $p_D = \sum_{i,j \text{ s.t. } i \geq 0, j \geq 0, i+j \leq d} p_{P_1}^i p_{P_2}^j$. If s is the number of allowable separation distances, the score for a dyad D, s_D is defined to be $s_D = \frac{o_D - s n_T p_D}{\sqrt{s n_T p_D (1 - p_D)}}$. For a minimum score threshold t, we set minimum number of occurrences for a dyad pattern D as $k_D = s n_T p_D + t\sqrt{s n_T p_D (1 - p_D)}$.

Two other types of information is the distribution of separation distances between the dyad signals and the distribution of the positions of instances of the signal. Many dyads which correspond to a binding site, have a peak in the histogram of separation distances at a certain separation distance such as in Figure 1(a) for $B.subtilis$. Similarly, many binding sites tend to occur in a similar position relative to the transcription start of the gene such as in Figure 1(c) $B.subtilis$. At the same time, for most regulatory signals the situation is more difficult, for example, the same histograms for $E.Coli$ 1(b,d) show less pronounced peaks.

We incorporate this information by assessing the statistical significance of the distribution of both the separation distance (between the two parts of the dyad) and of the position of the signal (relative to the estimated transcription start site). In the first case our null hypothesis is that every instance of the dyad is independently equally likely to fall in one of the $s_2 - s1 + 1$ possible bins: one for each possible separation distance. For the positional histogram, since there is often some flexibility in the position of the transcription factor relative to the transcription start site, we group the positions using bins of 30 bp.

We assess the statistical significance of the observed data using the statistic M which equals the maximal number of instances that fall in one bin. Under our null hypothesis M is distributed as the maximum multinomial bin which is given by [24]:

$$P(M \leq i) = \frac{N!}{N^N e^{-N}} \left[F_{\text{Pos}}[N/t](i) \right]^t P(W_i = N) \qquad i = 0, 1, \ldots,$$

where N is the total number of instances, t is the number of bins, $F_{\text{Pos}}[\lambda](i)$ is the cdf of a Poisson random variable with parameter λ evaluated at i, and W_i is a sum of t iid Poisson random variables ($\lambda = N/t$), each of which is subject to truncation at i.

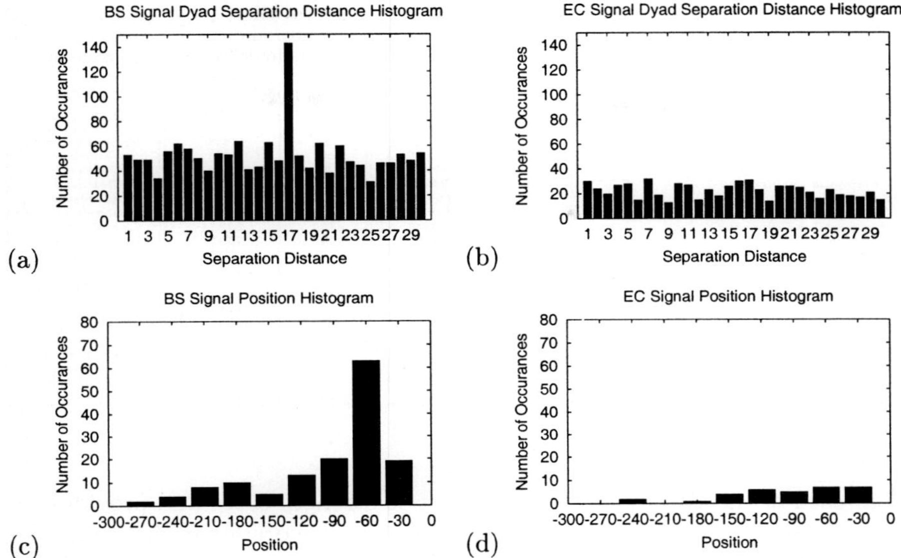

Figure 1: Histogram of Separation Distances and Positions for the Pribnow-Gilbert dyad signal TTGACA-17-TATAAT. Separation Distances in (a) BS genome and (b) EC genome. Positions in (c) BS genome (d) EC genome. We bin the positions of the signal instances into buckets (30 bp by default) since positions rarely exactly match and the exact transcription start position often unknown.

In practical terms $P(W_i = N)$ can be computed by a t-fold convolution of the truncated Poisson distribution. Note that the distribution of a Poisson (λ) random variable which is truncated at i is given by $P_{\mathrm{Pos}}[\lambda](j)/F_{\mathrm{Pos}}[\lambda](i)$ for $j = 0, 1, \ldots, i$, where $P_{\mathrm{Pos}}[\lambda](j) = e^{-\lambda}\lambda^j/j!$ is the standard Poisson probability mass function. Also, by Stirling's approximation, for a reasonably sized N we can approximate $\frac{N!}{N^N e^{-N}} \sim \sqrt{2\pi N}$. For both of these scores we compute the P-value of the statistic and report its negative log.

4 Finding Putative Regulatory Elements in Bacterial Genomes

We performed blind experiments over 20 bacterial genomes (Table 1) and extracted the top dyad signals for each genome. We searched for dyads consisting of two conserved regions of length 6 with separation distances from 3 to 23 bases. To validate our results, we checked the found signals against the known regulatory elements in bacterial genomes. Table 2 shows the top signals

Table 1: Genome Intergenic Region Statistics. The first column is name of the genome. The ID is an abbreviation for the genome. The next columns list the number and the lengths of intergenic regions for divergent and convergent samples. The last column describe genome taxonomy.

Genome Name	Genome ID	Div. Regions	Div. Nucleotides	Conv. Regions	Conv. Nucleotides	Genome Taxonomy
Bacillus subtilis	BS	552	132145	244	47087	*Bacillus* group
Campylobacter jejuni	CJ	168	25837	36	4638	ϵ-proteobacteria
Chlamydia muridarum	CMU	120	29176	55	5751	*Clamidiales*
Chlamydophila pneumoniae	CPX	136	33274	67	8805	*Chlamidiales*
Chlamydia pneumoniae	CQ	141	36973	68	11460	*Chlamidiales*
Synechocystis sp.	CY	525	138494	330	50311	*Cyanobacteria*
Escherichia coli	EC	589	148936	374	61262	γ-proteobacteria
Haemophilus influenzae	HI	228	48346	181	23944	γ-proteobacteria
Helicobacter pylori	HP	169	33123	109	26091	ϵ-proteobacteria
Lactococcus lactis	LLX	217	53342	139	23081	*Bacillus* group
Mycoplasma genitalium	MG	30	5013	7	1404	*Mycoplasmales*
Mycoplasma pneumoniae	MP	51	9535	41	10235	*Mycoplasmales*
Mycobacterium tuberculosis	MT	530	93471	290	43742	*Mycobacteria*
Neisseria meningitidis ser. B	NX	287	55651	266	45793	β-proteobacteria
Pseudomonas aeruginosa	PA	790	194787	450	80089	γ-proteobacteria
Rickettsia prowazekii	RP	104	42597	99	62895	α-proteobacteria
Streptococcus pyogenes	ST	169	47300	159	33165	*Bacillus* group
Thermotoga maritima	TM	163	25332	29	4542	*Thermotogales*
Ureaplasma urealyticum	UU	55	12728	40	10623	*Mycoplasmales*
Pasteurella multocida	VK	267	58281	238	30640	γ-proteobacteria

from each of the 20 genomes with respect to strength score and provides the strength score, the dyad score and the positional score for each signal. We do not report a signal if (i) it is a slight variation of a higher scoring signal, (ii) if it is a shifted variant of a higher scoring signal, or (iii) if it is a reverse complement of a higher scoring signal.

The set of putative signals identified by our algorithm contains a number of known signals and several promising candidates for more detailed analysis. Among the known signals are the classical promoter signals consisting of the standard Gilbert and Pribnow boxes. They have been found in all Gram-positive bacteria from the *Bacillus* group: *B. subtilis*, *S. pyogenes*, and *L. lactis*, as well as in alpha-proteobacterium *R. prowazekii*.

For nine genomes (BS, CY, EC, HI, LLX, NX, PA, ST, and VK) we discovered the particularly strong signals with high dyad and strength scores. For six of these genomes, these signals correspond to known biological signals or to variants of known signals.

In *B. subtilis. (BS)*, one of the strongest dyad signals that we recovered was the classical Pribnow-Gilbert promoter consensus $TTGACA - 17 - TATAAT$. This signal is over-represented in the divergent intergenic regions relative to the convergent intergenic regions, has a very strong distance peak at the distance 17 as well as a strong positional peak at distance in the range -90 to -60. The distribution of separation distances is shown in Figure 1(a) and the positional distribution is shown in 1(c). In *E. coli. (EC)*, the found signal in *E. coli* perfectly matches the binding signal of the transcription factor CRP (TGTGAT-4-ATCACA). In *H. influenzae. (HI)*, all three signals found in *H.*

influenzae are interesting. The first found dyad TGCGGT-12-CGTTTT signal has a strongly conserved region around the dyad represented by the longer dyad AAAAGTGCGGTNA-10-CGTTTT. The second found signal is the binding signal of the transcription factor CRP. In addition, the third signal has an additional interesting feature. Although it looks like an *AT* rich signal it tends to occur in non-*AT* rich regions which suggests that it is a real binding site. In *L. lactis. (LLX)*, the found dyad corresponds to the canonical Pribnow-Gilbert promoter consensus. In *S. pyogenes. (ST)*, the found dyad corresponds to the canonical Pribnow-Gilbert promoter consensus. In *P. multocida. (VK)*, the found dyad is a slightly shifted form of the binding signal of the transcription factor CRP. In addition, there are two lower-scoring but still interesting palindromes AATGTG-10-CACATT and AATTTG-12-CAAATT that may be the binding signals for yet unknown transcription factors. We plan to do detailed analysis of these signals in order to determine the corresponding regulons.

Many of the other signals detected also correspond to known binding sites. In *N. meningitidis (NX)* and *R. prowazekii (RP)*, we detect canonical Pribnow-Gilbert boxes. Among other identified signals, there are modified forms of the Pribnow-Gilbert promoter consensuses: TTGACA-19-ATAATT in *C. pneumoniae (CPX)* (the Pribnow box is shifted by 1 bp to the right; the spacer length is longer than in other species), TTAATC-21-TATAAT in *H. pylori (HP)*, identified earlier in [21] (unusual Gilbert box and longer spacer), TTGACC-17-TAGAAT in *P. aeruginosa (PA)* (modified boxes), and TTGCCA-17-TACAAT in *P. multocida (VK)* (modified boxes). Many of the other signals that we find are palindromic signals (Table 2).

The known promoter signal of *E. coli* was too weak to be discernible in this analysis as well as previous genome-wide analyses[21]. The promoter signal in *Mycoplasmas* is also very weak; even given a sample of mapped promoters it is not possible to derive a good consensus [25]. The signals identified in *M. tuberculosis* do not resemble the promoter consensus of a closely related bacterium *M. paratuberculosis* [26]. We also did not find any signal corresponding to the suggested consensus TTTAAGT-(15-19)-TATAAT of *C. jejuni* [27].

Some signals may still be artifacts. In particular, the AT-rich signals of *Mycoplasmas* represent neither promoters identified in experimental study [25], nor can they be binding signals of transcription factor HrcA (the CIRCE box TTAGCACTC-9-GAGTGCTAA) identified in [28]. Despite the method's inability to find these signals, the fact that we were able to identify many of the promoters and transcription factor binding signals demonstrates the power of the method and indicates that at least some of the identified candidates deserve closer look.

38

Table 2: Top scoring dyad signals in 20 bacterial genomes. Underlined signals are particularly strong (strength scores greater than 10 and either a dyad score or position score greater than 7). The Signal Class column labels if the signal falls into a known biological signal. Classes are defined as (PB) Pribnow-Gilbert signal, (PB*) variant Pribnow-Gilbert signal, (CRP) CRP signal, (CRP*) variant CRP signal, (PU) palindromic signal for a possibly unknown factor.

Genome ID	Signal Pattern	Number Occurrences	Strength Score	Dyad Score	Position Score	Signal Class
BS	TTGACA-17-TATAAT	143	20.10	8.86	7.42	PB
BS	CCTCCT-16-CATTAT	62	13.15	4.07	3.53	
BS	TATAAT-5-TATAAT	151	11.25	5.81	7.48	
CJ	TTCCCT-10-AAATTT	54	4.79	3.86	2.76	
CJ	TACCAT-8-TAAAAT	58	2.71	8.19	6.32	
CJ	TTTAAC-11-TAGAAT	71	4.19	8.30	6.74	
CMU	AATTAT-6-ATAATT	39	2.61	7.67	0.37	PU
CMU	AATATA-18-TATATT	37	1.21	7.51	2.37	PU
CMU	CATTGT-12-TCTTCT	21	1.81	6.75	0.09	
CMU	GCAACA-21-AAATAA	22	0.86	3.49	0.62	
CPX	TTGACA-19-ATAATT	27	3.43	7.16	1.92	PB*
CPX	GTGCAA-11-TTTTTC	25	3.81	6.95	0.30	
CPX	ATTAAT-12-ATTAAT	42	2.78	2.17	0.74	PU
CPX	ATTATT-6-ATTAAT	57	3.21	3.35	4.11	
CQ	AAAATT-5-ATAATG	49	3.17	7.76	1.57	
CQ	ATTATT-6-ATTAAT	57	3.91	3.25	5.98	
CQ	ATTAAT-14-ATTATT	54	4.42	4.48	2.76	
CY	ATTGTA-11-AATTTT	76	7.67	3.75	3.43	
CY	TGTTAA-4-TGTTAC	57	12.77	7.83	3.30	
EC	TGTGAT-4-ATCACA	108	24.04	8.01	2.99	CRP
HI	TGCGGT-12-CGTTTT	58	5.84	7.80	3.14	
HI	TGTGAT-4-ATCACA	62	17.64	7.11	6.44	CRP
HI	AAAATT-6-AATTTT	342	13.21	9.85	8.26	PU
HP	ATTATA-10-TATAAT	89	7.17	7.81	6.96	PU
HP	ATTTTA-18-TATGCT	49	6.04	7.97	2.51	
HP	TTAAGC-21-TATAAT	51	6.62	7.79	6.41	PB*
HP	GTATAA-7-ATTATA	56	8.78	8.09	6.49	
LLX	TTGACA-17-TATAAT	125	19.87	8.62	7.29	PB
LLX	TTATAA-5-TTATAA	203	8.41	9.54	7.76	PU
MG	AGTAAA-10-TTTACT	20	0.11	5.99	2.16	PU
MG	AATCAA-11-ACTTTT	21	0.78	6.70	0.09	
MP	TCCAAA-14-TTTTTA	23	2.91	6.14	0.49	
MP	TTGTAA-15-TAATTA	20	4.33	4.18	0.92	
MP	TTAAAA-17-TTAGTA	23	2.88	6.46	0.49	
MP	TTATTA-18-CAAACA	20	3.91	6.08	3.58	
MT	CGGCCC-10-CGGGCC	80	6.47	8.53	1.37	
MT	CGATAC-12-CGCCGC	51	4.84	7.57	2.22	
MT	GGCCCG-8-CCAGGC	66	7.37	8.31	4.86	
NX	CCGCCG-12-CGGCGG	33	10.86	7.40	3.13	PU
NX	TTGACA-17-TATAAT	43	12.87	7.71	6.24	PB
NX	CTTCAG-3-GCATAG	50	19.61	6.68	4.45	
NX	TATAGT-6-ACTATA	30	7.78	7.17	0.50	PU
PA	TTGACC-17-TAGAAT	31	9.99	7.18	5.91	PB*
PA	TGTCAC-5-TGTCAC	34	9.51	7.23	5.18	
PA	TATAAT-6-CAATTT	30	11.77	7.21	5.88	
PA	TATAAT-3-CGGCCT	50	8.63	7.50	6.39	
RP	TTGACA-17-TATAAT	42	6.55	7.56	4.80	PB
RP	CTTTAA-21-TTAAAG	44	6.99	7.32	1.32	PU
RP	AATTAT-22-TTTCCC	20	6.56	6.56	0.92	
RP	TAATTA-9-AACCAT	39	5.46	7.52	1.07	
ST	TTGACA-17-TATAAT	92	18.56	8.24	6.99	PB
ST	AATTAT-3-ATAATA	123	10.80	8.85	5.97	
TM	TTGACA-17-TATAAT	33	4.34	7.43	5.98	PB
UU	TATAAT-13-TACAAT	39	1.12	0.37	3.34	
VK	AATGTG-10-CACATT	58	12.36	7.76	3.14	PU
VK	AATTTG-12-CAAATT	79	8.97	8.57	3.94	PU
VK	ATTGTA-12-AAATTT	78	6.72	7.02	6.67	
VK	TTGCCA-17-TACAAT	30	7.85	7.13	5.88	PB*
VK	GTGATC-4-TCACAA	70	27.03	7.33	6.72	CRP*

5 Conclusion

We presented an approach for fully automatic discovery of putative regulatory signals in bacterial genomes. The approach emphasizes the interplay of three processes: sample generation, signal finding, and scoring.

We applied our MITRA algorithm to 20 bacterial genomes and detected signals that correspond to known binding sites in 12 of the 20 genomes. The majority of the strong signals detected by MITRA, do in fact correspond to known biological binding sites. Of the 14 particularly strong signals detected by MITRA, 4 correspond to a Pribnow-Gilbert signal or one of its variants and 3 correspond to a CRP signal. A very promising direction is to further examine the remaining 7 strong signals to determine whether or not they correspond to actual binding sites.

6 Acknowledgments

We are grateful to Alexei Dolgopolov for assistance with generating the data samples from genomic sequences. MG was partially supported by grants from INTAS (99-1476), HHMI (55000309) and RFBR (00-15-99362).

1. C.K. Stover et al. Complete genome sequence of pseudomonas aeruginosa pa01, an opportunistic pathogen. *Nature*, 406:959–964, 2000.
2. A. Vanet, L. Marsan, and M. Sagot. Promoter sequences and algorithmical methods for identifying them. *Research in Microbiology*, 150:779–799, 1999.
3. T. L. Bailey and C. Elkan. Unsupervised learning of multiple motifs in biopolymers using expectation maximization. *Machine Learning*, 21:51, 1995.
4. G. Hertz and G. Stormo. Identifying DNA and protein patterns with statistically significant alignments of multiple sequences. *Bioinformatics*, 15:563–577, 1999.
5. C. E. Lawrence, S. F. Altschul, M. S. Boguski, J. S. Liu, A. F. Neuwald, and J. C. Wootton. Detecting subtle sequence signals: a gibbs sampling strategy for multiple alignment. *Science*, 262:208–214, 1993.
6. A. Neuwald, J. Liu, and C. Lawrence. Gibbs motif sampling: Detection of bacterial outer membrane repeats. *Protein Science*, 4:1618–1632, 1995.
7. J. Buhler and M. Tompa. Finding motifs using random projections. In *Proceedings of the Fifth Annual International Conference on Computational Molecular Biology (RECOMB01)*, pages 69–76, 2001.
8. P. A. Pevzner and S. Sze. Combinatorial approaches to finding subtle signals in DNA sequences. In *Proceedings of the Eighth International Conference on Intelligent Systems for Molecular Biology*, pages 269–278, 2000.
9. D. GuhaThakurta and G. D. Stormo. Identifying target sites for cooperatively binding factors. *Bioinformatics*, 17:608–621, 2001.
10. M.S. Gelfand, E.V. Koonin, and A.A. Mironov. Prediction of transcription regulatory sites in archaea by a comparative genomic approach. *Nucleic Acids Research*, 28:695–705, 2000.
11. J. van Helden, A. F. Rios, and J. Collado-Vides. Discovering regulatory elements in non-coding sequences by analysis of spaced dyads. *Nucleic Acids Research*, 28:1808–1818, 2000.

12. X. Liu, D.L. Brutlag, and J.S. Liu. Bioprospector: Discovering conserved dna motifs in upstream regulatory regions of co-expressed genes. In *Proceedings of the 2001 Pacific Symposium on Biocomputing*, volume 6, pages 127–138, 2001.

13. M. Sagot. Spelling approximate or repeated motifs using a suffix tree. *Lecture Notes in Computer Science*, 1380:111–127, 1998.

14. G. Pavesi, G. Mauri, and G. Pesole. An algorithm for finding signals of unknown length in DNA sequences. *Bioinformatics*, 17:S207–S214, July 2001. Proceedings of the Ninth International Conference on Intelligent Systems for Molecular Biology.

15. E. Eskin and P. Pevzner. Finding composite regulatory patterns in dna sequences. *Bioinformatics*, Supplement 1:S354–63, 2002. Proceedings of the Tenth International Conference on Intelligent Systems for Molecular Biology (ISMB-2002).

16. L. Marsan and M. Sagot. Algorithms for extracting structured motifs using a suffix tree with applications to promoter and regulatory site consensus identification. *Journal of Computational Biology*, 7:345–360, 2000.

17. S. Karlin, F. Ost, and B. E. Blaisdell. Patterns in dna and amino acid sequence and their statistical significance. In M. S. Waterman, editor, *Mathematical Methods for DNA Sequences*, pages 133–158, Boca Raton, France, 1989. CRC Press.

18. A. Denise, M. Regnier, and M. Vandenbogaert. Assessing the statistical significance of overrepresented oligonucleotides. In *Proceedings of the 1st Workshop on Algorithms in BioInformatics*, Denmark, 2001.

19. Y. Barash, G. Bejerano, and N. Friedman. A simple hypergeometric approach for discovering putative transcription factor binding sites. In *Proceedings of the 1st Workshop on Algorithms in BioInformatics*, 2001.

20. T. Washio, J. Sasayama, and M. Tomita. Analysis of complete genomes suggests that many prokaryotes do not rely on hairpin formation in transcription termination. *Nucleic Acids Res.*, 26:5456–5463, 1998.

21. A. Vanet, L. Marsan, A. Labigne, and M. Sagot. Inferring regulatory elements from a whole genome. an analysis of Helicobacter pylori σ^{80} family of promoter signals. *Journal of Molecular Biology*, 297:335–353, 2000.

22. R. Overbeek, N. Larsen, G.D. Pusch, M. D'Souza, E.Jr. Selkov, N. Kyrpides, M. Fonstein, N. Maltsev, and E.T. Selkov. Wit: integrated system for high-throughput genome sequence analysis and metabolic reconstruction. *Nucleic Acids Res.*, 28:123–125, 2000.

23. M. H. DeGroot. *Optimal Statistical Decisions*. McGraw-Hill, New York, 1970.

24. B. Levin. A representation for multinomial cumulative distribution functions. *Ann. Statist.*, 9(5):1123–1126, 1981.

25. J.Weiner, R. Herrmann, and G.F. Browning. Transcription in mycoplasma pneumoniae. *Nucleic Acids Res.*, 28:4488–4496, 2000.

26. J.P. Bannantine, R.G. Barletta C.O. Thoen, and R.E. Andrews. Identification of mycobacterium paratuberculosis gene expression signals. *Microbiology.*, 143:921–928, 1997.

27. M.M. Wosten, M. Boeve, M.G. Koot, A.C. van Nuene, and B.A. van der Zeijst. Identification of campylobacter jejuni promoter sequences. *J Bacteriol.*, 180:594–599, 1998.

28. R. Segal and E.Z. Ron. Regulation and organization of the groe and dnak operons in eubacteria. *FEMS Microbiol Lett*, 138:1–10, 1996.

MOPAC: Motif Finding by Preprocessing and Agglomerative Clustering from Microarrays

R. GANESH[1], DEBORAH A. SIEGELE[2] and THOMAS R. IOERGER[1]

Department of Computer Science[1], and Department of Biology[2]
Texas A&M University, College Station, TX 77840, USA

We propose a novel strategy for discovering motifs from gene expression data. The gene expression data in our experiments comes from DNA Microarray analysis of the bacterium *E. coli* in response to recovery from nutrient starvation. We have annotated the data and identified the upregulated genes. Our interest is to find common regulatory motifs that are responsible for the upregulation of these specific genes. We assume that a common motif that a regulatory protein can bind to will be present in the upstream region of the upregulated genes and will not be present in the upstream regions of genes that showed a constant level of expression over time. Our objective is to find the common motifs that are present in at least some of the upstream sequences of upregulated genes and not present in the control set, which is the set of genes whose expression remained the same. Because it is possible that there could be several subsets of co-regulated genes under different control mechanisms among the co-expressed genes, we do not want to require motifs to be present in all upregulated sequences. Therefore, we propose a new algorithm for finding such motifs through stages of pre-processing, denoising, agglomerative clustering and consensus checking. Through this process, we have found some motifs that are good candidates for further validation.

1 Introduction

Analyzing gene expression data from DNA Microarrays is a well-studied problem. There are several ways that gene expression data can be used, from profiling of genes to inferring gene regulatory networks. Microarray experiments reveal genes that are co-expressed and this is a good starting point to find co-regulated genes among the co-expressed genes. The most common work that has been done in analyzing the microarray data is by using different clustering techniques[1, 5]. We augment these analyses by searching for motifs that are shared by the upstream sequences of the genes that are co-expressed. There could exist short sequences that are shared by these co-regulated genes, which serve as the binding sites for those proteins that initialize transcription of these genes. We are interested in finding motifs from genes that have similar expression profiles. Previous work has been done in this area using probabilistic and Bayesian approaches[13].

Many of the methods that have been proposed to solve this problem are based on local search techniques like Gibbs sampling and Expectation Maximization. Thijis *et al*[20] have classified the computational methods that are used to identify regulatory motifs into two categories, viz. string analysis methods and methods based on probabilistic sequence models. The former method is based on frequency analysis of nucleotides in the upstream sequences of co-expressed genes and the latter develops a probabilistic model as a position-specific probability matrix.

McGuire *et al* [13] have worked on identifying regulatory elements from yeast and *E. coli* using AlignACE, which uses Gibb's sampling algorithm. Hu *et al* [9] have used constructive induction to analyze potential motif combinations. Pevzner and Sze [15] have proposed a challenge problem to find a signal in a sample of sequences each containing an unknown signal of length 15 with 4 mismatches and have come up with two algorithms WINNOWER and SP-STAR by interpreting the problem as a maximum-clique graph problem.

Working with prokaryotes in general is hard, as we have to take into consideration the operon relationships in finding upstream sequences and with respect to sharing the common signal. We cannot expect all upstream sequences to share the common signal and there could be many signals, which means many regulatory proteins can control the expression of a trait. The signals aren't expected to be identical in the genes that share them, and the proteins need just a partial consensus, tolerating some noise to bind to these signals [11]. The motifs are expected to be of variable length and they are typically between 5 and 15 nucleotides [10].

Most of the approaches that have been tried so far either do not use a background model or use a probabilistic background model. A probabilistic model would rule out motifs that have a high probability of occurring elsewhere in a large genome based on product of nucleotide frequencies. We define background empirically based on upstream sequences of other genes that did not show differential regulation, i.e. the genes whose expression profiles remained the same. A good background model would help reduce spurious signals being recognized as motifs. Another approach from Thijs *et al* [21] is INCLUSive, which is an integrated tool for clustering, retrieving upstream sequences and sampling motifs using Gibbs sampling. They have reported that using an organism-dependent background model can enhance the outcome of their motif finder. Another paper from the same authors [19] suggests using a higher-order background model that would update the probabilities of finding a motif at a certain position in the sequence. They have found that overall recovery of the motifs in the presence of a higher-order model has been significantly improved, and also the program better handles noisy data. YEBIS is another tool that was developed based on hidden Markov models using a weight matrix method designed for high-speed computation [24]. There has also been work that has been carried on removing artifacts from real motifs after finding the motifs first using a greedy approach. [3] Use of clustering techniques to solve the DNA motif problem has been rarely attempted. Guralnik and Karypis [7] have tried hierarchical and k means clustering for protein sequences.

In short our approach requires the motif to be present only in a subset of the differentially regulated genes and absent in the genes that do not show a significant level of upregulation or downregulation.

2 Input Processing

2.1 Annotation of Input Data

The gene expression data has signal levels measured for each spot in the Microarray chip. For the nutrient starvation response, signals were recorded from the microarrays for RNA samples of *E. coli* culture in exponential growth phase, starved, 5 minutes and 15 minutes after recovery from starvation[12,18]. We assigned 1 or 0 or −1 as a regulation index for every gene based on whether they show a significant level of upregulation, no change, or downregulation, relative to growth phase (starved, recovery after 5 min, recovery after 15 min). We used 0.03 as a threshold to determine whether the signal is significant compared to the background, based on the signal levels of the genes that were known to be deleted from the genome. We used a threshold of two-fold increase or decrease in signal intensity compared to the exponential phase culture in order to determine whether it is an upregulation or downregulation; most internal variations of the signals of a same gene were less than this threshold. We classified six patterns as upregulation viz. (0,1,1), (0,1,0), (0,0,1), (-1,1,1), (-1,1,0), (-1,0,1) and another six patterns as downregulation viz. (0,-1,-1), (0,-1,0), (0,0,-1), (1,-1,-1), (1,-1,0), (1,0,-1) based on the relationships among the expression levels at the three time points, wherein the three numbers denote the regulation indices at time points 0, 5, and 15 minutes after recovery from starvation. For making the control (non-regulated) set, we grouped the genes that showed a constant level of expression in the three different time intervals viz. (0,0,0), (1,1,1) and (-1,-1,-1).

2.2 Extraction of Upstream Sequences

The next challenge in annotating the data was to extract the upstream nucleotide regions for these genes, since that is where common putative regulatory elements are likely to occur. In order to retrieve the upstream sequences, we needed the coordinates of the genes in the whole genome of *E. coli*. Like most prokaryotes, *E. coli* has operons, wherein more than one gene shares a common upstream sequence, which is located before the start of the first gene in the operon. In order to correctly extract the upstream sequence, we need to know its position in the operon and the upstream sequence of the first gene in the operon. We used the Linkage Map of *E.coli*[2] to determine the operon relationships. We used the NCBI ftp site[25] for information about the orientation and starting point of the ORFs of interest. We used the complete *E.coli* genome from the GOLD[26] database. We used the start of the predicted protein coding sequence as the boundary for extracting upstream sequence. We extracted 600 bp ahead of the start of the gene for upstream region[22]. If the gene were part of an operon, we used 300 bps before the start of the gene and

300 bps before the start of the first gene in operon to suit the total size. We also took the orientation of the gene into consideration in extracting the upstream sequences and we took the reverse complement of the sequences that were known to be transcribed in the counterclockwise direction.

3. Algorithm

An exhaustive approach of screening all potential patterns up to a certain length k by depth first search, especially with wildcards, is computationally intractable. Therefore, we decomposed the problem to a smaller problem of finding motifs of fixed length that are frequent in the upregulated set and not present in the non-regulated set. We can describe the functioning of our algorithm in the following steps.

3.1 Preprocessing

First, we preprocess the experimental (upregulated) set and control (non-regulated) set by extracting all possible sub-sequences which could form motifs. We list all the over-lapping fragments in the experimental set, which are windows of length k nucleotides. If the new motif that we are adding is already in the list and is from a different gene, we increment the count of the motif. If it is a part of an operon, we make an additional check to ensure that we don't increment the count for the same motif from two genes belonging to the same operon. Similarly, we pre-process the control set and, for every motif in control set, we check whether the list of putative motifs contains it and remove them from the list, since we want the motif not to show up the in the control set.

a) Phase I

List = ϕ

For an upstream sequence Z of every gene G_z in the experimental set

      For every subsequence $S_i \in Z$ of length L

          If $(S_i \notin$ List) List = List $\cup$ S_i

          Else Let S_j be the other element in the List that matches S_i

            if $(G_z \in$ operon$)$ and $($ operon$(S_i) \neq$ operon $(S_j))$

               Increment count for S_j

            Else if $($not $(S_i \in G_z$ and $S_j \in G_z))$

               Increment count for S_j

b) Phase II

For an upstream sequence Z of every gene G_z in the control set
    For every subsequence $S_i \in$ Z of length L
        If $(S_i \in$ List) List = List $- S_i$

3.2 De-noising

After pre-processing, there might be a lot of nucleotide patterns that we need to find consensus for. We can easily identify patterns that do not have much similarity to the other patterns in the whole set. It is important to remove these outliers, as they can interfere with the accuracy of the clustering method, and also they add to the complexity of the clustering algorithm. Thus, for every pattern, we compute the ratio of sum of distances of that pattern to other patterns that do not belong to the same gene or operon to the number of such occurrences. The distance between two patterns is scored by the number of mismatches from lexicographic comparison. It gives a negative score for the nucleotides that match and a positive score for every mismatch. The total score is computed by adding the scores of the individual nucleotides. Then we use a threshold, which is based on the distribution of the scores, to exclude the patterns that are far away from the other patterns. The distance is computed by a scoring function that compares two strings lexicographically.

$$\text{Score}_p = \frac{\Sigma \text{ distances of S to } S_i \text{ where S and } S_i \text{ do not share the same gene or operon}}{\text{number of such occurences}}$$

3.3 Distance Graph

Next, we compute the similarity matrix of all the patterns we have identified. Our goal is to group the similar ones together and derive a wildcard motif representation for each cluster. We get a dense graph wherein each edge has a weight associated with it, which represents the distance between the two patterns. The distance is computed based on matches and mismatches between any two patterns. The weight associated with the edge between two patterns would be less if the patterns are more similar. This is an undirected graph, since the distance between each pair is the same in either direction.

3.4 Agglomerative Clustering

Now, we want to group the most similar patterns, which we can do by applying a clustering algorithm.[8] We start by coloring the edges in the graph from the lowest cost to highest cost. Initially, we represent every point in the space as a set. For every edge in our graph, if the sets that each vertex belongs to are different, we make a union of them. If we iterate this we will get a Minimum Spanning Tree[4] (equivalent to single-linkage clustering [6]). However, if we set a threshold called Critical Cost over which we stop coloring edges, we will get a disjointed forest. Critical Cost is the value of the score that represents the similarity between two patterns at which we don't consider the two patterns to be similar enough to go in a single cluster.

    Compute distance for each edge (pair of patterns)
    Sort the edges of the graph in non-decreasing order
    For every vertex v
        Do MakeSet(v)
    For every edge m = [u_i, v_i]
        If distance(u_i, v_i) < θ (critical cost)
            R_1 = set-of(u_i);
            R_2 = set-of(v_i);
            If (R_1 != R_2) then
                Union(u_i, v_i);

3.5 Consensus Checking

The last step is to compute a consensus sequence for each cluster. We represent each pattern as a bit string. The valid nucleotides that can be represented in wild cards would be 2^4 and we keep turning on the bits as we encounter nucleotides that are observed to differ in a position. This approach is similar to the Find-S algorithm[14] for symbolic concept generalization.

    For every pattern in a cluster
        Represent Pattern as bit string
        Make bitwise disjunction (OR) of its members

To construct the consensus, we relax the pattern. Though the individual members of the pattern are guaranteed to be absent in the control set, other instances matching the consensus pattern might show up sometimes. We discard the clusters that show hits in the control sequences with their consensus. We drop the most remote member within a cluster if the average distance to other points is about twice more than the next closest distance not involving that member.

4. Results

After annotating the input data (microarray signals) using the above-mentioned criteria, we identified 22 genes that were upregulated for recovery from nutrient starvation. We also arrived at a control set by pulling out the genes that did not show an appreciable upregulation or downregulation for this environmental condition. We retrieved the upstream sequences for the upregulated and downregulated genes using the operon relationships. Also, we retrieved the upstream sequences for the entire control set of 1361 genes. We did not use operon information to find upstream sequences from control set because the control set is just used to remove the spurious signals.

Given this well-defined experimental set and control set, we then ran our pre-processing program, which produced many patterns. We then ran the De-noising algorithm with a threshold of 5.87 to remove the outliers and keep the ones that are closely related. After de-noising, we ran the clustering algorithm by creating a similarity graph using a matrix representation and doing agglomerative clustering as explained above. We arrived at an optimum threshold of -6 to stop the clustering based on our expectation of similarity within a cluster (any two patterns can be atmost 25% dissimilar), so that we get a reasonable number of good quality clusters (at least 3 members that belong to different genes/operons). We constructed a consensus string for every cluster. We examined the results and identified the patterns that did not show any hits in the control set. We excluded some motifs whose hits in the control sets were too far from transcription start (more than 450 bp upstream) or downstream of transcription start. Table 1 lists the motifs found and the genes that share the motif, and Figure 2 displays the sequence logo[17] for these motifs.

Table 1: Results of MOPAC – Motifs discovered and their genes
(represented by IUB nucleotide symbols)

ID	Motif	Gene name
1	AAsAAwTTmAwA	CmtB, ygjR, cysD
2	CmwTTkTTyTTC	CysH, B3914, MetR
3	TTCTwHTgAwAT	B1587, MetF, FliY
4	wTVAACwThCAA	B1587, asnB, cysA,P,W
5	rAkTTTwTTCAT	B3914, MetR, MetF
6	CAArTwTTTwTr	CmtB, yhaV, cysD
7	ATwAATAATksw	B1587, yhaV, CmtB
8	ACsdTTTTTmTw	CmtB, asnB, b3914, ygjR
9	rAAwTTmATAAT	MetF, CmtB, ygjR
10	vwTTAATAATkC	CmtB, b1587, yhaV, MetF
11	ATwTTGAATTww	AsnB, metR, metF
12	yTTTkhGATATT	YfiA, cysD, fliY
13	AkTTTwTTCATy	B3914, metR, metF

5. Validation and Discussion

We believe that biological experiments such as mutating the motifs and looking for changes in expression would be the best way to validate our results. However, in this paper, we use some heuristics to evaluate the likelihood of these patterns to be a motif.

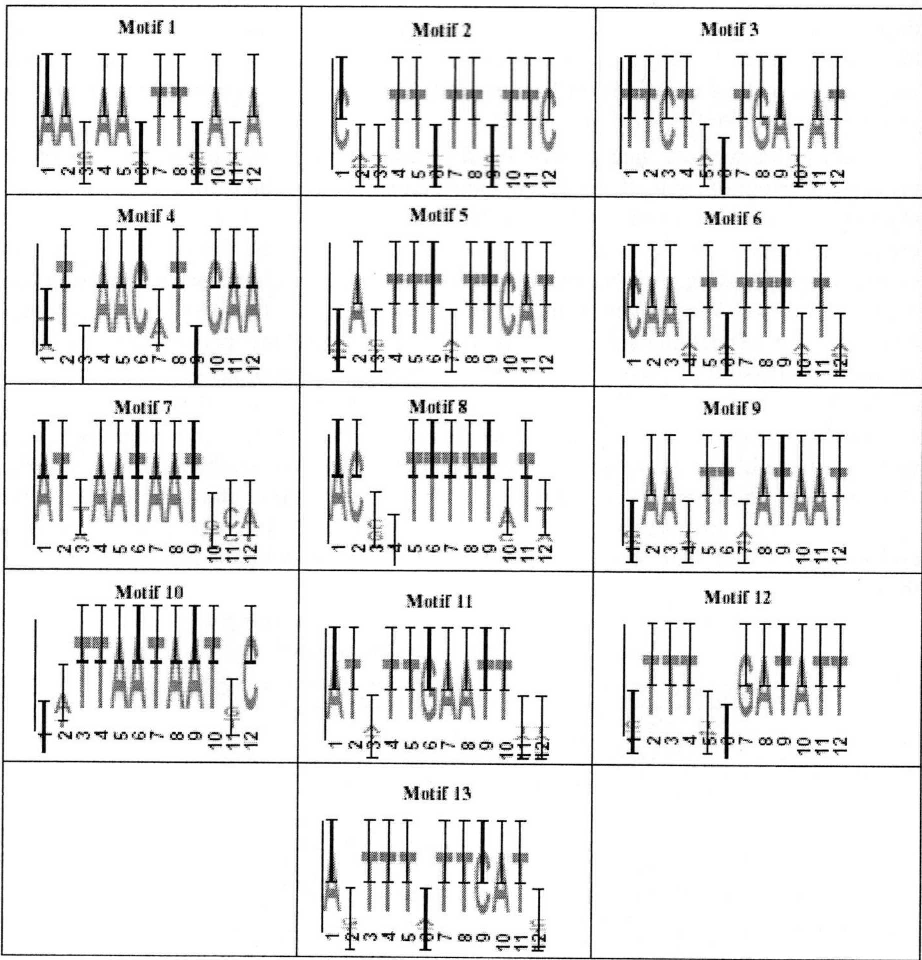

Figure 1: Sequence Logo[17] of motifs discovered using MOPAC

Also, we ran AlignACE[13] and MotifSampler[19] with our input data (upstream sequences of upregulated genes from starvation recovery response). AlignACE tried to find motifs that were shared by almost all the genes. It was difficult to compare the results of AlignACE because it didn't output a consensus, and when we tried to construct one, it seemed too general (with many wildcards, i.e. low specificity).

MotifSampler finds as many motifs as we want with a fixed length and gives the consensus too. The motifs identified by MOPAC were unique.

5.1 Distance from Transcription Start

We found most of the motifs to be close to and upstream of transcription start, which strengthens our belief that the candidate motifs are biologically significant. Since activators are often associated with upregulation, we hypothesize that they tend to bind upstream say between −30 to −300 and the strongest signals could be found within 100 base pairs from the transcription start. We used RegulonDB[16] to determine the transcription start for each gene (most of them were predictions). Figure 2 shows the distribution of position of motifs found using MOPAC.

We calculated the distances to the transcription start site based on the occurrence of the best motif found by AlignACE and MotifSampler. In MOPAC, motifs were typically found between 0-300 bp upstream. The motifs found by MotifSampler were mostly between −200 and −400 and in the case of AlignACE more than 60% of the upstream sequences that share a motif had their motifs concentrated around −300 to −500 bp upstream.

5.2 Palindromicity

Regulatory motifs are often observed to be palindromes[11]. So, we reverse complemented each motif and compared the resultant pattern with our motif and recorded their degree of palindromicity, which is defined as the ratio of the characters that match when the pattern is compared with itself after it is reverse complemented. Table 2 shows that the motifs found using MOPAC have average to high palindromicity.

5.3 Probability

Since the motifs that we have discovered show up in 3 or more genes out of 22 upregulated genes and do not show up at all in the 1361 control genes, we believe that these are significant signals. To quantify this, the probability P of finding a pattern of length n in a sequence of length L, allowing y positions in n to have x number of wildcards, is approximately $P = 1 - [1- \{1/4\}^{n-y}. \{x/4\}^{y}\}]^{L}$. For example, the probability of finding a pattern of length 12 with 4 wildcard positions in regions of size 600 bases with each of them having 2 possible wildcards is 0.000057. The probability of finding the motif in 3 sequences out of 22 sequences is $Q(3,22) = P^{3}.(1-P)^{19}.C^{22}_{3}$ which evaluates to approximately 10^{-10}. We can also see that the probability of finding it in 3 out of 22 sequences would be further reduced. Hence the chance of these motifs to be random occurrences is very minimal.

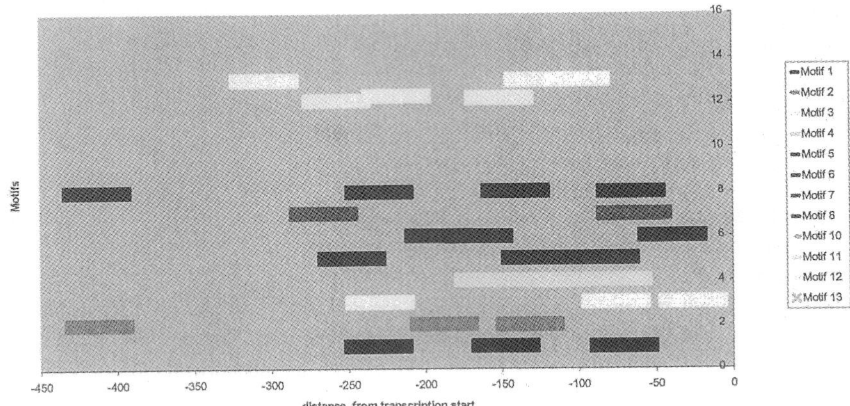

Figure 2: Relative locations of motifs upstream of transcription start [MOPAC]

Table 2: Palindromicity of motifs

MOPAC		MotifSampler	
Motif	Palindromicty	Motif	Palindromicity
1	0.5	1	0.5
2	0.33	2	0.33
3	0.5	3	0.33
4	0.83	4	0.5
5	0.5	5	0.66
6	0.75	6	0.33
7	0.66	7	0.33
8	0.5	8	0.58
9	0.66		
10	0.58		
11	0.66		
12	0.42		
13	0.5		

5.4 Relationship with Known Transcription Factors

Using the prokaryotic part of the TRANSFAC database[23], we searched for sites with 2 or 3 mismatches from our patterns. We got several hits from the database. For example, the motif "CmwTTkTTyTTC" from MOPAC, which is shared by metF, picked up the consensus of the MetJ-MetF site. The other known sites that matched some of the motifs are OmpR-ompC-bc, BlaI-P(p)1, BlaI-P(p)2, ArgRcarAB arg-box-1, AlgR1-algD, IHF-L1, IHF-L2, cI-Op72a, deoR-deoO(E), CRP-galO, XylR-UBS2, Nod box and CAP/CRP-lac. Many of the motifs predicted by MotifSampler had close homology to the binding site of lambda repressor.

5.5 Hits in the Control Set

The motifs discovered by MOPAC had zero hits in the control set in most cases. In some cases the motifs were still considered in spite of control hits, when the location of the motifs is far upstream or downstream of the transcription start, making them unlikely to be involved in regulation. When we tested for the occurrences of the consensus sequences of MotifSampler, they were found to occur in the control set occasionally. [Table 3] MOPAC, by design does not have any hits in the control set. If we were to make a consensus for AlignACE, it would pick up several hits because of its low specificity. MOPAC and MotifSampler found two different motifs for the genes asnB, metR and metF. The motif found by MOPAC lies between −150 and −250 in all three genes but the one that is found by MotifSampler lies between −300 and −450.

Table 3: Occurrences of motifs in upstream sequences using MotifSampler

Motif	Consensus	Expt hits	Control hits
1	mCGCwkCCGGCr	11	15
2	CGCCrGCGGwrA	10	4
3	GwCGTsnyTGAn	10	8
4	GCnTCTGsTnGs	7	7
5	wsCCGCkGyrCT	6	2
6	CCrCGCmGGAAr	5	2
7	TGTAGGCCGGAT	4	8
8	CGATATCnACCG	3	0

6. Conclusion

We have presented an algorithm, MOPAC that couples analysis of gene expression data with motif prediction in upstream sequences. Our approach is based on the assumption that motifs would be over represented in the genes that are upregulated for an environmental condition and not present in the genes that do not show any significant change in their expression for the same environmental condition. We also do not require the motif to be present in all the co-regulated genes, as some algorithms do. We feel that use of a real (empirically-defined) background sequence yields more biologically meaningful results, by the way it filters out spurious motifs unrelated to the environmental condition. Though our validations were purely computational, the motifs we discovered appear promising. Verifying these signals biologically is beyond the scope of this research. Our algorithm is not as fast as

AlignACE or MotifSampler as it uses an extensive background search and hierarchical clustering, which is slow and memory intensive. This method works only when we have results of a microarray experiment wherein we can clearly identify the genes that are affected and the genes that are not affected for a specific environmental condition. Several improvements can be made in reducing the complexity of the algorithm. Also, it might be better to associate a weight to every nucleotide in the wildcard instead of attaching equal importance when relaxing a pattern.

References

1. A. Ben-Dor and Z. Yakhini, in *RECOMB '99* (1999).
2. M. K. Berlyn, *Microbiology and Molecular Biology Rev*, **62**, 814 (1998).
3. M. Blanchette and S. Sinha in *ISMB 2001* (2001).
4. T.H. Cormen *et al.*, *Introduction to Algorithms* (MIT Press, Cambridge, 1990).
5. M.B.Eisen *et al.*, Proc. Natl. Acad. Sci. USA, **95**, 14863 (1998).
6. B.S. Everitt, *Cluster Analysis* (Heineman Educational Books Ltd., London, 1980).
7. V. Guralnik and G.Karypis in *KDD-2001* (2001).
8. J. Han and M. Kamber, *Datamining: Concepts and Techniques* (Morgan Kaufmann Publishers, San Francisco, 2001).
9. Y. J. Hu *et al.*, *Bioinformatics*, **16(3)**, 222 (2000).
10. S. Keles *et al.*, Bioinformatics **18**, 1167 (2002).
11. B. Lewin, *Genes VII* (Oxford University Press Inc., New York, 2000).
12. J. McEthanon and D.A. Siegele, *Pers. Comm.*
13. A.M. McGuire *et al.*, *Genome Research* **10(6)** 744 (2000).
14. T. M. Mitchell, *Machine Learning* (McGraw-Hill, 1997).
15. P.A. Pevzner and S. –H. Sze in *ISMB'2000* (2000).
16. H. Salgado *et al.*, *Nucleic Acids Res.* **29(1)**,72 (2001).
17. T.D. Schneider and R.M. Stephens, *Nucleic Acids Res.* **18**, 6097 (1990).
18. D.A. Siegele and L.J.Guynn, J. Bacteriology, **178(21)**, 6352 (1996).
19. G. Thijis *et al.*, *Bioinformatics* **17(12)**, 1113 (2001).
20. G. Thijis *et al.*, in *RECOMB'2001* (2001).
21. G. Thijis *et al.*, *Bioinformatics* **18(2)**, 331 (2002).
22. J. van Helden *et al.*, *J. Mol. Biol.* 281, 827 (1998).
23. E. Wingender *et al.*, *Nucleic Acids Res.* **24**, 238 (1996). http://transfac.gbf.de/TRANSFAC/.
24. T. Yada *et al.*, *Bioinformatics* **14(4)**, 317 (1998).
25. ftp://ftp.ncbi.nih.gov/genomes/Bacteria/Escherichia_coli_K12/
26. http://wit.integratedgenomics.com/GOLD/completegenomes.html

IMPROVED GENE SELECTION FOR CLASSIFICATION OF MICROARRAYS

J. JAEGER[*], R. SENGUPTA[*†], W.L. RUZZO[*†]

[*]Department of Computer Science & Engineering
University of Washington
114 Sieg Hall, Box 352350
Seattle, WA 98195, USA

[†]Department of Genome Sciences
University of Washington
1705 NE Pacific St, Box 357730
Seattle, WA 98195, USA

In this paper we derive a method for evaluating and improving techniques for selecting informative genes from microarray data. Genes of interest are typically selected by ranking genes according to a test-statistic and then choosing the top k genes. A problem with this approach is that many of these genes are highly correlated. For classification purposes it would be ideal to have distinct but still highly informative genes. We propose three different pre-filter methods — two based on clustering and one based on correlation — to retrieve groups of similar genes. For these groups we apply a test-statistic to finally select genes of interest. We show that this filtered set of genes can be used to significantly improve existing classifiers.

1   Introduction

Even though the human genome sequencing project is almost finished the analysis has just begun. Besides sequence information, microarrays are constantly delivering large amounts of data about the inner life of a cell. The new challenge is now to evaluate these gigantic data streams and extract useful information.

Many genes are strongly regulated and only transcribed at certain times, in certain environmental conditions, and in certain cell types. Microarrays simultaneously measure the mRNA expression level of thousands of genes in a cell mixture. By comparing the expression profiles of different tissue types we might find the genes that best explain a perturbation or might even help clarify how cancer is developing.

Given a series of microarray experiments for a specific tissue under different conditions we want to find the genes most likely differentially expressed under these conditions. In other words, we want to find the genes that best explain the effects of these conditions. This task is also called feature selection, a commonly addressed problem in machine learning, where one has class-labeled data and wants to figure out which features best discriminate among the classes. If the genes are the features describing the cell, the problem is to select the features that have the biggest impact on describing the results and to drop the features with little or no effect. These features can then be used to classify unknown data. Noisy or irrelevant attributes make the classification task more complicated, as they can contain random correlation. Therefore we want to filter out these features.

Typically, informative genes are selected according to a test statistic or p-value rank according to a statistical test such as the t-test. The problem here is that we might end up with many highly correlated genes. Besides being an additional computational burden, it also can skew the results and lead to misclassifications. Additionally, if there is a limit on the number of genes to choose we might not be able to include all informative genes. Our approach is to first find similar genes, group them and then select informative genes from these groups to avoid redundancy.

Besides t-like-statistics, there are many different techniques applicable. There are non-parametric tests like TNoM[1] (which calculates a minimal error decision boundary and counts the number of misclassifications done with this boundary) or Wilcoxon rank sum/Mann-Whitney[2] (which test statistic is identical[3] to the Park[4] score). It creates a minimal decision boundary too, but incorporates the distance from the boundary into the score. T-like statistics such as Fisher[5] and Golub[6] put different weights in the variance and number of samples. The Mutual Information score results from entropy and information theory, and the B-score[7] comes from Bayesian decision theory. Vapnik[8] describes an interesting method to optimize feature selection while generating support vector boundaries for SVMs.

In this paper we will compare classification done with five different test statistics: Fisher,[5] Golub,[6] Wilcoxon,[2] TNoM,[1] and t-test[2] on three different publicly available datasets, Golub,[6] Notterman[16] and Alon[9]. We will propose two algorithms based on clustering and one based on correlated groups to find similar genes. We then show that these prefiltering methods yield consistently better classification performance than standard methods using similar numbers of genes.

The rest of the paper is organized as follows: Section 2 will review important methods needed for our proposed approach. Section 2.1 will introduce Microarrays. Section 2.2 describes a method for evaluating different feature selection sets using support vector machines and a technique called leave-one-out cross-validation. In section 2.3 we will review clustering techniques used in this paper. In section 2.4 we will discuss how reducing redundancy in a dataset can help with the final classification process and elucidate why redundancy can cause problems for classification tasks. In section 2.5 we propose new methods to select genes from clusters using correlation, clustering and statistical information about the genes. In section 3 we will present results using our proposed approach on three publicly available data sets. Section 4 contains conclusions and future research directions.

2   Methods and Approach

2.1   Microarrays

A typical microarray holds spots representing several thousand to several tens of thousands of genes or ESTs (expressed sequence tags). After hybridization the microarray will be scanned and converted into numerical data. Finally the data should be normalized. The purpose of this step is to counter systematic variation (e.g. difference in labeling efficiency for different dyes, compensation for signal spill over from neighboring spots[10]) and to allow a comparison between different microarrays[11]. The data we work with is already background-corrected and normalized, and we do not address these problems in this paper.

2.2   Validation, Classification

As we are proposing new gene selection schemes we want to measure their performance and allow a comparison. All schemes provide us with a set of informative genes that will be used for future classification. Assume we have n samples. Leave-one-out cross-validation (LOOCV) is a technique where the classifier is successively learned on n-1 samples and tested on the remaining one. This is repeated n times so that every sample was left out once. To build a classifier for the n-1 samples, we extract the most revealing genes for these samples, and use a machine learner. With this classifier we try to classify the remaining (left out) sample. Repeating this procedure n times gives us n classifiers in the end. Our error score is the number of mispredictions. We use support vector machines (SVMs[12]) as the classification method as these are very robust with sparse and noisy data.

2.3   Support Vector Machines

Supports Vector machines (SVMs) expect a training data set with positive and negative examples as input (i.e., a binary labeled training data set). Then they create a decision boundary (the maximal-margin separating hyperplane) between the two groups and select the most relevant examples involved in the decision process (the so-called support vectors). The construction of the hyperplane is always possible as long as the data is linearly separable. If this is not the case, SVMs can use 'kernels', which provide a nonlinear mapping into a higher dimensional feature space. If a separating hyperplane in this feature space is found, it can correspond to a nonlinear decision boundary in the input space. If there is noise or inconsistent data a perfectly separating hyperplane may not exist. Soft-margin SVMs[12] attempt to separate the training set with a minimal number of errors. In this paper we used Bill Noble's SVM implementation 1.3 beta now called gist[13].

2.4 Clustering

Cluster analysis is a technique for automatically grouping and finding structures in a dataset. Clustering methods partition the dataset into clusters, where similar data are assigned to the same cluster whereas dissimilar data should belong to different clusters. Fuzzy clustering[14] deals with the problem that there is often no sharp boundary between clusters in real applications. Instead of assigning an element to one specific cluster there is a membership probability for each cluster. In doing so an element can be a member of several clusters. Fuzzy clustering can be seen as a generalization of k-means[15] clustering. We used the FCMeans Clustering MATLAB Toolbox V2-0.

2.5 Reducing Redundancy

Now that we have reviewed all the methods we need, we will return to the problem of feature selection. Table 1 shows a list of 7 genes from Notterman's Adenoma[16] dataset sorted by increasing p-value. For gene M18000 the expression value is generally higher in Adenoma than in Normals with the exception of Adenoma 1 and Normal 2. Looking at X62691 the same is true. Both genes have a very low p-value and would be pulled out by conventional methods, which focus on genes with the lowest p-values. Biologists are often interested in a small set of genes (for financial, personal workload or experimental reasons) that describes the perturbation as well as possible. Therefore we are limited in the number of genes to extract and e.g. assuming we could only extract 2 genes we would pull out the first two genes as they have the lowest p-value. However we would not get much additional information using the second gene as it shows the same overall pattern. It would be better to include a gene that provides us with extra information.

Table 1: Expression values for 7 selected genes of Adenoma and normal tissues, sorted by p-value

Accession Number	Adenoma				Normal				t-test p-value
	1	2	3	4	1	2	3	4	
M18000	705.41	1227.27	959.35	951.56	359.83	711.08	485.33	431.19	0.014
X62691	387.91	577.57	578.45	546.54	227.26	436.65	306.94	239.33	0.016
M82962	91.85	16.27	12.61	61.62	187.44	76.90	181.38	186.53	0.017
U37426	0.47	7.05	6.30	3.40	-3.88	1.58	-2.99	-2.91	0.018
HG2564	2.33	0.54	1.58	3.82	-2.91	-2.11	1.00	-2.91	0.019
Z50853	35.43	26.03	51.49	41.22	27.68	15.80	12.46	15.99	0.022
M32373	-48.02	-28.20	-64.62	-56.95	-15.05	-16.86	-7.97	-34.88	0.022

Table 2: Correlation between Adenoma genes from table 1

	M18000	X62691	M82962	U37426	HG2564	Z50853	M32373
M18000	1.000						
X62691	0.961	1.000					
M82962	-0.944	-0.971	1.000				
U37426	0.973	0.975	-0.983	1.000			
HG2564	0.592	0.653	-0.553	0.529	1.000		
Z50853	0.514	0.616	-0.633	0.597	0.614	1.000	
M32373	-0.509	-0.590	0.602	-0.580	-0.619	-0.874	1.000

Looking at the correlation values in table 2 we can see that the first four genes have an absolute correlation greater than 0.94. Not surprisingly highly correlated genes show the same misclassification pattern and in fact we find that the first four genes also have the same pattern of consistent outliers in Adenoma 1 and Normal 2. In order to increase the classification performance we propose to use more uncorrelated genes instead of just the top genes. We expect the phenomenon illustrated by this example to be a general one. By just using the k best ranking genes according to a test-statistic we would select highly correlated genes. Correlation can be a hint that the two genes belong to the same pathway, are coexpressed or are coming from the same chromosome. In general we expect high correlation to have a meaningful biological explanation. If, e.g., genes A and B are in the same pathway it could be that they have similar regulation and therefore similar expression profiles. If gene A has a good test score it is highly likely that gene B will, as well. Hence a typical feature selection scheme is likely to include both genes in a classifier, yet the pair of genes provides little additional information than either gene alone. Of course we could just select more genes in order to capture all relevant genes. But not only would more genes involve higher computational complexity for classification but it also can skew the result if we have a lot more genes from one pathway. Furthermore if there are several pathways involved in the perturbation but one pathway has the main influence, we will probably select all genes from this pathway. If we then have a limit for the number of genes we might end up with genes only from this pathway. If many genes are highly correlated we could describe this pathway with fewer genes and reach the same precision. Additionally, we could replace correlated genes from this pathway by genes from other pathways and possibly increase the prediction accuracy. The same issue might be true when selecting a lot of genes as well, but it is more compelling when we have a limited budget of genes and can only select a few genes.

Our method for gene selection will therefore be to prefilter the gene set and drop genes that are very similar. For the remaining genes we will apply a common test statistic and pull out the highest-ranking genes. One way to find correlated genes would be to calculate the correlation between all genes. In our first method we selected from the best genes (best according to a test statistic) those that have a pairwise correlation below a certain threshold. A simple greedy algorithm accomplishes this selection – the k-th gene selected is the gene with highest p-value among all genes whose correlation to each of the first k-1 is below the specified threshold. This method is called "Correlation" in the figures below. Greedy algorithms are of course simple, but often give results of poor overall quality due to their myopic decision-making. As an alternative allowing a more global view of the data, we also consider clustering algorithms. Clustering is very versatile as it can use different distance functions (Euclidean, L_k, Mahalanobis, and correlation), and different underlying models, shapes and densities, which are not captured by just correlation. In this paper we compared clustering and correlation methods. We used a fuzzy clustering algorithm because it assigns a membership probability for a cluster for each gene and may therefore capture the fact that some genes are involved in several pathways. Although a cluster does not automatically correspond to a pathway it is a reasonable approximation that genes in the same cluster have something to do with each other or are directly or indirectly involved in the same pathway. Our basic approach is to cluster the genes, and then to select one or more representative genes from each cluster. The details how many genes from which cluster depend on the "quality" of each cluster and will be discussed below.

2.6 Assigning Quality to Cluster

Once we have done the clustering we know that genes in a cluster show similar expression profiles and might be involved in the same pathway. Since we want to have as many pathways as possible involved in our list of significant genes, we would like to sample from each cluster/pathway. But it would not be fair to treat each cluster and gene equally. The size of the clusters as well as the quality of a cluster play a role, i.e. how close together are the genes, how far away are they from the cluster center. If a cluster is very tight and dense it can be assumed that the members are very similar. On the other hand if a cluster has wide dispersion the members of the cluster are more heterogeneous. To capture the biggest possible variety of genes, it would therefore be favorable to take more genes from a cluster of bad quality than from a cluster with good quality. To determine the quality for the fuzzy clustering algorithm we used the membership probability for a gene. We said that an element belongs to the cluster to which it has the highest membership probability. The cluster quality is then assessed by looking at the average membership probability of its elements.

A high cluster quality means low dispersion, and the closer the quality gets to 0 the more scattered the cluster becomes. In our first clustering algorithm we decided that no matter how bad the quality and how small the size of the cluster we should get at least one element from each cluster. Our reasoning is as follows. First, if the cluster size is very small but there is a very good gene in it, we do not want to miss that cluster. Second, by eliminating a cluster we lose all the information of that pathway, so getting at least one representative plays a role like pseudocounts. Third, if we have a cluster that is extremely correlated, we would have a very high quality score and therefore may not pick any gene from that cluster. But this one gene from that cluster might have had a very good contribution to the discrimination process.

The drawback is that a cluster might represent a pathway that is totally unrelated to the discrimination we look for. If the cluster then has a bad quality we might pick a lot of genes from that cluster even though they are not informative. To counteract this problem we implemented the possibility to mask out and exclude clusters that have an average bad test statistic p-value (this method is called "Masked out Clustering" in the figures, whereas "Clustering" refers to the method where we look at all clusters and do not mask out any). Lastly we want to have genes that have a high discriminatory power, i.e. can explain the symptoms. This can be achieved by using an appropriate test statistic.

3   Results and Discussion

For our experiments we selected three different publicly available microarray datasets, Alon[9](40 Adenocarcinoma and 22 normal samples), Golub[6](47 ALL and 25 AML leukemia samples) and Notterman[16](18 tumor and 18 normal samples). We compared five different test statistics: Fisher[5], Golub[6], Park[4], TNoM[1], and t-test and ran our three different filtering algorithms described above: Correlation, Clustering, Masked out Clustering. The performance of the feature selection was calculated using SVM and LOOCV scores.

For each possible combination of test statistic and clustering algorithm we evaluated the performance varying the number of clusters between 1 and 30 and the number of selected features between 2 and 100. We used the Euclidean distance metric and a fuzzy clustering softness of 1.2 (where 1 would be hard clustering and infinity is everything belonging to all clusters). For SVM we chose an RBF kernel function and used data normalization in the feature space.

We also calculated the LOOCV performance using all available data (i.e. not reducing the number of genes but using all of them). The result was that LOOCV produced 6 false classifications in the Alon's colon dataset, resulting in an error of 9.7%. In Golub's leukemia dataset LOOCV produced 2 false classifications, resulting in an error of 2.8% and in Notterman's carcinoma dataset LOOCV made 1 false classification, resulting in an error of 2.8%.

Figure 1 shows a 3d plot of the LOOCV performance varying the number of clusters selected between 1 and 10 and the number of features chosen between 10 and 100 in steps of 10. The plot shows the clustering algorithm without masking out.

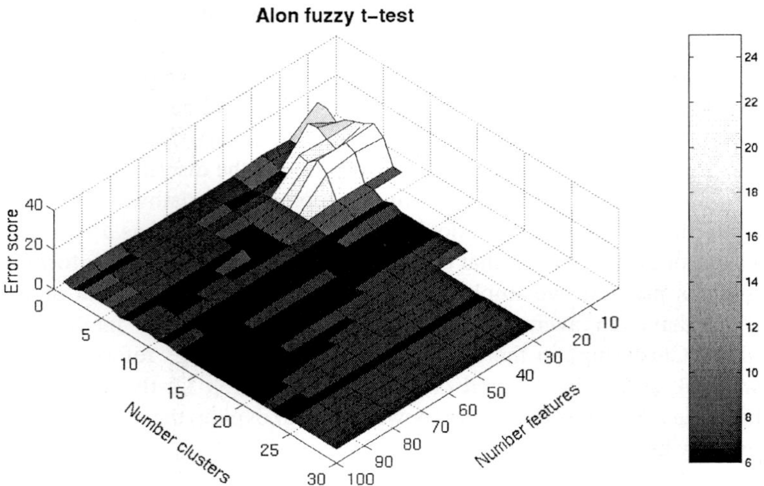

Figure 1: LOOCV performance for Alon's data set using clustering and conventional methods

There are no values for 10 features and more than 10 clusters, as well as 20 features and more than 20 clusters, as one of our constraints is that each cluster has at least one member in the feature set. So we can never have more clusters than features selected.

In the leftmost ribbon (starting in the lower left corner and going up to the top left corner) we can see the performance varying the number of features and using just one cluster (i.e. this is our standard comparison line, as this reflects just selecting features by test-statistic). The whole plot seems very flat once we have more than 30 features. But notice the darker spots in the middle (between 10 and 20 clusters), that reflect very low LOOCV scores. One reason that there is a high peak for less than 30 features is that the t-test selects highly correlated and therefore redundant genes, which makes it hard for the underlying SVM to learn a good classifier. The average correlation of the top 10 genes selected with the t-test is 0.85.

Figure 2 compares different test statistics on a given data set using the first clustering algorithm. Notice that Fisher and Golub behave very similarly as do Park and TNoM, but t-test has (besides the big bulk at only a few features) a very flat and robust behavior. Fisher and Golub seem to have a higher variance in classification but their best classification performance is similar to t-test. They achieve their best results with 6-25 clusters. TNoM and Park achieve their best results for fewer clusters (in the range of 1-6) and in fact they seem not to benefit from the clustering as much as t-test, Fisher or Golub do.

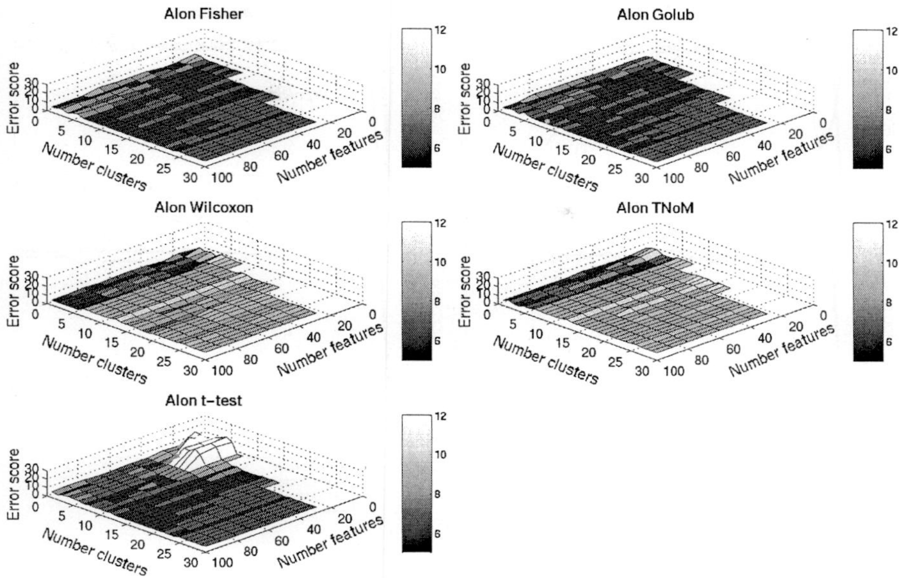

Figure 2: Comparison of different test statistics

For Notterman's carcinoma dataset the standard classification process already achieves 0% error when using more than 10 features but we can still improve the classification when using 10 features or less. Using only that few features we still manage to have a 0% error with most of the test statistics. Due to space restrictions figures are not shown here but can be accessed online[17].

Now consider how the LOOCV performance of our clustered result compares to the conventional methods (depicted as normal). In figure 3 we plot the normal score, the clustered scores (the minimum error score over all trials with cluster size from 2 to 30), the clustered scores with masking out and the correlation scores for the LOOCV performance for each of the five test-statistics. Here we did not plot the number of errors (plots for this are available online[17]), that refiect the efficiency of the classification but a ROC[18] (receiver operator curves) score (i.e., the area under the ROC graph, which takes both false negative and false positive errors into account and reflects the robustness of the classification). We can see that almost always the filtered performance is better than the conventional method. Another noticeable fact is that without clustering, TNoM would have on average the best LOOCV performance of the five scores for Alon's colon dataset. An explanation might be that TNoM, as a nonparametric test, extracts less correlated genes and therefore already does a good job in selecting different genes.

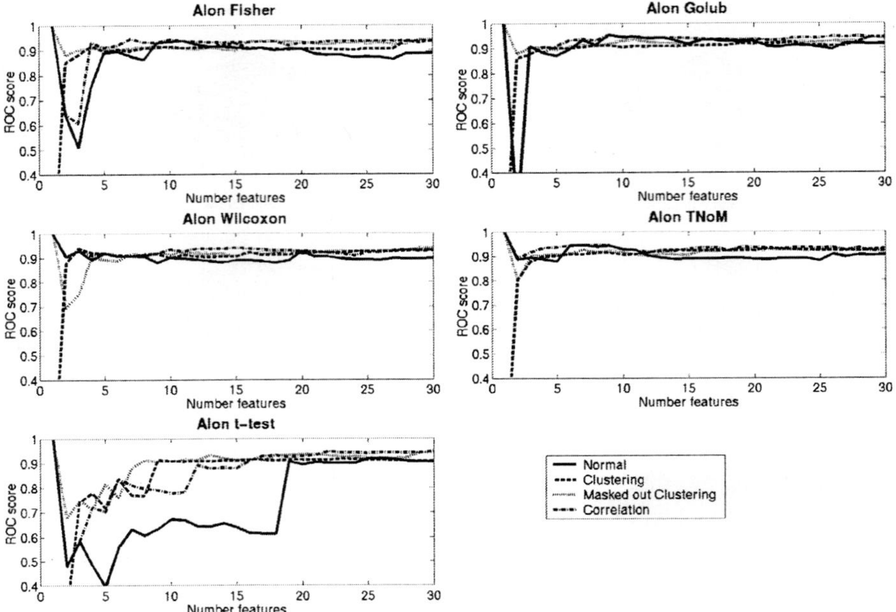

Figure 3: Comparison of test-statistics and different prefilter methods for Alon's data set

In the top 30 t-test genes in Alon's dataset, we have an average correlation of 0.79, whereas in the top 30 TNoM genes, we have only an average correlation of 0.56. Wilcoxon (also a nonparametric test) achieves the best result of all the tests (when comparing 30 features) and can reduce the absolute LOOCV error to 5.[17] The average correlation of the top 30 Wilcoxon genes is 0.43.

Doing the same comparison on Golub's dataset yields figure 4. In Alon's dataset it seemed that TNoM was on average the best test statistic whereas in Golub's leukemia dataset Wilcoxon performs best. We still see that clustering can lower the error in most of the cases. A remarkable fact is that with clustering we achieve 0% error with the t-test when using more than 50 features. Notice that we had 2 errors when using all the data. Here, not only can feature selection reduce the number of genes to find, but it can also decrease the error.

Although clustering for feature selection generally seems to improve the LOOCV error, the least improvements were obtained using it in conjunction with the Wilcoxon rank sum test and the best performance improvement was achieved using it together with t-tests. As illustrated above the reason for that could be that the t-test generally finds more correlated genes. The nonparametric tests do not take values into account and calculate their scores purely based on rank information what seems to have a positive effect on selecting fewer correlated genes.

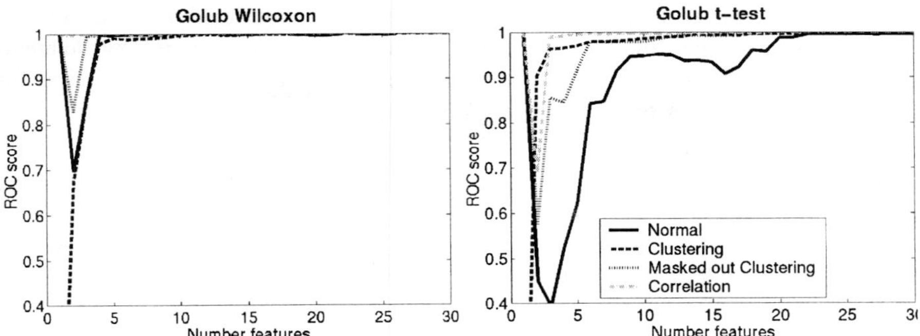

Figure 4: Comparison of different prefilter methods for Golub's data set

4 Conclusion and Further Research

In this paper we presented three novel prefilter methods to increase classification performance with microarray data. There is no clear winner between the three proposed methods and it depends largely on the dataset and parameters used. All the proposed feature selection methods find a subset that has better LOOCV performance than the currently used approaches. One question not addressed here is how to find the correct number of clusters. It is pretty expensive to try all possible numbers for clusters to find a setting that provides us with a good LOOCV performance. One direction for future work would be to estimate the number of clusters using a BIC[19] (Bayesian Information Criterion) score or switching over to model based clustering[20].

We addressed the problem of feature selection and outlined why feature selection has to be done and how it can be done without losing crucial information. The question not answered here is how many features/genes should be chosen in the end. One could argue to choose exactly that many genes as necessary to achieve the lowest LOOCV error. But in the end it comes down to a tradeoff between false positives and false negatives. The more genes we have in our set of interest (the feature set) the more genes might also be real candidates. The extreme would be to take all genes. Then we definitely have all candidates but also a very high percentage of false positives. The other extreme would be to take no gene at all. Then we would not mispredict anything to be a real gene but would have a very high false negative rate. The final answer on how many genes to select can only be answered by the biologists who must judge how much time they can invest in examining these genes further and which false positive/negative rate they will accept.

We feel, however, that for any fixed size the methods outlined here are likely to identify sets of genes that are stronger predictors than sets found by standard methods, which should be of significant value for diagnostic purposes as well as for guiding further exploration of the underlying biology.

Acknowledgment

Thanks to Bill Noble for his helpful suggestions. JJ and RS were supported in part by NIH grant NHGRI/K01-02350. WLR and JJ were supported in part by NSF DBI 62-2677.

References

1   A. Ben-Dor, L. Bruhn, N. Friedman, I. Nachman, M. Schummer, and Z. Yakhini. Tissue classification with gene expression profiles. *RECOMB*, (2000).
2   J.L. Devore, Probability and Statistics for Engineering and the Sciences, 4th edition, Duxbury Press, (1995).
3   Own unpublished results
4   P.J. Park, M. Pagano, M. Bonetti: A nonparametric scoring algorithm for identifying informative genes from microarray data. *PSB*:52-63, (2001).
5   C.M. Bishop: Neural Networks for Pattern Recognition, *Oxford University Press*, (1995)
6   T.R. Golub, D.K. Slonim, et al. Molecular classification of cancer: Class discovery and class prediction by gene expression monitoring. *Science*, **286**:531-537, (1999).
7   I. Lonnstedt and T. P. Speed. Replicated Microarray Data. *Statistical Sinica*, (2002).
8   J. Weston, S. Mukherjee, O. Chapelle, M. Pontil, T. Poggio, and V. Vapnik. Feature selection for SVMs, *Advances in Neural Information Processing Systems* **13**. MIT Press, (2001).
9   U. Alon, N. Barkai, D.A. Notterman, K. Gish, S. Ybarra, D. Mack and A.J. Levine: Broad patterns of gene expression revealed by clustering analysis of tumor and normal colon tissues probed by oligonucleotide arrays **PNAS 96**:6745-6750, (1999).
10   Y.H. Yang, M.J. Buckley, S. Dudoit, T.P. and Speed: Comparison of methods for image analysis on cDNA microarray data. *Technical report* (2000)
11   Y. H. Yang, S. Dudoit, P. Luu and T. P. Speed. Normalization for cDNA Microarray Data. *SPIE BiOS*, (2001).
12   C. Cortes and V. N. Vapnik. Support vector networks. *Machine Learning*, **20**:273-297, (1995).
13   http://microarray.cpmc.columbia.edu/gist/
14   J.C. Dunn, "A fuzzy relative of the ISODATA process and its use in detecting compact well-separated clusters", *Journal of Cybernetics*, **3**:32--57, (1973).
15   A.K. Jain, R.C. Dubes. *Algorithms for Clustering Data*. Prentice Hall, (1988).
16   D.A. Notterman, U. Alon, A.J. Sierk, A.J. Levine: Transcriptional Gene Expression Profiles of Colorectal Adenoma, Adenocarcinoma and Normal Tissue Examined by Oligonucleotide Arrays, *Cancer Research* **61**:3124-3130, (2001).
17   http://www.cs.washington.edu/homes/jj/psb
18   C.E. Metz, Basic principles of ROC analysis, *Seminars in Nuclear Medicine*, Vol **8**, No. **4**, 283-298, (1978).
19   G. Schwarz: Estimating the dimension of a model. *Annals of Statistics*, **6**:461-464 (1978).
20   K. Y. Yeung, C. Fraley, A. Murua, A. E. Raftery and W. L. Ruzzo, Model-based clustering and data transformation for gene expression data, *Bioinformatics* **17**:977-987 (2001).

KERNEL COX REGRESSION MODELS FOR LINKING GENE EXPRESSION PROFILES TO CENSORED SURVIVAL DATA

HONGZHE LI, YIHUI LUAN

Department of Statistics and Medicine, University of California, Davis,
CA 95616, USA

In functional genomics, one important problem is to relate the microarray gene expression profiles to various clinical phenotypes from patients. The success has been demonstrated in molecular classification of cancer in which gene expression data serve as predictors and different types of cancer are the binary or multi-categorical outcome variable. However, there has been less research in linking gene expression profiles to other types of phenotypes, in particular, the censored survival data such as patients' overall survival or cancer relapse times. In the paper, we develop a kernel Cox regression model for relating gene expression profiles to censored phenotypes in the framework the penalization method in terms of function estimation in reproducing kernel Hilbert spaces. To circumvent the problem of censoring, we use the negative partial likelihood as a loss function in the estimation procedure. The functional combinations of the original gene expression data identified by the method are highly correlated with the patients' survival times and at the same time account for the variability in the gene expression levels. We apply our method to data sets from diffuse large B-cell lymphoma, lung adenocarcinoma and breast carcinoma studies to verify its effectiveness. The results from these analysis indicate that the proposed method works very well in identifying subgroups of patients with different risks of death or relapse and in predicting the risk of relapse or death based on the gene expression profiles measured from the tumor samples taken from the patients.

1 Introduction

The recent development of DNA microarrays, which permits the simultaneous measurements of the expression levels of thousands of genes, raised the possibility of a powerful, genome-wide approach to the genetic basis of different types of tumors in the area of molecular classification of cancers, different levels of drug responses in the area of pharmacogenomics, and different patients' clinical outcomes in the area of clinical phenotype prediction. The problem of cancer class prediction using the gene expression data has been studied extensively in recent years. [1,2,3,4,5,6] This problem can be formulated as predicting binary or multi-category outcomes using gene expression data. However, there has been less development in relating gene expression profiles to other phenotypes such as quantitative continuous phenotypes, or the censored survival phenotypes. Relationship between gene expression profiles with other phenotypes such as quantitative phenotype or survival phenotypes is also important in clinical applications. For example, correlating gene expression profiles with to the drug responses of thousand of potential drug compounds in 60 human cell lines, researchers were able to identify sets of genes whose expressions are highly related to drug response of a set of compounds, which will eventually help development of new drugs.[7] Correlating gene expression profiles obtained from tumor samples prior to treatment with the time to cancer relapse or death due to cancer can be very important in clinical practice.

The goal of this paper is to develop new statistical methods for relating gene expression profiles to censored survival data such as time to cancer recurrence or death. From the statistical point of view, one challenge is that the time to cancer recurrence or death is often right censored because during the course of followups, some patients may still be cancer-free or alive. Another challenge is that the microarray gene expression data are often measured with great deal of noise, and that the sample size of tissues or cell lines is usually very small compared to the number of genes from expression arrays. The problem of censoring make the problem difficult compared to binary or continuous phenotypes. One popular approach is to first cluster tumor samples into several clusters based on their gene expression patterns across many genes, and then to use the Kaplan-Meier curves or the log-rank tests to test whether there is a difference in survival times among different tumor groups. One drawback of this approach is that the phenotype information is completely ignored in the clustering step and therefore may result in loss of efficiency. Another approach is to cluster genes first based on their expression across different samples, and use the sample averages of the gene expression levels in a Cox model[8] for survival outcome. Both methods of course depend on the methods of clustering used. Hastie *et al*[9] proposed a tree-harvesting approach in which a stepwise regression approach is used to select genes or clusters of genes that are related to the phenotypes using the Cox proportional hazards model. This approach still requires the intermediate results of the hierarchical clustering analysis. Nyuyen and Rock[10] proposed to generalize the idea of the partial least square in the framework of the Cox model by using the residuals. However, their method is limited to linear function of the gene expression levels. In addition, use of residuals in the estimation of the parameters in the Cox model is not well-established in the survival analysis literature since there are many different ways of defining residuals.[11]

In the paper, we develop methods for relating gene expression profiles to censored phenotypes in the framework of the support vector machines (SVMs) by using kernels.[12]. The SVMs technique is a relatively new but popular methods in machine learning, and was applied successfully in the problem of tumor classification.[13] The robustness of the methods with respective to the sparse and noisy data is making them the methods of choice for many problems in bioinformatics. As demonstrated by Wahba *et al* [14], the SVM can be reformulated as a penalization method in terms of function estimation in reproducing kernel Hilbert spaces, where the objective function can be written as "loss+penalty". In this paper, we formulate a kernel Cox regression model for censored survival data by using the negative partial likelihood in a generalized Cox model as the loss function.

The rest of the paper is organized as the following. We first present a general Cox model for relating gene expression profiles to the censored survival phenotypes. We then present methods for estimating the parameter of the model in the framework of the kernel Cox regression model and penalization methods. In Section 3, we present results of analysis of three different cancer survival data sets. We conclude in Section 4 with a discussion of the results presented in this paper and offer some directions for future work.

2   Statistical Methods

2.1   A general Cox model for censored survival data

Suppose that we have collected n patients with a particular cancer. For the ith patients, let (t_i, δ_i) be the observed phenotype, where t_i is the survival time (e.g., time to cancer relapse after treatment) when $\delta_i=1$, and is the censoring time (e.g., time of last known being cancer-free) when $\delta_i = 0$. Let $x_i = (x_{i1}, \cdots, x_{ip})$ be the vector of the gene expression levels of p genes for the ith sample taken from the ith patient. We assume the following general Cox model, where the hazard function for the ith patient is modeled as

$$\lambda_i(t|x_i) = \lambda_0(t) \exp(f(x_i)), \tag{1}$$

where $\lambda_0(t)$ is the unspecified baseline hazard function, $f(x_i)$ is an arbitrary function of the gene expression data x_i. In this model, the gene expression profile measured over p genes is related to the risk of death or cancer relapse through the score function $f(x)$.

2.2   Estimation of the model

Since the dimension of x_i vector is usually much larger than the sample size n, standard methods such as the Cox partial likelihood [8] for estimating the unspecified function f is unfeasible. To overcome this problem, a regularized formulation of the Cox regression is considered as a variational problem in a reproducing kernel Hilbert space [15] (RKHS) H_K generated by the kernel K,

$$minR_{reg}(f) = \frac{1}{n}\sum_{i=1}^{n} V(t_i, \delta_i, f(x_i)) + \xi||f||^2_{H_K} \tag{2}$$

where $V(t_i, \delta_i, f(x_i))$ is the loss function which is a functional of f depending on only the values of $f(x)$ at the data points, $\{f(x_i)\}_{i=1}^n$, $||f||^2_{H_K}$ is the norm defined in H_K, and $f = b + h$ with $h \in H_K, b \in R$, and $\xi > 0$ is a tuning parameter. For the general Cox model (1), we propose to use the negative log partial likelihood [8] as the loss function and reformulate the problem as finding function $f(x)$ such that

$$R_{reg}(f) = -\frac{1}{n}\sum_{i=1}^{n} \delta_i[f(x_i) - \log\{\sum_{j \in R_i} \exp(f(x_j))\}] + \xi||f||^2_{H_k} \tag{3}$$

is minimized, where $R_i = \{j : t_j \geq t_i, j = 1, \cdots, n\}$ is the set of individuals who were at risk at time t_i. The solution to this problem was given by Kimeldorf and Wahba [15], and is known as the representer theorem. By this theorem, the optimal $f(x)$ has the form:

$$f(x) = b + \sum_{i=1}^{n} a_i K(x, x_i), \tag{4}$$

where K is the reproducing kernel of H_K. Since b can be absorbed into the baseline hazard function in model (1), we omit b in the following discussion. For the simplest case of natural inner product kernel with $K(x_i, x_j) =< x_i, x_j >$, the function $f(x)$ can be expressed as

$$f(x) \quad = \quad \sum_{i=1}^{n} a_i K(x, x_i) = \sum_{j=1}^{p}(\sum_{i=1}^{n} a_i x_{ij})x^{(j)} = \sum_{j=1}^{p} \beta_j x^{(j)} \tag{5}$$

where $x = (x^{(1)}, \cdots, x^{(p)})$ the vector of the gene expression levels over p genes. Here the parameter β_j can be used as a measurement on how the gene expression level of gene j affects the risk of death or tumor recurrence. Because of this nice interpretation, this paper uses only this kernel in our analysis of real data sets. In the case when the data are not linearly separable, one can choose more general kernel such as the polynomial kernels with $K(x_i, x_j) = (< x_i, x_j > +1)^d$ or the Gaussian kernels with $K(x_i, x_j) = \exp(||x_i - x_j||^2/\sigma_d^2)$, where d and σ_d^2 are the kernel parameters.

From the representer formula (4), it can be shown that minimizing equation (3) is equivalent to minimizing over vector a the finite dimensional form:

$$R_a = -\delta'(K_a a) + \delta' \log\{\sum_{j \in R_i} \exp(K_a a)\}] + \xi a' K_a a \tag{6}$$

where $a' = (a_1, \cdots, a_n\}$, the regressor matrix $K_a = [K(x_i, x_j)]_{n \times n}$. Here the matrix K_a is called the kernel matrix.

For a fixed ξ, to find a, we set the derivative with respective to a in R_a to zero, and use the Newton-Raphson methods to iteratively solve the score equation. In case of singularity problem in the Newton-Raphson iterations, we used the downhill simplex algorithm to minimize the function (6). [16]

We propose to use the leave-one out method for selecting the tuning parameter ξ in equation (6). One approach is to choose ξ which minimizes the cross-validated sum of the square residuals. Another approach is to choose ξ which maximizes

$$\prod_{i=1, \delta_i=1}^{n} \frac{\exp(\hat{f}_{-i}(x_i; \xi))}{\sum_{j \in R_i} \exp(\hat{f}_i(x_j; \xi))}$$

where $f_{-i}(x; \xi)$ denote the kernel estimate of the function for a given ξ computed under model when δ_i is changed from one (uncensored) to zero (censored). This procedure essentially leaves only the uncensored individuals out in the cross validation procedure and has been demonstrated to work well for the problem of subset selection for the Cox regression model. [17]

3   Application to real data sets

We applied the proposed methods to three published data sets of three different cancers to illustrate the methods and to demonstrate that the proposed methods work well in separating patients into different risk groups and in predicting patient's risk of cancer relapse or overall survival. In all the analysis, we used the natural inner product kernel and chose the tuning parameter ξ by residual cross-validation.

3.1   Application to diffuse large B-cell lymphoma data set

Alizadeth *et al* [1] reported a genome-wide gene expression profiling analysis for diffuse large B-cell lymphoma (DLBCL), in which a total of 96 normal and malignant lymphocites samples were profiled over 17,856 cDNA clones. Details can be found in Alizadeth *et al*[1]. None of the patients included in the study has been treated before obtaining the biopsy samples. After biopsy, the patients were treated at two medical centers using comparable standard chemotherapy regimens. Among 42 patients, 40 of them had followup information, including 22 death with death time ranging from 1.3 to 71.3 months (median 10.6) and 18 being still alive with the follow-up times ranging from 51.2 to 129.9 months (median 74.7).

Alizadeth *et al*[1] first identified 4026 genes which showed large variations across all the samples. We further selected 319 genes with the p-value less than 0.05 by using the Cox model for each of these 4026 genes. We used the inner product kernel to build a general Cox model for the time to death of the 40 patients using the gene expression levels of the 319 genes as predictors. Figure 1 (a) shows the estimated survival curves for patients in two different groups defined by the scores $f(x) > 0$ or $f(x) < 0$, indicating large difference in overall survival in the two groups. One group of 15 patients with $f(x) < 0$ were all alive during the followups, another group of 25 patients with $f(x) > 0$ includes 22 deaths. The group with $f(x) < 0$ includes 12 germinal centre B cell-like DLBCL, and the group with $f(x) > 0$ includes 18 activated B cell-like DLBCL, indicating much better overall survival in germinal centre B cell-like DLBCL patients. Figure 1 (b) shows the estimated coefficient for each of the 319 genes that were used in estimating the score $f(x)$. Higher expression levels of the genes with positive coefficients increase the risk of death due to lymphoma, and higher expression levels of the gene with negative coefficients decrease the risk of death due to lymphoma.

To further examine how the model predicts the survival time of future new patient, we performed a leave-one-out cross validation analysis. We left one patient out each time, and estimated the function $f(x)$ in model (1) with the rest of the data. We then estimated the score for the patient who was left out from model building process. Figure 1 (c) shows the plot of the cross-validated score versus the observed times to death or times at censoring, indicating the higher the score, the high risk of death from lymphoma. Figure 1 (d) shows the estimated survival curves based on these cross-validated scores being negative or positive. This plot suggests that the proposed method works quite well in predicting future patient's risk of death.

3.2 *Application to lung cancer data set*

Garber *et al.*[4] reported global gene expression profiles for 67 human lung tumors representing 56 patients whose clinical course was followed for up to 5 years. Among these tumors, there were 41 Adenocarcinomas (ACs), of those, 22 patients had survival information available, including 12 death ranging from 0 to 36 months (median 12.5 months) and 10 death-free during the 5 year

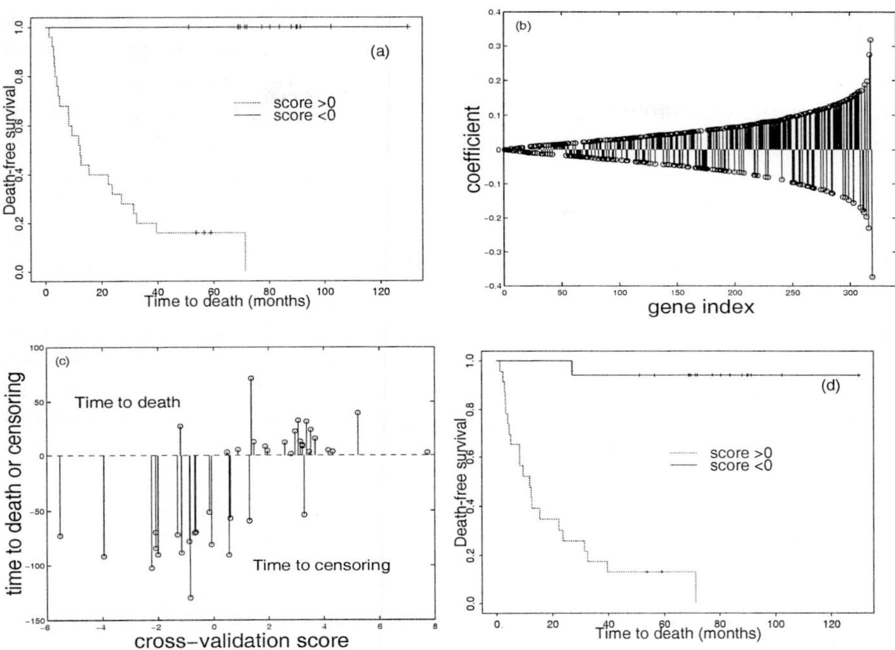

Figure 1: Analysis of lymphoma data set. (a) Survival curves for patients in two groups based on the estimated scores using gene expression profiles; (b) Coefficients of the gene effects sorted by the absolute values; (c) Plot of cross-validated scores versus observed data, where the death times are plotted as positive values on the y-axis, and the censored times are plotted as negative values on the y-axis; (d) Survival curves for patients in two groups based on cross-validated scores.

followups (followup times ranged from 18 to 54 months with median of 48 months). The original gene expression data include 23,100 cDNA clones representing 17,108 unique genes. Garber *et al* further reduced this list of gene to 918 cDNA clones representing 835 unique genes with large variabilities cross samples in different clusters.

There are 131 genes with the Cox score statistic larger than 3.84 (corresponding p-values less than 0.05). Using our proposed method with inner produce kernel, we obtained the estimate of $f(x)$ in the model (1) based on the 22 patients's survival information during the followups and the gene expression levels of 131 genes. Based on the estimated values of $f(x) > 0$ or $f(x) < 0$, we divided the 22 patients into two groups. Figure 2 (a) shows the overall survival curves for these two groups, indicating that large difference in lung cancer patients' survival can be identified based on their gene expression profiles. Figure 2 (b) shows the estimates of the β coefficients in model (5), indicating the expression levels of different genes have different effects on patients's survival. It is important to note that it is the linear combination of these gene expression levels that is used for separating patients into different groups with different risks of death due to lung cancer.

We also performed similar cross-validation analysis as we did for the lymphoma example. Figure 2 (c) shows the cross-validated scores for each patient, plotted against the observed time to death or time to last followups. The plot indicates that the patients with larger cross-validated scores tend to have higher risk of being death. Figure 2 (d) show the overall survival curves for patients with the cross-validated scores greater or less than zero. This example further demonstrates that the proposed method can effectively predict the patient's risk of death from lung cancer.

3.3 Application to breast cancer data set

Sorlie $et\ al$[5] demonstrated the use of the gene expression profiles of breast carcinomas to distinguish tumor subclass with clinical implications. In the following analysis, only 49 of the patients from the prospective study of locally advanced disease and with no distant metastases were used. The gene expression levels of the 456 cDNA clones (427 unique genes) in the intrinsic gene list were obtained for tumor samples prior to chemotherapy. The followup information including the time to cancer relapse and overall survival was available for all the 49 patients. During the followups, 24 patients had cancer relapse at various time ranging from 0 month to 59 months after the treatment, and 25 patients had no relapse during the followup period ranging from 22 to 92 months.

Based on the univariate Cox regression, we first selected 68 genes with the scores greater than 3.85 (significant at 0.05 level). Based on the gene expression data of these genes and the patients' time to relapse data, we fitted the model (1) and obtained the estimated $f(x)$ function and the score for each patient. Figure 3 (a) shows the plot of the estimated scores versus the time to relapse or time to last followups. In general, we observed that the higher the score, the higher the risk of having relapse after treatment. Based on these scores, we can divide the patients into high (if $f(x) > 0.85$), medium (if $-1.45 < f(x) \leq 0.85$) and low risk (if $f(x) \leq -1.45$) groups with different risks of relapse. Figure 3 (b) give the relapse-free survival curves for these three groups. Therefore, based on the scores derived from the gene expression profiles, we are able to

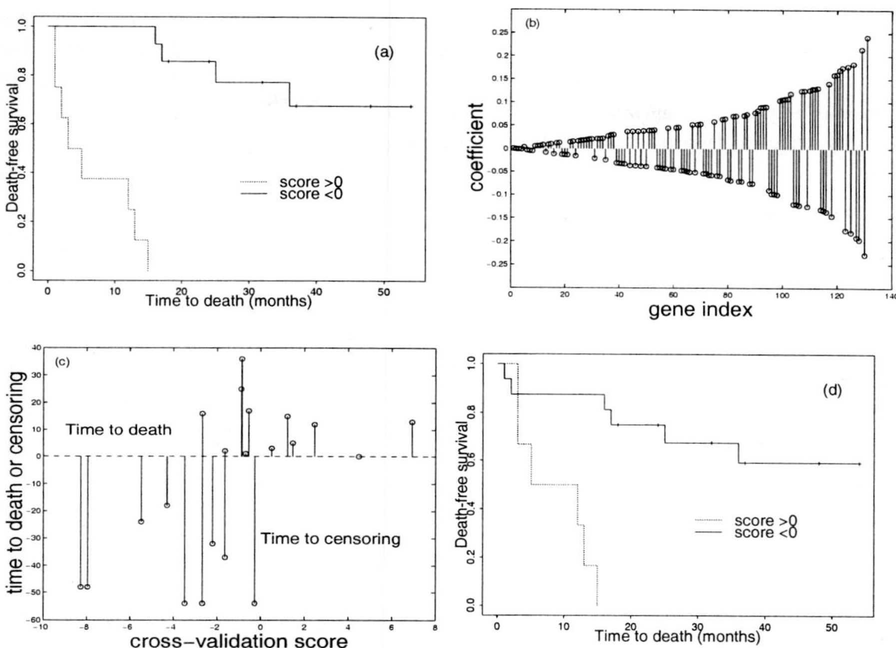

Figure 2: Analysis of lung cancer data set. (a) Survival curves for patients in two groups based on the estimated scores using gene expression profiles; (b) Coefficients of the gene effects sorted by the absolute values; (c) Plot of cross-validated scores versus observed data, where the death times are plotted as positive values on the y-axis, and the censored times are plotted as negative values on the y-axis; (d) Survival curves for patients in two groups based on cross-validated scores.

identify subgroups of the patients with different risks of breast cancer relapse after the chemotherapy.

In order to examine how well the model performs for predicting the risk of relapse for a new patient, we performed leave-one out cross-validation analysis, similar to the previous two examples, in which we left one patient out, and fitted the model with the rest of the patients, and estimated the score for the left-out patient. Figure 3 (c) shows the cross-validated scores versus the observed times to relapse/censoring. Comparing to the estimated scores in Figure 3 (a), it is less clear that there are three different risk groups for relapse. However, clear differences in risk of relapse can still been seen for those with very high or low scores. Figure 3 (d) shows the relapse-free survival curves for the three groups defined by the cut-off values of -1.45 and 0.85. The cross-validated results still identified one group of patients with large scores who has higher risk of cancer relapse. However, the difference of the other two groups was not significant, which suggests that for predicting cancer relapse risk, there might only be two distinct groups with different risks.

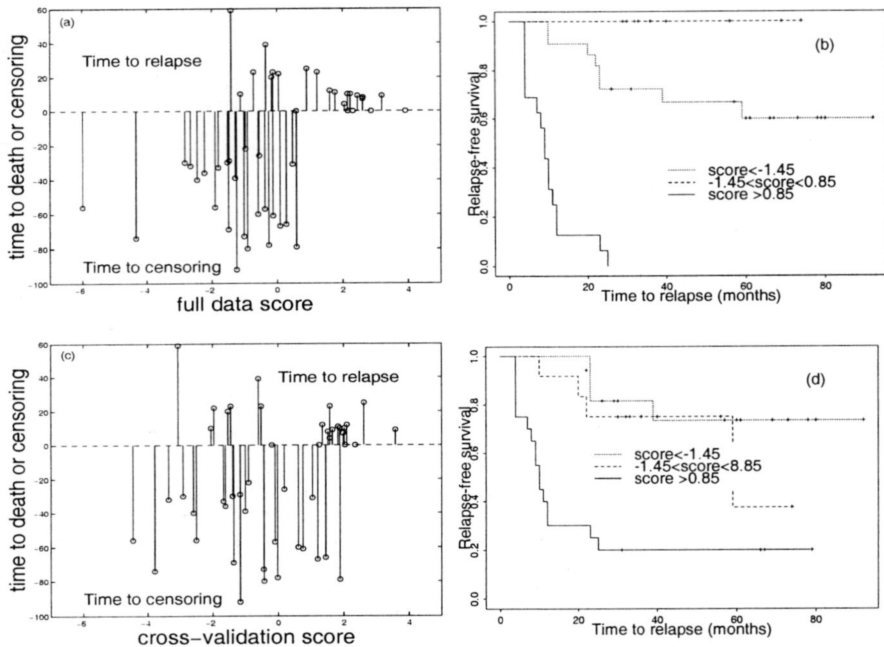

Figure 3: Analysis of breast cancer data set. (a) Estimated scores versus the observed times to cancer relapse or times at censoring, where the relapse times are plotted as positive values on the y-axis, and the censored times are plotted as negative values on the y-axis; (b) Survival curves for patients in three groups based on the estimated scores using gene expression profiles; (c) Cross-validated scores versus the observed times to cancer relapse or times to censoring; (d) Survival curves for patients in three groups based on cross-validated scores.

4   Discussion

We have introduced the kernel Cox regression models for relating gene expression profiles to censored survival data. The method generalizes the idea of the support vector machine for binary or multi-categorical data to censored survival data. The model automatically searches for the genes whose expression levels are related to survival phenotypes and identifies the optimal combination of the gene expression data in predicting the risk of cancer recurrence or death. Since the risk of cancer recurrence of death due to cancer might be due to interplay of many genes in certain way, methods such as our proposed one are expected to show better performance in predicting the risks. We demonstrated the applicability of our methods by analyzing time to death or tumor recurrence of large cell lymphoma, lung carcinoma, and breast carcinoma. In all the analysis, we used the simple inner cross product kernels and the linear combination of the gene expression levels as our scores, and obtained satisfactory results. However, it is important to emphasize that our proposed method is not limited to obtaining only linear scores. One advantage of the method is that it can handle nonlinear scores easily by using alternative kernels without introducing any computational difficulty.

Another advantage of the proposed method is that there is no computational or methodological limitation in term of the number of genes that can be used in the prediction of patient's overall survival or time to cancer recurrence. One important future research is to examine how different number of genes used in model affects the results of prediction. Since not all genes will be important in predicting censored survival phenotypes, we would expect better prediction results using only genes that are related to the phenotypes. For all the three cancer data sets we analyzed, we selected these genes by using the univariate Cox score statistic. An alternative would be to iteratively select important genes and build predictive models. One approach for this problem is to iteratively delete those genes with β coefficient smaller than a preset cutoff values and build predictive model until no such genes can be deleted. We plan to study this approach in details in the future.

In summary, we formulated the problem of linking gene expression profiles to the censored survival data in the framework of the kernel Cox regression model. As demonstrated by applications to several real cancer data sets, the methods can potentially be useful for identifying important genes that are related to clinical outcomes and for predicting time to cancer relapse or death to cancer based on the tumor molecular gene expression profiles.

Acknowledgments

This work was supported by an NIH grant (ES09911) and an UC Davis Health System Research Award grant.

References

1. A. Alizadeh *et al.*, *Nature* **403**, 503 (2000).
2. T. R. Golub *et al.*, *Science* **286**, 531 (1999).
3. U. Alon *et al.*, *Proc Natl Acad Sci USA*, **96(12)**, 6745 (1999).
4. M. E. Garber *et al.*, *Proc Natl Acad Sci USA* **98**, 13784 (2001).
5. T. Sorlie *et al.*, *Proc Natl Acad Sci USA* **98**, 10869 (2001).
6. J. E. Staunton *et al.*, *Proc Natl Acad Sci USA* **98**, 10787 (2001).
7. U. Scherf *et al.*, *Nat Genet* **24**, 236 (2000).
8. D. Cox, *J. Roy Stat. Sco B* **74**, 187 (1972).
9. H. Trevor *et al.*, *Genome Biology* **2**, research0003.1 (2001).
10. D. Nyuyen, D. M. Rocke, *Technical report*, UC Davis (2001).
11. W. E. Barlow, R. L. Prentice, *Biometrika* **75**, 65 (1988).
12. G. Wahba, *Technical report*, UW Madison (1998).
13. T. S. Furey *et al.*, *Bioinformatics* **16**, 906 (2000).
14. G. Wahba, (SIAM, Philadelphia, 1990).
15. G. Kimeldorf, G. Wahba, *J. Math. Anal. Applic.*, **33**, 82 (1971).
16. J. A. Nelder, R. Mead, *Computer Journal* **7**, 308 (1965).
17. A. Y. C. Kuk, *Biometrika*, **71**, 587 (1984).

EXTRACTING CONSERVED GENE EXPRESSION MOTIFS FROM GENE EXPRESSION DATA

http://genomics10.bu.edu/murali/xmotif

T. M. MURALI

Bioinformatics Program, 48 Cummington St., Boston University, Boston MA 02152

SIMON KASIF

Bioinformatics Program and Department of Biomedical Engineering,
48 Cummington St., Boston University, Boston MA 02152

Abstract

We propose a representation for gene expression data called conserved gene expression motifs or xMOTIFs. A gene's expression level is conserved across a set of samples if the gene is expressed with the same abundance in all the samples. A *conserved gene expression motif* is a subset of genes that is simultaneously conserved across a subset of samples. We present a computational technique to discover large conserved gene motifs that cover all the samples and classes in the data. When applied to published data sets representing different cancers or disease outcomes, our algorithm constructs xMOTIFs that distinguish between the various classes.

1 Introduction

Gene expression plays an important role in cell differentiation, development, and pathological behaviour. DNA microarrays[1,2] offer biologists the remarkable ability to monitor the expression levels of thousands of genes in a cell simultaneously. High-throughput gene expression analysis promises to produce new insights into cell function as well as stimulate the development of new therapies and diagnostics.

When a gene's expression level is measured across a variety of samples, the expression values usually span a wide range. Biologically, these values correspond to a small number of distinct states that the gene is in, e.g., up-regulated or down-regulated. Since the up-regulation of a gene is a temporal process, it is often difficult to determine a gene's state based on a set of noisy expression values. However, on average, it might be possible to differentiate the expression level of an up-regulated gene in one tissue type (e.g., a cancer tissue) from its level in another type of tissue (e.g., a healthy tissue) where the gene is not expressed with the same abundance.

78

Motivated by this observation, we say that a gene's expression level is *conserved* across a subset of samples if the gene is in the same state in each of the samples in this subset. A *conserved gene expression motif* or XMOTIF is a subset of genes whose expression is simultaneously conserved for a subset of samples; we say that each of these samples *matches* the motif. In this paper, we use a range of expression values to represent a gene's state. If we map each gene to a dimension, each sample to a point, and each expression value to a coordinate value, an XMOTIF is identical to a multi-dimensional hyper-rectangle that is bounded in the dimensions corresponding to the conserved genes in the motif and unbounded in the other dimensions. See Figure 1 for examples. Mathematically, the expression values of a sample that matches a motif satisfy a conjunction of inequalities, two for each gene in the motif.

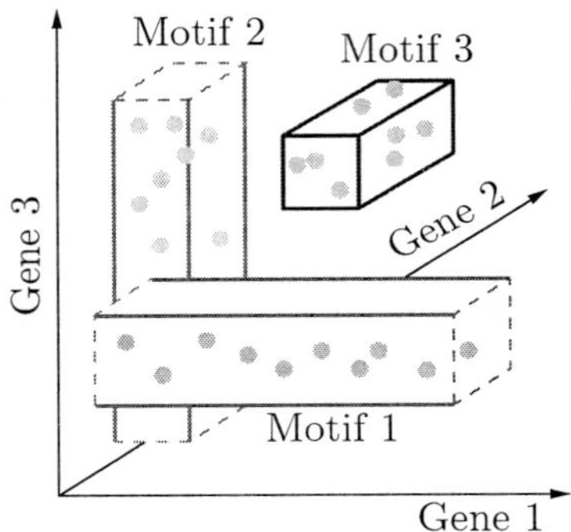

Figure 1: Example of XMOTIFs. Genes 2 and 3 are conserved in Motif 1, Genes 1 and 2 are conserved in Motif 2, and all three genes are conserved in Motif 3. A dashed face of a box indicates the dimension along which the box is unbounded.

In this work, we address the task of identifying XMOTIFs in gene expression data. We concentrate on data sets where each sample belongs to a particular class, e.g., different types of cancer,[3] cancerous and healthy tissues,[4] and patients with different survival rates.[5]

We believe that this work is the one of the first to suggest representing gene expression data concisely in the form of XMOTIFs. Such a representation has several potential biological advantages and applications. First, by comparing and contrasting the gene motifs for different classes, we can identify genes that are conserved in multiple classes but are in different states in different classes. If the classes correspond to different diseases or to diseased and normal tissues, such genes are possible drug targets. Second, if the genes in a motif are believed to interact in a pathway, the information present in the motif about which genes are highly expressed and which are suppressed could be used to deduce and refine the structure of the pathway. Third, by requiring that multiple genes be simultaneously conserved across the samples matching a motif, we might be able to characterise sub-classes in the data that no gene on its own provides high-quality evidence for.

Before attempting to develop an algorithm for computing xMOTIFs, it is useful to consider the properties that an xMOTIF should have. Computing one motif for each sample makes the representation over-specific. Therefore, we desire that each motif for a class should be matched by a large fraction of the samples in that class, if not all the samples in that class. Each motif should contain as many conserved genes as possible. While a motif that contains one or two genes is biologically feasible, it may not be statistically significant since such a motif could appear with high probability in randomly-generated data. On the other hand, a motif that contains too many genes may be too restrictive since no sample may match the motif. Motivated by these observations, we propose the following formal definition of an xMOTIF:

Definition 1.1 *Given a set of genes whose expression levels are measured across a set of samples and user-defined parameters $0 < \alpha, \beta < 1$, a conserved gene expression motif or* xMOTIF *is a pair (C, G), where C is a subset of the samples and G is a subset of the genes, that satisfies the following conditions:*

Size: *the number of samples in C is at least an α-fraction of all the samples,*

Conservation: *every gene in G is conserved across all the samples in C, i.e., the gene is in the same state in all the samples in C, and*

Maximality: *for every gene not in G, the gene is conserved in at most a β-fraction of the samples in C.*

The maximality condition enforces a balance between the number of genes in the motif and the number of samples matching the motif. If we add a gene to the motif, then the number of samples matching the new motif will decrease by a fraction of at least β, a cost we may not be willing to pay.

Given this definition, the gene expression data may contain many xMOTIFs. Among all xMOTIFs, we are interested in the largest xMOTIF, the one that contains the maximum number of conserved genes. In order to cover all the classes completely using xMOTIFs, we adopt the following iterative algorithm: find the largest xMOTIF in the data, remove the samples that satisfy this motif from the data, find the largest motif in the remaining data, and continue in this manner until all samples satisfy some motif.

Our approach has several desirable features. (i) We allow a gene to appear in more than one motif and in motifs representing different classes, modelling the possibility that the gene's expression level may be regulated in multiple conditions. (ii) By not deleting the samples that match a computed XMOTIF, we can allow samples to appear in different motifs. This property may be useful when a sample belongs to multiple classes or when we are interested in discovering new classifications of the samples. (iii) The system need not be told beforehand how many motifs to compute.[a] (iv) Using this approach, we can find XMOTIFs with vastly different numbers of genes; we are likely to discover motifs with many genes in earlier iterations and motifs with fewer genes in later iterations.

Our definition of an XMOTIF is based on the notion of "projective cluster" developed by Procopiuc et al.[6] in the context of problems in computer databases and computer vision. It can be shown that the problem of computing the largest XMOTIF is NP-complete by transforming the problem of computing the maximum-edge bipartite clique in a bipartite graph[7] to the motif computation problem. In Section 4, we present a probabilistic algorithm that exploits the mathematical structure of XMOTIFs to compute the largest XMOTIF efficiently. This algorithm extends the technique developed by Procopiuc et al. to compute projective clusters.

2 Previous Research

Previous work on the computational analysis of gene expression data falls into two broad categories. When the samples belong to distinct classes, researchers have used well-known techniques such as nearest-neighbour rules, support vector machines and feature selection to build a diagnostic tool that can distinguish between the various classes based on the expression profiles of predictive genes.[3,8] When the goal is to analyse the data in an unsupervised manner, techniques such as k-means clustering and hierarchical clustering[9] are common. See the paper by Tibshirani et al.[10] for a survey of such clustering methods. Unfortunately, it appears difficult to modify techniques like k-means clustering or hierarchical clustering to compute XMOTIFs since these techniques attempt to cluster in the space spanned by all the genes. Bayes networks[11] and graph-based clustering algorithms[12] have also been used to model and analyse gene expression data.

Recently, some researchers have developed techniques that simultaneously cluster genes and samples, an idea that seems closest to XMOTIFs. Since this approach to analysing gene expression data is relatively new, different papers use different criteria for simultaneously clustering genes and samples. As a

---

[a]While the user provides the parameter α, we may compute far less than $1/\alpha$ XMOTIFs, since α is a lower bound on the number of samples that match the largest XMOTIF.

result, it is difficult to compare these techniques. Here, we attempt to characterise in what terms our approach is different from other similar ideas.

Hartigan's block clustering algorithm[13,10] repeatedly rearranges the rows and columns of the matrix of gene expression values so that the rearranged matrix contains several disjoint blocks (contiguous sub-matrices) of highly correlated values. This technique requires that a gene be present only in one cluster, a condition that may not always be useful biologically (consider a situation when a gene's expression is regulated in different diseases). Cheng and Church adopt an approach called "biclustering"[14]. They compute sub-matrices or biclusters that have small "mean squared residue," a measure of the variance in the sub-matrix. Tanay, Sharon, and Shamir[15] adopt a graph-theoretic approach to biclustering. They represent gene expression data as a bipartite graph, whose nodes and edges are assigned weights under a graph model that they define. In this framework, a bicluster is a dense bipartite subgraph of the original graph. They develop algorithms to compute biclusters with large weights. Getz, Levine, and Domany apply a technique called "coupled two-way clustering."[16] They repeatedly apply a hierarchical clustering algorithm to different subsets of genes and samples. The sub-matrices they output correspond to stable clusters generated by this process.

All these techniques partition or cover the expression matrix by sub-matrices such that the expression values in each sub-matrix are highly coherent according to a suitable measure. The key property of an XMOTIF is that each gene in the XMOTIF is conserved across all the samples in that sub-matrix. It does not appear that the other techniques can capture this property. Finally, Hastie et al. [17] present a technique called "gene shaving" that tries to extract coherent and small clusters of genes that vary as much as possible across the samples; in essence, their goals are complementary to ours.

3 Determining Gene States

In this section, we describe our technique for computing the states corresponding to a gene. In our approach, a state is simply a range of expression values that is statistically significant. Similar ideas have been adopted by other researchers.[18] Thus, if we have n samples, there are $\binom{n}{2}$ possible states for each gene, each corresponding to one of the sub-intervals spanned by the expression values. Clearly, not all these states are biologically interesting. Intuitively, a state is interesting if it contains far more expression values than we would expect the state to contain if the expression values were generated at random. Formally, as a null hypothesis, we assume that gene expression values are generated by a uniform distribution. We define a state $[a, b]$ to be "interesting" if the expression values in it are unlikely to have been generated by a uniform distribution. We compute the p-value of the decision that $[a, b]$ is interesting,[b] order states by p-value, and consider only those states whose p-value is less than a user-defined parameter. When the samples belong to different classes, we also adopt a "supervised" version of this idea. We define a state $[a, b]$ to be interesting if there is a class such that the set of expression values of the samples from that class that lie in $[a, b]$ are unlikely to have been generated by a uniform distribution. For each class, we calculate a p-value and assign the smallest p-value to $[a, b]$.

In practice, we also discard those intervals that contain more than a user-specified number of expression values. The rationale for this step is that even if an interval containing a large number of expression values is statistically significant, it may not be biologically interesting since it is unlikely to help us in distinguishing between the various classes.

4   Algorithm

We are now ready to describe our algorithm for computing the largest XMOTIF. The input to the algorithm is a set of genes, a set of samples, an expression value for each gene-sample pair, and for each gene, a list of intervals representing the states in which the gene is expressed in the samples.

To determine an XMOTIF, we have to compute the set G of conserved genes, the states that these genes are in, and the set C of samples that match the motif. We observe that if we are given (i) the set G of conserved genes, (ii) the states of the conserved genes, and (iii) one sample c that matches this motif, we can compute the remaining samples in C simply by checking for each sample c' whether the genes in G are in the same state in c and c'. Informally, c is a "seed" from which we can compute the entire motif.

This observation is the starting point of our algorithm. Suppose we know a sample c that matches the largest XMOTIF. Given such a sample c, how can we compute the genes in the largest motif and the states they are in? Suppose we are given a set D of samples with the following properties: (i) for every sample c' in D and for every gene in the largest motif, there is exactly one state such that the gene is in that state in samples c and c' and (ii) for every gene g that is not in the largest motif, there exists a sample c' in D such that gene g is not in the same state in samples c and c'. We call D a *discriminating*

---

[b]If k values lie in the interval $[a, b]$ and if the gene's expression values fall in the range $[0, 1]$, then the probability that an expression value falls inside $[a, b]$ is $b - a$. Therefore, the p-value of the decision that $[a, b]$ is interesting (rejecting the null hypothesis) is the sum $\sum_{k \leq i \leq n} \binom{n}{i} (b - a)^i (1 - (b - a))^{n-i}$.

set. The key property of a discriminating set is that given the seed sample c and such a set D, we can compute the largest xMOTIF by including exactly those gene-states that satisfy these conditions and exactly those samples that agree with c on all these gene-states.

Algorithm 1 describes the steps of our probabilistic algorithm. We assume that for each gene, the intervals corresponding to that gene's states are disjoint.[c] It proceeds by selecting n_s samples uniformly at random from the set of all samples. These samples act as seeds. For each random seed, we select n_d sets of samples uniformly at random from the set of all samples; each set has s_d elements. These sets serve as candidates for the discriminating set. For each seed-discriminating set pair, we compute the corresponding xMOTIF as explained above. We discard the motif if less than an α-fraction of the samples match it. Finally, we return the motif that contains the largest number of genes. Suppose our data consists of n samples and m genes. We can extend the arguments made by Procopiuc et al. to prove that by choosing $n_s = O(1/\alpha)$, $s_d = O(\log m/\log(1/\beta))$, and $n_d = O(1/\alpha^{s_d})$, we can compute the largest xMOTIF with probability greater than $1/2$ in time $O(nn_s n_d) = nm^{O(\log(1/\alpha)/\log(1/\beta))}$. We can increase the probability of success by repeatedly executing this algorithm.

Algorithm 1 FINDMOTIF(): algorithm for computing the largest xMOTIF.

1: **for** $i = 1$ to n_s **do**
2:     Choose a sample c uniformly at random.
3:     **for** $j = 1$ to n_d **do**
4:         Choose a subset D of the samples of size s_d uniformly at random.
5:         For each gene g, if g is in the state s in c and all the samples in D, include the pair (g, s) in the set G_{ij}.
6:         C_{ij} = set of samples that agree with c in all the gene-states in G_{ij}.
7:         Discard (C_{ij}, G_{ij}) if C_{ij} contains less than αn samples.
8: **return** the motif (C^*, G^*) that maximises $|G_{ij}|, 1 \leq i \leq n_s, 1 \leq j \leq n_d$.

---

5   Results

We have implemented this algorithm in C++ under the Linux operating system. We present our analysis of three data sets: an ALL-AML data set,[3] a

---

[c]In practice, since the intervals for a gene often overlap, we modify Step 5 of the algorithm as follows: For each gene g, let I_g be the set of all states s such that g is in state s in sample c and all the samples in D. We pick a state s in I_g uniformly at random and include the pair (g, s) in the motif G_{ij}.

colon cancer data set,[4] and a B-cell lymphoma data set.[5] A detailed description of our results is available at http://genomics10.bu.edu/murali/xmotif.

For each data set, we computed 50 genes that were most informative about the class distinctions in the data.[d] For each gene in the data set, we computed its score using the "twoing" rule[19] and selected the genes with the 50 best scores. For each of these genes, we computed all the states as described earlier. Finally, we executed the XMOTIF-finding algorithm on each data set.

In all these experiments, we set the number of seeds $n_s = 10$ and the number of determinants $n_d = 1000$. The size s_d of the discriminating set varied from 7 to 10. The quality of our results did not change much if we varied these parameters slightly. For each data set, our algorithm took about 2–5 minutes to compute all the XMOTIFs when running on a computer equipped with an 800 MHz Intel Pentium III processor.

Our overall algorithm uses information about which class each sample belongs to only to compute the set of informative genes. The XMOTIF algorithm does not explicitly take class labels into account. As a result, samples from different classes may match a computed XMOTIF.

5.1　ALL-AML data

The ALL/AML data set[3] consists of the expression levels of roughly 6,800 human genes measured using an Affymetrix oligonucleotide array from bone marrow samples collected from 47 patients suffering from acute lymphoblastic leukaemia (ALL) and 25 patients suffering from acute myeloid leukaemia (AML).

For each gene, we considered only those states that had p-values at most 10^{-10} and contained at most 50 expression values. Our algorithm computes five XMOTIFs that cover the data. The samples matching four of these XMOTIFs are almost exclusively ALL patients. The fifth motif is matched almost exclusively by AML patients. The number of conserved genes in ALL-related motifs ranges from 11 to 21 while the AML-related motif contains 7 conserved genes. Table 1 displays information on each motif in the order they were computed.

The algorithm is able to compute motifs that distinguish between the two types of leukaemia almost perfectly. A total of 30 distinct genes appear in ALL-related motifs. Only one gene, TCF3 Transcription factor 3 (E2A immunoglobulin enhancer binding factors E12/E47), is conserved both for ALL patients and for AML patients. In all the motifs, almost every gene is in an under-expressed state (relative to the total expression range of the gene).

---

[d]The number 50 is somewhat arbitrary. We chose it because Golub et al. build a classifier using 50 genes for the ALL-AML data set.

To obtain another view of the degree of similarity between the samples that match a motif, we computed average intra- and inter-motif distances. To compute the distance between two samples that match a motif, we used the standard Manhattan (also called rectilinear or L_1) distance, except that we ignored genes that were not conserved in that motif and we divided the distance by the number of conserved genes in that motif. We defined the distance between two samples in different motifs similarly. We observed that all the average intra-motif distances ranged between 250 and 350. All the average inter-motif distances between ALL-related motifs ranged between 310 and 515 but the average distance between ALL motifs and the AML motif ranged from 1675 to 3115. These results indicate that the ALL-related and AML-related motifs captured distinct regions of the gene expression space.

ALL-AML motifs			
#genes	#samples	#ALL	#AML
18	19	18	1
11	17	15	2
7	20	1	19
21	9	8	1
19	5	5	0

Colon cancer motifs			
#genes	#samples	#tumour	#normal
11	15	14	1
13	18	2	16
12	11	9	2
6	10	9	1
1	8	6	2

B-cell lymphoma motifs			
#genes	#samples	#alive	#dead
15	14	11	3
17	14	0	14
3	10	10	0

Table 1: Motifs computed for the data sets.

5.2 Colon cancer data

Alon et al.[4] present a data set containing 62 samples of colon epithelial cells collected from patients suffering from colon cancer; 40 of these samples are collected from tumours and 22 from normal colon tissues of the same patients. They measured the absolute expression levels of about 6500 human genes using an Affymetrix oligonucleotide array.

We use the supervised algorithm for assigning p-values to gene states and discarded all states that contained more than 40 values or had p-value greater than 10^{-10}. Our algorithm computed five xMOTIFs. At least 75% of the samples matching four of these xMOTIFs were tumorous tissues. Almost 90% of the samples matching the fifth xMOTIF were normal tissues. Only one gene was conserved in the fourth tumour-related xMOTIF but the other xMOTIFs contained between 6 and 13 genes. Only one gene, P03001 Transcription Factor IIIA, was conserved in both types of motifs. If we ignore the fourth tumour-related xMOTIF (since it contains only one conserved gene), distances between tumour-related xMOTIFs ranged from 90 to 265, while all the distances between the normal xMOTIF and the tumour xMOTIFs were greater than 610.

5.3   B-cell lymphoma data

Alizadeh et al. [5] describe a data set of 96 normal and malignant lymphocyte samples whose expression levels they measured using a specialised "Lymphochip." Alizadeh et al. had information of the survival rates of the patients suffering from diffuse large B-cell lymphoma (DLBCL). Of the 40 such patients, 22 survived (data available at http://llmpp.nih.gov/lymphoma/data.shtml). We applied the supervised version of our algorithm to this classification (using only gene states with p-value at most 10^{-3}) and obtained three motifs, with two motifs being conserved across patients who survived and one motif conserved across patients who died. Only three genes were conserved across both classes. One of these is a NF-κB family member that is frequently amplified in diffuse large cell lymphoma. The other two are ESTs of unknown function. In each motif, the conserved genes appear in states ranging from relatively under-expressed to relatively over-expressed. The separation between the xMOTIFs in terms of distance was not as clear as for the other data sets.

6   Conclusions

We have introduced a useful and concise representation of gene expression data in the form of conserved gene expression motifs or xMOTIFs. These motifs capture the degree of conservation in the gene expression profiles of the samples belonging to a class at two levels: (i) each gene in the motif is similarly-expressed in each of the samples and (ii) all the genes in the motif are simultaneously conserved in all these samples. We believe that this representation has the potential to capture many key biological properties implicitly present in gene expression data. We have implemented a system to compute large xMOTIFs that cover all the classes in the data. Our analysis of three publicly-available data sets shows that our algorithm can compute xMOTIFs that appear to distinguish between the classes in these data sets quite well. Our technique has the potential to find clinically and biologically relevant subdivisions in gene expression data.

In our current formulation, we require that each gene in a motif be expressed in the same state across all samples matching the motif. A useful generalisation of this concept is the requirement that the gene's expression levels in a motif obey a specified distribution. Extending our algorithm to this case appears to be a challenging problem.

Acknowledgements   We would like to than Geoff Cooper and Michael Schaffer for useful discussions. We thank Yang Su for providing us with the gene-selection software. An NSF-KDI grant partly funded this research.

References

1. S Fodor, R Rava, X Huang, A Pease, C Holmes, and C Adams. Multiplexed biochemical assays with biological chips. *Nature*, 364(6437):555–6, 1993.
2. M Schena, D Shalon, R Davis, and P Brown. Quantitative monitoring of gene expression patterns with a complementary dna microarray. *Science*, 270(5235):467–70, 1995.
3. T.R. Golub, D.K. Slonim, P. Tamayo, C. Huard, M. Gaasenbeek, J.P. Mesirov, H. Coller, M.L. Loh, J.R. Downing, M.A. Caligiuri, C.D. Bloomfield, and E. S. Lander. Molecular classification of cancer: class discovery and class prediction by gene expression monitoring. *Science*, 286:531–537, 1999.
4. U Alon, N Barkai, D A Notterman, K Gish, S Ybarra, D Mack, and A J Levine. Broad patterns of gene expression revealed by clustering analysis of tumor and normal colon tissues probed by oligonucleotide arrays. *Proc Natl Acad Sci U S A*, 96(12):6745–50, 1999.
5. A A Alizadeh, M B Eisen, R E Davis, C Ma, I S Lossos, A Rosenwald, J C Boldrick, H Sabet, T Tran, X Yu, J I Powell, L Yang, G E Marti, T Moore, J Hudson, L Lu, D B Lewis, R Tibshirani, G Sherlock, W C Chan, T C Greiner, D D Weisenburger, J O Armitage, R Warnke, and L M Staudt. Distinct types of diffuse large b-cell lymphoma identified by gene expression profiling. *Nature*, 403(6769):503–11, 2000.

6. Cecilia Magdalena Procopiuc, Michael T. Jones, Pankaj K. Agarwal, and T. M. Murali. A monte-carlo algorithm for fast projective clustering. In *Proceedings of the 2002 International Conference on Management of Data*, 2002.

7. M. R. Garey and D. S. Johnson. *Computers and Intractability: A Guide to the Theory of NP-Completeness.* W. H. Freeman, New York, NY, 1979.

8. Robert Tibshirani, Trevor Hastie, Balasubramanian Narasimhan, and Gilbert Chu. Diagnosis of multiple cancer types by shrunken centroids of gene expression. *Proc Natl Acad Sci U S A*, 99(10):6567–72, 2002.

9. M.B. Eisen, P.T. Spellman, P.O. Brown, and D. Botstein. Cluster analysis and display of genome-wide expression patterns. *Proc. Natl. Acad. Sci. USA*, 95:14863–14868, 1998.

10. R. Tibshirani, T. Hastie, M. Eisen, D. Ross, D. Botstein, and P. Brown. Clustering methods for the analysis of dna microarray data. Technical report, Department of Statistics, Stanford University, 1999. Available at http://www-stat.stanford.edu/ tibs/ftp/jcgs.ps.Z.

11. N Friedman, M Linial, and I Nachman. Using bayesian networks to analyze expression data. *J Comput Biol*, 7(3-4):601–20, 2000.

12. R Sharan and R Shamir. Click: a clustering algorithm with applications to gene expression analysis. *Proc. Int. Conf. Intell. Syst. Mol. Biol.*, 8:307–16, 2000.

13. J. A. Hartigan. Direct clustering of a data matrix. *J. Amer. Statis.*, 67:123–129, 1972.

14. Y. Cheng and G. M. Church. Biclustering of expression data. In *Proc. ISMB*, 2000.

15. Amos Tanay, Roded Sharon, and Ron Shamir. Biclustering gene expression data. In *Proceedings of ISMB 2002*, pages S136–S144, 2002. Available at http://www.math.tau.ac.il/ roded/publications.html.

16. G. Getz, E. Levine, and E. Domany. Coupled two-way clustering of dna microarray data. *Proc. Natl Acad. Sci. USA*, 97:12079–12084, 2000.

17. T Hastie, R Tibshirani, M B Eisen, A Alizadeh, R Levy, L Staudt, W C Chan, D Botstein, and P Brown. "Gene shaving" as a method for identifying distinct sets of genes with similar expression patterns. *Genome Biol*, 1(2), 2000.

18. Jinyan Li and Limsoon Wong. Identifying good diagnostic gene groups from gene expression profiles using the concept of emerging patterns. *Bioinformatics*, 18(5):725–34, 2002.

19. S. K. Murthy, S. Kasif, and S. Salzberg. A system for induction of oblique decision trees. *Journal of Artificial Intelligence Research*, 2:1–33, 1994.

DECOMPOSING GENE EXPRESSION INTO CELLULAR PROCESSES

E. SEGAL , A. BATTLE AND D. KOLLER
Computer Science Department
Stanford, CA 94305-9010
E-mail: eran@cs.stanford.edu,ajbattle@stanford.edu,koller@cs.stanford.edu

Abstract

We propose a probabilistic model for cellular processes, and an algorithm for discovering them from gene expression data. A *process* is associated with a set of genes that participate in it; unlike clustering techniques, our model allows genes to participate in multiple processes. Each process may be active to a different degree in each experiment. The expression measurement for gene g in array a is a sum, over all processes in which g participates, of the activity levels of these processes in array a. We describe an iterative procedure, based on the EM algorithm, for decomposing the expression matrix into a given number of processes. We present results on Yeast gene expression data, which indicate that our approach identifies real biological processes.

1 Introduction

A living cell is a complicated system that performs multiple functions and has to respond to a variety of signals. To organize this complex web of activity, the cell tends to compartmentalize its activity into distinct *processes*, or modules. This global organization cannot be discerned by studying the properties of isolated components. Genome-wide measurements of mRNA expression level across multiple experimental conditions provide us with a global picture of the cell's activities, and provide the potential for a high-level understanding of its behavior.

Clustering techniques are the most common approaches to identifying functional groups in gene expression data. These approaches generate clusters of genes that have similar expression profiles over a range of experimental conditions [6, 4, 14]. However, these approaches group genes into mutually exclusive clusters, and are thus limited in their ability to represent the true underlying biological system: many genes are known to be multi-functional, and thus should belong to more than one functional group.

In this paper, we introduce a probabilistic framework for discovering biological processes from expression data. Each process is associated with a set of genes that participate in it; unlike clustering methods, our model allows genes to participate in multiple processes. Each process might be more active in some conditions and less active in others. Thus, our model defines for each experiment the extent to which each process is active in that experiment. In our model, the expression measurement for gene g in array a is a sum, over all processes in which g participates, of the activity levels of these processes in array a. Thus, we decompose the entire expression matrix as a sum of the expression levels of all the active processes.

Our model resembles the Plaid model proposed by Lazzeroni and Owen [11]. The Plaid model also decomposes the expression data as a sum of overlapping "layers"; each layer is associated with a set of genes and experiments that define the expression of that layer. There are several differences between our model and Plaid. Of these, the

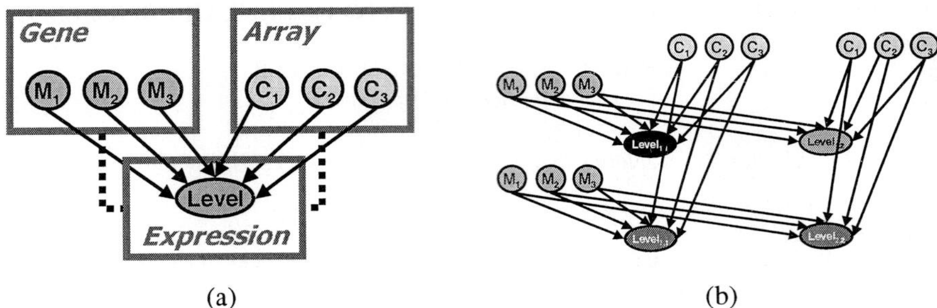

(a)   (b)

Figure 1: (a) PRM for the process model. (b) An instantiation of the PRM to a particular dataset with 2 genes, 2 arrays and 3 processes.

most important is that Plaid uses a greedy sequential approach to perform the decomposition, attempting, in each step, to explain as much of the unexplained expression data as possible. Once a layer has been learned, it remains unchanged and is subtracted from the expression data. In contrast, our model is trained as a unified whole, allowing the association of genes with processes to change as the process models become more refined, and the process models to change as the set of assigned genes changes. As we demonstrate in our experimental results, our approach discovers much "cleaner" processes than does Plaid: The set of genes associated with a process by our approach often contains a very high fraction of genes that are known to share a functional role. By contrast, Plaid layers are much larger and more heterogeneous, and do not correspond as neatly to a biological process.

We provide an iterative algorithm for learning the model from gene expression data, based on the *expectation maximization (EM)* algorithm [5]. Once a model has been learned, we can read the processes directly from it. For each process, we read both the genes that participate in it as well as the levels of activity of each process in all experiments. We describe encouraging results on real data, providing evidence that our approach identifies real biological processes. Specifically, we show a high correlation between the gene sets constructed and known biological processes. We also show significant DNA binding sites in the promoter regions of the genes in the process. Finally, we show cases where our learned activity levels for processes had extremely high correlations with the expression levels of known regulators of those processes; importantly, these regulators were *not* part of the input data given to our program, indicating that our program reconstructed the levels of activity of these processes.

2 Probabilistic Model

In this section we present our probabilistic model. Our approach is based on the language of *probabilistic relational models (PRMs)*, as described in [10, 7]. For lack of space, we do not review the general PRM framework, but focus on the details of the model, which follow the application of PRMs to gene expression in [14]. A simplified version of our model is presented in Fig. 1(a); we now describe its elements.

The PRM framework represents the domain in terms of the different biological entities that interact in it: genes, arrays, expression measurements, and biological processes. Also, each object may be associated with a set of attributes that are relevant to the interactions in the domain. Specifically, our model includes a set $\mathbf{G}$ of n gene objects $\mathbf{G} = \{g_1, \ldots, g_n\}$, a set $\mathbf{A}$ of k array objects $\mathbf{A} = \{a_1, \ldots, a_k\}$, and a set $\mathbf{E}$ of expression objects $\mathbf{E} = \{e_{1,1}, \ldots, e_{n,k}\}$, one for each gene in each array. Each expression object e is associated with a gene object $e.Gene = g$, an array object $e.Array = a$, and a real-valued attribute $e.Level$ denoting the mRNA expression level of $e.Gene = g$ in $e.Array = a$.

In addition, we include a set of j *process* objects. Our model makes explicit the notion that genes participate in biological processes and that processes are active to varying degrees in arrays. Thus, for each gene object g we define a set of binary attributes, $g.M_1, \ldots, g.M_j$, where $g.M_p$ represents the gene's *M*embership in process p. To represent process activity levels, we associate with each array object a, a set of continuous attributes $a.C_1, \ldots, a.C_j$, where $a.C_p$ represents the a*C*tivity level of process p in array a.

Each expression measurement e associated with gene $g = e.Gene$ and array $a = e.Array$ is assumed to be a (stochastic) function of the processes in which g participates and of the activity level of those processes in the array a. More precisely, let $g.\mathbf{M}$ be the set of all g's membership variables, and $a.\mathbf{C}$ be the set of all a's activity level variables. We assume that $e.Level$ is normally distributed with mean $\mu_e = \sum_p g.M_p \cdot a.C_p$ and standard deviation σ_a. More precisely:

$$P(e.Level) = \exp\left(\frac{(e.Level - \sum_p g.M_p \cdot a.C_p)^2}{2\sigma_a^2}\right) \tag{1}$$

where σ_a is the standard deviation of all expression measurements in array a. Thus, the expression measurement for gene g in array a can be viewed as the sum over expression components, with each component being the result of the activity in array a of some process to which g belongs.

To complete the description of our probabilistic model, we associate with each process p a prior probability $P(g.M_p) = q_p$, which is the prior probability with which any gene participates in p. We also associate with the continuous attribute $a.C_p$ a uniform distribution (over some appropriately bounded range).

Although the description of our model is compact, its instantiation to a particular data set is quite large. In a specific instantiation of the PRM model we might have 10 processes, 1000 genes and 100 arrays. Thus, we have as many as 1000×100 expression objects (if all expressions are observed), so the instantiation of our model to a particular dataset contains a large number of objects and variables that interact probabilistically. The resulting probabilistic model is a *Bayesian network* [12], where the local probability models governing the behavior of nodes of the same type (e.g., all nodes $g.M_p$ for different genes g in process p) are shared. Fig. 1(b) shows a small instantiation of such a network, for two genes, two arrays, and three processes.

Putting everything together, an instantiation of the PRM model specifies the membership $g.M_p$ for all genes $g \in \mathbf{G}$ and all processes p, the activities $a.C_p$ for all arrays $a \in \mathbf{A}$ and all processes p, and the expression level $e.Level$ for all $e \in \mathbf{E}$. The joint

92

distribution over all possible instantiations is given by:

$$P(\mathbf{G}.\mathbf{M}, Arrays.\mathbf{C}, \mathbf{E}.Level) =$$

$$\left(\prod_p \left(\prod_{g \in \mathbf{G}} P(g.M_p) \right) \left(\prod_{a \in \mathbf{A}} P(a.C_p) \right) \right) \left(\prod_{e \in \mathbf{E}} P(e.Level \mid g.\mathbf{M}, a.\mathbf{C}) \right) \quad (2)$$

where $a = e.Array$ and $g = e.Gene$.

Our model has several desirable properties. First, processes are represented explicitly making it easy to "read" off processes from the model: the $g.M_p$ variables tell us which genes participate in process p and the $a.C_p$ variables tell us the activity level of each process p in each array. Second, the model allows for genes to participate in more than one process (since we could have $g.M_p = 1$ and $g.M_q = 1$ for different processes p and q), which enables us to model multi-functional genes. Finally, as we will see in the next section, the probabilistic model allows us to utilize statistical optimization techniques to learn the models from data over several processes jointly.

3 Learning the Model

In the previous section, we described the different components of our probabilistic model. We now consider how we learn this model from data. We assume that the only information given is the expression data itself, and the number of processes we wish to identify. We do not know which genes participate in which processes nor the levels of activity of arrays processes. Thus, all attributes $g.M_p$ and $a.C_p$ are hidden. From the perspective of the model parameters, the prior probabilities of the different processes — $P(g.M_p) = q_p$ are also unknown. We assume that the expression level model is given, as in Eq. (1), and that the distribution $P(a.C_p)$ is uniform and fixed for all a, p.

The learning problem we have here is quite complex, as it involves a large number of hidden variables. The main technique we use to address this issue is the *Expectation Maximization (EM)* algorithm [5], which allows parameter estimation with incomplete data. The EM algorithm is an iterative method. Starting from an initial setting for the parameters, it repeatedly performs two steps. In the *E-step*, it computes the distribution over the unobserved attributes $g.M_p, a.C_p$ (for all g, a, p), given the observed expression data and the current estimate of the parameters. It uses this distribution to "fill in" each missing attribute. Two variants of the EM algorithm are *soft EM*, where the completion explicitly accounts for the probability over the values of each missing attribute, and *hard EM*, which simply selects the single most likely assignments to each attribute. The *M-step* re-estimates the parameters, using the completion of the missing attributes as if it were real, using a standard maximum likelihood estimation procedure. The process then repeats, using the new parameters, until convergence. Soft EM is guaranteed to converge to a local maximum of the likelihood of the observed data $P(\mathbf{E}.Level)$; hard EM is guaranteed to converge to a local maximum of the joint likelihood of the observed data *and the completion* — $P(\mathbf{G}.\mathbf{M}, \mathbf{A}.\mathbf{C}, \mathbf{E}.Level)$.

In applying either variant of EM to our model, we must deal with the complexities of the E-step, which require that we compute the distribution over all assignments to both $\mathbf{G.M}$ and $\mathbf{A.C}$. As we discussed, our model induces a complex set of interactions. In the experiments we describe below, we have 1010 genes, 173 arrays, and 10 processes. This results in a model with $11,830$ hidden variables. Moreover, the nature of the data is such that we cannot treat genes (or arrays) as independent samples. Instead, any two hidden variables are dependent on each other given the observations (see [7] for an elaboration of this point). For example, consider two genes g and h; h's assignment to processes influences our estimates of the $a.C_p$ variables, which in turn influence our membership probabilities for g. Due to these dependencies, the exact computation of the E-step is intractable for large domains.

However, our model is such that if we were given the values of $a.C_p$ for all a, p, then the assignments of the different genes to processes are rendered independent. Likewise, given a fixed assignment of genes to processes ($g.M_p$), the activity levels for each array a can be estimated from these assignments and from the expression data in a alone, without knowing the activity levels in other arrays.

This key observation suggests the use of hard assignments to the hidden attributes, rather than a soft assignment. Thus, we use a variant of hard EM. In this case, our goal in the E-step is to find the maximum probability assignment to the variables $\mathbf{G.M}, \mathbf{A.C}$, given the current parameter setting. This problem is still intractable, but we can use our observation to find a very good local maximum.

Specifically, starting from an initial assignment of genes to processes (which could come from standard clustering methods), we find the most likely activity levels $\mathbf{A.C}$. We then fix these activity levels, and find the most likely assignment to $\mathbf{G.M}$. Each step increases the joint likelihood $P(\mathbf{G.M}, \mathbf{A.C}, \mathbf{E}.\textit{Level})$ given the current parameters, and thus the process is guaranteed to converge. The resulting assignment to these variables is a fairly strong local maximum: No step that adapts only the gene memberships or the array activity levels can improve the likelihood; however, a step that adapts both gene memberships and array activities might.

At convergence, the E-step is complete, and we can use the final assignment $\mathbf{G.M}^*$, $\mathbf{A.C}^*$ to estimate the parameters q_p, using standard maximum likelihood estimation. More precisely, we compute the *expected* sufficient statistics:

$$N_{\mathbf{G}.M_p}[v] = \sum_{g \in \mathbf{G}} \eta(g.M_p^* = v) \tag{3}$$

where η is an indicator variable that takes value 1 when its argument holds and 0 otherwise. We then compute the probabilities q_p as:

$$q_p = \frac{N_{\mathbf{G}.M_p}[1]}{N_{\mathbf{G}.M_p}[0] + N_{\mathbf{G}.Member_p}[1]}. \tag{4}$$

The algorithm as a whole is shown in Fig. 2. As we can see, it defines two separate optimization tasks: finding the most likely memberships of genes in processes given activity levels (step 2(a)i), and finding the most likely activity levels given the memberships of genes in processes (step 2(a)ii). We discuss the implementation of these steps in the subsequent subsections.

A critical part of our approach is that our algorithm does not learn the membership and activity of each process in isolation. Rather, our model is learned over all

---

1. Initialize **G**.**M** using standard clustering techniques.

2. Repeat until convergence

   (a) **E-step** Repeat until convergence

      i. Find the assignment to each activity level $a.\mathbf{C}$ that maximizes $P(a.\mathbf{C} \mid \mathbf{E}.Level, \mathbf{G}.\mathbf{M})$.

      ii. Find the assignment to each gene membership $g.\mathbf{M}$ that maximizes $P(g.\mathbf{M} \mid \mathbf{E}.Level, \mathbf{A}.\mathbf{C}$.

   (b) **M-step** Estimate the parameters q_p as in Eq. (3) and Eq. (4).

Figure 2: Full Learning Algorithm

processes simultaneously, allowing information and (probabilistic) conclusions from one process to propagate and influence our conclusions about another. For instance, assume that our learning process places a gene g into process p at some step, and that this membership explains g's expression data very accurately. In this case, g will be less likely to be a member of other processes, allowing other genes assigned to the process to have a stronger influence on the activity level profile of the process.

One of our two tasks is to find the most likely activity levels given the memberships of genes in processes. Here, we assume that we are given, for each gene g, all the processes p in which it participates. Thus, we now need only to find the most likely assignment to the activity levels of arrays in processes, i.e., $\mathrm{argmax}_{\mathbf{A}.\mathbf{C}} P(\mathbf{A}.\mathbf{C} \mid \mathbf{E}.Level, \mathbf{G}.\mathbf{M})$. Using Bayes rule, our assumption of uniform prior over each $a.C_p$, and the model in Eq. (1), we can reformulate our maximization task as

$$\mathrm{argmax}_{\mathbf{A}.\mathbf{C}} P(\mathbf{E}.Level \mid \mathbf{A}.\mathbf{C}, \mathbf{G}.\mathbf{M}) =$$

$$\mathrm{argmax}_{\mathbf{A}.\mathbf{C}} \sum_{a \in \mathbf{A}} \sum_{e \in \mathbf{E}._a} \log \left(\frac{1}{\sqrt{2\pi}\sigma_a} \exp \left(\frac{(e.Level - \mu_e)^2}{2\sigma_a^2} \right) \right)$$

where $\mathbf{E}._a$ is the column of the expression matrix that corresponds to the array a. Simple algebraic reformulation shows that this problem is, in fact, a standard least squares problem $E \approx GC^T$, where: E is the standard $n \times k$ expression matrix; G is an $n \times j$ 0-1 matrix such that $G_{g,j}$ contains a 1 if gene g is a member in process j and C is a $k \times j$ matrix such that $C_{a,p}$ represents the activity level of array a in process p.

The matrices E and G are both fixed, and our goal is to find the matrix C that minimizes the squared-error for $E \approx GC^T$. It is well-known [9] that a least-squares solution to this system exists, and can be found effectively using standard methods.

We now turn to our second optimization problem, where we are given the activity levels of all processes in all arrays and our task is to learn the assignments of genes to processes. Thus, the attributes $g.M_p$ are hidden for all genes g and all processes p. However, with all activity levels given, assignments of genes to processes are independent across genes and we can find the most likely assignment for each gene separately.

To perform this maximization, we maximize $P(g.\mathbf{M} \mid \mathbf{E}_g, \mathbf{A}.\mathbf{C})$ separately for each gene g, where $\mathbf{E}_g$ is the row in the expression matrix corresponding to the gene g. More precisely, this computation can be done as follows:

$$\mathbf{v}^* = argmax_{\mathbf{v}'} P(g.\mathbf{M} = \mathbf{v}' \mid \mathbf{E}_g, \mathbf{A}.\mathbf{C}), \tag{5}$$

where

$$P(g.\mathbf{M} = \mathbf{v} \mid \mathbf{E}_g, \mathbf{A}.\mathbf{C}) = \alpha \prod_p P(g.M_p = v_p) \prod_{e \in \mathbf{E}_g} P(e.\mathit{Level} \mid g.\mathbf{M} = \mathbf{v}, \mathbf{A}.\mathbf{C}),$$

where $v_p \in \{0, 1\}$, and α is a normalization constant. The expression inside the final term in the product is simply the Gaussian model for the expression level given its parents, as in Eq. (1).

For models that include a large number of processes, we cannot perform this maximization over $g.\mathbf{M}$ exactly. The number of calculations required for each gene is exponential in the number of processes, since every possible joint assignment to $g.\mathbf{M}$ must be considered. In these cases, we use an approximation. Instead of considering every possible assignment to $g.\mathbf{M}$, we *include* only a subset $g.\mathbf{M}_I$ of processes, and *exclude* all others $g.\mathbf{M}_E$, forcing their value to 0. To select our subset, we relax the problem and allow each $g.M_p$ to be any real value between 0 and 1. We then maximize Eq. (5) subject to this relaxation. This problem reduces to a bounded least squares problem, which we can solve exactly [3]. We then select $g.\mathbf{M}_I$ from $g.\mathbf{M}$, by choosing those variables whose relaxed assignments are closest to 1 (in practice the majority of the variables are assigned to 0 in the relaxed solution). Finally, we find the most likely $\{0, 1\}$ assignment to $g.\mathbf{M}_I$ with all other variables fixed to 0 by maximizing:

$$P(g.\mathbf{M}_I = \mathbf{v} \mid \mathbf{E}_g, \mathbf{A}.\mathbf{C}, g.\mathbf{M}_E = 0) =$$
$$\alpha \prod_{p \in I} P(g.M_p = v_p) \prod_{e \in \mathbf{E}_g} P(e.\mathit{Level} \mid g.\mathbf{M}_I = \mathbf{v}, g.\mathbf{M}_E = 0, \mathbf{A}.\mathbf{C})$$

4 Model Evaluation

We first evaluated our approach on synthetic data. These experiments test whether we recover structure known to be present in the data. We generated a synthetic data set by sampling from a PRM model. To make the data realistic, we used PRM models learned from real biological data [8]. Specifically, we first learned a model with 7 processes. We then sampled data for 500 genes and 173 experiments (the original data contained 173 experiments) from the model: assignments of genes to processes were sampled from the distribution our model had for the **G.M** variables, and expression data was then derived by computing the expected expression levels (according to our model of expression) from the sampled assignment of genes to layers and the $a.\mathbf{C}$ means which were part of the learned model.

We then hid the true assignments of genes to processes and activity levels in arrays, as well as the original model parameters q_p, and learned a model with 7 processes from the synthetic expression data using the algorithm described in Section 3. To test the robustness of our learning algorithm to noise, we also learned models using various levels of perturbations, where a perturbation level of π corresponds to shuffling $\pi\%$ of the expression data across all genes and experiments. To gain statistical confidence, we generated five data sets for each perturbation π, and learned a model from each.

All models were evaluated by their ability to recover the "true" assignments of genes to processes (true assignments are the assignments in the sampled data) by performing a pairwise consistency test: we extracted all gene pairs appearing in the same

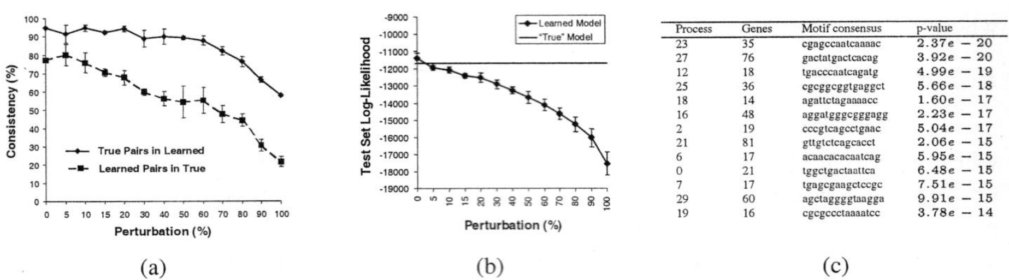

Process	Genes	Motif consensus	p-value
---	---	---	---
23	35	cgagccaatcaaaac	$2.37e - 20$
27	76	gactatgactcacag	$3.92e - 20$
12	18	tgacccaatcagatg	$4.99e - 19$
25	36	cgcggcggtgaggct	$5.66e - 18$
18	14	agattctagaaaacc	$1.60e - 17$
16	48	aggatgggcgggagg	$2.23e - 17$
2	19	cccgtcagcctgaac	$5.04e - 17$
21	81	gttgtctcagcacct	$2.06e - 15$
6	17	acaacacacaatcag	$5.95e - 15$
0	21	tggctgactaattca	$6.48e - 15$
7	17	tgagcgaagctccgc	$7.51e - 15$
29	60	agctaggggtaagga	$9.91e - 15$
19	16	cgcgccctaaaatcc	$3.78e - 14$

(a)	(b)	(c)

Figure 3: (a) Fraction of learned pairs appearing in the true data and fraction of true pairs in the learned model for various levels of perturbations. (b) Log-Likelihood on test data achieved for learned models for various levels of perturbations. (c) Motifs learned by searching for commonalities in the upstream regions of the genes in each process.

process in our learned model, and computed the fraction of these pairs appearing in the true data. We also tested the reverse, extracting all the true pairs and computing the fraction of these pairs appearing in a learned model. The results are summarized in Fig. 3(a), indicating that our algorithm reconstructs the true structure with very high accuracy even if 30% of the data is perturbed: gene pairs assigned to the same process in the true data, are likely to appear in our learned model and vice versa. Note that in fully randomized data (100% perturbation), a high fraction of the pairs in the learned model were indeed present in the true data ($58 \pm 0.4\%$). This occurs since the randomized data contains much weaker patterns and the total number of pairs learned is small, as can be seen by the poor coverage ($21.7 \pm 2.8\%$) of true pairs in these models.

As another evaluation, we measured the ability of our learned models to predict unseen data, by computing the likelihood that each model assigns to held out data. Specifically, we randomly partitioned the data into five equally sized sets of 100 genes and learned five models from all five possible combinations of four sets. For each such model we computed the likelihood it assigned to the held out subset. We compared these results to the likelihood that the "true" model from which the data was sampled assigned to the held out test data. These experiments were also performed in the presence of varying levels of perturbations. The results are summarized in Fig. 3(b). As can be seen, the test set likelihood is comparable (and even better with very little noise) for up to 30% perturbations, dropping sharply as more noise is added.

Recall that when the number of processes is large, we resort to the approximation described in Section 3. To evaluate our approximate algorithm, we learned a 12 process model, where we could apply the exact algorithm and compare the results. In our results, 77.2% of the genes had the same assignment to processes in the approximation and exact algorithm. However, the training and test set likelihoods of both models were practically the same, implying that the errors made by the approximation had little effect.

5 Biological Analysis

We now consider the data set of Gasch *et al.* [8], who characterized the genomic expression patterns of yeast genes in 15 different experimental conditions. We selected 1010 genes that had significant changes in gene expression (eliminating the ESR genes for which clustering is trivial), and the full set of 173 arrays.

We used the model discussed above, with 30 processes. Overall, our model predicted that 24 genes do not participate in any process, 552 genes participate in only one process, 257 in two, 119 in three, and 58 in four or more processes. As a comparison, we also tested a Plaid model with 30 processes, learned from the same data. (We obtained the Plaid software from http://www-stat.stanford.edu/~owen/plaid/.) The Plaid model assigned many more genes to layers than our model did, with 0 genes in no processes, 1 gene in one process, 4 genes in two, 10 genes in three, and 995 genes in four or more processes. According to Plaid, almost *all* genes participate in four or more processes, a situation not supported by current biological understanding. We note that the running time of our algorithm was 30 minutes on a 700MHZ Pentium 4, compared to 1 minute for running Plaid on the same machine.

To evaluate whether our assignments are biologically plausible, we checked whether the genes associated with each process showed any enrichment for known annotations. To do so, we used the GO [1] and KEGG [2] databases which assign genes to a diverse set of functional categories and biological pathways, respectively. For each process and each annotation, we counted the number of genes from the process with that annotation, and compared that to the total number of genes in our dataset with that annotation. If a process we learned indeed corresponds to known biological processes, then we expect the learned process to contain a high fraction of the genes with the corresponding annotation. For each combination of process p and annotation α, we can use the hyper-geometric distribution and assign a statistical significance (p-value) measure, corresponding to the probability that a randomly selected group of genes of the same size have similar enrichment for α. We performed this evaluation for our processes, the layers found by the Plaid model, and clusters from a standard clustering procedure.

The web supplement to this paper (http://cs.stanford.edu/~eran/psb03) lists the 30 processes, along with all annotations that were significant with a p-value of $1e-3$ or lower in either our model or in Plaid, where we removed some repetitive annotations from GO. Overall, we discovered highly significant processes relating to a variety of cellular functions. These included oxidative phosphorylation, various transport processes, protein folding, glycolysis, lipid metabolism, amino acid metabolism, carbohydrate metabolism, protein membrane targeting, ribosomal biogenesis, and cell cycle control. Some of the stronger active processes we identified were also present as Plaid layers, but Plaid layers typically included many extraneous genes, rendering the patterns less clear. For example, neutral lipid metabolism appears as a process of 17 genes with a p-value of $1.26e-21$ in our model, while in Plaid it appeared in a layer of 317 genes with a p-value of $1.22e-5$. Also, protein folding appeared as a process of 14 genes with a p-value of $1.46e-16$ in our model, while the corresponding Plaid layer had 254 genes with a p-value of $2.65e-7$. Fig. 4(a) shows a scatter plot com-

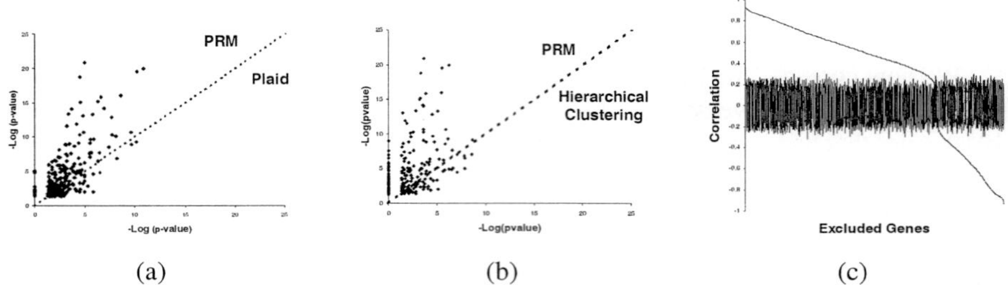

(a) (b) (c)

Figure 4: (a) Scatter plot of the negative log p-value of different GO and KEGG annotations for layers in Plaid on the one hand (X axis) and processes in our framework on the other (Y axis). Each point corresponds to one annotation. (b) Scatter plot of the negative log p-value of different GO and KEGG annotations for clusters from Pearson clustering on the one hand (X axis) and processes in our framework on the other (Y axis). (c) Correlation between all genes not included in the analysis and the learned activity levels. For each gene, the best correlation (or anti-correlation) is plotted as well as the best correlation achieved for that gene after permuting its expression measurements. The genes are sorted by best correlations.

paring the p-value for the GO and KEGG annotations that came up. We can see that in most cases (122 of 135 cases for pvalue of $1e-4$ or lower), the p-value achieved by our approach was always better and often much better than that achieved by Plaid. We performed a similar comparison to a standard hierarchical clustering algorithm [6], where we cut the hierarchy at 30 clusters to allow for a comparison to our model. The results are shown in Fig. 4(b), where again the majority of annotations appeared with greater significance in our model.

If genes assigned to the same process indeed participate together in a biological process, then the cell must have some regulatory mechanism by which it can coordinate their activity. One such mechanism is a shared DNA binding site (or multiple sites) recognized by a transcription factor (or several). To test whether genes that we associated with a process share DNA binding sites, we extracted the promoter regions of all genes (500bp upstream of translation start site) and applied a discriminative motif finder [13], searching for motifs of length 15. The result of the search is a standard position specific scoring matrix (PSSM) which can then be used to compute which genes have the binding site defined by the PSSM and which do not. From this we can derive a statistical significance measure assessing the uniqueness of the binding site to the set of genes associated with the process relative to the entire genes in the dataset. The consensus sequence of the best PSSMs learned along with the statistical significance of the PSSM to the process are summarized in Fig. 3(c). Overall, we were able to find unique binding sites in the set of genes in each process (see web supplement for full list), often with striking significance, consistent with significant annotations identified for each process using GO or KEGG. For example, we found a highly unique DNA binding site (p-value $2.37e-20$), occurring in the promoter region of 28 of the 35 genes in the oxidative phosphorylation pathway (process 23), compared to 102 appearances in the remaining 976 genes in the dataset, suggesting possible regulation of the pathway by this binding site. Indeed, the core element of this binding site is *ccaat*, which is the known target for the transcription factor regulator of oxidative phosphorylation, HAP4.

In addition to the assignments of genes to processes, our approach attempts to reconstruct the activity levels of each process p in each array a, as captured by the posterior mean of $a.C_p$. For each process, we can thus construct a vector $\mathbf{A}.C_p$ of the activity levels of p across all arrays $a \in \mathbf{A}$. We examined these activity levels, and found that they were biologically plausible for their respective processes. For instance, the process associated with protein folding (process 18) had high activity levels during heat shock and exposure to diamide, and low activity levels during amino acid and nitrogen depletion, reflecting accurately the biological function of the process.

Also, genes associated with process p should have high correlations between $\mathbf{A}.C_p$ and their average expression across all experiments. Indeed, a large number of genes were either highly correlated (286 genes with correlation 0.8 or above) or highly anti-correlated (13 genes with correlation -0.8 or below).

Much more exciting is to measure the correlation between the $\mathbf{A}.C_p$ vectors for all processes p and the 5147 genes that were *not* included in our analysis. Due to the way in which we selected the 1010 genes for our analysis, the genes included are likely to contain only a fraction of the genes associated with each process. If our model learned activity levels that indeed correspond to activity levels of real processes, then we expect to see high correlations between some of the left out genes and our learned activity levels. Indeed, there were many such genes: 614 of correlation above 0.8 and 252 of correlation below -0.8. To test whether this phenomenon could have happened by chance, we permuted the vector of expression measurements for each gene and recomputed the correlations. The results are summarized in Fig. 4(c), demonstrating that it is highly unlikely that our computed correlations could have resulted by chance, as the most significant correlation achieved for any of the 5147 permuted genes was -0.32. Surprisingly, the distributions of the correlation measurements were identical between the genes included in the analysis and those not included (data not shown).

Interestingly, there were several cases where the learned process activity levels had high correlation to the expression of known regulators (e.g., transcription factors) not included in the analysis. The web supplement lists all regulators with high correlation (or anti-correlation) to any process. Overall, we had 37 unique regulators with correlation above 0.7, of which 10 had correlation above 0.8, and 8 unique regulators with correlation below -0.7, of which 5 had correlation below -0.8. For 12 of the 30 processes, we learned activity levels that had extremely high correlation with known regulators. When information about the regulator was available in the literature, we could verify that the regulator that was highly correlated to a process, indeed was known to regulate the genes associated with that process. For example, CLB2, a G2/M phase specific cyclin, had correlation 0.88 with process 12, which in turn has significant cell cycle related annotations. Even when information was not available to verify our proposed regulation relationships, the regulators were known to be related to glucose starvation, cell wall stress, cell growth, cyclic AMP, ribosome synthesis, nitrogen starvation and mating, all processes known to be affected by the conditions in the Gasch [8] dataset.

6 Conclusions

In this paper, we have presented a probabilistic framework for extracting biological processes from gene expression data. Unlike most clustering methods, our approach does not attempt to associate each gene with a single process. Rather, it attempts to explain each gene's expression level as a sum of of the activity levels of the processes to which is belongs. For each process, we learn a set of genes that are associated with the process, and the extent to which the process is active in each array.

We compared our approach to the Plaid model of [11], that uses a related decomposition, and showed that our approach extracts processes that are more clearly identified with biological functions. In general, we showed that our approach provides a coherent global picture of biological processes.

An important advantage of our approach is that it is part of a general probabilistic framework for biological processes, as described in [14, 13]. Thus, it provides a mechanism by which we can integrate heterogeneous data sources, such as array annotations, clinical outcomes, or promoter region sequences, into a single coherent framework. Thus, for example, following [13], we could try to directly identify processes based not only on the expression data, but also on the existence of shared motifs in the promoter region. We intend to explore this extension and others in future work.

References

[1] Go: Gene ontology. In *http://www.geneontology.org*.

[2] Kegg: Kyoto encyclopedia of genes and genomes. In *http://www.genome.ad.jp/kegg*.

[3] A. Bjorck. *Numerical methods for least squares problems*. Siam, 1996.

[4] Y. Cheng and G.M. Church. Biclustering of expression data. In *ISMB'00*, 2000.

[5] A. P. Dempster, N. M. Laird, and D. B. Rubin. Maximum likelihood from incomplete data via the EM algorithm. *Journal of the Royal Statistical Society*, B 39:1–39, 1977.

[6] M. B. Eisen, P. T. Spellman, P. O. Brown, and D. Botstein. Cluster analysis and display of genome-wide expression patterns. 95(25):14863–8, 1998.

[7] N. Friedman, I. Nachman, and D. Peér. Learning of Bayesian network structure from massive datasets: The "sparse candidate" algorithm. Submitted, 1999.

[8] A.P. Gasch, P.T. Spellman, C.M. Kao, O.Carmel-Harel, M.B. Eisen, G.Storz, D.Botstein, and P.O. Brown. Genomic expression program in the response of yeast cells to environmental changes. *Mol. Bio. Cell*, 11:4241–4257, 2000.

[9] T. Kailath, A.H. Sayed, and B. Hassibi. *Linear Estimation*. Prentice Hall Information and Systems Sciences Series, 2000.

[10] D. Koller and A. Pfeffer. Probabilistic frame-based systems.

[11] L. Lazzeroni and A. Owen. Plaid models for gene expression data. Technical report, Stanford, 1999.

[12] J. Pearl. *Probabilistic Reasoning in Intelligent Systems*. Morgan Kaufmann, 1988.

[13] E. Segal, Y. Barash, I. Simon, N. Friedman, and D. Koller. From sequence to expression: A probabilistic framework. In *Proc. RECOMB*, 2002.

[14] E. Segal, B. Taskar, A. Gasch, N. Friedman, and D. Koller. Rich probabilistic models for gene expression. *Bioinformatics*, 17(Suppl 1):S243–52, 2001.

GENOME, PATHWAY, AND INTERACTION BIOINFORMATICS

PETER KARP, PEDRO R. ROMERO
Bioinformatics Research Group, SRI International
333 Ravenswood Avenue, Menlo Park, CA 94025
{pkarp,promero}@ai.sri.com

ERIC NEUMANN
Beyond Genomics
40 Bear Hill Road, Waltham, MA
eneumann@beyondgenomics.com

ALEXANDER J. HARTEMINK
Department of Computer Science and
Center for Bioinformatics and Computational Biology
Duke University, Box 90129
Durham, NC 27708-0129
amink@cs.duke.edu

The complete sequencing of the genomes of hundreds of organisms has fundamentally transformed both the focus and the practice of modern biology. While these efforts have largely succeeded in revealing the vocabulary of the "Book of Life", the syntax and semantics of the book remain in large measure undiscovered—the book reads more like Lewis Carroll's *Jabberwocky* than Shakespeare's *Hamlet*. Fathoming the meaning of the genomic vocabulary is the critical next step in better understanding biological systems and their complexity.

At the heart of this new challenge lies the need to understand the interplay of genes and their protein products. From the various interactions between genes and proteins, biological pathways and networks arise. These causal pathways and networks are responsible for the development, maintenance, regulation, and responsiveness of all living systems. Only when we go beyond the study of individual biological molecules to the analysis of their relationships does the complexity of biological systems come into full view. This degree of complexity calls for more powerful models and representations, new mathematical and computational frameworks, and possibly even altogether different analytical paradigms. The Genome, Pathway, and Interaction Bioinformatics session at PSB 2003 is dedicated to presenting novel research that attempts to answer this call by addressing fundamental issues in the recovery and analysis of the interactions, pathways, and networks governed by the genomes of all organisms.

This field has been significantly transformed by the emergence of new high-throughput functional genomics technologies that have enhanced our ability to observe the functioning of complex biological systems. Experimental data elucidating various aspects of bioloigical interactions, pathways, and networks is being generated at an ever-increasing rate. This large ensemble of information contains patterns that reflect pathway dynamics, and thus can be used to deduce causal pathway structures. Progress in this field will depend at least partially on the intelligent analysis and mining of this high-throughput functional genomics data in order to infer pathways and their regulation.

Specific relevant types of high-throughput data include gene expression profiles collected from high-density nucleotide arrays, protein expression profiles collected from 2D gels and newer array-based assays, gene-protein interaction data describing the binding location along the genome of DNA-binding proteins collected using chromatin immunoprecipitation, protein-protein interaction data collected using various hybrid systems, and the immense quantity of both gene and protein sequence information accumulating in publicly available databases. Each of these types of data presents a different perspective on the structure of biological pathways and networks, and the most effective analytical methods will likely draw their conclusions not from just a single type of data, but from the combined evidence of data of many different types.

The papers in the session this year examine a number of different approaches to the problem of network inference and modeling but retain some common themes. The point of widest disagreement among the papers seems to be the appropriate framework with which to model complex biological systems, the choices ranging from S-systems and extensions thereof, to Bayesian and dynamic Bayesian networks, to hybrid Petri nets, to full-blown systems of differential equations. The theme of greatest commonality among the papers seems to be the need for simulation to effectively evaluate the ability of these models to represent and recover properties of complex biological systems.

Beyond the issues addressed in the papers of this session, a number of other challenges in this field remain. Future work in the field will likely address such thorny questions as: How do we compare and align individual pathways or entire networks once uncovered? Can we infer qualitative network properties when the immense number of quantitative parameters that govern their behavior is not known? How can we effectively collect, organize, and integrate pathway information? Are there new experimental techniques that can be developed to reveal even more insights into the functioning of biological pathways? Finally, we note that in addition to the functional aspects of networks of interaction, spatial and temporal factors must also be taken into account when modeling complex biological systems, the difficulty of which is only beginning to be appreciated.

The field of genome, pathway, and interaction bioinformatics still includes many challenges and holds much promise; a better understanding of the interaction networks in complex biological systems will enable numerous advances in biotechnology, not the least of which is an enhanced ability to target therapeutics appropriately in diseased cells.

SUITABILITY AND UTILITY OF COMPUTATIONAL ANALYSIS TOOLS: CHARACTERIZATION OF ERYTHROCYTE PARAMETER VARIATION

R. E. ALTENBAUGH[*], K. J. KAUFFMAN[*], J. S. EDWARDS

Dept of Chemical Engineering, University of Delaware
Newark, DE 19716 USA
[*]Contributed equally

Systems engineering can provide insights into multivariate regulatory networks and pooling in complex biological networks that cannot be fully interpreted through experiments alone. Herein, we analyzed the use of phase planes, modal and time-lagged correlation (TLC) analyses of the human erythrocyte to explore the utility of these techniques for understanding the effect of single parameter changes on the behavior of a metabolic network. Specifically, several parameters in key regulatory steps in erythrocyte glycolysis, Rapoport-Leubering bypass, pentose phosphate pathway, adenosine metabolism, and membrane transport were perturbed. The most sensitive parameters were identified based on the steady-state metabolite concentration changes and were explored further. Modal analysis identified relevant time scales for each parameter change. These time scales were further explored using phase plane and TLC analyses. Phase plane and TLC both inferred pooling changes, while TLC also identified changes in the regulatory network structure that resulted from various parameter changes. Each method has strengths and weaknesses for exploring and gaining insight into complex biological networks.

1. Introduction

Biological system models can serve as suitable benchmarks for the development and comparison of analysis tools to study metabolic and regulatory networks. The modified Joshi and Palsson model of human Red Blood Cell (RBC) metabolism[1-6] is the most complete dynamic model of a cell that is currently available. Furthermore, the RBC model is relatively simple compared to other cellular networks, consisting of 44 enzymatic reactions and membrane transport systems describing the dynamics of 34 metabolites and ions. Although the RBC metabolic model is not complete in every aspect, it entails many of the issues expected to arise in the normal behavior of complex cellular systems such as nonlinear enzyme kinetics, multilayer feedback mechanisms, competing time scale effects, and pathway regulation.

Biological network models are primarily analyzed using purely linear techniques such as metabolic control analysis (MCA). In MCA, sensitive enzymes are identified by the steady-state impact of minute enzyme parameter changes.[7] Such techniques miss nonlinear feedback and regulatory effects that can significantly alter the underlying dynamics. Time-lagged correlation (TLC) analysis is a more general technique that measures nonlinear dynamic effects through random system disturbances selected to excite those dynamics. Arkin and Ross used TLC as the basis for a correlation metric construction analysis which takes the strongest correlations between metabolites, converts those correlations into a distance metric, and then projects those distances into an optimized 2D representation. The idea behind this work was that the 2D projection gave insight into the relative differences between metabolites.[8,9] However, in creating a simplified 2D projection, much of the TLC information is ignored. This necessitates alternative techniques.

Kauffman and coworkers explored the use of modal analysis, physiological pooling, phase plane analysis, and TLC for improved understanding of the dynamics of RBC metabolism.[10] These techniques are data intensive, and, at present, a full *in vivo* comparison is not feasible for large scale systems. Therefore, the RBC model provided sample data sets that allowed for a comparison of the techniques, evaluation of requirements and limitations of each technique, and exploration of experimental issues that may arise. However, Kauffman *et al.* did not explore the sensitivity of the results to variations in the kinetic parameters in the model— corresponding to discrete differences in metabolic networks.

The present work explores the ability of phase plane analysis and TLC to identify and characterize differences arising from single parameter changes in a complex metabolic network (corresponding to SNPs). Modal analysis was used to identify dynamic intervals of interest, thus focusing the data analysis techniques on time-scales of interest.

The RBC model used in this study modeled each of 34 metabolites using a series of differential-algebraic equations coded in FORTRAN. The resulting equations were solved using D02NDF and associated routines from the NAG™ Library.[11] This solver is ideally suited for sparse, stiff systems that undergo a series of dramatic environmental changes, including biological systems such as the RBC. The rate parameters were as previously published.[3,10] The complete code is available for download at http://epicurus.che.udel.edu/downloads/.

1.1 Parameter Sensitivity and Alterations

RBC model parameters were altered systematically one parameter at a time. Individual parameters were increased and decreased (ΔP_j) until the steady-state (SS^Δ) of the new system differed significantly from the nominal case (SS^o), as described by the normalized sum squared differences between the two steady-states. Parameter changes were selected so the sum squared difference between the steady-states approached 1. For most changes, this allowed the system to be altered without risking instability when energy loads or random disturbances were applied.

The sensitivity (Ψ) was numerically defined as the sum squared fractional difference of the steady-states (Γ) scaled by the corresponding parameter change:

$$\Psi_j^{\pm} = \frac{1}{\Delta P_j^{\pm}} \sum_{i=1}^{34} \left(\frac{SS_i^o - SS_i^{\Delta}}{SS_i^o} \right)^2$$

1.2 Analysis Techniques

<u>Phase Plane Analysis:</u> Phase planes plotted the concentrations of two metabolites, each parameterized with respect to time, against each other.[12] The phase plane analysis worked well for monitoring the changes in response to a step change in the environment of the metabolic network. In this study, one of three zero order step changes were made, simulating nonspecific changes in the usage efficiency of cellular currencies: 2.0 mM/h ATP load, 0.15 mM/h redox load, and a combination of 1.0 mM/h ATP load and 0.075 mM/h redox load. The strengths of the combined loads were smaller than the individual loads to prevent physiologically insignificant steady-states.[5] The effects of the three loads were followed for up to 310 h as the modeled cells responded from the steady-state identified following the parameter change to a new steady-state corresponding to the presence of the load(s). Phase planes constructed over time scales of interest were analyzed to identify pools of metabolites which had strong linear tendencies by computing a regression coefficient and the slope of the resulting best-fit line. Species with high regression coefficients and conserved, positive slope values were pooled together as equilibrium pools where the equilibrium constant was related to the slope of the best-fit line.[10]

<u>Time-Lagged Correlation Analysis:</u> A correlation array R, was constructed:

$$\mathbf{R}(\tau) = \left(r_{ij}(\tau) = \frac{\frac{1}{n}\sum_{k=1}^{n-\tau}\left(x_i(k)-\overline{x}_i\right)\left(x_j(k+\tau)-\overline{x}_j\right)}{\left[S_{ii}S_{jj}\right]^{0.5}} \right)$$

where S_{ii} was the variance and $\overline{x}_i$ was the mean value of the uniformly sampled dataset $x_i(t)$. Values of $r_{ij}(\tau)$ range from -1 to 1, where values approaching -1 indicate two datasets perfectly anti-correlated with a delay of τ time steps; values approaching 1 indicate two datasets perfectly correlated with a delay of τ time steps; and values approaching 0 indicate two datasets statistically unrelated at a delay of τ time steps. Pools of metabolites were identified by correlations stronger than 0.985 and peaking at zero lag.[10] Other peaks in correlation were used to infer potential connectivity and regulation.

TLC required that appropriate perturbations were applied to ensure persistent excitation of the network. Persistent excitation was achieved by application of truncated Gaussian perturbations for two extracellular concentrations (mM): glucose (GLU, $\mu=5$, $\sigma^2=2$), hypoxanthine (HX, $\mu=10^{-5}$, $\sigma^2=10^{-5}$) and three currency loads (mM/h): NADH ($\mu=0.1$, $\sigma^2=0.033$), GSH ($\mu=0.1$, $\sigma^2=0.033$), and ATP ($\mu=0.1$,

$\sigma^2=0.033$). Each input sequence was truncated to prevent negative concentrations or generation terms. Data were collected with a sampling frequency of 2 samples per perturbation and a fixed frequency of random perturbations for 2001 measurements. The resulting data were analyzed with TLC, which provided similar results to phase plane analysis with three key distinctions. First, TLC directly accounted for delayed effects whereas phase plane analysis did not. Second, TLC identified poolings and correlations based on deviations from the mean value whereas phase planes analyzed the absolute concentrations. Finally, whereas phase planes provided insight into how the metabolite moved from one steady-state to the next, TLC analyzed behavior with respect to persistent excitation.

<u>Modal Analysis</u>: A third technique used for analysis of the effects of various alterations was modal analysis. Modal analysis identified inherent time scales within the system and whether those time scales contained oscillatory contributions. Modal analysis first calculates the Jacobian matrix by multiplying the stoichiometric matrix with the vector of the gradient of the rate equations linearized about a desired steady-state. The Jacobian matrix was diagonalized as $J=M\Lambda M^{-1}$. The actual concentrations were transformed into a vector of independent modes, defined as $m=M^{-1}x$. Each mode moved on a characteristic time constant, defined as the negative inverse of the eigenvalue corresponding to the eigenvector of the Jacobian.[13] Significant changes in the time constants focused the TLC and phase plane analyses to interesting dynamic regions. While the modal analysis saved a significant amount of effort and time, it was not required for the other analyses (phase planes and TLC) which are data driven techniques and do not require an underlying model.

2. Results

Table I summarizes the resulting sensitivities for each parameter alteration. To prevent physiologically insignificant steady-states, parameters were decreased no more than 90% and increased no more than 1000 fold from their nominal values. When parameters changes within this region led to insignificant Ψ, the parameters were classified as "insensitive". Each of the remaining parameter changes were analyzed using the three methods described above. Parameter alterations were classified into four categories, depending on which component of RBC metabolism was affected: glycolysis, pentose-phosphate pathway (PPP), adenosine metabolism, and membrane transport. Each category is explored in greater detail.

2.1 Glycolytic Pathway

The glycolytic pathway is thought to be rate controlled by five enzymes: hexokinase (HK), phosphofructokinase (PFK), pyruvate kinase (PK), diphosphoglycerate (DPG) mutase (DPGM) and DPG phosphatase (DPGase).[3] The first three enzymes in this group are part of glycolysis, while the last two constitute the Rapoport-Leubering bypass, a feature specific to the RBC. For each of the core glycolytic enzymes, three parameters were changed: activity constant (V_{max}), ATP inhibition constant ($K_{m,ATP}$) or MgATP inhibition constant ($K_{m,MgATP}$) as appropriate, and substrate inhibition constant ($K_{m,Substrate}$).

Glycolytic parameter alterations manifested themselves as one of three phenotypes: increased ATP levels, decreased ATP levels, no significant effect (see Table II). TLC provided the most insight into glycolytic changes, typically manifested as pooling changes (Table III). Both the identified pools and the ATP phenotype were the same for changes in the V_{max} parameter of HK and PK. Increasing either of these two parameters led to sedoheptulose-7-phosphate (S7P) moving independently from the ribulose-5-phosphate (RU5P) pool. This common behavior was observed for several other parameter changes as well. Decreasing V_{max} for either HK or PK combined the 5-phosphoribosyl-1-pyrophosphate (PRPP) and RU5P pools. Almost all of the explored changes to the model parameters dissolved HX and inosine (INO) from the adenosine (ADO)/AMP/HX/INO pool. Under glucose limiting conditions (i.e., $K_{m,GLC}$ increased for HK), the RBC became more sensitive to perturbations, particularly ATP loads. Interestingly, this increased load sensitivity does not correspond to a significant change in the system steady-state. The largest $K_{m,GLC}$ change permitting stable application of the loads was $\Delta P^+=8.0$ ($\Gamma=0.27$). No other parameters explored demonstrated this type of sensitivity.

Other alterations to HK inhibition resulted in unique system changes. TLC revealed that increasing $K_{m,MgATP}$ or decreasing $K_{m,GLC}$ caused 6-phosphoglyconate (GO6P) to no longer pool with glucose-6-phophate (G6P) and fructose-6-phosphate (F6P). Under either parameter change, GO6P became less correlated with glycolysis while remaining highly correlated with the $NADP^+$/GSH pool in the PPP (Fig. 1). This resulted from PPP moving independent of glycolysis on the time scale explored.

2.2 *Pentose Phosphate Pathway (PPP)*

Three enzymes were perturbed in the PPP: G6P dehydrogenase (G6PDH), 6-phosphogluconolactonate (GL6P) dehydrogenase (GL6PDH), and glutathione oxidase (GSHR). The first two reactions were reversible: their equilibrium constants were shifted by changing the ratios of their forward and reverse kinetic constants. Changes in G6PDH and GL6PDH parameters only significantly affected levels of GO6P and $NADP^+$. Both G6PDH and GL6PDH were found to be insensitive to changes in their respective inhibition constants. G6PDH was found to be highly sensitive to shifts in equilibrium whereas GL6PDH was not. G6PDH equilibrium shifts were analogous to strong changes in V_{max} (G6PDH). Decreasing G6PDH activity decreased $NADP^+$ and GO6P without affecting 6-phosphogluconolactone

Table I: Parameter alterations and observed sensitivity. (A) glycolysis, (B) pentose phosphate pathway, (C) adenosine metabolism, and (D) transport in the RBC. Ψ=IS denotes insensitive parameters, IC denotes intracellular, XC denotes extracellular. *Γ=0.28 was used to prevent metabolic crash for HK.

(A)	HK	PFK	PK
V_{max}	Ψ^-=11.11, Ψ^+=10.00	Ψ^-=IS, Ψ^+=0.22	Ψ^-=2.86, Ψ^+=0.88
$K_{m,ATP}$	Ψ^-=IS, Ψ^+=0.95 (MgATP)	Ψ^-=1.67, Ψ^+=0.91	Ψ^-=1.89, Ψ^+=0.17
$K_{m,Substrate}$	Ψ^-=IS, *Ψ^+=0.034	Ψ^-=1.56, Ψ^+=IS	Ψ^-=2.04, Ψ^+=1.79
(B)	G6PDH	GL6PDH	GSHR
V_{max}	Ψ^-=1.30, Ψ^+=IS	Ψ^-=2.08, Ψ^+=IS	Ψ^-=1.79, Ψ^+=1.54
K_m	Ψ^-=IS, Ψ^+=0.01	Ψ^-=IS, Ψ^+=0.26	Not applicable
$\rightleftharpoons$	($\rightarrow$G6P), ($\rightarrow$GL6P) Ψ^-=9.10, Ψ^+=8.33	($\rightarrow$GO6P), ($\rightarrow$RU5P) Ψ^-=IS, Ψ^+=0.17	Not applicable
(C)	Adenosine Kinase (ADK)	ATPase	AMPase
V_{max}	Ψ^-=IS, Ψ^+=IS	Ψ^-=4.54, Ψ^+=4.00	Ψ^-=2.70, Ψ^+=2.00
$\rightleftharpoons$	($\rightarrow$ ATP), ($\rightarrow$ ADP) Ψ^-=2.50, Ψ^+=2.08	Not applicable	Not applicable
(D)	AEX	K^+ Leak	Na^+ Leak
V_{max}	Ψ^-=1.67, Ψ^+=0.01	Ψ^-=1.72, Ψ^+=IS	Ψ^-=2.08, Ψ^+=1.82
$K_{m, XC}$	Ψ^-=2.70, Ψ^+=1.11	Ψ^-=1.45, Ψ^+=IS	Ψ^-=IS, Ψ^+=0.20
$K_{m, IC}$	Ψ^-=2.38, Ψ^+=2.38	Ψ^-=IS, Ψ^+=0.05	Ψ^-=1.22, Ψ^+=0.50
	Na^+/K^+ Pump		
K^+ Inhibitor	Ψ^-=IS, Ψ^+=IS		
$K_{m,Na}$	Ψ^-=IS, Ψ^+=3.65		
$K_{m,ATP}$	Ψ^-=IS, Ψ^+=2.38		

Table II: Summary of glycolytic parameter manipulations resulting in observed phenotype..

Phenotype	Changes of parameters in glycolysis pathway
ATP Decrease	$\downarrow V_{max}$ (HK), $\uparrow K_{m,MgATP}$ (HK), $\downarrow K_{m,ATP}$ (PFK), $\downarrow V_{max}$ (PK), $\uparrow K_{m,PEP}$ (PK), $\downarrow K_{m,ATP}$ (PK)
ATP Increase	$\uparrow V_{max}$ (HK), $\downarrow K_{m,MgATP}$ (HK), $\downarrow K_{m,F6P}$ (PFK), $\uparrow V_{max}$ (PFK), $\uparrow V_{max}$ (PK), $\downarrow K_{m,PEP}$ (PK), $\uparrow K_{m,ATP}$ (PK), $\uparrow K_{m,ATP}$(PFK)
No noticeable effects	$K_{m,MgATP}$ (PFK), $K_{m,GLC}$ (HK)

Table III: Time-lagged correlation of glycolytic pathway reveals changing pools.

	Pools	
Nominal	G6P/F6P/G06P FDP/GAP/DHAP PG3/PG2/PEP R5P/RU5P/X5P/S7P/R1P	AD0/AMP/HX/INO NADPH/GSH PYR|NADH ADE|PRPP
	Decreased ATP	
$\downarrow V_{max}$ (HK,PK)	ADE|PRPP combine with (R5P/RU5P/X5P/S7P/R1P)	AD0/AMP/~~HX/INO~~
$\downarrow K_{m,ATP}$ (PFK)	R5P/RU5P/X5P/R1P/~~S7P~~	AD0/AMP/~~HX/INO~~
	Increased ATP	
$\uparrow V_{max}$ (HK,PK)	R5P/RU5P/X5P/R1P/~~S7P~~	AD0/AMP/~~HX/INO~~
$\downarrow K_{m,PEP}$ (PK)	R5P/RU5P/X5P/R1P/~~S7P~~	AD0/AMP/~~HX/INO~~
	No effect on system	
$\uparrow K_{m,MgATP}$ (HK)	G6P/F6P/~~G06P~~	AD0/AMP/~~HX/INO~~

Table IV: Major changes in steady-state metabolite concentrations arising from altered transport kinetics.

Parameter change	Steady-state levels
ADE Tx: $\downarrow V_{max}$, $\downarrow K_m$(ADE ext), $\uparrow K_m$(ADE int)	$\downarrow$ADE (38%), $\uparrow$PRPP(100%)
K^+ Leak: $\downarrow V_{max}$, $\downarrow K_m$(K^+ ext), $\uparrow K_m$(K^+ int)	$\uparrow K^+$(103%)
Na^+ Leak: $\downarrow V_{max}$, $\uparrow K_m$(Na^+ ext), $\downarrow K_m$(Na^+ int)	$\downarrow Na^+$(58%), $\downarrow K^+$(40%), $\downarrow$NADP$^+$(19%), $\uparrow$G6P(30%), $\uparrow$ATP(10%), $\uparrow$RU5P(13%), $\uparrow$PRPP(28%)
Na^+/K^+ ATPase Pump: $\uparrow K_m$(Na^+), $\uparrow K_m$(ATP)	$\downarrow K^+$(68%),$\uparrow Na^+$(78%), $\uparrow$G6P(23%), $\uparrow$ATP(8%),

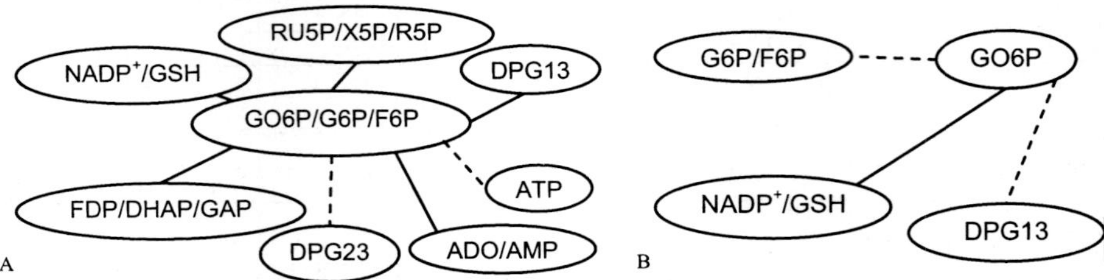

Figure 1: System connectivity to GO6P metabolite using TLC. (—) strong correlation (r_{ij} >0.80); (--) weak correlation ($0.05 \leq r_{ij} \leq 0.80$); A. Normalized system, B. Increased $K_{m,GLC}$ in HK.

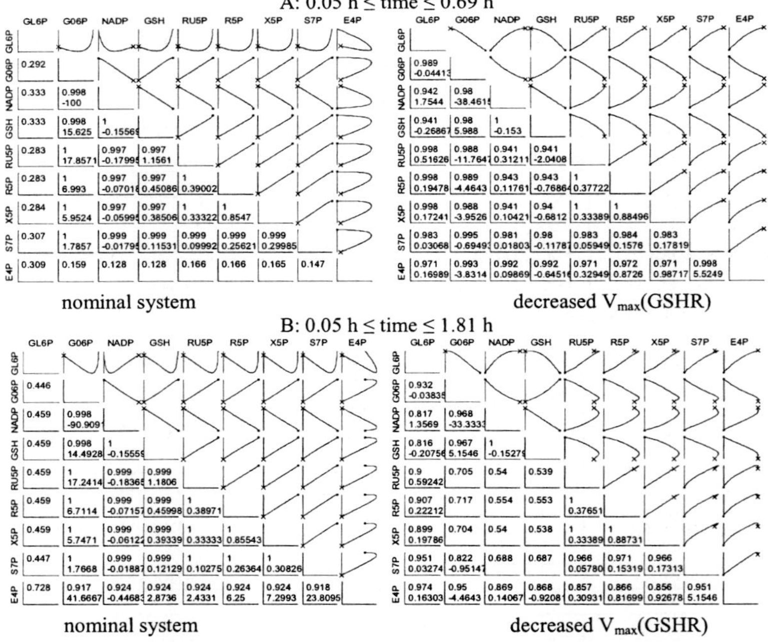

Figure 2: Phase planes for nominal and perturbed case. Manipulation of maximum rate of GSHR: (x) denotes starting point, (.) denotes ending point. The numbers in the lower left refer to the linear regression coefficient (upper number) and the slope of the straight line fit (lower number, for highly linear fits).

(GL6P) levels. GL6PDH parameter changes affected GO6P without noticeably changing the RU5P pool concentrations. Decreasing V_{max} (GL6PDH) increased GO6P levels two fold. GO6P levels were insensitive to GL6PDH inhibition and equilibrium constants.

Of the enzymes explored, GSHR affected the greatest number of metabolite concentrations. Decreasing V_{max} (GSHR) decreased NADPH (increased $NADP^+$) by 60%, GL6P by 57%, RU5P by 15%, and erythrose 4-phosphate (E4P) by 20% while increasing GO6P by 30%. These widespread changes represented the indirect effect of changing NADPH levels on the activity of several enzymes.

TLC analysis for 10 h perturbations demonstrated few changes in metabolite pooling with these parameter changes. All major pools were consistent over each parameter change in PPP kinetics. Modal analysis, however, demonstrated significant differences. Decreased GSHR activity dissolved of one pair of imaginary time constants ($\tau = 0.14 \pm 0.01i$ h) into two real constants ($\tau = 0.14$ h, $\tau = 0.36$ h). This change represents a loss of underdamped, oscillatory behavior and certain dynamic features such as overshoot. S7P and E4P made up the largest contributors to the underlying modes. TLC analysis over the time scale of interest (0.74 h perturbations) showed GO6P correlated less strongly with G6P and F6P. HX and INO no longer showed strong correlation.

Phase planes were analyzed with a combination of ATP and redox loads for both the nominal and decreased V_{max} (GSHR) systems. Two time scales were selected to observe the impact of the changes in dynamics predicted by the modal analysis on the observed phase plane evolution of the concentration profiles. Fig. 2A shows the phase plane evolutions over time ranges from 0.05 h to 0.69 h, allowing the 0.14 h mode to approach steady-state. Fig. 2B shows the concentration phase plane evolutions over time ranges of 0.05 h to 1.81 h, allowing the 0.36 h mode to approach steady-state. Three significant differences were observed between the nominal and altered systems: GL6P levels changed relative to the other PPP metabolites; E4P moved on a slower time scale; and many of the slopes for consistent linear relations changed (see lower left corner, each panel Fig. 2), potentially indicating changes in the equilibrium relationships for the altered system. Nominally, GL6P levels decreased under the loads. Under the altered case, GL6P levels increased 5 fold from steady-state and GL6P pooled with the RU5P pool. The nominal case showed E4P levels increased initially as NADPH levels increased and then eventually decreased (Fig. 2A). The altered case did not demonstrate this behavior until a later time scale, as shown in Fig. 2B and predicted by the modal analysis. Both systems showed strong linear correlations. However, there were several slope changes. Decreased V_{max} (GSHR) showed GO6P changed with respect to $NADP^+$, GSH, and S7P with slopes of ~1/3 the nominal values. Other GSHR changes had less pronounced effects.

112

2.3 Transport Mechanisms

The four transport mechanisms explored affected the RBC steady-state differently (Tables I and IV). Of the three leaks, parameter changes in Na^+ leak had the most system-wide effect due to resulting effects on Na^+/K^+ ATPase activity which led, via propagation, to widespread ATP effects. When the Na^+ leak decreased (internal Na^+ decreased), less ATP was consumed exporting Na^+. HK became more active and increased both RU5P and PRPP (Table IV). When the Na^+ leak increased, ATP levels decreased less dramatically, although G6P levels decreased slightly. Major poolings from TLC analysis for the Na^+ leak showed that when the Na^+ leak increased ADE pooled negatively with pentose phosphate metabolites. This type of pooling was also observed in the analysis of HK parameter changes.

Two of the three Na^+/K^+ ATPase Pump inhibition constants significantly affected the system. Increasing $K_{m,Na+}$ caused intracellular K^+ levels to fall while intracellular Na^+, ATP and G6P levels increased. The Na^+/K^+ ATPase pump was not as interconnected with the rest of the cell as other systems in the RBC. In this model, intracellular levels of Na^+ and K^+ only affected levels of other metabolites through the ATP required for the pump. When pump activity decreased, intracellular Na^+ increased and the additional ATP phosphorylated more glucose. Levels of other metabolites did not change significantly.

2.4 Adenosine Metabolism

Three key adenosine metabolism enzymes were analyzed: adenylate kinase (ADK), ATP phosphohydrolase (ATPase), and AMP phosphohydrolase (AMPase). Parameter changes in each of these enzymes demonstrated sensitivity. In particular changes in the ATPase parameters (Ψ^-=4.54, Ψ^+=4.00) influenced several RBC metabolite levels.

Decreased ATPase activity corresponded to a decreased glycolytic flux, leading to build up of the early glycolytic and PPP metabolites, including G6P, ADP, RU5P, PRPP, and NADPH. Modal analysis showed the loss of the pair of complex conjugate modes (τ=0.14 $\pm$ 0.01i h) to (τ=0.21 h, τ=0.16 h). Phase plane analysis comparing the nominal system to decreased ATPase activity over the time scale of 0.5-1.5 h showed a dramatic change in the dynamics of NADH. Over this time scale, NADH was nominally linear ($r^2\approx$0.96) with all other metabolites, but highly nonlinear in the case of the ATPase altered case. TLC analysis for perturbations every 0.21 h showed few differences between the systems except in the correlation between HX and INO. The maximum correlation between the two was 0.96 for the nominal case and 0.77 in the altered ATPase case. A similar trend was seen for many of the parameter changes studied. TLC analysis with 10 h perturbations did not show significant differences between the nominal and altered ATPase system.

Other parameter changes showed slight pooling changes. Decreased V_{max} (AMPase) and an ADK equilibrium shift towards ATP gave similar pooling due to net decreases in the concentrations of the glycolytic species. Shifting ADK equilibrium towards ATP/AMP caused levels of ADP, ATP and G6P to drop. Levels of DPG13, DPG23, fructose-1,6-bisphosphate (FDP) and $NADP^+$ increased. This behavior was examined through a flux analysis of the kinetic expressions. Under deficits in ADP, a larger fraction of the flux passed through the Rapoport-Leubering bypass and less ADP was used. However, this also caused ATP levels to decrease because a generation step was bypassed. Alternatively, decreasing AMPase activity led to unique behavior not seen by the other parameter changes explored. G6P and ADO levels decreased while AMP, ADP, and ATP levels increased.

3. Discussion

The most sensitive enzymatic parameters explored were V_{max} (HK) (Ψ^-=11.11, Ψ^+=10.00), G6PDH equilibrium (Ψ^-=9.10, Ψ^+=8.33), and V_{max} (ATPase) (Ψ^-=4.54, Ψ^+=4.00). Each of these significantly changed the steady-state of most metabolites in the system. Conversely, many parameter alterations only affected a few metabolite levels. In general, the model was sensitive to equilibrium constants for reversible reactions and relatively insensitive to inhibition constants.

An unexpected trend of the system occurred when HK was more active due to parameter changes resulting in increased overall ATP in the cell. In these cases, PRPP and the RU5P pool rose more than other metabolites. Nominally, ADE transport appeared limiting relative to PRPP production in the formation of AMP. This was not the case in systems with decreased ATP levels. Analysis of several parameter changes concluded that increased levels of the RU5P pool indicated increased fluxes through either PPP or glycolysis. GL6P is well-regulated and only directly affected by changes in the regulation by $NADP^+$.

Each method studied possessed distinct advantages and disadvantages. TLC depended on the frequencies of perturbation and measurement. With 10 h perturbations, only the most sensitive parameter changes caused new poolings. Generally, TLC offered more information than either of the other two analyses studied. However, as TLC analysis was more involved, it was helpful when suitable time scales for perturbing the system were known. The major TLC poolings revealed that most parameter alterations resulted in one or two pooling changes. Many of these changes represented moderate differences in correlation, indicating strong interaction was still present. In the nominal case, the correlations between HX, INO, and ADO were very strong, but in most of the perturbed cases studied, $r_{ij} \leq 0.5$. These metabolites were not as tightly controlled with respect to one another as other RBC poolings. This needs to be explored further as it may provide potential insight into the RBC model. GL6P never correlated strongly with GO6P, indicating distinct

114

regulation of the metabolites. TLC provided insight into interconnectivity between pools and how those connections changed in response to parameter changes. However, a priori identification of interesting time scales was helpful.

Modal analysis revealed time scale sensitivity to network parameter changes. Single parameter alterations typically caused significant changes in one or two time constants while the other time scales remained the same. Imaginary pairs of modes tended to show more interesting behavior: the addition or subtraction of a pair indicated significant dynamic changes. Analysis of the modes of the system determined where and how system dynamics changed with respect to the nominal system. Phase planes constructed for these regions, gave insight into the specific dynamic changes that occurred as a result of changing the parameters. For cases where a full model is not available, phase plane analysis could be used to identify the time scale where differences are most apparent. These differences could then be characterized more fully with either additional phase plane or TLC analyses.

These analyses demonstrate that common systems techniques are able to identify differences in metabolic networks that result from changes in single parameters. Although not presented here, introducing noise up to four percent into the system for time-lagged correlation analysis produced similar maximum correlation results, even though the peaks were not as pronounced. Using triplicate measurements effectively halved the effect of noise. The changes studied are representative of various pathological conditions such as infections, genetic variability, and environmental effects. Analyses such as this also provide insight into how several distinct pathological conditions can give rise to the same phenotypic symptoms.

4. Conclusions

Each of the analyses offered useful insight into cellular regulation. Modal analysis was most constructive as a starting point to determine the time scales on which alterations manifested. These time scales were then studied further with phase planes and correlation analysis. Rarely did more than one or two of the 34 time scales change in the modal analysis for the parameter changes observed. Phase planes identified major dynamic differences including changes in relationships. Phase planes visually demonstrated time scale differences manifestations better than either of the other techniques. As a result, phase planes could potentially be used to identify time scales of interest when modal analysis wasn't available due to lack of a complete model. TLC provided insight into changes in interaction and was better suited to decompose effects than phase plane analysis. A priori identification of the most significant time scale enhanced the utility of TLC. TLC showed subtle pooling differences under parameter changes, even though the analysis was more complex. Further work is needed to classify system responses for each possible single parameter alteration, explore the effects of multiple parameter alterations, examine changes to the fundamental kinetics of the system, and explore scalability issues regarding the sensitivity and utility of TLC, phase plane, and modal analyses.

Acknowledgements

Financial support was provided in part by a NASA Graduate Student Researcher's Program Fellowship (KJK, NGT5-50367), NIH BRIN Fellowship (REA, 1P20 RR16472-01) and NIH Grant 5P20 RR15588-02.

References

1. A. Joshi, B. O. Palsson, "Metabolic dynamics in the human red cell. Part I--A comprehensive kinetic model." *J Theor Biol* **141**, 515 (1989).
2. A. Joshi, B. O. Palsson, "Metabolic dynamics in the human red cell. Part II--Interactions with the environment." *J Theor Biol* **141**, 529 (1989).
3. A. Joshi, B. O. Palsson, "Metabolic dynamics in the human red cell. Part III--Metabolic reaction rates." *J Theor Biol* **142**, 41 (1990).
4. A. Joshi, B. O. Palsson, "Metabolic dynamics in the human red cell. Part IV--Data prediction and some model computations." *J Theor Biol* **142**, 69 (1990).
5. J. Edwards, B. Palsson, "Multiple steady states in kinetic models of red cell metabolism." *J Theor Biol* (2000).
6. N. Jamshidi, J. S. Edwards, T. Fahland, G. M. Church, B. O. Palsson, "Dynamic simulation of the human red blood cell metabolic network." *Bioinformatics* **17**, 286 (2001).
7. D. Fell, *Understanding the Control of Metabolism* (Portland Press, London, 1996).
8. A. Arkin, J. Ross, "Statistical Construction of Chemical Reaction Mechanisms From Measured Time-Series." *J Phys Chem* **99**, 970 (1995).
9. A. Arkin, P. D. Shen, J. Ross, "A test case of correlation metric construction of a reaction pathway from measurements." *Science* **277**, 1275 (1997).
10. K. J. Kauffman, J. D. Pajerowski, N. Jamashidi, B. O. Palsson, J. S. Edwards, "Description and Analysis of Metabolic Connectivity and Dynamics in the Human Red Blood Cell." *Biophys J* **83**, 646 (2002).
11. "FORTRAN Library Documentation, Mark 20" (Numerical Algorithms Group, 2002, pp. D02NDF.1).
12. J. Hubbard, B. H. West, *Differential equations: a dynamical systems approach*, Texts in applied mathematics; 5,18 (Springer-Verlag, New York, 1991).
13. A. Varma, M. Morbidelli, *Mathematical Methods in Chemical Engineering* (Oxford University Press, New York, 1997).

FOUNDATIONS OF A QUERY AND SIMULATION SYSTEM FOR THE MODELING OF BIOCHEMICAL AND BIOLOGICAL PROCESSES

M. ANTONIOTTI† F. PARK* A. POLICRITI+ N. UGEL† B. MISHRA†‡

† *Courant Institute of Mathematical Sciences, NYU, New York, NY, U.S.A.*

**Seoul National University, Seoul, S. Korea*

+ *Università di Udine, Udine (UD),* ITALY

‡ *Watson School of Biological Sciences, Cold Spring Harbor, NY, U.S.A.*

The analysis of large amounts of data, produced as (numerical) *traces* of *in vivo*, *in vitro* and *in silico* experiments, has become a central activity for many biologists and biochemists. Recent advances in the mathematical modeling and computation of biochemical systems have moreover increased the prominence of *in silico* experiments; such experiments typically involve the simulation of sets of Differential Algebraic Equations (DAE), *e.g.*, Generalized Mass Action systems (GMA) and S-systems [1,2,3]. In this paper we reason about the necessary theoretical and pragmatic foundations for a query and simulation system capable of analyzing large amounts of such trace data. To this end, we propose to combine in a novel way several well-known tools from numerical analysis (approximation theory), temporal logic and verification, and visualization. The result is a preliminary prototype system: simpathica/xssys. When dealing with simulation data simpathica/xssys exploits the special structure of the underlying DAE, and reduces the search space in an efficient way so as to facilitate any queries about the traces. The proposed system is designed to give the user possibility to systematically analyze and simultaneously query different possible timed evolutions of the modeled system.

1 Introduction

The emerging fields of system biology, and its sister field of bioinformatics, focuses on creating a finely detailed and "mechanistic" picture of biology at the cellular level by combining the part-lists (genes, regulatory sequences, other objects from an annotated genome, and known metabolic pathways), with observations of both transcriptional states of a cell (using micro-arrays) and translational states of the cell (using proteomics tools).

After more than a decade of progress, it has become evident that the mathematical foundation of these systems needs to be explored accurately, and that their computational models should be implemented in software packages faithfully while exploiting the potential trade-offs among usability, accuracy, and scalability dealing with large amounts of data. The work described in this paper is part of a much larger project still in progress, and thus only provides a partial and evolving picture of a new paradigm for computational biology.

We assume the following scenario. Imagine a biologist seeking to test some hypotheses against a corpus of data produced by several *in vitro*, *in vivo*, and *in silico* experiments regarding the behavior of a given biological system, *e.g.*, a regulated metabolic pathway in a given organism. A (graphical) metabolic map of the biochemical system under study, together with a specific associated S-system, is assumed available, and that the number or quantities recorded is large. The biologist can access one or both of the following items:

- Raw data stored somewhere about the temporal evolution of the biological system. This data may have been previously collected by *observing* an in vivo or an in vitro system, or by *simulating* the system in silico.

- Some mathematical model of the biological system[a].

The biologist will want to formulate *queries* about the evolution encoded in the data sets. For example, the biologist may ask: *will the system reach a "steady state"?*, or *will a temporary increase in the level of a certain protein repress the transcription of another?* Clearly the set of numerical *traces* of very complex systems rapidly becomes unwieldy to wade through for increasingly larger numbers of variables.

We first discuss what are the building blocks needed to construct a comprehensive and well-founded system capable of answering such queries. We implemented a prototype system called simpathica/xssys that builds on this foundation. The proposed computational tool derives its expressiveness, flexibility and power by integrating in a novel manner many commonly available tools from numerical analysis, symbolic computation, temporal logic, model-checking, and visualization.

Finally we show how we applied our system to a sizable example: the purine metabolism pathway as described in [3,4,5].

2   Related Works

The modeling and qualitative simulation of complex genetic regulatory networks of large dimension has received considerable attention in the literature,

---

[a]We note that simulating a system *in silico* actually requires a mathematical model. However, we want to consider the case when such mathematical model is unavailable to both the biologist and the software system.

118

(see, *e.g.*, [6] and the references cited therein). Most of these works treat situations in which the lack of quantitative information forces simulation in a qualitative way. These qualitative methods differ markedly from the types of qualitative modeling we propose, which are formulated in a *functional approximation* setting; the survey by DeVore [7] in the context of numerical analysis is particularly illuminating. Our approach also differs from systems such as *Cellerator*, [8] a Mathematica package for biological modeling. While this package bears many similarities with our approach, our focus is on a single collection of biochemical reactions rather than a hierarchy. Moreover the temporal ingredient introduced by our integration with a temporal language enhances the reasoning abilities of our proposed tool. Our temporal logic query language also differs from that of [9] (which can be viewed as an extension of the *Qualitative Reasoning* approach of *cf.* [10]) by relying on a full numerical simulation trace or a (sampled) trace of a physical experiment.

The kind of formalization and tools we are proposing in this paper could, in general, be used to study in a more systematic way hypotheses on properties of complex systems of biochemical reactions. The research by Bhalla et al. in [11], for example, aims at proving that a sort of "learned behavior" of biological systems is in fact stored within the mechanisms regulating intracellular biochemical reactions constituting signaling pathways. For this kind of studies, following [12], both qualitative and quantitative features of the system under study should be taken into account.

Finally, the recent control theory literature contains several relevant works on, *e.g.*, the general problem of constructing an automaton from a differential equation models (Brockett [13]), hybrid systems models for biochemical reactions (Alur *et al.* [14]). With respect to the former, our approach takes advantage of the specific mathematical models (S-systems) under consideration, and relates the numerical integration of the S-system with the construction of the automaton. With respect to the latter, the idea of using the discrete part of the (hybrid) automaton to switch from one mode to another when, for example, the number of molecules grows over a certain threshold, can be replicated in our framework by "gluing" different simulations (corresponding to different sets of parameters) in a bottom-up fashion.

3   Mathematical and Computational Models

3.1   Canonical Forms of Biological Systems Models

In the following we build on ideas introduced in [1,2,3] (from which we will borrow also most of the notation) and, e.g., [15]. Central to our discussions will be the notion of S-system (or of a GMA system). The basic ingredients of an S-system are n dependent variables to be denoted $X_1, \ldots, X_n$, m independent variables $X_{n+1}, \ldots, X_m$, each of which has domain $D_1, \ldots, D_{n+m}$, respectively. We augment the S-system form with a set of *algebraic constraints* that serve to characterize the conditions under which a given set of equations is derived from a set of maps. The justification for this construction is beyond the scope of this paper and it appears elsewhere. The basic differential equations constituting the system then take the following *power law* form:

$$\dot{X}_i = \alpha_i \prod_{j=1}^{n+m} X_j^{g_{ij}} - \beta_i \prod_{j=1}^{n+m} X_j^{h_{ij}} \tag{1}$$

$$C_j(X_1(t), \ldots, X_m(t)) = \sum \left(\gamma_j \prod_{k=1}^{n+m} X_k^{f_{jk}} \right) = 0 \tag{2}$$

where the α_i's and β_i's are called *rate constants* and govern the positive or negative contributions to a given substance (represented by X_i as a function of time) with other variables entering in the differential equation with exponents to be denoted as g_{ij}'s and h_{ij}'s. The γ_j are called *rate constraints* acting concurrently with the exponents f_{jk} to delimit the evolution of the system over a specified manifold embedded in the $n + m$-dimensional surface. Note that we have $\alpha_i \geq 0$ and $\beta_i \geq 0$ for all i's. S-systems can be integrated numerically, and symbolically in certain special cases (we address the symbolic case in a much more general way in a future work—see [16] for a preliminary treatment of the topic). The particular S-system "canonical" form allows for very efficient computations of both the function flows X_i and their derivative fields $\dot{X}_i$ [17].

3.2 XS-Systems: S-Systems Extended with Automata

The following properties of biochemical metabolic systems and corresponding S-systems serve as our starting point:

- The value of the dependent and independent variables uniquely characterize the state of the system when *normalized* with respect to time (and possibly other values);

- The transitions from one state to another are not necessarily encoded in the metabolic map of the system, and are instead *parametric* with respect to the value of constants in the S-system;

- There exists an *exogenous* set of "functions" that represent special perturbations of the system that the user may decide to include in the trace generation, *e.g.*, *ramps*, *impulses*, and *oscillations* (possibly with parametric amplitude and frequency).

We start with snapshots of the system variables' values constituting the possible states of the automaton. Transitions will be inferred from *traces* of the system variables' values evolution.

Definition 1 *Given an S-system S, the S-system automaton $\mathcal{A}_S$ associated to S is 4-tuple $\mathcal{A}_S = (S, \Delta, S_0, F)$, where $S \subseteq D_1 \times \cdots \times D_{n+m}$ is a (finite or infinite) set of states, $\Delta \subseteq S \times S$ is the transition relation, and $S_0, F \subset S$ are the initial and final states, respectively.*

Definition 2 *A* trace *of an S-system automaton $\mathcal{A}_S$ is a (finite or infinite) sequence $s_0, s_1, \ldots, s_n, \ldots$, such that $s_0 \in S_0$, $\Delta(s_i, s_{i+1})$ for $i \geq 0$. Equivalently, a trace can be defined as:*

$$\mathsf{trace}(\mathcal{A}_S) = \langle \langle X_1(t) \ldots X_n(t) \rangle \mid t \in \{t_0 + k\,\mathbf{step} : \; k \in \mathbb{N}\} \rangle.$$

Notice that, for fixed values of the independent variables, a unique trace is obtained when a simulation is performed. Moreover, while studying a trace of a system, it can be useful to focus on one or more specific variables, which justifies the following definition:

Definition 3 *Given any set of variables $U \subseteq \{X_1, \ldots, X_{n+m}\}$, the sequence:*

$$\mathsf{trace}(\mathcal{A}_S|_U) = \langle \langle X_i(t) \mid X_i \in U \rangle : \; t \in \{t_0 + k\,\mathbf{step} : \; k \in \mathbb{N}\} \rangle,$$

is called the trace of U. *If U consists of a single variable X_i the trace is called the* trace of X_i.

Multiple traces arise as we start varying the values in the primary parameter sets. Collection of such traces, in general, allows one to study the different instances of the simulated metabolic pathway evolution. Such a collection will give rise to an automaton with corresponding transitions once a suitable *equivalence relation* on states has been defined.

3.3 Construction of a Collapsed Automaton

Given a trace $\mathsf{trace}(\mathcal{A}_S)$ we want to construct an automata that can be used in conjunction with a TL based query system. The construction we present in the following corresponds to finding a partition of the real line (time) into a set I of non-overlapping intervals where the functions $X_i(t)$ are approximated in a piecewise-linear fashion. This is a very simple case of approximation as defined in much greater detail in [7]. Here we show a very simple automaton construction method that essentially is an *adaptive approximation* for the given $\vec{X}(t)$; more sophisticated construction algorithms can be constructed based on, *e.g.*, adaptive step-size integration methods for ordinary differential equations, and other types of segmentation algorithms. We will describe an idea about how to improve the construction of such *collapsed automata* when there is more information available about the kind of *in silico* simulation experiment. Essentially we will use a "dictionary" of base functions to guide the collapsing method in a more adaptive way. We will present a fully developed treatment of this approach in a subsequent paper.

A *linear* automaton can be constructed simply by associating a different state to each tuple $\langle X_1(t)\ldots X_n(t)\rangle$ as time is incremented according to a given fixed time step. To illustrate, consider the function $X_i(t)$ at times $t_i, t_{i+1}, \ldots t_{i+5}$ as depicted in Figure 1. In this case we have $\mathbf{step} = t_{i+1} - t_i$ as a result of (fixed) sampling or numerical integration. We associate the automata $\mathcal{A}_{\mathcal{S}}$ to the trace of X_i simply by taking into account each time step.

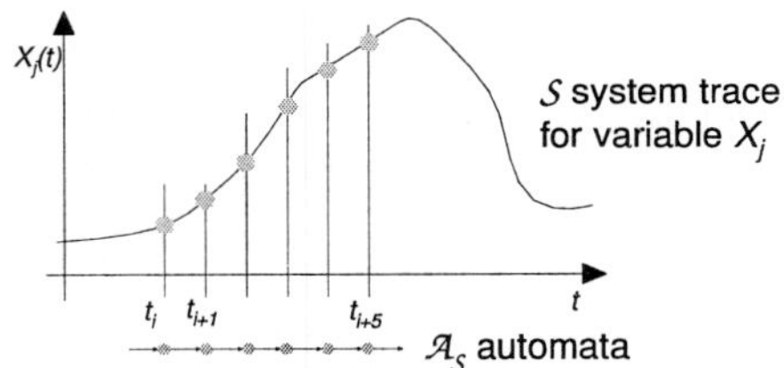

Figure 1. Simple one-to-one construction of the "trace" automata $\mathcal{A}_{\mathcal{S}}$ for a S-system $\mathcal{S}$.

We now suggest a simple method for collapsing the states of a linear automaton into piecewise-linear equivalence classes. Given a numerical solution for our S-system, two states $X_i(t + k\,\mathbf{step})$ and $X_i(t + (k + j)\,\mathbf{step})$, $j > 0$, are said to be equivalent under the relation R_{i,δ_i} if

$$| \dot{X}_i(t + k\,\mathbf{step}) - \dot{X}_i(t + (k - 1)\,\mathbf{step}) | \leq \delta_i.$$

The above construction is easily extended to the full collection of variables (see the following definition) and allows us to consider equivalent two states if the difference between the rate of growth of the corresponding variables is sufficiently small (with respect to the parameter δ_i).

Definition 4 *The relation $R_{\vec{\delta}}$ holds between two states $s_k = \vec{X}(t + k\,\mathbf{step})$ and $s_{k+j} = \vec{X}(t + (k + j)\,\mathbf{step}))$ with $j > 0$, if and only if, for each $i \in$*

$\{1, \ldots, n+m\},$

$$|\dot{X}_i(t + k\,\mathbf{step}) - \dot{X}_i(t + (k+j)\,\mathbf{step})| \leq \delta_i.$$

The collection $\{\delta_i \mid 1 \leq i \leq n+m\}$ is denoted by $\vec{\delta}$.

The above relation turns out to be an equivalence relation collapsing states whose variables' values are sufficiently similar, *if* computed *over* the numerical approximation of the solution. The (simple) idea is to choose as representative in each equivalence class, the element corresponding to the minimum time in the class. It is easy to build an iterative algorithm based of the above relation that determines the *knot* points in the state space that can be used as *state set representatives* in the collapsed automata. The algorithm is presented in [18] and has $O(hk)$ time complexity, where h is the size of the trace and k is the number of variables in the system.

In the following, given a collection of linear automata (traces), we will propose to "glue" them together in a unique automaton capable of modeling various possible behaviors of the system.

Example 5 *Consider again the function $X_i(t)$ depicted in Figure 1. In Figure 2 the effects of applying the collapsing algorithm are shown. With respect to $X_i(t)$ we obtain an automata $\mathcal{A}_{\mathcal{S}i}$ which has the reduced number of states*

$$\mathsf{states}(\mathcal{A}_{\mathcal{S}i}) = \langle \ldots (t_i, X_j(t_i), \dot{X}_j(t_i)), \quad (t_{i+2}, X_j(t_{i+2}), \dot{X}_j(t_{i+2})),$$
$$(t_{i+5}, X_j(t_{i+5}), \dot{X}_j(t_{i+5})), \ldots \rangle$$

Now suppose we have a different function $X_k(t)$. We associate to $X_k(t)$ the collapsed automata $\mathcal{A}_{\mathcal{S}k}$, such that

$$\mathsf{states}(\mathcal{A}_{\mathcal{S}k}) = \langle \ldots (t_i, X_k(t_i), \dot{X}_k(t_i)), \quad (t_{i+4}, X_k(t_{i+4}), \dot{X}_k(t_{i+4})), \ldots \rangle$$

i.e. the "landmark" times are t_i and t_{i+4} in this case. In order to construct a useful automata for the analysis tool we construct the merged automata $\mathcal{A}_{\mathcal{S}jk}$ *such that*

$$\mathsf{states}(\mathcal{A}_{\mathcal{S}k}) = \langle \ldots (t_i, X_j(t_i), \dot{X}_j(t_i)),$$
$$(t_{i+2}, X_j(t_{i+2}), \dot{X}_j(t_{i+2})),$$
$$(t_{i+4}, X_k(t_{i+4}), \dot{X}_k(t_{i+4})),$$
$$(t_{i+5}, X_j(t_{i+5}), \dot{X}_j(t_{i+5})), \ldots \rangle$$

i.e. automata $\mathcal{A}_{\mathcal{S}jk}$ is an ordered merge of the two automata $\mathcal{A}_{\mathcal{S}j}$, $\mathcal{A}_{\mathcal{S}k}$.

3.3.1 Normalization and Projection

In order to capture the notion of state-equivalence *modulo* normalization we begin with the following definition:

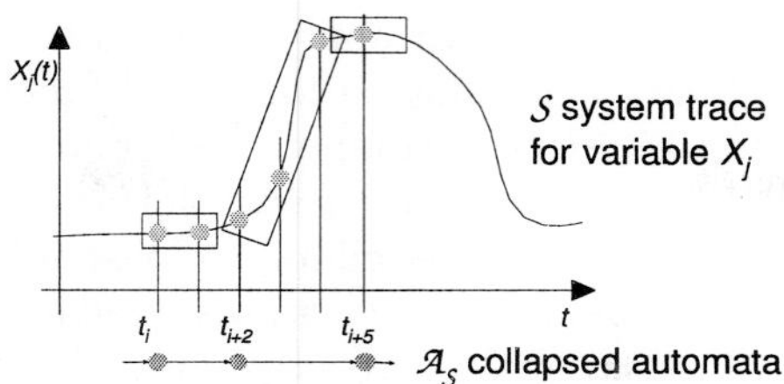

Figure 2. The effects of the *collapsing* construction of the "trace" automata $\mathcal{A}_S$ for a S-system $\mathcal{S}$.

Definition 6 *Given a subset V of the set $\{X_1, \ldots, X_{n+m}\}$ of variables, we define the set of states normalized with respect to V as the following set of tuples:*

$$S_{\backslash V} = \left\{ \left\langle \frac{X_i(t_0 + k\,\mathbf{step})}{\nu_i(V, t_0 + k\,\mathbf{step})} \right\rangle \; : \; k \in \mathbb{N} \right\},$$

with the ν_i's are normalizing functions.

More complex forms of normalization can be obtained when other variable contribute into play. Moreover, notice that when we normalize with a collection of normalizing functions defined as follows:

$$\nu_i(U, t) = \begin{cases} X_i & \text{if } X_i \notin U, \\ 1 & \text{otherwise,} \end{cases}$$

then the normalization corresponds to projecting with respect to the set of variables U.

3.3.2 Adaptive Collapsing by Dictionary Functions

Suppose we know that the user is trying to see what is the response of a simulated system to an impulse (*cf.* the example in Section 4.1). This is a piece of information that we can use to build a collapsed automaton more effectively. We assume that our tool comes equipped with a dictionary **D** of predefined basic functions: *ramps, impulses, sigmoids*, etc. Suppose that the user specifies that the simulation will include an impulse $I(t; \tau)$ at time τ. The collapsing algorithm will then use this information to compute the partition from time τ to $\tau + \Delta\tau$ using the function $I(t; \tau)$ as a component of the approximation of the solution.

This is just a special form of an approximation scheme. The computational complexity will vary according to the set of functions in **D**, e.g. it is known that when using *wavelets* as components, finding an *optimal* approximation to a function using a dictionary of wave-forms is an NP-hard problem [19]. We conjecture that, since we are willing to loose information in the collapsing operation, we will be able to use a greedy adaptive approach as well that will keep the computational complexity low. A full treatment of this topic will appear in a forthcoming paper.

4 Pathways Simulation Query language

In this section we briefly outline a language that can be used to inspect and formulate queries on the simulation results of XS-systems.

Our aim is to provide the biologists with a tool to formulate various queries against a repository of *simulation traces* and a set of *known pathways specifications*. We propose a *Temporal Logic* language (*cf.* [20]) with a specialized set of predicate variables whose aim is to make it easy to formulate queries on numerical quantities. For instance, all our discussions can be carried out in the standard **CTL**, Computation Tree Logic: a branching-time propositional temporal logic.

We can also augment the standard **CTL** language with a set of *domain dependent* queries. Such queries may be implemented in a more efficient way and express typical questions of interest to biologists.

4.1 An Example: Purine Metabolism

We now revisit in detail the example of purine metabolism described in [3] Chapter 10 and fully analyzed in [4,5]. The pathway for purine metabolism is presented in Figure 3. A brief description of the key reactions follows, and the reader is invited to examine the more detailed summaries contained in the literature referenced in [3,4,5].

The main metabolite in purine biosynthesis is *5-phosphoribosyl-α-1-pyrophosphate* (PRPP). A linear cascade of reactions converts PRPP into *inosine monophosphate* (IMP). IMP is the central branch point of the purine metabolism pathway. IMP is transformed into AMP and GMP. Guanosine, adenosine and their derivatives are recycled (unless used elsewhere) into *hypoxanthine* (HX) and *xanthine* (XA). XA is finally oxidized into *uric acid* (UA). In addition to these processes, there appear to be two "salvage" pathways that serve to maintain IMP level and thus of *adenosine* and *guanosine* levels as well. In these pathways, *adenine phosphoribosyltransferase* (APRT)

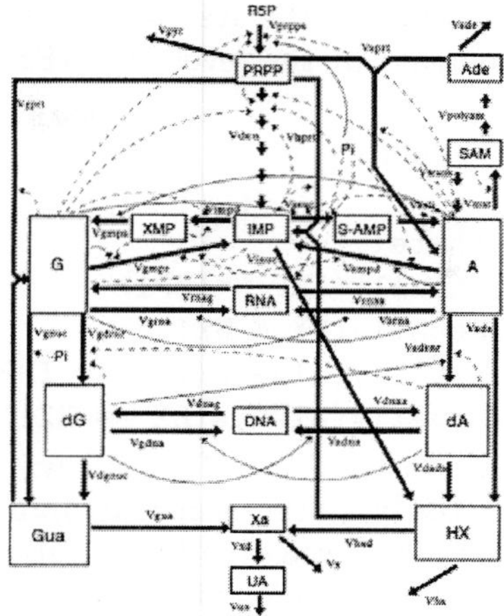

Figure 3. The metabolic scheme of purine metabolism in human. (Reprinted from [5], where a full description and further references may be found.)

and *hypoxanthine-guanine phosphoribosyltransferase* (HGPRT) combine with PRPP to form ribonucleotides.

The consequences of a malfunctioning purine metabolism pathway are severe and can lead to death. The entire pathway is quite complex and contains several feedback loops, cross-activations and reversible reactions, and thus an ideal candidate for reasoning with the computational tools we have developed.

In [3], a sequence of models for purine metabolism is presented alongside an analysis of how to identify discrepancies with physically observed data, and how to amend the current model in order to explain these discrepancies.

We show how to formulate queries over the simulation traces to express various desirable properties (or absence of undesirable ones) that the model should possess. Should any of these queries "fail," the model will be marked for further examination, experimentation and correction.

As an example consider the "Final" model for purine metabolism presented in [3]. The in silico experiment shows that when an initial level of PRPP is increased by 50-fold, the steady state concentration is quickly absorbed by the system. The level of PRPP returns rather quickly to the expected steady state values. IMP concentration level also rises and HX level falls before returning to predicted steady state values.

Suppose that we wanted to ask the system how it will respond to a temporary (instantaneous) increase in the level of PRPP. Such request can be formulated as follows:

```
always(PRPP > 50 * PRPP1
        implies
          (steady_state()
            and eventually(IMP > IMP1)
            and eventually(HX < HX1)
            and eventually(always(IMP == IMP1))
            and eventually(always(HX == HX1))
```

an (instantaneous) increase in the level of PRPP will not make the system stray from the predicted steady state, even if temporary variations of IMP and HX are allowed. A detailed discussion in the full paper shows how xssys responds to the query and helps the biologist to acquire the needed intuition.

5 Concluding Remarks

We have presented a novel framework under which we bring together well known tools from numerical analysis (approximation theory), temporal logic and verification, and visualization. This is a work in progress whose final aim is to construct an effective tool to aid biologists analyze experimental results and design new ones[b].

There are several open questions in our work that we need to address. We are now considering how to extend our automata construction by using notions from *approximation theory* that will allow us to take into consideration time and frequency domain aspects of a trace. Of course more theoretical treatments of the subject are possible as well (*cf.* [13]).

References

1. M. A. Savageau. *Biochemical System Analysis: A Study of Function and Design in Molecular Biology*. Addison-Wesley, 1976.
2. E. O. Voit. *Canonical Nonlinear Modeling, S-system Approach to Understanding Complexity*. Van Nostrand Reinhold, New York, 1991.
3. E. O. Voit. *Computational Analysis of Biochemical Systems A Practical Guide for Biochemists and Molecular Biologists*. Cambridge University Press, 2000.
4. R. Curto, E. O. Voit, A. Sorribas, and M. Cascante. Mathematical models of purine metabolism in man. *Mathematical Biosciences*, 151:1–49, 1998.

---

[b]A more complete version of this paper is available in our web site `bioinformatics.cims.nyu.edu`.

5. R. Curto, E. O. Voit, A. Sorribas, and M. Cascante. Analysis of abnormalities in purine metabolism leading to gout and to neurological dysfunctions in man. *Biochemical Journal*, 329:477–487, 1998.

6. H. de-Jong, M. Page, C. Hernandez, and J. Geiselmann. Qualitative simulation of genetic regulatory networks: methods and applications. In B. Nebel, editor, *Proc. of the 17th Int. Joint Conf. on Art. Int.*, San Mateo, CA, 2001. Morgan Kaufmann.

7. R. A. DeVore. Nonlinear approximation. *Acta Numerica*, 7:51–150, 1998.

8. B. E. Shapiro and E. D. Mjolsness. Developmental simulation with cellerator. In *Proc. of the Second International Conference on Systems Biology (ICSB)*, Pasadena, CA, November 2001.

9. B. Shults and B. J. Kuipers. Proving properties of continuous systmes: qualitative simulation and temporal logic. *Artificial Intelligence Journal*, 92(1-2), 1997.

10. B. Kuipers. *Qualitative Reasoning*. MIT Press, 1994.

11. U. S. Bhalla and R. Iyengar. Emergent properties of networks of biological signaling pathways. *SCIENCE*, 283:381–387, 15 January 1999.

12. D. Endy and R. Brent. Modeling cellular behavior. *Nature*, 409(18):391–395, January 2001.

13. R. W. Brockett. Dynamical systems and their associated automata. In U. Helmke, R. Mennicken, and J. Saurer, editors, *Systems and Networks: Mathematical Theory and Applications—Proceedings of the 1993 MTNS*, volume 77, pages 49–69, Berlin, 1994. Akademie-Verlag.

14. R. Alur, C. Belta, F. Ivančić, V. Kumar, M. Mintz, G. Pappas, H. Rubin, and J. Schug. Hybrid modeling and simulation of biological systems. In *Proc. of the Fourth International Workshop on Hybrid Systems: Computation and Control*, LNCS 2034, pages 19–32, Berlin, 2001. Springer-Verlag.

15. A. Cornish-Bowden. *Fundamentals of Enzyme Kinetics*. Portland Press, London, second revised edition, 1999.

16. B. Mishra. A symbolic approach to modeling cellular behavior. In *Proceedings of HiPC 2002, Bangalore,* INDIA, December 2002.

17. D. H. Irvine and M. A. Savageau. Efficient solution of nonlinear ordinary differential equations expressed in S-System canonical form. *SIAM Journal on Numerical Analysis*, 27(3):704–735, 1990.

18. M. Antoniotti, A. Policriti, N. Ugel, and B. Mishra. XS-systems: extended S-Systems and Algebraic Differential Automata for Modeling Cellular Behaviour. In *Proceedings of HiPC 2002, Bangalore,* INDIA, December 2002.

19. G. Davis, S. Mallat, and M. Avellaneda. Adaptive Greedy Approximations. *Constructive Approximation*, 13:57–98, 1997.

20. E. A. Emerson. Temporal and Modal Logic. In J. van Leeuwen, editor, *Handbook of Theoretical Computer Science*, volume B, chapter 16, pages 995–1072. MIT Press, 1990.

INCORPORATING BIOLOGICAL KNOWLEDGE INTO EVALUATION OF CAUSAL REGULATORY HYPOTHESES

LONNIE CHRISMAN,[a][†] PAT LANGLEY,[†] STEPHEN BAY,[†]
and ANDREW POHORILLE[‡]

[†] *Institute for the Study of Learning and Expertise*
2164 Staunton Court, Palo Alto, CA 94306
[‡] *Center for Computational Astrobiology and Fundamental Biology*
NASA Ames Research Center, M/S 239-4, Moffett Field, CA 94035

Biological data can be scarce and costly to obtain. The small number of samples available typically limits statistical power and makes reliable inference of causal relations extremely difficult. However, we argue that statistical power can be increased substantially by incorporating prior knowledge and data from diverse sources. We present a Bayesian framework that combines information from different sources and we show empirically that this lets one make correct causal inferences with small sample sizes that otherwise would be impossible.

1 Introduction and Motivation

There is a growing interest in the development and application of new computational methodologies for analyzing genomic and proteomic data, ranging from clustering techniques[1] to algorithms for inferring regulatory networks.[2,3,4,5,6] However, most such methods concentrate on discovering regularities in individual data sets and operate in a knowledge-lean manner. This contrasts sharply with the strategies of most biologists, who focus on testing specific hypotheses formulated in the context of biological knowledge and previous studies.

This observation suggests that biologists would benefit from better computational aids for hypothesis evaluation. Many such tools already exist, but their statistical power remains generally weak because, like most computational discovery techniques, they focus on data collected from a single study and typically ignore available knowledge. In this paper, we demonstrate how one can utilize prior biological knowledge to substantially increase the statistical power of causal hypothesis evaluation. Along the way, we address a number of challenges that this idea raises, including the facts that knowledge may come from different sources under different experimental conditions, have varying levels of uncertainty, and involve quantities that are not measured directly.

In the section that follows, we provide a motivating example that describes a biological hypothesis and relevant background knowledge. We then present a computational framework and associated algorithm that lets us calculate the evidence in favor of such a hypothesis given both prior knowledge and data.

---

[a] Corresponding author: lonnie@apres.stanford.edu.

We take a Bayesian approach to hypothesis evaluation, since this paradigm provides ready mechanisms for combining data and knowledge from multiple sources. After this, we report experimental studies with the algorithm on synthetic data, to determine its robustness, and a specific biological hypothesis, to ensure its relevance. In closing, we review related work on causal models in biology and suggest some directions for future research in this area.

2 A Motivating Example

Mitogen-activated protein kinase signal transduction pathways process a wide range of extracellular stimuli to determine a cell's transcriptional response to environmental changes or inter-cellular messages. One example, the c-Jun NH_2-terminal kinase[7] (JNK/SAPK) pathway, responds to growth factors (e.g., TGF-β and EGF), cytokines (e.g., TNF and IL-1), and forms of environmental stress (e.g., osmotic and radiation). It terminates in the phosphorylation and activation of JUN-family transcription factors, which dimerize with FOS, ATF, or other JUN factors to form AP-1 leucine-zipper transcription factor complexes,[8] which in turn enhance or repress transcription of many immediate-early genes. The JNK pathway has been implicated in many cellular processes and pathologies, including embryonic morphogenesis, cancer, immune system response, apoptotic signaling, cardiac hypertrophic response, neurodegenerative disease, and diabetes complications.[9] The JNK pathway is also considered a promising intervention point for many pathological conditions.

In humans and mice, the JNK family of kinases is derived from three genes, each of which elicits distinct responses under distinct conditions.[10,11] Jnk3 is found almost exclusively in brain, heart, and testes, whereas Jnk1 and Jnk2 are present in all tissues.[10] JNKs are known to phosphorylate several components of AP-1 complexes, including c-Jun, JunD, and Atf2,[9] although the different JNKs differ in their ability to phosphorylate each target.[12,13,10] The JUN family of transcription factors consists of c-Jun, JunB, JunD, and the viral oncogene v-Jun. They differ in their activation conditions, the AP-1 complexes in which they participate, and their transcriptional targets. To date, most laboratory studies involving JNK pathways have studied the involvement of JNK or JUN as a group, rather than looking at specific JNK or JUN variants. However, understanding the interactions between specific variants is essential to untangling functional roles of these pathways. We use as our motivating example the hypothesis that c-Jun is uniquely activated by Jnk2, and therefore is not activated by Jnk1.[b]

---

[b]For example, Kallunki et al.[13,11] found that Jnk2 binds to c-Jun 25 times more efficiently than Jnk1, and Gupta[10] found that Jnk2 isoforms tend to have higher affinities for c-Jun and Atf2 than do Jnk1 isoforms. Until recently, no kinase other than JNKs had been found

3 Representing Background Knowledge, Hypotheses, and Data

A computational system that evaluates causal biological hypotheses in the context of background knowledge and data must first represent such knowledge, data, and hypotheses. We encode these in terms of relations among discrete variables that can take on the values + (up-regulated), − (down-regulated), or 0 (unchanged) relative to a control condition. Facts and experimental data are divided into *scenarios*, with $X_{i,j}$ denoting the value of variable $\mathbf{x}_i$ in scenario j. When referring to a specific variable by name, we need only the scenario subscript, e.g., c-Jun$_j$ for the level of phosphorylated c-Jun in scenario j.

Consider the background knowledge that increasing TPA causes IL-11 expression to increase[16] [with other factors held constant]. To encode this, we assign this to a scenario, say $j = 1$, then define *causal conditions*, $C_1 = \{\mathrm{TPA}_1 = +, \mathrm{TNF}\text{-}\alpha_1 = 0\}$ and *known effects*, $E_1 = \{\mathrm{IL}\text{-}11_1 = +\}$. Encoding of experimental data is done in a similar manner. Consider an experiment in which cells are exposed to an increased amount of TNF-α with TPA exposure unchanged, and in which expressions of nur77, FL1, and IL-11 increase, whereas expressions of p19 and p53 decrease. Assigning this experiment to scenario $j = 2$, we encode the *experimental conditions* or interventions as $I_2 = \{\mathrm{TNF}\text{-}\alpha_2 = +, \mathrm{TPA}_2 = 0\}$ and the *observed effects* as $D_2 = \{\mathrm{EL\_nur77}_2 = +, \mathrm{EL\_p19}_2 = -, \mathrm{EL\_FL1}_2 = +, \mathrm{EL\_IL}\text{-}11_2 = +, \mathrm{EL\_p53}_2 = -\}$. If Jnk1 were also knocked out, the experimental conditions would be expressed as $I_3 = \{\mathrm{TNF}\text{-}\alpha_3 = +, \mathrm{TPA}_3 = 0, \mathrm{Jnk1}_3 = 0\}$, where $\mathrm{Jnk1}_3 = 0$ states that Jnk1 is controlled to be unchanged.

Background knowledge also indicates which direct causal links are plausible. We say that variable $\mathbf{x}_1$ is a causal *parent* of $\mathbf{x}_2$ when externally changing $\mathbf{x}_1$ can effect a change in $\mathbf{x}_2$ when all other variables are held unchanged. We specify *plausible parent relationships*, since there may be uncertainty over any specific causal link. In our example, we specify that each stimulus variable (TNF-α or TPA) is a plausible parent of every kinase (Jnk1 or Jnk2), that each kinase is a plausible parent of every transcription factor (JunD or c-Jun), and that each transcription factor is a plausible parent of each expression variable. The hypothesis that Jnk2 uniquely activates c-Jun is represented by the assertion that there is no link from Jnk1 to c-Jun, which may be true or false in any particular hypothetical causal model. Additional background knowledge takes the form of numeric assessments, α, that encode subjective beliefs about model parameters, including link likelihoods, reliability of causal statements, expected noise levels, and beliefs that particular links should be positive or negative influences.

---

capable of phosphorylating c-Jun,[12] but some evidence has emerged that an ERK kinase may phosphorylate c-Jun in specific cell types and developmental conditions.[14,15]

4 From Background Knowledge to Prior Probabilities

Background knowledge consists of plausible causal links between variables, causal conditions C, known effects E, and numeric assessments, α. Because we are working in a Bayesian framework, we must transform all this background knowledge into prior probabilities. Also, we must be able to express causal relations like "TNF-α (directly or indirectly) up-regulates JunD" and the results of experimental manipulations like applying TPA and knocking out Jnk1. Traditional notations for conditional probability do not capture the distinction between an estimate when a variable is observed and an estimate when a variable is manipulated. Thus, we introduce the notation $P(X|Y:Z)$ for the probability of X when Y is believed or observed to be true and Z is exogenously forced to be true. With our background knowledge consisting of α, E, and C, we define the *prior causal probability* as $P(\cdot|E,\alpha:C)$.

A causal model structure, M, is an acyclic subset of directed links between variables. Each variable $\mathbf{x}_i$ has an associated conditional probability, $P(\mathbf{x}_i|par_M(\mathbf{x}),M,\theta_M)$, which specifies its probability distribution conditioned on possible values of its parents, provided $\mathbf{x}_i$ is not exogenously controlled. The conditional probability is parameterized by a set of parameters, θ_M, and does not depend on the scenario, so that together an instance of (M,θ_M) defines a Bayesian network. For example, when M contains a link from Jnk1 to JunB, θ_M includes the probabilities that JunB is up-regulated, stays the same, or is down-regulated given that Jnk1 is up-regulated. When JunB has multiple parents according to M, our parameterization combines the contribution from each incoming link as a weighted mixture. We provide a detailed elucidation of our specific parameterization in a supplement.[17] We handle causal intervention by setting the local model to be $P(X_{i,j} = z|par_M(\mathbf{x}),M,\theta : X_{i,j} = z,\ldots) = 1$ whenever a variable is exogenously controlled, effectively severing the influence from the variable's parents.[18,3] Following the Bayesian network product rule and combining scenarios, the joint distribution over all variables under the causal interventions in C is the product of the conditional probabilities:

$$P(X|M,\theta_M:C) = \prod_i \prod_j P(X_{i,j}|X_{par_M(\mathbf{x}_i),j},M,\theta_M:C_{i,j}) \qquad (1)$$

Knowing M and θ_M is sufficient for estimating probabilities in situations that involve hypothetical causal interventions, such as $P(\text{c-Jun} = +|\text{EL\_IL-11} = 0,M,\theta_M : \text{TNF-}\alpha = +)$. In terms of M and θ, we rewrite the prior causal probability as

$$P(X,M,\theta_M|E,\alpha:C) = P(M)P(\theta_M|\alpha)P(X|E,M,\theta_M:C) , \qquad (2)$$

132

where $P(M)$ expands to $k \cdot exp(-|M|)$ to provide a simplicity bias and $P(\theta_M|\alpha)$ consists of products of Dirichlet priors. The final term is encoded by the Bayesian network (M, θ_M) from Equation (1).

5 Evaluating Causal Biological Hypotheses

As we have noted, biologists are often concerned with evaluating whether experimental data, encoded by experimental conditions I and observed effects D, support some particular causal hypothesis, H. It is important to distinguish between the degree to which (a) the data alone support the hypothesis, (b) the data and prior knowledge together support the hypothesis, or (c) the data support the hypothesis in the context of the prior knowledge. Classical statistical hypothesis testing generally focuses on (a) and Bayesian analysis usually focuses on (b), whereas we focus on (c). To this end, we first combine the prior probability of a causal model M with experimental data to determine the posterior probability that H is true, as in (b), then we utilize this to define a p value that codifies (c). For example, the scientist may which to publish results demonstrating how significantly his new data supports (or refutes) a hypothesis in the context of what was previously known.

In our framework, a causal hypothesis is an expression that is either true or false as a function of the model M, its parameters θ_M, and the values of latent variables X.[c] The hypothesis used in our example, $(Jnk1, \text{c-Jun}) \notin M$, is a function only of M. If we let

$$
\begin{aligned}
prior(H) &= P(H|E, \alpha : C) \\
posterior(H|D) &= P(H|D, E, \alpha : I, C)
\end{aligned}
$$

the latter expression, the *posterior causal probability*, can be rewritten as

$$
P(X, M, \theta_M|D, E, \alpha : I, C) = \xi P(M)P(\theta_M|\alpha)P(X, D, E|M, \theta_M : I, C) \quad (3)
$$

$$
\text{where} \qquad \xi = 1/ \sum_{M \in \mathcal{M}} \int P(M)P(\theta_M|\alpha)P(D, E|M, \theta_M : I, C)d\theta_M
$$

and where $\mathcal{M}$ is the set of possible causal models.

A posterior probability may be an optimal metric to employ in decision making contexts, but it often does not measure the real concerns in scientific data analysis. The posterior can be hard to interpret due to subjective assessments in prior knowledge and does not specifically reveal the data's support for the conclusions. A typical data analysis question is whether the experimental

---

[c]We also let H depend on a hypothetical control condition, which lets us evaluate the outcomes of hypothetical causal interventions, but we have omitted this case here for the sake of simplicity.

data support the hypothesis in the context of prior knowledge, which, as we have explained, differs from whether the data and prior knowledge together support the hypothesis. Many scientists are more familiar with the frequentist notion of a p value, which is conventionally not applicable to such knowledge-rich contexts. However, by combining the p value with the posterior, we obtain

$$p(H) = Pr\,[\,posterior(H|D') > posterior(H|D) \mid D' \sim prior(\cdot|\neg H)\,]$$

where $D' \sim prior(\cdot|\neg H)$ denotes that D' are hypothetical observations drawn at random from $prior(\cdot|\neg H)$. As in classical statistics, the p value gives the probability that random observations would appear to support H as much as the actual data even when H does not hold, i.e., the probability that one can be misled by the data into thinking H is true when it is actually false. A p value near zero indicates that the data provide strong support for H. If $p(\neg H) \approx 0$, then they provide strong evidence against the hypothesis.

We utilize a Metropolis-Hastings sampler[19] to compute $posterior(H|D)$, based directly on Equation 3. The basic strategy, which is an instance of a Markov chain Monte Carlo algorithm,[20] rapidly samples as many plausible models as it can in a short time. The method samples each model with a probability that asymptotically approaches its posterior probability, and uses the resulting counts to estimate how often the hypothesis H is true. We sample M, θ, and X simultaneously, as opposed to most other approaches to Bayesian network induction,[2,3] which usually use one method to search over possible model structures, M, a separate nested method (often based on EM) to fit parameters, θ, and yet different algorithms for inference over X.

Our algorithm begins with a starting model structure M, parameters θ, and value assignments to all latent variables, X. One point in the sample space, (M, θ, X), can be conceptualized as one possible model of the underlying biological system and how its latent variables respond in each situation. The method then proceeds to sample new values for $z_t = (M_t, \theta_t, X_t)$, based on z_{t-1}, to converge on the posterior distribution given by Equation 3.

At each step a new point, $z' = (X', M', \theta')$, is randomly proposed according to a proposal probability $g(\cdot|z_{t-1})$, and the Hastings acceptance probability

$$a(z_{t-1}, z') = \min\left\{1, \frac{\pi(z')}{\pi(z_{t-1})}\,\frac{g(z_{t-1}|z')}{g(z'|z_{t-1})}\right\}$$

is computed, where $\pi(z)$ is the right hand side of Eq. 3, and where z_t is set to z' with probability $a(z_{t-1}, z')$ and otherwise to z_{t-1}. The Hastings update rule guarantees *detailed balance*,[20] and thus that $P(z_t) \to \pi(z)$ as $t \to \infty$, provided that g can reach all points of the sample space.

We will not specify in detail our choice for g, but note that it involves a mixture of strategies, one for altering M by adding or deleting links, one for updating θ, one for changing X, and a few more specialized methods. When a new sample changes only a few links, parameters, or values, $\pi(z')$ can be updated incrementally, so that computational complexity depends on the number of changes rather than on the size of the model. Moreover, the normalization factors for π and g are not relevant and thus are not computed. Finally, we follow the standard practice for Markov chain Monte Carlo algorithms and begin with a 'burn-in' period before tallying statistics (usually 1/10th the total number of iterations), and we usually tally only every 100th sample in a 100,000 sample run. Our Java-based implementation typically explores about 2,000 proposals per second on a 1.5 GHz Pentium IV, with a typical proposal acceptance rate around 35 percent.

To compute the p value for a hypothesis H, we first utilize the above algorithm to draw a sample (M, θ_M, X), enforcing $\neg H$ during the process and treating all variables in D as latent (i.e., ignoring their original observed values). We then extract D' from the sampled X, use D' in place of D, and run the above procedure to compute $posterior(H|D')$. We repeat this N times, tallying the mean and variance of $posterior(H|D')$ across the different choices of D'. With small N, the number of times that $posterior(H|D') > posterior(H|D)$ is a poor estimate of the p value, so we instead assume $posterior(H|D')$ is normally distributed and use the area under the normal curve with the tallied mean and variance to estimate the probability that $posterior(H|D') > posterior(H|D)$.

6 Experimental Evaluation

Our basic claim is that incorporating prior knowledge into causal biological hypothesis evaluation enhances statistical power, therefore making it possible to infer causal effects that would otherwise be undetectable in small sets of experimental data. To support this claim, we utilized the hypothetical, but biologically plausible, model in Figure 1 to generate synthetic data under six different experimental conditions. These experiments involved one wild type organism and two knockout conditions, each under stimulation by TPA or TNF-α. The 'true' model takes the same form as those described earlier, that is, a causal Bayesian network in which each random variable has the domain $\{+, 0, -\}$ and each causal influence is stochastic (with 10% of the observed values being altered by noise). We generated two sets of data for our evaluation: the first assumed a model in which Jnk1 does not influence c-Jun, whereas the second assumed this influence does occur.

Our first study aimed to demonstrate that prior background knowledge lets one obtain statistical support for a causal hypothesis even when some of

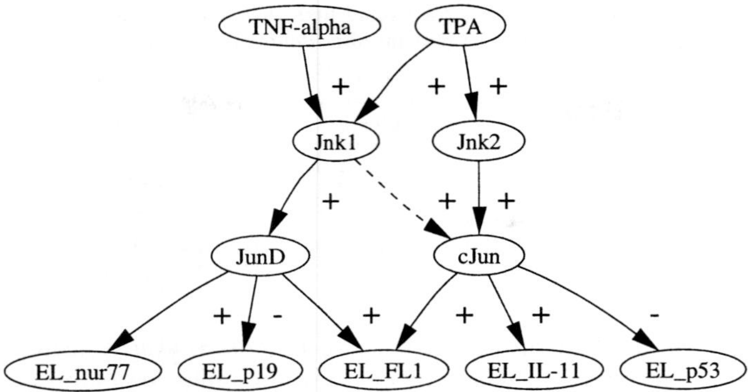

Figure 1: A hypothetical regulatory system used to generate synthetic data. The variables TNF-α and TPA are extracellular stimulus levels controlled by experimental conditions. Jnk1 and Jnk2 denote activation levels for the respective kinases, each of which are knocked out in two of six experiments. JunD and c-Jun, which denote activation levels of the respective transcription factors, are unobserved in the experiments. Variables prefaced with EL denote the observed expression levels of specific genes. Enhancement and suppression influences are indicated with pluses and minuses, respectively.

the relevant variables are latent. We used background knowledge about the JNK/JUN signaling pathways collated from published literature, subdivided into 12 facts: c-Jun is a transcription factor (i.e., parent) of IL-11[16] and p53,[21] JunD is a transcription factor of nur77,[22] TPA up-regulates IL-11,[16] JunD,[23] c-Jun,[24] and nur77,[22] JunD up-regulates nur77,[25] c-Jun up-regulates IL-11[16] and down-regulates p53,[21] and Jnk1 and Jnk2 are kinases, so they positively influence their targets. We used this knowledge in evaluating two hypotheses

H_1 : There is no direct causal link from Jnk1 to c-Jun.

H_2 : There is a direct causal link from Jnk1 to c-Jun.

against the two data sets. We performed each evaluation twice, once using the entire corpus of background knowledge, and once using no prior background knowledge (except for the set of variables and plausible links). Table 1 shows the posterior probabilities and p values that resulted from these runs.

Statistically significant support for the correct hypothesis occurs at the 0.05 level only for the two cases that utilize background knowledge. From the data alone, neither hypothesis is supported at a significant level. Since c-Jun and JunD are both latent, the knowledge is necessary to relate them to the observed data; otherwise, the system finds no support for the hypotheses that involve those variables.

Table 1: Statistical significance levels for causal hypotheses with and without the incorporation of background knowledge. The first number denotes the posterior probability of each hypothesis, whereas the second denotes the p value.

True Model	All Prior Knowledge		No Prior Knowledge	
	H_1 : no link	H_2 : link	H_1 : no link	H_2 : link
No Influence	0.99 / 0.003	0.01 / 0.77	0.42 / 0.46	0.58 / 0.51
Influence	0.03 / 0.500	0.97 / $< 10^{-7}$	0.44 / 0.55	0.56 / 0.44

Figure 2 shows the results of a more extensive study, using only data generated by the model in which Jnk1 does not influence c-Jun, that relates statistical power to the amount of background knowledge available. In these runs, we ordered the 12 facts randomly and evaluated the hypothesis H_1 (that the link is not present) while varying the number of facts given to the system from 0 to 12. The graphs reveal that a certain amount of background knowledge, about half the corpus, must be available before strong support for the hypothesis becomes evident. The sixth fact, which happened to be that c-Jun influences EL_IL-11, was a major clue the system needed to disambiguate between two main competing systems-level scenarios that appeared possible prior to this fact. With nine or more facts, the system detects that the data sets indeed support the hypothesis statistically at the 0.05 level.

Without this ample biological knowledge, it would have been impossible to validate or refute the hypotheses from experimental data alone. However, when combined with such prior information, the data reveal clear support for the hypothesis at statistically significant levels. These results are consistent with our central thesis — that utilizing preexisting background knowledge can be crucial for effective hypothesis testing when only small samples are available or the variables are only partially observable.

7 Discussion

Our approach to interpreting biological data contrasts sharply with most computational work in this area. The most common methodology passes available data to an induction algorithm, which extracts regularities that, hopefully, are biologically meaningful. However, work in this paradigm typically assumes data are plentiful and makes little use of background knowledge about biology. Our framework instead takes advantage of prior knowledge and previous experimental results in analyses of new observations, increasing their statistical power even when few samples are available.

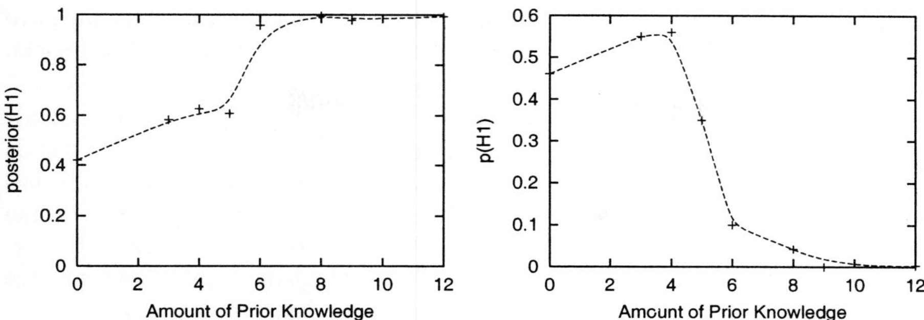

Figure 2: Evidence that statistical power increases with available background knowledge, as reflected by the number of biological facts provided to the system. The data set supports the (correct) hypothesis at a significant level ($p(H_1) < 0.05$) only when at least nine of the 12 domain facts are utilized during evaluation.

Despite this crucial difference, our approach has clear links to earlier research in computational biology. The strongest connection is to methods for learning Bayesian networks,[2,26,27,28] which also search a space for causal models that match the experimental data. For the special case in which each possible link is a hypothesis, we can view these systems as evaluating causal hypotheses with respect to their support by the data. Previous efforts within this framework have also dealt with the technical complications that arise with latent variables[29,27] and causal interventions,[3,18] both addressed in our own work.

Nevertheless, most research on inducing Bayesian networks of biological systems has taken a knowledge-lean, data-mining approach. Despite a few exceptions that incorporate knowledge about promoter sequences,[4,30,31] typical work in this paradigm attempts to construct a causal model from scratch, rather than evaluate particular causal relations in the context of background knowledge. Our own previous research on computational methods for revising causal biological models[5] comes much closer, but still emphasizes model discovery rather than hypothesis evaluation. A few researchers[32,33] have focused on model evaluation as opposed to model discovery. In particular, the JustAid system[32] also supports the use of experimental data to evaluate qualitative causal hypotheses, although its underlying algorithms are quite different and it addresses neuroendocrinology rather than gene regulation.

At a computational level, our approach draws heavily on Markov chain Monte Carlo techniques for hypothesis testing from the statistical literature.[34] However, our use of p values in a Bayesian context appears novel, in that the standard approach to Bayesian hypothesis testing involves comparing the *Bayes Factors*[34] or *Bayesian Information Criteria*[35] for alternative models.

Both approaches are legitimate from a statistical perspective, but we have chosen to utilize p values because they are generally more familiar to biologists.

Despite our encouraging results, we must extend our computational framework along a number of dimensions before it can become a useful tool for biologists. For example, we should explore other representations that let us encode qualitative causal knowledge about biological systems, especially notations that make stronger contact with established biological concepts like phosphorylation and dimerization. Moreover, we should incorporate this extended formalism into a user interface that lets biologists visualize and manage their background knowledge, hypotheses, and experimental data.

Another limitation of our current implementation concerns efficiency, in that its sampling strategy does not scale well to very large corpora of background knowledge. Also, since our posterior distributions often exhibit isolated regions of high probability, achieving a workable mixing rate from the Hastings algorithm is challenging. In future work, we plan to make our inference methods more efficient by incorporating additional ideas from the literature on Markov chain Monte Carlo and Metropolis-Hastings algorithms.[36,20] Our approach would also benefit from computational methods that generate plausible hypotheses automatically by reasoning over biological knowledge.[37] Finally, we must demonstrate the utility of our framework on data from actual biological experiments and on hypotheses they were designed to test.

In summary, we have presented a computational approach to evaluating causal hypotheses that takes advantage of background knowledge and previous experimental results to increase statistical power. Our framework encodes biological knowledge, hypotheses, and data in terms of qualitative relations between variables, and it utilizes Bayesian inference to calculate the evidence for and against each candidate hypothesis. We illustrated this approach to hypothesis evaluation in the context of knowledge and data about the JNK/JUN signaling pathways, and we demonstrated the increase in statistical power that background knowledge provides in this setting. We believe that the techniques we have described will make future tools for computational biology more robust and let them use available data more effectively.

8   Acknowledgements

We thank Jessica Ross, Jeff Shrager, and Hilary Spencer for their input and contributions to this project. This research was supported in part by Grants NCC 2-1202, NCC 2-5471, and NCC 2-1335 from NASA Ames Research Center and in part by NTT Communication Science Laboratories, Nippon Telegraph and Telephone Corporation.

9 References

1. M. B. Eisen *et al.* *Proc.Natl.Acad.Sci.*, 95(25):14863–8, 1998.
2. N. Friedman *et al.* *Comp.Biol.*, 7:601–20, 2002.
3. D. Pe'er *et al.* *Bioinformatics*, pages 1–9, 2001.
4. A. J. Hartemink *et al.* In *Pacific Symp.Biocomp.*, pages 437–449, 2002.
5. J. Shrager *et al.* In *Pacific Symp.Biocomp.*, pages 486–97, 2002.
6. T. Akutsu *et al.* *Bioinformatics*, 16:727–34, 2000.
7. C. R. Weston *et al.* *Curr.Op.Gen.Dev.*, 12(1):14–21, 2002.
8. M. Z. L. Karin *et al.* *Cur.Op.Cell Biol.*, 9(2):240–6, 1997.
9. R. J. Davis. *Cell*, 103(2):239–52, 2000.
10. S. Gupta *et al.* *J. EMBO*, 15(11):2760–70, 1996.
11. F. Bost et al. *Mol.Cell Biol.*, 19(3):1938–49, 1999.
12. A. Minden *et al.* *Biochemica et Biophysica Acta*, 1333:F85–F104, 1997.
13. T. Kallunki *et al.* *Genes Dev.*, 8(24):2996–3007, 1994.
14. S. Leppa *et al.* *J. EMBO*, 17:4404–13, 1998.
15. Y. Levkovitz *et al.* *J.Neurosci.*, 22(10):3845–54, 2002.
16. W. Tang *et al.* *J.Biol.Chem.*, 273(10):5506–13, 1998.
17. http://cll.stanford.edu/~lonnie/papers/psb2003sup.ps.
18. C. Yoo *et al.* In *Pacific Symp.Biocomp.*, pages 498–509, 2002.
19. W. K. Hastings. *Biometrica*, 57:97–109, 1970.
20. R. M. Neal. Tech.Rep. CRG-TR-93-1, Dept. C.S., Univ. Toronto, 1993.
21. M. Schreiber *et al.* *Genes Dev.*, 13:607–19, 1999.
22. C.Y. Young *et al.* *Oncol.Res.*, 6(4–5):203–10, 1994.
23. T. Li *et al.* *J.Biochem.*, 277:32668–76, 2002.
24. Zhou H. *et al.* *J.Biochem.*, 275(30):22868–75, 2000.
25. C. O. Stocco *et al.* *J.Biochem.*, 277(5):3293–3302, 2002.
26. D. Heckerman. Tech.Rep. MSR-TR-95-06, Microsoft Res., 1996.
27. P. Spirtes *et al.* *Causation, Prediction and Search.* MIT Press, 2001.
28. G. Cooper *et al.* *Machine Learning*, 9:309–47, 1992.
29. N. Friedman. In *UAI*, pages 129–38, 1998.
30. I. M. Ong *et al.* *Bioinformatics*, 18:241–8, 2002.
31. E. Segal *et al.* In *RECOMB*, 2002.
32. A. Mahidadia *et al.* In *Proc. Discovery Science*, pages 214–27, 2001.
33. A. J. Hartemink *et al.* In *Pacific Symp.Biocomp.*, pages 422–33, 2001.
34. A. Raftery. In *[36]*, pages 163–188. 1996.
35. H. Akaike. *Trans.Auto.Control*, 19:716–27, 1974.
36. W. R. Gilks *et al.* *Markov Chain Monte Carlo in Practice.* Chapman & Hall, 1996.
37. P. Romero *et al.* *Pacific Symp.Biocomp.*, pages 471–82, 2001.

ASSESSMENT OF THE RELIABILITY OF PROTEIN-PROTEIN INTERACTIONS AND PROTEIN FUNCTION PREDICTION

MINGHUA DENG,   FENGZHU SUN,   TING CHEN

Molecular and Computational Biology Program, Department of Biological Sciences
University of Southern California
1042 West 36th Place, Los Angeles, CA 90089-1113, USA
Contact: fsun@hto.usc.edu or tingchen@hto.usc.edu

Abstract

As more and more high-throughput protein-protein interaction data are collected, the task of estimating the reliability of different data sets becomes increasingly important. In this paper, we present our study of two groups of protein-protein interaction data, the physical interaction data and the protein complex data, and estimate the reliability of these data sets using three different measurements: (1) the distribution of gene expression correlation coefficients, (2) the reliability based on gene expression correlation coefficients, and (3) the accuracy of protein function predictions. We develop a maximum likelihood method to estimate the reliability of protein interaction data sets according to the distribution of correlation coefficients of gene expression profiles of putative interacting protein pairs. The results of the three measurements are consistent with each other. The MIPS protein complex data have the highest mean gene expression correlation coefficients (0.256) and the highest accuracy in predicting protein functions (70% sensitivity and specificity), while Ito's Yeast two-hybrid data have the lowest mean (0.041) and the lowest accuracy (15% sensitivity and specificity). Uetz's data are more reliable than Ito's data in all three measurements, and the TAP protein complex data are more reliable than the HMS-PCI data in all three measurements as well. The complex data sets generally perform better in function predictions than do the physical interaction data sets. Proteins in complexes are shown to be more highly correlated in gene expression. The results confirm that the components of a protein complex can be assigned to functions that the complex carries out within a cell. There are three interaction data sets different from the above two groups: the genetic interaction data, the in-silico data and the syn-express data. Their capability of predicting protein functions generally falls between that of the Y2H data and that of the MIPS protein complex data. The supplementary information is available at the following Web site: http://www-hto.usc.edu/~msms/AssessInteraction/.

1   Introduction

The development of high-throughput bio-techniques for functional genomic analysis generated a large amount of protein-protein interaction data. These include the yeast two-hybrid assay [1,2,3] and mass spectrometry [4,5]. Several databases have been developed to collect different sources of protein interaction data, including the Munich Information Center for Protein Sequences (MIPS) [6], the Database of Interacting Proteins (DIP)[7], and the Biomolecular Interaction Network Database (BIND)[8]. However, even using the same experimental technique of the yeast two-hybrid assay, Ito's data and Uetz's data share few overlaps[3], which suggest that errors are present in these data sets. In this paper, we estimate the reliability of these data sets using three different measurements: (1) the distribution of gene expression correlation coefficients, (2) the reliability based on gene expression correlation coefficients, and (3) the accuracy of protein function predictions.

Interacting proteins are more likely to be involved in similar biological processes and functions and thus they are more likely to be co-expressed. Based on this observation, Grigoriev (2001)[9] first showed that the mean correlation coefficient of gene expression profiles for interacting protein pairs is higher than that for random protein pairs. Therefore, the significance of a set of protein interaction data can be tested by comparing the mean correlation coefficient of gene expression profiles with that of random protein pairs[10,11,12]. Using the same idea, Ge et al. (2001) [13] showed that interacting protein pairs are more likely to be in the same cluster of gene expression data than random pairs. Although this idea can be used to study the significance of a set of protein interaction data, it cannot be used to estimate the *reliability* of this set. Here the *reliability* is defined as the fraction of real interactions over the observed protein interactions.

Two methods have been proposed to estimate the reliability of a set of putative protein interactions. Mrowka et al. (2001)[14] used the protein physical interactions in MIPS as the reference of real interactions and estimated the reliability of a putative set by comparing the distributions of correlation coefficients with those of random pairs. They used a bootstrap method to count how many random pairs needed to be added to the reference data to create the same statistical behavior of gene expression correlation coefficients as the putative interaction data, and then they computed the reliability from this sampling data. Deane et al. (2002)[15] used INT, a subset of DIP interactions that are derived from small-scale experiments, as the reference for real interactions. They assumed that the distribution of the square of the Euclidian distance between expression profiles of putative interacting pairs is a mixture of distributions for real interacting pairs and for random pairs. They then used a least-square approach to estimate the reliability of the putative protein interaction data. Although both ideas are novel and interesting, their estimation methods do not optimally correspond to their models. Other methods have also been proposed to assess the reliability by looking at protein functions [16] and "interaction generality"[17].

In this paper, we develop a maximum likelihood estimation(MLE) method for estimating the reliability of several interaction data sets based on the model of Deane et al. (2002). We studied two groups of interaction data. The first group included the MIPS, Uetz's and Ito's interaction data, all of which contain pairwise physical interactions. In this group, the interactions in MIPS were treated as real interactions, and we estimated the reliability of Uetz's data, Ito's data and Ito's data with multiple IST hits (Ito1IST, $\cdots$, Ito8IST). We show that the reliability of Uetz's interaction data (53%) is much higher than the reliability of Ito's data (17%) and is comparable to that of the Ito2IST data with at least two IST hits (56%). The reliability of Ito's data generally increases with the number of IST hits, which is consistent with our intuition that multiple IST hits could reduce the false positives significantly. The second group contained protein complexes: the MIPS complexes, the TAP data, and the HMS-PCI data. They provide protein components in a complex but do not give the physical interactions. The two groups are different, although a relationship does exist, since proteins in the same complex are more likely to physically interact with one another. The two groups have to be treated differently. For the protein complex data, the meaning of reliability is the fraction of protein pairs that are in the same protein complex in the putative complex data. We used the same MLE approach as above to estimate the reliability of the protein complex data using the MIPS complexes as the true complexes. The reliability of the TAP data and the HMS-PCI data are 58% and 25%, respectively.

Several methods have been developed to predict functions using protein-protein interaction data [18,19,20,21]. The basic assumption behind these studies is that proteins involved in similar functions are more likely to be interacting. In this paper, we consider two methods: the neighborhood-counting method and the chi-square method for the prediction of protein function. We use a leave-one-out method to estimate the accuracy of predictions, and we compare the results from different data sets. As expected, the reference data sets show higher accuracy than others in predicting functions. The MIPS complex data have the highest accuracy (70% sensitivity and specificity), while Ito's data have the lowest accuracy (15% sensitivity and specificity). The results are consistent with the estimation of the reliability of the protein interaction data.

2   Method

2.1   Estimating the reliability of a putative protein interaction data set

The reliability of a set of putative protein interactions is defined as the fraction of real protein interactions over all the putative protein interactions. Let α be the reliability of a given set of putative interactions. Let $O(\cdot)$, $T(\cdot)$ and $R(\cdot)$ be the distributions of the correlation coefficients of the gene expression profiles for the given set of putative interaction pairs, the true interacting pairs and the random pairs, respectively. Then we should have

$$O(\cdot) = \alpha T(\cdot) + (1 - \alpha)R(\cdot).$$

Suppose we split the values of correlation coefficients into K bins. Let n_k be the number of observed interaction pairs in the kth bin. Let p_k and q_k be the fractions of real interactions and random pairs in the kth bin, respectively. Then the likelihood function can be defined as:

$$L(\alpha) = \prod_{k=1}^{L} (\alpha p_k + (1 - \alpha)q_k)^{n_k}. \tag{1}$$

$L(\alpha)$ is a convex function and we can use a classical gradient algorithm to estimate the parameter α, $\widehat{\alpha}$, by maximizing $L(\alpha)$.

To find the precision of the estimation, we use the following formula to calculate the variance of $\widehat{\alpha}$,

$$Var(\widehat{\alpha}) = \frac{1}{\sum_{k=1}^{K} n_k \frac{(p_k - q_k)^2}{(\widehat{\alpha}p_k + (1 - \widehat{\alpha})q_k)^2}}.$$

2.2   Estimating the reliability of protein interactions from different experiments

If we have protein interaction data from several experiments, how do we estimate the reliability of the different sets of putative interactions? For example, we have two sets of putative protein interactions, E_1 and E_2, with M_1 and M_2 pairs, respectively. How do we estimate the reliability of the putative interacting pairs in $E_1/E_1 \cap E_2$, $E_2/E_1 \cap E_2$, and $E_1 \cap E_2$? If the number of protein pairs in a set is large, the above approach can be applied. On the other hand, if the number of protein pairs is not too large, we propose the following method.

As in Deng et al. [10], we define the false positive rate (fp) and the false negative rate (fn) for a specific data set, where the *false positive* rate is the probability that two proteins do not interact in reality but are observed to be interacting in the experiment. The *false negative* rate is the probability that two proteins interact in reality but are not observed to be interacting in the experiment. Let O_{ij} and P_{ij} be the variables for the observed and the real interaction for proteins P_i and P_j, respectively, with value 1 for interaction and 0 for no interaction. Then

$$fp = \Pr(O_{ij} = 1 \,|\, P_{ij} = 0), \qquad fn = \Pr(O_{ij} = 0 \,|\, P_{ij} = 1). \tag{2}$$

Thus, the probability for the observed protein-protein interaction is

$$\Pr(O_{ij} = 1) = \Pr(P_{ij} = 1)(1 - fn) + \Pr(P_{ij} = 0)fp.$$

As in Mrowka et al.[14] and Deane et al. [15], the reliability of a protein-protein interaction data set is measured by the fraction of real interactions in the data set, denoted by α. Let K_r and K_n be the sizes of *real interactions* and *non-interactions*, respectively, and let M be the size of the data set. We assume that the observed interactions and non-interactions are random samples from the real interaction set and the non-interaction set, respectively. The false positive rate and false negative rate can be estimated as

$$fp = \frac{M(1 - \alpha)}{K_n}, \qquad fn = 1 - \frac{\alpha M}{K_r}. \tag{3}$$

Using the above formula, we can estimate the reliability of a putative protein-protein pair given two protein-protein interaction sets. We use the following notation for the kth data set:

$O_{ij}^{(k)}$: observed interaction result for P_i and P_j,
$fp^{(k)}$ and $fn^{(k)}$: false positive rate and false negative rate,
α_k: reliability,
M_k: the number of interaction pairs.

From equations 2 and 3, we have,

$$\Pr(P_{ij} = 1 | O_{ij}^{(1)} = 1, O_{ij}^{(2)} = 1)$$
$$= \frac{\Pr(P_{ij} = 1)(1 - fn^{(1)})(1 - fn^{(2)})}{\Pr(P_{ij} = 1)(1 - fn^{(1)})(1 - fn^{(2)}) + \Pr(P_{ij} = 0)fp^{(1)}fp^{(2)}} \tag{4}$$
$$= \frac{\alpha_1 \alpha_2}{\alpha_1 \alpha_2 + \frac{K_r}{K_n}(1 - \alpha_1)(1 - \alpha_2)}.$$

$$\Pr(P_{ij} = 1 | O_{ij}^{(1)} = 1, O_{ij}^{(2)} = 0)$$

$$= \frac{\Pr(P_{ij} = 1)(1 - fn^{(1)})fn^{(2)}}{\Pr(P_{ij} = 1)(1 - fn^{(1)})fn^{(2)} + \Pr(P_{ij} = 0)fp^{(1)}(1 - fp^{(2)})} \qquad (5)$$

$$= \frac{\alpha_1(1 - \frac{\alpha_2 M_2}{K_r})}{\alpha_1(1 - \frac{\alpha_2 M_2}{K_r}) + (1 - \alpha_1)(1 - \frac{(1 - \alpha_2)M_2}{K_n})}.$$

Therefore, if $K_r << K_n$ (which is true for yeast), the reliability for the intersection of both interaction sets will be very high, close to 1. If $M_k << K_r$ and $M_k << K_n$ such as in the yeast two-hybrid assays of Ito's and Uetz's, the reliability for $E_1/E_1 \cap E_2$ is very close to α_1, and the reliability for $E_2/E_1 \cap E_2$ is very close to α_2.

2.3 *Protein function prediction using protein-protein interaction data*

We used two simple methods to predict protein functions based on protein-protein interaction data. One is referred to as the "neighborhood-counting method"[18], which assigns k functions to a protein with the k largest frequencies in its interacting partners. The other, referred to as the "chi-square method"[19], assigns k functions to a protein with the k largest chi-square scores. The chi-square score for a function j and a protein P_i is defined as

$$S_i(j) = \frac{[n_i(j) - e_i(j)]^2}{e_i(j)},$$

where $n_i(j)$ is the number of interaction partners of protein P_i having function j, $e_i(j) = n_i(j) \times p_j$ is the expected number of partners having function j, and p_j is the fraction of proteins having function j among all the proteins.

3   Results

Our study assessed the reliability of two different groups of protein-protein interaction data: the protein physical interaction data and the protein complex data. The protein physical interaction data included two yeast two-hybrid data sets, by Uetz et al.[1] and Ito et al.[2,3], and DIP (http://dip.doe-mbi.ucla.edu), a collection of protein interactions from the literature and the yeast two-hybrid assays. A special feature of Ito's protein-protein interactions is that each interaction is accompanied by an IST number, indicating how many times the interaction was observed. We compared these data with the experimentally determined MIPS physical interaction data set (http://mips.gsf.de). The protein complex data included two data sets, the TAP data[4] and the HMS-PCI data[5],

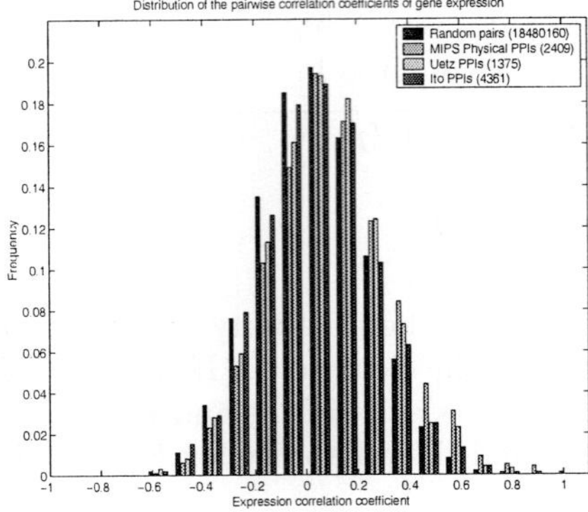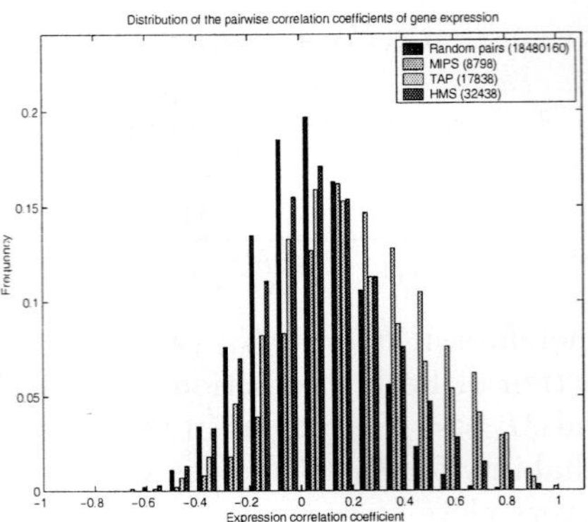

Figure 1: Distribution of expression coefficients for Uetz, Ito, MIPS physical interactions and random protein pairs (left), and that for TAP, HMS-PCI, MIPS complex data and random protein pairs (right).

obtained by systematic purification of protein complexes and protein identification via mass spectrometry. We compared them with a set of experimentally determined protein complexes called "MIPS Complex" [6]. For completeness, we also included the data obtained by other methods: co-expressed proteins measured at mRNA levels ("SynExpress"), computationally predicted interactions ("In-silico") and genetic interactions [16] ("Genetic"). We used the YPD [22] protein names in all data sets.

3.1 Distribution of gene expression correlation coefficients

We computed the correlation coefficient for every interacting protein pair using the cell cycle gene expression data [23], which contains 6,080 genes with 77 data points (2 cln3, 2 clb, 18 alpha, 24 cdc15, 17 cdc28, and 14 elut). Figure 1 shows the distributions of pairwise correlation coefficients. One is for all gene pairs, the MIPS physical interactions, Uetz's data and Ito's data. The other is for all gene pairs, the MIPS complex protein pairs, the TAP data and the HMS-PCI data. The distributions for the other three sets: "SynExpress", "In-silico", and "Genetic", are not included, as they cannot be easily classified as being in either of the above two groups.

The statistical significance for the difference between the mean expression correlation coefficient of a putative interaction set and that of random pairs is measured by the T-score and the P-value for the null hypothesis of no difference between the sample mean and the mean of random gene pairs. The T-scores are calculated as the standard two sample T-test statistic.

Table 1 gives some descriptive statistics for the distributions of correlation coefficients for different data sets in the three groups: the protein physical interaction data, the protein complex data, and the data obtained by other methods. As expected, the MIPS complex data has the largest mean and the largest T-score among all the data sets. In the first group, the MIPS physical interaction data has the largest mean, while Ito's data have the smallest mean and the smallest T-score. Generally, as the IST number increases, the mean for Ito's data increases as well. In the second group, the TAP data shows a higher mean than the HMS-PCI data.

Data	#Pairs	Mean	Variance	T-score	P-value
Random	18480160	0.0305	0.200	0.00	—
Physical Interactions					
MIPS Physical	2409	0.0985	0.224	16.71	6.31e-063
DIP	14351	0.0852	0.236	32.78	5.63e-236
Uetz	1375	0.0692	0.210	7.18	3.70e-013
Ito1IST	4361	0.0410	0.209	3.47	2.64e-004
Ito2IST	1408	0.0714	0.214	7.69	7.82e-015
Ito3IST	751	0.0833	0.223	7.23	2.42e-013
Ito4IST	541	0.0941	0.217	7.40	6.85e-014
Ito5IST	442	0.0979	0.223	7.09	6.96e-013
Ito6IST	351	0.0821	0.210	4.84	6.75e-007
Ito7IST	291	0.0883	0.217	4.94	4.04e-007
Ito8IST	257	0.0938	0.223	5.08	1.95e-007
Protein Complex					
MIPS Complex	8798	0.2560	0.250	105.90	0.00
TAP	17838	0.1642	0.270	89.31	0.00
HMS-PCI	32438	0.0801	0.245	44.69	0.00
Other Methods					
SynExpress	16063	0.1650	0.238	85.28	0.00
In-silico	7152	0.1111	0.234	34.10	3.89e-255
Genetic	878	0.0990	0.240	10.16	1.67e-024

Table 1: Statistics of distributions of gene expression correlation coefficients for different protein-protein interaction data sets.

3.2 Reliability of the different data sets

We estimated the reliability of the different data sets using maximum likelihood estimation on the distribution of gene expression correlation coefficients. As in equation 3, we needed to specify the distribution for real interactions and that for non-interactions. We used the distribution for random pairs as that for the non-interactions, since it is believed that the size of real interactions is much smaller than the size of the non-interaction pairs. We chose the MIPS physical interactions as the reference for the physical interaction data and the MIPS complex data as the reference for the TAP and the HMS-PCI complex data.

Data	Pairs	PairsExp	α	Variance
Physical interactions				
Uetz	1436	1375	0.529	0.0843
DIP	14454	14351	0.815	0.0244
Ito1IST	4443	4361	0.167	0.0383
Ito2IST	1469	1408	0.558	0.0831
Ito3IST	802	751	0.753	0.1144
Ito4IST	584	541	0.895	0.1436
Ito5IST	476	442	0.964	0.1567
Ito6IST	379	351	0.676	0.1768
Ito7IST	312	291	0.791	0.1942
Ito8IST	276	257	0.878	0.2054
Protein Complex				
TAP	17962	17838	0.585	0.0081
HMS-PCI	32667	32438	0.248	0.0053

Table 2: Reliability of the physical interaction data (Uetz's, DIP, and Ito's with different IST hits) and the protein complex data (TAP and HMS-PCI).

The results are listed in Table 2. The table shows that the Ito5IST data, with ≥ 5 IST hits, are the most reliable, with $\alpha = 0.96$, while the Ito1IST data are the least reliable with $\alpha = 0.17$. The Ito2IST data are as reliable as the Uetz data. The TAP data are more reliable than the HMS-PCI data. Again, the maximum likelihood approach cannot be applied to the other three data sets because they cannot be easily classified into either of the above two groups.

3.3 Cellular role prediction based on different data

We applied the neighborhood-counting method and the chi-square method[20] to predict protein functions based on the protein-protein interaction data. Both methods assign functions to a protein based on the functions of its immediate interaction proteins. The functional annotations were obtained from YPD, which assigns protein to three functional categories: "cellular role", "subcellular localization", and "biochemical function". Here, we considered the functional annotation based on the cellular role. Up to April 8, 2002, YPD included 6,416 proteins, among which 3,894 proteins have been assigned to one or more functions of 43 cellular roles, and the rest 2,522 proteins are unknown.

We used a leave-one-out method to measure the accuracy of the predictions. The leave-one-out method randomly selects a protein with known functions, assuming its functions as unknown, and then uses the neighborhood-counting method or the chi-square method to predict its functions. Finally, the predictions were compared with the actual functions of the protein. We repeated the leave-one-out experiment for K known proteins, $P_i, \cdots, P_K$. Let n_i be the number of functions for protein P_i in YPD, m_i be the number of $predicted$ functions for protein P_i, and k_i be the overlap between them. The

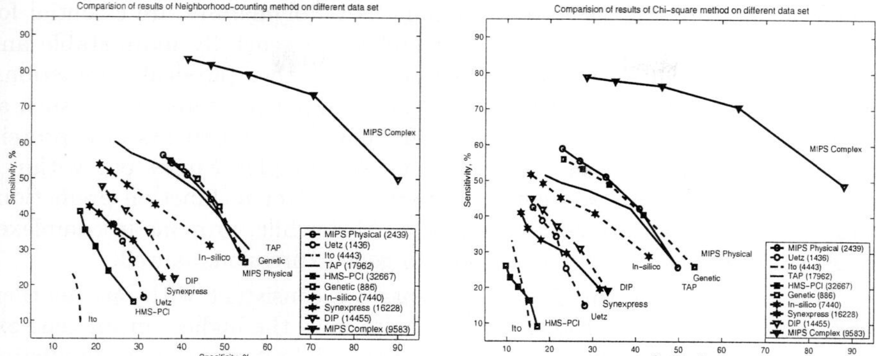

Figure 2: Sensitivity and specificity of functional predictions for different protein-protein interaction data sets using the neighborhood-counting (left) and the chi-square (right) methods.

specificity (SP) and the sensitivity (SN) can be defined as

$$SP = \frac{\sum_i^K k_i}{\sum_i^K m_i}, \qquad SN = \frac{\sum_i^K k_i}{\sum_i^K n_i} \qquad (6)$$

Figure 2 shows the relationship between specificity and sensitivity for the neighborhood-counting method and the chi-square method on different protein interaction data sets. As expected, the MIPS protein complex data have the best performance, while the Ito1IST data do the worst. Excluding the MIPS protein complex data, a group of three data sets (the TAP data, the MIPS Physical data and the Genetic Interaction data) shows better performance than the others. Overall, the results are very consistent with those of the distributions of expression correlation coefficients and the reliability estimation.

4 Discussions

We studied two groups of protein-protein interaction data, the physical interaction data and the protein complex data, and estimated the reliability of these data sets using three different measurements: the distribution of gene expression correlation coefficients, the reliability based on gene expression correlation coefficients, and the accuracy of protein function prediction. We separated protein complex data from protein physical interaction data because of their obvious difference: not all protein pairs in a complex interact with one another, and not all physically interacting protein pairs are in the same complex. Many protein complexes such as ribosomes and RNA Polymerases are essential for a cell, and the interactions within a complex are generally more stable and stronger and have a longer life span than most other physical interactions, while other physical interactions include other important interactions such as signal transductions. Our results confirm that the components of a protein complex can be assigned to functions that the complex carries out within a cell. The complex data sets generally perform better in function predictions than do the physical interaction data sets. Meanwhile, proteins in complexes are shown to be more highly correlated in gene expression, as well.

The results of the three measurements are consistent with one another. For example, the MIPS protein complex data have the highest mean gene expression correlation coefficient (0.256) and the highest accuracy in predicting protein functions (70% sensitivity and specificity), while Ito's Y2H data have the lowest mean (0.041) and the lowest accuracy (15% sensitivity and specificity); Uetz's data are more reliable than Ito's data in all three measurements, and the TAP data are more reliable than the HMS-PCI data in all three measurements. The Ito1IST data containing many interactions with only 1 IST hit are believed to contain many false positives and are the least reliable. However, the Ito2IST data containing interactions with at least 2 IST hits have better performance in all three measurements. This confirms that multiple ISTs can reduce false-positives in Y2H assays significantly.

There are three interaction data sets different from the above two groups: the genetic interaction data, the In-silico data, and the Synexpress data. Their capability of predicting protein functions generally falls between that of the Y2H data and that of the MIPS protein complex data. It should be noted that these interactions contain not only real physical interactions but also other protein pairs that are functionally associated. These data are important in understanding protein functions on a global scale.

We used three different ways to assess protein-protein interaction data. Although they show consistency with one another, there are some limitations to the reliability estimation. First, we assumed that the MIPS interactions and complexes were unbiased real interactions. However, the MIPS data sets may contain errors and may be biased to certain functions, cell compartments, mRNA expression levels, and so on, and thus the measurements may be inaccurate. Second, we used the gene expression profiles as a measurement. Ideally, a good measurement should itself be unbiased for assessing the reliability of interactions, and the data collection process should be unbiased as well. In this study, the gene expression data we used are certainly biased. It is also well-known that only a small set of interacting protein pairs are correlated at the mRNA expression levels. Another limitation is the assumption that known proteins have all their functions annotated. Based on this assumption, we can estimate the accuracy. In reality, the proteins may have un-discovered functions.

Acknowledgments

This research was partially supported by National Institutes of Health Grant DK53392, National Institutes of Health Grant 1-R01-RR16522-01, National Science Foundation EIA-0112934, and the University of Southern California.

References

1. P. Uetz, et al. *Nature* **403**, 623 (2000).
2. T. Ito, K. Tashiro, S. Muta, R. Ozawa, T. Chiba, M. Nishizawa, K. Yamamoto, S. Kuhara and Y. Sakaki. *Proc. Natl. Acad. Sci. USA* **97**, 1143 (2002).
3. T. Ito, T. Chiba, R. Ozawa, M. Yoshida, M. Hattori and Y. Sakaki. *Proc. Natl. Acad. Sci. USA* **98**, 4569 (2001).
4. A. Gavin, M. Böche, R. Krause, P. Grandi,M. Marzioch, A. Bauer, J. Schultz, J.M. Rick, A. Michon, C. Cruciat *et al.. Nature* **415**, 141 (2002).
5. Y. Ho, *et al.. Nature* **415**, 180 (2002).
6. H.W. Mewes, D. Frishman, U. Guldener,G. Mannhaupt, K. Mayer, M. Mokrejs, B. Morgenstern, M. Munsterkotter, S. Rudd and B. Weil. *Nucleic Acids Research* **30**, 31 (2002).
7. I. Xenarios, L. Salwinski, X.J. Duan, P. Higney, S. Kim and D. Eisenberg. *Nucleic Acids Research* **30** 303 (2002).
8. G.D. Bader, I. Donaldson, C. Wolting, B.F. Ouellette, T. Pawson, C.W. Hogue. *Nucleic Acids Research***29**, 242 (2001).
9. A. Grigoriev. *Nucleic Acid Res.* **29**, 3513 (2001).
10. M. Deng, S. Mehta, F. Sun and T. Chen. *Proceedings of the Sixth International Conference on Computational Molecular Biology (RECOMB2002)*, 117 (2002).
11. R. Jansen, D. Greenbaum and M. Gerstein. *Genome Research* **12**, 37 (2002).
12. P. Kemmeren, N.L.V. Berkum, J. Vilo, T. Bijma, R. Donders, A. Brazma and F.C.P. Holstega. *Molecular Cell* **9**, 1133 (2002).
13. H. Ge, Z. Liu, G.M. Church and M. Vidal. *Nature Genetics*, **29**, 482 (2001).
14. R. Mrowka, A. Patzak and H. Herzel. *Genome Research* **11**, 1971 (2001).
15. C.M. Deane, L. Salwinski, I. Xenarios and D. Eisenberg. *Molecular and cellular proteomics* (2002)
16. C.V. Mering, R. Krause, M. Snel, S.G. Oliver, S. Fields and P. Bork. *Nature* **417**, 399 (2002).
17. Saito, R., Suzuki, H. and Hayashizaki, Y.. 2002. *Nucleic Acids Research* **30:** 1163-1168.
18. B. Schwikowski, P. Uetz and S. Fields. *Nature Biotechnology* **18**, 1257 (2000).
19. H. Hishigaki, K. Nakai, T. Ono, A. Tanigami and T. Takagi. *Yeast* **18**, 523(2001).
20. M. Deng, K. Zhang, S. Mehta, T. Chen and F. Sun. *Proceeding of IEEE Computer Society Bioinformatics Conference*, Page 197-206 (2002).
21. M. Fellenberg, K. Albermann, A. Zollner, H.W. Mewes, and J. Hani. In *Proc. of the Eighth Int. Conf. on Intelligent System for Molecular Biology (ISMB2000)*, 152 (2000).
22. M.C. Costanzo, et al. *Nucleic Acids Research* **29:**, 75 (2001).
23. P.T. Spellman, G. Sherlock, M.Q. Zhang, V.R. Iyer, K. Anders, M.B. Eisen, P.O. Brown, D. Botstein and B. Futcher. *Molecular Biology of the Cell* **9**, 3273 (1998).

BOUNDARY FORMATION BY NOTCH SIGNALING IN DROSOPHILA MULTICELLULAR SYSTEMS: EXPERIMENTAL OBSERVATIONS AND GENE NETWORK MODELING BY GENOMIC OBJECT NET

HIROSHI MATSUNO,* RYUTARO MURAKAMI*, RIE YAMANE,
NAOYUKI YAMASAKI, SACHIE FUJITA, HARUKA YOSHIMORI

*Faculty of Science, Yamaguchi University,
1677-1 Yoshida, Yamaguchi 753-8512, Japan*

SATORU MIYANO

*Human Genome Center, Insititute of Medical Science, University of Tokyo,
4-6-1 Shirokanedai, Minato-ku, Tokyo, 108-8639, Japan*

The Delta-Notch signaling system plays an essential role in various morphogenetic systems of multicellular animal development. Here we analyzed the mechanism of Notch-dependent boundary formation in the *Drosophila* large intestine, by experimental manipulation of Delta expression and computational modeling and simulation by Genomic Object Net. Boundary formation representing the situation in normal large intestine was shown by the simulation. By manipulating Delta expression in the large intestine, a few types of disorder in boundary cell differentiation were observed, and similar abnormal patterns were generated by the simulation. Simulation results suggest that parameter values representing the strength of cell-autonomous suppression of Notch signaling by Delta are essential for generating two different modes of patterning: lateral inhibition and boundary formation, which could explain how a common gene regulatory network results in two different patterning modes in vivo. Genomic Object Net proved to be a useful and flexible biosimulation system that is suitable for analyzing complex biological phenomena such as patternings of multicellular systems as well as intracellular changes in cell states including metabolic activities, gene regulation, and enzyme reactions.

1   Introduction

Pattern formation of multicellular organisms includes intracellular regulatory events such as gene activation/repression, enzymatic reactions generating/degrading various kinds of biomolecules, as well as cell-to-cell interactions that coordinate intracellular events of individual cells.

The Delta-Notch signaling pathway plays an essential role in various morphogenetic systems of multicellular animal development. Lateral inhibition through Delta-Notch signaling pathway was examined during the emergence of ciliated cells in *Xenopus* embryonic skin [1]. Ghosh and Tomlin [2] modeled

---

*THESE AUTHORS EQUALLY CONTRIBUTED TO THIS PAPER.

the Delta-Notch signaling pathway as a hybrid system and presented results in both simulation and reachability analysis of this hybrid system. Bockmayr and Courtois [3] gave an another hybrid system approach, hybrid concurrent constraint programming, for modeling dynamic biological systems including the Delta-Notch signaling pathway. In their methods, to analyze a biological system, it has to be translated into a complete set of mathematical formulas for simulation. However, this task is in practice difficult for most biologists. In contrast, we have proposed a new method for modeling biological phenomena, which is based on the hybrid Petri net (HPN) [4]. In this approach, we only need to diagrammatize known or hypothetical biological pathways, without writing mathematical formulas, and define simple rules for the kinetics of each biomolecular component.

We have been developing a biosimulation system, Genomic Object Net (GON), whose architecture is essentially based on the hybrid functional Petri net (HFPN) and XML technology. The HFPN [5] was introduced by extending the notion of HPN [6] so that various aspects in biopathways can be modeled smoothly while inheriting good traditions from the the research on Petri net [7]. With GON, we have modeled and simulated many biopathways including the gene switch mechanisms of λ phage [4,8,18], the gene regulation for circadian rhythm in *Drosophila* [9,18], the signal transduction pathway for apoptosis induced by the protein Fas [9,18], and the glycolytic pathway in *E.coli* with the *lac* operon gene regulatory mechanism [5,18].

As a next step for exploiting GON, we here present a method to model a multicellular patterning system by HFPN with a novel visualizing function suitable for monitoring the simulation process of multicellular systems. In this paper, we analyzed the mechanism of Notch-dependent patterning events in the *Drosophila* large intestine, by combining experiments on live materials and computational modeling by GON.

2   Outline of the Delta-Notch Signaling Pathway

Cell-to-cell interactions mediated by Notch signal transduction pathway play essential roles in development of a multicellular organism [10]. Both of Delta and Notch proteins are initially expressed as membrane proteins. In canonical Notch signaling pathway, Delta binds to inactive Notch protein of adjacent cells, and triggers activation of Notch. After a few steps of activation, the intracellular domain of Notch is released by proteolytic cleavage and becomes active. The active form of Notch causes a change in gene expression pattern, including a down-regulation of *Delta* gene expression [10]. In addittion to the activation of Notch protein of adjacent cells, Delta has a suppressive effect

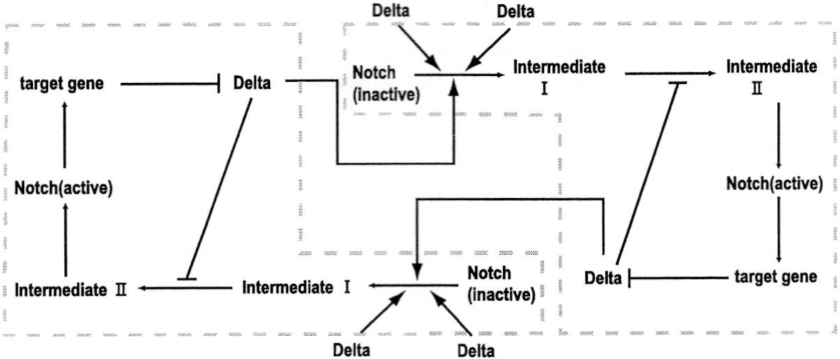

Figure 1. Two cells model of Delta-Notch Signaling pathway. Notch activation process by Delta of adjacent cells as well as the cell-autonomous suppression of Notch signaling by Delta in Delta-positive cells is included. Arrows and bars in the pathway represent activation and suppression, respectively.

on Notch signaling within Delta-positive cells [11,12]. This is a core feature of the Delta-Notch pathway common to various patterning systems of the multicellular animal development (Figure 1). These feedback loops, an essential feature of the Delta-Notch system, often make it difficult to predict how the system works in vivo. Furthermore, in spite of a seemingly simple core regulatory pathway, Notch signaling causes various types of pattern formation, depending on the developmental system in which the Notch pathway is working. One of the most well-known example of Notch-dependent patterning is the lateral inhibition [10], in which one cell is singled out from a group of equivalent precursor cells. The other example is the boundary formation between two different fields of cells [10]. In the present study, we analyzed the mechanism of the boundary cell formation in the *Drosophila* large intestine, in which a single row of boundary cells is induced between dorsal and ventral domains [13,14,15,16].

3 Hybrid Functional Petri Net Modeling of the Delta-Notch Pathway

3.1 Basic Elements

In general, a *Petri net* [7] is a network consisting of the following four kinds of elements: place, transition, arc, and token. A *place* holds *tokens* as its content. A *transition* has arcs coming from places and arcs going out from

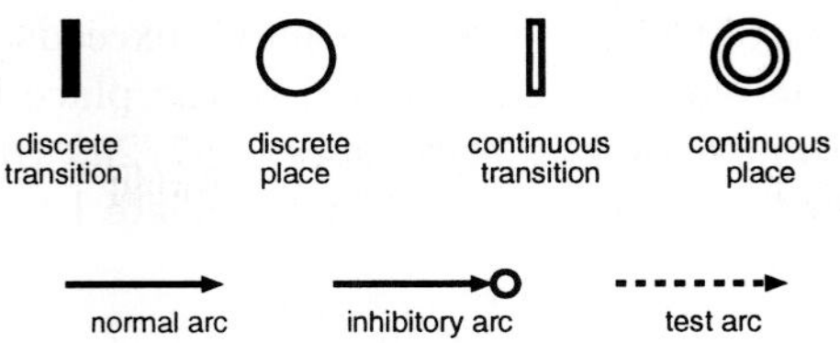

Figure 2. Basic elements of hybrid (functional) Petri net.

the transition to some places. A transition with these arcs defines a *firing rule* in terms of the contents of the places where the arcs are attached.

In *hybrid Petri net* (HPN) model [6], two kinds of places and transitions are used, *discrete/continuous places* and *discrete/continuous transitions*. A discrete place and a discrete transition are the same notions as used in the traditional *discrete Petri net* model [7]. A continuous place holds a nonnegative real number as its content. A continuous transition fires continuously in the HPN model and its firing speed is given as a function of values in the places in the model. The graphical notations of a discrete transition, a discrete place, a continuous transition, and a continuous place are drawn in Figure 2, together with three types of arcs. The same basic elements in Figure 2 are used in HFPN. Refer to reference [5] for the details of HFPN.

Specific values are assigned to each arc as a weight. When a *normal arc* with weight w is attached to a discrete transition, w tokens are transferred through the normal arc. On the other hand, when a normal arc is attached to a continuous transition, the amount of token that flows is determined by the firing speed of the continuous transition. An *inhibitory arc* with weight w enables the transition to fire only if the content of the place at the source of the arc is less than or equal to w. For example, an inhibitory arc can be used to represent repressive activity in gene regulation. A *test arc* does not consume any content of the place at the source of the arc by firing. For example, test arcs can be used to represent enzyme activity, since the enzyme itself is not consumed.

3.2 *Modeling the Delta-Notch Signaling Pathway*

The Delta-Notch pathway depicted in Figure 1 is modeled by an HFPN, which includes the intracellular regulatory circuit as well as cell-to-cell interactions (Figure 3). In Figure 3, an HFPN model of the complete intracellular circuit of a single Cell A, with interactions with adjacent Cells B and C, is illustrated.

When the amount of Delta in Cell B (Cell C) exceeds level 1, token value is transferred from the place Notch(inactive) to the place Intermediate I. This token value is determined by the firing speed $m7/200$ ($m8/200$). To define the repression level of the processing of Intermediate I to Intermediate II, we use the following formula;

$$\frac{\alpha \times m2}{\beta \times m6 + m2},\tag{1}$$

which is assigned to the transition T_a. This formula describes the following two functions;

- the firing speed of the transition T_a becomes faster as the amount $m2$ in the place Notch(inactive) increases, and

- the firing speed of the transition T_a becomes slower as the amount $m6$ in the place Delta increases.

Note that the firing speed of the transition T_a can be manipulated by changing the two parameters α and β.

The production rate of Delta is defined by the parameter d at the transition T_b. The forced-expression rate of Delta can be also set to the parameter d_m at the transition T_c.

4 Experimental Results

The large intestine of *Drosophila* embryo occupies a major middle portion of the hindgut, and is subdivided into dorsal and ventral domains (Figure 4 (a)). A one-cell-wide boundary cell strand forms between the dorsal and ventral domains [13] (Figure 4 (a), (b)). Delta is expressed exclusively in the ventral domain, and essenial for the activation of Notch signaling in abutting Delta-negative dorsal cells [14,15,16]. In Delta mutant embryo, in which no Delta protein is produced, boundary cells failed to form (Figure 4 (c)). When Delta protein is expressed throughout large intestine by the GAL4-UAS system, an established method for forced gene expression [17], boundary cell formation was strongly affected. In about 60% of large inestines examined, only a few boundary cells formed randomly (Figure 4 (e)). About 20% of large intestines failed to form boundary cells (Figure 4 (d)). Large intestines with many boundary cell clusters were occasionally found (about 20%, Figure 4 (f)).

These results are summarized as follows:

- Boundary cell strand forms between the ventral and dorsal domains when Delta is expressed only in the ventral domain, i.e., in the case of normal large intestine.

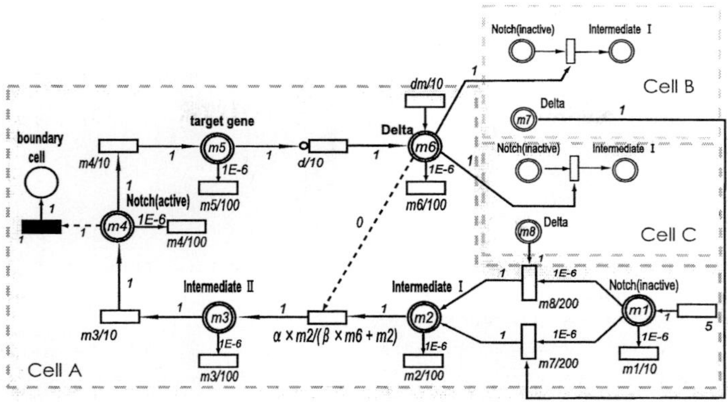

Figure 3. The HFPN model of the Delta-Notch signaling pathway. An HFPN model of complete intracellular circuit of a single Cell A with interactions with adjacent Cells B and C is illustrated. Continuous places represent concentrations of the molecules depicted in Figure 1. Production rates and degradation rates are assigned to the continuous transitions. When the discrete place **boundary cell** gets token(s), the corresponding cell becomes a boundary cell. Test arcs are used at the reactions where no substances are consumed. Inhibitory arcs are used for modeling repressive activity. The weight *1E-6* which is assigned at some transitions represents 10^{-6}. This means that if the token value of the relevant place becomes over 10^{-6}, the transition begins to fire.

- Boundary cells fail to form in the absence of Delta.

- Forced expression of Delta throughout the large intestine suppresses boundary cell formation, with ectopic induction of a small number of boundary cells.

5 Simulation by Genomic Object Net

Genomic Object Net (GON) is a biosimulation system which is developed based on the hybrid functional Petri net (HFPN) and XML architecture [8].

Genomic Object Net (GON) consists of two tools, GON Assembler and GON Visualizer. GON Assembler allows us to model target biopathways without complicated mathematical formulas, and to perform simulations easily by manipulating parameters directly and smoothly using its GUI [4,9,18]. GON Visualizer was developed based on XML technology [8]. Users can realize visualization of simulation results of biological phenomena by describing it as an XML document, in which CSV files produced by GON Assembler are included

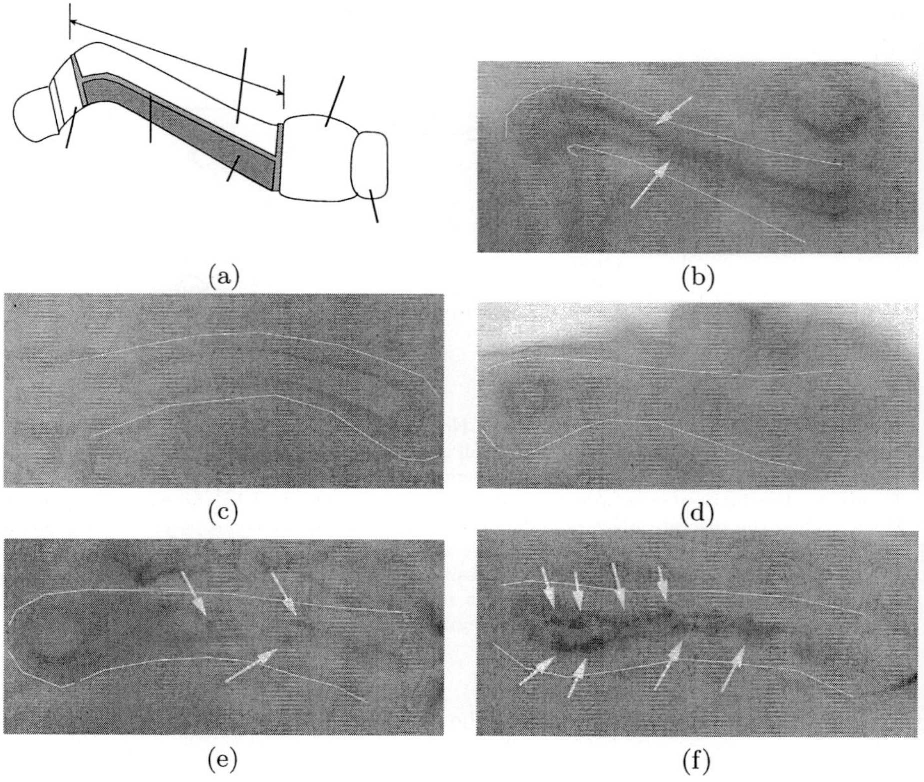

Figure 4. (a): A diagrammatic illustration of the hindgut of *Drosophila* embryo. The large intestine is a major middle portion of the hindgut, and subdivided into the dorsal and ventral domains. A one-cell-wide strand of boundary cells develops between the two domains. (b): Boundary cell strand in the large intestine of a wild-type embryo. Boundary cell strands in the right and left sides of the large intestine are indicated by arrows. Outline of the large intestine is marked with white lines. Staining of boundary cells (in brown color) was performed by use of anti-Crumbs antibody. (c): Boundary cells fail to differentiate in *Delta* mutant embryo. (d), (e), (f): Forced-expression of Delta caused suppession of boundary cell differentiation. In (d), no boundary cells have developed. In (e), a few boundary cells have formed ectopically (arrows). Most of the large intestines examined showed these two patterns. (f) In fewer cases (less than 20% of large intestines examined) many clusters of boundary cells were induced (arrows).

as basic data for visualization. With GON Visualizer, users in biology and medicine can design a personalized visualization for simulation suitable for the purpose of their studies [8]. By combining these two tools, GON provides an efficient environment for biopathway simulations.

5.1 Simulation Model

We carried out simulations of the patterning of boundary cells by using GON. Figure 5 shows the simulation model consisting of 60 cells. Each cell has the HFPN model illustrated in Figure 3. Refer to the website [19] for the full connection model.

For representing cell-to-cell interactions, arcs are drawn from the place Delta of (up to 6) adjacent cells to the transitions between the places Notch(inactive) and Intermediate I. Since the whole HFPN model constructed in this way is very complicated and messy, it is actually difficult to monitor progress of the simulation on GON Assembler.

To address this issue, we wrote an XML document for GON Visualizer which realizes a model of 60 cells (Figure 6 (a), (b), Figure 8 (a)). In this model, the color of each cell can be changed according to the token value in the places which we want to observe.

5.2 Simulation Results

Figures 6 presents the simulation results of boundary cell formation. Parameters used in the simulations are summarized in Table 1. We choose parameter values 0.7 and 49 for α and β, respectively. Initial condition for Delta level (d) is: 0 for cells 1-36 (dorsal cells) and 10 for cells 37-60 (ventral cells). This condition represents a prepattern of Delta expression in normal large intestine, in which Delta is expressed only in the ventral cells.

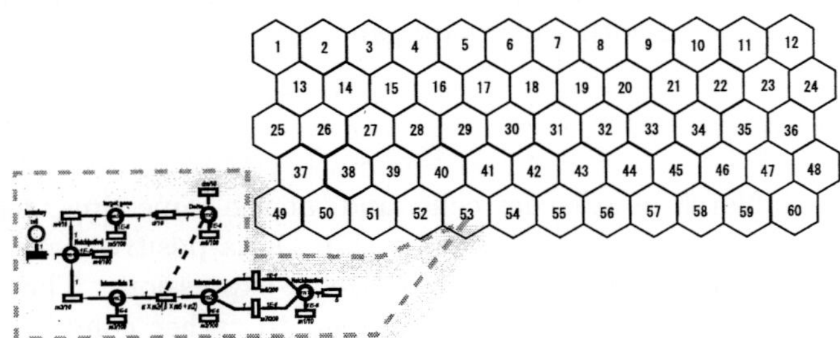

Figure 5. 60 cells model for simulation by GON. Each cell has the HFPN model presented in Figure 3.

Simulation with this condition generated a single strand of boundary cells that are abutting ventral cells (a). The values obtained for Delta ($m6$) and Notch(active) ($m4$) of the cells marked with bold lines in (a) are shown in Figure 7 (a). For the condition of *Delta* mutant, the parameter value of $m8$ was fixed at 0, resulting in no boundary cell formation (b). For the simulation of forced expression of Delta, parameter values 6 and 30 are chosen for d_m.

In the condition of forced expression of Delta at $d_m = 30$, no boundary cells were generated (c), while a few boundary cells were generated ectopically at $d_m = 6$ (d). The values obtained for Delta ($m6$) and Notch(active) ($m4$) of the cells marked with bold lines indicated in (d) are shown in Figure 7 (b). Note that ratios of Delta ($m6$) to Notch(active) ($m4$) of the cells around boundary cells are slightly higher than those of boundary cells in this case. These results correspond well to the experimental results described above (Figures 4), though many boundary cell clusters could not be generated in the present condition.

We also tried a simulation with an initial condition of a uniform Delta level ($d = 3$) for all the cells, in order to represent a situation of lateral inhibition, in which specified cells are singled out from equivalent precursors. When the parameter β was reduced to 5, a regular distribution pattern of specified cells (with a high Delta level) was obtained (Figure 8 (a)), a pattern of which is considered to correspond to the patterning event regulated by lateral inhibition, such as neural cell determination and ciliated cell differentiation in *Xenopus* embryo [1]. The values obtained for Delta ($m8$) and Notch(active) ($m4$) of the cells in the area indicated in Figure 8 (a) are shown in Figure 8 (b).

It should be emphasized that all these patterns were obtained only by changing a few parameter values and initial conditions of a common HFPN model, which is a reasonable approach because all the cells of a multicellular organism are equipped with a common genome.

6   Discussion

Delta activates Notch of adjacent cells, and, at the same time, represses autonomously Notch signal transduction within Delta-positive cells. Activated Notch, in turn, autonomously represses Delta expression. The ambivalent nature of Delta on Notch signaling may lead to rather complicated results when expressed ubiquitously. Forced expression of Delta largely suppressed boundary cell formation, and, at the same time, induces a few boundary cells ectopically. Similar patterns of boundary cell formation were also generated computationally by GON. Obtained ratios of Delta ($m6$) to Notch(active) ($m4$)

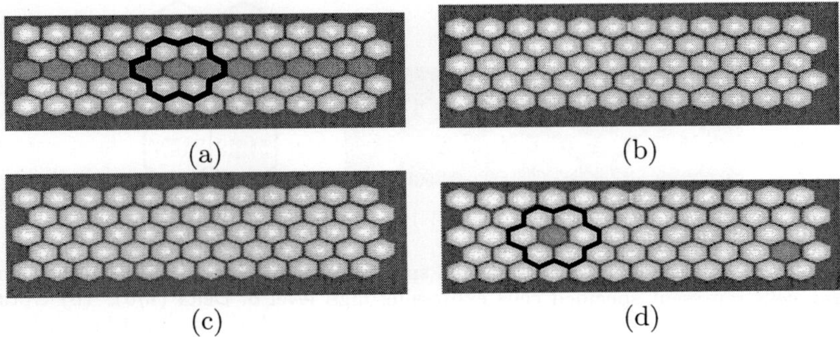

Figure 6. Simulation results of boundary cell formation. Gray cells represent boundary cells. Area marked with bold lines correspond to those illustrated in Figure 7. (a) wild type. (b) *Delta* mutant (realized by removing arc from the transition T_b to the place Delta) (c) and (d) Forced-expression of Delta. Parameter values 30 and 6 were chosen for d_m of (c) and (d), respectively.

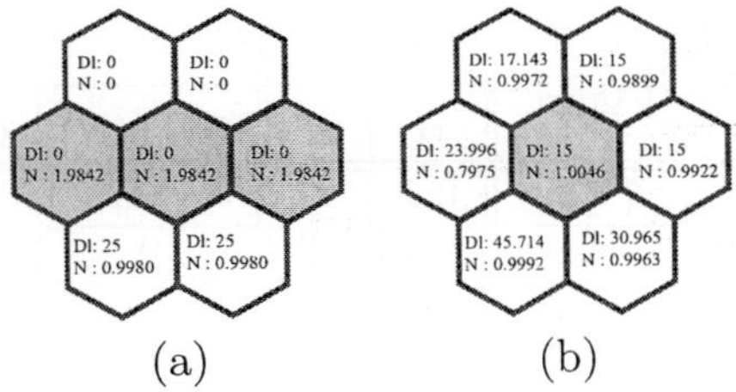

Figure 7. Values of Delta ($m6$) and Notch(active) ($m4$) generated by the simulation in the cells marked with bold lines in Figure 6. Note that ratios of Delta ($m6$) to Notch(active) ($m4$) in cells around a boundary cell are slightly higher than those of the boundary cell.

in cells around boundary cells are slightly higher than those of boundary cells in these cases. Local differences in ratios of the contents of places Delta to Notch(active) occurring among cells is considered to induce boundary cells.

In addition, by reducing parameter value β to 5, with a modification of the initial condition d, GON simulation brought about another type of Notch-dependent patterning, the lateral inhibition. This suggest that two distinct types of Notch-dependent patterning, boundary formation and lateral inhibition, is a consequence of different β values, which represent the susceptibility to autonomous suppressive activity of Delta on Notch signaling.

GON has been demonstrated to be a useful tool for modeling and simulating intracelluar biological phenomena through several examples [5,8,9,18]. In

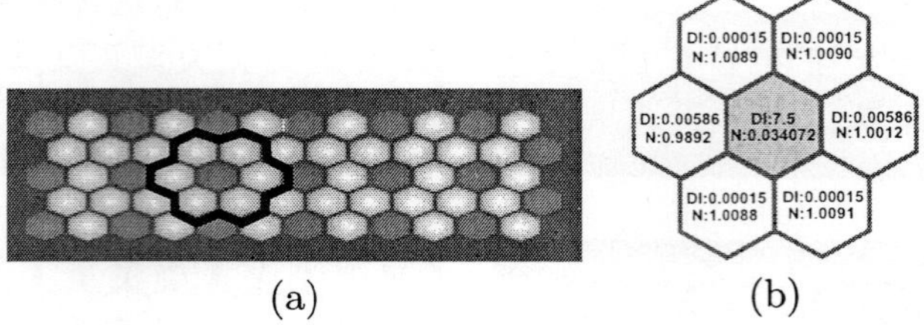

Figure 8. Simulation result representing cell specification by lateral inhibition mechanism. (a) Gray cells represent specified cells expressing high level of **Delta** ($m6$). (b) Obtained values of **Delta** ($m6$) and **Notch(active)** ($m4$) in the cells marked with bold line in (a).

Table 1. Parameters in HFPN model of Figure 3 used in the simulation. α and β represent the firing speed of the transition T_a in the formula (1). In the case of boundary cell, different initial production rates of Delta d at the transition T_b are set to dorsal cells:1-36 and ventral cells:37-60, while the same initial value 49 is set to all 60 cells in the case of lateral inhibition. d_m is the forced-expression rate of Delta assigned to the transition T_c.

phenomenon	Figure	(1)		d		d_m
		α	β	cell:1-36	cell:37-60	
boundary cell	7 (a)	0.7	49	0	10	0
	7 (c)	0.7	49	0	10	30
	7 (d)	0.7	49	0	10	6
lateral inhibition	9 (a)	0.7	5	3	3	0

the present study, we tried to model and simulate multicelluar phenomena by using GON, and succeeded in obtaining results corresponding to experimental observations. It is expected that variable multicelluar phenomena can be computationally analyzed by GON based on the technique demonstrated in this paper.

Although GON Assembler has an excellent GUI which allows us to tune up parameters smoothly, it is still difficult to obtain intuitive observations of simulation results, since GON Assembler can present only time-course graph representations. As is demonstrated in this paper, with the support of GON Visualizer, we can realize a more effective and creative environment for biopathway simulations.

We are currently developing a new version of GON Assembler which has the scalability in modeling and simulating more complex biological systems such as the development mechanism of *C. elegans* embryo. For this purpose, we extended HFPN architechture by allowing more "types" of data (integer, real, boolean, string) with which more complex information such as localization, multicellular process, etc. can be handled smoothly.

Acknowledgments

This work was partially supported by the Grand-in-Aid for Scientific Research on Priority Areas "Genome Information Science" and Grand-in-Aid for Scientific Research (B) (No.12480080) from the Ministry of Education, Culture, Sports, Science and Technology in Japan.

References

1. Marnellos, G., Deblandre, G.A., Mjolsness, E. and Kintner, C., *Proc. Pacific Symposium on Biocomputing 2000*, 329–340 (2000).
2. Ghosh, R. and Tomlin, C.J., *Proc. 4th International Workshop on Hybrid Systems:Computation and Control*, LNCS 2034, 232–246 (2001).
3. Bockmayr, A. and Courtois, A., *Proc. 18th International Conference on Logic Programming*, LNCS2401, 85–99 (2002).
4. Matsuno, H., Doi A., Nagasaki M. and Miyano, S., *Proc. Pacific Symposium on Biocomputing 2000*, 338–349 (2000).
5. Matsuno, H., Fujita, S., Doi, A., Hirata, Y., and Miyano, S., *submitted*, (2002). *http : //genome.ib.sci.yamaguchi-u.ac.jp/~matsuno/papers/lac_operon.pdf*
6. Alla, H. and David, R., *Journal of Circuits, Systems, and Computers* 8(1), 159–188 (1998).
7. Reisig, W. Petri Nets, *Springer-Verlag*, (1985).
8. Matsuno, H., Doi, A., Hirata, Y., and Miyano, S., *Genome Informatics* 12, 54–62 (2001).
9. Matsuno, H., Tanaka, Y., Aoshima, H., Doi, A., Matsui, M., and Miyano, S., *submitted*, 2002. *http : //genome.ib.sci.yamaguchi-u.ac.jp/~matsuno/papers/circadian_apoptosis.pdf*
10. Artavanis-Tsakonas, S., Rand, M. D., and Lake, R. J., *Science* 284, 770-776 (1999).
11. Jacobsen, T. L., Brennan, K., Martinez-Arias, A. and Muskavitch, M. A. T., *Development* 125, 4531–4540 (1998).
12. Baker, N. E., *Bioessays* 22, 264–273 (2000).
13. Takashima, S. and Murakami, R., *Mech. Dev.* 101, 79–90 (2001)
14. Fusse, B. and Hoch, M., *Curr. Biol.* 12, 171–179 (2002).
15. Iwaki, D.D. and Lengyel, J.A.A., *Mech. Dev.* 114, 71–84 (2002).
16. Takashima, S., Yoshimori, H., Yamasaki, N., Matsuno, K., and Murakami, R., *Dev. Genes Evol.* 212, in press, 2002.
17. Brand, A. and Perrimon, N., *Development* 118, 401-415 (1993).
18. *http : //www.GenomicObject.Net/*
19. *http : //genome.ib.sci.yamaguchi-u.ac.jp/~fujita/Notch/*

INFLUENCE OF NETWORK TOPOLOGY AND DATA COLLECTION ON NETWORK INFERENCE

V. ANNE SMITH, ERICH D. JARVIS

Department of Neurobiology, Duke University Medical Center, Box 3209
Durham, NC 27710, USA

ALEXANDER J. HARTEMINK

Department of Computer Science, Duke University, Box 90129
Durham, NC 27708, USA

We recently developed an approach for testing the accuracy of network inference algorithms by applying them to biologically realistic simulations with known network topology. Here, we seek to determine the degree to which the network topology and data sampling regime influence the ability of our Bayesian network inference algorithm, NETWORKINFERENCE, to recover gene regulatory networks. NETWORKINFERENCE performed well at recovering feedback loops and multiple targets of a regulator with small amounts of data, but required more data to recover multiple regulators of a gene. When collecting the same number of data samples at different intervals from the system, the best recovery was produced by sampling intervals long enough such that sampling covered propagation of regulation through the network but not so long such that intervals missed internal dynamics. These results further elucidate the possibilities and limitations of network inference based on biological data.

1 Introduction

Large amounts of data on the expression of genes have recently become available due to advances in gene expression array analysis. Functional network inference algorithms have begun to be used to draw causal information from this correlational data, producing networks representing putative regulatory interactions of hundreds to thousands of genes (for an early review, see [1]). However, a difficulty arises in evaluating the success of these algorithms, because no high throughput method of assessing genetic regulatory influences yet exists. To experimentally validate all links in even a single putative network, using antisense blocking or other methods, could take impractically long given current technology.

In Smith et al. [2] and Jarvis et al. [3] we developed an approach to circumvent this problem, by creating a biologically realistic computer simulation of a system in which we make and know all the rules. We run the simulation representing the behaving organism and sample data from it as one would in a real biological experiment. We then present this sampled data to network inference algorithms. We compare putative networks produced by the algorithms to the known truth underlying the simulation, and thus evaluate the success of the algorithms. Additionally, because we sample data from the simulation, we can also use our framework to evaluate the effect of differing sampling regimes on the output of the inference algorithms. This ability has implications for designing biological experiments for use with network inference algorithms[1].

In this paper, we use our approach to evaluate the degree to which the recovery performance of our Bayesian network inference algorithm depends on the topology of the network being recovered. In particular, we examine various network topologies exhibiting features commonly found in genetic regulatory pathways like feedback loops, multiple regulators of a single gene, and multiple targets of a single regulator. We also evaluate its success on two complex pathways of biologically realistic topology, a gene cascade and a convergence network. In addition, we use our approach to further examine the effect of sampling regime on network inference results. Previously, we showed that sampling at an interval that matches underlying system dynamics produces considerably better results than sampling with an interval twice as long [2]. Here, we investigate the issue more closely to determine the extent of the usable range of sampling intervals.

2 Approach and Methods

2.1 Simulator

Our simulator, BRAINSIM, is based on the vocal communication system of the songbird brain [3]. Songbirds are vocal learners, along with only five other animal groups, including humans, and thus provide a model system for understanding human language [5]. Songbird vocal communication has been studied extensively at the behavioral, electrophysiological, neuroanatomical, and molecular levels of organization [6]. This simulator is part of an integrative project that aims to synthesize information across multiple levels of biological organization [3]. As vocal learning represents an integration of motor (singing) and sensory (hearing) systems, such synthesis in the songbird is expected to have implications for a wide variety of brain systems.

BRAINSIM is programmed in C++ and simulates behavior, electrophysiological activity in distinct brain regions, and gene expression in these brain regions for six subject birds (Fig. 1). A detailed account of BRAINSIM's implementation is found in Smith et al. [2]. Briefly, the simulation progresses in discrete time steps, each step representing approximately one minute. Behavior of the animal takes one of two values, 0 or 1, which correspond to off-on states of a behavior, such as silence and singing. At each time step, activity in four brain regions ranges from 0-400 Hz to simulate recordings from extracellular multi-unit electrodes [7], and is correlated

---

[1]After this paper was submitted, an independent generation of a similar approach by Zak et al. [4] was brought to our attention, using a different type of simulation. Zak et al. evaluate linear, linear-log, and linear-squashing network inference algorithms.

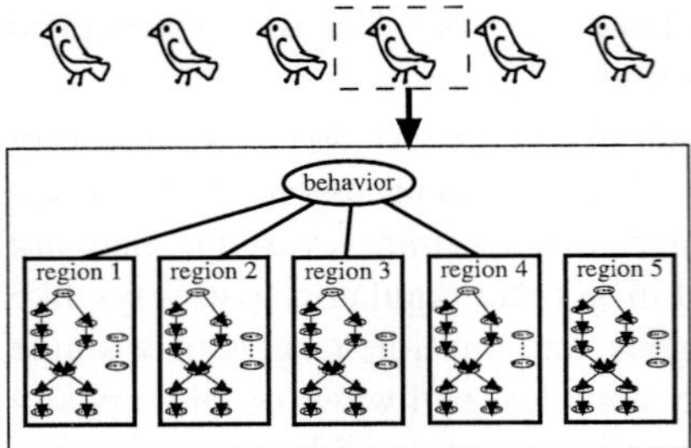

Figure 1. Overview of BRAINSIM. Each simulated data set consists of behavior, activity in five brain regions, and expression levels of 100 genes in each region, for six subjects. A small fraction of the 100 genes are regulated as part of a pathway; the remainder wander randomly. Specific pathway topologies we considered are found in Fig. 2. Figure modified from [2].

either positively or negatively with behavior. Activity in the fifth region wanders randomly in the same range. Three factors influence gene expression level, whose regulatory network is assumed to be the same in all brain regions. First, a returning function moves each gene towards its constitutive expression level, representing degradation of mRNA or return to transcription after suppression. Second, for genes in a pathway, gene regulation is implemented by adding (for up-regulation) or subtracting (for down-regulation) a proportion of the regulator's expression level in the previous time step. This proportion is set at 0.2 but can be varied, and is the **influence** of a regulator on its target. Genes not in a pathway add or subtract a random amount from their previous expression level to simulate influence from other unmeasured processes. Third, a final random amount is added or subtracted from all genes to simulate stochasticity in gene expression [8]. All genes have a minimum possible expression level of 0 and a maximum of 50, in arbitrary units.

For this paper, the simulation began with behavior at 0 and was run for 250 time steps to stabilize the system. Behavior then switched to 1 for 50 time steps, and back to 0 for 100 more time steps. This represents a likely experimental situation where an animal would perform a behavior, such as singing, for a set amount of time in response to some stimulus, such as presentation of a potential mate.

2.2 Network inference algorithm

Our inference algorithm, NETWORKINFERENCE, described in detail previously [2,9], is written in C and uses simulated annealing to identify a high scoring network from data using a discrete Bayesian network framework. A Bayesian network is a graph-based representation of a joint probability distribution, with nodes connected by links that represent conditional dependence. A link from one node to another indicates that the child node is conditionally dependent on the parent node. Unlike many other network inference frameworks, Bayesian networks are capable of representing combinatorial, nonlinear, and stochastic relationships, such as are often found in biological systems. Additionally, due to their probabilistic nature, Bayesian networks can handle noisy data [10]. While static Bayesian networks are limited to having no cycles, regulatory cycles like those found in biological systems can be handled using dynamic Bayesian networks (DBN). A DBN represents all elements at time t and $t+\Delta t$, with controlling elements at time t having links to controlled elements at time $t+\Delta t$. For example, a cyclic interaction in time between nodes A and B can be represented acyclically in a DBN using links from $A(t)\rightarrow B(t+\Delta t)$ and $B(t)\rightarrow A(t+\Delta t)$ [11]. NETWORKINFERENCE is designed to recover DBNs from pairs of data points, disallowing links that flow backwards in time, i.e., from time $t+\Delta t$ to time t [2].

2.3 Genetic regulatory network topology

We wished to examine the performance of our algorithm on three types of topologies found in genetic regulatory pathways: (1) multiple regulator genes of a single target gene, (2) regulatory loops, and (3) multiple targets of a single regulator. We simulated networks to test each topology, and also tested multiple regulators and multiple targets in two large, more biologically realistic topologies.

We considered two types of multiple regulators, one where two inputs to one gene were governed by pathways originating from a common variable and were thus **correlated**, and another where the two inputs were governed by **uncorrelated**

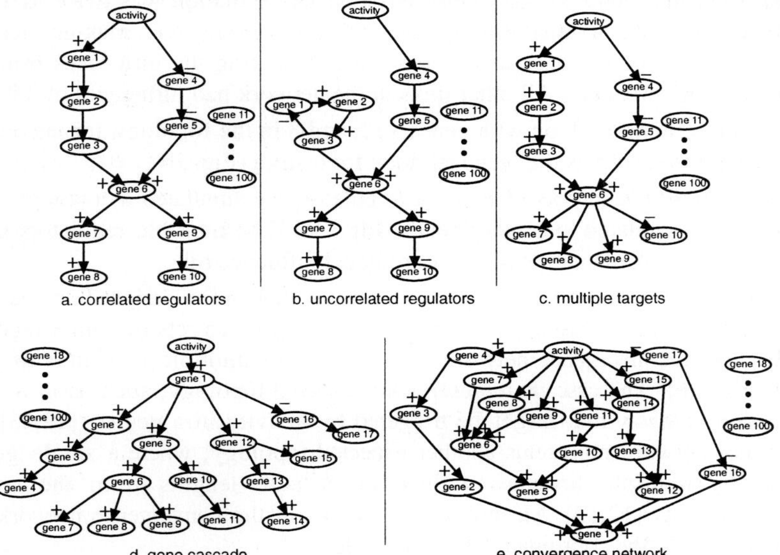

Figure 2. Genetic regulatory network topologies. Different topologies for electrical-genetic regulation pathways. All topologies contain 100 genes, with those not in the pathway wandering randomly. **a.** Genes 3 and 5 serve as correlated regulators of gene 6. **b.** Genes 3 and 5 serve as uncorrelated regulators of gene 6, and genes 1, 2, and 3 from a regulatory loop. **c.** Genes 7-10 serve as multiple targets of gene 6. **d.** A regulatory cascade, with increasing branches of influence from gene 1. **e.** A convergence network, with regulatory paths of varying lengths of regulation converging on a single target.

168

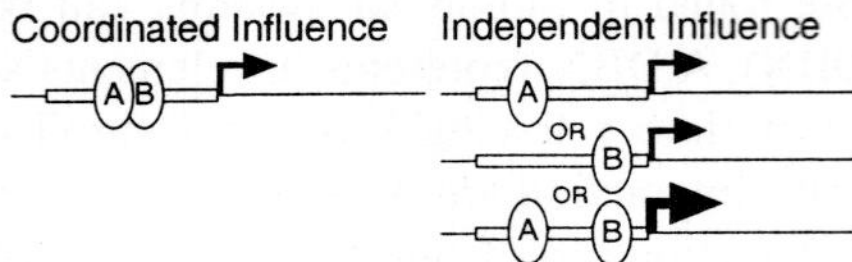

Figure 3. Coordinated versus independent influence. In the case of coordinated influence, both regulators A and B must bind to the promoter region (boxed area) to activate transcription of the gene (right of boxed area, transcription noted by arrow). In the independent case, either A or B alone is capable of activating transcription but transcription is elevated with both present (additive effect, thicker arrow).

pathways. The correlated regulators were produced using the topology we developed previously [2], where gene 6's regulators, 3 and 5, are both downstream of activity (Fig. 2a). Uncorrelated regulators of gene 6 were created by modifying the topology of this network so that genes 1, 2, and 3 formed a self-generating loop uncorrelated with activity (Fig. 2b). With multiple regulators, the manner of combining their influences on the target needs to be considered. We simulated **coordinated influence**, which models a situation where regulation of the target gene requires a molecular complex formed from all regulators, and **independent influence**, which models a situation where regulation of the target gene requires any of the regulators singly, and their influence is additive (Fig. 3) [12]. For independent influences, we tested two situations: one where the two influences on gene 6 were equal to the other influences in the network, i.e. 0.2, such that each link was as "strong" as any other, but also such that gene 6's maximum possible up-regulation was twice as fast as any other gene; and another where these influences were 0.1, making gene 6's dynamics equal to other genes in the network but making the influences from each of its regulators "weaker". All other links in the network had influences of 0.2.

The self-generating loop with genes 1, 2, and 3 in the topology for uncorrelated multiple parents served as the feedback loop for testing (Fig. 2b).

To test multiple targets of a single regulator, we simulated a topology where gene 6 directly regulated four other genes (Fig. 2c). The multiple regulators in this topology were implemented with the coordinated influence rule.

Finally, we tested two complex pathways that were designed to be more biologically realistic: a cascade, which contains multiple targets of single regulators (Fig. 2d), and a convergence network, which contains multiple regulators of single targets (Fig. 2e). The cascade represents an expected topology, such as in a signal transduction pathway that might be triggered by activity in a neuron [12,13]. The convergence network represents another expected topology, where a single gene is regulated by multiple branches with different path lengths from the original triggering activity [12,13]. All multiple regulators in the convergence network were implemented with the coordinated influence rule.

2.4 Sampling regime

These network topologies were sampled using the regime described in [2], sampling at an interval that intuitively matched the system, every 5 time steps, following the reasoning that a 0.2 influence each time step leads to a full influence in 5 time steps. Sampling began just before the animal started its behavior (value switched to 1), and covered the period of the behavior and an equal amount of time after its cessation (value 0), for a total of 21 samples. We also investigated the effect of varying the sampling interval, using the original regulatory topology (Fig. 2a). We sampled at intervals of 1-4, 6-9, 15, 20, and 30 time steps, and examined these along with previous results for intervals of 5 and 10. In order to maintain the same number of data points across the behavior and its cessation, behavior length was adjusted to the sampling interval. For example, when sampling at an interval of 3, instead of behavior lasting for 50 time steps, behavior lasted 30 time steps; when sampling at an interval of 15, behavior lasted 150 time steps. The simulation was run long enough to collect 21 data points in each situation. This is experimentally reasonable, as it possible to adjust the duration of stimulus presentation to an animal or allow an animal to continue to perform its behavior.

2.5 Data generation and discretization

Each data set consisted of sampled data from a simulation containing six subjects. This represents a biologically reasonable maximum sample size for gene expression studies (21 time points x 6 animals per time point = 126 animals). Although not expected to be carried out experimentatally, for statistical purposes we generated multiple data sets to evaluate the robustness of the algorithm to random variation in data with the same structure. Ten data sets were collected for each genetic network topology and sampling interval, and presented to the network inference algorithm independently.

Since NETWORKINFERENCE searches for discrete Bayesian networks, we must present it with data that has been discretized. A smaller number of discrete states reduces computational time and smooths out random noise; a larger number retains more of the information present in the original data. We chose to use four discretization levels as a balance. As in Smith et al. [2], we discretized activity and gene expression levels into quartiles, with the lowest 25% of values being state 1, the next 25% state 2, and so on.

2.6 Search settings and parameters

Because in this study we were interested in evaluating the ability of the NETWORKINFERENCE algorithm to infer different types of genetic network structures, we only presented the inference algorithm the task of determining relationships between activity and gene expression within brain regions and not between activity and behavior as in [2]. To infer a dynamic Bayesian network, we presented to the algorithm pairs of sampled data points at time t and time $t+\Delta t$ for the activity and values of all 100 genes. The 21 sampled points thus corresponded to 20 sets of pairs. For the five brain regions of the six birds in each data set, this resulted in 600=20x5x6 observations for 202 variables (activity plus 100 genes each at two time points). NETWORKINFERENCE had no prior information about which genes were regulated and which wandered randomly. NETWORKINFERENCE was run approximately 160 minutes on each dataset on a 1GHz Pentium III CPU or 100 minutes on a 1.6GHz Athlon CPU, both running Linux. The algorithm searched an average of 23.6 million structures (range 20.5-24.7 million) and found the best scoring structure within the first 3.8 million on average (range 0.6-23.1 million).

3   Results

3.1 Influence of genetic regulatory network topology

We tested the ability of NETWORKINFERENCE to recover a network with coordinated influence of correlated regulators in our previous report [2] and these results are shown for comparison (Fig. 4, top left). As noted in [2], NETWORKINFERENCE recovered nearly 100% of the topology, only missing one of two parents of gene 6. When we simulated a network with independent influence of correlated regulators, NETWORKINFERENCE performed well, but still did not recover the two parents and instead linked a gene upstream of one of gene 6's parents (gene 2) as the sole parent for over half of the data sets, and its other parent (gene 5) as the sole parent in the rest, for both influence amounts tested (Fig. 4, top right two graphs).

When testing coordinated influence from uncorrelated regulators, no parents were recovered for gene 6 in most data sets (Fig. 4, bottom left). With independent influences of genes 3 and 5 on gene 6 set at a weight of 0.2, gene 5 was found as the parent in all 10 data sets. However, many unregulated genes were incorrectly linked to the downstream genes in over half the data sets, a phenomenon that did not occur in any other situation (Fig. 4, bottom center). With independent influences set at 0.1, gene 5 was linked as a parent in 4 of 10 data sets (Fig. 4, bottom right).

The loop structure was recovered completely in 100% of the data sets tested (Fig. 4, bottom graphs). Recovery of multiple targets was also highly successful (Fig. 5).

NETWORKINFERENCE performed well in recovering the gene cascade, finding most links with reasonable consistency (Fig. 6). It performed less well in recovering the convergence network, only finding one parent of each gene with multiple parents. This was usually the gene with the shortest path length from the original influence at activity. When path length of multiple regulators was the same, it found each in reasonably equal proportions of the data sets (Fig. 6).

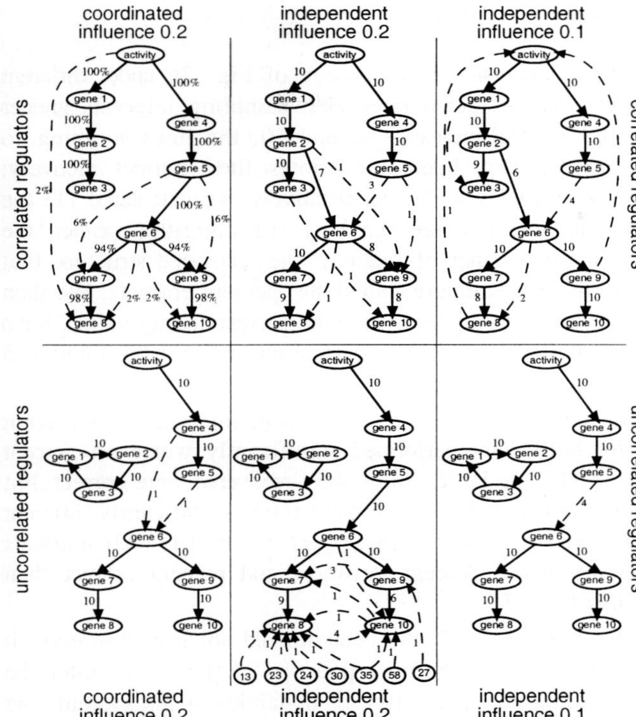

Figure 5. Multiple target recovery. Numbers next to links represent the number of data sets in which that link was found, as described in Fig. 4. Compare with the true structure in Fig. 2c.

Figure 4. Multiple regulator and feedback loop recovery. The number of data sets in which a particular link was recovered is shown next to the link. Links discovered in at least half of the data sets are shown as solid lines; those in less than half are dashed. The upper left graph shows the result of the original structure from [2], with percentages of data sets instead of absolute numbers reported. In all cases except for the lower center, all 90 unregulated genes were correctly identified as having no part in any pathway. In the lower center, incorporated unregulated genes are shown as numbers in circles. All graphs depict a causal translation of the recovered DBN. Compare with the true structure in Fig. 2a for the top graphs; Fig 2b for the bottom.

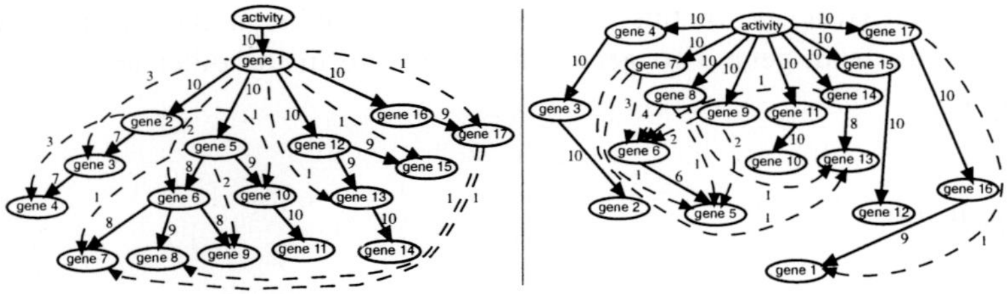

Figure 6. Cascade and convergence recovery. Recovery of the cascade structure is shown on the left; the convergence on the right. Numbers next to links represent number of data sets in which that link was found, as described in Fig. 4. Compare with the true structures in Fig. 2d and 2e.

3.2 Influence of sampling regime

Fig. 7 shows recovery of the genetic network topology of Fig. 2a under different sampling intervals. Recovery success deteriorates with sampling intervals greater than 5, and gets worse as the interval increases beyond 10, the links jumping to genes further and further downstream. Interval 1 shows little correct recovery; however, 2, 3, and 4 perform similarly to 5. Quantitatively, we calculated (1) the percentage of links of the true structure recovered by the algorithm, called the "percent recovered", and (2) the percentage of links in the recovered structure that correspond to correct links of the true structure, called the "percent correct". We then compared these values from each sampling interval to the corresponding values for a sample interval of 5, using a one-tailed student's t-test, assuming that the interval 5 represented the best recovery possible.

Intervals 1 and 2 were not significantly different from interval 5 on percent correct (t_{58}=1.17, P=0.12 for both), but performed significantly worse on percent recovered ($t_{58} \geq 16.77$, P<0.0001). That is, the links they recovered were correct, but they did not recover as many links. Intervals 3 and 4 were not significantly different from 5 on either percent recovered or percent correct ($t_{58} \leq 1.27$, P$\geq$0.1). All intervals 6 and greater had significantly lower percent recovered and percent correct than interval 5 ($t_{58} \geq 5.04$, P<0.001) (Fig. 7).

With the two shortest intervals, 1 and 2, the behavior did not last long enough for influences to propagate throughout the entire pathway. A failure to infer the structure in this situation could be due to the lack of dynamics in the system, and not lack of ability of the algorithm to find the structure. When tested in a situation where the behavior persisted throughout the sampling period and influences were able to propagate through the entire pathway, interval 2 performed similarly to 5 on both measures ($t_{58} \leq 0.58$, P$\geq$0.3); however, interval 1 recovered significantly worse than 5 on both measures ($t_{58} \geq 13.02$, P<0.001) (Fig. 7). For interval 1, the sampling period lasted only 21 time steps, and the latter elements in the pathway did not receive their regulation until 30 or 40 time steps after a behavioral switch, thus it is possible that it simply still did not observe enough of the network dynamics. In order to determine whether an interval of 1 could recover the network given a long enough sampling time, we created a single data set sampling 61 data points over two behavioral switches. In this situation, all the links in the original network, except the link from gene 3 to gene 6 not recovered in the other tests, were found (not shown).

4 Discussion

We found that our inference algorithm, NETWORKINFERENCE, is successful at recovering both multiple targets of a regulator and regulatory loops. Our algorithm

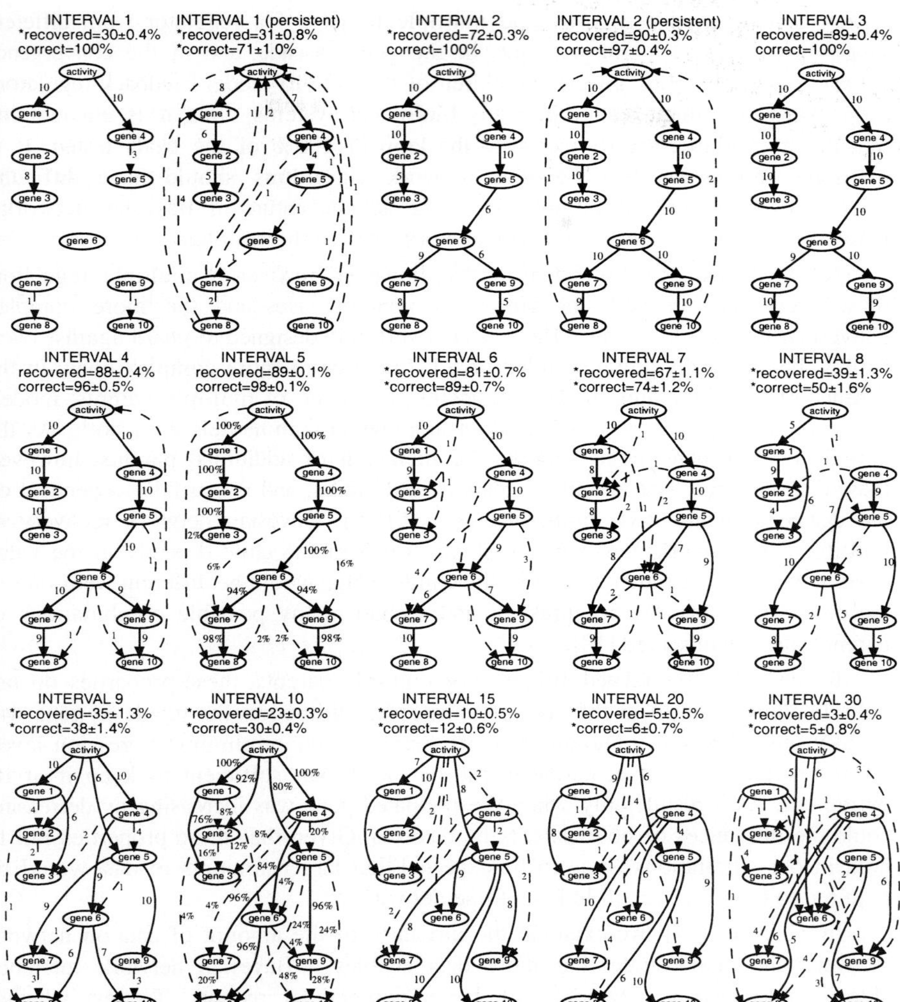

Figure 7. Sampling recovery. Recovery for various sampling intervals from 1 to 30. Numbers next to links represent number of data sets in which that link was found, as described in Fig. 4. Graphs with intervals 5 and 10 are from [2], and show percentages instead of number of trials. Intervals 1 and 2 (persistent) represent the situation when the behavior persisted, in order to allow influence to propagate through the entire pathway. Mean±SE of percent recovered and correct are shown for each sampling type. Asterisks preceding the percent recovered and percent correct indicate that the value is significantly different from those of interval 5. Compare with the true structure in Fig. 2a.

repeatedly proved its ability to find multiple targets of a regulator from different topologies, in our 10 gene network, in the gene cascade, and in the convergence network where it found all eight children of the activity node. Feedback regulatory loops are a prominent feature in many biological systems, thus it is encouraging that our algorithm correctly recovered the loop in 100% of the trials tested. It is often the impression that Bayesian network algorithms should have difficulty discovering loops in a structure, due to the acyclic nature of Bayesian networks. However, use of a dynamic Bayesian network negated this problem.

NETWORKINFERENCE had difficulty, however, recovering multiple regulators of a target gene in both the simple two-parent cases and the more complex convergence network. Because Bayesian networks are designed to guard against over-fitting, they naturally prefer simpler model structures to more complex ones. In the absence of large amounts of data suggesting inclusion of multiple parents, models with only one regulator as a parent often score better than models with both. As the amount of data increases to support the inclusion of additional parents, however, these edges appear. The amount of data needed varies, and is partially dependent on the degree to which two properties are satisfied: Bayesian network recovery of multiple parents works best when first, the value of the child depends on the value of each parent independently, that is, the child's value would be different were one or the other of the parents not present; and second, many possible combinations of parent states are observed [10].

In the case of correlated influence of multiple parents, these properties do not hold. In addition, even in the case of independent influence, the regulated gene was usually pushed in our simulations to its maximum or minimum expression level, where regulation by either parent would be more than sufficient to keep it at its extremum. As a result, only when the regulated gene was increasing or decreasing could the influences of both parents be detected. Given that these properties fail to hold in our simulations, it is to be expected that larger amounts of data would be required to fully recover all links in these networks.

In Yu et al. [14], we examine the influence of the amount of data on network recovery and demonstrate that more data enables a Bayesian network inference algorithm to recover networks with larger numbers of parents. To test whether NETWORKINFERENCE was capable of identifying multiple parents given larger amounts of data in this context, we ran BRAINSIM with uncorrelated regulators (Fig. 2b) of independent influence weight 0.1 through 20 behavioral transitions, resulting in 401 sampled data points per bird region and thus 12,000 observations of the 202 variables, unrealistic experimentally. NETWORKINFERENCE successfully recovered all links in the underlying structure, including gene 3 and gene 5 as joint parents of gene 6. We have not yet determined the minimum amount of data required for successful recovery, and it will be interesting to see if the required amount is close to the realm of experimental feasibility.

Our simulation framework enabled us to test in fine detail another important experimental consideration, sampling regime. It is clear that sampling with an interval too long for the system's dynamics produces poor results. Even the sampling interval of 6, whose recovery graph did not visually look much worse than the intuitive choice of 5 (Fig. 7), performed significantly more poorly than the shorter intervals in terms of both proportion recovered and proportion correct. It is encouraging that the intervals of 2, 3, and 4 all performed as well as interval 5. Interval 1 only observed enough of the network dynamics to recover the entire system when the period of data sampling was lengthened. In Fig. 7, it can be seen how those links most often recovered with the shorter sampling period were the first links in the pathway. Thus an interval of 1 is not too short to recover the dynamics of the system; however, it is too short to observe the entire system with a biologically realistic number of data samples.

Our results suggest a strategy for finding the correct sampling regime for a biological system by slowly increasing the interval. With increasing sampling interval, first more distal elements of the pathway are recovered as the samples expand to cover the dynamics of the entire system. Then there is a plateau when the recovered graph does not change. When the sampling interval exceeds the internal dynamics of the system, recovery begins to skip nodes in the graph with more and more regularity, and links genes to further and further downstream elements. Thus an interval somewhere on the plateau should be chosen.

In summary, our network inference evaluation approach is showing us the promises and limitation of this algorithm, and thus, areas that need improvement. Our approach also is informative on how to sample data from real biological systems for use with network inference algorithms.

Reference

1. P. D'Haeseleer, S. Liang, and R. Somogyi, *Bioinformatics* **16**, 707 (2000).
2. V.A. Smith, E.D. Jarvis, and A.J. Hartemink, *Bioinformatics* **18**, S216 (2002).
3. E.D. Jarvis, et al., *J Comp Physiol A* in press (2002).
4. D.E. Zak, et al., *ICSB* **2**, 231 (2001).
5. E.D. Jarvis, et al., *Nature* **406**, 628 (2000).
6. E.A. Brenowitz, D. Margoliash, and K.W. Nordeen, *J Neurobiol* **33**, 495 (1997).
7. N.A. Hessler and A.J. Doupe, *J Neurosci* **19**, 10461 (1999).
8. H.H. McAdams and A. Arkin, *PNAS* **94**, 814 (1997).
9. A.J. Hartemink, et al., *PSB* **7**, 437 (2002).
10. D. Heckerman, D. Geiger, and D.M. Chickering, *Mach Learn* **20**, 197 (1995).
11. N. Friedman, K. Murphy, and S. Russell, *UAI* **14**, 139 (1998).
12. A. West, et al., *Proc Natl Acad Sci U S A* **98**, 11024 (2001).
13. K. Miyazono, K. Kusanagi, and H. Inoue, *J Cell Physiol* **187**, 265 (2001).
14. J. Yu, et al., *Genome Research* (submitted).

INTRODUCTION TO INFORMATICS APPLICATIONS IN STRUCTURAL GENOMICS

SEAN D. MOONEY

Stanford Medical Informatics
Department of Genetics, Stanford University
Stanford, CA 94305

PATRICIA C. BABBIT

Departments of Biopharmaceutical Sciences and Pharmaceutical Chemistry,
University of California
San Francisco, CA 94143

The goal of structural genomics is to determine representative three dimensional structures of all proteins and macromolecules found in nature. Current efforts in protein structure determination have a major focus in the development of high-throughput methods. This session presents progress and achievements in solving the computational challenges in this field. Topic areas included are (1) Determination of all common scaffolds found in naturally evolved proteins, (2) Structure-based prediction and classification of function and (3) Elucidating structurally defined function and phenotype.

1. Structural Genomics

1.1 Introduction

Structural genomics initiatives aim to determine all of the naturally evolved macromolecular scaffolds. The large numbers of structures resulting from these projects are stored in publicly available databases such as the PDB or the NDB. While these projects are clearly far from finished, researchers have made great strides toward achieving their primary goals. This session focuses on the applications that use the large datasets created by these discovery projects. These datasets can be used to infer function, identify targets and to understand the underlying physical properties that dictate how proteins fold. This introduction is not meant to be a review of the field; see Chance, *et al.*[1] or Norin and Sundstrom[2] for a more thorough treatment.

The three dimensional structure of a protein is typically determined by one of three methods: X-ray crystallography, NMR spectroscopy or modeling based on inferred similarities to homologous proteins or other macromolecules. Crystallographic methods continue to be the gold standard for protein structure determination and the vast majority of experimentally determined structures are solved using these methods. Structures determined by NMR spectroscopy are equally useful in characterizing in-solution protein behavior, yet their representation in the PDB is still relatively low. Homology modeling methods are becoming more popular, but limitations in the technology continue to hinder widespread adaptation.

Currently there are less than 18,000 structures in the protein databank (PDB)[3]. This comprises a small subset of the more than 500,000 characterized protein sequences and the millions of structures derived from variants in populations. Methods for characterizing the protein structure universe are being developed. Equally important is the structural classification and functional annotation of these protein structures.

The number of protein sequences far outnumbers the number of protein scaffolds. Many sequences share the same fold, with current estimates putting the number of folds between 500 and 2000. A complete set of protein folds will greatly advance our ability to build and store large numbers of protein structures based on comparative models. Progress in this field is excellent, in 2001, nearly 4,000 new structures were deposited in the protein databank and ModBase[4], a database of theoretical models, now contains over 500,000 reliable protein structures. SCOP, the resource for the structural classification of proteins[5], has identified and annotated over 600 folds and nearly one thousand superfamilies comprising over 31,000 domains.

A nearly complete set of scaffolds is a powerful research tool that raises many future challenges and opportunities. These include, 1) macromolecular structure prediction, 2) identification of functionally and structurally important motifs, 3) understanding the relationship between structure and molecular phenotype, and 4) understanding the physical principles that specifiy the structure and dynamics of macromolecules. Here we present a brief summary of the problems and approaches addressed in this session.

1.2 Macromolecular structure prediction

Comparative modeling of protein structures becomes an increasingly important problem as more naturally evolved scaffolds are determined and characterized. Comparative modeling often consists of a four step process: fold identification, alignment between target and scaffold, model building and model evaluation and refinement. Several contributions to this process are presented here. First, Xu and Li present a linear programming method for threading, the process of building an alignment between the target and scaffold. This rapid method can be used for both fold identification and building a high-resolution alignment suitable for model building. Second, Edgar and Sjolander describe a method for building accurate alignments using hidden Markov models. Third, Kersting, *et al.* describe a different application of hidden Markov models to find specific structural motifs in protein sequences. Finally, Ohsen *et al.* demonstrate their use of profile-profile alignments to achieve accurate alignments with highly similar sequences as well as more distant relatives.

1.3 Identifying important motifs from a database of macromolecular structure

An important problem is to understand both the structure and function of many solved or modeled protein (or nucleic acid) structures. Singh and Saha describe a method for identifying known motifs from a set of protein structures. Liang *et al.* introduce a new method for describing and categorizing structural motifs based on characteristic sequence motifs within them. They show that their automatically generated motifs perform similarly to manually determined motifs in sequence alignments and yield better alignments than those based on simple sequence motifs.

1.4 Understanding the underlying physical properties of proteins

Understanding the underlying physical principles that dictate the folding and dynamics of proteins is required for understanding macromolecular function. Song *et al.* present a method that gives sophisticated insight into the folding energy landscape of a protein. Using homologous model proteins, protein G and L, their method captures folding differences between the structurally similar proteins.

Radivojac *et al.* describe a method for predicting the boundaries between intrinsically structured and "disordered" regions in protein sequences. This work continues their previous research suggesting that lack of structure may play a functional role and quantitatively determines where these regions are likely to occur.

2. Conclusions

The underlying theme of our discussion is to translate structural information now becoming available into a functional understanding of a macromolecule's purpose. This process of linking structure to function involves organizing and recognizing structural information at many levels. New methods of recognizing sequence and structural motifs within protein scaffolds, aligning these motifs and finding similarities between homologous folds is required for understanding protein structure. Inferring function from structure requires such sophisticated methods as those that are being developed now both in industry and academia.

References

1. Chance, M., et al. "Structural Genomics: A Pipeline For Providing Structures For The Biologist." Protein Science, 11(4), 723-738 (2002).
2. Sundstrom, M. and M. Norin "Structural Proteomics: Developments in Structure-to-Function Predictions." Trends Biotechnology, 20(2), 79-84 (2002).
3. Berman, H., et al. "The Protein Data Bank." Nucleic Acids Research, 28(235-242 (2000).
4. Sanchez, R. and A. Sali "ModBase: A Database Of Comparative Protein Structure Models." Bioinformatics, in press, (2001).
5. Murzin, A., et al. "SCOP: A Structural Classification Of Proteins Database For The Investigation Of Sequences And Structures." Journal of Molecular Biology, 247(536-540 (1995).

SIMULTANEOUS SEQUENCE ALIGNMENT AND TREE CONSTRUCTION USING HIDDEN MARKOV MODELS

ROBERT C. EDGAR

195 Roque Moraes Drive
Mill Valley, California 94941, USA
bob@drive5.com

KIMMEN SJÖLANDER

Department of Bioengineering
University of California, Berkeley
California 94720, USA
kimmen@uclink.berkeley.edu

We present a new algorithm (SATCHMO) that simultaneously estimates a tree and generates a set of multiple sequence alignments given a set of protein sequences. Alignments are constructed for each node in the tree. These alignments predict the structurally conserved elements of the sequences in a subtree and are therefore of different lengths, and represent different amino acid preferences, at different nodes. Hidden Markov Models (HMMs) are also generated for each node and are used to determine branching order, to align sequences and to predict structurally alignable regions. In experiments on the BAliBASE benchmark alignment database, SATCHMO is shown to perform comparably to ClustalW and the UCSC SAM HMM software. Results using SATCHMO to identify protein domains are demonstrated on potassium channels, with implications for the mechanism by which tumor necrosis factor alpha affects potassium current.

1 Introduction

In the words of David Jones, "There are really only three things that govern the overall accuracy of comparative modeling: alignment quality, alignment quality, and...alignment quality" [1]. Comparative modeling is not the only application for which alignment quality is critical: multiple sequence alignments are used for profile construction, detection of critical residues, prediction of functional subfamilies, and a host of other tasks. Because of its central importance, the construction of multiple sequence alignments is a focus of the computational biology community. When sequences are similar to each other, virtually any alignment method will produce good results. However, evolutionary divergence in multi-gene families can result in family members with pairwise similarity so low as to be indistinguishable from chance. Even when sequence similarity is detectable, local changes in structure between members can be significant and represent a great challenge.

Methods for multiple sequence alignment can be broadly classified into two groups: Hidden Markov Model (HMM) methods [2-4], and those that optimize other scoring schemes, such as ClustalW [5]. Both approaches have essential limitations when applied to highly variable protein sequences. HMM methods tend to be successful at detecting and aligning critical motifs and conserved core structure of protein families, but may not correctly align regions between these conserved motifs. Other methods are often superior to HMMs at correctly aligning sequences within similar subgroups; however, subgroups with significant divergence may not be correctly aligned to the consensus structure, causing misalignment of family-defining conserved motifs. Here we present a new method designed to overcome these limitations by constructing a hierarchical tree and identifying conserved structural elements at each level in the tree, as illustrated in Figure 1.

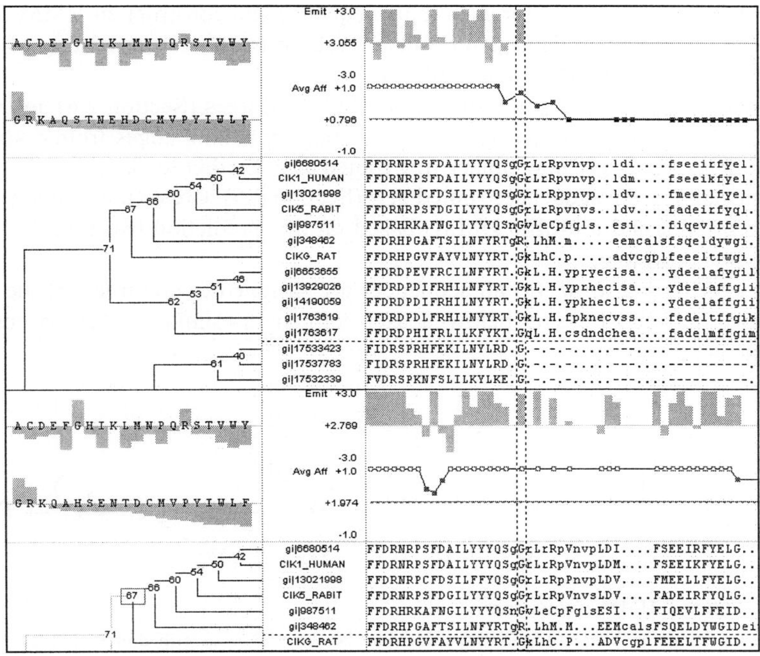

Figure 1. SATCHMO graphical interface.

Here we show results using Satchmo to align two groups of sequences which share a common domain: voltage-gated potassium channels (above) and TNF-alpha induced protein B12 homologs (below). The lower-left pane in each view shows the tree, clicking on a node displays the alignment at that node. Edge lengths in this view of the tree are chosen for convenient display and are uninformative. In the upper view the root node is selected, showing a region that is aligned across all sequences. Below, an internal node is selected to show the same region aligned across K$^+$ channels only. The upper-right pane shows the affinity at each position (histogram) and smoothed affinity (see Section 2.5). The upper-left pane shows the affinity by residue type at the selected position, sorted alphabetically and by value. The lower-right pane shows the sequence alignment. Upper-case letters are aligned, lower-case letters are unaligned. See Section 3.5 for further discussion.

2 The SATCHMO algorithm

2.1 Algorithm

We call our algorithm SATCHMO, for Simultaneous Alignment and Tree Construction using Hidden Markov mOdels. SATCHMO is an agglomerative algorithm that uses HMMs to model classes produced in each iteration, to choose which two subtrees to join, and to generate an alignment of those two subtrees.

Input: A set of unaligned protein sequences.

Step 1: Create a node for each input sequence and construct an HMM from the sequence (Section 2.7). This step results in a set of trees, each consisting of a single node containing an HMM and a sequence.

Step 2: Measure the similarity S_{ij} of all pairs of trees (Section 2.4) and identify a pair ab with highest similarity. Merge this pair by adding edges from a and b to a new node g, forming a new binary tree rooted at g. Predict the conserved structural elements among the sequences rooted at g (Section 2.5). Create a multiple sequence alignment Aln_g and a profile HMM_g based on this prediction (Section 2.7), assign HMM_g and Aln_g to g.

Repeat Step 2 until: (a) all sequences are assigned to one tree, (b) the highest similarity between trees is below a user-defined threshold, or (c) no conserved elements are predicted.

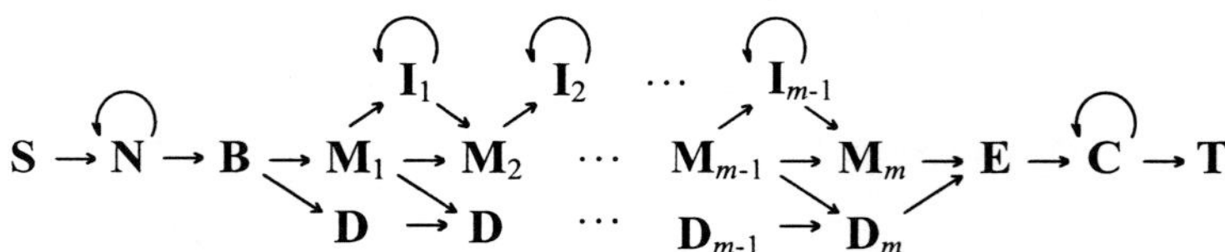

Figure 2. HMM architecture.

Our model architecture follows the HMMer Plan 7 model, as described in Section 2.2.

2.2 HMM architecture

We chose to make our HMMs compatible with the Plan 7 architecture defined by version 2.2 of the HMMer package [6], as shown in Figure 2.

The *k*th node in the model ($k = 1...m$) has match, delete and insert states (M_k, D_k and I_k). The insert state in the last node (I_m) is disabled; terminal insert states N and C emit according to the null model, allowing N- and C-terminal extensions respectively; they emit on N→N and C→C transitions only. Note that D→I and I→D transitions are forbidden. This configuration is designed to encourage alignments that are global to the model and local to a sequence.

2.3 Aligning an alignment to an HMM

A key element of the SATCHMO algorithm is the alignment of two multiple sequence alignments to each other by aligning one (the "target") to a profile HMM constructed from the other (the "template"). This is done in such a way that columns of both alignments appear intact in the combined alignment; sequences in the target are not permitted to align individually to the profile HMM. The motivation for this constraint is the assumption that closely related sequences are already accurately aligned; hence any change that would be induced by re-aligning to a more distant profile is more likely an error than an improvement. This is implemented by allowing emitter (match and insert) states to generate multiple residues. If there are n sequences in the alignment, an emitter state is required to emit n times, producing one column. Weighting (Section 2.6) is accounted for by including transition probabilities N times (see Equation 6) and by calculating the emission probability for a column as $\prod_i p_i^{n_i}$, where p_i is the residue probability calculated from the template counts using Equation 7, and n_i are the weighted counts of the target column.

Columns in the input alignment that have both gaps and residues ("mixed columns") require special consideration. We allow emitter states to emit a gap (or, equivalently, not to emit) with a given probability. This allows the model to generate mixed columns without requiring sequences to take different paths. Gaps from emitter states (as opposed to gaps produced by delete states) are emitted with a background probability, so in log-odds terms contribute zero to the column score after subtraction of the null model. Columns in the input alignment which are predicted not to be alignable are assigned a zero probability of being emitted by a match state, thus forcing them into insert states.

We use the Viterbi algorithm [7] to determine the most probable path through the model. The implementation does not require an inner loop over sequences; it suffices to pre-calculate the total number of residues of each type found in each column of the alignment being scored against the HMM.

2.4 Relationship score and similarity measure

We define a relationship score r_{ij} between two trees i and j from the log-odds score per sequence of the alignment of i (Aln$_i$) to the profile HMM of tree j (HMM$_j$):

$$r_{ij} = (1/N) \log_2(P(\mathrm{Aln}_i \mid \mathrm{HMM}_j) / P(\mathrm{Aln}_i \mid \mathrm{Null})) / \mathrm{length}(\mathrm{HMM}_j). \qquad (1)$$

Here, N is the estimated number of independent observations (Equation 6). Dividing by N gives the score per sequence, dividing by the model length corrects a bias towards longer models. We also define a symmetrical similarity measure,

$$S_{ij} = (r_{ij} + r_{ji})/2. \tag{2}$$

A large positive value of S_{ij} indicates a close relationship between the sequences in i and j; zero indicates a degree of relationship indistinguishable from chance.

2.5 Prediction of structurally alignable regions

The structurally alignable regions between two trees i and j are estimated as follows. One tree is chosen to be the target and one the template by calculating r_{ij} and r_{ji} using Equation 1. If $r_{ij} \geq r_{ji}$, then j is the template, otherwise j is the target. The target is then aligned to the template as described in Section 2.3, and affinities of template nodes to target columns are calculated. The affinity $A(k)$ of the kth model node is defined to be the log-odds score of the target column $c(k)$ generated by its match state:

$$A(k) = (1/N) \log_2(P(c(k) \mid \mathrm{HMM}) / P(c(k) \mid \mathrm{Null})). \tag{3}$$

If the most probable path passed through its delete state, the node is assigned an affinity value of zero. Affinity is a noisy signal; we therefore define a smoothed affinity $a_w(k)$ over a window of length w, an odd integer ≥ 1:

$$a_w(k) = |W(k,w)|^{-1} \sum_{q \in W(k,w)} A(q). \tag{4}$$

Here, $W(k,w)$ is the set of nodes found in a window of length w centered on the kth node. $|W(k,w)|$ is the number of nodes in that set, $|W(k,w)|=w$, $w \leq k \leq (m-w)$. Window slots beyond an end of the model are empty. For example, if $w=5$, then the window at the first node $k=1$ has only three nodes, i.e. $|W(1,5)|=3$.

The kth node is predicted to be an alignable position if and only if the following condition is met:

$$a_w(k) \geq Z. \tag{5}$$

Z is described as the minimum smoothed affinity threshold; it is a parameter of the algorithm.

2.6 Sequence weighting and amino acid probability estimation

Following standard practice, we employ relative weights to compensate for correlation among the sequences, using the scheme described by Gerstein *et al.* [8].

With Bayesian methods such as ours, the number of observations may swamp the contribution of priors to posterior amino acid estimates. We estimate the number of independent observations to be:

$$N = n^{1-C}. \tag{6}$$

Here, n is the number of sequences in the alignment and C is the average over all columns of the fractional occurrence of the most common residue in that column. Hence C ranges from 1 (all positions identical) to 1/20 (all amino acids having equal frequency) and N ranges from 1 to approximately n. When calculating measured counts in model construction, sequence weights are scaled to N rather than to n. We introduce a maximum permitted value U for N as a second line of defense against under-weighting the priors; U is a parameter of the algorithm.

Estimated residue probabilities p_i, $i=1...20$ are calculated at each position in the alignment by combining the weighted observed residue counts n_i with a Dirichlet mixture density [9], as follows (for simplicity, we refer to the jth component by its hyperparameters α_j):

$$p_i = \sum_j P(\alpha_j \mid counts)\, (n_i + \alpha_{ji}) \, / \, (N + |\alpha_j|). \tag{7}$$

2.7 Constructing an HMM that models the consensus structure of two alignments

When two nodes are joined to make a new tree, their alignments are aligned to each other (Section 2.3) and alignable regions are estimated (Section 2.5). A profile HMM that models the consensus structure of the combined alignment is constructed by creating a node for each position that meets the minimum smoothed affinity condition (Equation 5); i.e., for each position that is predicted to be alignable across both alignments. The nodes in the new model thus correspond to a subset of the nodes in the template profile HMM. This guarantees that model lengths are monotonically non-increasing (i.e. can get shorter but not longer) along a path from a leaf towards its root, reflecting the expectation that increasingly diverged proteins have progressively shorter mutually alignable regions. In the special case of constructing an HMM from a single sequence in Step 1 of the algorithm (see Section 2.1), a node is created for each residue.

Match state emission probability distributions are estimated by combining weighted residue counts with a Dirichlet mixture prior (Equation 7). In the preliminary implementation described here we use the HMMer default priors for transition and insert state emission probabilities; we do not attempt to estimate these parameters from observations (see Section 4 for further discussion).

2.8 Complexity

Given n sequences of length L, the space complexity of SATCHMO is dominated by the dynamic programming matrix used by the Viterbi algorithm, which is $O(L^2)$. The total time complexity is $O(L^2 n^2 + L n^3)$. On representative current hardware (a PC with one 2.5 GHz Pentium 4 processor), our implementation of SATCHMO is typically able to align 100 sequences of length 100 in under two minutes.

3 Experimental results

3.1 Reference alignments

We used version 1 of the BAliBASE benchmark alignment database [10]. BAliBASE is divided into five reference sets. Ref1 contains alignments of a small number (< 6) of equidistant sequences, meaning that the percent identity between two sequences is within a specified range. These Ref1 alignments contain sequences of similar length, with no large insertions or extensions. Alignments in Ref2 add up to three distantly related sequences ($< 25\%$ identical) from Ref1 with a family of at least 15 closely related sequences. Ref3 contains alignments of up to four subgroups, with $< 25\%$ identity between sequences from different groups. Ref4 contains alignments with long N/C-terminal extensions of up to 400 residues. Ref5 has long insertions of up to 100 residues. Ref1, 2 and 3 are divided into groups with short, medium and long sequences. Ref1 is further subdivided by percent identity.

3.2 Alignment quality scoring

BAliBASE provides a module (*BaliScore*) that defines two scores. SP is the ratio of the number of correctly aligned pairs of positions in the test (predicted) alignment to the number of aligned pairs in the reference (structurally informed) alignment. TC is the ratio of the number of correctly aligned columns in the test alignment to the number of aligned columns in the reference alignment. Both SP and TC range from 1.0 for perfect agreement to 0.0 for no agreement. The designers of BAliBASE recommend SP as the best quality score for Refs1, 2 and 3, TC as the best score for Refs4 and 5 [11]. We wrote our own module to compute SP and TC as the published *BaliScore* software module produces incorrect results on some inputs: specifically we found that *BaliScore* would report scores that were less than the correct value for alignments with gapped positions. Using the published *BaliScore* gave very similar relative rankings of the tested methods to our own scoring module, but reduced median scores.

We felt that while the SP and TC scores are useful, they have limitations as measures of alignment quality. Neither measure penalizes columns in the test alignment that are not structurally alignable, i.e. over-alignment. They thus fail to distinguish between algorithms that predict alignable regions, such as SATCHMO and SAM, from those that do not, such as ClustalW. Moreover, the "correct" alignment is sometimes ambiguous as experts may disagree on the identification of homologous positions; however SP and TC give no credit for positions with small shifts.

As a complementary measure of alignment quality, we used the Cline shift score (CSS) [12]. CSS penalizes both over- and under-alignment and gives positive, though reduced, scores for positions with small shifts. CSS is controlled by a parameter ε which determines how long a shift must be to contribute a negative value to the score. We follow Cline's recommendation and set $\varepsilon = 0.2$, which gives a negative score to shifts of more than five positions (approximately one turn in a helix). Cline defines the score on a pair-wise alignment; we take the average over all pairs of sequences present in both the test and reference alignments. CSS ranges from 1.0 in the case of perfect agreement between the test and reference alignments to $-\varepsilon$ as a lower limit.

3.3 Algorithm parameters

In addition to Dirichlet mixture priors, SATCHMO has the following parameters: U, the maximum number of independent observations, Z, the minimum smoothed affinity, and w, the window length for affinity smoothing. Two additional parameters g_1 and g_2 are defined to isolate tuning of gap-related transitions: g_1, multiplies the M$\rightarrow$D (gap-open) probability, g_2 multiplies the D$\rightarrow$D (gap-extend) probability. The transition distributions from M and D are of course re-normalized after the multiplications have been applied.

For the BAliBASE reference alignments, we found that the SP, TC and CSS scores were optimized by setting $Z = -\infty$, causing all candidate positions to be aligned irrespective of w. This parameter setting is not optimal when aligning sequences that may share only a single domain, or are otherwise more variable structurally than those in BAliBASE (as in Figure1). For such inputs, we find setting Z close to 0, and using a window size w of 5 or 7, to be optimal. There was some sensitivity to g_1 and g_2: we found that $g_1 = g_2 = 0.75$ gave a small but significant improvement over the HMMer defaults induced by $g_1 = g_2 = 1$. Very little variation was found with U, which is not surprising given that BAliBASE alignments have small numbers of sequences; we therefore set $U = \infty$. We performed some experiments where we separated the reference alignments into training sets used for parameter optimization and test sets. The parameters found to be optimal for these smaller training sets were identical to those found to be optimal for all of BAliBASE.

Table 1. Alignment quality scores for BAliBASE reference sets.
For each program and each scoring method, the median score is shown for each BAliBASE reference set. In the case of Refs 1, 2 and 3, the median score in each sub-category is also shown. Finally the median score over all BAliBASE alignments is given for each program. We show here SP results for Refs 1, 2 and 3, and TC for Refs 4 and 5, following the suggestion of the BAliBASE authors and for consistency with their published analyses.

	SP / TC			CSS		
	ClustalW	SAM	Satchmo	ClustalW	SAM	Satchmo
Ref1 short <25% identity	0.72	0.40	0.40	0.39	0.26	0.42
Ref1 medium <25% identity	0.68	0.61	0.73	0.52	0.22	0.51
Ref1 long <25% identity	0.64	0.60	0.59	0.47	0.10	0.33
Ref1 short 20-40% identity	0.92	0.95	0.93	0.70	0.59	0.73
Ref1 medium 20-40% identity	0.96	0.97	0.96	0.78	0.56	0.80
Ref1 long 20-40% identity	0.96	0.96	0.95	0.77	0.52	0.75
Ref1 short >35% identity	0.99	0.99	0.98	0.90	0.94	0.89
Ref1 medium >35% identity	0.98	0.99	0.99	0.93	0.90	0.92
Ref1 long >35% identity	0.99	0.99	0.99	0.89	0.92	0.89
All Ref1	*0.94*	*0.97*	*0.94*	*0.77*	*0.59*	*0.77*
Ref2 short	0.88	0.00	0.76	0.68	0.81	0.67
Ref2 medium	0.86	0.89	0.78	0.71	0.85	0.73
Ref2 long	0.88	0.90	0.80	0.54	0.83	0.51
All Ref2	*0.88*	*0.77*	*0.78*	*0.66*	*0.82*	*0.69*
Ref3 short	0.72	0.00	0.79	0.65	0.58	0.66
Ref3 medium	0.74	0.76	0.66	0.75	0.62	0.77
Ref3 long	0.91	0.90	0.88	0.64	0.77	0.66
All Ref3	*0.84*	*0.76*	*0.60*	*0.71*	*0.71*	*0.74*
Ref4	0.52	0.32	0.74	0.24	0.00	0.25
Ref5	0.58	0.75	0.71	0.23	0.47	0.23
All BAliBASE	**0.88**	**0.90**	**0.88**	**0.70**	**0.67**	**0.69**

3.4 Comparison with ClustalW and SAM

We chose to compare the performance of SATCHMO with two tools, ClustalW [5] and the UCSC SAM "tuneup" algorithm based on their SAM-T99 clustering and alignment method [13]. These tools are representative of the non-probabilistic and

HMM categories of alignment methods respectively. ClustalW has been shown to have excellent performance against BAliBASE, ranking behind only PRRP [14] in a comparison of ten algorithms [11]. We used ClustalW version 1.81 with default parameters. Following the procedure recommended Karplus and Hu for evaluating tuneup performance on BAliBASE alignments [13], we assigned zero scores to the 18 reference sets where tuneup failed to produce a complete alignment of the test sequences owing to rejection of one or more sequences deemed to be too distantly related.

Median scores for the three programs on BAliBASE reference sets are shown in Table 1. In our preliminary implementation we do not handle undetermined residues; the following reference alignments that include letters X or B were therefore excluded: 1bbt3_ref1, 1havA_ref1, 1ppn_ref1, 5ptp_ref1, 9rnt_ref1 2trx_ref2, 1ajsA_ref2, 2myr_ref2, 4enl_ref2, 1ajsA_ref3, 2myr_ref3, 4enl_ref3, 1lkl_ref4 and 2abk_ref4.

3.5 Domain identification

Our preliminary experiments with SATCHMO suggest that it is effective at identifying protein domains. In Figure 1, we show the tree constructed by SATCHMO for two sets of proteins: TNF-alpha-induced protein B12 and homologs, and voltage-gated potassium channels. The surprising homology between these two groups was discovered by one of us (Sjölander) while scoring the NR database with an HMM constructed for voltage-gated potassium channels [15], where these B12 proteins received weak but significant scores. SATCHMO assigns these two groups to separate subtrees, and identifies a common domain. Our analysis shows this region to be the tetramerization, or T1, domain of potassium channels, for which a solved structure exists. This allows us to predict the fold of TNF-alpha-induced protein B12 and homologs. Even more intriguingly, tumor necrosis factor alpha is known to affect potassium current, but the precise mechanism is unknown [16, 17]. Since TNF-alpha induces the B12 protein and its homologs, and these B12 proteins share homology with the tetramerization domain, we predict that at least one of the mechanisms by which TNF-alpha affects potassium current is by inducing the B12 proteins which tetramerize with potassium channels.

4 Discussion

The test results suggest that SATCHMO performs competitively with state-of-the-art algorithms while providing more insight into common structural elements and variations of those elements among the input sequences. We find this encouraging, especially given that we regard the current implementation of SATCHMO as preliminary and expect that significant improvements are possible. We plan to investigate several areas. We will pursue a more rigorous treatment of gaps and better estimation of transition probabilities and insert state emission probabilities. We will also enable local-local alignment. As presented here, SATCHMO is a progressive algorithm, meaning that once a pair of sequences has been aligned, this part of the alignment is fixed. We may find that improvements are possible by keeping regions of high confidence fixed and re-aligning marginal regions using iterative methods when more sequences have been added. We also plan to investigate different smoothing heuristics and alternative methods for predicting alignable regions.

We also plan to investigate the performance of SATCHMO on input that is more representative of typical applications than the small number of carefully screened sequences found in the BAliBASE alignments. Adding homologs of the BAliBASE reference sequences may enable us to improve upon the scores reported hereby providing intermediates that interpolate between, and therefore guide the alignment of, those sequences.

We will further explore the use of SATCHMO in phylogenetic tree construction through experiments based on functionally characterized proteins. Other work in this area suggests that this approach can produce trees and alignments of high quality [18]. We will evaluate the trees constructed using experimental information on molecular function, binding pocket positions, and other biological data.

Finally, preliminary results (as shown in Figure 1) suggest that SATCHMO is effective at identifying domain-level similarities between proteins, and this aspect of its functionality will be examined further.

Acknowledgements

The authors thank Sean Eddy for making the HMMer package and its source code readily available, Melissa Cline for helpful discussions, and Wayne Christopher for providing a subroutine for efficient calculation of the log-gamma function.

References

1. Jones, D.T., *Progress in protein structure prediction*. Curr Opin Struct Biol, 1997. **7**(3): p. 377-87.
2. Krogh, A., M. Brown, I.S. Mian, K. Sjolander, and D. Haussler, *Hidden Markov models in computational biology. Applications to protein modeling*. J Mol Biol, 1994. **235**(5): p. 1501-31.

3.   Eddy, S.R., *Hidden Markov models.* Curr Opin Struct Biol, 1996. **6**(3): p. 361-5.

4.   Karplus, K., K. Sjolander, C. Barrett, M. Cline, D. Haussler, R. Hughey, L. Holm, and C. Sander, *Predicting protein structure using hidden Markov models.* Proteins, 1997. **Suppl 1**: p. 134-9.

5.   Thompson, J.D., D.G. Higgins, and T.J. Gibson, *CLUSTAL W: improving the sensitivity of progressive multiple sequence alignment through sequence weighting, position-specific gap penalties and weight matrix choice.* Nucleic Acids Res, 1994. **22**(22): p. 4673-80.

6.   Eddy, S.R.; *HMMER: Profile hidden Markov models for biological sequence analysis.* http://hmmer.wustl.edu/

7.   Forney, G.D., *The Viterbi Algorithm.* Proc. IEEE,, 1973(61): p. 268-278.

8.   Gerstein, M., E.L. Sonnhammer, and C. Chothia, *Volume changes in protein evolution.* J Mol Biol, 1994. **236**(4): p. 1067-78.

9.   Sjolander, K., K. Karplus, M. Brown, R. Hughey, A. Krogh, I.S. Mian, and D. Haussler, *Dirichlet mixtures: a method for improved detection of weak but significant protein sequence homology.* Comput Appl Biosci, 1996. **12**(4): p. 327-45.

10.   Thompson, J.D., F. Plewniak, and O. Poch, *BAliBASE: a benchmark alignment database for the evaluation of multiple alignment programs.* Bioinformatics, 1999. **15**(1): p. 87-8.

11.   Thompson, J.D., F. Plewniak, and O. Poch, *A comprehensive comparison of multiple sequence alignment programs.* Nucleic Acids Res, 1999. **27**(13): p. 2682-90.

12.   Cline, M., R. Hughey, and K. Karplus, *Predicting reliable regions in protein sequence alignments.* Bioinformatics, 2002. **18**(2): p. 306-14.

13.   Karplus, K. and B. Hu, *Evaluation of protein multiple alignments by SAM-T99 using the BAliBASE multiple alignment test set.* Bioinformatics, 2001. **17**(8): p. 713-20.

14.   Gotoh, O., *A weighting system and algorithm for aligning many phylogenetically related sequences.* Comput Appl Biosci, 1995. **11**(5): p. 543-51.

15.   Sjolander, K. *Automated domain identification in proteins using HMMs.* Presented at Genome Sequencing and Analysis Conference, Miami, FL; September, 1999

16.   Soliven, B., S. Szuchet, and D.J. Nelson, *Tumor necrosis factor inhibits K+ current expression in cultured oligodendrocytes.* J Membr Biol, 1991. **124**(2): p. 127-37.

17.   McLarnon, J.G., M. Michikawa, and S.U. Kim, *Effects of tumor necrosis factor on inward potassium current and cell morphology in cultured human oligodendrocytes.* Glia, 1993. **9**(2): p. 120-6.

18.   Mitchison, G.J., *A probabilistic treatment of phylogeny and sequence alignment.* J Mol Evol, 1999. **49**(1): p. 11-22.

TOWARDS DISCOVERING STRUCTURAL SIGNATURES OF PROTEIN FOLDS BASED ON LOGICAL HIDDEN MARKOV MODELS

K. KERSTING[1], T. RAIKO[1,2], S. KRAMER[1], L. DE RAEDT[1]

[1]*Institute for Computer Science*
Machine Learning Lab
University of Freiburg
Georges-Koehler-Allee 079
79112 Freiburg, Germany

[2]*Helsinki University of Technology*
Laboratory of Computer and
Information Science,
P.O. Box 5400,
02015 HUT, Finland

With the growing number of determined protein structures and the availability of classification schemes, it becomes increasingly important to develop computer methods that automatically extract structural signatures for classes of proteins. In this paper, we introduce and apply a new Machine Learning technique, Logical Hidden Markov Models (LOHMMs), to the task of finding structural signatures of folds according to the classification scheme SCOP. Our results indicate that LOHMMs are applicable to this task and possess several advantages over other approaches.

1 Introduction

In recent years, the number of proteins with determined structure has been growing rapidly due to large-scale structural genomics projects. Consequently, the Protein Data Bank (PDB) is growing at high rates. In parallel, researchers have developed classification schemes of proteins based on their sequence, structure and function. The development of classification schemes is a common scientific activity to make sense and gain a deeper understanding of experimental data. Given the determined structures and classification schemes, the discovery of structural characteristics of protein classes becomes an important topic. The primary interest is to gain insights into structural characteristics of fold classes, but ultimately structural signatures should also be useful for the prediction of protein folds. In fact, some successful approaches in the CASP predictive exercises made use of knowledge about structural signatures. So far, most signatures have been discovered by human experts based on extensive manual/visual inspection of the data. However, few experts in the world are in a position to find/provide these signatures, and few systematic attempts exist to catalog known signatures. So, there is a need to develop computer methods that automatically extract structural signatures in a systematic way [11].

Recently, Hidden Markov Models (HMM) have been used to analyze classes in SCOP [6]. HMMs are among the most widely and successfully used tools for the analysis of sequence data in bioinformatics. Despite their successes, however, it is well-known that HMMs have a number of weaknesses. One of the major weaknesses is that HMMs handle only flat sequences, i.e. sequences of unstructured symbols. In this paper we will overcome this weakness by introducing Logical Hidden Markov Models (LOHMMs).

This paper is organized as follows. In Section 2, we present the task and the dataset. In Section 3, we introduce LOHMMs. Section 4 describes experiments with LOHMMs for the discovery of structural signatures. Subsequently, we discuss related work and conclude.

2 Task and Dataset

In this section, we describe the task of finding structural signatures of protein folds and the dataset used. The basis of our study is the SCOP (Structural Classification of Proteins) database due to A. Murzin and maintained by the MRC Laboratory of Molecular Biology. Our goal was to find structural characteristics of the domains at the second level of the SCOP hierarchy, i.e., the level of folds. In our study, we focused on alpha and beta proteins (a/b), a class consisting of domains with mainly parallel beta sheets (beta-alpha-beta units). From this class, we chose the five most populated subclasses, that is, folds: TIM beta/alpha-barrel, NAD(P)-binding Rossmann-fold domains, Ribosomal protein L4, glucosamine 6-phosphate deaminase/isomerase and and leucine aminopeptidas. The overall set-up is quite similar to the one by Turcotte *et al.* [12,11] The data have been extracted automatically from the PDB release #96 and SCOP version 1.57.

Information for domains from the above five folds was extracted in the form of "logical sequences" of secondary structure elements. Logical sequences are sequences of logical atoms. An example of such a sequence (corresponding to a Ribosomal protein L4) is:

$$st(null, 2), he(h(right, alpha), 6), st(plus, 2), he(h(right, alpha), 4), st(plus, 2),$$
$$he(h(right, alpha), 4), st(plus, 3), he(h(right, alpha), 4),$$
$$st(plus, 1), he(h(right, alpha), 6).$$

There are two predicates *he* and *st*. Atoms *he(Type, Length)* model helices of a certain type and length, whereas atoms *st(Orientation, Length)* model strands of a certain orientation and length. The helix types are: $h(left, alpha), h(right, alpha), h(left, gamma), h(right, gamma),$ $h(left, omega), h(right, omega), h(right, pi), h(right, 3to10), 27ribbon$ and *polyproline*. The orientation of strands can be *null* (the beginning of a sheet), *plus* (a parallel strand of a sheet), or *minus* (an anti-parallel strand of

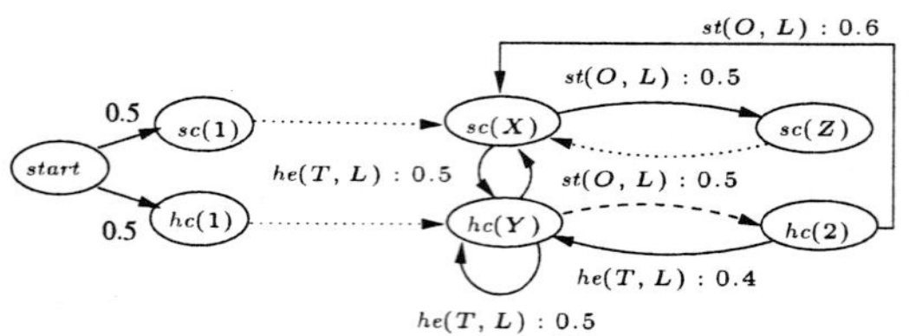

Figure 1: A logical hidden Markov model encoding default reasoning. The dashed edge represents a *more general than* relation.

a sheet). The length is defined as the number of acids and was quantized in the experiments (see below).

For each of the above five folds, we are modeling their domains in terms of their secondary structure using a logical variant of HMMs. So, what can be expected from an application of LOHMMs to this task? First of all, it should be clear that we do not obtain structural signatures for each of these classes immediately. What we obtain instead, is a model for each fold. Each model provides a precise probabilistic and logical characterization of the respective fold. Structural signatures can then be found by a comparison of models. As will be shown below, it is quite easy to find characteristics upon inspection of the trained models.

3   Logical (Hidden) Markov Models

Logical (hidden) Markov models (LOHMM) extend the unstructured model representation of HMMs [10,9] by incorporating complex, internal structure into the specification of transitions (and therefore of emissions) between states.

Sets of states are summarized by *abstract states*, which are represented by logical atoms. A logical atom then represents all states that can be obtained by instantiating the atoms (i.e. by replacing the variables by terms). E.g. the abstract state $hc(X)$, where X is a variable, could represent the set of states $\{hc(1), hc(2)\}$ depending on the terms (1 and 2 in this case) in the LOHMM. If the logical atom does not contain any variables such as $hc(1)$, it represents a singleton set. Abstract states are connected by *abstract transitions*, which summarize sets of transitions between states. When a transition is made, a state is sampled from the encompassing abstract state. Subsequently an observation symbol is generated in the same manner. We will explain these concepts on an example. For more details, we refer to a technical report [7].

3.1 An Example of a LOHMM

Fig. 1 shows an example of a LOHMM. The vertices in the model represent abstract (hidden) states where the predicate $hc(ID)$ (resp. $sc(ID)$) represents a block *ID* of consecutive helices (resp. strands). In such models, we find three different types of edges:

Solid edges between abstract states specify the abstract transitions. Transition probabilities and emission symbols are associated to them. An example transition from Fig. 1 is $sc(X) \xleftarrow{st(O,L):0.5} hc(Y)$. Such a solid edge expresses that if one is in one of the states represented by $hc(Y)$ one will go to one of the states in $sc(X)$ with probability 0.5 while emitting a symbol in $st(O, L)$.

Dotted edges indicate that two abstract states behave in exactly the same way. If we follow a transition to an abstract state with an outgoing dotted edge, we will automatically follow that edge. Consider the dotted edge going from $sc(Z)$ to $sc(X)$ in Fig 1. The two abstract states are identical. The dotted edge is needed in this case because the variables appearing in the abstract states are different. We could not have written this using solid edges alone as the meaning of the solid edge $sc(X) \xleftarrow{st(O,L):0.5} sc(X)$ is different from that of $sc(Z) \xleftarrow{st(O,L):0.5} sc(X)$. Whereas the first transition only allows a transition between the same state, say $sc(1)$ (because the X is identical), the second one allows transition between different states such as $sc(1)$ and $sc(2)$. In a logical sense, dotted edges implement a kind of recursion.

Dashed edges represent a kind of default reasoning. This is often used to model exceptions. Consider the dashed edge in Fig. 1 connecting $hc(Y)$ and $hc(2)$. This dashed edge denotes that $hc(2)$ is a more specific state than $hc(Y)$. This implies that the set of states represented by the more specific (abstract) state is a subset of that represented by the more general one. Logically speaking, the more specific state $hc(2)$ can be obtained by substituting Y by 2 in the more general state $hc(Y)$. Dashed edges and default reasoning are useful because they represent exceptions. Indeed, in our current example, the outgoing probability labels associated to $hc(2)$ are different from those for $hc(Y)$. This actually implies that the $hc(2)$ acts as an exception to the states represented by $hc(Y)$. So for $Y = 2$ we employ the transitions from $hc(2)$ and for $Y \neq 2$ we follow those indicated by $hc(Y)$.

Let us now explain how the model in Fig. 1 generates the sequence of observations $he(h(right, 3to10), 10), st(plus, 10), st(plus, 15), he(h(right, alpha), 9)$, (cf. Fig. 2). Starting from the artificial state *start*, it chooses an initial abstract state, say $hc(1)$. Forced to follow the dotted edge, it enters the abstract state $hc(Y)$. In each abstract state, the model samples values for all variables that

196

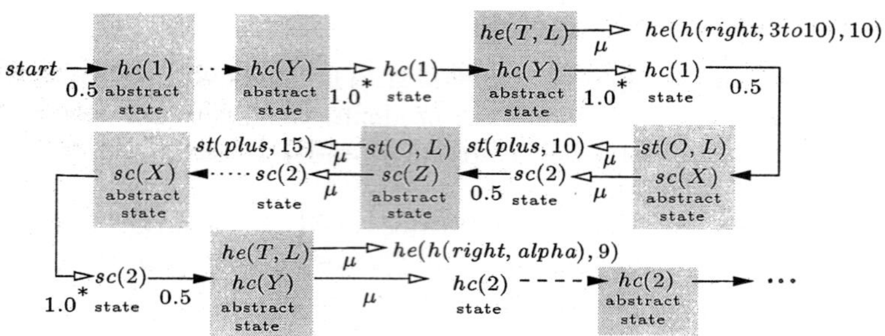

Figure 2: Generating the observation sequence $he(h(right, 3to10), 10)$, $st(plus, 10)$, $st(plus, 15)$, $he(h(right, alpha), 9)$ by the LOHMM in Fig. 1 (* $\mu = 1.0$ due to unification).

are not instantiated yet according to a *selection distribution* μ.

The function μ specifies for each abstract state a distribution over the possible instantiations of the abstract state. E.g. $\mu(he(h(right, alpha), 4) \mid he(h(T, alpha), 4)) = 0.5$ says that the model samples $he(h(right, alpha), 4)$ with probability 0.5 from $he(h(T, alpha), 4)$ whereas $\mu(he(h(right, alpha), 4) \mid he(h(T, A), 4)) = 0.05$ specifies that $he(h(right, alpha), 4)$ is sampled with probability 0.05 from $he(h(T, A), 4)$. In general, any probabilistic representation such as Markov chains or Bayesian networks might be used to represent μ. In our experiments, we followed a naïve Bayes approach, i.e. each argument of a predicate is assumed to be independent of the other arguments. E.g., to compute $\mu(he(h(right, alpha), 4) \mid he(T, L))$, we compute the product of $P_T(h(right, alpha))$ and $P_L(4)$.

Since the value of Y was already instantiated in the previous abstract state $hc(1)$, the model samples with probability 1.0 the state $hc(1)$. It selects the transition to $hc(Y)$ observing $he(T, L)$. Since Y is shared among the head and the body, the state $hc(1)$ is selected with probability 1.0. The observation $he(h(right, 3to10), 10)$ is sampled from $he(T, L)$ using the selection distribution μ. Now, the model goes over to the abstract state $sc(X)$, emitting $st(plus, 10)$ which in turn was sampled from $st(O, L)$. Variable X in $sc(X)$ is not yet bound; so, a value, say 2, is sampled using μ. Next, we move on to abstract state $sc(Z)$, emitting $st(plus, 15)$. The variable Z is sampled to be 3. The dotted edge brings us back to $sc(X)$ and automatically unifies X with Z, which is bound to 3. Emitting $he(h(right, alpha), 9)$, the model returns to abstract state $hc(Y)$. Assume that it samples 2 for variable Y, it has to follow the dashed outgoing edge to $hc(2)$, which represents an exception to $hc(Y)$. This process is similar to unrolling dynamic Bayesian networks [2] and to grounding logic programs [8].

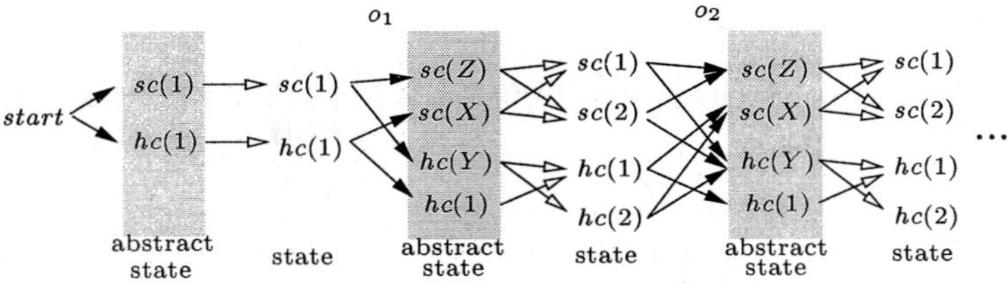

Figure 3: Illustration of the trellis induced by the LOHMM in Fig. 1. In contrast with HMMs, there is an additional layer where the states are sampled from abstract states.

3.2 Semantics and Evaluation

Each HMM is a LOHMM consisting of propositional, logical transitions only. Having the described grounding/unrolling process in mind, it is clear that a LOHMM defines a HMM given a *selection distribution* μ. There is a finite set of abstract transitions, and each domain associated to an argument of a predicate is finite. Thus, the set of states and, therefore, the set of (ground) transitions is finite. To summarize, there are two primary differences to HMMs. First, transition probabilities are defined by a product of abstract transition probabilities and the selection probability. Second, the set of states represented by an abstract state can vary with the domains associated to predicates.

A trellis can be built as follows: After selecting an abstract transition, μ generates the relevant states from the head of the abstract transition (cf. Fig. 3). Based on the trellis, it is easy to adapt the *forward- backward*, the *Viterbi* and the *Baum-Welch* algorithms for HMMs to LOHMMs. E.g. in the forward-backward procedure, the probabilities α and β are computed for each reachable state (sets S_t) recursively. The $\alpha_t(s)$ is the probability of the partial observation sequence $o_1, \ldots, o_{t-1}$ and state s at time t given the LOHMM. The $\beta_t(s)$ is the probability of the partial observation sequence $o_t, \ldots, o_T$ given a state s at time t and the LOHMM. Set $\alpha_0(start) = 1.0$ and $\beta_T(s) = 1.0$ for every $s \in S_T$. Recursive formulae are $\alpha_t(h) = \sum_{cl} \sum_{b \in S_{t-1}} \alpha_{t-1}(b) p_{cl} p_\mu \delta(cl, b, h, o_t)$ and $\beta_t(b) = \sum_{cl} \sum_{h \in S_{t+1}} \beta_{t+1}(h) p_{cl} p_\mu \delta(cl, b, h, o_t)$, where cl is a transition in the LOHMM, p_{cl} is the transition probability and p_μ is the selection probability given by μ. The indicator function $\delta(cl, b, h, o_t) = 1$ whenever transition cl can take from state b to h observing o_t and the transition cl has the most specific body for b. The other algorithms can be adapted analogously.

4   Experiments

The aim of the experiments described below is to put the following hypotheses to test:

H1 LOHMMs are capable of distinguishing between different folds based on a logical representation of the secondary structure of domains.

H2 The inspection of LOHMMs reveals distinguishing features of folds.

H3 LOHMMs can be applied to real-world problems.

H4 In some applications in computational biology, LOHMMs are by at least an order of magnitude smaller than their instantiations which are HMMs.

We implemented the EM algorithm (with pseudocounts) using the Prolog system Sicstus-3.8.6. The experiments were ran on a Pentium-III-600 MHz machine. Our task was to classify sequences representing protein secondary structures into one of five folds. To do so, we followed the standard approach to classification based on HMMs. We chose a LOHMM (see Fig. 5), fixed its structure, and randomly generated for each fold a set of initial abstract transition probabilities and domain distributions. From each fold dataset [a] described in Section 2, we randomly sampled a training set consisting of 200 sequences. The remaining sequences were used as a test set. Then, we trained these five LOHMMs, one per fold. We used a simple, but common stopping criterion: EM stops if a change in log-likelihood is less than 10^{-1} from one iteration to the next. To evaluate the learned models, we computed the log-likelihood that each model gave to a sequence in the test sets. If the i-th model was the most likely one, then we classified the sequence as a member of class i.

The used LOHMM structure is given in Fig. 5. The hidden states are modeled using $hc(ID, T, L)$ and $sc(ID, O, L)$ representing blocks of consecutive helices and strands. Being in a block ID of consecutive helices (resp. strands), the model will remain in the block or transition to a new block $s(ID)$ of strands (resp. helices). This model takes into account type T, length L and orientation O information. Moreover, there are specific abstract transitions for helices of types $h(right, alpha)$ and $h(right, 3to10)$, and for parallel and anti-parallel strands, and for being at the beginning of a sheet. This enabled us to model the "process" within blocks of consecutive helices quite detailed, and of transitions from blocks of consecutive helices to strands and vice versa. The ID enables

---

[a] For the extraction of the Prolog facts from the PDB, we adapted the program `secondary.c` made available by the Learning and Planning group of the University of Texas at Arlington (`http://cygnus.uta.edu/subdue/databases/db/proteins.tar.gz`).

Table 1: Confusion matrix showing actual vs. predicted fold classification.

actual \ predicted	fold1	fold2	fold23	fold37	fold55
fold1	736	61	51	62	30
fold2	49	291	53	31	11
fold23	18	23	166	11	15
fold37	55	44	27	282	19
fold55	0	1	1	3	147

Table 2: Precision and recall for each fold rounded to second decimal.

	fold1	fold2	fold23	fold37	fold55
Precision	0.86	0.69	0.56	0.72	0.66
Recall	0.78	0.67	0.71	0.66	0.96

us to have general directed transitions from one block to exactly one successor block.

4.1 Results

Our implementation of EM took at most five iterations and approximately 5 minutes to estimate the maximum-likelihood parameters per fold. Given our quantization of the helix and strand lengths, the LOHMM consisted of 74 abstract transition and 46 domain distribution probabilities, whereas the corresponding HMM would consist of over $62,000$ transition probabilities. So, the abstract representation of states and transitions in LOHMMs achieves, by design, a remarkable compression of the model, which supports hypothesis H4.

The classification results are summarized by the confusion matrix in Table 1. In this section, the TIM beta/alpha-barrel fold will be denoted as $fold1$, the NAD(P)-binding Rossmann-fold as $fold2$, the Ribosomal protein L4 fold as $fold23$, the glucosamine 6-phosphate deaminase/isomerase fold as $fold37$, and the leucine aminopeptidas fold as $fold55$. In total, 74% (1622 out of 2187) sequences were correctly classified. This result is in the same range as the one reported by Turcotte et al.[11] (75%). However, we have to emphasize that the datasets are not completely comparable. In contrast to this result, a learner predicting always the majority class would achieve an predictive accuracy of 43%. These results suggest that hypothesis H1 holds. In Table 2, we also give our results in terms of the *recall* and *precision*. Recall is defined as the sum of true positives divided by the sum of true positives and false negatives. Precision is defined as the sum of true positives divided by the sum of true positives and false positives. As can be seen, the recall and precision figures vary among the folds, but within the folds recall and precision are well balanced. In other

200

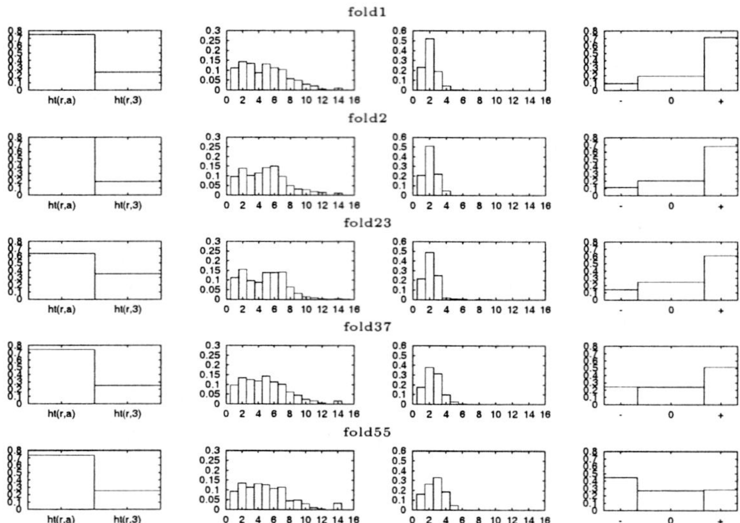

Figure 4: Estimated selection distributions for the five folds (from left to right: helix types, helix lengths, strand lengths, and strand orientations). The distributions specify the probability that a state is sampled from an abstract state (if needed) using a naïve Bayes scheme. The distribution over helix types shows, that only the types $ht(right, alpha)$ (shortly $ht(r, a)$) and $ht(right, 3to10)$ (shortly $ht(r, 3)$) occurred in the data. Due to pseudocounts, no probability value is zero.

words, a good precision is not bought at the expense of a good recall, and vice versa. The relatively low precision values for *fold23* and *fold55* are explained by a smaller number of test examples for these two folds. Finally, we inspected the trained LOHMM for characteristic differences. More precisely, we plotted for each of the five estimated LOHMMs the probability distributions implicitly defining μ (see Fig. 4) following a naïve Bayes scheme. Please note that μ defines the probability of sampling a state from an abstract state taking variable binding into account, i.e. that μ and therefore the distributions in Fig. 4 depend on the logical structure of the LOHMM. Upon visual inspection, differences can be found as follows (hypothesis H2):

- Regarding the helix types, $fold23$ differs from the others in that the probability of selecting right-handed alpha helices seems to be lower. Also, the probability of right-handed 3to10 helices to be selected seems to be higher than for the other folds.

- The first three and last two folds can be grouped w.r.t. the strand lengths.

- As for strand orientations, we have uniform "patterns" for $fold1$, $fold2$ and $fold23$, but characteristic patterns for $fold37$ and $fold55$.

To summarize, we believe that the results obtained in our experiments are quite promising also for what concerns the application domain. Therefore, they indicate that the answer to hypothesis H3 should be positive.

5   Related Work

Gough *et al.*[6] presented an approach to sequence annotation based on profile HMMs trained on the primary structure of domains for each superfamily in SCOP. The present study is at a different level of abstraction: Firstly, we are working with a secondary structure representation, and not a primary structure representation. Secondly, we are dealing with SCOP folds, not SCOP superfamilies. It might be interesting to apply our approach also at the (more detailed) superfamily level. The goal in the work by Gough *et al.*[6] was to annotate sequences based on a library of HMMs that represent all proteins of known structure. In contrast, our short-term goal was to give a proof of the principle, with the intermediate-term goal of providing a tool that helps to gain insights into structural characteristics. In further work we are planning to predict the fold based on the primary structure, with the stepping stone of secondary structure prediction.

Turcotte *et al.*[12,11] applied the Machine Learning and Inductive Logic Programming (ILP) tool Progol to a similar task as the one tackled in this paper. The task there was also to predict SCOP folds based on a high-level logical representation. The difference is that we are working with a larger, more recent dataset, a different representation, and that we are applying a different Machine Learning approach based on probability theory.

HMMs have been extended in a number of different ways e.g. hierarchical HMMs[3], factorial HMMs[5] and based on tree automata[4]. None of them utilize logical representations. *Relational Markov Models* (RMMs), as recently introduced and applied to web navigation by Anderson *et al.*[1], are an exception. RMMs do not allow for variable binding, unification nor hidden states.

6   Conclusion

In this paper, we have introduced Logical Hidden Markov Models (LOHMMs) and applied them to the task of finding structural signatures of protein folds. LOHMMs offer the possibility to specify states and transitions at an abstract level, and thereby offer a significant reduction in model size compared to regular

HMMs. Our experiments show that the learning performance of LOHMMs is good. We have also shown that it is easy to extract characteristic patterns from the learned models. In the future, we will conduct further experiments with more folds in SCOP. Current work includes the development of algorithms for learning the (logical) structure of LOHMMs.

Acknowledgements This research was partly supported by the European Union IST programme under contract number IST-2001-33053, *APrIL*. T. Raiko was supported by a Marie Curie fellowship at DAISY, HPMT-CT-2001-00251.

1. C. R. Anderson, P. Domingos, and D. S. Weld. Relational Markov Models and their Application to Adaptive Web Navigation. In *Proceedings of KDD-2002*, July 2002.
2. T. Dean and K. Kanazawa. Probabilistic temporal reasoning. In *Proceedings of AAAI-88*, 1988.
3. S. Fine, Y. Singer, and N. Tishby. The hierarchical hidden markov model: analysis and applications. *Machine Learning*, 32, 1998.
4. P. Frasconi, G. Soda, and A. Vullo. Hidden markov models for text categorization in multi-page documents. *Journal of Intelligent Information Systems*, 18(2/3):195–217, 2002.
5. Z. Ghahramani and M. Jordan. Factorial hidden Markov models. *Machine Learning*, 29:245–273, 1997.
6. J. Gough, K. Karplus, R. Hughey, and C. Chothia. Assignment of homology to genome sequences using a library of hidden markov models that represent all proteins of known structure. *Journal of Molecular Biology*, 313(4):903–919, 2001.
7. K. Kersting, T. Raiko, S. Kramer, and L. De Raedt. Towards discovering structural signatures of protein folds based on logical hidden markov models. Tech. Rep. 175, University of Freiburg, June 2002.
8. J. W. Lloyd. *Foundations of Logic Programming*. Springer, 1989.
9. L. R. Rabiner. A Tutorial on Hidden Markov Models and Selected Applications in Speech Recognition. *Proceedings of the IEEE*, 77(2), 1989.
10. L. R. Rabiner and B. H. Juang. An Introduction to Hidden Markov Models. *IEEE ASSP Magazine*, January:4–16, 1986.
11. M. Turcotte, S. Muggleton, and M. J. E. Sternberg. Discovery of structural signatures of protein fold and function. *Journal of Molecular Biology*, 306(3):591–605, 2001.
12. M. Turcotte, S. Muggleton, and M. J. E. Sternberg. The effect of relational background knowledge on learning of protein three-dimensional fold signatures. *Machine Learning*, 43(1/2):81–95, 2001.

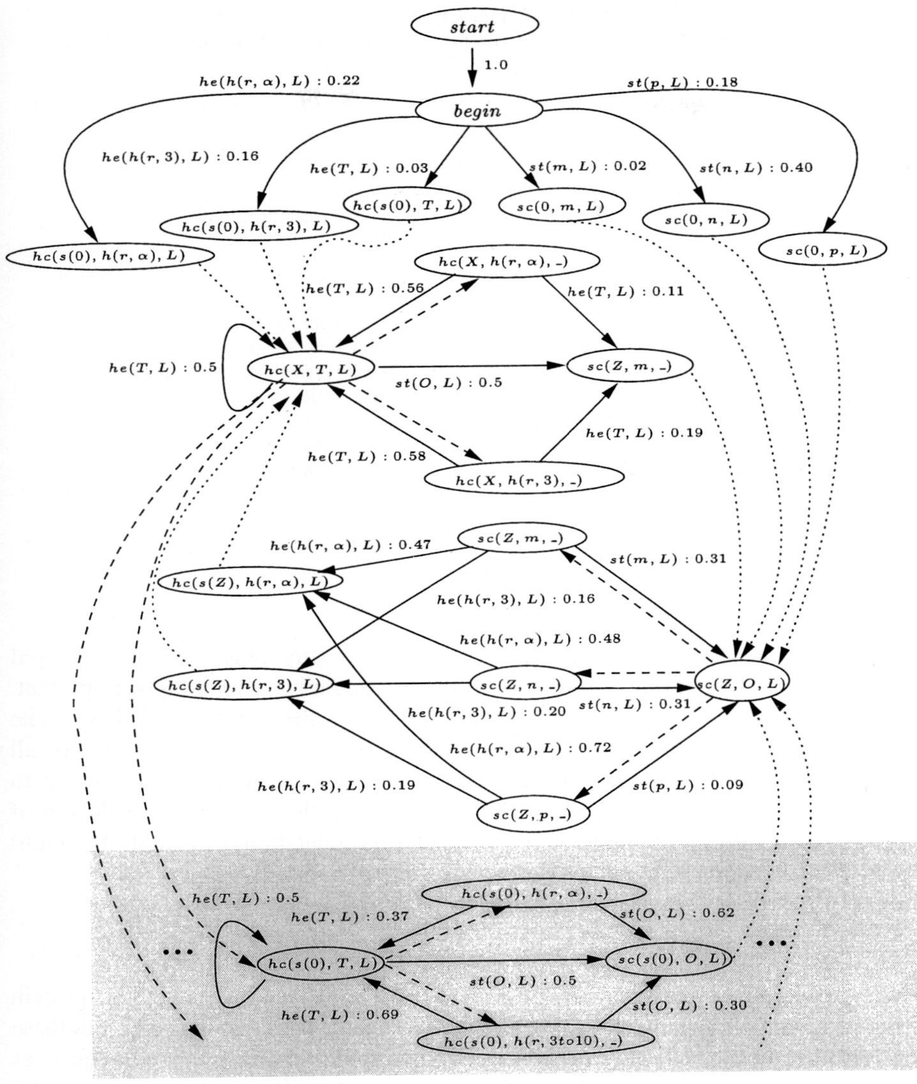

Figure 5: The estimated logical (hidden) Markov model of fold 1. The *end* state is omitted. If probabilities do not sum to 1.0, then there is a transition to *end*. The symbol _ denotes anonymous variables which are read and treated as distinct, new variables each time they are encountered. There are copies cf the shaded part for $hc(s^2(0), T, L), \ldots, hc(s^7(0), T, L)$. Terms are abbreviated using their starting alphanumerical and α for *alpha*.

AUTOMATED CONSTRUCTION OF STRUCTURAL MOTIFS FOR PREDICTING FUNCTIONAL SITES ON PROTEIN STRUCTURES

M.P. LIANG, D.L. BRUTLAG, R.B. ALTMAN

Stanford University, 251 Campus Drive X-215,
Stanford, CA 94305-5479, USA
E-mail: {mliang, brutlag, altman}@smi.stanford.edu

Structural genomics initiatives are beginning to rapidly generate vast numbers of protein structures. For many of the structures, functions are not yet determined and high-throughput methods for determining function are necessary. Although there has been extensive work in function prediction at the sequence level, predicting function at the structure level may provide better sensitivity and predictive value. We describe a method to predict functional sites by automatically creating three dimensional structural motifs from amino acid sequence motifs. These structural motifs perform comparably well with manually generated structural motifs and perform better than sequence motifs. Automatically generated structural motifs can be used for structural-genomic scale function prediction on protein structures.

1 Introduction

1.1 Structural genomics

With the sequencing of the human genome and many other organisms, rapid determination of gene and protein function is becoming increasingly important. Structural genomics initiatives are helping elucidate the function of these gene products by developing high-throughput methods for determining structures for all unique protein folds. These new structure targets are specifically selected not to have sequence similarity to existing proteins[1-4]. With the anticipated explosion of available structures, it is imperative to develop computational methods for high-throughput function prediction on protein structures.

1.2 Sequence motifs

There has been extensive work in identifying conserved residues in protein sequences with similar function. Amino acid sequence patterns that represent these conserved residue positions can be created from multiple alignments of sequences with similar function. These patterns, or sequence motifs, can be used to assign function to sequences that contain the pattern.

Numerous sequence motif databases have been established with different methods for creating sequence motifs. Some of the databases are manually curated by experts, while others are automatically derived. The BLOCKS+ database provides an integration of many sequence motif databases by generating blocks of ungapped alignment of sequences from families in the databases[5, 6].

The eMotif database uses sequence alignments from the BLOCKS+ database to create sequence motifs (eMotifs) at various specificities for a family of protein sequences. By providing motifs at different specificities for a block of sequences, eMotifs can represent families with high specificity and yet provide sensitivity by combining multiple motifs[7, 8].

Although sequence motifs can provide insight into protein function, when new proteins do not have significant sequence similarity with known proteins, using only sequence information fails. Proteins that do not have high sequence similarity may still have similar function because of conservation of physicochemical properties at the structural level[9, 10].

1.3   Structural motifs

Analogous to sequence motifs, structural motifs provide a description of conserved properties in the three dimensional structure of proteins sharing molecular function. Investigators have devised different techniques to construct and define structural motifs; each technique emphasizes different conserved properties.

Wallace et al have developed a system, PROCAT, for identifying catalytic sites by geometric orientation of residues with known functional importance. By using previous knowledge of the critical residues involved in the catalytic activity, a structural motif representing the conserved relative positions of those residues is constructed. This motif can be used to scan a new protein structure for occurrence of the catalytic site using a geometric hashing algorithm[11, 12].

Fetrow and Skolnick have developed Fuzzy Functional Forms (FFF) for representing distances between interesting residues. The critical residues involved in a functional site are identified by careful examination of the literature. Examples of known structures containing these residues are used to find mean distance and variance between the residues. The structural motif representing the conserved distance and variance of the residues is used to identify functional sites on protein structures[13, 14].

Wei and Altman have developed the FEATURE system that describes the physicochemical environment around functional sites. The environment, characterized by observing the frequency of physicochemical properties in radial shells around the site of interest, represents a structural motif used for predicting the functional sites[15, 16].

Unlike PROCAT and FFF, FEATURE does not require the conserved properties to be known in advance, but can discover them automatically given a training set. Once a set of examples of the functional site and a background control set is provided, FEATURE can automatically build the structural motif without having to manually identify which residues are considered important.

We introduce a new method, SeqFEATURE, that automatically creates structural motifs around functional sites by characterizing the structural environment around sequence motifs using FEATURE. Since the three dimensional structures of protein are more conserved than the sequence of amino acids[10], focusing on the protein structure should provide better sensitivity in identifying protein function.

2   Methods

2.1   *SeqFEATURE overview*

SeqFEATURE is a method for automatically building a structural motif from sequence motifs (Figure 1). The sequence motifs can be obtained from any of the sequence motif databases; here we use the eMotif database. Protein structures that contain the sequence motif are identified and a training set for the FEATURE system is automatically generated. FEATURE then uses the training set to construct a structural motif describing the physicochemical environment around the sequence motif.

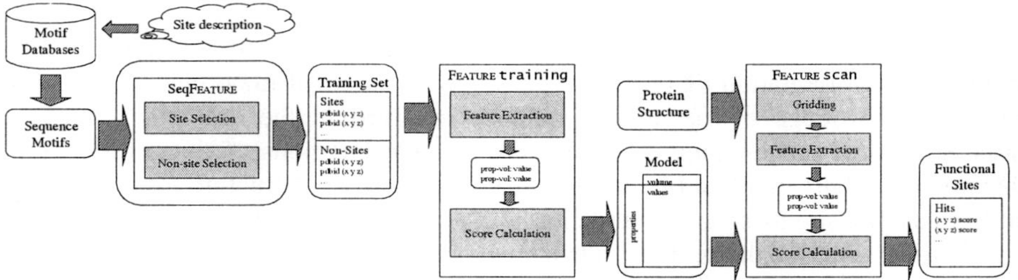

Figure 1: Data flow diagram for SeqFEATURE. Sequence motifs are selected by querying a sequence motif database. These motifs are fed into SeqFEATURE for automatic selection of sites and non-sites to create a training set. The training set is used by FEATURE for creating a model of the microenvironment around the functional site. The model is used to predict functional sites on new protein structures.

2.2   *FEATURE system*

The FEATURE system is detailed and evaluated elsewhere[16]; here we describe it briefly. The FEATURE system is used to build structural motifs and predict functional sites. FEATURE builds a physicochemical model of the environment around functional sites. The model represents the statistical distribution of physicochemical properties at radial distances from the site of interest. The physicochemical properties range from atom names and atom properties to residue names and secondary structure (Table 1). By using properties at the atomic level

Table 1: List of FEATURE's physicochemical properties

AtomName	C	N	O	S
	ANY	OTHER		
ChemicalGroup	Hydroxyl	Amide	Amine	Carbonyl
	RingSystem	Peptide		
AtomProperties	VDWVolume	Charge	NegCharge	PosCharge
	ChargeWithHis	Hydrophobicity	Mobility	SolventAccessibility
ResidueName	ALA	ARG	ASN	ASP
	CYS	GLN	GLU	GLY
	HIS	ILE	LEU	LYS
	MET	PHE	PRO	SER
	THR	TRP	TYR	VAL
	HOH	OTHER		
ResidueProperties	Hydrophobic	Charged	Polar	NonPolar
	Basic	Acidic		
SecondaryStructure	3Helix	4Helix	5Helix	Bridge
	Strand	Turn	Bend	Coil
	Het	Unknown		

and not just at the residue level, FEATURE provides a closer view of the chemistry necessary for function.

FEATURE requires a training set composed of positive examples of the functional site and negative background control examples. Each example is a specific 3D coordinate in a protein structure locating the site of interest. The environment around each site of interest is divided into radial shell volumes. FEATURE then creates a feature vector v for each site by counting the occurrence of physicochemical properties within each volume around the site.

The structural motif of the functional site is constructed by comparing the distribution of feature vectors from the positive examples (sites) with those from the negative examples (non-sites). Properties at each volume are evaluated as statistically more present or absent in the functional site by using the Wilcoxon rank sum test.

Finally, using a Bayesian inference method, FEATURE predicts functional sites at specific locations on protein structures. It places a 3D grid over the new protein structure and calculates the feature vector described above from the environment around each grid point. The grid point is given a score proportional to the likelihood that it is a functional site given its feature vector v.

$$\text{Score} = \log \frac{p(\text{Site} \mid v)}{p(\text{Site})} \tag{1}$$

By assuming that the components of the feature vector are independent, i.e. the property-volume values v_i are independent, the score can be broken down to summing the probability that the grid point is a site given each vector component v_i.

$$\text{Score} = \sum_i \log \frac{p(\text{Site} \mid v_i)}{p(\text{Site})} \qquad (2)$$

2.3 *eMotif database*

Here we use the eMotif database[8] as the source of sequence motifs for automatic creation of structural motifs, although any of the other sequence motif databases would be acceptable. The eMotif database provides a broad collection of sequence motifs, each representing functional signatures or domains built from BLOCKS+.

eMotif takes advantage of predefined conserved residue substitution groups based on shared chemical or physical properties of the amino acids, such as size or hydrophobicity. The resulting motifs represent variation in motif positions that are biologically meaningful.

In addition, multiple eMotifs are built for a block of sequences with different levels of specificity and sensitivity. These properties make eMotif particularly amenable to predicting functional sites at a proteomic level.

2.4 *Training set selection*

We select sequence motifs related to the functional site from the eMotif database using a keyword search. A training set of positive example sites and negative control non-sites is automatically generated from the sequence motif.

Structures from the Protein Data Bank[17] (PDB) whose sequence contains the sequence motifs are selected. Bennett et al. have created a database of all structures containing eMotifs (3MOTIF) along with accompanying tools for visualizing the eMotifs on the 3D structures[18, 19].

To pick a positive site example, the residues matching the sequence motif are extracted from the structure and the geometric center of their alpha carbons is selected.

To pick a control non-site example, the atom density around the selected positive sites is calculated and a random point on the same protein structure with similar atom density is selected. Points within a particular distance from the positive site are excluded.

2.5 *Evaluation metrics*

The performance of a structural motif or sequence motif is measured by its sensitivity and positive predictive value.

Sensitivity is the true positive rate, representing how many of the true sites are detected by the motif. It is calculated by taking the ratio of the number of sites that are predicted correctly (TP_{sites}) to the total number of sites ($TP_{sites} + FN_{sites}$).

$$\text{Sensitivity} = \frac{\text{Number of Sites Predicted Correctly}}{\text{Total Number of Sites}} = \frac{TP_{sites}}{TP_{sites} + FN_{sites}} \qquad (3)$$

Positive predictive value (PPV) measures how often the positive predictions are correct. Positive predictions (hits) are predictions that a functional site occurs at a particular location. Positive predictive value is calculated by taking the ratio of the number of hits that are near true sites (TP_{hits}) to the total number of hits ($TP_{hits} + FP_{hits}$).

$$\text{PPV} = \frac{\text{Number of Hits near True Sites}}{\text{Total Number of Hits}} = \frac{TP_{hits}}{TP_{hits} + FP_{hits}} \qquad (4)$$

It is important to point out that the units for sensitivity and PPV are not interchangeable. Sensitivity is a ratio of sites, whereas PPV is a ratio of hits. Because hits detect the area around a single point representing the site, there could be more than one predicted location per actual site.

3 Results

3.1 *EF-Hand calcium binding sequence motif*

Calcium ions have a spectrum of essential biological roles, affecting a myriad of regulatory processes, enzymatic activity, and protein stability. The EF-Hand is the most common motif for binding calcium in proteins[20, 21].

Sequence motifs for the EF-Hand family of calcium binding sites were selected from eMotif by keyword search. Thirteen motifs were found with specificity ranging from 10^{-3} to 10^{-9} (Table 2). The motifs were 12 to 13 residues in length.

Table 2: EF-Hand family eMotifs at varying specificities

eMotif	specificity
`d.[dn]........[de]`	10^{-3}
`[dn].[dn]....[ilmv]...[de][filvy]`	10^{-4}
`d........[filmv].e[fwy]`	10^{-4}
`d.[dn]..g.[ilmv]...[de]`	10^{-5}
`[dn]...d....[fly].e[fwy]`	10^{-5}
`d.[dn]..g.[ilmv]...e[fly]`	10^{-6}
`d.[dn].d..[iv]...[de][fly]`	10^{-6}
`d.[dn].[dn]....[filmv].e[fwy]`	10^{-6}
`d.[dn].dg.[ilv]...[de][fly]`	10^{-7}
`d.[dn]..g.[ilmv].[fly].e[fwy]`	10^{-7}
`d.[dn].d..[ilv].[filmv].e[fwy]`	10^{-7}
`d.[dn].dg.[ilv].[filmvy].e[fwy]`	10^{-8}
`d.[dn].dg.[iv]..[de]ef`	10^{-9}

3.2 Automated construction of structural motif

The 13 EF-Hand motifs had a total of 1374 hits on 62 structures. These were automatically selected and fed into the training set generator. The positive and negative examples were selected as described in the methods generating a total of 220 sites and 119 non-sites. Figure 2 shows the automatically constructed structural motif.

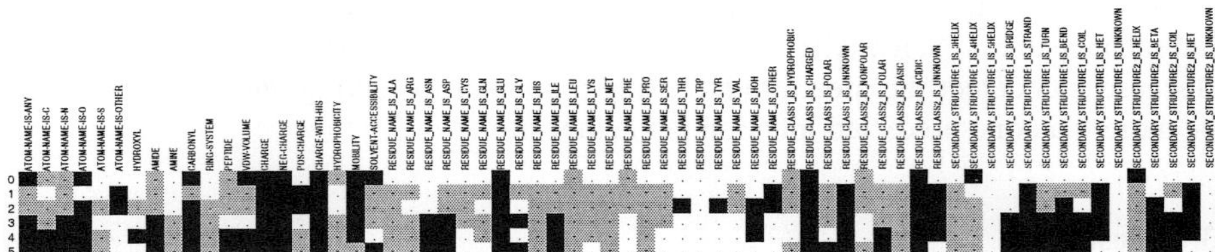

Figure 2: SeqFEATURE model of calcium binding automatically generated from EF-Hand sequence motifs. This model describes the 3D environment around the sequence motif. Each row represents the distribution of properties in increasingly larger spherical shell around the site. Properties significantly present are shaded darker and properties significantly absent are shaded lighter.

3.3 Manual construction of general calcium binding motif

A general calcium binding structural motif was created by manual generation of a training set for FEATURE. Forty protein structures from the PDB with unique folds for binding calcium were manually selected. The calcium ions in the protein structures were used as sites for the training set. In addition, random backbone atoms in 14 structures known not to have calcium binding activity were used as non-sites.

3.4 *Performance on general calcium binding proteins*

In the following performance figures, the presence of calcium ion in the crystal structure is used as the gold standard for identifying true sites.

The general calcium binding site test set had 54 structures containing 91 true sites. The number of hits from FEATURE is varied by adjusting the cutoff threshold for the score. The eMotif sequence motifs predicted 81 of 98 hits near sites, detecting 14 unique sites (15% sensitivity, 83% PPV). At similar sensitivity, SeqFEATURE with score cutoff at 115 had 23 of 24 hits near sites, detecting 14 unique sites (15% sensitivity, 96% PPV). The manual FEATURE with score cutoff at 65 had 17 of 18 hits near sites, detecting 16 unique sites (18% sensitivity, 94% PPV). Figure 3 shows the performance over varying sensitivity.

General Calcium Binding

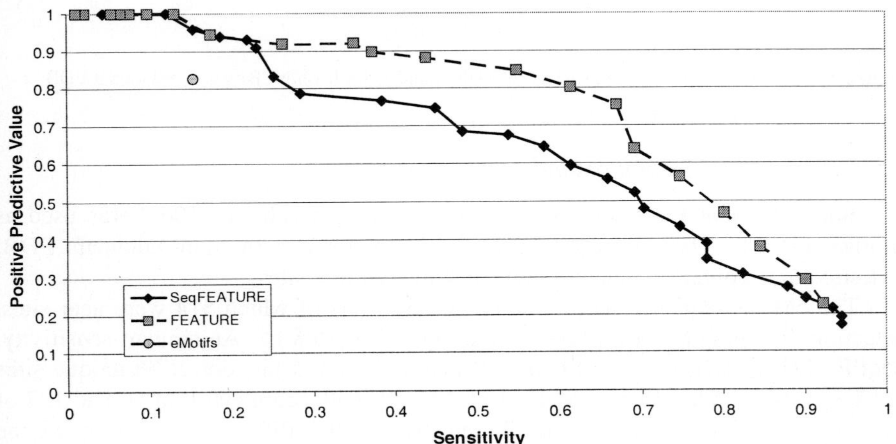

Figure 3: Performance of SeqFEATURE, FEATURE, and eMotifs on general calcium binding sites.

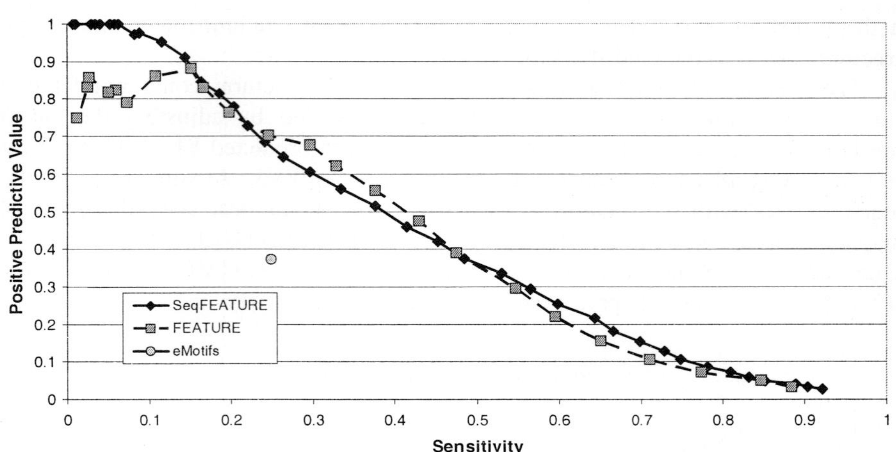

Figure 4: Performance of SeqFEATURE, FEATURE, and eMotifs on NCBI's non-redundant PDB.

3.5   *Performance on non redundant database*

The non-redundant PDB set from NCBI with Blast e-value of 10e-7 was used as another test set.   There were a total of 2122 structures in the non-redundant PDB. 131 structures contained calcium ions, making 398 true sites.

The eMotif sequence motifs predicted 1376 hits of which 514 were near sites, detecting 99 unique sites (25% sensitivity, 37% PPV).   At similar sensitivity, SeqFEATURE with score cutoff at 110 had 284 of 415 hits detect 96 unique sites (24% sensitivity, 68% PPV).   Manual FEATURE with score cutoff at 60 had 137 of 195 hits detect 98 unique sites (25% sensitivity, 70% PPV).   Figure 4 shows the performance over varying sensitivity.

4   Discussion

In both the general calcium binding sites and the non-redundant PDB test sets, the automatically generated EF-Hand structural motif from SeqFEATURE had comparable performance to manual construction of a calcium binding site from FEATURE.   The FEATURE calcium binding site performed better in the general calcium binding test set, with higher positive predictive value across the sensitivity range.   This is expected since the SeqFEATURE model was created from only examples of EF-Hand calcium binding whereas the manual FEATURE model was trained on a variety of different calcium binding structures.

The SeqFEATURE model of the EF-Hand was able to have better PPV and sensitivity than the EF-Hand sequence eMotifs in both test cases. This demonstrates how observing the three dimensional environment around sequence motifs can provide more information in predicting function than just using the sequence information.

SeqFEATURE and FEATURE use a discriminative method for learning the model for a functional site. By taking positive examples and negative control examples, these methods identify what shared attributes in the positive examples make them different from the control examples and filter out attributes and noise common to both sets. The eMotifs use a generative method for representing the conservation of residues. They only use a set of positive examples, identifying the properties conserved in the set and filtering out those that have too much variance. If the eMotifs are created using discriminative methods, they may provide more power in predicting function than the current generative methods.

In the evaluation of the structural and sequence motifs for calcium binding, we used the presence of calcium ion in the crystal structure as a gold standard. This standard may be subject to errors as some calcium ions may have been missed in the structure determination due to different experimental conditions. By taking a large set of examples, we hope these errors will be averaged out.

5   Conclusion

The three dimensional structure of the proteins provide more information for predicting functional sites than does sequence homology. A structural motif for calcium binding was automatically created from sequence motifs of the EF-Hand calcium binding family. By using structural information, the structural motif had better performance in detecting calcium binding than the sequence motif. This method for automatic construction of structural motifs from sequence motifs can be used to build a library of models for performing genomic-scale prediction of functional sites. Structural motifs for other types of functional sites, such as catalytic sites and small molecule binding sites, will be tested in the future.

Acknowledgments

This work was supported by the National Institute of Health (NIH) grants LM-05652, LM-07033, GM-63495, and HG-02235. We thank Rey Banatao, Liping Wei and Soumya Raychaudhuri for helpful discussions.

References

1. Brenner, S.E., "A tour of structural genomics". *Nat Rev Genet.* **2**, 10 (2001) pp. 801-9.
2. Burley, S.K., et al., "Structural genomics: beyond the human genome project". *Nat Genet.* **23**, 2 (1999) pp. 151-7.
3. Baker, D. and A. Sali, "Protein structure prediction and structural genomics". *Science.* **294**, 5540 (2001) pp. 93-6.
4. Skolnick, J., J.S. Fetrow, and A. Kolinski, "Structural genomics and its importance for gene function analysis". *Nat Biotechnol.* **18**, 3 (2000) pp. 283-7.
5. Henikoff, J.G., et al., "Increased coverage of protein families with the blocks database servers". *Nucleic Acids Res.* **28**, 1 (2000) pp. 228-30.
6. Henikoff, S., J.G. Henikoff, and S. Pietrokovski, "Blocks+: a non-redundant database of protein alignment blocks derived from multiple compilations". *Bioinformatics.* **15**, 6 (1999) pp. 471-9.
7. Nevill-Manning, C.G., T.D. Wu, and D.L. Brutlag, "Highly specific protein sequence motifs for genome analysis". *Proc Natl Acad Sci U S A.* **95**, 11 (1998) pp. 5865-71.
8. Huang, J.Y. and D.L. Brutlag, "The EMOTIF database". *Nucleic Acids Res.* **29**, 1 (2001) pp. 202-4.
9. Chothia, C. and A.M. Lesk, "The relation between the divergence of sequence and structure in proteins". *Embo J.* **5**, 4 (1986) pp. 823-6.
10. Chothia, C. and A.M. Lesk, "The evolution of protein structures". *Cold Spring Harb Symp Quant Biol.* **52**, (1987) pp. 399-405
11. Wallace, A.C., N. Borkakoti, and J.M. Thornton, "TESS: a geometric hashing algorithm for deriving 3D coordinate templates for searching structural databases. Application to enzyme active sites". *Protein Sci.* **6**, 11 (1997) pp. 2308-23.
12. Wallace, A.C., R.A. Laskowski, and J.M. Thornton, "Derivation of 3D coordinate templates for searching structural databases: application to Ser-His-Asp catalytic triads in the serine proteinases and lipases". *Protein Sci.* **5**, 6 (1996) pp. 1001-13.
13. Fetrow, J.S. and J. Skolnick, "Method for prediction of protein function from sequence using the sequence-to-structure-to-function paradigm with application to glutaredoxins/thioredoxins and T1 ribonucleases". *J Mol Biol.* **281**, 5 (1998) pp. 949-68.
14. Fetrow, J.S., A. Godzik, and J. Skolnick, "Functional analysis of the Escherichia coli genome using the sequence- to-structure-to-function paradigm: identification of proteins exhibiting the glutaredoxin/thioredoxin disulfide oxidoreductase activity". *J Mol Biol.* **282**, 4 (1998) pp. 703-11.

15. Bagley, S.C., et al., "Characterizing oriented protein structural sites using biochemical properties". *Proc Int Conf Intell Syst Mol Biol.* **3**, (1995) pp. 12-20

16. Wei, L. and R.B. Altman, "Recognizing protein binding sites using statistical descriptions of their 3D environments". *Pac Symp Biocomput.*, (1998) pp. 497-508.

17. Berman, H.M., et al., "The Protein Data Bank". *Nucleic Acids Res.* **28**, 1 (2000) pp. 235-42.

18. Bennett, S.P., C.G. Nevill-Manning, and D.L. Brutlag, "3MOTIF: Visualizing Conserved Protein Sequence Motifs in the Protein Structure Database". Submitted for Publication (2002)

19. Bennett, S.P. and D.L. Brutlag, "Protein Sequence Motifs in the Protein Structure Database". Personal Communication (2002)

20. Donato, R., "Functional roles of S100 proteins, calcium-binding proteins of the EF- hand type". *Biochim Biophys Acta.* **1450**, 3 (1999) pp. 191-231.

21. Lewit-Bentley, A. and S. Rety, "EF-hand calcium-binding proteins". *Curr Opin Struct Biol.* **10**, 6 (2000) pp. 637-43.

PREDICTION OF BOUNDARIES BETWEEN INTRINSICALLY ORDERED AND DISORDERED PROTEIN REGIONS

PREDRAG RADIVOJAC, ZORAN OBRADOVIĆ

Center for Information Science and Technology, Temple University, U. S. A.

CELESTE J. BROWN, A. KEITH DUNKER

School of Molecular Biosciences, Washington State University, U. S. A.

Using proteins with both disordered and ordered regions collected through literature searches and database scanning, we assembled a set of 24-residue long segments centered on their order/disorder boundaries as well as a larger set of non-boundary segments consisting of either order or disorder. We analyzed position-specific amino acid compositions around the order/disorder boundaries and found more than thirty significant ($p < 0.05$) compositional differences between boundary and non-boundary data. From this analysis, we constructed several logistic regression predictors of order/disorder boundaries using slightly different data modeling approaches. Exact boundary prediction accuracies were estimated to be in the range from 74% to 80% for the different predictors.

1 Introduction

Intrinsically disordered protein is gaining increased attention in the biological community.[1-4] Following the prior work of Ptitsyn and Uversky[5], we proposed that *native* proteins may exist in any of three forms: ordered (fully folded), collapsed (molten globule-like) or extended (random coil-like). These three forms can occur in localized regions or over entire sequences. Furthermore, we proposed that protein function may arise from any of the three forms or from inter-form structural transitions resulting from changes in environmental or cellular conditions such as ligand binding.[2, 6] The collapsed and extended forms correspond to intrinsic disorder while the fully folded, ordered form is generally comprised of three secondary structure types: α-helix, β-sheet, and other. Collapsed disorder can have secondary structure, and so presence or absence of secondary structure is not a feature that distinguishes ordered and disordered proteins. Instead, disordered proteins or regions have coordinates and Ramachandran angles that vary significantly over time while ordered protein ensembles are distinguished by relatively fixed coordinates and the same canonical set of Ramachandran angles that are invariant over time. Disordered proteins have also been called natively unfolded[7] and intrinsically unstructured.[1]

Disordered regions can be detected by X-ray crystallography as stretches of missing electron density corresponding to both backbone and side chain atoms. Disordered regions can also be identified especially by several features of their NMR spectra, such as the longitudinal relaxation rate, the transverse relaxation rate, and the heteronuclear Overhauser effect between the amide proton and its attached nitrogen.[8, 9] Other disorder indicators include an extended hydrodynamic radius and a random coil-type circular dichroism spectrum, both of which can be usefully coupled with limited, time-resolved proteolysis.[2]

Disordered proteins are found in all kingdoms, but are predicted to be more common in eukaryotes than in archaea or eubacteria.[10] Intrinsic disorder is apparently necessary for certain critical functions. A recent survey[11] classified the functions of approximately 100 disordered regions into four categories: molecular recognition, molecular assembly/disassembly, protein modification, and entropic chains (i.e. flexible linkers and entropic clocks, bristles and springs). Many disordered proteins have yet-to-be-determined functions, and it is doubtful that all of the functions associated with intrinsic disorder have been identified.

Previous studies indicated that disordered regions are compositionally distinct from ordered proteins. Compared to ordered proteins, disordered proteins have, among other distinguishing attributes, a higher average flexibility index value[12], a lower sequence complexity[13, 14] as estimated by Shannon's entropy[15], and a lower aromatic content.[16, 17] We noticed the importance of charge and hydrophobicity for distinguishing between order and disorder,[16, 17] while others have emphasized the combined use of these two attributes.[18, 19]

In accordance with the hypothesis that a protein's structure and function are determined by its amino acid sequence[20], long stretches of 30 or more consecutive disordered residues are predictable from primary structure, with accuracies in the 70-75% range on a per-residue basis.[21-24] These studies employed ordinary least squares regression[25], feed-forward neural networks[26] or ensembles thereof.[27, 28]

The various order/disorder predictors just described were based on the premise that different types of disordered sequences are more similar to each other than to ordered sequences and *vice versa*. Prediction accuracies well above those expected by chance support this two-state approximation. However, this is not to suggest that either the ordered or disordered class has homogeneous flexibility. Thus, on the ordered side we are developing sequence-based predictors of high versus low B-factors (submitted for publication), and on the disordered side, we are developing predictors of different types, or flavors, of disorder (submitted for publication). Overall, then, we are first predicting order/disorder and then investigating the two predicted subsets for further separations.

Per-residue prediction accuracies were improved to about 82% by using input data from longer segments and from post-processing to eliminate isolated errors.[29] Although these modifications led to overall gains in accuracy, especially for very long ordered and disordered regions, both of these improvements led to localized decreases in accuracies around order/disorder boundaries. Prediction of order/disorder boundaries would therefore have the potential of reversing such localized losses in accuracy.

Methods for predicting boundaries of secondary structure classes were first reported by Presta & Rose[30] and Richardson & Richardson.[31] They analyzed amino acid compositions around helix/non-helix boundaries and reported regularities that were used to improve predictions of helical secondary structure. A similar approach was reported by Blom and coworkers for successfully predicting cleavage[32] and phosphorylation[33] sites. Following the approaches of Richardson & Richardson and Blom *et al.*, we studied position-specific amino acid compositions around order/disorder boundaries. Then, using machine-learning techniques, we constructed order/disorder boundary predictors and evaluated their prediction accuracies. Our eventual goal will be to combine order/disorder boundary predictors and standard order/disorder predictors to give improved estimations of intrinsic protein disorder.

2 Methods

2.1 Dataset

From a set of 154 proteins containing intrinsically disordered regions 30 or more residues in length (available at http://disorder.chem.wsu.edu), we extracted a set of non-redundant, 24-residue long sequences centered at order/disorder boundaries. The number of assembled order/disorder boundaries is much smaller than the potential of $154 \times 2 = 308$ because some of the proteins in the dataset are completely disordered (e.g. acidic ribosomal protein, gi:133069) and because many of the disordered regions start from the N-terminus (e.g. APEX nuclease, gi:299037) or end at the C-terminus (e.g. antibacterial protein, gi:1706745). In addition, many of the proteins in our datasets consist of fragments rather than whole proteins. If an isolated polypeptide fragment is completely disordered and the structure of the rest of the protein is unknown, an ordered boundary could not be inferred (e.g. CFTR, gi:14753227). The positive set of order/disorder boundaries resulting from this extraction contained 123 sequences.

Using the same set of disordered proteins, we singled out 1,691 24-residue long completely disordered fragments that were at least 12 residues from the disorder boundary. All the segments were selected at random and with caution to avoid over-representation of very long disordered regions. This set comprised half of the negative control set. To form the other half of the negative control set, we used a dataset of 290 completely ordered proteins selected from the Protein Data Bank[34] by Smith *et al.*[35] All the proteins from this set have a resolution better than 2Å, an R-factor lower than 20%, and at least 80 residues. None of these proteins have missing backbone or side-chain atoms. We randomly selected 1,691 24-residue long ordered fragments from these proteins and included them in the negative control set, henceforth denoted as $\mathcal{N}$. The balanced composition of ordered and disordered segments in $\mathcal{N}$ was designed to prevent a trained predictor from simply adapting to disorder anywhere in an input sequence. Overall, the complete dataset consisted of 123 positive and 3,382 negative segments.

In order to ensure non-redundancy within the data sets, sequence identity was analyzed for the segments collected in both sets using a series of pairwise alignments among the sequences. For this purpose, we used the Smith-Waterman algorithm[36] and the BLOSUM62 scoring matrix[37] with a gap-opening penalty of 10 and a gap-extension penalty of 0.6, as optimized in Vogt *et al.*[38] The average sequence identity for the set of positive examples was 15% with a maximum of 50%. The examples with 50% sequence identity correspond to the order/disorder boundaries of neurofilament triplet proteins (sp:P08553, sp:P02548, gi:128127), which are highly similar on the ordered side and significantly different on the disordered side. We therefore decided to keep these three sequences. The average sequence identity of the negative set was 16%.

The original set of positive sequences consists of 51 boundaries with the disordered region on the right (order/disorder) and 72 boundaries with disorder on the left (disorder/order). In order to study and visualize the true determinants of the boundaries, we constructed another set of positive sequences where all segments had the disordered region on the right by reversing the sequence of the disorder/order sequences. A practical importance of this dataset arises from the fact that a boundary predictor trained on it can be used in combination with currently existing order/disorder predictors[29, 39] towards better overall prediction of intrinsic disorder. A simple scheme that would guarantee good performance would use the outputs of the standard order/disorder predictor and reverse the sequences fed to a boundary predictor only around the right ends of putative disorder regions.

Although there might exist a difference in left and right ends of disordered regions, the number of sequences in the original set available at this moment is insufficient to separately characterize and predict them. We refer to the original set of positive sequences as $\mathcal{P}_U$ (unidirectional) and the set with reversed disorder/order segments as $\mathcal{P}_B$ (bidirectional).

2.2 Sequence logos and statistical significance of differences in distributions

A sequence logo for the $\mathcal{P}_B$ set is shown in Figure 1. Sequence logos were introduced by Schneider & Stephens[40] in order to visualize the deviation of position specific amino acid compositions at sites of interest from the uniform distribution. Deviations are measured in bits per amino acid and range from 0 if the observed distribution is uniform to $\log_2(20) \approx 4.32$ when only one completely conserved amino acid is observed. Amino acids are plotted with the size proportional to their contribution in overall deviation for a particular position.

To estimate the statistical significance of differences in the observed relative frequencies of $\mathcal{P}_U$ over $\mathcal{N}$, and $\mathcal{P}_B$ over $\mathcal{N}$, we performed Fisher's permutation test[41], with 10,000 permutations, for each position in the range ±12 residues around the boundaries. For $\mathcal{P}_U$, we observed 282 (out of 24 positions × 20 amino acids = 480) compositional differences greater than 0.01; of these 282 attributes, 33 had p-values lower than 0.05, and another 37 had p-values below 0.1. In the case of $\mathcal{P}_B$, 43 attributes out of the 302 with compositional differences above 0.01 exhibited p-values below 0.05, while an additional 32 had p-values below 0.1. The analysis corresponding to the sets $\mathcal{P}_B$ and $\mathcal{N}$ is presented in Table 1. The statistical significance of some position specific amino acid compositions enabled us to construct an order/disorder boundary predictor.

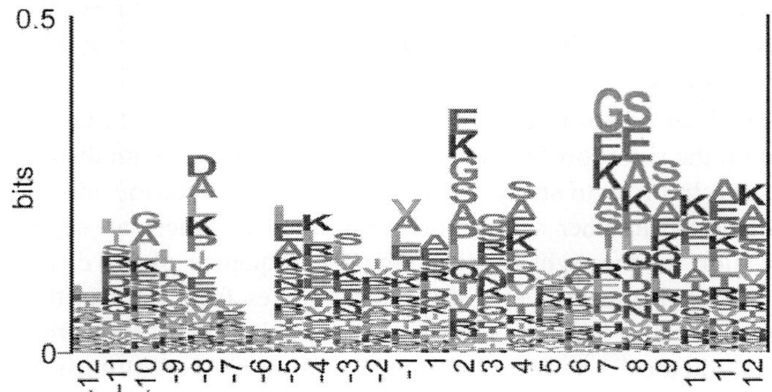

Figure 1. *Sequence logos for the set $\mathcal{P}_B$.*

Table 1. *Differences of position specific amino acid compositions between two sets of data, $\mathcal{P}_B$ and $\mathcal{N}$. The first and the third column represent a position relative to the order/disorder boundary. The second and fourth columns are in the following format: (amino acid, p-value, compositional difference between sets $\mathcal{P}_B$ and $\mathcal{N}$). The compositional difference is positive if the set $\mathcal{P}_B$ is enriched in a given amino acid at a specific position, as compared to $\mathcal{N}$; if an amino acid is depleted in $\mathcal{P}_B$, the difference is negative. Entries that are significant at the 0.05 level are denoted in bold, those significant at the 0.1 level are not bold.*

Pos.	(amino acid, p-value, compositional difference)	Pos.	(amino acid, p-value, compositional difference)
−12	(I, 0.02, 0.04) (L, 0.04, 0.05)	1	(R, 0.06, 0.04)
−11	(A, 0.00, −0.06) (I, 0.03, 0.04) (L, 0.07, 0.04) (N, 0.07, 0.03) (P, 0.04, −0.04) (R, 0.08, 0.03) (V, 0.04, −0.04) (Y, 0.04, −0.03)	2	(H, 0.04, −0.03) (I, 0.00, −0.04)
−10	(G, 0.05, −0.05) (T, 0.06, −0.04) (V, 0.06, −0.04)	3	(N, 0.03, 0.04)
−9	(I, 0.03, 0.05) (K, 0.05, −0.04) (L, 0.02, 0.06) (M, 0.01, 0.04) (Q, 0.06, −0.03)	4	(H, 0.06, −0.02)
−8	(D, 0.03, 0.05) (G, 0.04, −0.05) (H, 0.06, −0.02) (M, 0.02, 0.03) (N, 0.01, −0.04) (Y, 0.01, 0.04)	5	(A, 0.03, −0.05) (L, 0.08, -0.03) (N, 0.07, 0.03) (S, 0.10, −0.04)
−7		6	(F, 0.05, 0.03) (G, 0.06, 0.05) (M, 0.01, 0.03)
−6	(A, 0.02, −0.06) (E, 0.01, −0.06) (M, 0.01, 0.04) (R, 0.09, 0.03) (W, 0.05, 0.02)	7	(G, 0.00, 0.10) (L, 0.02, −0.05) (Q, 0.00, −0.05)
−5	(D, 0.02, −0.04) (G, 0.03, −0.05) (L, 0.06, 0.04)	8	(F, 0.03, −0.03) (I, 0.04, −0.03) (S, 0.01, 0.07) (V, 0.09, −0.03)
−4	(G, 0.08, −0.04) (N, 0.08, −0.02) (R, 0.07, 0.03)	9	(I, 0.05, −0.04) (N, 0.05, 0.04) (P, 0.00, −0.06) (Q, 0.02, 0.04) (S, 0.04, 0.05)
−3	(A, 0.00, 0.07) (S, 0.05, 0.04) (V, 0.09, 0.04)	10	(I, 0.10, 0.03) (K, 0.04, 0.05) (M, 0.08, 0.02) (V, 0.02, −0.05)
−2	(I, 0.00, 0.06) (K, 0.08, −0.04) (R, 0.10, 0.03)	11	(A, 0.09, 0.04) (I, 0.02, −0.04)
−1	(F, 0.05, 0.03) (H, 0.06, −0.02)	12	(G, 0.06, −0.04) (K, 0.05, 0.05)

2.3 Attribute construction and selection

The input to a predictor of order/disorder boundaries is a 24-residue long sequence. For each position of the input sequences, we constructed a 20-dimensional vector of 0s with the only 1 for the residue actually observed at that position. In total, considering all positions, we constructed examples consisting of $24 \times 20 = 480$ binary attributes. A binary target variable (1 = boundary, 0 = non-boundary) was then added to each example. Consequently, three matrices were constructed: matrix $\mathbf{P_U}$ consisting of examples based on the set of sequences $\mathcal{P_U}$, matrix $\mathbf{P_B}$ based on $\mathcal{P_B}$, and matrix $\mathbf{N}$ corresponding to the negative control set $\mathcal{N}$.

The number of positive examples available for model training is small and the data for each sample is highly dimensional and sparse (the majority of the data is zeros with 24 ones in each row). We, therefore, kept only those position specific amino acids with significant compositional differences (p-value < 0.1). After reducing the number of attributes, however, the sample was still dominated by zeros which can cause collinearity problems during model training.[25] To deal with collinearity, we applied principal component analysis[26] (PCA) to further reduce the dimensionality of the sample.

Finally, the sample construction resulted in 3 matrices: $\mathbf{P_U}$, of size $123 \times (m + 1)$, $\mathbf{P_B}$, also of size $123 \times (m + 1)$, and $\mathbf{N}$ of size $3,382 \times (m + 1)$. Here, m represents the number of attributes after principal component analysis. We use term $m + 1$ to indicate that the target variable was added to each row of the sample. Although the matrices are denoted by the same symbols before and after dimensionality reduction, we believe that this actually simplifies further discussion. In sections 3 and 4, we show and analyze prediction accuracy of the order/disorder boundary predictor versus sample dimension m after applying PCA.

2.4 Model training and testing

We combined each set of positive examples $\mathbf{P_U}$ and $\mathbf{P_B}$ with the set of negative examples $\mathbf{N}$ to construct a linear predictor based on logistic regression[25], a maximum likelihood technique suited to classification problems. To further prevent overfitting, model building and testing were performed using the leave-one-out method[26], which is appropriate for very small datasets. More specifically, the first example from the positive set and 25% of the data selected at random from the negative set were removed to form a test set. The remaining 122 positive examples and a set of 122 negative examples randomly chosen from the remaining 75% of negative examples were included in a balanced training set, and a linear predictor was trained. The training was repeated for I random selections of 122 negative examples, and the prediction on the test set was made by averaging raw outputs from all I models. The whole procedure was then repeated for all other positive examples. To remove the dependence of prediction results on the choice of examples forming the test set the whole procedure was repeated 30 times and performance results were reported. This algorithm is presented in Fig. 2. In machine learning approaches, usually, balanced sets of positive and negative examples are constructed and predictor performance is evaluated on a test set. However, in this case the abundance of negative examples allows us to randomly select negative examples more than once ($I \geq 1$), with the potential of improving prediction results.

Input:
  $\mathbf{P}$ = matrix of $|\mathbf{P}|$ = 123 positive examples ($\mathbf{P_U}$ or $\mathbf{P_B}$);
  $\mathbf{N}$ = matrix of $|\mathbf{N}|$ = 3,382 negative examples;
repeat 30 times
  for every example $\mathbf{x} \in \mathbf{P}$
    randomly select subset $\mathbf{N_{25\%}} \subset \mathbf{N}$; $\mathbf{N_{75\%}} = \mathbf{N} - \mathbf{N_{25\%}}$
    construct a test set $\mathbf{Ts} = \{\mathbf{x}\} \cup \mathbf{N_{25\%}}$
    for i = 1 to I
      randomly select $|\mathbf{P}| - 1$ examples from $\mathbf{N_{75\%}}$ making $\mathbf{N_{|P|-1}}$
      train predictor p_i using training set $\mathbf{Tr} = \mathbf{P} - \{\mathbf{x}\} \cup \mathbf{N_{|P|-1}}$
      make a raw prediction $p_i(\mathbf{Ts})$ on test set $\mathbf{Ts}$ using predictor p_i
    end
    make final prediction i.e. $p(\mathbf{Ts}) = 1/I \cdot \sum_{i=1}^{I} p_i(\mathbf{Ts})$
    quantize $p(\mathbf{Ts})$ and calculate sensitivity and specificity
  end
  average sensitivity and specificity over all $|\mathbf{P}|$ iterations
end
Output:
  sensitivity and specificity averaged over all 30 iterations;
  95% confidence intervals;

Figure 2. *The process of model building and testing*

2.5 Performance evaluation

In order to evaluate the performance of different predictors, we measure sensitivity and specificity for a given set of parameters. This approach is commonly used when the sizes of prediction classes are not equal. Sensitivity (denoted as *sn*) is defined as the percentage of positive examples i.e. order/disorder boundaries, correctly predicted, while specificity (denoted as *sp*) is the percentage of negative examples correctly predicted. Assuming that the class sizes are equal, the accuracy of prediction (denoted as *ac*) can be expressed as the arithmetic mean of sensitivity and specificity. This sets the results of a prediction at random to 50% accuracy. Since all experiments were repeated n = 30 times, together with sensitivity and specificity we also report 95% confidence intervals calculated as $\pm 2 \cdot \sigma / \sqrt{n}$. Here, σ is the standard deviation of the estimated parameter (*sn* or *sp*).

3 Experiments and Results

3.1 Poor order/disorder boundary prediction with standard methods

The first experiment was to test a standard order/disorder predictor near the boundaries of protein disorder. Applying a linear order/disorder predictor[29] to the set of proteins used to extract order/disorder boundaries, we estimated the degree of deviation of the predicted boundary from the true boundary. After predicting intrinsic disorder on all proteins and discarding outliers, we found that 27.0% of predicted disorder boundaries were within ±10 residues from the true boundary, 42.7% were within ±20, while 79.8% were within ±50 residues from the true boundary. Using the expectation-maximization algorithm[42] with 100 random starts and suitably chosen convergence parameters, we approximated these data with the Gaussian distribution with a mean $\mu = 0.9$ and standard deviation $\sigma = 40.9$ residues. The predicted boundaries were inside the true disordered regions 66.3% of the time, and outside of the disordered region 33.7% of the time. This coincides with the lower sensitivity of the linear predictor reported in the study of Vucetic *et al.*[29] Therefore, although the order/disorder predictor has high overall prediction accuracy, its performance drops significantly near the true order/disorder boundaries.

3.2 Prediction of boundaries between order and disorder

Using the procedure described in Methods, we trained and tested the performance of the three types of order/disorder boundary predictors. The first type of boundary predictor p_U was built using datasets $\mathbf{P_U}$ and $\mathbf{N}$. The second predictor p_B was built using $\mathbf{P_B}$ and $\mathbf{N}$, while the third predictor p_C combined the first two and its output values were the arithmetic mean of the soft predictions outputted from p_U and p_B. Since the linear predictor outputs soft values in the (0, 1) interval, the output values obtained after predictions were then quantized using a threshold of 0.5 and prediction accuracy was reported.

We report performance results of all three predictors for $I \in \{1, 5, 10, 30, 50\}$ and for different dimensionalities $m \in \{5, 10, 15, 20\}$, after the principal component analysis was applied. The results are shown in Table 2. Prediction results are reported with 95% confidence intervals.

4 Discussion

The primary goals of our research were to examine statistical properties of amino acid compositions around order/disorder boundaries and then to use these properties to design predictors of such boundaries. A major uncertainty in this research relates to the quality of the input data. In contrast to many helix/non-helix boundaries, a sharp boundary may not exist between ordered and disordered structure. A gradation in flexibility as the chain transitions between the ordered and the disordered states seems possible. Furthermore, the NMR and X-ray data used to assign ordered and disordered categories were each not interpreted by standardized protocols and the two methods could give different results on the same protein. All of these uncertainties could lead to variable boundary determination for essentially identical circumstances. Despite these uncertainties, the analysis performed in section 2 did reveal statistically significant differences in amino acid compositions in the boundary regions as compared to non-boundary ordered or disordered regions. As described in sections 2 and 3, the observed statistical differences enabled the development of order/disorder boundary predictors with accuracies between ~74-80%.

Table 2. *Percent sensitivity (sn), specificity (sp) and accuracy (ac) ± 95% confidence intervals for three order/disorder boundary predictors (p_U, p_B, p_C). Output dimensionality after PCA indicated by m. I is the number of random selections of 122 negative examples. Confidence intervals for specificity (sp) are lower than 0.001 in all cases and are denoted as 0.0 in the table.*

m	I	p_U			p_B			p_C		
		sn	sp	ac	sn	sp	ac	sn	sp	ac
5	1	68.7 ± 1.2	64.7 ± 0.0	66.7	71.7 ± 1.2	69.1 ± 0.0	70.4	73.7 ± 1.1	72.4 ± 0.0	73.1
	5	71.7 ± 0.8	68.9 ± 0.0	70.3	74.8 ± 0.9	71.5 ± 0.0	73.2	77.2 ± 0.8	75.5 ± 0.0	76.4
	10	72.7 ± 0.9	70.4 ± 0.0	71.6	76.3 ± 0.8	72.8 ± 0.0	74.6	78.4 ± 0.6	76.6 ± 0.0	77.5
	30	73.6 ± 0.9	72.0 ± 0.0	72.8	78.2 ± 0.5	73.8 ± 0.0	76.0	79.2 ± 0.6	77.7 ± 0.0	78.5
	50	73.1 ± 0.8	72.4 ± 0.0	72.8	78.4 ± 0.5	74.0 ± 0.0	76.2	79.5 ± 0.5	77.9 ± 0.0	78.7
10	1	71.3 ± 1.2	67.2 ± 0.0	69.3	73.6 ± 1.2	70.8 ± 0.0	72.2	75.6 ± 0.9	74.6 ± 0.0	75.1
	5	72.4 ± 1.1	70.5 ± 0.0	71.5	77.0 ± 0.6	72.8 ± 0.0	74.9	78.8 ± 0.8	77.1 ± 0.0	78.0
	10	73.0 ± 0.9	71.4 ± 0.0	72.2	77.9 ± 0.5	73.6 ± 0.0	75.8	79.2 ± 0.5	77.8 ± 0.0	78.5
	30	74.3 ± 0.6	72.6 ± 0.0	73.5	78.9 ± 0.6	74.3 ± 0.0	76.6	79.9 ± 0.7	78.6 ± 0.0	79.3
	50	74.6 ± 0.5	72.9 ± 0.0	73.8	79.4 ± 0.5	74.4 ± 0.0	76.9	80.0 ± 0.6	78.8 ± 0.0	79.4
15	1	70.7 ± 1.0	68.3 ± 0.0	69.5	74.9 ± 0.9	71.7 ± 0.0	73.3	76.9 ± 1.0	75.5 ± 0.0	76.2
	5	73.7 ± 1.0	71.0 ± 0.0	72.4	76.2 ± 0.7	73.4 ± 0.0	74.8	79.0 ± 0.8	77.6 ± 0.0	78.3
	10	75.2 ± 0.7	71.8 ± 0.0	73.5	77.3 ± 0.7	74.0 ± 0.0	75.7	79.4 ± 0.6	78.2 ± 0.0	78.8
	30	75.0 ± 0.6	72.7 ± 0.0	73.9	79.3 ± 0.6	74.5 ± 0.0	76.9	80.0 ± 0.5	78.9 ± 0.0	79.5
	50	75.3 ± 0.6	72.8 ± 0.0	74.1	79.7 ± 0.6	74.7 ± 0.0	77.2	80.2 ± 0.4	79.1 ± 0.0	79.7
20	1	71.0 ± 1.2	69.2 ± 0.0	70.1	73.6 ± 0.9	71.8 ± 0.0	72.7	77.2 ± 0.9	75.9 ± 0.0	76.6
	5	73.7 ± 0.9	71.5 ± 0.0	72.6	77.0 ± 0.8	73.4 ± 0.0	75.2	78.8 ± 0.7	77.9 ± 0.0	78.4
	10	75.0 ± 0.7	72.1 ± 0.0	73.6	78.1 ± 0.9	73.9 ± 0.0	76.0	78.8 ± 0.7	78.3 ± 0.0	78.6
	30	75.4 ± 0.7	72.7 ± 0.0	74.1	79.2 ± 0.6	74.4 ± 0.0	76.8	79.1 ± 0.4	78.9 ± 0.0	79.0
	50	75.2 ± 0.6	72.8 ± 0.0	74.0	79.5 ± 0.5	74.5 ± 0.0	77.0	79.5 ± 0.5	79.0 ± 0.0	79.3

Since balanced sets were used, for which 50% accuracy would be expected by chance, the substantially greater than 50% prediction accuracies argue that reasonably sharp order/disorder boundaries exist and also that protocols for order/disorder data interpretation are not too dissimilar using the same or different methods in different laboratories. Sharp order-disorder boundaries fit with the idea that protein folding is highly cooperative so that, for the most part, individual residues either have stable packing, which leads to order, or they don't have stable packing, which leads to disorder. Even structured regions of low stability typically exhibit two-state equilibria between order and disorder.

Identifying the determinants of transitions between ordered and disordered regions may provide improved understanding of protein structure. As mentioned in the introduction, our previous research identified numerous sequence characteristics that distinguish order and disorder. Here we report that there appear to be differences between internal regions of order or disorder and regions at the boundaries between order and disorder as discussed below.

Isoleucines, leucines, and valines occur significantly more often on the ordered side of boundaries and significantly less often on the disordered side (Table 1). Aromatic hydrophobic groups appear to be better than the aliphatic hydrophobic groups for the general stabilization of order[16, 17]; therefore it is all the more interesting that the aliphatics appear to be preferred over the aromatics on the ordered sides of the boundaries. Perhaps the smaller size of the aliphatics is an advantage near boundaries where burial within an ordered core is unlikely, and perhaps the complete absence of hydrogen bonding potential is especially important for preventing the incursion of water into the ordered side of the boundaries.

On the disordered side, there are significantly more glycines and asparagines (Table 1). In our previous studies, to our surprise, glycine was not indicated to be a strong promoter of disorder, perhaps because its configurational pliability helps to stabilize tight packing within ordered cores.[16, 17] Glycine's importance in disorder near boundaries might be for a similar reason: to help free the locally disordered segment from steric constraints, especially from constraints arising from the nearby ordered structure. As for the asparagine, this was the only hydrophilic amino acid for which disordered regions were depauperate rather than enriched compared to ordered regions; we speculated that asparagine might be disfavored in disordered regions because hydrogen bonding with the backbone would tend to increase the order of the chain.[16, 17] On the disordered side of boundaries, however, transient hydrogen bonding to the backbone could both stabilize the ends of ordered regions and help to disrupt backbone hydrogen bonding between ordered and disordered segments.

The small number of training segments that straddle order/disorder boundaries is a limitation of the current work. A larger set of order/disorder boundaries will allow the identification of different types of order/disorder boundaries, especially whether the boundary determinants depend on chain direction, that is, whether order/disorder boundaries are different from those for disorder/order. Thus, a continuing research goal is to identify more proteins with disordered regions. To this end, we are developing a website for the deposition of disordered protein information (http://disprot.wsu.edu).

Our disorder predictions provide support for the re-assessment of the current protein structure-function paradigm.[1,8] Combining order/disorder boundary prediction with standard disorder prediction has the potential for significantly improving the prediction accuracy, which in turn should be useful in helping to relate disorder to protein function.

Acknowledgements

NIH Grant 1R01 LM06916 awarded to AKD and ZO, NSF Grant CSE-IIS-9711532 awarded to ZO and AKD, and generous funding from Molecular Kinetics, Inc. are gratefully acknowledged. We also thank to Gary Daughdrill and Timothy R. O'Connor for important discussions and help in manuscript preparation.

References

1. Wright, P. E., and Dyson, H. J., *J. Mol. Biol., 293*, 321 (1999).
2. Dunker, A. K., Lawson, J. D., Brown, C. J., Williams, R. M., Romero, P., Oh, J. S., Oldfield, C. J., Campen, A. M., Ratliff, C. M., Hipps, K. W., Ausio, J., Nissen, M. S., Reeves, R., Kang, C., Kissinger, C. R., Bailey, R. W., Griswold, M. D., Chiu, W., Garner, E. C., and Obradovic, Z., *J. Mol. Graph. Model., 19*, 26 (2001).
3. Demchenko, A. P., *J. Mol. Recognit., 14*, 42 (2001).
4. Uversky, V. N., *Eur. J. Biochem., 269*, 2 (2002).
5. Ptitsyn, O. B., and Uversky, V. N., *FEBS Lett., 341*, 15 (1994).
6. Dunker, A. K., and Obradovic, Z., *Nat. Biotechnol., 19*, 805 (2001).
7. Weinreb, P. H., Zhen, W., Poon, A. W., Conway, K. A., and Lansbury, P. T., Jr., *Biochemistry, 35*, 13709 (1996).
8. Dyson, H. J., and Wright, P. E., *Curr. Opin. Struct. Biol., 12*, 54 (2002).
9. Bracken, C., *J. Mol. Graph. Model., 19*, 3 (2001).
10. Dunker, A. K., Obradovic, Z., Romero, P., Garner, E. C., and Brown, C. J., *Genome Inform., 11*, 161 (2000).
11. Dunker, A. K., Brown, C. J., Lawson, J. D., Iakoucheva, L. M., and Obradovic, Z., *Biochemistry, 41*, 6573 (2002).
12. Vihinen, M., Torkkila, E., and Riikonen, P., *Proteins, 19*, 141 (1994).
13. Romero, P., Obradovic, Z., and Dunker, A. K., *FEBS Lett ., 462*, 363 (1999).
14. Romero, P., Obradovic, Z., Li, X., Garner, E. C., Brown, C. J., and Dunker, A. K., *Proteins, 42*, 38 (2001).
15. Wootton, J. C., and Federhen, S., *Methods Enzymol ., 266*, 554 (1996).
16. Xie, Q., Arnold, G. E., Romero, P., Obradovic, Z., Garner, E., and Dunker, A. K., *Genome Inform., 9*, 193 (1998).
17. Williams, R. M., Obradovic, Z., Mathura, V., Braun, W., Garner, E. C., Young, J., Takayama, S., Brown, C. J., and Dunker, A. K., *Pac. Symp. Biocomput., 6*, 89 (2001).

18. Williams, R. J. P., *Biol. Rev. Camb. Philos. Soc., 54*, 389 (1979).
19. Uversky, V., Gillespie, J., and Fink, A., *Proteins, 41*, 415 (2000).
20. Anfinsen, C. B., *Science, 181*, 223 (1973).
21. Romero, P., Obradovic, Z., and Dunker, A. K., *Genome Inform., 8*, 110 (1997).
22. Romero, P., Obradovic, Z., Kissinger, C. R., Villafranca, J. E., and Dunker, A. K., *IEEE Int. Conf. Neural Netw., 1*, 90 (1997).
23. Li, X., Romero, P., Rani, M., Dunker, A. K., and Obradovic, Z., *Genome Inform., 10*, 30 (1999).
24. Li, X., Obradovic, Z., Brown, C. J., Garner, E. C., and Dunker, A. K., *Genome Inform., 11*, 172 (2000).
25. Davidson, R., and MacKinnon, J., *Estimation and Inference in Econometrics*, Oxford University Press, New York 1993.
26. Haykin, S., *Neural networks: a comprehensive foundation*, Prentice Hall, Upper Saddle River, N.J. 1999.
27. Breiman, L., *Mach. Learn., 24*, 123 (1996).
28. Schapire, R. E., in *MSRI Workshop on Nonlinear Estimation and Classification* (2002).
29. Vucetic, S., Radivojac, P., Obradovic, Z., Brown, C. J., and Dunker, A. K., in *IEEE Int. Conf. Neural Networks, 4*, 2718 (2001).
30. Presta, L. G., and Rose, G. D., *Science, 240*, 1632 (1988).
31. Richardson, J. S., and Richardson, D. C., *Science, 240*, 1648 (1988).
32. Blom, N., Hansen, J., Blaas, D., and Brunak, S., *Protein. Sci., 5*, 2203 (1996).
33. Blom, N., Gammeltoft, S., and Brunak, S., *J. Mol. Biol., 294*, 1351 (1999).
34. Berman, H. M., Westbrook, J., Feng, Z., Gilliland, G., Bhat, T. N., Weissig, H., Shindyalov, I. N., and Bourne, P. E., *Nucleic Acids Res., 28*, 235 (2000).
35. Smith, D. K., Radivojac, P., Obradovic, Z., Dunker, A. K., and Zhu, G., (Submitted).
36. Smith, T. F., and Waterman, M. S., *J. Mol. Biol., 147*, 195 (1981).
37. Henikoff, S., and Henikoff, J. G., *Proc. Natl. Acad. Sci. USA, 89*, 10915 (1992).
38. Vogt, G., Etzold, T., and Argos, P., *J. Mol. Biol., 249*, 816 (1995).
39. Romero, P., Obradovic, Z., and Dunker, A. K., *Artificial Intelligence Rev., 14*, 447 (2000).
40. Schneider, T. D., and Stephens, R. M., *Nucleic Acids Res., 18*, 6097 (1990).
41. Efron, B., and Tibshirani, R., *An introduction to the bootstrap*, Chapman & Hall, New York 1993.
42. McLachlan, G. J., and Peel, D., *Finite Mixture Models*, Wiley, New York 2000.

IDENTIFYING STRUCTURAL MOTIFS IN PROTEINS

ROHIT SINGH[a,b]

Accelrys Inc, San Diego

MITUL SAHA

Department of Mechanical Engineering, Stanford University

In biological macromolecules, structural patterns (motifs) are often repeated across different molecules. Detection of these common motifs in a new molecule can provide useful clues to the functional properties of such a molecule. We formulate the problem of identifying a given structural motif (pattern) in a target protein (example) and discuss the notion of complete matches vis-a-vis partial matches. We describe the precise error criterion that has to be minimized and also discuss different metrics for evaluating the quality of partial matches. Secondly, we present a new polynomial time algorithm for the problem of matching a given motif in a target protein. We also use the sequence and (if available) secondary structure information to annotate the different points in motif and the target protein, thus reducing the search space size. Our algorithm guarantees the detection of a perfect match, if present. Even otherwise, the algorithm computes very good matches. Unlike other methods, the error minimized by our algorithm directly translates to root mean square deviation (RMSD), the most commonly accepted metric for structure matching in biological macromolecules. The algorithm does not involve any preprocessing and is suitable for the detection of both small and large motifs in the target protein. We also present experiments exploring the quality of matches found by the algorithm. We examine its performance in matching (both full and partial) active sites in proteins.

1 Introduction

Before we can go very far in modeling and manipulating proteins *in-silico*, we need to develop quantitative measures of structural similarities between (parts of) two proteins. In the simplest case— comparing two conformations of the same protein— one usually calculates the root mean squared deviation (RMSD) between the two structures. Suppose, however, we would like to know if some part of the given protein matches a particular sub-structure. For example, finding a particular *active site* inside a protein can provide insights into the protein's function. Most of the current approaches for identification of active-sites rely on building sophisticated sequence-based models to build a 'consensus' representation of the active site [1,2]. These sequence based representations, however, are only an approximation to the underlying structural information. A method based on structure-based matching would be much more useful. The usefulness of such a technique is not restricted to small motifs ($\simeq$ 3-10 amino acids) only. Often, even larger sub-structures ($\geq$ 15 amino acids) are conserved across different proteins. These larger motifs often form *sub-domains* in a large protein. Proteins sharing such common sub-domains often share similar structural and functional properties [3,4].

The problem of finding a good match for a pattern (motif) in an example (protein) has two parts. The first part is finding the best match for the pattern in the example.

---

[a]This work was done while the author was at the Department of Computer Science, Stanford University

[b]Corresponding Author: `rohitsi@cs.stanford.edu`

The second part is to evaluate if this match is significant enough: e.g., finding a close match for a pattern of 10 amino acids is more significant than finding a close match for a pattern for 2 amino acids. The process of evaluating the quality of a match, however, is somewhat subjective and depends on the goal of the biologist. There has only been a little work on this subject[5,6]. In the rest of this chapter, we shall restrict our focus to the first sub-problem: finding the best solution for the given input. We first formulate the problem of finding an optimal match for a pattern in a given example. We then discuss a method for solving the problem and evaluate it.

2 Problem Description

First, we introduce the notion of a *multipoint*.

Definition 1 *A multipoint* $\mathbf{a} \in \mathcal{M}$ *is an abstraction of a collection of points in* $\Re^3$ *defined as* $\mathbf{a} = \langle \vec{p}, \langle \hat{u}, \hat{v}, \hat{w} \rangle \rangle$ *where* $\vec{p}, \hat{u}, \hat{v}, \hat{w} \in \Re^3$*;* $\langle \hat{u}, \hat{v}, \hat{w} \rangle$ *define a right-handed reference frame with its origin at* $\vec{p}$ *(also referred to as the* anchor*);* $\mathcal{M}$ *is set of all multipoints and* $\Re$ *is the set of real numbers. Every multipoint* $\mathbf{a}$ *has a label* $l_{\mathbf{a}} \in \mathcal{L}$ *where* $\mathcal{L}$ *is the set of all possible labels. Let* $\mathcal{T}$ *be the set of all rigid-body transformations (rotations and/or translations) in* $\Re^3$*. Applying* $T \in \mathcal{T}$ *on* $\mathbf{a}$ *is the same as applying* T *on the points which* $\mathbf{a}$ *represents:* $T(\mathbf{a}) = \mathbf{a}' = \langle T(\vec{p}), \langle T(\hat{u}), T(\hat{v}), T(\hat{w}) \rangle \rangle$.

Definition 2 *With a multipoint* $\mathbf{a} \in \mathcal{M}$*, a* **distance measure** *,* $D_{\mathbf{a}}(\mathbf{b})$*, can be defined as providing its distance from some other multipoint* $\mathbf{b}$ *which has the <u>same label</u> as* $\mathbf{a}$*. The calculation of this distance may depend on the internal representation of the two multipoints. The distance measure associated with* $\mathbf{a}$*, if any, should be preserved under transformations i.e.,* $D_{\mathbf{a}}(\mathbf{b}) = D_{T(\mathbf{a})}(T(\mathbf{b}))$

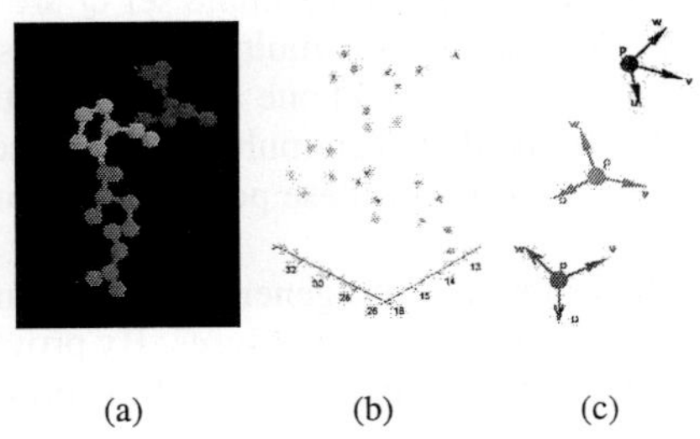

(a)   (b)   (c)

Figure 1: **Multipoints**
The above figures show how a multipoint would typically be constructed. (a) a protein structure (b) can be broken into a collection of 'significant' points. For each group of points that belong to one feature, (c) we can construct an anchor and a reference frame.

A *pattern set* $P \subset \mathcal{M}, |P| = m$, is a set of m multipoints. Each multipoint $\mathbf{p_i} \in P$ has an associated distance measure $D_{\mathbf{p_i}}()$. An *example set* $Q \subset \mathcal{M}$, where $|Q| = n$ and $n \geq m$, is a set of n multipoints. Given P and Q, we can define the set of possible correspondences between their multipoints:

$$\mathcal{C}_{PQ} = \{\langle r_1, r_2, \ldots, r_m \rangle | r_i \in \{1 \ldots n\}; \ r_i \neq r_j \forall i, j; \ and \ label(\mathbf{p_i}) = label(\mathbf{q}_{r_i})\}$$

Notation: Henceforth, we shall use $X[i]$ to refer to the ith element of the ordered set X.

Each correspondence $C \in \mathcal{C}_{PQ}$ induces an *optimal* rigid-body transformation T_C such that

$$T_C = \operatorname*{argmin}_{T \in \mathcal{T}} \sum_{k=1}^{m} D_{T(\mathbf{p_i})}(\mathbf{q}_{C[i]}) \tag{1}$$

We can now state the matching problem as follows:

Problem 1 (MATCH) *Given the above definitions, find the correspondence $C^\star$ and the optimal transformation induced by it, $T^\star$, such that*

$$\langle C^\star, T^\star \rangle = \operatorname*{argmin}_{C \in \mathcal{C}_{PQ}, T \in \mathcal{T}} \sum_{i=1}^{m} D_{T(\mathbf{p_i})}(\mathbf{q}_{C[i]}) \tag{2}$$

Intuitively, a multipoint can be thought of as a model of a collection of points (e.g. atoms in an amino acid) that behaves as a rigid body i.e., the relative orientation of the points stays the same. Labels capture the intuition that an amino acid of feature X can only match another acid of the same feature. In simple cases, the feature could just indicate the residue type or the hydrophobicity of the amino acid. For larger motifs, the label could indicate the secondary structure (helix, loop, or strand) the amino acid is part of. Features can be aggregated ($\alpha = \{X, Y, Z\}$) for greater flexibility (α matches X, Y, or Z).

Thus, we have formulated MATCH as the problem of finding the optimal correspondence (and alignment) of the multipoints in the pattern set to the multipoints in the example set. The pattern set P and the example set Q are abstractions of a motif and a protein, respectively. Typically, each multipoint models one amino acid. This abstraction might depend only on the backbone atoms or on the side-chain atoms as well. Henceforth, we shall assume that each multipoint represents a list of points and the number (and the relative ordering) of these points is the same for two multipoints if they have the same label.

As it stands, however, the problem is too general. We need more information about how to model multipoints and their distance measures. By providing more information about these, we can formulate different special cases of the problem, each of which has a special biological relevance.

Problem 2 (COMPLETE-MATCH) *Solve MATCH under the following constraint: given any two multipoints $\mathbf{a}$ and $\mathbf{b}$, both of which represent sets of k points each, the distance between them is*

$$D_{\mathbf{a}}^{\circ}(\mathbf{b}) = \sum_{i=0}^{k} \|\mathbf{a}(i) - \mathbf{b}(i)\|^2$$

where $\mathbf{a}(i)$ is the ith point in the list which the multipoint $\mathbf{a}$ represents and $\|x - y\|$ is the Euclidean distance between $x, y \in \Re^3$.

Problem 3 (PARTIAL-MATCH) *Solve MATCH under the following constraint: given any two multipoints $\mathbf{a}$ and $\mathbf{b}$, both of which represent sets of k points each, the distance between them is of the form*

$$D_{\mathbf{a}}^{+}(\mathbf{b}) = \sum_{i=0}^{k} \rho(\|\mathbf{a}(i) - \mathbf{b}(i)\|)$$

where $\rho(x)$ is a monotonically non-decreasing function such that $0 \leq \frac{d\rho}{dx} \leq x$, and ρ is the same across all multipoints with a common label.

COMPLETE-MATCH abstracts the problem of looking for *well-preserved* matches of a given active site or sub-domain in a protein. PARTIAL-MATCH, on the other hand, would be of interest when we expect that the matching region (e.g., active site) may not be well-preserved. The distance measure used in COMPLETE-MATCH is equivalent to the Root Mean Squared Distance (RMSD) criteria. Under this distance criterion, it is relatively easy to find the optimal transformation $T^\star$ that minimizes $D^o_{T(\mathbf{a})}(\mathbf{b})$[7,8]. The problem with this metric is that the influence of each pairwise distance is quadratic in the size of the distance. This creates problems when we expect a partial match between the two point-sets. In PARTIAL-MATCH, we use an *influence function, $\rho(x)$*, designed so that the larger inter-point distances do not overshadow the smaller ones. This idea is borrowed from the notion of *M-estimators* which are often used in Statistics and Computer Vision for robust estimation. The choice of a particular ρ depends on the situation and the biological motivation. For example, we might only be interested in matches where the distance between two matching points is less than a given threshold. The Tukey estimator captures that intuition. Similarly, this formulation can also be used in the case where the structure of the pattern (e.g., active site) is ambiguous.

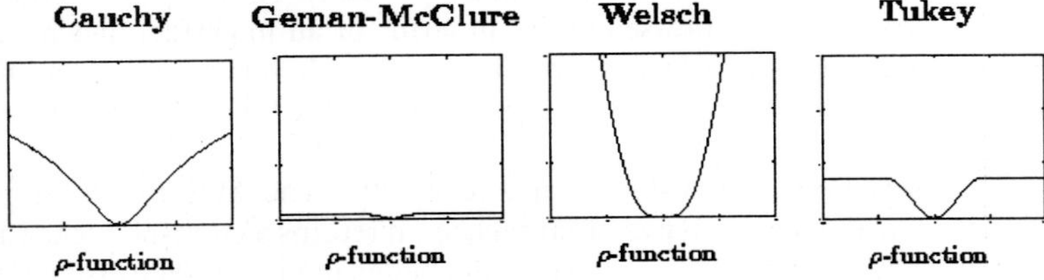

Figure 2: **M-estimators**

The above figure shows the plots of 4 important M-estimators: (a) Cauchy: $\rho(x) = \frac{c^2}{2} log(1 + (x/c)^2)$, (b) German-McClure: $\rho(x) = \frac{x^2/2}{1+x^2}$, (c) Welsch: $\rho(x) = c^2[1 - exp(-(x/c)^2)]$, (d) Tukey: $\rho(x) = \frac{c^2}{6}(1 - [1 - (x/c)^2]^3)$; where c is a constant.

2.1　Previous Work

The problem of identifying motifs in a protein has strong parallels to the object recognition problem in computer vision. Before we proceed, it is worth noting that many of the methods mentioned below rely on a well-known technique for the finding the optimal transformations— rotations and translations— for aligning two sets of n points i.e. the rotation and translation that result in the least squared-error in the alignment of the two point sets. The method, initially proposed by Faugeras[7] and Horn[8], assumes that the correspondences between the points in the two sets are known and boils down to finding the largest eigenvalue of a 4×4 matrix.

Kleywegt[9] et al. have released programs (SPASM, RIGOR) that look for small motifs (e.g., active sites) in a given protein using a brute-force approach. Their method, based on the idea that inter-point distances are invariant w.r.t. rotation and translation, performs an *exhaustive* search of sets of matching inter-point distances in the two point-sets. Their method can take advantage of the labeling information available with the motif and the protein and use it to prune the search space (e.g. a carbon atom should only match another carbon atom). Exhaustive searching, however, limits the size of motifs this method can be easily applied to.

Wolfson and Nussinov [10] proposed a technique (*Geometric Hashing*) also based upon the idea that distances are invariant under rotations and translations. An ordered triplet of points can be used to define a reference frame. For each ordered triplet of points in the point-set, coordinates of the other points (in the ref frame defined by the triplet) are stored in a hash-table. For points from the matching region, the key for the protein's hash-table will also be found in the motif's hash-table (and vice-versa) so that the appropriate transformation between the two ref frames (for the two triplets) can be found. The most popular transformation, across all triplets, wins.

Geometric hashing is a powerful approach, but it has the same kind of problem as the approach of Kleywegt et al: it is hard to establish a link between the output of this algorithm and the desired optimum because they handle the matching problem in an indirect fashion. When two sets of n points match perfectly, the $\binom{n}{2}$ pairwise distances corresponding to each set will also match perfectly. However, when the match is imperfect, it can not be expressed easily in terms of an imperfect match between the corresponding pair-wise distances (see Fig 3). As such, it is not necessary that the final answer returned by this approach will be optimal in the LSE (least sum of squared errors) sense.

Iterative Closest Point Algorithm, proposed by Besl and McKay [11], computes a locally optimum match between two point sets i.e., it returns a correspondence and the induced optimal transformation between two unlabelled point sets. The distance metric is the same as the RMSD metric or the one used in COMPLETE-MATCH. The basic idea is that at any point, the (currently) best estimate of the optimal transformation can be used to improve the (currently) best estimate of the optimal set of correspondences and vice-versa (see Fig 4). By iterating over these operations, the method converges to a locally optimal solution. The algorithm is guaranteed to converge because in each step of the iteration, the least-mean-squared-error can only decrease. However, the minima which it reaches may well be a local minima and the algorithm is highly dependent on the starting placement for the pattern set w.r.t. to the example set.

3   Method

Here are some of the observations that guided our approach:

- In ICP, if we replace the Euclidean distance function by any other monotonically non-decreasing function of the Euclidean distance (see the defn of PARTIAL-MATCH), the algorithm is still guaranteed to converge to a local optimum because each step in each iteration of the algorithm can only reduce the error. Of course, the optimal transformation to map the pattern set to the example set must be optimal with respect to the specific distance measure and the methods of Faugeras et al may not be

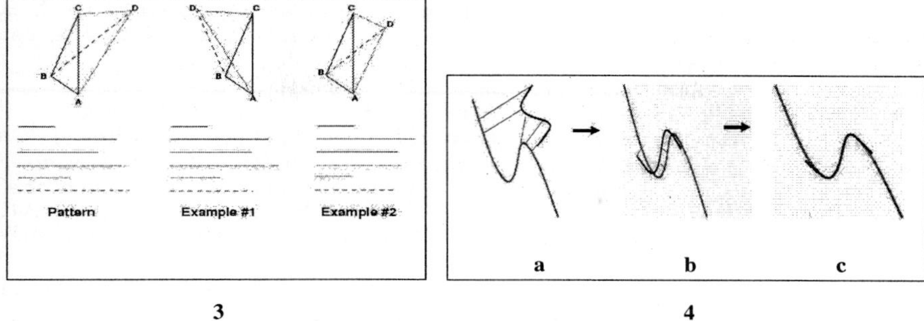

3   4

Figure 3: Matching Pairwise Distances Is Not Always a Good Idea

The above figure shows a typical case where matching pairwise distances can lead to a sub-optimal result. Example #2 is more similar to the pattern than Example #1. However, only one line (BD) in Example #2 does not have the same length as the corresponding line in the pattern. In Example #2 there are three such lines (AD, BD, CD). Thus, Kleywegt's algorithm would choose Example #1 over Example #2.

Figure 4: ICP

The goal is to align the pattern (short) curve with the example (long) curve. (a): For each point of the pattern (short curve) set, find the closest point on the example (long curve) set and build the list of corresponding points. (b): Find the optimal transformation that aligns these corresponding points. (c): After many iterations the curves have been optimally aligned.

applicable.

- The quality of the final match returned by ICP depends on the initial placement of the pattern relative to the example. Suppose that we can identify, in the example, a few (small) *regions of interest* one of which, with high probability, also contains the optimal match. For each such region, we can *seed* ICP by placing the pattern near this region of the example.
- In our formulation, only multipoints of the same label can be matched. The proper choice of label for this purpose can help us in reaching the optimum quickly by reducing the size of the search space. The frequency of occurrence of different features among the set of amino acids is often uneven. For example, if one were to label amino acids just by their residue type, the most frequent amino acid, Leucine, is 6.85 times more abundant than the least frequent amino acid, Tryptophan [12].

Our method to solve MATCH is described in Algorithm 1. The initial step of our method consists of identifying regions of interest by using a small set of multipoints from the pattern as *pivots*. We find sets of multipoints from the example such that these multipoints could correspond to the pivoting multipoints in an optimal match. For each such possible correspondence, we transform the pattern so that the pivoting multipoints are aligned with their (guessed) counterparts. We then use an ICP-like algorithm to find the best match in that region. Two functions in the algorithm need further elaborations:

RegionsOfInterest This function returns the regions in the example set, Q, where a

Input: Given a pattern $P = \{\mathbf{p_1}, \mathbf{p_2}, \ldots, \mathbf{p_m}\}$ and an example $Q = \{\mathbf{q_1}, \mathbf{q_2}, \ldots, \mathbf{q_n}\}$ where $\mathbf{p_i}, \mathbf{q_j} \in \mathcal{M}$ and $m \leq n$. Recall that $\mathcal{M}$ is the set of multipoints in $\Re^3$

Goal: Define the set of possible correspondences between P and Q: $\mathcal{C}_{PQ} = \{\langle r_1, r_2, \ldots, r_m \rangle \mid r_i \in \{1 \ldots n\},\ r_i \neq r_j \forall i, j\ and\ label(\mathbf{p_i}) = label(\mathbf{q}_{r_i})\}$. A transformation T is a rotation and translation in $\Re^3$. Since this operation is analogous for both points and multipoints, we shall use the symbol T for both the meanings, i.e., $T: \Re^3 \to \Re^3$ and also, $T: \mathcal{M} \to \mathcal{M}$. Find the optimal correspondence $C^\star \in \mathcal{C}_{PQ}$ and the corresponding transformation $T^\star$ such that

$$\langle C^\star, T^\star \rangle = \operatorname*{argmin}_{\langle C, T \rangle} \sum_{i=1}^{m} D_{T(\mathbf{p_i})}(\mathbf{q}_{C(i)})$$

Algorithm:

1. Generate seeds: choose some multipoints $\mathbf{p}_\alpha, \mathbf{p}_\beta, \ldots \in P$ such that $RegionsOfInterest((\mathbf{p}_\alpha, \mathbf{p}_\beta, \ldots), Q) = S$ returns only a few sets of possible matches.
2. Initialize $d_{best} = \infty$
3. For each $(\mathbf{q}'_\alpha, \mathbf{q}'_\beta, \ldots) \in S$, do

    (a) Find Initial Transform: $T_1 = OptimalTransform((\mathbf{p}_\alpha, \mathbf{p}_\beta, \ldots), (\mathbf{q}'_\alpha, \mathbf{q}'_\beta, \ldots))$

    (b) Initialize: $P_1[i] = T_1(P[i]), i = 1, \ldots, m$.

    (c) Initialize: $r = 1, d_0 = \infty, d_1 = LARGE\text{-}NUMBER$

    (d) While ($r < MAX\text{-}ITERATIONS$ and $d_r < d_{r-1}$)

        i. Set $r = r + 1$

        ii. Find correspondences: set correspondences so that each multipoint in P_r corresponds to the closest multipoint in Q:

    $$C_r[i] = \operatorname*{argmin}_{y \in \{1, \ldots, n\}} D_{P_{r-1}[i]}(Q[y])$$

    where $X[i]$ is the ith multipoint in ordered set X

        iii. Find the optimal transformation for these correspondences: $T_r = OptimalTransformation((P[1], P[2], \ldots, P[m]), (Q[C_r[1]], Q[C_r[2]], \ldots, Q[C_r[m]]))$

        iv. Set $P_r[i] = T_r(P[i]),\ i = 1, \ldots, m$

        v.

    $$d_r = \sum_{i=1}^{m} D_{P_r[i]}(Q[C_r[i]])$$

    (e) Update: if $d_{best} > d_r$ then set $C^\star = C_r$, $T^\star = T_r$, $d_{best} = d_r$

4. Return $C^\star, T^\star$

Algorithm 1 *General algorithm for solving MATCH*

globally optimum match can be found. It takes as input a set of *pivoting* multipoints in the pattern P and returns a list of possibly optimal correspondences that this set can have in Q. This function can be implemented by choosing some multipoints from the pattern and looking, in the example set, for sets of multipoints whose labels and relative orientations are consistent with those of the chosen multipoints from the pattern. Our aim is to get only a few regions of interest. As the simplest choice, one could pick just one pivoting multipoint from the pattern and, from the example, choose all multipoints that have the same label as this multipoint. In a more sophisticated approach, one can pick a collection of multipoints from the pattern. The matching constraints would also be based on the relative orientation of these multipoints. Such constraints can be efficiently implemented, say, by using *kd*-trees.

OptimalTransform Given a set of corresponding multipoints $X = \{\mathbf{x_1}, \mathbf{x_2}, \ldots, \mathbf{x_k}\}$ and $Y = \{\mathbf{y_1}, \mathbf{y_2}, \ldots, \mathbf{y_k}\}$, this function returns the transformation $T^\star$ that results in the minimum total error:

$$T^\star = \underset{T}{\operatorname{argmin}} \sum_{i=1}^{k} D_{T(\mathbf{x_i})}(\mathbf{y_i})$$

In the case of COMPLETE-MATCH, the distance function is just the sum of the squares of the Euclidean distances between the points making up the two multi-points. As mentioned before, there are well known methods that can compute the optimal transformation (in a least squared error sense) such that final error is minimum[7,8]. In the case of PARTIAL-MATCH, our method is based on the following observation: the fraction of 'bad' data-points (or outliers), w, is expected to be low— otherwise the match won't have any biological significance. On picking a random sample from the data, there is a low probability that some the points in the sample will not be 'good'. This probability can be reduced further by taking samples many times. Then we can use the method by Faugeras and Horn,[7,8], to calculate an optimal transformation for this subset, *using the Euclidean distance measure*. Since our distance function gives lower weight to the bad data-points, this transformation should be a pretty good choice for the whole data too. Our method is similar to the RANSAC[13] method and the method of Venkatsubramanian et al[14,15]. Due to a lack of space, we can not provide a formal specification of our algorithm in this paper.

4   Analysis

The speed of Algorithm 1 depends on our ability to generate a small list of regions of interest in the example (protein) while ensuring that the global optimum will be one of these. There is a trade-off between the complexity of this step and the quality of its results. Still, the time it takes will, typically, be much smaller than the time taken by the rest of the algorithm.

It should be clear that if a perfect match (exact replica of the pattern in the example) exists, Algorithm 1 will always detect it– one of the regions of interest will result in a perfect alignment of the pattern and the example.

In the second part of the algorithm, we iterate over each of the regions of interest discovered earlier and start our search by aligning the pivoting multipoints. The basic intuition is that once we have placed the pattern near the region of interest in the example, the algorithm can take us to the locally best match which could also be the global minimum. Of course, the algorithm might still go wrong and stray towards another local optimum even if the global optimum was indeed present in that region. We shall now try to provide some intuition for why something like this is unlikely i.e., once the algorithm has gotten on the trail towards the optimal match, it is unlikely that it will stray.

We analyze the case corresponding to COMPLETE-MATCH. Given a pattern set $\mathcal{U} \subset \Re^3$, we construct an example set $\mathcal{V} \subset \Re^3$ as follows: first, we apply a random rotation and translation to $\mathcal{U}$ and randomly add points around it such that the probability of a point being present in any given unit volume is γ. This is the test set $\mathcal{V}$. γ indicates the density of points in the test set– if the test set is densely packed, there should be a greater chance for any method to make mistakes, i.e., infer incorrect correspondences between points of the pattern set and those of the test set, even if it is close to the global optimum. Also, for notational convenience, we shall use $\mathcal{U}' \subset \mathcal{V}$ to refer to the subset of points which exactly match $\mathcal{U}$, i.e., we want the algorithm to output $\mathcal{U}[i] \leftrightarrow \mathcal{U}'[i]$ as the final correspondences. We also introduce the term β which is the ratio of the largest inter-point distance in $\mathcal{U}$ to the smallest inter-point distance in $\mathcal{U}$. β captures the shape of the pattern: skinny, cylindrical patterns will have a larger β than fat, cuboidal patterns.

Lemma 1 *If, at the end of Step   3.(d).iii in Algorithm 1, T_r maps at least 3 points from $\mathcal{U}$ within an ϵ-neighborhood of their correct matches i.e. $\|T_r(\mathcal{U}[s_i]) - \mathcal{U}'[s_i]\| < \epsilon$, $i = 1, 2, 3$, then the probability that, for any point $\mathcal{U}[i] \in \mathcal{U}$, $T_r(\mathcal{U}[i])$ is not the closest point to $\mathcal{U}'[i]$ is less than $\frac{4}{3}\pi\gamma(\beta\epsilon)^3$*

Proof: Look at the Fig 5.

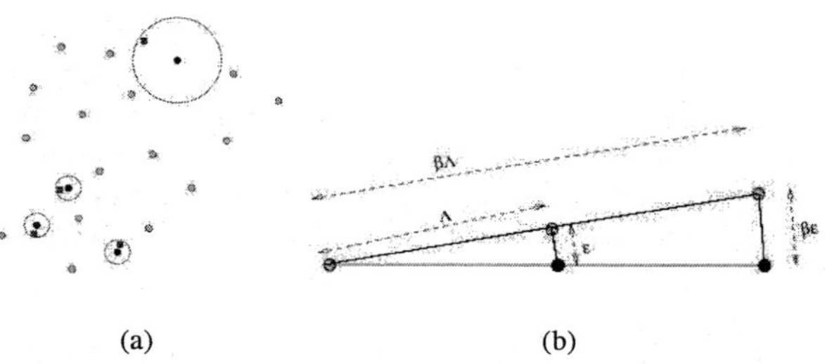

(a)   (b)

Figure 5: **Algorithm 1 Won't Stray Too Often**

The filled squares represent points belonging to $\mathcal{U}$. The filled circles represent points belonging to $\mathcal{V}$. Of those, the lighter circles are randomly selected and the darker circles belong to $\mathcal{U}'$. If at least 3 points in $\mathcal{U}$ are within ϵ of the matching points from $\mathcal{U}'$, then any other point can only be $\beta\epsilon$ away from its corresponding point in $\mathcal{U}'$.

The probability that one or more randomly chosen points will be present within that neighborhood is less than $\frac{4}{3}\pi\gamma(\beta\epsilon)^3$ ∎

5 Results

To test the quality of the matches returned by our algorithm (for COMPLETE-MATCH), we downloaded two sets of protein structures from the Protein Data Bank . We chose about 42 different variants of the protease trypsin and about 37 different types of kinases. Choosing the trypsin (consensus) active site[16] (3 amino acids) as our pattern, we ran our matching algorithm against the trypsins and then against the kinases.The results of our experiments are summarized in the plots shown below (Fig 6). For 32 of the 42 trypsin-like molecules, we were able to achieve matches with RMSDs less than 0.5 Å, indicating a near perfect match. In some other cases, the structural information in the PDB was incomplete and hence a good match could not be found. Finally, in the remaining 3 cases, our algorithm had not converged on to the optimal match.

When we tried to match the trypsin active site against the kinases, we expected to get low-quality matches. Most of the matches returned alignments where the RMSD was more than 2 Å– a low score considering that active site of trypsin is relatively small and hence a rough match for it can be found in many proteins. The interesting observation was that there were some kinases in which we could obtain really high quality matches. Some of these turned out to be buried inside the protein and hence could not be an active site. Such spurious matches, however, are a problem faced by almost all structure based techniques. Finally, there were 2 kinase-type molecules which matched the trypsin active site on their surface. It would be an interesting biological problem to determine if these kinases share any functional similarity with trypsins.

To evaluate **partial matches**, we distorted the trypsin active site by displacing two of the amino acids and then changing their orientation randomly. One of these amino acids was given a large displacement ($\simeq 8\mathring{A}$) while the other was given a smaller displacement ($\simeq 1.5\mathring{A}$). Our aim was to find an alignment that would *ignore the most outlying amino acid* and align (a part of) the protein with the rest of the motif. First, we used the same distance measure as in COMPLETE-MATCH i.e. the *Least-Sum-of-Squared-Error* criterion. We then replaced the the distance measure with one of the M-estimator (we used the Tukey Estimator) based criteria (PARTIAL-MATCH) and ran our algorithm. In one set of experiments, we used a gradient descent based approach to find the best transformation that minimized the error (in *OptimalTransform()*). In the other set of experiments, we used the method mentioned before (based on repeated sampling of a subset of points from the set). Fig 7 summarizes the results of a comparison of these three implementations with a distorted active site.

6 Conclusion

In this paper, we have formulated the problem of detecting motifs in proteins and have presented a general method for it. Our formulation of the problem provides for a rigorous measure of the best fit between a given pattern and an example. At the same time, there is a certain flexibility in the choice of an appropriate distance measure— biological context of the particular problem instance will be important in

238

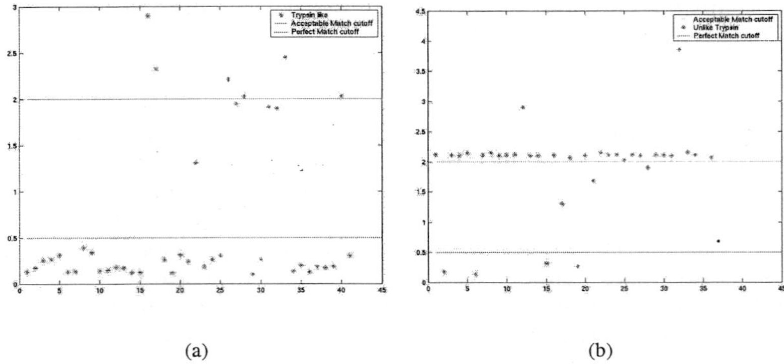

(a)   (b)

Figure 6: **COMPLETE-MATCH Results**

(a) Trypsin-like Proteins: **Y axis**: RMSDs between motif (trypsin active site) and (matching part of protein) in $\overset{\circ}{A}$. **X axis**: 42 proteins belonging to the Trypsin family. Observe that the quality of the match is near-perfect for most of the proteins. This confirms that all these proteins share the same structural and functional properties. (b) Kinase-like Proteins: **Y axis**: RMSDs between motif (trypsin active site) and (matching part of protein) in $\overset{\circ}{A}$. **X axis**: 37 proteins belonging to the Kinase family. Observe that most molecules don't have good matches.

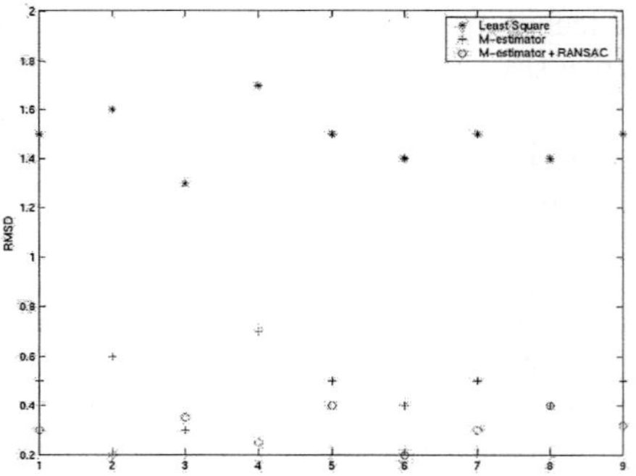

Figure 7: **Comparison of partial matching methods**

Y axis: RMSD between the aligned pattern and the matching sub-structure in the protein, *after removing the outlier pair from the match*. The pattern is a distorted version of the trypsin active site. **X axis** 9 different proteins from the Trypsin family. Here we compare the minimum alignment error after excluding the 'bad' amino acid. Ideally, this should be close to zero. Using the normal Least Square Error criterion gives the poorest performance. Using M-estimator based error criteria and gradient-descent based optimization technique, we can get better performance. If we use the RANSAC-like method mentioned before for finding the optimal transforms, we get the best results.

choosing this. Our method for solving the matching problem is fast and we also provided some intuition for why its results should be near-optimal most of the time. The algorithm, as presented, leaves significant scope in the choice of various parameters.

For example, the appropriate choice of features for labeling purposes will depend on the pattern/example at hand. In this context, we are in the process of conducting experiments to ascertain the performance of our algorithm under different choices of features. We are also testing our algorithm on much bigger data-sets and motifs of varying sizes. Another area that needs further work is related to finding the optimal transformation under PARTIAL-MATCH conditions. Though our current algorithm for *OptimalTransform( )* works well, there is scope for improvement.

References

[1] K. Hofmann, P. Bucher, L. Falquet, A. Bairoch *The PROSITE database, its status in 1999* Nucleic Acids Res. 27:215-219, 1999

[2] A. Bateman, E. Birney, L. Cerruti, R. Durbin, L. Etwiller, S.R. Eddy, S. Griffiths-Jones, K.L. Howe, M. Marshall, and E.L.L. Sonnhammer *The Pfam protein families database* Nucleic Acids Research, 30(1):276-280, 2002

[3] L Lo Conte et al.*SCOP database in 2002: refinements accommodate structural genomics* Nucl. Acid Res. 30(1), 264-267, 2002.

[4] F.M.H. Pearl, et al. *Assigning genomic sequences to CATH* Nucleic Acids Research. Vol 28. No 1. 277-282, 2002

[5] M. Levitt and M. Gerstein *A Unified Statistical Framework for Sequence Comparison and Structure Comparison* Proc. Natl. Acad. Sci., 95, 5913-5920, 1998

[6] P. Bradley, P. S. Kim, B. Berger *Trilogy: Discovery of Sequence-Structure Patterns Across Diverse Proteins* International Conference on Research in Computational Biology (RE-COMB), 2002

[7] O. D. Faugeras and M. Hebert, *The representation, recognition, and locating of 3-D objects*, Int. J. Robotic res. vol. 5, no. 3, pp. 27-52, Fall 1986.

[8] B. K. P. Horn, *Closed-form solution of absolute orientation using unit quaternions*, J. opt. Soc. Amer. A vol. 4, no. 4, pp. 629-642, Apr. 1987.

[9] GJ Kleyweg *Recognition of spatial motifs in protein structures* J Mol Biol. 29;285(4):1887-97, Jan 1999

[10] R. Nussinov, H.J. Wolfson *Efficient Detection of Three - Dimensional Motifs In Biological Macromolecules by Computer Vision Techniques*, Proceedings of the National Academy of Sciences, U.S.A., 88, 10495-10499, 1991

[11] P.J. Besl and N.D. Mckay, *A Method for Registration of 3-D Shapes*, IEEE Transactions on PAMI, Vol. 14. No.2, Feb. 1992.

[12] CRC Handbook of Chemistry and Physics (ISBN 0-8493-0458-X, CRC Press, Inc., Cleveland, Ohio)

[13] W. Forstner *Robust Estimation Procedures in Computer Vision* In Third Course in Digital Photogrammetry, 1998

[14] S. Venkatasubramanian *Geometric Shape Matching and Drug Design* Ph.D. Thesis, Stanford University, 1999

[15] P. Finn, L. Kavraki, R. Motwani, J.C. Latombe, C. Shelton, S. Venkatasubramanian and A. Yao *RAPID: Randomized Pharmacophore Identification in Drug Design* Proc. 13th ACM Symposium on Computational Geometry, 1997

[16] SF Russo, DN Morris *A fluorescent probe for the active site of bovine trypsin* Physiol Chem Phys Med NMR. 1983;15(3):223-7.

A PATH PLANNING-BASED STUDY OF PROTEIN FOLDING WITH A CASE STUDY OF HAIRPIN FORMATION IN PROTEIN G AND L

GUANG SONG,[†] SHAWNA THOMAS,[†] KEN A. DILL,[‡] J. MARTIN SCHOLTZ,[§] NANCY M. AMATO[†]

We investigate a novel approach for studying protein folding that has evolved from robotics motion planning techniques called *probabilistic roadmap* methods (PRMs). Our focus is to study issues related to the folding process, such as the formation of secondary and tertiary structure, *assuming* we know the native fold. A feature of our PRM-based framework is that the large sets of folding pathways in the roadmaps it produces, in a few hours on a desktop PC, provide global information about the protein's energy landscape. This is an advantage over other simulation methods such as molecular dynamics or Monte Carlo methods which require more computation and produce only a single trajectory in each run. In our initial studies, we obtained encouraging results for several small proteins. In this paper, we investigate more sophisticated techniques for analyzing the folding pathways in our roadmaps. In addition to more formally revalidating our previous results, we present a case study showing our technique captures known folding differences between the structurally similar proteins G and L.

1 Introduction

There are large and ongoing research efforts whose goal is to determine the native structure of a protein from its amino acid sequence.[1,2] A protein's 3D structure is important because it affects the protein's function. In this work, we assume the native structure is known, and our focus is on the study of protein folding mechanisms. That is, instead of performing fold prediction, we aim to study issues related to the folding process, such as the formation of secondary and tertiary structure, and the dependence of the folding pathway on the initial denatured conformation. Such questions have taken on increased practical significance with the realization that mis-folded or only partially folded proteins are associated with many devastating diseases.[3] Moreover, increased knowledge of folding mechanisms may provide insight for protein structure prediction. Despite intensive efforts by experimentalists and theorists, there are major gaps in our understanding of the behavior and mechanism of the folding process.

[†]{gsong,sthomas,amato}@cs.tamu.edu Department of Computer Science, Texas A&M University, College Station, TX 77843-3112, USA.

[‡]dill@maxwell.ucsf.edu. Department of Pharmaceutical Chemistry, University of California, San Francisco, CA 94143-1204.

[§]jm-scholtz@tamu.edu. Department of Medical Biochemistry & Genetics, Texas A&M University, College Station, TX 77843-1114, USA.

Table 1: A comparison of protein folding models.

Comparison of Models for Protein Folding					
Approach	**Folding Landscape**	**#Paths Produced**	**Path Quality**	**Compute Time**	**Need Native**
Molecular Dynamics	No	1	Good	Long	No
Monte Carlo	No	1	Good	Long	No
Statistical Model	Yes	0	N/A	Fast	Yes
PRM-Based	Yes	Many	Approx	Fast	Yes
Lattice Model	Not used on real proteins				

In previous work,[4] we proposed a technique for computing protein folding pathways that is based on the successful *probabilistic roadmap* (PRM)[5] method for robotics motion planning. We were inspired to apply this technique to protein folding based on our success in applying it to folding problems such as carton folding and paper crafts.[6] We obtained promising results for small proteins ($\sim$60 amino acids) and validated our pathways by comparing the secondary structure formation order with known experimental results.[7]

A major feature of our PRM-based framework is that in a few hours on a desktop PC it produces roadmaps containing large sets of unrelated folding pathways that provide global information about the protein's energy landscape. In this paper, we investigate more sophisticated techniques for analyzing the pathways in our roadmaps. In addition to more formally revalidating our previous results, we present a case study showing our technique captures known folding differences between the structurally similar proteins G and L.

1.1 Comparison to Related Work

Table 1 provides a summary comparison of various models for protein folding. Both Monte Carlo simulation and molecular dynamics provide a single, usually high quality, folding trajectory. Each run is computationally intensive because they attempt to simulate complex kinetics and thermodynamics. Statistical mechanical models, while computationally efficient, assume extremely simplified molecular interactions and are limited to studying global averages of folding kinetics. They also cannot detect multiple kinetics behavior such as the two-state and three-state kinetics exhibited by hen egg-white Lysozyme.[8,9] Lattice models[10] have been well studied and possess great theoretical value but cannot be applied to real proteins.

Our PRM approach, by constructing a roadmap that approximates the folding landscape, computes multiple folding pathways in a single run and provides a natural way to study protein folding kinetics at the pathway level. What we sacrifice is path quality, which can be improved through bigger roadmaps, oversampling, or other techniques.

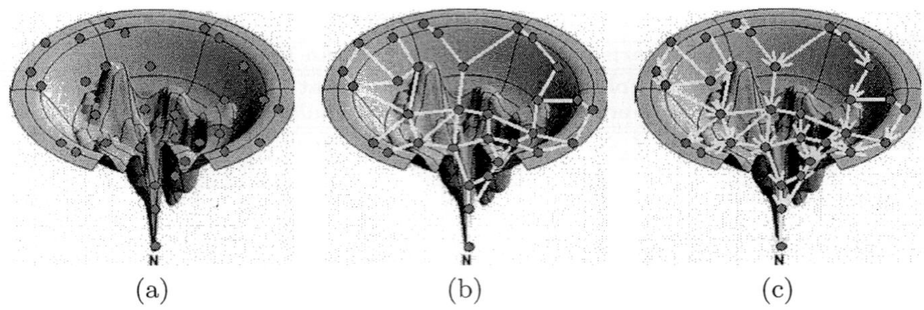

Figure 1: A PRM roadmap for protein folding shown imposed on a visualization of the potential energy landscape: (a) after node generation (note sampling is denser around **N**, the known native structure), (b) after the connection phase, and (c) using it to extract folding paths to the known native structure.

2 A Probabilistic Roadmap Method for Protein Folding

Our approach to protein folding is based on the *probabilistic roadmap* (PRM) approach for motion planning.[5] A detailed description of how the PRM framework can be applied to protein folding is presented in our previous work.[4] The basic idea is illustrated in Figure 1. We first sample some points in the protein's conformation space (Figure 1(a)); generally, our sampling is biased to increase density near the known native state. Then, these points are connected to form a graph, or roadmap (Figure 1(b)). Weights are assigned to directed edges to reflect the energetic feasibility of transition between the conformations corresponding to the two end points. Finally, folding pathways are extracted from the roadmap using standard graph search techniques (Figure 1(c)).

2.1 Modeling Proteins (C-Space)

The amino acid sequence is modeled as a tree-like linkage. Using a standard modeling assumption for proteins,[11] the only degrees of freedom (dof) in our model of the protein are the backbone's phi and psi torsional angles, which we model as revolute (rotational) joints taking values in $[0, 2\pi)$. Moreover, side chains are modeled as spheres and have zero dof.

Since we are not concerned with the absolute position and orientation of the protein, a *conformation* of an $n + 1$ amino acid protein can be specified by a vector of $2n$ phi and psi angles, each in the range $[0, 2\pi)$, with the angle 2π equated to 0, which is naturally associated with a unit circle in the plane, denoted by S^1. That is, the conformation space (C-space) of interest for a protein with $n + 1$ amino acids can be expressed as:

$$\mathcal{C} = \{q \mid q \in S^{2n}\}. \tag{1}$$

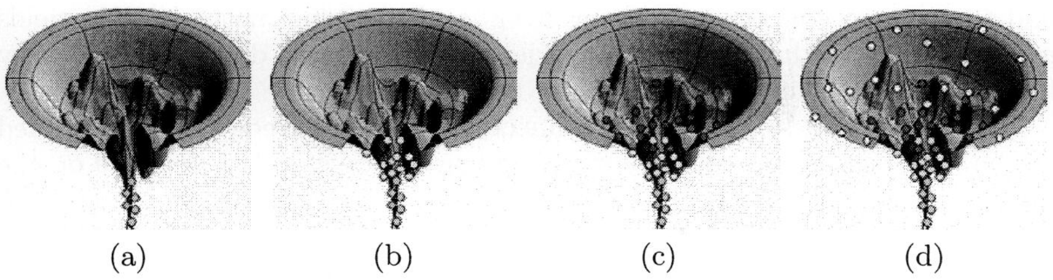

(a)   (b)   (c)   (d)

Figure 2: An illustration of our iterative perturbation sampling strategy shown imposed on a visualization of the potential energy landscape.

Note that $\mathcal{C}$ simply denotes the set of all possible conformations. The feasibility of a point in $\mathcal{C}$ will be determined by potential energy computations.

2.2 Node Generation

Recall that we begin with the known native structure and our goal is to map the protein-folding landscape leading to the native fold. The objective of the node generation phase is to generate a representative sample of conformations of the protein. Due to the high dimensionality of the conformation space, simple uniform sampling would take too long to provide sufficiently dense coverage of the region surrounding the native structure.

The results presented in this paper use a biased sampling strategy that focuses sampling around the native state by iteratively applying small perturbations to existing conformations.[4] The process is illustrated in Figure 2. A node q is accepted and added to the roadmap based on its potential energy $E(q)$ with the following probability:

$$P(\text{accept } q) = \begin{cases} 1 & \text{if } E(q) < E_{\min} \\ \frac{E_{\max} - E(q)}{E_{\max} - E_{\min}} & \text{if } E_{\min} \leq E(q) \leq E_{\max} \\ 0 & \text{if } E(q) > E_{\max} \end{cases}$$

We set $E_{\min} = 50000$ kJ/mol and $E_{\max} = 70000$ kJ/mol which favors configurations with well separated side chain spheres. This acceptance test, which retains more nodes in low energy regions, was also used in PRM-based methods for ligand binding[12,13] and in our previous work on protein folding.[14]

2.3 Connecting the Roadmap

Connection is the second phase of roadmap construction. The objective is to obtain a roadmap encoding representative, low energy paths. For each roadmap node, we first find its k nearest neighbors, for some small constant k, and then try to connect it to them using local planning method. This yields a connectivity roadmap that can be viewed as a net laid down on the energy landscape (see Figure 1(b)).

When two nodes q_1 and q_2 are connected, the directed edge (q_1, q_2) is added to the roadmap. Each edge (q_1, q_2) is assigned a weight that depends on the sequence of conformations $\{q_1 = c_0, c_1, c_2, \ldots, c_{n-1}, c_n = q_2\}$ on the straight line in $\mathcal{C}$ connecting q_1 and q_2. For each pair of consecutive conformations c_i and c_{i+1}, the probability P_i of moving from c_i to c_{i+1} depends on the difference between their potential energies $\Delta E_i = E(c_{i+1}) - E(c_i)$.

$$P_i = \begin{cases} e^{\frac{-\Delta E_i}{kT}} & \text{if } \Delta E_i > 0 \\ 1 & \text{if } \Delta E_i \leq 0 \end{cases} \tag{2}$$

This keeps the detailed balance between two adjacent states, and enables the weight of an edge to be computed by summing the logarithms of the probabilities for consecutive pairs of conformations in the sequence.

$$w(q_1, q_2) = \sum_{i=0}^{n-1} -log(P_i), \tag{3}$$

In this way, we encode the energetic feasibility of transiting from one conformation to another in the edge connecting them.

2.4 Extracting Folding Pathways

The roadmap is a map of the protein-folding landscape of the protein. One way to study this landscape is to inspect and analyze the pathways it contains.

An important feature of our approach is that the roadmap contains *many* folding pathways, which together represent the folding landscape. We can extract many such paths by computing the single-source shortest-path (SSSP) tree from the native structure (see Figure 1(c)).

3 Potential Energy Calculations

The way in which a protein folds depends critically on the potential energy. Our PRM framework incorporates this bias by accepting conformations based on their potential energy (Section 2.2) and by weighting roadmap edges according to their energetic feasibility (Section 2.3).

While our framework is flexible enough to use any method for computing potential energies, our current work uses a very simplistic potential.[4] Briefly, we use a step function approximation of the van der Waals potential component. Our approximation considers only the contribution from the side chains. Additionally, in our model of each amino acid, we treat the side chain as a single large 'atom' R located at the C_β atom. For a given conformation, we calculate the coordinates of the R 'atoms' (our spherical approximation of the side chains) for all residues. If any two R 'atoms' are too close (less then 2.4 Å during node generation and 1.0 Å during roadmap connection), a very high potential is returned. If all the distances between all R 'atoms' are larger than 2.4 Å, then we proceed to calculate the potential as follows:

$$U_{tot} = \sum_{restraints} K_d\{[(d_i - d_0)^2 + d_c^2]^{1/2} - d_c\} + E_{hp}, \tag{4}$$

The first term represents constraints which favor the known secondary structure through main-chain hydrogen bonds and disulphide bonds and the second term is the hydrophobic effect.

Finally, we note that in our case, the minimum potential is not necessarily achieved by the native structure, and thus our energy model does not yield true funnel landscapes as are shown in the figures.

4   Timed Contact Analysis

Contact analysis provides us with a formal method of validation and allows for detailed analysis of the folding pathways. We first identify the *native contacts* by finding all pairs of C_α atoms in the native state that are at most 7 Å apart. If desired, attention can be restricted to *hydrophobic contacts* between hydrophobic residues. To analyze a particular pathway, we examine each conformation on the path and determine the time step on the path at which each native contact appears. Although these time steps cannot be associated with any real time, they do give a temporal ordering and produce a *timed contact map* for the given pathway, see Figures 3 and 4.

The timed contact map provides a formal basis for determining secondary structure formation order along a pathway. Here, structure formation order is based on the formation order of the native contacts.[15] We have looked at several metrics to determine when a secondary structure appears: average appearance time of native contacts within the structure, average appearance of the first $x\%$ of the contacts, average appearance ignoring outliers, etc. We can also focus our analysis on smaller pieces of secondary structure such as β-turns (instead of the entire β-sheet). This is especially helpful when looking for fine details in a folding pathway.

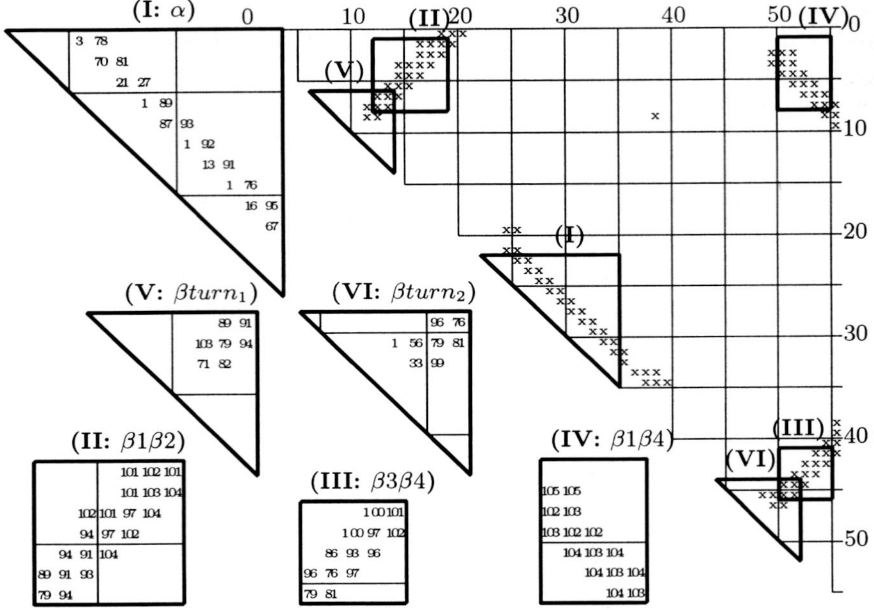

Figure 3: Timed Contact Map for Protein G. The full contact matrix (right) and blow-ups (left) showing the time steps when contacts appear on a path. The blow-ups: I - alpha helix contacts, II - beta 1-2 contacts, III - beta 3-4 contacts, IV - beta 1-4 contacts, V - turn 1 (beta 1-2) contacts, and VI - turn 2 (beta 3-4) contacts.

Because the roadmap contains multiple pathways, we can estimate the probability of a particular secondary structure formation order occurring. If the roadmap maps the potential energy landscape well, then the percentage of pathways in the roadmap that contain a particular formation order reflects the probability of that order occurring.

5 Experimental Validation and Discussion

In this section, we present results obtained using our PRM-based approach. For each protein studied, we construct a roadmap, extract the folding pathways as described in Section 2.4, and analyze the pathways as described in Section 4.

We study several small proteins^a in detail, see Table 2. The structures for all the proteins were obtained from the Protein Data Bank.[17] Protein A is an

---

^{a}Abbreviations of proteins: G, B1 immunoglobulin-binding domain of streptococcal protein G; L, a 62-residue variant[16] of B1 immunoglobulin-binding domain of peptostreptococcal protein L; A, B domain of staphylococcal protein A; CTXIII, Cardiotoxin analogue III.

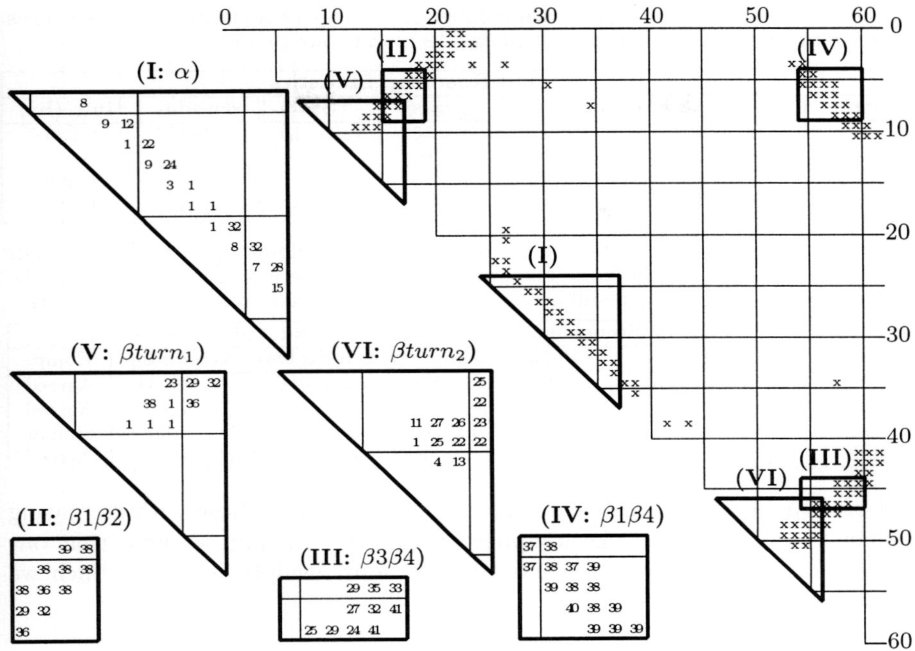

Figure 4: Timed Contact Map for Protein L. The full contact matrix (right) and blow-ups (left) showing the time steps when contacts appear on a path. The blow-ups: I - alpha helix contacts, II - beta 1-2 contacts, III - beta 3-4 contacts, IV - beta 1-4 contacts, V - turn 1 (beta 1-2) contacts, and VI - turn 2 (beta 3-4) contacts.

all alpha protein composed of three alpha helices. Protein G and protein L are mixed proteins which are both composed of one α-helix and a 4-stranded β-sheet. CTXIII is an all beta protein composed of a 2-stranded β-sheet and a 3-stranded β-sheet.

As discussed in Section 1.1, our PRM-based method sacrifices accuracy in favor of rapid coverage. As can be seen from the running time and roadmap statistics shown in Table 2, our roadmaps containing thousands of folding paths are computed in just a few hours on a desktop PC. In contrast, traditional methods such as molecular dynamics, compute a single trajectory, have tremendous computational requirements, and are subject to local minima.

Contact analysis was performed on the pathways for proteins A, G, L and CTXIII. Timed contact maps for proteins L and G are shown in Figure 3 and Figure 4. The dominant formation order found for each protein is shown in Table 3. It is clearly seen that our results are in good agreement with the hydrogen-exchange experimental results described by Li and Woodward.[7]

Table 2: Proteins studied. Shown are number of residues (#res), number of α helices and β strands ($\alpha + \beta$), roadmap size (#nodes), and construction time.

Proteins and Roadmap Statistics						
name	pdb	brief description	#res	SS	#nodes	time (hr)
G	1gb1	Protein G, B1 domain	56	$1\alpha + 4\beta$	16407	6.985
A	1bdd	Protein A, B domain	60	3α	21917	11.325
CTXIII	2crt	Cardiotoxin III	60	5β	14532	6.386
L	2ptl	Protein L, B1 domain	62	$1\alpha + 4\beta$	17407	9.152

Table 3: The secondary structure formation order on dominant pathways in our roadmaps and some validations. The brackets indicate there was no clear order. The last column compares our results with those from hydrogen-exchange experiments.

Secondary Structure Formation Order and Validation				
pdb	Out-Exchange[7]	Pulse-Labeling[7]	Our SS Formation Order	Comp.
1gb1	$[\alpha,\beta1,\beta3,\beta4]$, $\beta2$	$[\alpha,\beta4]$, $[\beta1,\beta2,\beta3]$	α, $\beta3$-$\beta4$, $\beta1$-$\beta2$, $\beta1$-$\beta4$	Agreed
1bdd	$[\alpha2,\alpha3]$, $\alpha1$	$[\alpha1,\alpha2,\alpha3]$	$[\alpha1,\alpha2,\alpha3]$, $\alpha2$-$\alpha3$, $\alpha1$-$\alpha3$	Agreed
2crt	$[\beta3,\beta4,\beta5]$, $[\beta1,\beta2]$	$\beta5$, $\beta3$, $\beta4$, $[\beta1,\beta2]$	$\beta3$-$\beta4$, $[\beta1$-$\beta2,\beta3$-$\beta5]$	Similar
2ptl	$[\alpha,\beta1,\beta2,\beta4]$, $\beta3$	$[\alpha,\beta1]$, $[\beta2,\beta3,\beta4]$	α, $\beta1$-$\beta2$, $\beta3$-$\beta4$, $\beta1$-$\beta4$	Agreed

All proteins seem to form local contacts first, and then those with increasing sequence contact order, like a zipper process.[18,15] Finally, we note that our results for CTXIII may be affected by the four disulphide bonds which we model as hydrogen bonds.

5.1 *Protein G and Protein L: A Detailed Study*

Proteins G and L present a good test case for our technique because they are known to fold differently although they are structurally similar. In particular, although they have only 15% sequence identity,[19] they are both composed of an α-helix and a 4-stranded β-sheet. β strands 1 and 2 form the N-terminal hairpin (hairpin 1) and β strands 3 and 4 form the C-terminal hairpin (hairpin 2). Experimental results show that β-hairpin 1 forms first in protein L, and β-hairpin 2 forms first in protein G.

In native state out-exchange experiments for protein G and L, numerous NHs out-exchange very slowly, which makes it difficult to unambiguously identify the slowest out-exchange residues. It is found that the slow exchanging NHs are in β_1, β_3, β_4 and the helix for G, and the α helix, β_1, β_2, and β_4 for L. On the other hand, pulse-labeling experiments identify that the first NHs to gain protection during folding are in α and β_4 for G and α and β_1 for L. (See Li and Woodward [7] and references therein.) In summary, out-exchange and pulse-labeling experiments strongly suggest that the α and β_4 form first for G and that the α and β_1 form first for L. Furthermore, this is consistent with Φ-value analysis on G [19] and L [20] which indicates that, in the folding transition state, β-hairpin 2 is more formed than the rest of the structure for

Table 4: Comparison of analysis techniques for proteins G and L using roadmaps computed with energy thresholds $E_{\min} = 50,000$ kJ/mol and $E_{\max} = 70,000$ kJ/mol. For each combination of contact type (all or hydrophobic) and number of contacts (first $x\%$ to form), we show the percentage of pathways with a particular secondary structure formation order. Recall that β-hairpin 2 ($\beta3$-$\beta4$) forms first in protein G and β-hairpin 1 ($\beta1$-$\beta2$) forms first in protein L.

Comparison of Analysis Techniques – Helix and Hairpins							
			analyze first x% contacts				
Name	Contacts	SS Formation Order	20	40	60	80	100
Protein G	all	α, $\beta3$-$\beta4$, $\beta1$-$\beta2$, $\beta1$-$\beta4$	76	66	77	55	58
		α, $\beta1$-$\beta2$, $\beta3$-$\beta4$, $\beta1$-$\beta4$	23	34	23	45	42
	hydrophobic	α, $\beta3$-$\beta4$, $\beta1$-$\beta2$, $\beta1$-$\beta4$	85	78	77	62	67
		α, $\beta3$-$\beta4$, $\beta1$-$\beta4$, $\beta1$-$\beta2$	11	11	9	8	8
		α, $\beta1$-$\beta2$, $\beta3$-$\beta4$, $\beta1$-$\beta4$	4	10	14	29	24
Protein L	all	α, $\beta1$-$\beta2$, $\beta3$-$\beta4$, $\beta1$-$\beta4$	67	76	78	78	92
		α, $\beta1$-$\beta2$, $\beta1$-$\beta4$, $\beta3$-$\beta4$	15	4	4	4	4
		α, $\beta3$-$\beta4$, $\beta1$-$\beta2$, $\beta1$-$\beta4$	19	20	18	18	4
	hydrophobic	α, $\beta1$-$\beta2$, $\beta3$-$\beta4$, $\beta1$-$\beta4$	54	65	74	73	86
		α, $\beta1$-$\beta2$, $\beta1$-$\beta4$, $\beta3$-$\beta4$	9	3	3	2	2
		α, $\beta3$-$\beta4$, $\beta1$-$\beta2$, $\beta1$-$\beta4$	36	32	23	26	13

G and β-hairpin 1 is similarly more formed for L.

For both protein G and L, we use the same definition of the beta strands as is contained in the protein Data Bank. These definitions include all the observed residues that are found in the slowest exchange core in the native state out-exchange experiments and that are among the first gaining protection in the pulse labeling experiments, see Figures 3 and 4. This enables us to have a fair comparison of our results with those from these experiments.

Table 4 shows our results. For each protein, one roadmap was constructed and then its (thousands of) pathways were studied using the different analysis methods described in Section 4. When only the specified contacts were considered, the percentage of paths that had the given secondary structure formation order is shown. For example, for all contacts, and limiting our consideration to only the first 60% of the contacts to form, in 77% of the pathways for protein G β-hairpin 2 (β_3-β_4 contacts) formed before β-hairpin 1 (β_1-β_2 contacts), while in 82% of the pathways for protein L β-hairpin 1 formed before β-hairpin 2. Thus, the helix and β-hairpin 2 form first by a significant percentage for protein G, while for protein L, the helix and β-hairpin 1 consistently form first by a significant percentage. Both results agree very well with experimental observations. We also performed the study considering only the hydrophobic contacts, and obtained similar results, further confirming our findings.

We also study the formation order of β turns (see Figures 3 and 4 for

Table 5: β turn formation: comparison of analysis techniques for proteins G and L using the same roadmaps as in Table 4. Recall that turn 2 forms first in protein G and turn 1 forms first in protein L.

Comparison of Analysis Techniques – Helix and Turns							
			analyze first x% contacts				
Name	Contacts	SS Formation Order	20	40	60	80	100
Protein G	all	α, turn 2, turn 1	53	52	52	50	50
		turn 2, α, turn 1	15	9	17	22	22
		α, turn 1, turn 2	25	33	26	23	24
	hydrophobic	α, turn 2, turn 1	96	96	85	96	87
		α, turn 1, turn 2	4	4	12	2	11
Protein L	all	α, turn 1, turn 2	24	30	37	38	41
		1st turn, α, 2nd	3	4	4	4	6
		α, turn 2, turn 1	73	63	60	48	39
	hydrophobic	α, turn 1, turn 2	72	68	72	70	69
		turn 1, α, turn 2	5	9	5	7	15
		α, turn 2, turn 1	23	22	22	23	15

our definition). Remarkably, the results (see Table 5) are in good agreement with those obtained using the beta strands. For protein G, the second β turn forms consistently earlier than the first β turn, which confirms our results that the second hairpin forms first. For protein L, our results show that the second β turn forms first when considering all contacts. However, when only hydrophobic contacts are considered, then the first β turn forms first by a significant percentage. This indicates that some hydrophobic contacts form earlier in the first turn than in the second.

6 Conclusion and Future Work

We have shown that our PRM-based approach for studying protein folding pathways is able to correctly reproduce known folding differences between the structurally similar proteins G and L. This result gives confidence in our approach and implies it could be valuable for analyzing proteins whose structure is known but for which we lack experimental data on the folding pathway.

Acknowledgments

This research supported in part by NSF Grants ACI-9872126, EIA-9975018, EIA-0103742, EIA-9805823, ACR-0081510, ACR-0113971, CCR-0113974, EIA-9810937, and EIA-0079874. Song supported in part by an IBM TJ Watson PhD Fellowship and Thomas supported in part by an NSF Graduate Research Fellowship. We would like to thank Xinyu Tang for helping with the experiments,

Jean-Claude Latombe for pointing out the connection between box folding and protein folding, and Michael Levitt and Vijay Pande for useful suggestions.

References

1. Levitt, M., Gerstein, M., Huang, E., Subbiah, S., and Tsai, J. *Annu. Rev. Biochem.* **66**, 549–579 (1997).
2. Reeke, Jr., G. N. *Ann. Rev. Comput. Sci.* **3**, 59–84 (1988).
3. Lansbury, P. *Proc. Natl. Acad. Sci. USA* **96**, 3342–3344 (1999).
4. Amato, N. M. and Song, G. *J. Comput. Biol.* **9**(2), 149–168 (2002). RECOMB 2001 special issue.
5. Kavraki, L., Svestka, P., Latombe, J. C., and Overmars, M. *IEEE Trans. Robot. Automat.* **12**(4), 566–580 August (1996).
6. Song, G. and Amato, N. M. In *Proc. IEEE Int. Conf. Robot. Autom. (ICRA)*, 948–953, (2001).
7. Li, R. and Woodward, C. *Protein Sci.* **8**, 1571–1591 (1999).
8. C.M.Dobson, Sali, A., and Karplus, M. *Angew. Chem. Int. Ed.* **37**, 868–893 (1998).
9. Radford, S. E. and Dobson, C. M. *Phil. Trans. R. Soc. Lond.* **B348**, 17 (1995).
10. Bryngelson, J., Onuchic, J., Socci, N., and Wolynes, P. *Protein Struct. Funct. Genet* **21**, 167–195 (1995).
11. Sternberg, M. J. *Protein Structure Prediction.* OIRL Press at Oxford University Press, (1996).
12. Bayazit, O. B., Song, G., and Amato, N. M. In *Proc. IEEE Int. Conf. Robot. Autom. (ICRA)*, 954–959, (2001). This work was also presented as a poster at RECOMB'01.
13. Singh, A., Latombe, J., and Brutlag, D. In *7th Int. Conf. on Intelligent Systems for Molecular Biology (ISMB)*, 252–261, (1999).
14. Song, G. and Amato, N. M. In *Proc. Int. Conf. Comput. Molecular Biology (RECOMB)*, 287–296, (2001).
15. Fiebig, K. M. and Dill, K. A. *J. Chem. Phys* **98**(4), 3475–3487 (1993).
16. Yi, Q. and Baker, D. *Protein Sci* **5**, 1060–1066 (1996).
17. http://www.rcsb.org/pdb/.
18. Dill, K. A., Fiebig, K. M., and Chan, H. S. *Proc. Natl. Acad. Sci. USA* **90**, 1942–6 (1993).
19. McCallister, E. L., Alm, E., and Baker, D. *Nat. Struct. Biol.* **7**(8), 669–673 (2000).
20. Kim, D. E., Fisher, C., and Baker, D. *J. Mol. Biol.* **298**, 971–984 (2000).

PROFILE-PROFILE ALIGNMENT: A POWERFUL TOOL FOR PROTEIN STRUCTURE PREDICTION

NIKLAS VON ÖHSEN

FhI-SCAI - Fraunhofer Institute for Algorithms and Scientific Computing, Schloss Birlinghoven, 53754 Sankt Augustin, Germany

INGOLF SOMMER

Max-Planck-Institut für Informatik, Stuhlsatzenhausweg 85, 66123 Saarbrücken, Germany

RALF ZIMMER

Institut für Informatik, LMU München, Theresienstrasse 39, 80333 München, Germany

Abstract

The problem of computing the tertiary structure of a protein from a given amino acid sequence has been a major subject of bioinformatics research during the last decade. Many different approaches have been taken to tackle the problem, the most successful of which are based on searching databases to identify a similar amino acid sequence in the PDB and using the corresponding structure as a template for modeling the structure of the query sequence. An important advance for the evaluation of sequence similarity in this context has been the use of a frequency profile that represents a part of the protein sequence space close to the query sequence instead of the query sequence itself. In this paper, we present a further extension of this principle by using profiles instead of the template sequences, also. We show that, by using our newly developed scoring model, the profile-profile alignment approach is able to significantly outperform current state of the art methods like PSI-BLAST, HMMs, or threading methods in a fold recognition setup. This is especially interesting since we show that it holds for closely related sequences as well as for very distantly related ones.

Since the first use of alignment procedures for evaluation of similarity between two sequences, various successful advances in this concept have been proposed. One major improvement of the alignment procedure that made its way into many popular bioinformatics tools is the use of a frequency profile instead of a sequence which was first proposed by Gribskov[1] and is one of the major ingredients of the well-known PSI-BLAST program[2]. The aim of replacing a single sequence by a frequency profile representing its protein family is to discard part of the sequence information that is not conserved throughout this family. Therefore, the profile will be a better representation of strongly conserved features like the tertiary structure of the protein than the sequence itself. While this concept proved useful when replacing one sequence in an alignment by a profile, it has been shown recently by Rychlewski et al.[3] that using profiles on both sides of the alignment is even better when trying to establish relationships between distantly related proteins. Using this profile-profile approach in their FFAS method, they managed to reach the second rank in the CAFASP2 contest of fully automated protein structure prediction servers. Despite the straight forward idea of replacing both sequences by frequency profiles, it is far from obvious how the alignment score should be calculated in this case. Rychlewski et al. used the simple dot product for computing the score but also noted that a more sophisticated method might prove advantageous.

A major new contribution to the profile-profile alignment approach has recently been made by Yona and Levitt[4] who were the first to propose a scoring formula for profile-profile alignments that was constructed on a theoretically sound basis. Their scoring system is based on an information theoretic measure of difference between the two probability distributions represented by the profiles. Since their profile-profile score is not constructed along the lines of the common similarity score for sequences, there are two major drawbacks to the approach: First, they measure only the similarity of the two probability distributions provided by the profiles and do not take into account the similarities between amino acids. Ignoring these contributions which have been crucial to sequence alignment methods for the last decades will most likely limit the sensitivity of an alignment scoring system. Second, they have to construct an ad-hoc transformation that will make their score applicable in the case of local sequence alignment. Our approach is an extension of the usual amino acid similarity score to the profile-profile situation (the sequence-sequence score is a special case of our formula) Thereby, our alignment score avoids both of these drawbacks and can directly be used for local alignment without any changes. Therefore, we believe that the proposed log average score may have significant practical and performance advantages. Yona and Levitt compare their profile-profile tool with tools from the BLAST family like Gapped BLAST, IMPALA and PSI-BLAST, showing that they can detect more protein superfamily relationships using their method than using PSI-BLAST. Due to lack of availability at the time of setting up the benchmarks performed in this paper, we have not been able to include their **prof_sim** tool in this study.

In the following, we are presenting a formula for applying the popular similarity matrix scoring for sequences to the more general case of scoring two frequency profiles which has been introducede in more detail earlier[5] and compare the performance of the new alignment score with other popular alignment methods in terms of fold recognition performance.

1 Introducing Log Average Scoring

The popular similarity matrix alignment scores like PAM[6] or BLOSUM[7] have a strong foundation in statistical test theory. The alignment score of an alignment of two sequences without gap penalties is a direct measure of the statistical evidence supporting the hypothesis that the two sequences are related against the alternative that they are unrelated. The definition of the term "related" in this context is part of the used substitution model. The most interesting features of alignment scores are an immediate consequence of this property. A score of zero means that there is no evidence in the alignment whether the sequences are related or unrelated whereas a positive score indicates relatedness and a negative score indicates that the sequences are rather unrelated. If p_i is the background amino acid probability distribution and p_{rel} denotes the probability distribution of "related" amino acid pairs, the substitution matrix alignment score for a pair (i, j) of amino acids is calculated by

$$M(i,j) = \log\left(\frac{p_{\mathrm{rel}}(i,j)}{p_i p_j}\right) \tag{1}$$

This is usually scaled by a factor $\frac{10}{\log 10}$ in the Dayhoff models and $\frac{2}{\log 2}$ for the BLOSUM matrices. While it seems straightforward to extend this score to two profile vectors α and β by using the formula

$$\mathrm{score}_{\mathrm{average}}(\alpha, \beta) = \sum_{i=1}^{20}\sum_{j=1}^{20} \alpha_i \beta_j \log \frac{p_{\mathrm{rel}}(i,j)}{p_i p_j} \tag{2}$$

(called *average scoring*) we have shown earlier[5] that the original meaning of the alignment score can be extended to the profile-profile case by using the *log average score*:

$$\mathrm{score}_{\mathrm{logaverage}}(\alpha, \beta) = \log \sum_{i=1}^{20}\sum_{j=1}^{20} \alpha_i \beta_j \frac{p_{\mathrm{rel}}(i,j)}{p_i p_j} \tag{3}$$

The double sum occuring here can be interpreted as a bayesian probability for the profile vectors being related according to the substitution model, thus giving a meaning to the sum of this score over all alignment positions.

2   Benchmarks

In order to evaluate whether the proposed profile-profile alignment scores are useful for measuring the relatedness of two profiles, we performed several tests measuring fold recognition performance. The SCOP-1.50 database[8] was taken as gold standard for measuring the relatedness of proteins on different levels. Using the ASTRAL server[9], a subset of this protein domain database was selected such that every two domains in the database were showing a maximum homology level of 40% sequence identity. This domain set is referred to as PDB40D[10] and its members will be called templates in the following.

2.1   Frequency Profile Construction

For each template sequence, a multiple alignment was constructed by running PSI-BLAST[2] against the KIND database[11], a non redundant protein sequence database. A frequency profile was calculated from this multiple alignment using a sequence weighting algorithm that is a slightly modified version of an algorithm by Henikoff[12]. [a] The resulting frequency profile was cut down to the original template length by throwing away positions that correspond to a gap of the template sequence in the multiple alignment.

This template database of frequency profiles was used in a fold recognition setup: The objective is to find the most closely related protein domain in the PDB40D set for a given protein chain (in the following also called *target*). This is done by first constructing a frequency profile for the target using PSI-BLAST and sequence weighting methods exactly as described for the templates. Then, all the template frequency profiles are subsequently aligned against the query frequency profile and the template with the highest score is the best guess for the structure of a domain contained in this chain.

2.2   Data Set

The performance of some state-of-the-art algorithms for fold recognition were compared with the newly developed profile-profile alignment scoring formula using a modified "leave-one-out" benchmark. All 2232 protein chains from the PDB containing one complete domain from the PDB40D set were used as benchmark set. When performing the fold recognition for each chain, all the domains belonging to the chain itself were removed from the PDB40D template

---

[a]The extended version tries to minimize the relative entropy regarding the background amino acid distribution rather than to maximize the absolute entropy of the profile, leading to small differences in the profile as compared to the Henikoff version. See also Krogh et. al.[13] for the connection between sequence weighting and entropy.

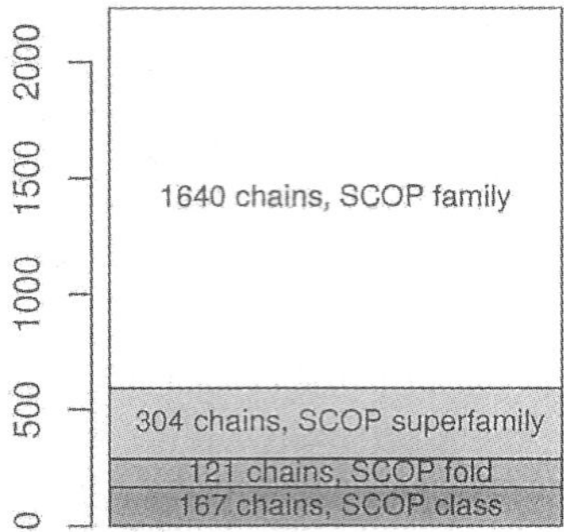

Figure 1: Test set composition: number of chains with the most closely related PDB40D domain belonging to the same SCOP level as indicated.

set. Each of the chains was aligned against the PDB40D template set except for the domains belonging to the current chain.

Figure 1 shows the composition of the benchmark set. Chains that contain a domain for which the PDB40D holds a member of the same SCOP family are likely to be quite easy fold recognition targets since SCOP family members ought to be quite closely related according to the definition of SCOP, making them tractable targets for fold recognition methods. Chains that have the most closely related domain on the SCOP superfamily level are harder to predict correctly, but according to the SCOP definition these domains probably have a probable common evolutionary origin. Hence it is possible that algorithms trying to find distant sequence similarities are successful in finding these relationships. The hardest level for fold prediction is the SCOP fold level since the templates in the PDB40D that share the same fold with a domain contained in the chain do not share more than a "major structural similarity" according to the SCOP definition. The SCOP class level finally is an impossible target for fold recognition by definition since all templates in our database have a SCOP fold different from the domains belonging to the chain. Nevertheless, these chains were taken into account when calculating the overall fold recognition percentages in order to get an unbiased estimation of the performance for a completely unknown target.

2.3 Algorithms, Implementations and Parameters

We used our Java implementation JProP of a dynamic programming engine using the two profile-profile alignment scores introduced in equations 2 and 3 to compare the performance of the profile-profile alignment with three other successful fold prediction approaches: HMMs, threading and PSI-BLAST. In addition, we used a plain sequence alignment program to serve as a lower bound in the evaluation procedure.

Sequence Alignment: We used the 123D threading program to produce global dynamic programming sequence alignments using Dayhoff's 250 PAM matrix.

PSI-BLAST: PSI-BLAST[2] was run against the KIND database[11] augmented by the PDB40D proteins and the first hit in the PDB40D set that did not belong to the chain itself was taken as the fold prediction.

HMMer: We chose HMMer[14,15] in its latest version 2.2g as a representative for the current state of the art in profile HMMs (see Lindahl et al.[16] for a comparison of HMMer with SAM-T98 and other programs). A HMM database was built from the PDB40D list containing HMMs trained from the same multiple alignments used for constructing the frequency profiles. The search parameters of the HMM database were calibrated using the `hmmcalibrate` tool from the HMMer package. The target sequence was used to search this database using the `hmmpfam` tool. The e-value output of the found templates was used as score.

Profile 123D: 123D[17] is a fast profile threading program based on contact capacity potentials and dynamic programming which has been in use for some years already, e.g. in CAFASP2 and CASP4. It has recently been subject to parameter optimization[18] and the basis for an analysis of confidence measures[10]. We chose 123D as a representative for the class of threading algorithms tractable by dynamic programming. We used the parameters from the Zien et al. paper on parameter optimization using a machine learning approach[18]. Frequency profiles on the target side were used. Information on the tertiary structure coded in contact capacity potentials, secondary structure information and the sequence itself were used on the template side to produce a global alignment. The resulting threading score was used for further analysis.

JProP profile-profile alignment: Our program JProP is a pure Java implementation of the dynamic programming alignment algorithm and can be configured to perform various alignment scoring schemes, average scoring and log average scoring being two of them. For the benchmarks we used gap cost parameters that were optimized using a machine learning approach on a small benchmark set described by Zien et al.[18]. The substitution model used for computing the scores was the BLOSUM62 model[7] and we applied it using an affine gap cost model and global dynamic programming alignment.

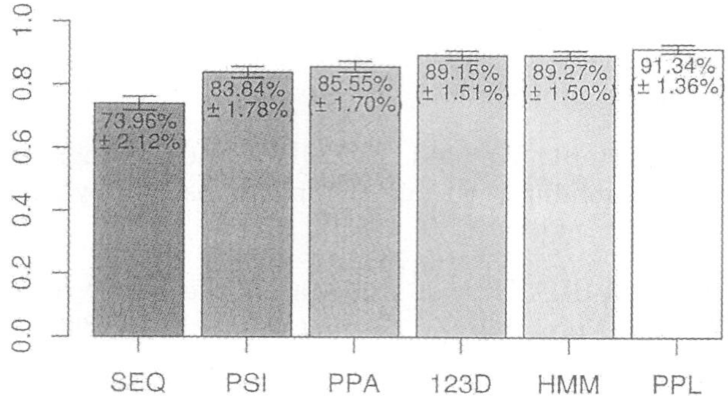

Figure 2: Fold recognition performance for the 1640 chains from the SCOP family level. Numbers indicating percentage of correctly predicted targets from this difficulty level, line segments indicating estimated 95% confidence intervals. SEQ: plain sequence-sequence alignment, PSI: PSI-BLAST, PPA: profile-profile alignment using average scoring, 123D: 123D profile threading, HMM: HMMer, PPL: JProP profile-profile alignment using log average scoring

3   Fold Recognition Results

3.1   SCOP Family Level

We performed the modified leave-one-out fold recognition benchmarks and analysed the results separately for the difficulty of the targets as described above. Figure 2 shows the results for the 1640 chains from the SCOP family level. The 95% confidence intervals indicated in the plot were estimated by using a normal approximation. It is remarkable that even on this level of close relationship (SCOP definition is "clear evolutionary relationship" with a general level of pairwise sequence identity greater than 30%) the sequence alignment is clearly outperformed by all other methods. PSI-BLAST is very good at detecting these close relationships but is already outperformed by the simple profile-profile average scoring approach and clearly left behind by the threading program 123D. As expected, the HMM is very good at detecting and precisely evaluating these close relationships with 89.27% correctly assigned targets. Thus it is very interesting to see that the newly introduced profile-profile alignment with log average scoring can still add more than 2%, yielding a total of 91.34% of fold recognition performance which is a quite significant lead due to the high performance level and the large sample size.

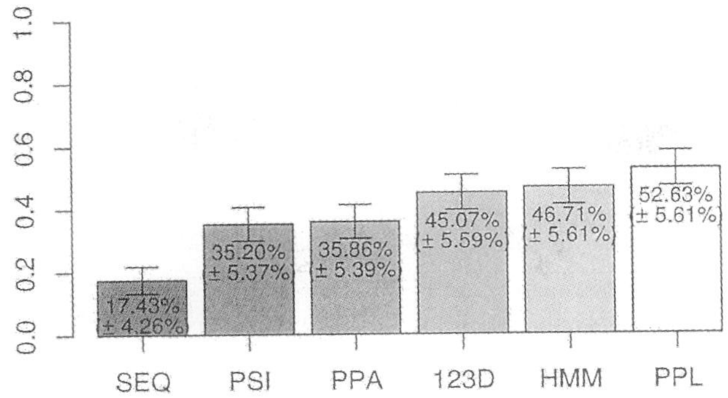

Figure 3: Fold recognition performance for the 304 chains from the SCOP superfamily level. See text or caption of figure 2 for details.

3.2 SCOP Superfamily Level

Since profile-profile alignment was originally designed to detect remote homology relationships we expect to see the largest performance gain on the SCOP superfamily level shown in Figure 3. The plain sequence alignment is clearly the worst when compared to the more sophisticated methods, getting less than half the performance of the next candidate PSI-BLAST. While PSI-BLAST is on par with the simple profile-profile approach, the performance gap to the threading program is already widening. The HMM is better than the threading approach on this level of relationship, predicting 46.71% of the targets correctly. The log average scoring profile-profile alignment clearly shows its strength in detecting weak sequence homology relationships here by outperforming the HMM approach by almost 6% getting a total of 52.63% correct recognition results.

3.3 SCOP Fold Level

On the fold level the relationships between the proteins to be recognized are fairly weak. Since the relations are weaker than the SCOP superfamily level it is not likely that the most closely related domain from the PDB40D set shares the same evolutionary origin with part of the chain. Only a major structural similarity is present. This is the setting for which threading approaches are designed, since they make use of tertiary and secondary structure information instead of relying on the sequence information alone.

Figure 4 shows the results for the 121 chains from this category. A slightly

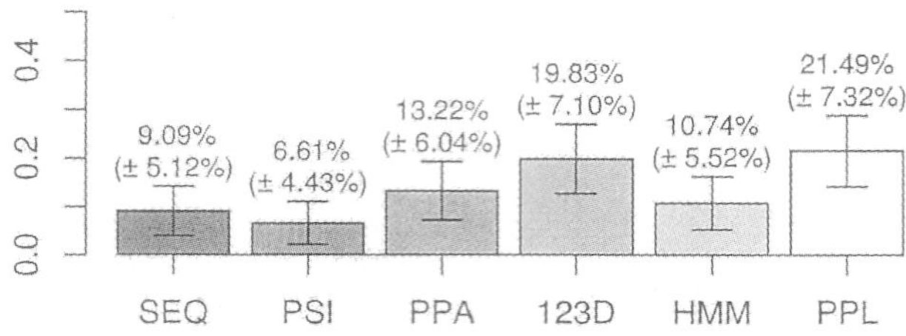

Figure 4: Fold recognition performance for the 121 chains from the SCOP fold level. See text or caption of figure 2 for details.

different picture shows up here. The worst performer is PSI-BLAST with only 6.6% followed by sequence alignment and HMMs with 9.09% and 10.74%, respectively. It should be noted that the results of the PSI-BLAST and the HMMer program on this level are probably hampered by the fact that these programs are the only ones to use significance cutoffs. Thus, sometimes no prediction at all is produced by these two programs, lowering their chance of producing "random" hits. The 123D profile threading programs performs competitively on this level, but again, even the threading approach on these hard targets at 19.81% is outperformed by the profile-profile alignment with the log average score leading with 21.49%. The confidence intervals indicate that these differences are not very significant due to the small sample size, but it is still intriguing to see the completely sequence homology based profile-profile alignment outperform the threading program which makes use of additional structural information. A closer look at the composition of the $\approx 20\%$ shares for 123D and log average profile-profile alignment revealed here that only about 10% of the recognized targets for these two candidates were identical. This stresses the usefulness of trying different algorithms when predicting folds in this very hard category. It also leaves room for speculations on an algorithm combining the strengths of these candidates being possibly capable of reaching 30% fold recognition performance on this SCOP level.

3.4 Overall Fold Recognition

Figure 5 shows a weighted average of the previous results combined with the 167 chains from the SCOP class level that cannot be correctly predicted by homology search. The theoretical maximum performance that can be reached in this plot is thus 92.52%. The results obey the pattern from the previous results. The threading approach and the HMM both at about 72.5% outper-

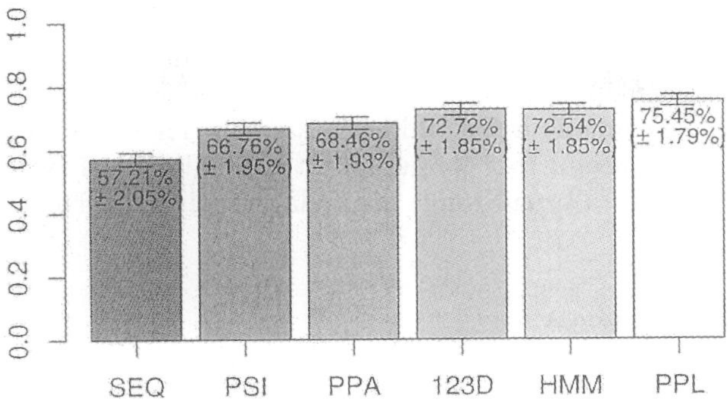

Figure 5: Fold recognition performance for all 2232 chains. See text or caption of figure 2 for details.

form PSI-BLAST and the average scoring profile-profile approach leaving the plain sequence alignment well behind. The log average scoring profile-profile alignment manages to increase the recognition rate by another 3%.

4   Discussion

The profile-profile alignment approach to fold recognition is basically a method for detecting very remote sequence similarity relationships. Sequence information that is not conserved in the most closely related sequences is thrown away by using the frequency profiles constructed by PSI-BLAST instead of the sequences themselves. Thus, a frequency profile is representing a part of the sequence space around the sequence rather than a single point in sequence space. In principle, this should allow for an improved detection of remote sequence homologies.

Nevertheless it is crucial to use a meaningful and sensitive approach to calculate the alignment score in order to receive best results. The effect of this can clearly be seen in the differing results between the mediocre performance of the average score and the superior performance of the log average score.

Our results show, that even simple profile-profile approaches like the average scoring perform competitively to PSI-BLAST. HMM and threading methods are already capable of outperforming PSI-BLAST in terms of fold recognition performance. Choosing the log average scoring, our profile-profile alignment tool outperforms these more advanced tools. On the superfamily level, which is by design the most suitable application scenario for profile-profile alignment, the log average profile-profile alignment leads the competition by 6% fold recognition performance. Perhaps even more interesting is the fact that it can outperform the applications HMMer, fine tuned on the family level, and 123D threading, fine tuned on the fold level, as well. Thus profile-profile alignment proves to be a useful tool for judging the similarity of two proteins by the alignment score for a broad range of similarity relations from very close to very remote.'

5   Further Developments

We are currently working on an extension of the profile-profile alignment score to incorporate a secondary structure component into the scoring system. Furthermore, it will be interesting to see whether the promising results of the alignment score when used for fold recognition also translate into a gain of alignment quality and reliability.

6   Acknowledgements

Part of this work was supported by the DFG priority programme "Informatik-methoden zur Analyse und Interpretation großer genomischer Datenmengen", grant ZI616/1-1. The authors thank Alexander Zien for the help with using his optimization tools and Daniel Hanisch for the collaboration on the program library underlying the JProP implementation.

1. Michael Gribskov and Stella Veretnik. Identification of sequence patterns with profile analysis. In *Methods in Enzymology*, volume 266, chapter 13, pages 198–212. Academic Press, Inc., 1996.

2. Stephen F. Altschul, Thomas L. Madden, Alejandro A. Schäffer, et al. Gapped BLAST and PSI-BLAST: a new generation of protein database search programs. *Nucleic Acids Research*, 25(17):3389–3402, 1997.

3. Leszek Rychlewski, Lukasz Jaroszewski, Weizhong Li, and Adam Godzik. Comparison of sequence profiles. Strategies for structural predictions using sequence information. *Protein Science*, 9:232–241, 2000.

4. Golan Yona and Michael Levitt. Within the twilight zone: A sensitive profile-profile comparison tool based on information theory. *J. Mol. Biol.*, 315:1257–1275, 2002.

5. Niklas von Öhsen and Ralf Zimmer. Improving profile–profile alignment via log average scoring. In Olivier Gascuel and Bernard M. E. Moret, editors, *Algorithms in Bioinformatics, First International Workshop, WABI 2001, Aarhus, Denmark, August 2001, Proceedings*, volume 2149 of *Lec-*

ture Notes in Computer Science, pages 11–26. Springer-Verlag Berlin Heidelberg New York, 2001.

6. Margaret O. Dayhoff, R.M. Schwartz, and B.C. Orcutt. A model of evolutionary change in proteins. In *Atlas of Protein Sequence and Structure*, volume 5, chapter 22, pages 345–352. National Biochemical Research Foundation, Washington DC, 1978.

7. Steven Henikoff and Jorja G. Henikoff. Amino acid substitution matrices from protein blocks. *Proc. Natl. Acad. Sci. USA*, 89(22):10915–9, 1992.

8. L. Lo Conte, B. Ailey, T. J. Hubbard, S. E. Brenner, A. G. Murzin, and C. Chothia. SCOP: a structural classification of proteins database. *Nucleic Acids Res*, 28(1):257–9., 2000.

9. S. E. Brenner, P. Koehl, and M. Levitt. The ASTRAL compendium for protein structure and sequence analysis. *Nucleic Acids Res*, 28(1):254–6, 2000.

10. Ingolf Sommer, Alexander Zien, Niklas von Öhsen, Ralf Zimmer, and Thomas Lengauer. Confidence measures for protein fold recognition. *Bioinformatics*, 18(6):802–812, 2002.

11. Yvonne Kallberg and Bengt Persson. KIND – a non-redundant protein database. *Bioinformatics*, 15(3):260–261, March 1999.

12. Steven Henikoff and Jorja G. Henikoff. Position–based sequence weights. *J. Mol. Biol.*, 243(4):574–578, 1994. 4. November.

13. Anders Krogh and Graeme Mitchison. Maximum entropy weighting of aligned sequences of protein or DNA. In C. Rawlings, D. Clark, R. Altman, L. Hunter, T. Lengauer, and S. Wodak, editors, *Proceedings of ISMB 95*, pages 215–221. AAAI Press, 1995.

14. S. R. Eddy. Profile hidden markov models. *Bioinformatics*, 14:755–763, 1998.

15. S. R. Eddy. Hmmer: Profile hidden markov models for biological sequence analysis. (http://hmmer.wustl.edu/), 2001.

16. Erik Lindahl and Arne Elofsson. Identification of Related Proteins on Family, Superfamily and Fold Level. *J. Mol. Biol.*, 295(3):613–625, January 2000.

17. Nick Alexandrov, Ruth Nussinov, and Ralf Zimmer. Fast protein fold recognition via sequence to structure alignment and contact capacity potentials. In Lawrence Hunter and Teri E. Klein, editors, *Pacific Symposium on Biocomputing'96*, pages 53–72. World Scientific Publishing Co., 1996.

18. Alexander Zien, Ralf Zimmer, and Thomas Lengauer. A simple iterative approach to parameter optimization. *Journal of Computational Biology*, 7(3):483–501, 2000.

PROTEIN THREADING BY LINEAR PROGRAMMING

JINBO XU, MING LI

Department of Computer Science, University of Waterloo,
Waterloo, Ont. N2L 3G1, Canada, {j3xu,mli}@math.uwaterloo.ca

GUOHUI LIN

Department of Computer Science, University of Alberta
Edmonton, Alberta, T6G 2E8, Canada, ghlin@cs.ualberta.ca

DONGSUP KIM, YING XU

Protein Informatics Group, Life Science Division and Computer Sciences and
Mathematics Division,
Oak Ridge National Lab, Oak Ridge, TN 37831, USA, {afg,xyn}@ornl.gov

Protein three-dimensional structure prediction through threading approach has
been extensively studied and various models and algorithms have been proposed.
In order to further explore ways to improve accuracy and efficiency of the threading
process, this paper investigates the effectiveness of a new method: protein threading via linear programming. Based on the contact map model of protein 3D structure, we formulate the protein threading problem as a large scale integer programming problem, then relax to a linear programming problem, and finally solve the
integer program by a branch-and-bound method. The final solution is optimal with
respect to energy functions incorporating pairwise interaction and allowing variable
gaps. The algorithm has been implemented as software package RAPTOR – RApid
Protein Threading predictOR. Experimental results for fold recognition show that
RAPTOR significantly outperforms other programs at the fold similarity level. The
RAPTOR webserver is at http://www.cs.uwaterloo.ca/~j3xu/RAPTOR_form.htm

1   Introduction

The Human Genome Project has led to the identification of over 30 thousand
genes in the human genome, which might encode, by some estimation, 100,000
proteins as a result of alternative splicing. To fully understand the biological
functions and functional mechanisms of these proteins, the knowledge of their
3-D structures is required. The ambitious *Structural Genomics Initiatives*,
launched by NIH in 1999, intends to solve these protein structures within
the next ten years, through the development and application of significantly
improved experimental and computational technologies.

A protein structure is typically solved using *x-ray crystallography* or *nuclear magnetic resonance spectroscopy (NMR)*, which are costly and very time-
consuming (ranging from months to years per structure) and is quite difficult
for high-throughput production. The overall strategy of the NIH Structural

Genomics Initiatives is to solve protein structures using experimental techniques like x-ray crystallography or NMR only for a small fraction of all the proteins and to employ computational techniques to model the structures for the rest of the proteins. The basic premise used here is that though there could be millions of proteins in nature, the number of unique structural folds is probably 2-3 (or even more) orders of magnitude smaller. Hence by strategically selecting the proteins with unique structural folds for experimental solutions, we can put the vast majority of other proteins "within the modeling distance" of these proteins. Model-based structure prediction techniques could play a significant role in helping to achieve the goal of the Structural Genomics Initiatives. *Protein threading* represents one of the most promising such techniques.

Protein threading can be used for both structure prediction and protein fold recognition, i.e., detection of homologous proteins. Numerous computer algorithms have been proposed for protein structure prediction, based on the threading approach. Based on the energy function models and computational methods, they can be grouped into three classes:

(1)The energy function does not include the pairwise interaction preferences explicitly. For this kind of model, a simple dynamic programming is employed to optimize the energy function. GenTHREADER[1] is a typical example. The prediction speed is fast, but theoretically, the prediction accuracy is worse than those incorporating pairwise interactions.

(2)The energy function includes the pairwise interaction preferences. However, it has been proved that this problem is NP-hard when variable gaps and pairwise interactions are considered simultaneously[2]. Some kinds of approximation algorithms are used to optimize the energy function. These methods include double dynamic programming[3], frozen approximation[4], and Monte Carlo sampling algorithm[5]. Unfortunately, T. Akutsu have proved that this problem is MAX-SNP-hard[6], which means that it cannot be approximated to arbitrary accuracy in polynomial time.

(3)The energy function includes the pairwise interaction preferences and an exact algorithm is designed to optimize the energy function. Xu *et al.* have proposed a divide-and-conquer method[7] which runs fast on simple protein template (interaction) topology but could take a long time for proteins with dense residue-residue interactions. In addition, this approach does not treat the following two special features explicitly: (i) interaction significance could be different from residues to residues; and (ii) interaction potentials could be heavily correlated with other non-pairwise scores such as mutation scores and fitness scores.

Our main focus in this paper is on the development of a globally optimal and practically efficient threading algorithm based on the alignment model incorporating the pairwise interaction preferences explicitly and allowing variable gaps by using the integer programming approach. Integer programming formulation can fully exploit the abovementioned two special features of the pairwise interaction preferences. It allows us to use the existing powerful linear programming packages together with some branch and bound algorithm to rapidly arrive at the optimal alignment. To our knowledge, this is the first time that integer programming is applied to protein threading.

2   Alignment Model

We represent the amino acid sequence, of length m, of a protein template by $t_1 t_2 \ldots t_m$ and the query sequence, of length n, by $s_1 s_2 \ldots s_n$. In formulating the protein threading problem, we follow a few basic assumptions widely adopted by the protein threading community[7]. We assume that:

(1) Each template consists of a linear series of cores with the connecting loops between the adjacent cores. Each core is a conserved segment of an α-helix or β-sheet secondary structure among the protein's homologs. Although the secondary structure is often conserved, insertion or deletion may occur within a secondary structure. So we only keep the most conserved part. Let $c_i = core(head_i, tail_i)$ denote all cores of one template, where $i = 1, 2, \ldots, M$ with M being the number of the cores, and $1 \leq head_1 \leq tail_1 < head_2 \leq tail_2 < \ldots < head_M \leq tail_M \leq m$. The region between $tail_i$ and $head_{i+1}$ is a loop. The length of c_i is $len_i = tail_i - head_i + 1$. Let loc_i denote the sum of the length of all cores before c_i, i.e., $loc_i = \sum_{j=1}^{i-1} len_j$.

(2) When aligning a query protein sequence with a template, alignment gaps are confined to loops, *i.e.*, the regions between cores or the two ends of the template. The biological justification is that cores are conserved so that the chance of insertion or deletion within them is very little.

(3) We consider only interactions between core residues. It is generally believed that interactions involving loop residues can be ignored as their contribution to fold recognition is relatively insignificant. We say that an interaction exists between two residues if the spatial distance between their C_α atoms is within $7A$ and they are at least 4 positions apart in the template sequence. *We say that an interaction exists between two cores if there exists at least one residue-residue interaction between the two cores.*

Our threading energy function consists of environment fitness score E_s, mutation score E_m, secondary structure compatibility score E_{ss}, gap penalty E_g and pairwise interaction score E_p. The overall energy function E has the following form:

$$E = W_m E_m + W_s E_s + W_p E_p + W_g E_g + W_{ss} E_{ss},$$

where $W_m, W_s, W_p, W_g, W_{ss}$ are weight factors to be determined by training.

Global alignment and global-local alignment methods are employed to align one template to one sequence. For the detailed description, please refer to Fischer's paper[8]. In the case that the template size is larger than the query sequence size, it is possible that some cores at the two ends of the template cannot be aligned to the sequence. But we can always extend the sequence by adding some "artificial" amino acids to the two ends of the sequence to make all cores are aligned to the (extended) sequence. All scores involving those extended positions are set to be 0.

3 Formulation

Definition 3.1 *We use an undirected graph $CMG = (V, E)$ to denote the contact map graph of a protein template structure. Here, $V = \{c_1, c_2, ..., c_M\}$ where c_i represents i^{th} core, and $E = \{(c_i, c_j) | there is an interaction between c_i and c_j, or $|i - j| = 1\}$.*

For simplicity, when we say that core c_i is aligned to position s_j, we always mean that core $c_i = (head_i, tail_i)$ is aligned to segment (s_j, s_{j+len_i-1}). In order to speed up the search, RAPTOR employs some knowledge-based filtering process proposed in PROSPECT[7] that indicates certain alignments as *invalid*.

Definition 3.2 *Let B denote the alignment bipartite graph of one threading pair. Each core of the template corresponds to one vertex in B, labeled as $c_i (i = 1, 2, \ldots, M)$, each residue in the query sequence corresponds to one vertex in B, labeled as $s_j (j = 1, 2, \ldots, n)$. The edges of B consist of all valid alignments (after initial filtering) between each core and each sequence position. The edges of B are also called the alignment edges.*

Definition 3.3 *For any two different edges $e_1 = (c_{i_1}, s_{j_1})$ and $e_2 = (c_{i_2}, s_{j_2})$ in an alignment bipartite graph B, if $(loc_{i_1} - loc_{i_2}) \times (s_{j_1} + loc_{i_2} - loc_{i_1} - s_{j_2}) \leq 0$, then we say e_1 and e_2 are in conflict.*

The proof of the following three lemmas is omitted due to space limit.

Lemma 3.1 *For any three different edges $e_r = (c_{i_r}, s_{j_r})$, $r = 1, 2, 3$ and $loc_{i_1} < loc_{i_2} < loc_{i_3}$, if e_1 conflicts with e_2 and e_2 conflicts with e_3, then e_1 conflicts with e_3.*

Lemma 3.2 *For any three different edges $e_r = (c_{i_r}, s_{j_r})$, $r = 1, 2, 3$ and $loc_{i_1} < loc_{i_2} < loc_{i_3}$, if e_1 does not conflict with e_2 and e_2 does not conflict with e_3, then e_1 does not conflict with e_3.*

Lemma 3.3 *For any three different edges $e_r = (c_{i_r}, s_{j_r})$, $r = 1, 2, 3$ and $loc_{i_1} < loc_{i_2} < loc_{i_3}$, if e_1 conflicts with e_3, then e_2 conflicts with e_3 or e_2 conflicts with e_1.*

Definition 3.4 *An alignment is called a valid alignment if: (1) each core of the template is aligned to some position of the (extended) sequence[a]; (2)For any two different cores c_1 and c_2, their two alignment edges do not conflict in the alignment graph.*

Let $x_{i,l}$ be a boolean variable such that $x_{i,l} = 1$ if and only if core c_i is aligned to position s_l. Similarly, for any $(c_{i_1}, c_{i_2}) \in E(CMG)$, let $y_{(i_1,l_1),(i_2,l_2)}$ indicate the pairwise interactions between x_{i_1,l_1} and x_{i_2,l_2} if the two edges $(c_{i_1}, s_{l_1}), (c_{i_2}, s_{l_2})$ do not conflict. $y_{(i_1,l_1),(i_2,l_2)} = 1$ if and only if $x_{i_1,l_1} = 1$ and $x_{i_2,l_2} = 1$. We say $y_{(i_1,l_1),(i_2,l_2)}$ is generated by x_{i_1,l_1} and x_{i_2,l_2}.

The x variables are called the alignment variables and y variables are called the interaction variables. Let $D[i]$ denote all valid query sequence positions that c_i could be aligned to. Let $R[i, j, l]$ denote all valid alignment positions of c_j given c_i is aligned to s_l.

Now the objective function of the protein threading problem can be formulated as follows:

$$\min W_m E_m + W_s E_s + W_p E_p + W_g E_g + W_{ss} E_{ss}, \tag{1}$$

$$E_m = \sum_{i=1}^{M} \sum_{l \in D[i]} [x_{i,l} \times \sum_{r=0}^{len_i-1} Mutation(head_i + r, l + r)], \tag{2}$$

$$E_s = \sum_{i=1}^{M} \sum_{l \in D[i]} [x_{i,l} \times \sum_{r=0}^{len_i-1} Fitness(head_i + r, j + r)], \tag{3}$$

$$E_{ss} = \sum_{i=1}^{M} \sum_{l \in D[i]} [x_{i,l} \times \sum_{r=0}^{len_i-1} SS(t_{head_i+r}, s_{j+r})], \tag{4}$$

$$E_p = \sum_{1 \leq i < j \leq M, (c_i,c_j) \in E(CMG)} \sum_{l \in D[i]} \sum_{k \in R[i,j,l]} y_{(i,l),(j,k)} P(i, j, l, k), \tag{5}$$

$$P(i, j, l, k) = \sum_{u=0}^{len_i-1} \sum_{v=0}^{len_j-1} \delta(t_{head_i+u}, t_{head_j+v}) Pair(l + u, k + v) \tag{6}$$

---

[a] As mentioned before, global and global-local alignment are employed.

$$E_g = \sum_{i=1}^{M} \sum_{l \in D[i]} \sum_{k \in R[i,i+1,l]} y_{(i,l),(i+1,k)} G(i,l,k), \tag{7}$$

where $\delta(t_u, t_v) = 1$ if there is an interaction between residues at position u and v in the template, otherwise 0. $G(i,l,k)$ is the gap potential between c_i and c_{i+1} when they are aligned to query sequence position l and k respectively. $G(i,l,k)$ could be computed by dynamic programming in advance given i, l, k.

The constraint set is as follows:

$$\sum_{j \in D[i]} x_{i,j} = 1, i = 1, 2, \ldots, M; \tag{8}$$

$$\sum_{l \geq l_0, l \in D[i]} x_{i,l} + \sum_{k \in D[i+1] - R[i,i+1,l_0]} x_{i+1,k} \leq 1, l_0 \in D[i], i = 1, 2, \ldots, M-1; \tag{9}$$

$$\sum_{k \in R[i,j,l]} y_{(i,l),(j,k)} \leq x_{i,l}, \forall l \in D[i], i, j = 1, 2, \ldots, M; \tag{10}$$

$$\sum_{l \in R[j,i,k]} y_{(i,l),(j,k)} \leq x_{j,k}, \forall k \in D[j], i, j = 1, 2, \ldots, M; \tag{11}$$

$$\sum_{k \in R[i,j,l]} y_{(i,l),(j,k)} \geq x_{i,l} + \sum_{k \in R[i,j,l]} x_{j,k} - 1, l \in D[i], i, j = 1, 2, \ldots, M; \tag{12}$$

$$\sum_{l \in R[j,i,k]} y_{(i,l),(j,k)} \geq x_{j,k} + \sum_{l \in R[j,i,k]} x_{i,l} - 1, k \in D[j], i, j = 1, 2, \ldots, M; \tag{13}$$

$$x_{i,j} \geq 0, j \in D[i], i = 1, 2, \ldots, M; \tag{14}$$

$$y_{(i,l)(j,k)} \geq 0, \forall l \in D[i], k \in D[j], i, j = 1, 2, \ldots, M. \tag{15}$$

Constraint 8 says that one core can be aligned to a unique sequence position. Constraint 9 forbids the conflicts between the adjacent two cores. Therefore, based on lemma 3.2, this constraint can guarantee that there are no conflicts between any two cores if variable x and y are integral. Constraints 10 and 11 say that at most one interaction variable can be 1 between any two cores that have interactions between each other. Constraints 12 and 13 enforce that if two cores have their alignments to the sequence respectively and also have interactions between them, then at least one interaction variable should be 1. Constraints 8,14 and 15 guarantee x and y to be between 0 and 1 when this problem is relaxed to linear program.

There is another set of more obvious constraints which can replace Constraints 9-13. They are:

$$x_{i,l} + x_{i+1,k} \leq 1, \forall k \in D[i+1] - R[i, i+1, l]; \tag{16}$$

$$y_{(i,l)(j,k)} \leq x_{i,l}, k \in R[i, j, l], (c_i, c_j) \in E(CMG); \tag{17}$$

$$y_{(i,l)(j,k)} \leq x_{j,k}, l \in R[j, i, k], (c_i, c_j) \in E(CMG); \tag{18}$$

$$y_{(i,l)(j,k)} \geq x_{i,l} + x_{j,k} - 1, (c_i, c_j) \in E(CMG); \tag{19}$$

Constraint 16 forbids the conflict between two neighboring cores and Constraints 17-19 guarantee that one interaction variable is 1 if and only if its two generating x variables are 1. Constraints 16-19 can be inferred from Constrains 9-13. Conversely, it is not true. Therefore, Constraints 16-19 are weaker than Constraints 9-13.

In order to improve running time, we found yet another set of Constraints 20 and 21 from which both 9-13 and 16-19 can be inferred (proofs omitted due to space limitation).

$$\sum_{k \in R[i,j,l]} y_{(i,l)(j,k)} = x_{i,l}, (c_i, c_j) \in E(CMG); \tag{20}$$

$$\sum_{l \in R[j,i,k]} y_{(i,l)(j,k)} = x_{j,k}, (c_i, c_j) \in E(CMG); \tag{21}$$

Constraints 20 and 21 imply that one x variable is 1 is equivalent to that one of the y variables generated by it is 1 . These two are the strongest constraints. Experimental results show that our algorithm with Constraints 20 and 21 (combining with Constraints 8,14 and 15) runs significantly faster. Our program RAPTOR uses 20 and 21 by default.

4 RAPTOR – Implementation

4.1 Scoring System

We calculated the averaged energy over a set of homologous sequences, as demonstrated in PROSPECT-II[9]. Given a query sequence of length n, an $n \times 20$ frequency matrix $PSFM$ is calculated by using PSI-BLAST[10] with maximum iteration number being set to 5. Each column of this matrix describes the occurring frequency of 20 amino acids at this position. Assume a template position i is aligned to the sequence position j. Then the mutation score and fitness score are calculated as follows.

$$Mutation(i, j) = \sum_a p_{j,a} M(t_i, a)$$

$$Fitness(i, j) = \sum_a p_{j,a} F(env_i, a)$$

where $p_{j,a}$ represents the occurring frequency of amino acid a at sequence position j, $M(a, b)$ represents the mutation potential between two amino acid a and b which is taken from PAM250 matrix[11], $F(env, a)$ denote the fitness potential when amino acid a is placed into environment env.

The 9 combinations of three secondary structure types (α-helix, β-strand and coil) and three solvent accessibility levels are used to define the local environments of a position in the template. The boundaries between thre solvent accessibility levels are at 7% and 37%. Secondary structure and solvent accessibility assignments are all taken from FSSP database[12].

The gap penalty function is assumed to be an affine function, i.e., a gap open penalty plus a length-dependent gap extension penalty. Gap open penalty is set at 10.6 and gap elongation penalty is 0.8 per single gap[13]. We use PSIPRED [14] to predict the secondary structure of the query sequence.

If the two ends of an interaction are aligned to j_1^{th} and j_2^{th} positions of the query sequence respectively, then the pair score for this interaction is given by:

$$Pair(j_1, j_2) = \sum_a p_{j_1,a} \sum_b p_{j_2,b} P(a, b)$$

where $P(a, b)$ denotes the pairwise interaction potential between two amino acids a and b. F, P are taken from PROSPECT-II[9].

4.2   Branch-and-Bound Method

We use a branch-and-bound algorithm to solve the above integer programming problem. First we relax the above integer program by allowing all x and y to be real between 0 and 1 and solve the resulting linear program. If the solution (x^*, y^*) of the linear program is integral, then we get the optimal solution. Otherwise, we select one non-integral variable according to some criterion, and generate two subproblems by setting it to 0 and 1 respectively. These two subproblems are solved recursively. More details on solving integer programming problem can be found in[15]. IBM OSL(Optimization and Solution Library) package is used to implement this process.

4.3   Weight Training

The weight factors $W_m, W_s, W_{ss}, W_g, W_p$ are chosen through optimizing the overall alignment accuracy. The optimal alignment accuracy does not necessarily imply the best fold recognition capability though. In the following subsection, an SVM (Support Vector Machine) method is used to carry out fold recognition. A set of 95 structurally-aligned protein pairs are chosen from Holm's test set[16] as the training samples, each of which only has fold-level similarity. The alignments generated by RAPTOR is compared with the structural alignments generated by SARF[17]. An alignment for a residue is regarded as correct if it is within 4 residue shift away from the correct structure-structure alignment by SARF. The overall alignment accuracy is defined as the ratio between the number of the correctly-aligned positions of all threading pairs and the number of the maximum alignable positions. Our objective is to maximize the overall alignment accuracy. A genetic algorithm plus a local pattern search method implemented in DAKOTA[18] is used to search for the optimal weight factors. We attained 56% alignment accuracy over this set of training pairs. A set of 1100 protein pairs which are in the fold-level similarity is also generated from Holm's test set[16] to test the weight factors and 50% alignment accuracy is attained. We have also selected 95 structurally-aligned protein pairs from Holm's test set, each of which is in superfamily-level or family-level similarity, 80% alignment accuracy is achieved when the same set of weight factors is used.

4.4   z-score and Fold Recognition

After threading one pair of sequence and template, z-score is calculated according to the method proposed in paper[19] to cancel out the composition bias. Let z_{raw} denote this kind of z-score. However, since the accurate z_{raw} is expensive to compute, we just approximate it by (i) fixing the alignment positions; (ii) shuffling the query sequence randomly; and (iii) calculating the alignment scores based on the existing alignment rather than doing optimal alignments again and again. The free software SVM light[20] with RBF kernel is employed to adjust the approximate z-score. Due to space limit, we refer the reader o Vapnik's book[21] for a comprehensive tutorial of SVM. A set of 60000 training pairs formed by all-against-all threading between 300 templates (randomly chosen from the FSSP database) and 200 sequences (randomly chosen from Holm's test set [16]) is used as the training samples of our SVM model. The relationship between two proteins is judged based on SCOP database[22]. If one pair is in at least fold-level similarity, then it is treated as a positive example, otherwise a negative example. Each of training samples consists of the following features: (1)z_{raw}; (2) the sequence size; (3) the template size; (4) the number of cores in the template; (5) the sum of the core size in the template; (6) the number of aligned cores; (7) the number of aligned positions; (8) the number of identical residues; (9) the number of contacts with both ends on the aligned cores; (10) the number of cut contacts with one end on the aligned cores and the other on the unaligned cores; (11) the total score; (12) mutation score; (13) singleton fitness score; (14) gap score; (15) secondary score; (16) pair score. Given one threading result, SVM outputs a real value. The value greater than 0 means this threading pair is in at least fold-level similarity. We do not use this directly due the abundance of the false negatives. We calculate the final z-score for each query sequence. For all threading pairs of one given sequence, let $o_1, o_2, ..., o_q$ denote the outputs from SVM model. The final z-score is calculate by $\frac{o_i - u(o)}{std(o)}$, where $u(o)$ is the mean value of o_i and $std(o)$ is the standard deviation of o_i. Daniel Fischer's benchmark[8] is used to fix the parameters of the model.

5   Preliminary Experimental Results

Fischer's benchmark consists of 68 target sequences and 301 templates. RAP-TOR ranks 56 pairs out of 68 pairs as top 1, achieving 82% prediction rate, while the previous best was 76.5%.

The fold recognition performance of RAPTOR was further tested on Lindahl's benchmark set consisting of 976 protein sequences [23]. By threading them all against all, there are totally 976×975 pairs. We measure RAPTOR's performance in three different similarity levels: fold, superfamily and family. The results are shown in Table 1. The results of other methods are taken from Shi et al's paper[24].

As shown in Table 1, the performance of RAPTOR at the fold level is much better than the others. At the superfamily level, RAPTOR performs a little bit worse than FUGUE [24], the best method (for superfamily and family level) listed in this table. However, at the family level, RAPTOR performs better than only THREADER, which means that RAPTOR is superior in recognizing fold-level similarity but bad in doing homology detection. RAPTOR-np is a variant of RAPTOR without considering pairwise interactions when doing optimal alignment, but the pairwise score is still calculated based on the non-pairwise alignment. The corresponding weight factors and SVM model are optimized separately using the same sets of training samples. Compared with RAPTOR-np, RAPTOR is better in fold level and superfamily level and same in family level. Thus, we may conclude that a strict treatment of the pairwise interactions is necessary for fold level recognition or even superfamily level.

method	Family		Superfamily		Fold	
	Top 1	Top 5	Top 1	Top 5	Top 1	Top 5
RAPTOR	**75.2**	**77.8**	**39.3**	**50.0**	**25.4**	**45.1**
RAPTOR-np	**68.9**	**72.8**	**34.0**	**49.7**	**19.0**	**36.6**
FUGUE	82.2	85.8	41.9	53.2	12.5	26.8
PSI-BLAST	71.2	72.3	27.4	27.9	4.0	4.7
HMMER-PSIBLAST	67.7	73.5	20.7	31.3	4.4	14.6
SAMT98-PSIBLAST	70.1	75.4	28.3	38.9	3.4	18.7
BLASTLINK	74.6	78.9	29.3	40.6	6.9	16.5
SSEARCH	68.6	75.7	20.7	32.5	5.6	15.6
THREADER	49.2	58.9	10.8	24.7	14.6	37.7

Table 1: The performance of RAPTOR at three different similarity levels

6 Computing Efficiency Issues

An outstanding advantage of our algorithm is that the memory requirement is just about $O(M^2 n^2)$ and, at most of time, the computing time does not increase exponentially with respect to the sequence size. Figure 1 shows the CPU time of threading 100 sequences (chosen randomly from Lindahl's benchmark) with size ranging from 25 to 572 to a typical template 119l of length 162 (with topological complexity[7] 3 and 12 cores). According to Xu *et al.*'s paper[7], the computing time of PROSPECT is $O(Mn^5)$ and its memory usage is $O(Mn^4)$. The observed memory usage of RAPTOR is $100 \sim 200M$ for most of threading pairs. Figure 1 shows that the computing time of our algorithm increases very slowly with respect to the sequence size. In fact, we found out that our relaxed linear programming gave the integral solutions most of time or generated only a few branch nodes when the solution was not integral.

References

1. D.T. Jones. *J. Mol. Biol.*, 287:797–815, 1999.
2. R.H. Lathrop. *Protein Engineering*, 7:1059–1068, 1994.
3. D.T. Jones, W.R. Taylor, and J.M. Thornton. *Nature*, 358:86–98, 1992.
4. A.Godzik, A.Kolinski, and J.Skolnick. *J. Mol. Biol.*, 227:227–238, 1992.
5. S. Bryant. *Proteins: Struct. Funct. Genet.*, 26:172–185, 1996.
6. T. Akutsu and S. Miyano. *Theoretical Computer Science*, 210:261–275, 1999.
7. Y. Xu *et al. Journal of Computational Biology*, 5(3):597–614, 1998.
8. D. Fischer *et al.* pages 300–318. PSB96, 1996.

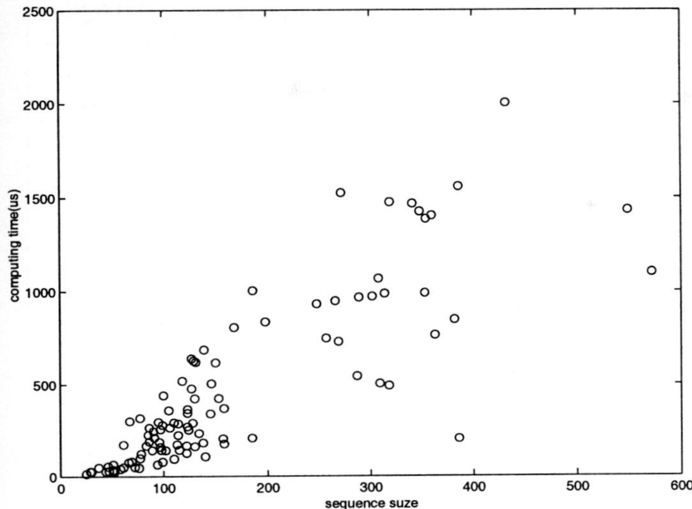

Figure 1: computing time of threading 100 sequences to template 1191.

9. D. Kim, D. Xu, J. Guo, K. Ellrott, and Y. Xu. 2002. Manuscript.

10. S.F. Altschul *et al.* *Nucleic Acids Research*, 25:3389–3402, 1997.

11. R.M. Schwartz and M.O. Dayhoff. pages 353–358. Natl. Biomed. Res. Found., 1978.

12. L. Holm and C. Sander. *Science*, 273:595–602, 1996.

13. G.H. Gonnet *et al.* *Science*, 256:1443–1445, 1992.

14. D.T. Jones. *J. Mol. Biol.*, 292:195–202, 1999.

15. Laurenece A. Wolsey. John Wiley and Sons, Inc., 1998.

16. L. Holm and C. Sander. *ISMB*, 5:140–146, 1997.

17. N.N. Alexandrov. *Protein Engineering*, 9:727–732, 1996.

18. M.S. Eldred *et al.* Technical Report SAND2001-3796, Sandia, 2002.

19. S.H. Bryant and S.F. Altschul. *Curr. Opin. Struct. Biol.*, 5:236–244, 1995.

20. T. Joachims. MIT Press, 1999.

21. V.N. Vapnik. Springer, 1995.

22. A.G. Muzrin *et al.* *J. Mol. Biol.*, 247:536–540, 1995.

23. E. Lindahl and A. Elofsson. *J. Mol. Biol.*, 295:613–625, 2000.

24. J. Shi, L. B. Tom, and M. Kenji. *J. Mol. Biol.*, 310:243–257, 2001.

GENOME-WIDE ANALYSIS AND COMPARATIVE GENOMICS

LIPING WEI
Nexus Genomics, Inc.
229 Polaris Avenue, Suite 6
Mountain View, CA 94043
wei@nexusgenomics.com

INNA DUBCHAK
Lawrence Berkeley National Laboratory
MS 84-171, Berkeley, CA 94720
ildubchak@lbl.gov

VICTOR SOLOVYEV
Softberry, Inc.
victor@softberry.com

This is the second year at the Pacific Symposium on Biocomputing (PSB) that a session is devoted to genome-wide analysis and comparative genomics. In the past year, we have witnessed even greater growth of the amount of genomic sequence data. For example, the number of complete eucaryote genome sequences available at NCBI has nearly doubled in the past year, to a total number of eight as of September, 2002. The draft sequences of several major eucaryotes have been published, including mouse and rice. In addition, numerous procaryote genomes have been completely sequenced. The explosion in the amount of genomic sequence data has resulted in unprecedented opportunities for discoveries from computational genome analyses. At the same time, the large amount of data has proven challenging for computational scientists to develop accurate and efficient algorithms. The papers in this section represent excellent examples of new analysis and novel algorithms that are contributing greatly to our understanding of genome biology.

One of the most important and challenging steps in genome annotation is gene prediction. In eukaryotic genomes, which have long noncoding regions and large introns, ab initio gene finders such as Fgenesh and Genscan identify about 90% of the genes, but generate false positive predictions and even more often predict partially incorrect gene structures. Thus, incorporation of as much available evidence as possible for gene prediction is necessary. The paper by Yada at al. presents the DIGIT algorithm that predicts genes by combining the results from several existing gene finders. It was able to successfully discard many false positive exons predicted by the individual programs and showed remarkable improvements in sensitivity and specificity. Another approach to improve quality of gene prediction is to implement more accurate models of splice site recognition, which is highly nontrivial. An interesting attempt in this direction is presented in the paper by Ott et al. The paper suggests that the splicing of long introns might be facilitated by splicing inner parts of the intron prior to the splicing of the long intron itself.

One of the challenges of the postgenomic era is to understand the regulation of gene transcription. The combinatorial nature of regulation and the practically unlimited number of cellular conditions significantly complicate the experimental identification of transcription factor (TF) binding sites on a large scale. Therefore, computational approaches to reveal potential regulatory elements become very important. Cheremushkin and Kel described a new technique, a variant of phylogenetic footprinting implementation, to mine conserved noncoding regions between human and mouse, and compiled a database of ~60,000 predicted potential TF binding sites. Another paper by Olman at al. introduced a beautiful minimum spanning tree algorithmic solution to reveal regulatory binding sites in a set of similarly regulated genes. Such sets of genes are presumed to share common TF binding sites. They could be extracted from the increasing volume of microarray gene expression data. The paper by Phang et al. outlined a novel non-parametric method called trajectory clustering that was able to cluster groups of genes with known related function better than alternative approaches.

Completion of the sequencing of many genomes and the exponential growth of biological databases present new challenges to sensitive homology searches. The size of the sequences is perhaps the biggest hurdle, since many alignment algorithms were designed for comparing single proteins and are extremely inefficient when processing large genomic intervals. The paper by Kahveci and Singh addresses one of those problems. They proposed an efficient technique to align long genomic strings up to ~100 times faster than the BLAST algorithm.

Finally, three articles deal with different aspects of phylogenetic analysis of genomic data. Distance-based methods are widely used for inferring phylogenies. Recently introduced reversal distance based on the orders of genes is defined as the minimum number of signed/unsigned reversals needed to account for the difference in gene order between two genomes. Wu and Gu presented an effective algorithm to reconstruct optimal distance-based phylogenetic trees for genomes. Recent findings reinforced the view that while considering evolutionary relationships, we need to account for such genomic events as gene duplication, loss, convergence and lateral (horizontal) gene transfer. The paper by Addrario-Berry et. al. is concerned with evaluating the performance of the model and algorithm for detecting lateral gene transfer events. Such biological processes as hybridization of horizontal gene transfer require network rather than tree structure of relationships. The paper by Nakleh with co-authors reports the development of computational tools for evaluating phylogenetic network reconstruction methods.

The session co-chairs are grateful to all the authors who had submitted their work to this session, and to all the reviewers for their help in the difficult task of choosing the best contributions from a large number of excellent submissions.

TOWARDS IDENTIFYING LATERAL GENE TRANSFER EVENTS

L. ADDARIO-BERRY AND M. HALLETT

McGill Centre for Bioinformatics, McGill University, Montréal, Canada
E-mail: {laddar,hallett}@mcb.mcgill.ca

J. LAGERGREN

Stockholm Bioinformatics Center and Dept. of Numerical Analysis and Computer Science, KTH, Stockholm, Sweden
E-mail: jensl@nada.kth.se

This paper is concerned with evaluating the performance of the model and algorithm in [5] for detecting lateral gene transfers events. Using a Poisson process to describe arrival times of transfer events, a simulation is used to generate "synthetic" gene and species trees. An implementation of an efficient algorithm in [5] is used to estimate the minimum number of transfers necessary to explain disagreements between the generated gene and species trees. Our first result suggests that the algorithm can solve realistic size instances of the problem. Our second result suggests that the mean error and variance are low when saturation does not occur. Additionally, certain plausible evolutionary events allowed by our model of evolution used to generate gene and species trees but not detectable by the algorithm occur rarely implying the framework should work well in practice. Our third, surprising result suggests that the number of optimal scenarios is on average low for realistic input sizes.

1 Introduction

Recent findings have reinforced the view that evolutionary relationships between taxa (i.e. the species tree) cannot be inferred from a single gene family (i.e. a single gene tree) due to genomic events such as *gene duplication, gene loss, gene convergence,* and *lateral gene transfer* (a.k.a. horizontal gene transfer) [3,4,5,6,7,8,9,10]. In essence, these events cause gene trees to not be equal to the species tree; that is, they "disagree". To explain such disagreements, various models have been developed that assume a simplified evolutionary process restricted to a subset of these genomic events. A natural computational problem is to find the most parsimonious *scenario* that explains how, via these events, the disagreements between the gene tree and species tree arise.

A well studied model is the *duplication, loss model* [4,9]. Here a species tree S and a gene tree T are given. Via a (computationally easy) least common ancestor mapping from the vertices of T to the vertices of S, it is possible to identify all vertices in T that correspond to duplication events and locate where they occur in the evolution represented by S. In a manner analogous to the duplication, loss model, the authors of [5,6] construct a model for lateral gene transfer. Here we are given a (hypothetically correct) species tree S and a (hypothetically correct) gene tree T. The goal is to find a most parsimonious scenario that explains disagreements between the two trees using lateral gene transfer events. This work extends previous work on the *subtree transfer* [7,8] and *network* [10] models in that the resulting scenarios are biologically sound. The price of this biological realism is an increase in complexity of the model itself - it is difficult to study analytically the behavior of the system.

This paper is concerned with evaluating experimentally the performance of the model and algorithm in [5] for detecting lateral transfers. Our experimental technique is analogous to those commonly used by the phylogenetic community. We begin with a method for simulating evolution with lateral transfer events. Using a Poisson process with rate parameter λ to describe the arrival times of transfer events, we use a discrete event simulation to probabilistically generate "synthetic" gene trees w.r.t. the species tree and λ. Next, using an implementation of the algorithm in [5] for what is termed activity level 1 (at most one gene per gene family may exist in a genome at any point in the evolution of the taxa), the minimum number of transfers necessary to explain disagreement between the gene and species tree is estimated.

The simulation of evolution is pessimistic in the sense that biologically plausible evolutionary events can occur that are not detectable by the algorithm. The first such event termed a *useless* transfer is analogous to a "back substitution" in molecular sequence evolution. Useless transfers will not be detected in a parsimony-based framework and hence the algorithm from [5] will underestimate the true number of transfers. The second type of degenerate event is termed a *transfer-loss* event. In certain cases, a transfer event followed by a gene loss event can cause the algorithm from [5] to grossly under-estimate or over-estimate the true number of transfer events for a gene and species tree. It is conjectured that such events, although biologically plausible, occur with low frequency. Furthermore, the authors of [5] conjecture that any algorithm that does detect such transfer-loss events would be computationally infeasible for even small instances of the problem. One of the primary goals of this paper is to test the frequency of harmful transfer-loss events under a reasonable model of evolution.

We answer several questions concerning the model from [5]. Our first result suggests that, although the running time of the algorithm is high, it is fast enough to solve instances one expects to encounter in practice. On a desktop machine, we managed to compute minimum cost scenarios when the number of leaves of the species tree (taxa) n is 300 and the number of transfers τ is 20. The algorithm from [5] has a worst case running time of $O(2^{4\tau}n^2)$ and consists of two phases. Although it is possible to construct examples where both phases are required, in over $10,000$ experiments we did not find a single example of a scenario that required the second phase (cycle removal). This suggests that the first phase of the algorithm is sufficient for realistic data sets and implies a more optimistic $O(2^{2\tau}n^2)$ running time for such data sets. Our second result suggests that the mean error (actual number of transfers used minus minimum cost of scenario found) and variance are low for all realistic values of n, τ and λ. This indicates that the number of harmful transfer-lost and other degenerate events are negligible and the framework should work well in practice. When λ is sufficiently large as to cause low levels of *saturation*, the algorithm still gives reasonable estimates of the number of transfers. Our third, surprising result suggests that the number of valid minimum cost scenarios is on average low for realistic values of n, τ and λ.

2   Definitions

We consider rooted directed trees where the arcs are directed from the root towards the leaves and a vertex has out-degree at most 2. We call such a tree a *rooted* tree. For such a tree T, $V(T)$ denotes the set of vertices and $A(T)$ denotes the set of arcs. The *internal vertices* of T are $V(T) \setminus L(T)$. The *root* of a tree T is denoted $r(T)$. For $u \in V(T)$, any vertex v reachable from u by a directed path is a *descendant* of u (this means that u is a descendant of u). We denote this by $v \leq_T u$. We also say that u is an *ancestor* of v ($u \geq_T v$). We say that v is a *proper descendant* (*proper ancestor*) of u, if $v \leq_T u$ ($v \geq_T u$) and $v \neq u$ and denote this relationship by $v <_T u$ ($v >_T u$). Both a *gene tree* T and a *species tree* S are binary rooted directed trees. We assume that $n = |L(S)| \geq |L(T)|$. By a rooted forest we mean a union of disjoint rooted trees. The set of leaves of a rooted forest F is denoted $L(F)$. For a vertex $u \in V(F)$, let F_u be the rooted subtree of F consisting of the vertices of $V(F)$ reachable by directed paths from u. Let T be a rooted tree. For $X \subseteq L(T)$, the *least common ancestor* of X in T, written $lca_T(X)$, is defined as follows: if $X = \{v\}$, then $lca_T(X) = v$; otherwise, $lca_T(X)$ is the vertex v such that $X \subseteq L(T_v)$ but $X \not\subseteq L(T_u)$ for each proper descendant u of v. Let T be a gene tree and let F, $F \subset T$, be a forest. The mapping $\lambda_{F,S} : V(F) \to V(S)$ is defined as follows: $\lambda_{F,S}(v) = lca_S(L(F_v))$. A *mixed graph* G is a graph containing arcs as well as undirected edges. The arcs of G are denoted $A(G)$, the edges $E(G)$, and the vertices $V(G)$. If G is a mixed graph and A is a set of arcs, then $G \cup A$ is used to denote the mixed graph with arcs $A(G) \cup A$, edges

$E(G)$, and vertices $V(G)$. For a set of edges E, $G \cup E$ is defined similarly. A *directed mixed cycle* is a mixed graph where each vertex has total degree 2, which contains arcs and edges, and where the cycle can be traversed in a direction that respects the arcs but edges may be traversed in either direction. If A is a set of arcs, then $E(A)$ denotes the underlying undirected edges, i.e. $E(A) = \{(u,v) : \langle u,v \rangle \in A\}$. For $u,v \in V(T)$, let $P^T_{u,v}$ be the unique directed path between u and v in T. Let length of $P^T_{u,v}$, $|P^T_{u,v}|$, be the number of arcs in $P^T_{u,v}$.

3 Lateral Transfer Scenarios

The following section gives a simplified version of the model and an intuitive explanation of the algorithm that appeared in [5] for *lateral transfer scenarios*. A more detailed description of this work can be found in [6].

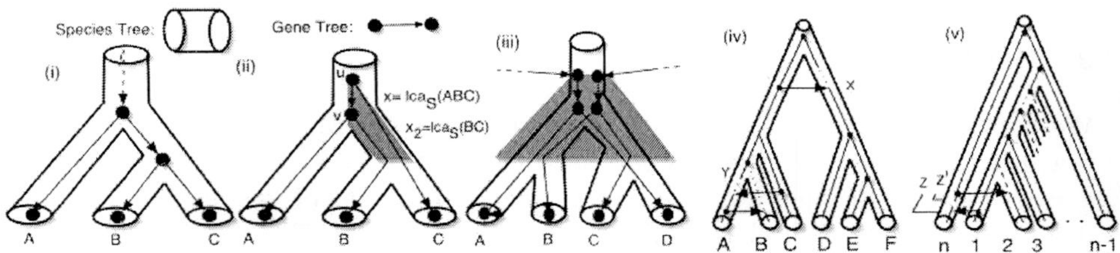

Figure 1. (i) The gene tree (thin lines and vertices) is drawn within the species tree (thick pipes). We do not explicitly direct the thick pipes of the species tree; however, note that the species tree is also rooted and directed. Here the gene tree and species trees agree. That is, the root of the species tree and the root of the gene tree both have A and BC (the ancestor of B and C) as their children. In (ii), the gene tree disagrees with the species tree since the root of the gene tree has C and AB as children. If we postulate a lateral transfer event from either v to child A or v to B, the resulting scenario would then be 1 active (see H-moves below). At any point during the evolution represented by the shaded region in (ii-iii), there exist two copies of the gene in the genome of these ancestral organisms. These examples are said to be 2 active. A second example of how activity levels > 1 can arise is depicted in (iii). Here two lateral transfer events have occurred prior to the root of $ABCD$. (iv) Two examples of useless transfers. (v) A harmful transfer-loss event.

The $\lambda_{T,S}$ mapping from a gene tree T to a species tree S places each vertex of the gene tree at its least common ancestor in the species tree. Figure 1 graphically depicts gene and species trees that agree (i) and trees that disagree (ii-iii). The model from [5] is built upon the assumption that: *Since A and B are siblings in the gene tree, either the ancestral gene AB must have been present in the ancestor of A and B in the species tree (i.e. the $lca_S(A,B)$)*

or a lateral transfer event has occurred from the A lineage to the B lineage (or vice versa). The postulated transfer thus explains why the gene tree has the sibling pair A and B. To capture this mapping of *gene trees into species trees,* we introduce the notions of *lateral transfer schemes* and *scenarios.*

Definition 1 *A* lateral transfer scheme *(or, simply* scheme*) for a species tree S is a pair (S', A') where S' is a subdivision of S and $A' \subseteq \{\langle x, y \rangle : x, y \in V(S') \setminus V(S), x \neq y\}$ such that: (1) the mixed graph $S' \cup E(A')$ does not contain a directed mixed cycle, (2) the tail of each arc in A' has in-degree 1 and out-degree 2 in $S' \cup A'$, and (3) the head of each arc in A' has in-degree 2 and out-degree 1 in $S' \cup A'$.*

Figure 2 gives an example of a scheme. In order for our *scenario* for a gene tree T and species tree S to be biologically meaningful, it must satisfy the following constraints.

Definition 2 *A* lateral transfer scenario *(or simply* scenario*) for a species tree S and a gene tree T is a triple (S', A', g) where (S', A') is a scheme for S and $g : V(S') \to V(T)$ such that: (1) $g(r(S')) = r(T)$; (2) if v_1 and v_2 are children of v_0 in T, then there exists x_0 with children x_1 and x_2 in $S' \cup A'$ (where $x_1 \neq x_2$) s.t. $v_i = g(x_i)$, for $i = 0, 1, 2$; (3) for each $v \in V(T)$, the vertices $\{x \in V(S') : g(x) = v\}$ induce a directed path in S'; (4) $g(l) = l$, for all $l \in L(S)$.*

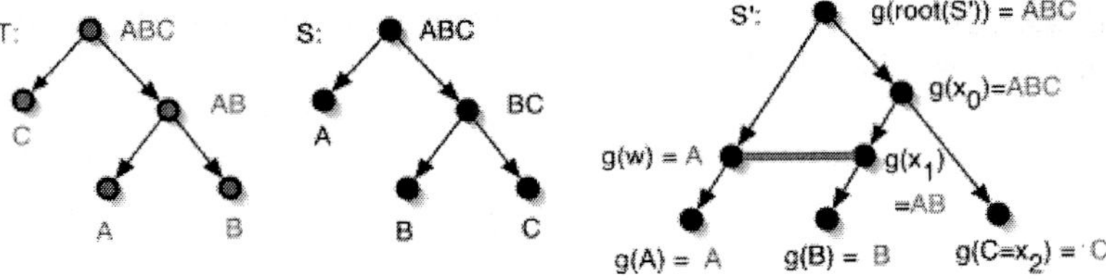

Figure 2. An example of a *scenario* for gene tree T and species tree S. S' represents a scheme for S. In S' there is only one transfer (between x_1 and w in S) and this arc is included in A'.

The *cost* of (S', A', g) w.r.t. T is simply $|A'|$. The *cost of S and T*, denoted $\tau(S, T)$, is the minimum cost of any scenario (S', A', g) for S and T. A detailed justification of each of these conditions along with a more in-depth discussion of mixed cycles in given in [5,6]. It is easy to verify that Figure 2 depicts a scenario for the gene tree T and species tree S with cost $|A'| = 1$. We note that Definition 2 is a simplification of a more general definition in [5,6]. In particular, the above definition holds only for what is termed *activity level* 1. Figure 1 (ii-iii) give an intuitive explanation of activity level.

The input to the τ-TRANSFER PROBLEM is a species tree S, a gene tree

T, and an integer τ. The output is a τ' lateral transfer scenario for S and T, $\tau' \leq \tau$. Let $T' \subseteq T$. The two basic operations the algorithm performs are H-moves and I-moves. In [6], the authors prove that these moves are sufficient to find all optimal scenarios.

Definition 3 (H-fat) *A vertex $x \in V(S)$ is H-fat for T' iff there exist $u, v \in \lambda_{T',S}^{-1}(x)$ such that: (1) v is $\leq_T$-minimal in $\lambda_{T',S}^{-1}(x)$, (2) u has out-degree 2 in T', and (3) $v <_T u$. If x is H-fat for T', then the two outgoing arcs of v (call them e_1 and e_2) are H-moves at x in T'.*

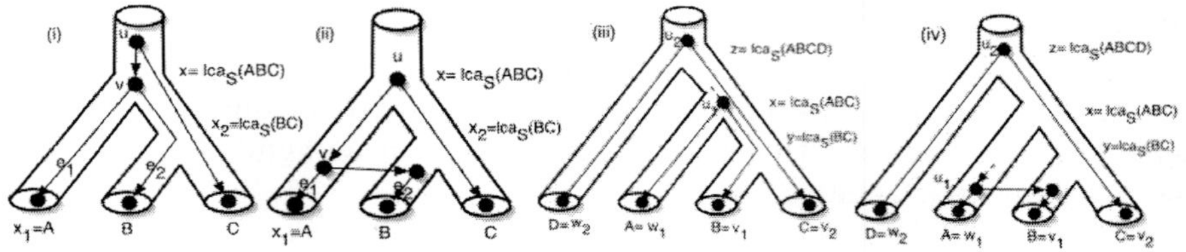

Figure 3. (i) An H-fat vertex x. (ii) one of 2 alternative H-moves at x. (iii) An I-fat vertex x, (iv) one of 4 alternative I-moves.

Definition 4 (I-fat) *A vertex $x \in V(S)$ is I-fat for T' iff, for a child y of x, there are arcs $\langle u_1, v_1 \rangle, \langle u_2, v_2 \rangle \in A(T')$ such that: $\lambda_{T',S}(v_i) \leq_S y$ for $i \in \{1, 2\}$, $\lambda_{T',S}(u_1) = x$, $x \leq_S \lambda_{T',S}(u_2)$, and u_i is $\leq_T$-minimal in $\lambda_{T',S}(\lambda_{T',S}^{-1}(u_i))$ for $i \in \{1, 2\}$. Notice that the latter implies that there are arcs $\langle u_1, w_1 \rangle, \langle u_2, w_2 \rangle \in A(T') \setminus \{\langle u_1, v_1 \rangle, \langle u_2, v_2 \rangle\}$. If x is I-fat for T', then the arcs $\langle u_1, w_1 \rangle, \langle u_2, w_2 \rangle, \langle u_1, v_1 \rangle$, and $\langle u_2, v_2 \rangle$, defined as above, are I-moves at x in T'.*

Figure 3 depicts graphically (i) an H-fat vertex, (ii) one of the two H-moves, (iii) an I-fat vertex, and (iv) one of the four I-moves.

A vertex is *fat* for T' iff it is H-fat for T' or I-fat for T'. A *candidate* F where $F \subseteq T$ is a directed forest without isolated vertices such that out degree ≤ 1 for all $v \in V(F)$.

The algorithm proceeds in two phases. Phase I (given below) consists of repeatedly picking a fat vertex and making the appropriate H- or I-moves, until there are no longer fat vertices. The resulting set of candidates is then examined for mixed cycles in phase II. Let C be a queue initially equal to one empty forest $F = \emptyset$ and let $X = \emptyset$. Phase I outputs X when the queue C is empty.

1. Dequeue F from queue C; Let $T' \leftarrow T \setminus F$. Compute $\lambda_{T',S}$. Pick a vertex x of S which is fat for T'.

2. If x is H-fat, let F_1 and F_2 be the candidates obtained from F by making the H-moves e_1 and e_2 respectively;

3. else if x is I-fat, let F_1, F_2, F_3, and F_4 be the candidates obtained from F by making the I-moves e_1, e_2, e_3, e_4 respectively.

4. If there does not exist a fat vertex in $T' \setminus F_i$, then let $X \leftarrow X \cup \{F_i\}$ else enqueue F_i in C.

Although each candidate $F \in X$ found in phase I is guaranteed to be 1-active, it may be the case that a mixed cycle is present. (In [6], an example of a gene and species tree is given where a cycle will exist after phase I). Phase II of the algorithm involves examining each candidate F and finding a minimal set of transfers to make the resultant candidate F' acyclic. The algorithm that enumerates these sets and tests if F' has no mixed cycle has time complexity $4^{|F|}n^2$. The overall running time of the algorithm is therefore $2^{4\tau}n^2$.

4 Simulations

We sketch how "synthetic" gene and species trees are generated for use in the experiments and detail a number of *degenerate* events that the simulation can generate.

Tree Generation. *Species Trees.* We begin by creating a random species tree S on n leaves as follows. Starting with a forest of n singleton vertices and stopping when only one tree exists, we remove two distinct trees x, y from the forest, create an internal vertex z with children x and y, and placed z back in the set. We verified experimentally that the trees had expected $\Theta(lg\ n)$ depth, as predicted analytically. Since species trees reflect the true evolutionary relationships between taxa, they are ultrametric and the weights on arcs correspond to time. The weight of a root to leaf path in S is always 1. Let an *ultrametric species tree* S be a binary rooted directed tree that is arc-weighted by $w : A \to [0..1)$. For vertex $u \in V(S)$, let $w(u) = \Sigma_{(p,p') \in P^S_{r(S),u}} w(p,p')$. In order to randomly assign weights between $[0..1)$ to the arcs of S s.t. every root to leaf path has total weight 1, we use the following routine in a root to leaf fashion on S:

1. Let $u \in V(S)$. Let l be a leaf in $L(S_u)$ that maximizes $|P^{S_u}_{u,l}|$ and such that no arc in $P^{S_u}_{u,l}$ has yet been assigned a weight. Let $P = P^{S_u}_{u,l}$ be the path $u = p_0, p_1, \ldots, p_{|P|} = l$.

2. Pick uniformly variates $\rho_1, \ldots, \rho_{|P|}$ from interval $[w(u)..1)$; sort $\rho_1, \ldots, \rho_{|P|}$ and assign the resulting weight $\rho'_{i+1} - \rho'_i$ to arc $\langle p_i, p_{i+1} \rangle$.

We experimented with several alternative approaches for generating ultrametric species trees and found that this produces trees that minimize the difference between arc weights under the L_∞ norm. This is important during the gene tree creation phase as it tends to "distribute" the lateral transfer events more evenly throughout the species tree.

Gene Trees. Using an ultrametric species tree S with n leaves generated as described above, we create a gene tree via a discrete event simulation. We assume that lateral transfer events occur according to a Poisson process with rate parameter λ. The four essential events in the simulation are *speciation, termination, transfer,* and *loss.* We start at the root of S (a speciation event at time 0) and work towards the leaves (termination events at time 1). For vertex $x_0 \in V(S)$ with children x_1, x_2, we generate the appropriate exponential variates for the arrival time of a transfer event along arc $\langle x_0, x_1 \rangle$. If the variate is less than $w(x_0, x_1)$, we schedule the transfer event and check for additional arrivals in the remaining interval along $\langle x_0, x_1 \rangle$. Otherwise, a speciation event for x_1 is scheduled. The process is repeated for arc $\langle x_0, x_2 \rangle$. When a transfer event is encountered at time t along an arc (call it $\langle x, y \rangle$), a second (distinct) arc is chosen uniformly randomly from all other arcs that exist at time t. In other words, only those arcs $\langle x', y' \rangle$ where $w(x') \leq t$ and $w(y') \geq t$ are considered.

Suppose that arc $\langle x, y \rangle$ is the tail and $\langle x', y' \rangle$ is the head of the lateral transfer. If there already exist events scheduled for time t', $t' \geq t$, along arc $\langle x', y' \rangle$, then all such events are aborted. The gene lineages associated with these events are *lost.* This corresponds to the foreign gene "knocking out" the resident gene in the genome of the organism. Such a protocol is necessary if we are to guarantee the 1-activity constraint of the scenario.

Degenerate Events. The simulation of evolution is pessimistic in the sense that the resulting scenarios may contain evolutionary events that are biologically plausible but not detectable by the algorithm. In this sense, the simulation is more general than our model for identifying transfer events. These events have varying effects on the ability of the algorithm to identify the correct number and location of lateral transfer events. We classify these events into two categories: *useless transfers* and *transfer-loss events.*

Useless Transfers. Consider Figure 1 (iv). At the point of evolution marked by X, there is a lateral transfer between two arcs in the species tree that share a common parent. Clearly, the gene tree is not changed by such transfers. In other words, the root of the species tree has children ABC and DEF and the root of the gene tree has children ABC and DEF even though a transfer has occurred at X. In the subtree labeled Y of the species tree, we show an example of two useless transfers that together do not cause the gene tree to disagree with the species tree. This subtree of the species tree has the ancestor AB and C as siblings. In the gene tree, A remains closer to B due to the "later" lateral transfer and C remains being a sibling with the ancestor AB via the "earlier" lateral transfer.

Transfer-Loss Events. Consider Figure 1 (v). At the point marked Z' in the diagram, a lateral transfer occurs from taxon n to taxon 2. Between point Z' and Z, this lineage is lost. Note that one child of the vertex of the gene tree at point Z' is a transfer event and one child is a loss event; we term this a *transfer-loss event*. Let T be a gene tree and S be a species tree and let τ be the true number of lateral transfer events that occurred during the period of evolution (the true number of lateral transfer events generated by our simulation of evolution). Let τ' be the minimum cost of a scenario for T w.r.t. S (the minimum cost scenario found by our algorithm). When a transfer-loss event occurs, it may be the case that $\tau' < \tau$, $\tau' = \tau$ or $\tau' > \tau$. We term these *helpful, harmless,* and *harmful* resp. The example in Figure 1 (v) shows that a single harmful transfer-loss event can cause the algorithm to require $\Omega(n)$ lateral transfers to explain the disagreement between the gene and species tree. It is easy to verify that the minimum cost scenario for this particular example requires $n - 2$ transfer events. It is equally easy to create examples of helpful and harmless transfer-loss events.

5 Experimental Results

For the remainder of this section, let n represent the number of leaves in the species tree, and Ω represent the number of repetitions performed for each experiment. Let τ represent the true number of lateral transfer events generated by a simulation and τ' represent the minimum cost scenario found by the algorithm. Let λ represent the rate parameter in our Poisson process. A *trial* is a species tree and gene tree pair generated by the simulation for a specified λ and n.

The largest gene and species tree for which we can compute the minimum cost scenario has $\tau = 20$ transfers and $n = 300$ leaves. The computation takes approximately 3 days on a standard desktop PC. For $\tau = 10$ and $n = 20$, the computation takes approximately 30 seconds on a standard PC.

In Figure 4 (a), we see that the number of transfers in a simulation rises linearly as a function of λ for a fixed n. This is consistent with a Poisson process and our species tree generation routine. As the rate parameter λ grows, the number of transfers τ will eventually become sufficiently large so that no further transfers will be detected be the algorithm. This is trivially the case if τ exceeds $n - 2$ for a species tree with n leaves. Figure 4 (b) plots the average estimated number of transfers τ' versus the λ. As τ' is consistently less than τ, we may conclude the majority of transfers are not harmful transfer-

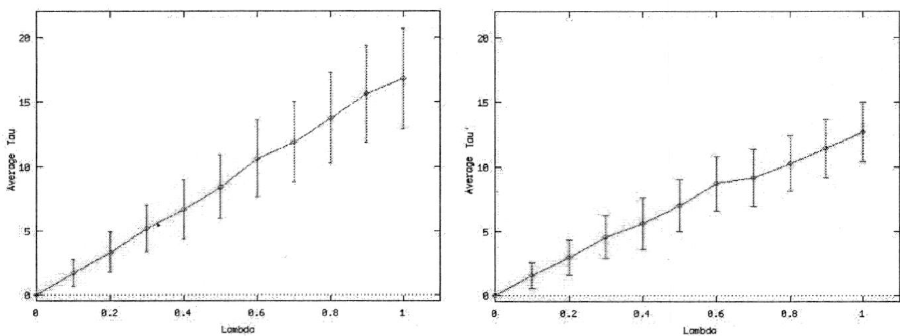

Figure 4. (a) Average τ versus λ. (b) Average τ' versus λ. $\Omega = 300$, $n = 80$.

loss events. When $\lambda = 0.6$, the average value for τ is approximately 11 and τ' is 8.6 with variance 2.11. Note that for $\lambda > 0.6$, one can see slight saturation occurring. To test the saturation point (defined informally as the point where average τ' stops increasing), we generated a large set of random trees. Using half of the set as species trees and the other half as gene trees, the average value of τ' is computed. This should give a very good estimate of the saturation point. For $n = 11$, the trial requires an average of 5.81 lateral transfers with variance 0.65. For $n = 21$, this average is 13.27 with variance 0.87. Trials for trees with larger n suggest that the saturation point is slightly above $n/2$ (graph not shown). At this time, our computational power does not give us an accurate estimate of where this converges. However, it allows us to say with some confidence that any scenario for a gene and species tree, where the cost of the scenario is $> n/2$ transfers, does not represent a meaningful explanation of disagreements between the trees.

When 10 transfers occur in a species tree of size $n = 20$, the algorithm under-estimated the number of transfers by 4.4 with variance 0.85. We note that such a 1 : 2 ratio of transfers to leaves is unrealistic for real-world data sets. We (informally) conjecture that a ratio of 1 : 10 might be more accurate. If such a ratio were true, our algorithm would tend to predict 9 transfer events in a 100 taxa tree where the true transfer number is 10.

Figure 5 and additional graphs available on-line reaffirm that harmful transfer-loss events are very rare and, when they do occur, their effect is negated by the occurrence of useless transfers and the effects of saturation or via the existence of alternative scenarios with approximately the same overall cost. In under 1% of all trials, the algorithm did not find a valid scenario with cost $\leq \tau$. Without exception, every trial had a scenario with cost $\leq \tau + 3$.

We also note that over some 10, 000 trials, we did not find a scenario that

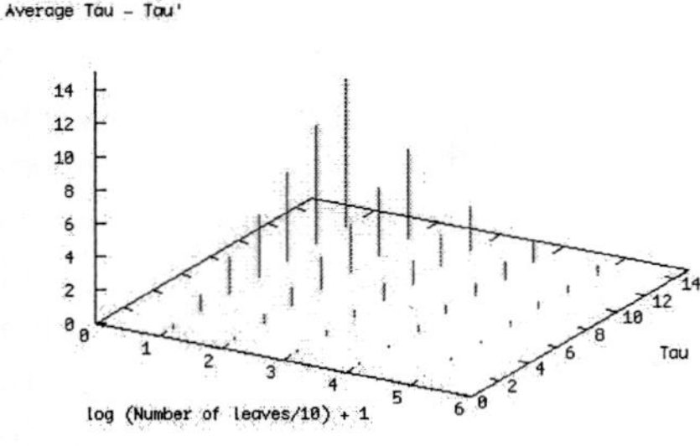

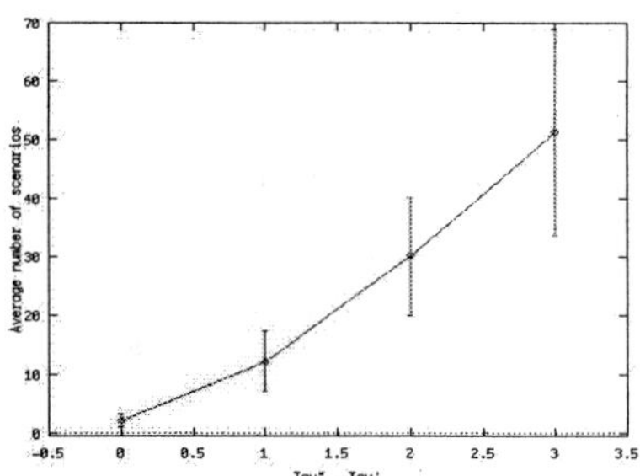

Figure 5. Error versus τ versus logarithm of the number of leaves.

Figure 6. Average number of minimum cost scenarios versus $\tau'' - \tau'$. $\Omega = 300$, $n = 20$. $\lambda = 0.3$

required cycle elimination (Phase II of the algorithm). Although it is possible to construct an example where the algorithm will require this phase, it appears that these scenarios are extremely rare or the rate of useless transfers and helpful transfer-loss events is sufficiently high that a scenario with cost $\tau' \leq \tau$ is created. Readers familiar with Hannehalli-Pevzner theory might note that this result is similar to that found for *hurdles* [1].

Figure 6 captures how many minimum and near minimum cost scenarios exists for a gene and species tree trial. Consider a gene and species tree where τ is the actual number of transfers that occurred during the simulation. Let τ' be the minimum cost over all scenarios found by the algorithm. For this graph, only trials where $\tau' \leq \tau$ were used for simplicity. Let k range from 0 to 3. The x-axis of this graph shows the number of scenarios (with cost $\tau'' = \tau' + k$) found on average. For $n = 20$ and $\lambda = 0.3$, we would expect that $\tau \approx 5$ (see Figure 4 (a)). When $k = 0$ ($\tau'' = \tau'$), there exist on average 2.09 scenarios with cost τ'. Approximately half of the time (probability 0.46), the scenario is unique. When $k = 3$, ($\tau'' = \tau' + 3$), the number of scenarios with cost τ'' is 51.32 with variance 34.52. It is surprising that so few scenarios exist given that is extremely easy to construct gene and species trees by hand that have exponentially many minimum cost scenarios. We repeated this experiment with various values of n and λ in such a way that the $\tau : n$ ratio was preserved, however the curve did not change significantly.

6 Conclusions and Open Problems

This paper demonstrates the feasibility of the model and algorithm presented in [5] and provides empirical evidence of the relative frequency of degenerate events. Our experiments suggest that transfer-loss events which cause the algorithm to over-estimate the true number of transfer events occur with small probability. Furthermore, the algorithm provides near-optimal scenarios when λ, the rate parameter for lateral transfers, is low enough as to not cause saturation. For realistic size instances of the problem, it was the case roughly half of the time that the minimum cost scenario was unique. In over $10,000$ trials, we did not find a single example of a scenario that required the cycle elimination phase of the algorithm. This suggests that the first phase of the algorithm is sufficient for realistic data sets and implies a $O(2^{2\tau}n^2)$ running time.

It is important to consider various extensions to this framework. It would be interesting to combine this model with *gene order*-based models such as those for *gene reversals*, since gene order will provide important clues as to where and when lateral transfers have occurred. In collaboration with L. Graur's group, we are now extending our framework to included species trees where the arcs are labelled with time. Lastly, for a fixed n and λ, it seems feasible that one could analytically compute the expected number of lateral transfer events needed for a random species tree and random gene tree. A first step would be to do this for, e.g., a balanced species tree and a random gene tree. An implementation of our algorithm and additional experimental results are available at www.cs.mcgill.ca/ ~laddar/lattrans.

References

1. A. Caprara, *J. of Comb. Opt.*, 3:149-18 (1999)
2. C. Delwiche and J. Palmer, *Mol. Biol. Evol.*, 13(6), pp. 873–882 (1996).
3. M. Goodman et. al. *Syst. Zool.*, 28 (1979)
4. R. Guigó et. al. *Molec. Phylogenet. and Evol.*, 6, 2, pp. 189-213 (1996)
5. M. Hallett and J. Lagergren *RECOMB '01*, Montreal, pp. 141-148 (2001)
6. M. Hallett and J. Lagergren Models and Algorithms for Lateral Gene Transfer Problems. Full journal version. Available at www.mcb.mcgill.ca/~hallett (2002)
7. J. Hein *J. Mol. Evol.*, 36, pp. 396-405 (1993)
8. J. Hein, T. Jiang, L. Wang, and K. Zhang *Disc. Appl. Math.*, 71, pp. 153-169 (1996)
9. R. D. M. Page and M. A. Charleston *Molecular Phylogenetics and Evolution*, 7, pp. 231-240 (1997)
10. A. von Haseler and G. A. Churchill *J. Mol. Evol.*, 37, pp. 77-85 (1993)

WHOLE GENOME HUMAN/MOUSE PHYLOGENETIC FOOTPRINTING OF POTENTIAL TRANSCRIPTION REGULATORY SIGNALS

E. CHEREMUSHKIN[1]; A. KEL[1,2]

[2]*Institute of Cytology & Genetics SB RAN, 10 Lavrentyev pr., 630090, Novosibirsk, Russia;*

[2]*BIOBASE GmbH, Halchtersche Strasse 33, 38304 Wolfenbuettel, Germany.*

Phylogenetic footprinting is an efficient approach for revealing potential transcription factor binding sites in promoter sequences. The idea is based on an assumption that functional sites in promoters should evolve much slower then other regions that do not bear any conservative function. Therefore, potential transcription factor (TF) binding sites that are found in the evolutionally conservative regions of promoters have more chances to be considered as "real" sites. The most difficult step of the phylogenetic footprinting is alignment of promoter sequences between different organisms (f.e. human and mouse). The conventional alignment methods often can not align promoters due to the high level of sequence variability. We have developed a new alignment method that takes into account similarity in distribution of potential binding sites (motif-based alignment). This method has been used effectively for promoter alignment and for revealing new potential binding sites for various transcription factors. We made a systematic phylogenetic footprinting of human/mouse conserved non-coding sequences (CNS). 60 thousand potential binding sites were revealed in human and mouse genomes. We have developed a database of the predicted potential TF binding sites. Availability: http://compel.bionet.nsc.ru/FunSite/footprint/; www.gene-regulation.com/

1. Introduction

Genes in genomes of higher eukaryotic organisms are regulated mainly by the means of multiple regulatory proteins - transcription factors (TF), acting through specific regulatory sequences (TF binding sites) that are located usually in the proximity of the genes when constituting a promoter, or at more remote locations when being a part of an enhancer. The unique pattern of regulation of every gene in the multiplicity of different cellular, tissue-specific and developmental conditions can be fully understood through identifying all functional TF binding sites in the regulatory regions of genes. New techniques of large scale analysis such as ChiP on chip (Ren et al, 2002) provide means for extensive identification of TF binding sites. But the large number, combinatory nature of regulation as well as practically unlimited verity of cellular conditions renders unlikely an experimental genome-scale identification of TF binding sites. Computational identification of potential TF binding sites in gene regulatory regions become very important for gene prediction, functional characterization of newly discovered genes (Bucher, 1999; Werner, 1999; Wasserman & Fickett, 1998) as well as for understanding of regulatory circuits in the cell. There are a number of pattern-based as well as matrix based methods for prediction of potential TF binding sites (Quant et al., 1995; Goessling et al, 2001) but the false positive prediction rates are pretty high (Tronche et al, 1997).

Phylogenetic footprinting (Gumucio et al., 1996; Duret and Bucher, 1997) is a very effective approach for identification of regulatory elements, which relies on the cross-species sequence comparison. Phylogenetic footprinting is based on the assumption that the parts of gene regulatory regions that are liable to no or little variations in the course of evolution have important functions (Duret & Bucher, 1997). The essence of the method is the alignment of the regulatory regions of the orthologous genes and finding of the most highly conserved regions.

Though, it is clear that this method works as far as the pattern of regulation of the particular gene and their control mechanisms are conserved across species, which can be true for rather similar organisms only. Sites will stay conserved when they provide basic regulatory function of a gene which is conserved between considered species. Species-specific regulatory nuances will not be recognized by this method. On the other side, sequence comparison of a closely related species does not help to distinguish functionally conserved features against a background similarity of recently evolved sequences. Therefore, human and mouse are generally accepted as a reasonably similar as well as pretty distant species to perform efficient phylogenetic footprinting of their genomes (Mueller et al, 2002).

In the last years, the approach of phylogenetic footprinting was used several times to reveal potential TF binding sites in different genes: in upstream region of the beta-like and ε-globine genes (Gumucio et al, 1993; Gumucio et al, 1996), in COX5B gene that encodes subunit Vb of cytochrome c oxidase (Bachman et al, 1996). Recently, phylogenetic footprinting helps to reveal new E2F sites and proposes a regulation in cell cycle for a number of new genes (Kel et al, 2001). Conservative combinations of two binding sites (composite elements) were revealed in many T-cell specific genes (Kel et al., 1999). Using principles of phylogenetic footprinting a number of novel binding sites were revealed in regulatory regions of 502 genes associated with a variety of different human diseases (Levy et al., 2001).

A few computational implementation of the idea of the phylogenetic footprinting have been developed. These are: FunSiteFootprint (http://compel.bionet.nsc.ru/FunSite/footprint/; Cheremushkin & Kel, 2002), Consite (http://forkhead.cgr.ki.se/cgi-bin/consite, Wasserman et al., 2000), rVISTA (http://pga.lbl.gov/rvista.html, Loots et al, 2002), TRES (http://bioportal.bic.nus.edu.sg/tres/info.htm), FootPrint2.0 (http://bio.cs.washington.edu/software.html, Blanchette et al, 2002). The most critical step of the phylogenetic footprinting is alignment of regulatory sequences such as promoters and enhancers. The conventional alignment methods often can not align promoters due to the high level of sequence variability. We have developed a new alignment method that takes into account similarity in distribution of potential binding sites. We call this method: motif-based alignment. This method has been used effectively for promoter alignment and for revealing new potential binding sites for various transcription factors. We made a systematic phylogenetic footprinting of human/mouse conserved non-coding sequences (CNS) that were downloaded from the Berkeley Genome Pipeline site (http://pipeline.lbl.gov/) of the global comparison of human and mouse genomes. 60 thousand potential binding sites were revealed in human and mouse genomes. We have developed a database of the predicted potential TF binding sites.

This database as well as the program for motif-based alignment and phylogenetic footprinting are available at: (compel.bionet.nsc.ru/FunSite/footprint).

2 Method

2.1 Search for TF binding sites

We developed a new method for alignment of regulatory sequences that includes information about TF binding sites. To search for the sites we apply position weight matrices (PWM) from TRANSFAC database (www.biobase.de) (Wingender et al., 2001). Every nucleotide in a sequence can potentially be belong to one or several TF binding sites. We estimate the probability $w_p(\overline{S},k)$ of k-th nucleotide of a sequence $\overline{S}$ to be belong to a binding site of a factor T_p ($p \in [1,P]$):

$$w_p(\overline{S},k) = \alpha \times \sum_{j=k-L+1}^{k} \exp(\beta \times s_p(\overline{S},j)), \quad \overset{\rho}{w}(\overline{S},k) = \langle w_1(\overline{S},k)..., w_P(\overline{S},k) \rangle$$

where $s_p(\overline{S},j)$ - score of p-th matrix at j-th position of sequence, L - length of $\overline{S}$, α and β are two normalization constants.

The corresponding scores for different weight matrices can be seen in the Figure 1. We use different smoothing functions that weight differently the core positions of the sites (Fig. 1 a and b). First smoothing function gives more weight to the core positions of the site, the second function gives similar weights to all positions of the site.

It is known that the library of weight matrices contains matrices that are similar to each other. These are different matrices for the same transcription factor or for the transcription factors that are very similar in their DNA binding signature. We consider a similarity matrix M that takes into account similarities between weight matrices. We use M to convert the probability to a new function:

$\overset{\rho}{\phi}(\overline{S},k) = \overset{\rho}{w}(\overline{S},k) \cdot M$, where M - $P \times Q$ similarity matrix. We will use $\overset{\nu}{\phi}(a)$ instead of $\overset{\rho}{\phi}(\overline{S},k)$, where $a \in \Sigma \times \Phi$ - sequence element, $\gamma(a) \in \Sigma$ - nucleotide for this element. The components of the vector $\overset{\nu}{\phi}(a)$ we will call TF belonging coefficients.

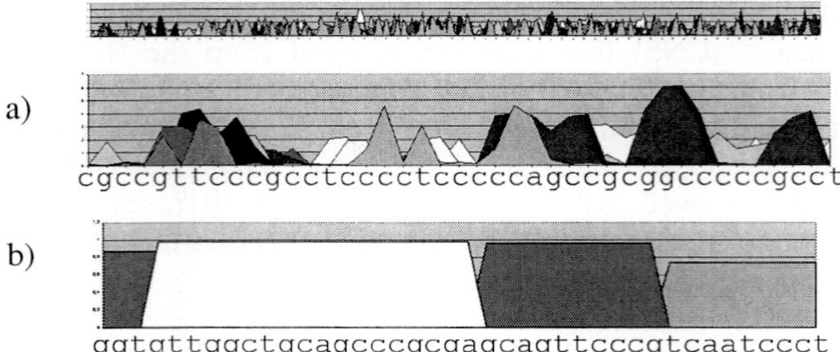

Figure 1. Distribution of TF belonging coefficients in a sequence. For each nucleotide in the sequence we compute a vector of weights that reflects the probability of the nucleotide to be belong to a TF binding site. Different gray colors correspond to different TFs. (a,b) – usage of two different smoothing functions.

2.2 Motif-based alignment algorithm

We have developed an alignment algorithm for pair-wise and multiple alignment of nucleotide sequences. The algorithm is similar to the generally accepted Needleman-Wunsch dynamic programming algorithm. A major modification is made in the way of calculating the nucleotide substitution weights and gap penalty. The PWM scores were considered at every sequence positions in order to compute the corresponding substitution weights and gap penalty (see Fig. 2).

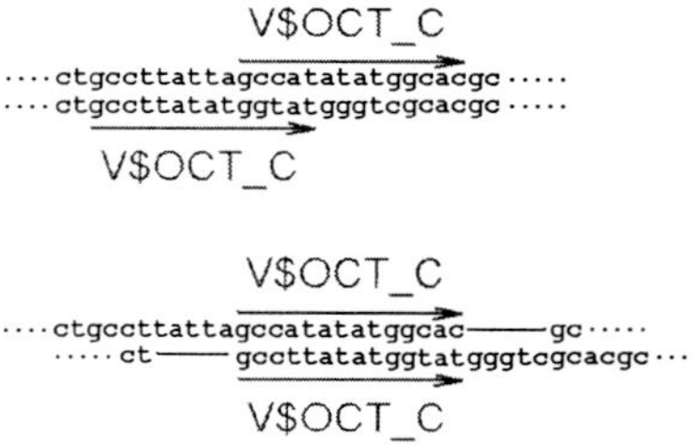

Figure 2. We consider alignment as a favorable one, if sites are aligned to each other.

For calculation of the gap penalty we construct a disjoining score function that can be applied to any two neighbor positions a and b in one nucleotide chain:

$$X_{gap}(a,b) = \left(\overset{0}{\phi}(a) + \overset{0}{\phi}(b) \right)^2 .$$

This score estimates how similar are the two TF belonging vectors for these two positions. If both these positions have high belonging coefficients for the similar sets of transcription factors then the score X_{gap} is high. It means that most of the predicted

TF binding sites in this region span over these two neighbor positions and disjoining of these positions by a gap will cause braking of these TF binding sites and is considered as an unfavorable event.

For N sequences in the alignment we use the following variant of the disjoining score function, where C_{gap}, W_{gap} are optimized constants:

$$Y(a,b) = C_{gap}/N + W_{gap} \cdot X_{gap}(a,b).$$

We use this function for calculation of the gap penalty, while inserting gap in $\overline{S}^1$ between $k-1$ and k under position l in $\overline{S}^2$:

$$GAP(\overline{S}^1,\overline{S}^2,k,l) = \frac{G(\overline{S}^1,k) + R(\overline{S}^2,l)}{2},$$

where

$$G(\overline{S}^1,k) = Y(s_{k-1}^1, s_k^1),$$

which describes the score of disjoining of positions *k-1* and *k* due to a possible insertion in the sequence $\overline{S}^2$ during evolution;

$$R(\overline{S}^2,l) = \frac{Y(s_{l-1}^2, s_l^2) + Y(s_l^2, s_{l+1}^2)}{2},$$

which describes the score of simultaneous disjoining of positions *l-1* and *l* as well as *l* and *l+1* due to a possible deletion in the sequence $\overline{S}^1$ during evolution. Both deletions and insertions are considered to be equally probable that is why the gap penalty is a mean value between G and R.

Substitution weight for two aligned positions k and l in the sequences $\overline{S}^1$ and $\overline{S}^2$ correspondingly:

$$SUB(\overline{S}^1,\overline{S}^2,k,l) = Z(s_k^1, s_l^2),$$

$$Z(a,b) = \frac{\Delta}{N} \cdot C_{sub} - W_{sub} \cdot \sum_{i=1}^{3} \lambda_i \cdot E_i(a,b) \Big/ \sum_{i=1}^{3} \lambda_i, \text{ where, } \Delta = \begin{cases} 1, \gamma(a) \neq \gamma(b) \\ 0, \gamma(a) = \gamma(b) \end{cases},$$

$$E_1(a,b) = \begin{cases} (\phi(a) + \phi(b))^2, \gamma(a) = \gamma(b) \\ \phi(a)^2 + \phi(b)^2, \gamma(a) \neq \gamma(b) \end{cases} \quad E_2(a,b) = \max_i(\varphi_i(a) \cdot \varphi_i(b)),$$

$$E_2(a,b) = \begin{cases} 0, m > C_{min} \\ (C_{min} - m)/C_{min}, m \leq C_{min} \end{cases}, \text{ where } m = \min_i |\varphi_i(a) - \varphi_i(b)|.$$

$\gamma(a) \in \Sigma$ - nucleotide, C_{sub}, C_{gap}, W_{sub}, W_{gap}, λ_i - constants.

In the Figure 3 we present an example of alignment of two sequences that is done by the motif-based algorithm. The score values of the aligned sequences are shown above and under the sequences correspondingly. One can see that the peaks of the TF belonging coefficients are aligned to each other.

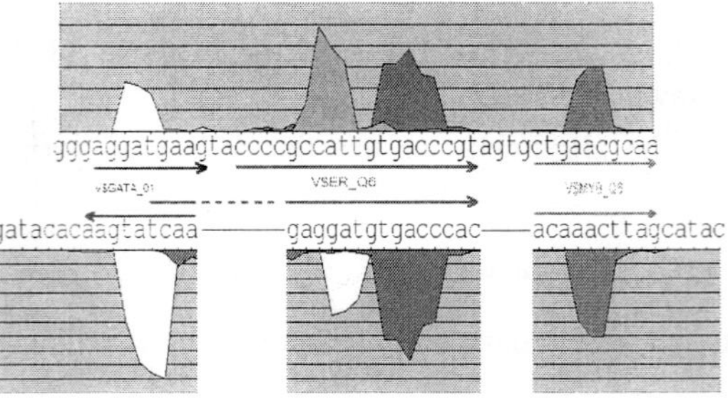

Figure 3. Example of alignment of a sequences. Graphical representation of the TF belonging coefficients.

3 Implementation and results

3.1 Implementation and availability

The motif-based alignment algorithm was implemented as a Java standalone program. It takes two sequences as input and align them. First it runs the Match program (Goessling et al., 2001) that finds potential TF binding sites in the sequences. Specific collection of weight matrices with predefined cut-off values for every matrix can be specified by the user: taxon-dependent collection, tissue or function specific, minimizing false positive or false negative error. User can build his own profile with the help of TRANSPLORER program (http://www.biobase.de/pages/products/transplorer.html).

3.2 Testing of the alignment using a model of orthologous promoter sequences

In order to validate the developed alignment algorithm we have constructed a computer model of evolution of promoter sequences. An ancestor sequence of a length L is randomly created. In this sequence we implant N_{sites} binding sites with $N_{sites} + 1$ spacers between them and on flanks. From this sequence we generate two descendant sequences by introducing R_{spacer} random mutations (insertions, deletions and substitutions) in the spacer regions and R_{site} substitutions in the sites. We require that after each iteration all sites should remain "functional". For that, we check the PWM score for each of them and discard cases when the score drops below a certain cut-off (CO_{site}). Then, these two sequences are aligned and positions of the alignment blocks are compared with the sites that were originally implanted. In the case of misalignment of one of the sites we report a failure.

We have compared the developed motif-based alignment algorithms with the ClustalW by counting the percentage of failures. Our algorithm shows much better performance in finding correct alignment. With the homology of sequences equals to homology between human and mouse, the failure rate of our algorithms was about 0.1% whereas ClustalW gives approximately 2.5% of failures. It is worth mentioning here that both alignment algorithms were always reporting a variant of alignment. Although, in the case of the failure the ClustalW alignments does not match the correct implanted sites but often matches wrong sites that were not implanted.

3.3 Phylogenetic footprinitng of human/mouse conserved non-coding sequences (CNS)

Evolutionary conserved non-coding regulatory sequences (CNS) could serve as good landmarks on genome to find functionally important promoters, enhancers or silencers (Duret & Bucher, 1997). Phylogenetic footprinting of CNS will help us to reveal TF binding sites and assign a regulatory function to the regulatory regions and to the adjacent genes. We use results of the Berkeley Genome Pipeline (http://pipeline.lbl.gov/) of the global comparison of human and mouse genomes. We have download the complete list of CNS and made the phylogenetic footprinting of all of them. Two types of alignment were used. First, we made an alignment using ClustalW, and second, using the developed motif-based alignment algorithm. An example of VISTA large scale analysis of the PTH gene located in the human chromosome 11 is presented in the Fig. 4. 3 conservative non/coding sequences

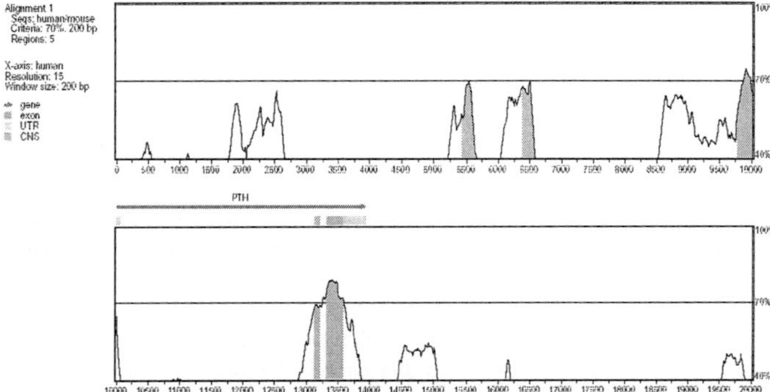

Figure 4. An example of large scale sequence comparison between orthologous PTH genes from human and mouse. The sequence comparison was done by the VISTA tool (http://www-gsd.lbl.gov/VISTA/). The program makes a global alignment and presents the result in a form of plot showing the percent of homology in the 200nt window sliding along the alignment. The position 10000 corresponds to the start of transcription of the human PTH gene. Regions of the highest homology are located in the coding regions of the second and third exons. 3 conservative non-coding sequences (homology > 70%) were found in the 5' sequence of the gene: CNS1 (5431-5639), CNS2 (6395-6599), CNS3 (9772-9999).

(CNS) are marked at the homology profile.

Two examples of detailed phylogenetic fotprinting are shown at the Figures 5 and 6. We can confirm most of the experimentally known sites. In addition a number of new sites are found.

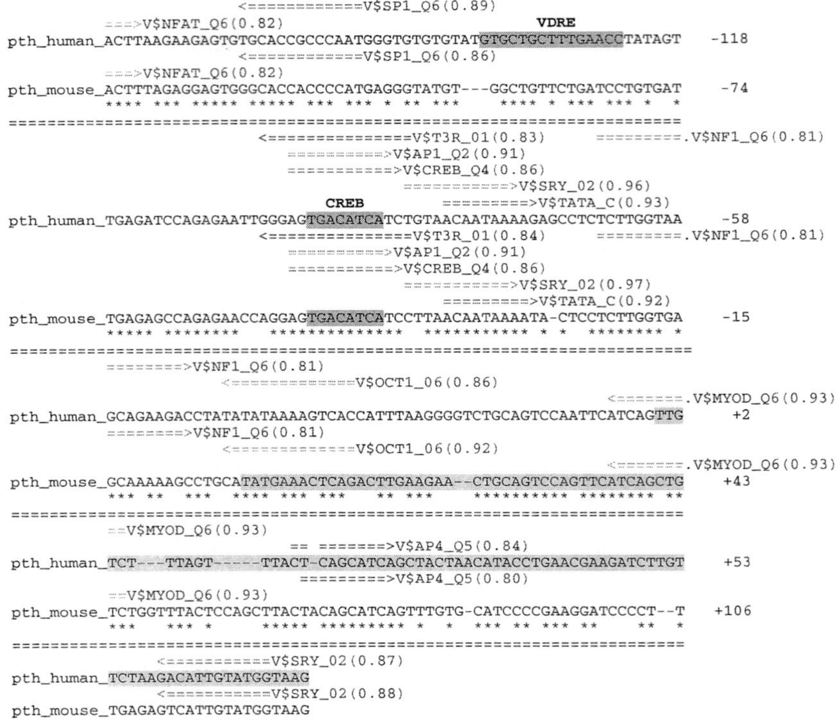

Figure 5. The result of applying of the phylogenetic footprinting tool to the proximal part of the human/mouse PTH CNS3. The beginning of the fist exon in both sequences is marked by a shadow. Two sites are marked (CREB and VDRE) that have been described previously for this region in human PTH gene. Conservative sites found by this analysis are shown by the arrows above each sequence. A number of new potential sites is revealed: CREB site, OCT site near start of transcription, a couple of NF-AT sites, SP-1 sites, GATA sites and some others.

Phylogenetic footprinting was done by the previously developed tool (http://compel.bionet.nsc.ru/FunSite/footprint/) that takes two or several aligned sequences, finds conservative binding sites and display them. Binding sites with the score exceeding a predefined cut-off, for transcription factors that belong to the same family and that have overlapping location on the alignment are considered as the positive match of the phylogenetic footprinting.

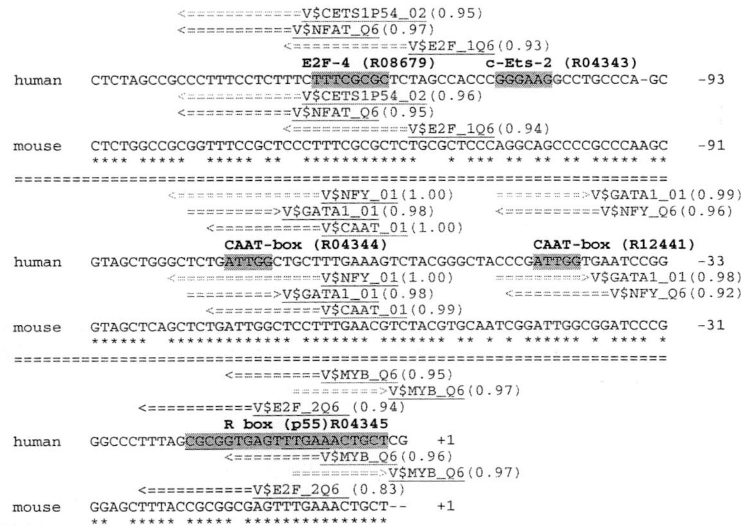

Figure 6. The result of phylogenetic footprinting of the promoter regions of human /mouse cdc2 gene. 5 sites annotated in TRANSFAC are marked. 3 of them are confirmed by the phylogenetic footprinting. On the bases of this analysis we can propose the structure of R box: it contains E2F and Myb sites. c-Ets site in human promoter is not conserved in mouse, although we revealed a new conserved c-Ets site 20bp upstream.

The list of 17117 CNS of the total length of alignment 2418267 bp was analysed. We applied a set of 240 weight matrices from TRANSFAC rel. 5.3 with the cut-offs optimized to minimize the sum of false positive and false negative errors. Using ClustalW alignments we found 54075 conservative TF binding sites. Using the motif-based alignment we found 58106 conservative TF binding sites. So, our algorithm that includes information about potential TF sites at the very early stage of analysis allows us to reveal 4031 more binding sites then using other alignment algorithm. In the figure 7 one can see the comparison of the number of revealed sites using ClustalW alignment versus our motif-based alignment.

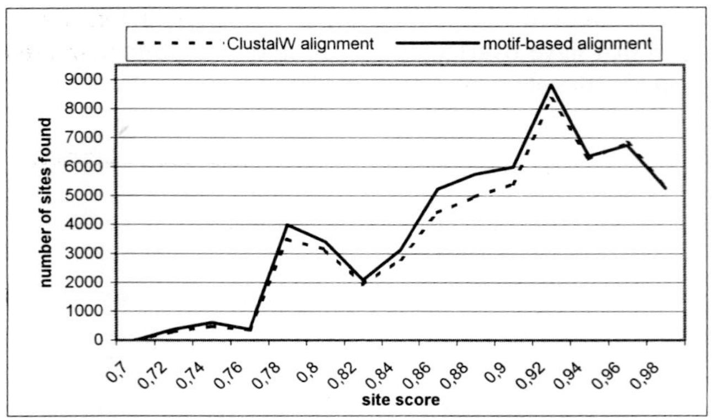

Figure 7. Comparison of the number of revealed sites using ClustalW alignment versus motif-based alignment algorithm. More sites with the score values from 0.78 to 0.92 can be revealed by the motif-based alignment algorithm.

It is interesting to observe that motif-based alignment algorithm helps to reveal more sites with the score values from 0.78 to 0.92, which are the most functionally relevant sites. Low scoring sites (lower then 0.72) and peak scoring (higher then 0.92) are revealed in the same amount as using the ClustalW alignment.

We made a comparison between two alignment algorithms whether they identify the same sites. The comparison shows that only about 80% of sites are found the same (data not shown). Due to the higher failure rate for ClustalW to reveal the correct sites (see the simulation results) we can assert that motif-based alignment not only finds more genomic sites but all sites reported by this alignment are expected to be more often correct then using the ClustalW algorithm.

An example of comparison of results of two alignments done by ClustalW and motif-based alignment algorithm is presented in the Fig. 8. One can see that our algorithm enables revealing an additional conservative potential TF binding site that was missed by ClustalW.

We have developed a database of predicted potential TF binding sites in human genome by analyzing the human/mouse CNS. Using this database user can retrieve all conservative sites for a selected chromosome or for a region at the chromosome and can visualize gene information for the nearest upstream and downstream genes, that can be targets for regulation through found TF binding sites. Using the developed database molecular biologists can plan their experiments for validation of found target genes and can make regulatory functional annotation of human and mouse genome.

a)

```
GGTAT-------------AATTCTGTATT--------TGTTAAAA--------------   1274
GATTTTTTTTAAAATATAAACTCAGTATTATCGATTATGTCAAAAATTTCTAGGTGGAC   1557
*** *                ** ** *****         *** ****
```

b)

```
                                     ============>V$CEBP_01(0.97)
GAT-----------GGTATA------ATT-CTGTATTTGTTAAAA--------------   1274
                                     ===========>V$CEBP_01(0.88)
GATTTTTTTTAAAATATAAACTCAGTATTATCGATTATGTCAAAAATTTCTAGGTGGAC   1557
***             ** *       ***     *   * *** ****
```

Figure 8 An example of two alignments done by a) ClustalW and b) motif-based alignment algorithm. First alignment is more optimal (it have 20 matching positions vs 18 in the second alignment), whereas second alignment allows to reveal a conservative C/EBP site.

Acknowledgments

The authors are indebted to Vadim Ratner, Edgar Wingender and Hubert Meier for fruitful discussion of the results. Parts of this work was supported by Siberian Branch of Russian Academy of Sciences, by grant of Volkswagen-Stiftung (I/75941)

References

1. Bachman N.J., Yang T.L., Dasen J.S., Ernst R.E., Lomax M.I. "Phylogenetic footprinting of the human cytochrome c oxidase subunit VB promoter." *Arch Biochem Biophys* **333**, 152-162 (1996)

2. Blanchette M., Schwikowski B., Tompa M. "Algorithms for phylogenetic footprinting." *J Comput Biol.* **9**, 211-223 (2002)

3. Bucher P. "Regulatory elements and expression profiles." *Curr Opin Struct Biol.* **9**,400-407 (1999)

4. Cheremushkin E., Kel A. "PromoterFootprint: A new method for alignment of regulatory genomic sequences. Phylogenetic footprinting of TF binding sites." In Currents in Computational Molecular Biology (edited by L.Florea, B.Walenz, S. Hannenhalli), 40-41 (2002)

5. Duret, L. & Bucher, P. "Searching for regulatory elements in human noncoding sequences." *Curr. Opin. Struct. Biol.* **7**, 399-406 (1997)

6. Goessling E., Kel-Margoulis O.V., Kel A.E. and Wingender E. "MATCH^{TM} - a tool for searching transcription factor binding sites in DNA sequences. Application for the analysis of human chromosomes." In: *Proceedings of the German Conference on Bioinformatics (GCB2001),* October 7-10, 2001, Braunschweig, pp.158-160 (2001)

7. Gumucio, D.L., Shelton, D.A., Zhu, W., Millinoff, D., Gray, T., Bock, J.H., Slightom, J.L., Goodman, M. "Evolutionary strategies for the elucidation of cis and trans factors that regulate the developmental switching programs of the beta-like globin genes." *Mol. Phylogenet. Evol.* 5, 18-32 (1996)

8. Gumucio D.L., Shelton D.A., Bailey W.J., Slightom J.L., Goodman M. "Phylogenetic footprinting reveals unexpected complexity in trans factor binding upstream from the epsilon-globin gene." *Proc Natl Acad Sci U S A.* **90**, 6018-6022 (1993)

9. Kel, A., Kel-Margoulis, O., Babenko, V., Wingender, E. " Recognition of NFATp/AP-1 Composite Elements within Genes Induced upon the Activation of Immune Cells" *J. Mol. Biol.* **288** , 353-376 (1999)

10. Kel A.E, Kel-Margoulis O.V., Farnham P.J., Bartley S.M., Wingender E., and Zhang M.Q. "Computer-assisted identification of cell cycle-related genes - new targets for E2F transcription factors." *J. Mol. Biol.* **309** , 99 – 120 (2001)

11. Loots G.G., Ovcharenko I., Pachter L., Dubchak I., Rubin E.M. *Genome Res.* **12**, 832-839 (2002)

12. Levy S., Hannenhalli S., Workman C. *Bioinformatics* **17**, 871-877 (2001)

13. Quandt, K., Frech, K., Karas, H., Wingender, E., and Werner, T. *Nucleic Acids Res.*, **23**, 4878-4884 (1995)

14. Ren B., Cam H., Takahashi Y., Volkert T., Terragni J., Young R.A., Dynlacht B.D. "E2F integrates cell cycle progression with DNA repair, replication, and G(2)/M checkpoints." *Genes Dev.* **15;16**, 245-256 (2002)

15. Tronche, F., Ringeisen, F., Blumenfeld, M., Yaniv, M. & Pontoglio, M. "Analysis of the distribution of binding sites for a tissue-specific transcription factor in the vertebrate genome." *J. Mol. Biol.* **266**, 231-245 (2002)

16. Muller F., Blader P., Strahle U. *Bioessays* **24**, 564-572 (2002)

17. Wasserman W.W., Palumbo M., Thompson W., Fickett J.W., Lawrence C.E. "Human-mouse genome comparisons to locate regulatory sites." *Nat Genet.* **26**, 225-228 (2000)

18. Wasserman, W. W., Fickett, J. W. "Identification of regulatory regions which confer muscle-specific gene expression." *J. Mol. Biol.* **278** , 167-181 (1998)

19. Werner T. "Models for prediction and recognition of eukaryotic promoters." *Mamm. Genome* **10**, 168-175 (1999)

20. Wingender, E., Chen, X., Fricke, E., Geffers, R., Hehl, R., Liebich, I., Krull, M., Matys, V., Michael, H., Ohnhäuser, R., Prüß, M., Schacherer, F., Thiele, S. and Urbach, S. "The TRANSFAC system on gene expression regulation." *Nucleic Acids Res.* **29**, 281-283 (2001)

MAP: SEARCHING LARGE GENOME DATABASES

TAMER KAHVECI   AMBUJ SINGH

Department of Computer Science
University of California
Santa Barbara, CA 93106
{tamer,ambuj}@cs.ucsb.edu

Abstract

A number of biological applications require comparison of large genome strings. Current techniques suffer from both disk I/O and computational cost because of extensive memory requirements and large candidate sets. We propose an efficient technique for alignment of large genome strings. Our technique precomputes the associations between the database strings and the query string. These associations are used to prune the database-query substring pairs that do not contain similar regions. We use a hash table to compare the unpruned regions of the query and database strings. The cost of the ensuing search is determined by how the hash table is constructed. We present a dynamic strategy that optimizes the random disk I/O needed for accessing the hash table. It also provides the user a coarse grain visualization of the similarity pattern quickly before the actual search. The experimental results show that our technique aligns genome strings up to 97 times faster than BLAST.

1 Introduction

The growth in the amount of genomic information has spurred increased interest in large scale comparison of genetic strings. Conditions such as skin and colon cancer can already be classified into much finer categories than before and soon the same approach may be possible for heart disease, schizophrenia and many other conditions. Using this kind of genetic information helps to target the treatments at the precise form of the illness that has been diagnosed, thus helping patients and doctors weigh the risk and benefits of different treatments.

One important emerging application, called *Comparative Genomics*, analyzes and compares the genetic material of different species. It is the most reliable way to identify genes and predict their functions. The functions of the higher level organisms, like humans, can be revealed by comparing to their counterparts in similar or lower level organisms. Such genome analysis involve comparison of huge strings, as large as the whole genome of a species.

Phylogenetics and evolutionary studies are other important applications that use complete genetic information of different species. Phylogenetics is used to infer the ancestral relationships among different species. This sort of relations can be captured by comparing the genetic code of the whole genomes.

The amount of biological data has been growing exponentially. This growth is on a collision course with current homology search and database query techniques and presents new challenges to biological database design. Queries (as mentioned above) will be large and complex, databases will be huge, some data will be on disk, and significant portions of datasets will be local because of networking bottleneck and proprietary data.

Fast and sensitive homology search algorithms are needed to 1) answer large queries, 2) handle huge databases stored on disk, and 3) interact with multiple datasets seamlessly.

In this paper, we propose an efficient algorithm for alignment of large genome strings. Our algorithm constructs a boolean *Match Table* for a given query string and database string with the help of the MRS index structure [1] (an index structure that stores the summary information for strings). The size of the MRS index structure is approximately 1-2% of that of database. Each entry of the Match Table corresponds to a query/database substring pair. An entry in the Match Table is marked as *True* if the corresponding query substring and database substring potentially contain similar patterns. It is marked as *False* otherwise. The size of the Match Table is negligible compared to that of database (typically 0.1% of the database.).

Once the Match Table is computed, we build hash tables on these strings as follows. First, we find the number of marked rows and columns by projecting all the *True* entries of the Match Table to its rows and columns. If the number of marked columns is more than the number of marked rows, we choose a vertical slice of the Match Table, and construct the hash table on the string that corresponds to rows of the Match Table. Otherwise, we choose a horizontal slice, and construct the hash table on the string that corresponds to columns of the Match Table. The width of these slices is restricted by available memory: the size of the hash table for a slice is no more than the available memory.

Once the hash table of a string for a slice is constructed, the marked substrings of the other string are read sequentially and exactly matching substrings (i.e. seeds) of the prespecified size (i.e. 11) are found using this hash table. The seeds are then extended in both directions to find better matches. Later, the results are reported in a descending score order. We call this technique *MAP* (MAtch table based Pruning).

The experimental results show that, MAP runs up to 97 times faster than BLAST [2] without decreasing the output quality. Furthermore, MAP can work well even for small memory sizes. This drastic reduction in CPU and I/O cost has the potential of making homology searches viable on desktop PCs. The filtering and scheduling techniques of MAP can easily be used to speedup and reduce the memory requirements of any of the current genome alignment tools. MAP also provides the user a coarse grain visualization of the similarity pattern between the strings prior to actual search.

2   Related Work

The dynamic programming solution to the problem of finding the best alignment between two strings of lengths m and n runs in $O(mn)$ time and space [3,4]. For large data and query strings, this technique becomes infeasible in terms of both time and space. Myers [5] improved the time and space complexity to $O(rn)$ by maintaining only the required part of the distance matrix, where r is the amount of allowed error. However, for large error rates r is $O(m)$, and hence the complexity is still $O(mn)$. SIM [6] uses dynamic programming technique to find all the alignments. However, it is extremely slow for large database/query strings. GLASS [7] accelerates dynamic programming solution by finding exactly matching long substrings first. However, the time and space complexity of GLASS is still high since it requires the extraction of k-mers.

Many heuristic based search tools are developed to perform string alignment faster. We consider these tools in two categories: 1) hash table based 2) suffix tree based.

Some of the important hash table based tools are FASTA[8], BLAST[2], MegaBLAST[9], BL2SEQ[10], WU-BLAST[11], SENSEI[12], FLASH[13], PipMaker (BLASTZ)[14], and PatternHunter[15]. These techniques are similar in spirit: they construct a hash table on either the query string, or the database string (or both) for all possible substrings of a prespecified size (say l). The value of l varies for these search tools and for different applications (e.g. BLAST uses $l = 11$ for nucleotides and $l = 3$ for proteins.). They start by finding exactly matching substrings of length l using this hash table. These exactly matching substrings are also called *seeds*. In the second phase, these seeds are extended in both directions, and combined if possible, in order to find better alignments. The main difference between these search tools is that they use different seed lengths and different seed extensions strategies. FLASH and PatternHunter also differs from them by choosing non-contiguous seeds. Current hash table based search tools handle short queries well, but become very inefficient in handling long queries in terms of both time and space.

There are also a number of homology search tools based on suffix trees (see [16] for suffix tree algorithms). These include MUMmer[17], QUASAR[18], and REPuter[19]. QUASAR builds a suffix tree on one of the strings, and counts the number of exactly matching seeds using this suffix tree. If the number of seeds for a region is more than a prespecified threshold, this region is searched using BLAST. REPuter builds a suffix tree on a string to find repetitions on the fly. MUMmer builds the suffix tree on both of the strings to find maximal unique matches. These tools are suitable for highly similar sequences. The main problems with suffix tree approach are twofold: (a) Suffix trees are inefficient at managing mismatches. This approach is perfect for highly similar sequences but fails to recognize more distant homologies. MUMmer and QUASAR implement various ways of linking up neighboring precisely matched blocks while REPuter lets user define a constant edit distance (default value is 3 at REPuter web site), the computation time increases very fast as the edit distance allowed increases. (b) Suffix trees have large space overhead. For example, the suffix tree in MUMmer costs $37n$ bytes of memory, where n is the input length[17], although with careful implementation, this can be reduced to $8n$.

All of the above mentioned tools require the use data structures larger than the database. Some of them are even larger than 200 times the database size. Extensive memory requirements make these tools infeasible for large scale genome comparison. Unlike these tools, the size of the MRS index structure is less than 2% of the database size. We will employ the MRS index structure in our technique.

3 Our contribution

3.1 MRS index structure

Let s be a string from an alphabet Σ, where $|\Sigma| = \sigma$. The *frequency vector*, $f(s)$, of s is defined as the σ-dimensional vector whose entries are the number of occurrences of the letters in Σ. The frequency vector of a DNA string has 4 dimensions. For example, if $s = $ CTACCCTTAG is a DNA string, then $f(s) = [2, 4, 1, 3]$. A single edit operation on a string has one of the following effects on the frequency vector of that string: 1) Increase the frequency of a character by one (insert operation). 2) Decrease the frequency of a character by one (delete operation). 3) Increase the frequency of a character by one and decrease another by one (modify operation). Based on this fact, the *frequency distance* $(FD(f(q), f(s)))$ between the frequency vectors of a query string q and a database string s is defined as the minimum number of increments and/or decrements in order to transform $f(q)$ to $f(s)$.

Each frequency vector defines an equivalence class of a set of strings whose frequency vectors are equal. The frequency distance between two frequency vectors is equal to the minimum possible edit distance between two strings in their equivalence classes. As a result, if $r < FD(f(q), f(s))$, then we can conclude that the edit distance between q and s is also greater than r. There are no false drops and indels (insertion/deletions) are also handled. The frequency distance for DNA strings can be computed in constant time using 8 integer additions and 4 integer comparisons (see [1]).

Let $S = \{s_1, s_2, ..., s_d\}$ be a database consisting of potentially long strings from alphabet $\Sigma = \{\alpha_1, \alpha_2, ..., \alpha_\sigma\}$. Let $w_1 = 2^a$ be the length of the shortest possible query string. The MRS index structure stores a grid of trees $T_{i,j}$, where i ranges from a to $a + l - 1$, and j ranges from 1 to d. The parameter l represents the number of resolution levels available in the index structure. Tree $T_{i,j}$ is the index structure for the jth string corresponding to window size 2^i.

Tree $T_{i,j}$ is obtained by sliding a window of length 2^i on string s_j starting from the leftmost point of s_j. For each possible placement of the window, the frequency vector of the corresponding substring of s_j is computed. Note that each substring corresponds to a point in the σ dimensional integer space. The frequency vectors of the c consecutive windows are covered using a *Minimum Bounding Rectangle* (MBR), where c is the *box capacity*. Typically, the box capacity ranges between 1000 and 10000. Note that, only the lower and the higher end points of the MBRs are stored along with the starting locations of the first substring contained in that MBR (i.e. the frequency vectors are not stored.). Since the index structure considers frequencies at different resolutions, this index structure is called as the *Multi Resolution String (MRS) Index Structure* [1]. The MRS index structure can be constructed using a single sequential scan on the database. Therefore, the time complexity for index construction is $O(|D|)$, where $|D|$ is the database size.

3.2 *Computing scores in the affine gap model*

The MRS index structure was originally used for string comparisons using edit distance [1]. In this paper, we extend it to work with scores and affine model for gaps. An upper bound to the score of the best alignment (in the affine gap model) of a query string q to the strings contained in an MBR, B, of the MRS index structure can be approximated efficiently. This algorithm is shown in Figure 1. The algorithm takes a frequency vector v and a box B as input. It starts by finding the number of mismatches (Steps 1 and 2). Later, the scores for matches and mismatches are computed (Steps 3 and 4), and the cost of insertions/deletions are added (Steps 5 and 6). Note that the algorithm assumes a single gap for all insertions and deletions in order to maximize the score. In the worst case, $3 \cdot \sigma + 7$ integer additions, 5 integer multiplications and $\sigma + 1$ integer comparisons are sufficient to compute $FS_w(v, B)$, where σ is the alphabet size. For DNA strings this number is 19 integer additions, 5 integer multiplications, and 5 integer comparisons regardless of the length of the strings. Hence the cost of computing FS_w is negligible compared to classic dynamic programming based algorithms with quadratic time complexity.

3.3 Match Table construction

The Match Table is the crucial element of the proposed search technique. It computes local similarities of a query string and a database string using the approximation function (FS_w) on the MRS index structure. Unlike the indexes of the genome search tools discussed above, the size of the MRS index structure is negligible compared to database size. Furthermore, index searching using our upper bounding function for scores in Figure 1 is extremely fast.

As shown in Figure 2, the Match Table M for a query string q and a database string s, is a boolean matrix. It is constructed by performing a search on the MRS index structure. Each column of the Match Table corresponds to the substrings contained in an MBR of the MRS index structure. For example, if the box capacity is 1000, then the first column corresponds to the first 1000 substrings of the database, the second column corresponds to the second 1000 substrings, and so on. For simplicity, we set the box capacity to the page size. A number of subqueries are constructed by sliding a window of length w on query q with a shift amount of Δ. Each row of M corresponds to a set of consecutive subqueries of total size equal to a page size. In the figure, seven subqueries q_1, q_2, $\cdots$, q_7, each having a size equal to a page size, have been shown. Entry $M_{i,j}$ is set to *True* if the MBR B_j (or equivalently s_j) potentially contains a similar substring to the query substring q_i. A row/column is called *marked* if at least one of the entries in that row/column is set to *True*. For example, first and second rows in this figure are marked, but third row is not marked.

Figure 3 presents the algorithm used to construct the Match Table. The algorithm takes a query string q, an error rate ϵ, and a shift amount Δ as input and constructs their match table M. The algorithm starts by initializing M to *False* (Step 1). Later, the largest resolution in the MRS index structure that is less than $|q|$ (Step 2) is determined. A *cutoff threshold*, τ is determined using this resolution and the error rate (Step 3). A window of the specified resolution, w, is then slid on q with a shift amount of Δ, and the match table is constructed by comparing the frequency distance between the query substring and the MBRs to the cutoff threshold (Step 4). The default value for ϵ is 0.1 and for Δ and w is 4K. The time complexity of the *FIND_MATCHES* algorithm is $O(\frac{|q|-w}{\Delta} \cdot \frac{|s|-w}{c})$, and the space complexity is $O(\frac{|q|-w}{P} \cdot \frac{|s|-w}{P})$, where c is the box capacity of MBRs, P is the page size, and $|s|$ is the total database size. This time complexity can be reduced to $O((\frac{|q|-w}{\Delta} + \frac{|s|-w}{c}) \cdot log(\frac{|q|-w}{\Delta} + \frac{|s|-w}{c}))$ using a plane-sweep algorithm. High values of c ($c = P$) make these negligible compared to other costs.

3.4 Scheduling disk I/O

Once the match table is constructed, only the query/database page pairs that are marked as *True* need to be searched. For example, since $M_{1,1}$ is marked (in Figure 2), the first

```
/* v is σ dimensional integer point. */
/* B is σ dimensional integer box of lower and
higher coordinates B.L and B.H. */
```
Procedure $FS_w(v, B)$

1. $inc := dec := sum := 0;$

2. for $i := 1$ to σ

   - if $v[i] < B.L[i]$ then
     $inc \mathrel{+}= B.L[i] - v[i];$
     $sum \mathrel{+}= B.L[i];$
   - else if $B.H[i] < v[i]$ then
     $dec \mathrel{+}= v[i] - B.H[i];$
     $sum \mathrel{+}= B.H[i];$
   - else
     $sum \mathrel{+}= v[i];$

3. $ScoreInc :=$
   $(min\{sum, w\} - inc) \cdot S_{match}$
   $+ inc \cdot S_{mismatch};$

4. $ScoreDec :=$
   $(min\{sum, w\} - dec) \cdot S_{match}$
   $+ dec \cdot S_{mismatch};$

5. if $w < sum$ then
   $ScoreInc \mathrel{+}=$
   $S_{gapopen} \cdot (sum - w - 1)$
   $+ S_{gap\_extend};$

6. else if $sum < w$ then
   $ScoreDec \mathrel{+}=$
   $S_{gapopen} \cdot (w - sum - 1)$
   $+ S_{gap\_extend};$

7. return $min\{ScoreInc, ScoreDec\};$

Figure 1: Procedure $FS_w(v, B)$ for computing the best score of the alignment between a string x and a set of strings $\mathcal{X}$, where v is the frequency vector of s and B is the MBR that covers the frequency vectors of the strings in $\mathcal{X}$.

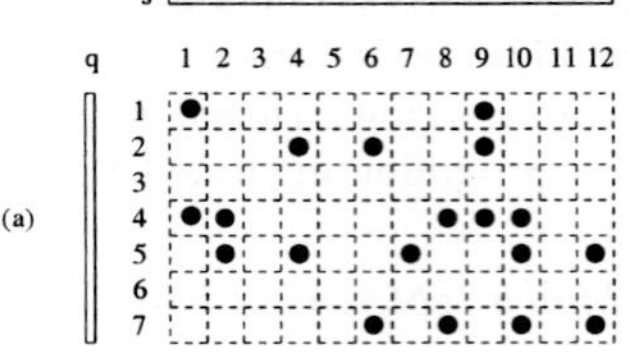

Figure 2: An illustration of Match Table on query q and database string s. The black dots correspond to *True* entries.

```
/* INPUT    q : query sequence
            ε : error rate
            Δ : shift amount
OUTPUT      M : match table      */
```

1. initialize M to *False*;

2. $w := 2^{r_{max}};$ /* the largest resolution in the index structure which is less than $|q|$. */

3. $\tau = w \cdot (1 - \epsilon);$ /* Compute cutoff threshold */

4. $start := 0; stop := |q| - w;$

5. While $start < stop$

   (a) $i = \lfloor (start + w)/pageSize \rfloor;$

   (b) $q' = q[start : start + w];$

   (c) For each MBR B_j

       - If $FS_{r_{max}}(f(q'), B_j) \leq \tau$ then
         $M_{i,j} :=$ True;

   (d) $start \mathrel{+}= \Delta;$

Figure 3: Construction of Match Table.

page of the query string (i.e. q_1) must be compared to the fist page of the database string (i.e. s_1). In order to do this, we need to find the seeds (i.e. exactly matching substrings of length 11) within s_1 corresponding to q_1 or vice versa.

One can simply consider constructing hash table on one of the strings and sequentially scanning the other string on the marked rows or columns. This is an improvement over currently available string search tools since the search is restricted to the marked entries. However, if both of the strings are very large, even the hash table of one of strings on marked rows/columns may not fit into the memory. This scheme may still cause excessive amount of random disk I/Os. In order to prevent this, we iteratively cut slices from the match table by splitting it either vertically or horizontally. Later, we construct a hash table on the marked pages of the unsplit string, and sequentially scan the other string.

Figure 4 illustrates the slices obtained from a sample Match Table by MAP. In this example, we assume that the available memory can store the hash table for 3 pages at most. The Match Table contains 9 marked columns and 5 marked rows. Since the number of marked columns is larger than the number of marked rows, MAP chooses a vertical slice in the first iteration (Figure 4(a)). This slice is shown using a rectangle. The slice has two properties: 1) It has at most 3 rows. 2) It is as wide as possible. The first restriction is imposed to ensure the hash table built on the rows fit into memory. Second restriction forces MAP to process as many entries as possible at each iteration. The first slice has rows r_1,r_4, and r_5, and columns c_1 and c_2. MAP constructs a hash table on the substrings in r_1,r_4, and r_5. Later, it sequentially scans c_1 and c_2, and searches their contents within r_1,r_4, and r_5 using the hash table. Once this search is completed, MAP removes this slice from the Match Table, and iterates on the rest of the Match Table. In the second iteration (Figure 4(b)), the match Table contains 7 marked columns and 5 marked rows. Therefore, MAP splits the table vertically. Note that, MAP does not need to consider columns c_3 and c_5 since they do not contain any marked entries. Match Table has 4 marked columns and 5 marked rows in the third iteration (Figure 4(c)). Therefore, it is a horizontal split in this step: hash table is constructed on the columns.

The decision for split direction is made as follows. Let r and c be the number of marked rows and columns of the match table. If $r < c$ then the Match Table is split vertically. Otherwise, the Match Table is split horizontally. For example, for the Match table in Figure 2, $r = 5$ and $c = 9$. Therefore, it is split vertically.

The size of a slice is determined based on memory size: The size of the hash table for the marked substrings of slice can not be more than the available memory. This can be determined by iteratively incrementing the split location. For example, consider the Match Table in Figure 2. Let memory size be 25 pages. Assume that the size of the hash table for a page is 8 times the size of a page (i.e. 2 words per letter). If the Match Table is split at the second column, then the slice contains three marked rows (r_1, r_4, and r_5). Hence, the memory requirement for this slice is $8 \cdot 3 + 1 = 25$ pages (here, the additional one page is reserved to store a page sequentially read from the other string). If we advance the cut to row r_4, the slice would contain four marked columns (r_1, r_2, r_4, and r_5), causing a memory requirement of $8 \cdot 4 + 1 = 33$ pages, which is more than available memory. Therefore, we decide to cut the Match Table at the second row.

Once a slice is cut from the Match Table, we iterate through the split string on an MBR by MBR basis. A page from the split string is read only if its corresponding row/column contains at least one *True* entry in that slice. The pages are read using an optimal disk read schedule [20]. The optimal disk read schedule guarantees that total disk I/O cost of reading candidate strings is never more than that of a sequential scan. For example, for the match table in Figure 2, once we cut a slice horizontally at third row, we iteratively read q_1 and q_3, and search them using the hash table constructed on s_1 and s_4. The optimality of our dynamic splitting algorithm can be proved by considering the expected number of marked entries in each slice.

Figure 5 presents the pseudocode for the MAP algorithm. The inputs to the program are Match Table, Match Table boundaries, and the size of the available memory. The algorithm starts by checking the boundaries. If the Match Table is all consumed, it quits (Step 1). If there are still unprocessed regions, the number of marked rows and columns are computed first (Steps 2 and 3). If the number of marked rows is less than the number of marked columns, the algorithm goes into vertical split mode (Step 4), otherwise it switches to horizontal split more (Step 5). In vertical split mode, the splitting point is iteratively advanced column wise unless the hash table for the slice fits into available memory (Step 4.a). The lower boundary for the columns of the Match Table is updated (Step 4.b). A hash table is then constructed on the marked rows of the slice (Step 4.c), and a search is performed by reading the marked columns of the slice using optimal disk scheduling (Step 4.d). Horizontal partitioning is performed similar to vertical partitioning. The MAP search algorithm is then recursively called on the remaining Match Table (Step 6).

4 Experimental Evaluation

For our experimental evaluation, we used human chromosomes 18 and 21, mouse chromosome 18, E.coli.K12, and E.coli.O157H7.EDL933 taken from NCBI. These chromosomes are composed of the alphabet $\Sigma = \{A, C, G, T\}$. The human chromosome 18 contains 4M characters, human chromosome 21 contains 33M characters, mouse chromosome 18 contains 2.3M characters, E.coli.O157H7.EDL933 contains 5.5M characters, and E.coli.K12 chromosome 18 contains 4.6M characters. We downloaded the source code of the BLAST program, and implemented the MRS index structure. We ran our experiments on a 400 MHz Pentium II computer with 1 GB memory. Page size is set to 4K in all experiments.

4.1 Quality comparison

Our first experiment set compares the quality of outputs. For this experiment, we generated five query sets from human chr18 dataset for $|q| = \{500, 1000, 2000, 4000, 8000\}$. Each of these query sets contains 100 queries. Later, we modified these queries with 0.05 mutation probability using three edit operations (i.e. insert, delete, and modify). We performed queries using these five query sets on human chr18 dataset using BLAST and MAP with quality cutoff values of $\epsilon = \{0.1, 0.2, 0.25\}$.

Figure 6 plots the score of the first 1000 outputs of BLAST and MAP for $\epsilon = \{0.1, 0.25\}$ for $|q| = 4000$, when w is chosen as the maximum possible resolution in

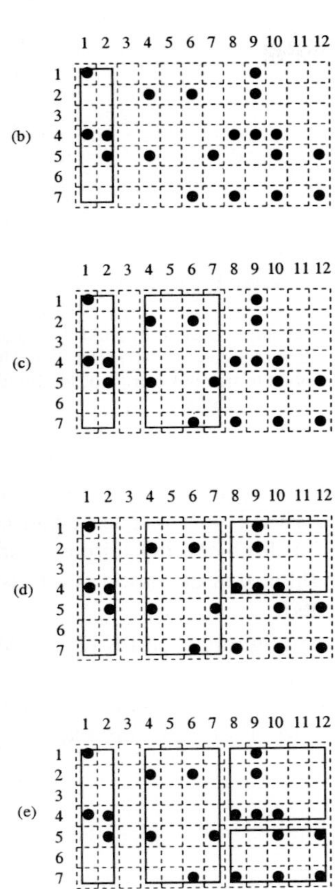

Figure 4: The slices determined by the MAP algorithm. We assume that the available memory can store the hash table for 3 pages at most.

```
/* INPUT   M : match table
           rowStart, rowStop : row boundaries
           colStart, colStop : column boundaries
           B : memory size */
MAP(M, rowStart, rowStop, colStart, colStop, B)

  1. if ((rowStart > rowStop) or
         (colStart > colStop)), then
     /* Match Table is consumed. */
         Return;

  2. numMarkedRow := number of marked rows in the
     range;

  3. numMarkedCol := number of marked columns in
     the range;

  4. if (numMarkedRow < numMarkedCol)
     /* Vertical partition */

     (a) For i := colStart to colStop
           - M' :=
               M[rowStart : rowStop][colStart : i];
           - If B < hashTableSize(M', rows) then
               If i > colStart then i--;
               Break;

     (b) colStart := i + 1;

     (c) Construct hash table on marked rows of M';

     (d) Search marked columns of M' on the hash table;

  5. else /* Horizontal partition */

     (a) For i := rowStart to rowStop
           - M' :=
               M[rowStart : i][colStart : colStop];
           - If
               B < hashTableSize(M', columns) then
               If i > rowStart then i--;
               Break;

     (b) rowStart := i + 1;

     (c) Construct hash table on marked columns of M';

     (d) Search marked rows of M' on the hash table;

  6. MAP(M, rowStart, rowStop, colStart, colStop, B);
```

Figure 5: MAP Search algorithm.

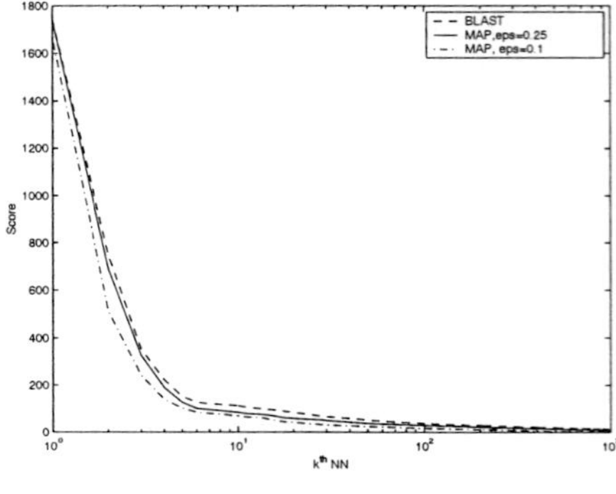

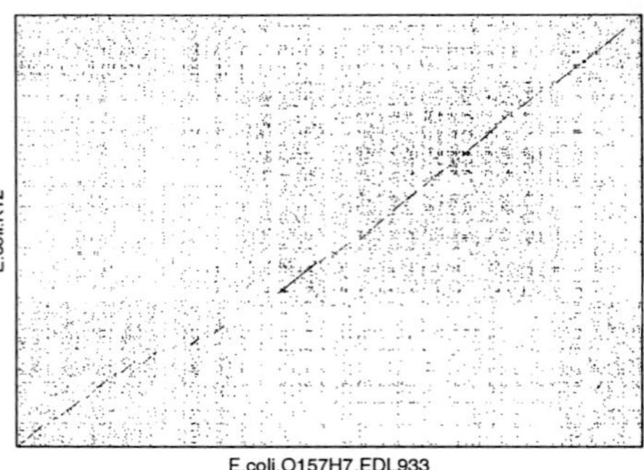

Figure 6: The average score for the first 1000 results for BLAST and MAP for different values of ϵ, for $|q| = 4000$ and $\Delta = w =$ maximum possible resolution.

Figure 7: Match Table created by MAP for aligning E.coli.K12 with E.coli.O157H7.EDL933. The points correspond to marked entries of the Match Table.

the index structure and $\Delta = w$. The results for other query lengths were similar. The quality of MAP outputs are close to those of BLAST even when $\epsilon = 0.1$. The deviation between BLAST outputs and MAP outputs are about 5% for high scoring outputs, and 10-30% for low scoring outputs when $\epsilon = 0.1$. For $\epsilon = 0.25$, this deviation is almost zero for high scoring outputs and 5% for low scoring outputs.

Figure 7 shows a part of the match table constructed for aligning E.coli.K12 (4.6 M base pairs) with E.coli.O157H7.EDL933 (5.5 M base pairs). The diagonal run of marked entries depicts the similarity patterns. This figure shows that MAP can visualize similar patterns in a coarse grain without aligning the sequences. MAP created this matrix in less than 4 seconds.

4.2 *Performance comparison*

Our second set of experiments compares the performance of MAP to BLAST. Figure 8(a) shows the total cost of BLAST and MAP for aligning human chromosome 18 with mouse chromosome 18 for memory sizes 100, 200, 400, 800, 1600, and 3200 pages. In this experiment, for BLAST's advantage, we set BLAST to construct the hash table on the shorter string (i.e. mouse chromosome 18). For the considered memory sizes, MAP performed 74 to 97 times faster than BLAST.

Figure 8(b) shows the total cost of BLAST and MAP for the comparison of human chromosome 18 with human chromosome 21 for memory sizes 100, 200, 400, 800, 1600, 3200, and 6400 pages. MAP performed 20 to 74 times faster than BLAST for these datasets.

We can summarize the experimental results as follows. 1) The output quality of MAP is very close to that of BLAST. 2) MAP is up to 97 times faster than BLAST. 3) MAP works well even for small memory sizes.

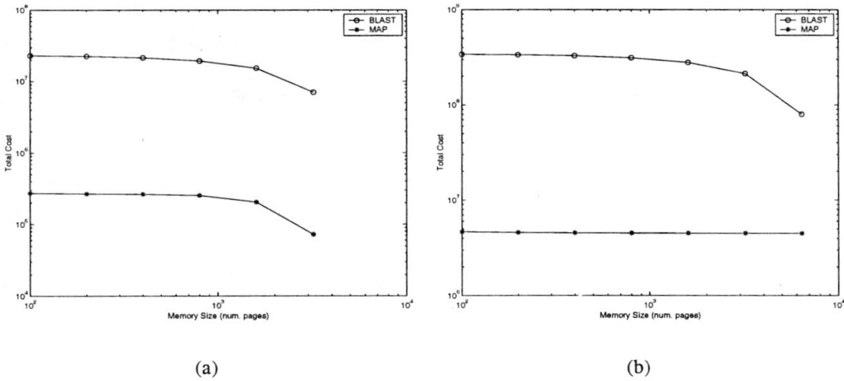

Figure 8: The total cost of BLAST and MAP in terms of page transfers for different memory sizes for aligning human chromosome 18 with (a) mouse chromosome 18, (b) human chromosome 21 when $\Delta = w = 4K$.

5   Discussion

In this paper, we considered the problem of aligning huge genome strings. We also made significant improvements to the MRS index structure by presenting an algorithm that computes scores with affine gaps in the frequency domain. We presented an algorithm that finds the local associations between a query string and a database string. The algorithm constructs a boolean *Match Table* which uses the MBRs of an index structure as its columns and substrings of the query as its rows. An entry in the Match Table is marked as *True* if the corresponding query substring and database MBR (which in turn represents a database substring) potentially contains a similar patterns. It is marked as *False* otherwise. The Match Table enables the user to visualize the similarity pattern between the strings prior to search in coarse grain.

The Match Table is split iteratively either vertically or horizontally minimizing the I/O and CPU costs. A hash table is constructed on the marked pages of the unsplit string. The marked pages of the split string are read sequentially and exactly matching substrings (i.e. seeds) of some certain prespecified size are found using this hash table. The seeds are then extended in both directions to find better matches. Later, the results are reported in a descending score order. The resulting search technique is called MAP. This MBR-based decomposition reduces the memory requirements and consequently the I/O cost. The CPU cost is also reduced since only the *True* entries of the Match Table need to be examined.

In our experiments, we used the BLAST [2] for comparison. According to our experimental results, MAP runs 20 to 97 times faster than BLAST. Furthermore, MAP can work well even for small memory sizes. Unlike the massive data structures used by current techniques, the size of the MRS index structure is only 1-2% of the database. MAP achieves these performance improvements while keeping quality of the resulting answer set very close to that of BLAST.

MAP is a very general technique, in the sense that its Match Table based pruning and dynamic splitting scheme can be used to improve any of the current string search tools. Hence, one can view MAP as a technique that improves the available techniques instead of as a competitor to these techniques.

Aligning large genome strings is an important emerging application. The explosive growth of these datasets and the complexity of computing matches makes it imperative that faster disk-resident techniques be devised. The techniques presented in this paper are an important step in this regard, and should be widely applicable. We are currently building a web-based server which makes the MAP software available. We are also extending MAP to align protein strings using alphabet specific scoring schemes.

References

1. T. Kahveci and A. Singh. In *VLDB*, pages 351–360, Roma, Italy, September 2001.
2. S. Altschul and W. Gish. Basic local alignment search tool. *J. Molecular Biology*, 1990.
3. S.B. Needleman and C.D. Wunsch. *J. Molecular Biology*, 48:443–53, 1970.
4. T.F. Smith and M.S. Waterman. *J. of Molecular Biology*, March 1981.
5. E. Myers. An O(ND) difference algorithm and its variations. *Algorithmica*, pages 251–266, 1986.
6. X. Huang and W. Miller. *Adv. Appl. Math*, 12:337–357, 1991.
7. S. Batzoglou, L. Pachter, J.P. Mesirov, B. Berger, and E.S. Lander. *Genome Research*, 10:950–958, 2000.
8. W.R. Pearson and D.J. Lipman. In *Proc. Natl. Acad. Sci*, volume 85, pages 2444–2448, 1988.
9. Z. Zhang, S. Schwartz, L. Wagner, and W. Miller. *J. of Computational Biology*, 7(1-2):203–214, 2000.
10. T.A. Tatusova and T.L. Madden. *FEMS Microbiol. Lett.*, pages 247–250, 1999.
11. W. Gish. WU-blast. http://blast.wustl.edu.
12. D.J. States and P. Agarwal. In *ISMB*, 1996.
13. A. Califano and I. Rigoutsos. FLASH: Fast look-up algorithm for string homology. In *CVPR*, 1993.
14. S. Schwartz, Z. Zhang, K.A. Frazer, A. Smit, C. Riemer, J. Bouck, R. Gibbs, R. Hardison, and W. Miller. *Genome Research*, 10(4):577–586, April 2000.
15. M. Ma, J. Tromp, and M. Li. *Bioinformatics*, 18(0):1–6, 2002.
16. D. Gusfield. Cambridge University Press, 1 edition, January 1997.
17. A.L. Delcher, S. Kasif, R.D. Fleischmann, J. Peterson, O. White, and S.L. Salzberg. Alignment of whole genome. *Nucleic Acids Research*, 27(11):2369–2376, 1999.
18. S. Burkhardt, A. Crauser, P. Ferragina, H.-P. Lenhof, 'E. Rivalsd, and M. Vingron. In *RECOMB*, Lyon, April 1999.
19. S. Kurtz and C. Schleiermacher. *Bioinformatics*, 15(5), 1999.
20. B. Seeger. *Information Systems*, 21(5):387–407, 1996.

TOWARDS THE DEVELOPMENT OF COMPUTATIONAL TOOLS FOR EVALUATING PHYLOGENETIC NETWORK RECONSTRUCTION METHODS

LUAY NAKHLEH, JERRY SUN, TANDY WARNOW

Dept. of Computer Sciences, U. of Texas, Austin, TX 78712

C. RANDAL LINDER

Section of Integrative Biology, U. of Texas, Austin, TX 78712

BERNARD M.E. MORET, ANNA THOLSE

Dept. of Computer Science, U. of New Mexico, Albuquerque, NM 87131

We report on a suite of algorithms and techniques that together provide a simulation flow for studying the topological accuracy of methods for reconstructing phylogenetic networks. We implemented those algorithms and techniques and used three phylogenetic reconstruction methods for a case study of our tools. We present the results of our experimental studies in analyzing the relative performance of these methods. Our results indicate that our simulator and our proposed measure of accuracy, the latter an extension of the widely used Robinson-Foulds measure, offer a robust platform for the evaluation of network reconstruction algorithms.

1 Introduction

Phylogenies, i.e., the evolutionary histories of groups of organisms, play a major role in representing the interrelationships among biological entities. Many methods for reconstructing such phylogenies have been proposed, but almost all of them assume that the underlying evolutionary history of a given set of species can be represented by a tree. While this model gives a satisfactory first-order approximation for many families of organisms, other families exhibit evolutionary mechanisms that cannot be represented by a tree. Processes such as hybridization and horizontal gene transfer result in networks of relationships rather than trees of relationships. Although this problem is widely appreciated, there has been comparatively little work on computational methods for estimating evolutionary networks.

A standard technique for assessing the performance of phylogenetic reconstruction methods is to use simulation studies. In such studies, a model topology (tree or network) is generated, after which a sequence is evolved (including bifurcations and non-treelike events) down the edges of the model topology according to some chosen model of sequence evolution. Finally, a phylogeny is reconstructed on the resulting set of sequences and its topology is compared to the model topology in order to assess the quality, or topological accuracy, of the reconstruction. While many simulation tools and accuracy measures are available for studying the performance of phylogenetic tree reconstruction methods, such tools and measures are lacking in the context of phylogenetic networks.

We describe a collection of such techniques and quality measures: (i) a technique for generating random phylogenetic networks and simulating the evolution of sequences on these networks, and (ii) measures to assess the topological accuracy of the reconstructed networks. We implemented those techniques and conducted a simulation study on one network reconstruction method, `SplitsTree` [1], and two tree reconstruction methods, *Neighbor-Joining* and *greedy Maximum Parsimony*. We assessed the performance of these methods on datasets generated in our simulation.

The rest of the paper is organized as follows. Section 2 provides some background on phylogenetic trees and describes the various steps involved in a simulation study. Sections 3 and 4 introduce our new techniques and measures, and review the existing reconstruction method used in our study. Section 5 describes our experimental setup, while Section 6 reports the results of our experiments and offers some remarks on the usefulness of our simulation flow.

2 Phylogenetic Trees

A *phylogenetic tree* on a set S of taxa is a rooted tree whose leaves are labeled by S. Such a tree represents the evolutionary history of a set of taxa, where the leaves of the phylogenetic tree correspond to the extant taxa and the internal nodes represent the (hypothetical) ancestors. Many algorithms have been designed for the inference of phylogenetic trees, mainly from biomolecular (i.e., DNA, RNA, or amino-acid) sequences [2]. Evaluating these algorithms cannot be done with real data alone, since we typically do not know with high confidence the details of the "true" evolutionary history. Thus the standard means of performance evaluation for phylogenetic reconstruction methods is the simulation study, in which "model" phylogenies are constructed using some chosen model of evolution, the "modern data" (found in the leaves of the model phylogeny) are fed to the reconstruction algorithms, and the output of the algorithms compared with the model phylogeny.

2.1 Model trees

Model trees are typically taken from some underlying distribution on all rooted binary trees with n leaves; commonly used distributions include the uniform distribution and the Yule-Harding distribution [3,4]. Although we generate networks rather than trees, we have based our network generation on the widely used model of birth-death evolution, which we now briefly review in the context of tree generation.

To generate a random birth-death tree on n leaves, we view speciation and extinction events as occurring over a continuous interval. During a short time interval, Δt, since the last event, a species can split into two with probability $b(t)\Delta t$ or become extinct with probability $d(t)\Delta t$. To generate a tree with n taxa, we begin this process with a single node and continue until we have a tree with n leaves. (With some nonzero probability some processes will not produce a tree of the desired size, since all nodes could go "extinct" before n species are generated; we then repeat the process until a tree of the desired size is generated.) Under this distribution, a natural length is associated with each edge, namely the time elapsed between the speciation event that gave rise to that edge and the (speciation or extinction) event that ended that edge. Thus birth-death trees are inherently ultrametric, that is, the branch lengths are proportional to time.

2.2 Tree reconstruction methods

In our experiments, we used one network reconstruction method and two tree reconstruction methods, neighbor-joining (NJ) and greedy maximum parsimony (MP).

- Neighor-joining[5] is the most popular distance-based method. For every pair of taxa, it determines a score based on the pairwise distance matrix, then joins the pair with the smallest score, building a subtree of two leaves whose root replaces the two chosen taxa in the matrix. Pairwise distances for this new "supertaxon" are then recalculated and the entire process is repeated until only three nodes remain; these are then joined to form an unrooted binary tree.
- Maximum parsimony[6] is one of the two main optimization criteria used in phylogenetic reconstruction. Because the problem is NP-hard, we use a simple greedy heuristic[7,8], which adds taxa to the tree one at a time in some random order—the placement of each new taxon is optimized locally.

2.3 Measures of accuracy

A commonly used measure of the topological accuracy of reconstructed trees is the *Robinson-Foulds (RF)* value[9]. Every edge e in an unrooted leaf-labeled tree T defines a bipartition π_e on the leaves (deleting e cuts the tree); we set $C(T) = \{\pi_e : e \in E(T)\}$, where $E(T)$ is the set of all internal edges of T. If T is a model tree and T' is a reconstructed tree, the *false positives* are the edges of the set $C(T') - C(T)$ and the *false negatives* are those of the set $C(T) - C(T')$.

- The *false positive rate (FP)* is $(|C(T') - C(T)|)/(n - 3)$.
- The *false negative rate (FN)* is $(|C(T) - C(T')|)/(n - 3)$.

Since $n - 3$ is the number of internal edges of an unrooted binary tree on n leaves, the false positive and false negative rates are values in the range $[0, 1]$. The RF distance between T and T' is simply the average of these two rates, $\frac{\text{FN}+\text{FP}}{2}$.

3 Phylogenetic Networks

3.1 Hybridization and gene transfer

Two of the mechanisms that can result in non-tree evolution are *hybridization* and *horizontal gene transfer*.

- In hybridization, two lineages recombine to create a new species, as symbolized in Figure 1.

318

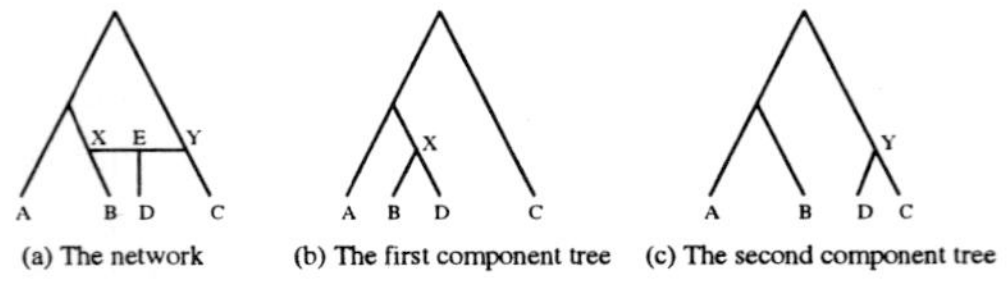

(a) The network   (b) The first component tree   (c) The second component tree

Figure 1: Hybridization: the network and its two induced trees

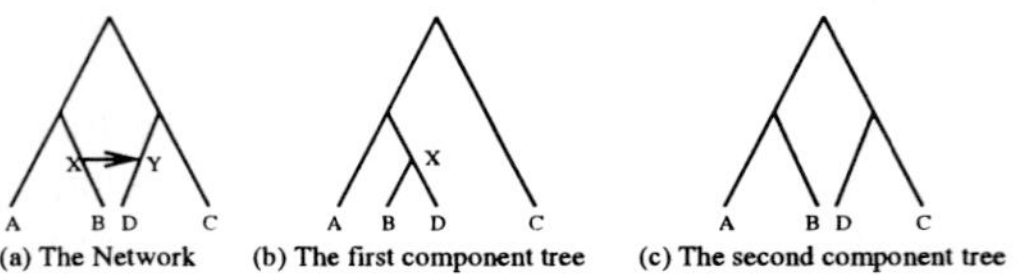

(a) The Network   (b) The first component tree   (c) The second component tree

Figure 2: Horizontal transfer: the network and its two induced trees

The new species may have the same number of chromosomes as its parent (*diploid hybridization*) or the sum of the numbers of chromosomes of its parents (*polyploid hybridization*).

- In horizontal gene transfer, genetic material is transferred from one lineage to another, producing a new lineage, as symbolized in Figure 2.

In these two cases, the true evolutionary history is best represented by a network, or *rooted directed acyclic graph*, rather than by a tree.

Consider how an individual site evolves down a network. For diploid organisms, each chromosome consists of a pair of homologs. In a diploid hybridization event, the hybrid inherits one of the two homologs for each chromosome from each of its two parents. Since homologs assort at random into the gametes (sex cells), each has an equal probability of ending up in the hybrid. In polyploid hybridization, both homologs from both parents are contributed to the hybrid. Prior to hybridization, each site on the homolog has evolved in a tree-like fashion, although due to meiotic recombination (exchanges between the parental homologs during gamete production), different strings of sites may have different histories. Thus each site in the homologs of the parents of the hybrid evolved in a tree-like fashion on one of the trees contained inside (or, induced by) the network representing the hybridization event; see Figures 1(b) and 1(c). Similarly, in an evolutionary scenario involving horizontal transfer, certain sites are inherited through horizontal transfer from another species, as in Figure 2(b), while all others are inherited from the parent, as in Figure 2(c). Thus, in each of these two scenarios, *each site evolves down one of the trees induced by the network*.

3.2 Representation

Phylogenetic networks can be represented by *rooted directed acyclic graphs*, where each node (except for the root) has indegree 1 or 2. Nodes of indegree 1 are called *tree nodes*, whereas nodes of indegree 2 are called *hybrid nodes*. A hybrid node typically takes its genetic material from both of its parents, whereas a tree node takes its genetic material from its sole parent. The leaves of a network represent the extant taxa, and the internal nodes represent the hypothetical ancestral taxa. Whereas phylogenetic trees have a standard representation, the Newick format (a form of pre-order traversal), no such representation exists for phylogenetic networks. We thus simply represent a network as a list of its edges, where each edge is defined by its two endpoints and its weight (the expected number of changes along that edge).

3.3   Model networks

We propose a model for generating random networks based on birth-death model for trees. A birth event in trees represents regular speciation; in networks, a birth event can be either regular speciation or hybrid speciation. We describe first the general model, then the restricted model used in our experiments.

Our model incorporates three types of events: regular speciation, hybrid speciation, and extinction. (Lateral gene transfers can easily be added as well.) The model supports several types of hybrid speciation. To generate a random network, we start with one node (the root) and a pair of sequences (the two homologs of the chromosome at the root) and initiate a regular speciation event, thus creating two lineages.

Each lineage is represented by an edge in the network. An edge e is defined by its two endpoint nodes u and v. Associated with each node u is a time-stamp, $t(u)$, and associated with each edge e is a positive real number, $w(e)$, indicating the expected number of changes along that edge. Whenever a new lineage is created at node u, the sequences at node u evolve along the newly created lineage, according to a given model of evolution and to the $w(e)$ value associated with that lineage. (The model can be easily extended to support nonuniform rates among segments of the sequences by associating with each edge a collection of values, one for each segment.)

At any time t, we consider all of the lineages that exist at that time. For each such lineage l, and based on certain probabilities, either nothing happens, which means the lineage l is continued, or one of three mutually exclusive events occurs:

Extinction: Lineage l becomes extinct and a leaf u is created with stamp t.

Speciation: A node u is created with stamp t and two new lineages are started from u.

Hybridization: Let H be the set of all lineages at time t and choose a lineage $l' \in H$ to hybridize with l. (The choice of l' depends on the ploidy level and number of chromosomes of l', as well as on the evolutionary distance between l and l'.) When l and l' hybridize, the two lineages are continued and a new, third lineage arises from the hybrid speciation event. We allow each lineage to hybridize only once at each point in time.

This process may generate a network with fewer than the desired number of leaves, since all lineages might go extinct before enough lineages are created; in such a case, we can repeat the process until we obtain a network of the desired size or we can conduct longitudinal studies in a population of networks of diverse sizes.

In our general model, the topology of the network and the sequences at each node are generated together, in an interdependent manner. Such dependence is important in networks, since the probability of a particular hybridization's taking place depends, among other things, on the actual genomes of the putative hybridization parents. In the generation of trees, in contrast, simulation studies generally generate a tree topology first, then use it to derive many different collections of sequences—the topology does not depend on the sequences. Our restricted model follows the outline of our general model, but assumes the same independence used in tree simulations and thus begins by generating a network and then evolves the sequences down the completed network. To generate a random network N with one hybrid, we start with a random birth-death tree T, with times associated with its nodes. Let the time at the root be 0 and that at the last generated leaf be t_l; let t_i be a uniform variate in the interval $[0, t_l]$. We identify all nodes in the tree with age not exceeding t_i, but whose children have ages exceeding t_i: these nodes are the potential parents of the hybrid. We calculate the pairwise evolutionary distances (the length of the tree paths) among all of these nodes and choose the two parents with a probability inversely proportional to the evolutionary distance between the two nodes. Hybridization thus occurs between nodes that coexist in time and is more likely between more closely related ancestral taxa. To generate p hybrid nodes, we repeat the process p times, thus allowing hybrids to hybridize again. The networks thus generated are ultrametric; for each edge, we use a uniform variate x in the range $[-\ln c, \ln c]$ and multiply the edge length by e^x to deviate the networks from ultrametricity. (We call c the *deviation factor*.)

3.4 Simulating sequence evolution on networks

We based our sequence generation on the popular Seq-Gen [10] tool, which takes a tree in the Newick format and simulates the evolution of sequences along that tree under a choice of models of evolution. Seq-Gen simulates the evolution of sequences on a tree by placing a random sequence of the desired length at the root of the tree and then evolving it down the tree using the specified model of evolution and other parameters, including scaling factor, codon-specific corrections, gamma rate heterogeneity, etc.

Our modified version, Seq-Gen2, takes a network as an input; it assumes diploid hybridization and uses networks produced by our restricted model. The main change is that Seq-Gen2 evolves a *pair* of sequences, A and B, corresponding to the two homologs of the chromosome. The two sequences A and B are evolved independently down the tree, using the specified model of evolution and the other parameters. In the case of tree nodes, both the A and B sequences are evolved in exactly the same manner as they evolve on a tree. In the case of a diploid hybrid node, the node inherits the A or B sequence from one parent and the A or B sequence from the other parent. There is no evolution on the edges between the parents of the hybrid and the node at the origin of the hybrid, since, at the scale of evolution, hybridization is essentially instantaneous. The output of Seq-Gen2 is a set of pairs of sequences at the leaves.

4 Measuring the Distance Between Two Phylogenetic Networks

We want to measure the error between a model network N_1 and an inferred network N_2; this measure, $m(N_1, N_2)$, must be nonnegative and symmetric and satisfy two properties:

- N_1 and N_2 are the same network exactly when $m(N_1, N_2) = 0$.
- If N_1 and N_2 are trees, then $m(N_1, N_2) = RF(N_1, N_2)$.

These are very mild requirements; in particular, they do not require that m be a metric, since we have not included any form of triangle inequality.

We propose two measures. Our first measure, based on a graph model, seeks to extend the RF measure by viewing the networks as extensions of trees. (Removing a chosen subset of edges from the network leaves a tree; the number of such subsets, however, can be very large.) This approach leads to a sophisticated view of the relationship between two networks, but the resulting measure is too expensive to compute exactly and tends to overemphasize the importance of guessing just the right number of non-tree events. Our second measure seeks to extend the RF measure by viewing the networks in terms of partitions. Just as the RF measure counts the number of compatible bipartitions, our new measure counts the number of compatible tripartitions. This second measure is easy to define and compute, is very closely related to the RF measure, and can also support a weighting scheme.

Let N_1 have p hybrid nodes and N_2 have q hybrid nodes. Then N_1 induces a set $\mathcal{T}_1$ of at most 2^p trees and N_2 induces a set $\mathcal{T}_2$ of at most 2^q trees. Figure 1 shows a network with one hybrid node and its two induced trees. Define the complete bipartite graph $G^{N_1,N_2} = (\mathcal{T}_1 \cup \mathcal{T}_2, E)$ (which has an edge $e = (u, v)$ between every $u \in \mathcal{T}_1$ and $v \in \mathcal{T}_2$) and assign to each edge $e = (u, v)$ a weight $w(e)$, the RF value between the two trees that correspond to nodes u and v. Our first measure can now be defined.

Definition 1 *The error rate between N_1 and N_2 is the weight of the minimum-weight edge-cover of G^{N_1,N_2}.*

The minimum-weight edge cover of a graph $G = (V, E)$ is a subset $E' \subseteq E$ such that E' covers V and the sum of edge weights, $\sum_{w(e) \in E'} w(e)$, is minimum. This measure clearly satisfies our two requirements. Although the minimum-weight edge-cover problem is solvable in polynomial time, the size of the bipartite graph is exponential in the number of hybrid nodes, so that computing the error rate may require exponential time. In practice, because hybridization is a rare event, the true network will have relatively few hybrid nodes, so that good reconstructions can be evaluated quickly. More damaging is the fact that this measure places a very strong emphasis on reconstructing networks that have the right number of hybridization events—small errors in that number dominate even very large errors in the choice of other edges.

This shortcoming leads us to design a measure based directly on partitions. Let N be a network, leaf-labeled by a set of taxa S, and e be an edge in N, where $e = (u, v)$, and u is the parent of v. Each edge $e \in N$ induces a tripartition of S, defined by the sets

- $A(e) = \{s \in S : s \text{ is reachable from the root of } N \text{ only via } v\}$.
- $B(e) = \{s \in S : s \text{ is reachable from the root of } N \text{ via at least one path passing through } v \text{ and one path not passing through } v\}$.
- $C(e) = \{s \in S : s \text{ is not reachable from the root of } N \text{ via } v\}$.

For each edge e, the three sets $A(e)$, $B(e)$ and $C(e)$ are weighted; the weight of an element in $A(e)$ or $C(e)$ is 0 (or any fixed constant) and the weight of an element $s \in B(e)$ is the maximum number of hybrid nodes on a path from v to s, where v is the target of edge e. Two weighted sets S_1 and S_2 are interchangeable, denoted by $S_1 \equiv S_2$, whenever they contain the same elements and each element has the same weight in both sets. Two edges e_1 and e_2 are *compatible*, denoted by $e_1 \equiv e_2$, whenever we have $A(e_1) \equiv A(e_2)$ and $B(e_1) \equiv B(e_2)$ and $C(e_1) \equiv C(e_2)$.

We define the *false negative rate* (FN) and *false positive rate* (FP) between two networks N_1 and N_2 as follows.

- $FN(N_1, N_2) = |\{e_1 \in N_1 : \nexists e_2 \in N_2 \text{ s.t. } e_1 \equiv e_2\}|/|N_1|$.
- $FP(N_1, N_2) = |\{e_2 \in N_2 : \nexists e_1 \in N_1 \text{ s.t. } e_1 \equiv e_2\}|/|N_2|$.

Definition 2 *The error rate between N_1 and N_2 is the average of $FN(N_1, N_2)$ and $FP(N_1, N_2)$.*

This measure clearly satisfies our two conditions and is computable in time polynomial in the size of the two networks.

5 Experimental Settings

In order to obtain statistically robust results [11,12], we used 30 *runs*, each composed of a number of *trials* (a trial is a single comparison), computed a mean outcome of each run, and studied the mean and standard deviation of these runs. The standard deviation of the mean outcomes in our studies was generally negligible for NJ and MP (except for very small numbers of taxa), and only rarely larger for SplitsTree. We graphed the average of the mean outcomes for the runs, but omitted the standard deviation from the figures for clarity.

We ran our studies on random networks generated using the technique described earlier. These networks had diameter 2; in order to obtain networks with other diameters, we scaled the edge lengths by factors of 0.01, 0.05, 0.1, 0.5, 1, and 2, producing networks with diameters of 0.02, 0.1, 0.2, 1, 2, and 4, respectively. To deviate the networks from ultrametricity, we used a deviation factor of 4. We generated networks with 0, 1, 2, 3, 4, and 5 hybrids, for 10, 20, 40 and 80 leaves—one network for each combination of diameter, number of taxa, and number of hybrids. We then evolved sequences on these networks using the K2P+Gamma [13] model of evolution (we chose $\alpha = 1$ for the shape parameter and set the transition/transversion ratio to 2). We used a fixed factor [14] of 1 for distance correction and used sequence lengths of 25, 50, 100, 250, and 500. We used our Seq-Gen2 to evolve DNA sequences down the network under the K2P+Gamma model of evolution. The datasets we generated consisted of pairs of sequences, as described; however, the phylogenetic methods under study take datasets with only a single sequence per taxon. Therefore, from each sequence dataset that we generated, we created two sets: one that consisted of the A sequence of each taxon, and one that consisted of concatenation of the A and B sequences of each taxon. Thus, the effective sequence lengths that we looked at were 25, 50, 100, 250, and 500, when the A sequences were used, and 50, 100, 200, 500, and 1000, when the concatenation of A and B was used. We used PAUP* [15] for the greedy MP method and the SplitsTree software package [1].

6 Results and Discussion

We present two sets of results: one set compares the output of SplitsTree with the true network in terms of the number of hybrid nodes created, while the second compares the topological error rate of the three methods. Our first finding is that SplitsTree tends to reconstruct more hybrid nodes than are present in the true network, as illustrated in Figure 3 (the correct answer is on the oblique line). The effect decreases as the number of true hybrids increases—indeed, the worst-case instances for SplitsTree are trees.

Our second set of figures compares the FN and FP error rates (as defined in Section 4) of SplitsTree, NJ, and MP as a function of the number of true hybrids (Figure 4), the sequence length (Figure 5), the number of taxa (Figure 6), and the scaling factor (Figure 7). The two tree reconstruction methods are nearly indistinguishable throughout our experiments (with a slight advantage of MP over NJ in some

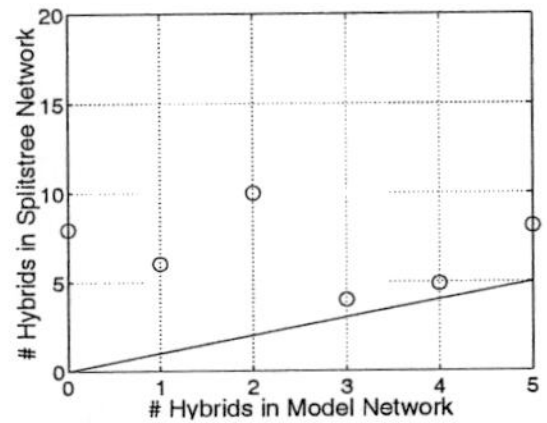

(a) single sequence (length 500)

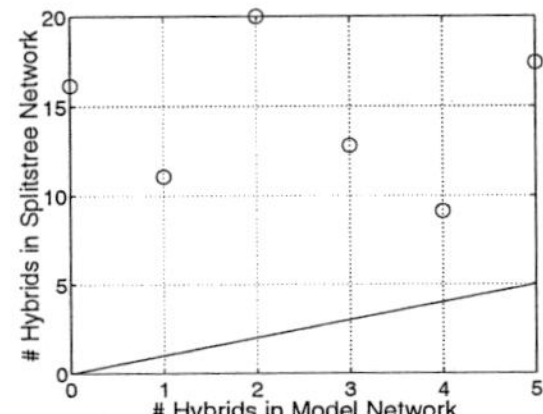

(b) concatenated sequences (length 1000)

Figure 3: The number of hybrids in the output of SplitsTree vs. the true number of hybrids, for 80 taxa at a scaling of 0.5

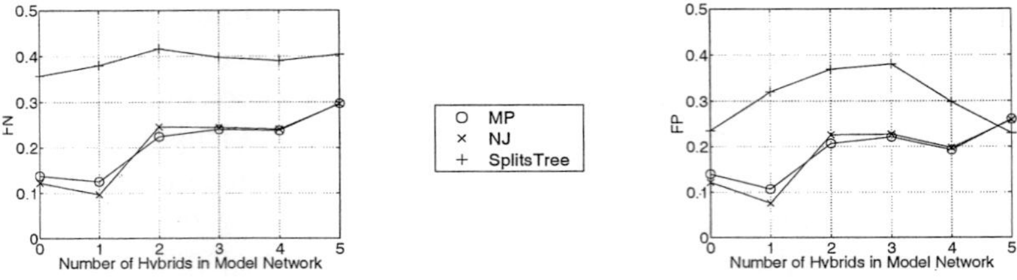

Figure 4: The FN and FP error rates of the three methods as a function of the number of hybrids in the model network. Scaling=0.5, concatenated sequences, 40 taxa, and sequence length=1000

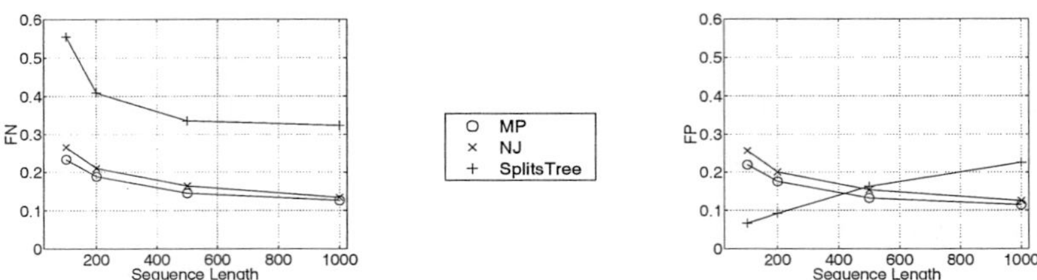

Figure 5: The FN and FP error rates of the three methods as a function of the sequence length. Scaling=0.5, concatenated sequences, 80 taxa, and 1 hybrid in the model networks

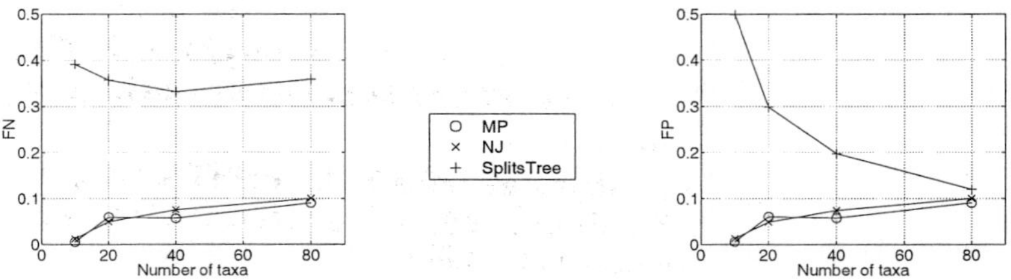

Figure 6: The FN and FP error rates of the three methods as a function of the number of taxa. Scaling=0.1, concatenated sequences, sequence length=1000, and no hybrids in the model networks

cases). As expected from the results of many previous experimental investigations, their error rate grows slowly with the number of taxa, decreases slowly with increasing sequence length, and grows very slowly with increasing scaling factor. Since these methods never create hybrid nodes, their error rate as a function of the number of true hybrids must grow linearly in p, where p is the number of true hybrids, a growth clearly demonstrated in Figure 4.

The accuracy of SplitsTree suffers from its overestimates of the number of hybrid nodes: every hybrid not present in the true network affects all edges from which the hybrid is reachable. Thus Figure 4 confirms our findings from Figure 3: as the number of true hybrid nodes increases, SplitsTree infers a more accurate

Figure 7: The FN and FP error rates of the three methods as a function of the scaling. Concatenated sequences, sequence length=1000, 40 taxa, and 3 hybrids in the model networks.

number of such nodes and its error rate does not noticeably increase with the number of hybrids. (Note that it is the FP rate that varies; the FN rate appears unaffected by the number of hybrids.) So far, our measure of topological accuracy behaves as desired. However, its behavior for `SplitsTree` in Figures 5 and 6 is, at first sight, surprising: one might expect the same trends as for the two tree reconstruction algorithms. Once again, however, this behavior accurately reflects the characteristics of the reconstructions. The mild increase of the FP rate in Figure 5 is due to the increasing number of incompatible splits (the decision criterion used by `SplitsTree`) due to an increasing number of characters. The sharp decrease in the FP rate in Figure 6 is due to the normalization by the increased number of taxa and the fact that the error is mostly attributable to excess hybrid nodes, whose number does not strongly depend on the number of taxa.

Overall, our proposed (second) measure of topological accuracy gives results that follow qualitative expectations, neither over- nor underemphasizing the importance of hybrid events.

7   Conclusions

We have presented a suite of tools that forms the basis of a simulation flow for the study of network reconstruction methods. The paucity of tools in this area, coupled with the universal recognition that tree models are too limited in many areas of the Tree of Life, makes the development of measures, algorithms, and tools an urgent task in phylogenetic research. While our test suite (generators and measures) is tailored for network reconstruction, our results at this point do not allow us to conclude much about network reconstruction methods—we could only test one existing method (others we tried would not run properly) and our range of experimental data in this study was somewhat limited. Yet, our test suite shows that it is possible to devise and use simulations for the assessment of reconstruction methods for phylogenetic networks. Our error measure has the advantage that it does not handle independently tree errors and network characteristics, avoiding the pitfall of having to assign relative weights or priorities to the two; if it does appear to favor trees slightly over networks, that is but a reflection of Occam's razor: hybridization events should be used only sparingly to explain data features otherwise explainable under a tree model.

8 Acknowledgments

This work is supported by National Science Foundation grants ACI 00-81404 (Moret), DEB 01-20709 (Linder, Moret, and Warnow), EIA 01-13095 (Moret), EIA 01-13654 (Warnow), EIA 01-21377 (Moret), and EIA 01-21680 (Linder and Warnow), and by the David and Lucile Packard Foundation (Warnow).

1. D. H. Huson. SplitsTree: A program for analyzing and visualizing evolutionary data. *Bioinformatics*, 14(1):68–73, 1998.
2. D. L. Swofford, G. J. Olsen, P. J. Waddell, and D. M. Hillis. Phylogenetic inference. In D. M. Hillis, B. K. Mable, and C. Moritz, editors, *Molecular Systematics*, pages 407–514. Sinauer Assoc., 1996.
3. G. L. Yule. A mathematical theory of evolution based on the conclusions of Dr. J.C. Willis, FRS. *Philosophical Transactions of the Royal Society*, 213:21–87, 1924.
4. E. F. Harding. The probabilities of rooted tree shapes generated by random bifurcation. *Adv. Appl. Prob.*, 3:44–77, 1971.
5. N. Saitou and M. Nei. The neighbor-joining method: A new method for reconstructing phylogenetic trees. *Mol. Biol. Evol.*, 4:406–425, 1987.
6. W. M. Fitch. On the problem of discovering the most parsimonious tree. *Am. Nat.*, 111:223–257, 1977.
7. O. Bininda-Emonds, S. Brady, J. Kim, and M. Sanderson. Scaling of accuracy in extremely large phylogenetic trees. In *Proc. 6th Pacific Symposium on Biocomputing (PSB01)*, pages 547–557. World Scientific, 2001.
8. L. Nakhleh, B.M.E. Moret, U. Roshan, K. St. John, J. Sun, and T. Warnow. The accuracy of fast phylogenetic methods for large datasets. In *Proc. 7th Pacific Symposium on Biocomputing (PSB02)*, pages 211–222. World Scientific, 2002.
9. D. F. Robinson and L. R. Foulds. Comparison of phylogenetic trees. *Mathematical Biosciences*, 53:131–147, 1981.
10. A. Rambaut and N. C. Grassly. Seq-gen: An application for the Monte Carlo simulation of DNA sequence evolution along phylogenetic trees. *Comp. Appl. Biosci.*, 13:235–238, 1997.
11. C. McGeoch. Analyzing algorithms by simulation: variance reduction techniques and simulation speedups. *ACM Comp. Surveys*, 24:195–212, 1992.
12. B. M. E. Moret and H. D. Shapiro. Algorithms and experiments: the new (and the old) methodology. *J. Univ. Comp. Sci.*, 7(5):434–446, 2001.
13. M. Kimura. A simple method for estimating evolutionary rates of base substitutions through comparative studies of nucleotide sequences. *J. Mol. Evol.*, 16:111–120, 1980.
14. D. Huson, K. A. Smith, and T. Warnow. Correcting large distances for phylogenetic reconstruction. In *Proceedings of the 3rd Workshop on Algorithms Engineering (WAE)*, 1999. London, England.
15. D. L. Swofford. PAUP*: Phylogenetic analysis using parsimony (and other methods), 1996. Sinauer Associates, Underland, Massachusetts, Version 4.0.

IDENTIFICATION OF REGULATORY BINDING SITES USING MINIMUM SPANNING TREES

VICTOR OLMAN, DONG XU, YING XU*

Protein Informatics Group, Life Sciences Division
Oak Ridge National Laboratory, Oak Ridge, TN 37831-6480, USA.
** To whom correspondence should be addressed. Email: xyn@ornl.gov.*

Recognition of protein-binding sites from the upstream regions of genes is a highly important and unsolved problem. In this paper, we present a new approach for studying this challenging issue. We formulate the binding-site recognition problem as a *cluster identification problem*, i.e., to identify clusters in a data set that exhibit significantly different features (e.g., density) than the overall background of the data set. We have developed a general framework for solving such a cluster identification problem. The foundation of the framework is a rigorous relationship between data clusters and subtrees of a minimum spanning tree (MST) representation of a data set. We have proposed a formal and general definition of clusters, and have demonstrated that a cluster is always represented as a connected component of the MST, and further it corresponds to a substring of a linear representation of the MST. Hence a cluster identification problem is reduced to a problem of finding substrings with certain features, for which algorithms have been developed. We have applied this MST-based cluster identification algorithm to a number of binding site identification problems. The results are highly encouraging.

1 Introduction

One of major challenging problems in systems biology is to understand the mechanisms governing the regulation of gene transcriptions. Microarray gene expression chips allow researchers to directly observe the dynamics of transcriptions of many genes simultaneously. A gene's transcriptional level is regulated by proteins (transcription factors), which bind to specific sites in the gene's promoter region, called *binding sites*[1]. Identification of genes that share common protein-binding sites can provide highly useful constraints in modeling of gene-transcriptional machineries[2].

Typically, a protein-binding site is a short (contiguous) fragment located in the upstream region of a gene. The binding sites by the same protein for different genes may not be exactly the same, rather they are similar on the sequence level. Computationally, the binding-site identification problem is often defined as to find short "conserved" fragments, from a set of genomic sequences, which cover many (or all) of the provided genomic sequences.

Because of the significance of this problem, many computer algorithms have been proposed to solve the problem[1,3]. Among the popular computer software for this problem are CONSENSUS[4] and MEME[5]. The basic idea among many of these algorithms/systems is to find a subset of short fragments from the provided genomic sequences, which show "high" information content [1] in their gapless multiple sequence alignments[a]. While many good algorithms have been proposed, this highly challenging problem remains not fully solved.

We have recently developed a new approach for the binding-site identification problem. Different than the previous methods, we have treated this problem as a clustering problem. Conceptually, we map all the fragments, collected from the provided genomic sequences, into a space so that similar fragments (on the sequence level) are mapped to nearby positions and dissimilar fragments to far away positions. Because of the relatively high frequency of the conserved binding sites appearing in the targeted genomic sequence regions, a group of such sites should form a "dense" cluster in a sparsely-distributed background. So the computational problem becomes to identify and extract such clusters from a "noisy" background. It is worth mentioning that this problem is different from classical clustering, which partitions all the elements of a data set into clusters.

Here we present a new framework for solving such a cluster identification problem. The foundation of this framework is a representation of a data set as a minimum spanning tree. We have demonstrated that no essential information is lost for the purpose of clustering with this representation. The simplicity of the minimum spanning tree structure allows us to deal with the clustering problem in a rigorous and efficient way. The main contribution of this work can be summarized as follows. We proposed a rigorous definition of a cluster, which we believe captures the essence of the concept of *clusters* that people frequently use but without a formal definition. This definition allows us to establish rigorous relationships between data clusters and subtrees of a minimum spanning tree, which further allow us to deal with the binding-site identification problem in a rigorous manner. Preliminary application results to three binding sites are highly encouraging. Though minimum spanning trees (MST) have long been used for data classification and clustering [6,7], we have not seen the kind of in-depth study of minimum spanning trees *versus* data clustering as what we present in this paper.

2 Minimum Spanning Trees *versus* Clusters

Let $D = \{d_i\}$ be a data set. We define a weighted (undirected) graph $G(D) = (V, E)$ as follows. The vertex set $V = \{d_i | d_i \in D\}$ and the edge set $E =$

---

[a]Very few closely related binding sites differ by an insertion/deletion due to the specific binding configuration of the transcrition factor on the DNA sequence.

$\{(d_i, d_j)|$ for $d_i, d_j \in D$ and $i \neq j\}$. Each edge $(u, v) \in E$ has a distance[b], $\rho(u, v)$, between u and v of V. Here *distance* is a binary relationship, which does not have to be a *metric*. A *spanning tree* T of a (connected) weighted graph $G(D)$ is a connected subgraph of $G(D)$ such that (i) T contains every vertex of $G(D)$, and (ii) T does not contain any cycle. A *minimum spanning tree* (MST) is a spanning tree with the minimum total distance. In this paper, any connected component of a MST is called a *subtree* of the MST.

Prim's algorithm represents one of the classical methods for solving the minimum spanning tree problem [8]. The basic idea of the algorithm can be outlined as follows: *the initial solution is a singleton set containing an arbitrary vertex; the current partial solution is repeatedly expanded by adding the vertex (not in the current solution) that has the shortest edge to a vertex in the current solution, along with the edge, until all vertices are in the current solution.* A simple implementation of Prim's algorithm runs in $O(\|E\| \log(\|V\|))$ time [9], where $\| \cdot \|$ represents the number of elements in a set.

Our first goal is to establish a rigorous relationship between a minimum spanning tree representation of a data set and clusters in the data set. To do this, we need to a formal definition of a cluster.

Definition 1: Let D be a data set and $\rho(u, v)$ denote the distance between any pair of data u, v in D. The **necessary condition** for any $C \subseteq D$ to be a cluster is that for any non-empty partition $C = C_1 \cup C_2$, the closest data point $d \in D - C_1$ to C_1 (measured by ρ) must be from C_2. $\square$

Note that the distance between a point to a data set means the distance between the point and its closest data point in the set. Here we provide only the necessary condition of a cluster since we believe that the sufficient condition of a cluster ought to be problem-dependent. We believe that this definition captures the essence of our intuition about a cluster: that is *distances among neighbors within a cluster should be smaller than any inter-cluster distances.*

Theorem 1: For any data set D and a distance measure ρ defined on D, let T be a MST representing D with its edge distances defined by ρ. If $C \subseteq D$ is a cluster by **Definition 1**, then C's data points form a subtree of T.

The proof of Theorem 1 can be found in our previous paper [10]. This theorem implies that a clustering problem can be rigorously reduced to a tree partitioning problem since each cluster is represented as a subtree of the MST representation of the data set. Note that a MST may not be unique for a given graph, and the non-uniqueness of MSTs does not affect the correctness of this theorem.

---

[b]For the simplicity of proofs and discussions, we assume that no two edges have the same distance. The proofs stay correct when edges are allowed to have the same distance.

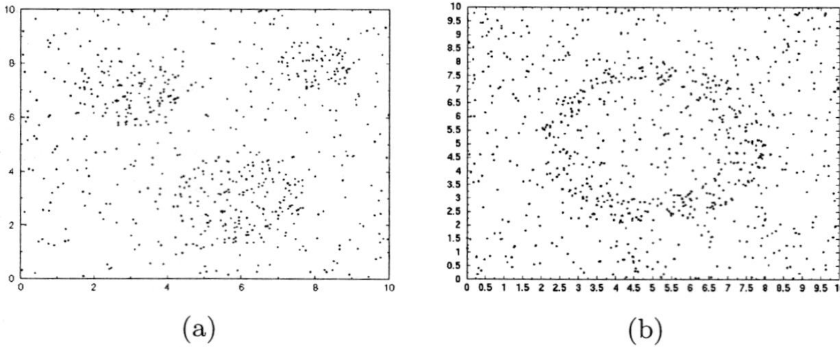

Figure 1: Examples of clusters in the an approximately uniformly-distributed background. (a) Three clusters with higher density than the background. (b) A ring-shaped cluster with higher density than the background.

3 Minimum Spanning Trees *versus* Cluster Identification

A *clustering problem* is typically defined as to partition a given data set into K clusters to optimize some objective function, that generally requires that the data points of the same cluster are "similar" and data points of different clusters are "dissimilar", for some $K > 1$. We have recently developed a set of MST-based clustering algorithms for solving this type of clustering problem [10,11]. A basic assumption for a clustering problem is that data points can be divided into clusters. However in real applications, it is often the case that data points of well-defined clusters may not appear in a vacuum but rather appear in a more general background. Figure 1 shows two such examples. Now we define a more general clustering problem. A *cluster identification problem* is defined as to partition a data set D as $D = B \cup D_1 \cup ... \cup D_p$ such that B is a data set that is approximately uniformly distributed, and each D_i forms a cluster in D, $i \in [1, p]$. A data set D is *uniformly distributed* in a bounded space G if for any region $A \subseteq G$, the number of data points of D in A is approximately proportional to the volume of A. We found that many biological data analysis problems can be modeled as cluster identification problems.

Before we present our algorithm for solving the cluster identification problem, we give an equivalent definition of a cluster. We continue to use D and ρ to denote a data set and the distance measure defined on D. For any $C \subseteq D$, we define an augmentation operation $\mathcal{A}(C) = C \cup \{c^*\}$, where $c^* \in D - C$ is the closest such element to C. We define $\mathcal{A}(D) = D$, and $\mathcal{A}^k = \mathcal{A}(\mathcal{A}^{k-1})$.

Definition 2: The **necessary condition** for a $C \subseteq D$ forms a cluster if for any $c \in C$, $\mathcal{A}^{\|C\|-1}(c) = C$. $\square$

This cluster definition is more closely related to the Prim's algorithm, which allows us to prove some of our results more easily.

Theorem 2: Definitions 1 and **2** are equivalent. That is, if $C \subseteq D$ is a cluster under **Definition 1**, then it is a cluster under **Definition 2**; and vise versa.

Proof: It is straightforward to show that **Definition 1** implies **Definition 2**. Hence we omit the proof. We now prove that **Definition 2** also implies **Definition 1**. It is not difficult to see that all we need to prove is that if C is a cluster under **Definition 2**, then for any $B \subseteq C$, $\mathcal{A}^{\|C\|-\|B\|}(B) = C$.

Let's assume that this is not true, and let $B^* \subset C$ be the subset that does not satisfy this equation. So there exists a $d_0 \in \mathcal{A}^{\|C\|-\|B^*\|}(B^*) \cap (D-C)$. We assume, without loss of generality, $d_0 \in \mathcal{A}(B^*)$ (if not, we can always expand B^*). Hence there exists a $b_0 \in B^*$ such that

$$\rho(b_0, d_0) < \min_{b \in B^*, c \in C - B^*} \rho(b, c). \tag{1}$$

Since C is a cluster under **Definition 2** and $b_0 \in C$, we know $\mathcal{A}^{\|C\|-1}(b_0) = C$. However this contradicts the inequality (1) since inequality (1) implies that d_0 will be added to $\mathcal{A}^{\|C\|-1}(b_0)$ before any other elements of $C - B^*$ can be added. However we know $d_0 \notin \mathcal{A}^{\|C\|-1}(b_0)$. This contradiction implies the correctness of the theorem. $\square$.

Clearly, $\mathcal{A}$ represents the basic selection operation in the Prim's algorithm. Let $(d_1, ..., d_{\|D\|})$ be the list of elements selected (in this order) by the Prim's algorithm when constructing a MST of the data set D and starting from element $d_1 \in D$. This list maps the data set D (of any dimensions) to a string $d_1....d_{\|D\|}$, which we call a *linear representation* of D.

Theorem 3: For any linear representation $L(D)$ of a data set D, any cluster C of D is represented as one substring of $L(D)$.

Proof: We need to show that after the Prim's algorithm selects the first element c_0 of C, it will not select any element outside of C until it has selected every element of C. We assume that this is not the case. Let d_0 be the first element of $D - C$ that gets selected after c_0's selection. We use C' to denote the subset of C that have been selected when d_0 is selected. By our assumption, $C' \subset C$. Let (c_0, d) be the MST edge that gets c_0 selected; similarly (d_0, d') be the corresponding edge of d_0 (note that $d' \in D - C$ by **Definition 2**; otherwise it would contradict the definition). Also let (c_1, c_2) be the first edge selected after d_0's selection, with $c_1 \in C'$ and $c_2 \in C - C'$ (note that such (c_1, c_2) must exist since elements of C form a subtree of the MST). By the order of their selections, we have

$$\rho(c_1, c_2) > \rho(d_0, d') > \rho(c_0, d).$$

332

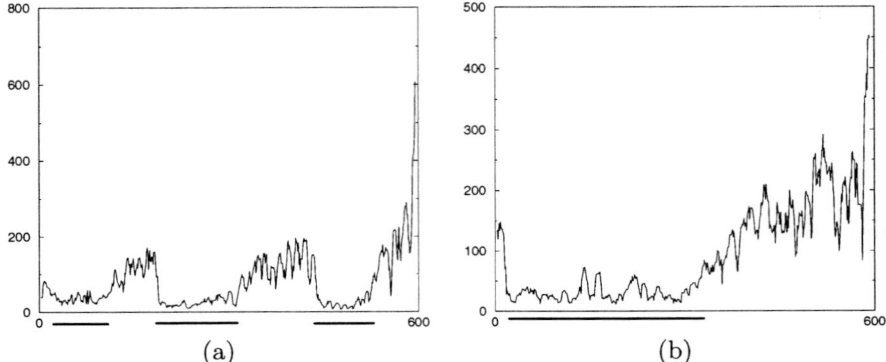

Figure 2: Plots of edge-distances in the order of their selection by the Prim's algorithm. Each valley is underlined. (a) Edge-distance plot of Figure 1 (a). (b) Edge-distance plot of Figure 1 (b).

This apparently contradicts the fact that $\rho(c_1, c_2) < \rho(c_0, d)$, implied by the proof of Theorem 2. So we have proved the theorem. $\square$

A direct application of Theorem 3 is that we can plot the edge distances in the selection order by the Prim's algorithm. In this plot, the x-axis is a linear representation $L(D)$ of data set D, and the y-axis represents the distance of the corresponding MST edge. By Theorem 3 and the definition of a cluster, each cluster should form a "valley" in this plot. Figure 2 shows two examples of such plot. In Figure 2(a), the three (visibly apparent) clusters of Figure 1 (a) are nicely represented by three valleys. Similarly, the one ring-shaped cluster of Figure 1 (b) is well represented as a valley in Figure 2(b). This suggests that we can find clusters from a noisy background through searching a linear string and finding the substrings with certain properties. The following result forms the foundation of our search algorithm. For a substring S of a linear representation $L(D)$ of a data set D, the *left-boundary edge* of S is defined to be the corresponding MST edge when the leftmost element of S was added into the MST. The *right-boundary edge* of S is defined to be the first MST edge, linked with the rightmost element of S, that gets selected after the rightmost element is selected. If the leftmost element of S is a first element of $L(D)$, the distance of the left boundary edge of S is infinity; similarly if the rightmost element of S is a last element of $L(D)$, the distance of the right boundary edge of S is infinity.

Theorem 4: A subtring S of $L(D)$ represents a cluster if and only if (a) S's elements form a subtree, T_S, of D's MST, and (b) S's both boundary edges have larger distances than any edge-distance of T_S.

Proof: Apparently, Theorem 3 implies the *only-if* condition. The *if* condition can be proved easily as follows. Since the right-boundary edge of S is larger than all tree edges of T_S, we know all other edges connecting S with $D - S$ are larger than the tree edges of T_S, possibly except for the left-boundary edge (because the right-boundary is the first selected such edge after the selection of the left-boundary edge). Since we also know that the left-boundary edge is also larger than T_S's edges, we conclude that S forms a cluster by the proof of Theorem 2. $\square$

We now present an algorithm for finding clusters in a noisy data set D. The algorithm goes through all substrings of $L(D)$ and checks if the (a) and (b) conditions of Theorem 4 are satisfied. We use K to represent the smallest cluster size we care to identify.

Procedure ClusterIdentification(D, K)

Construct a MST, T_D, of data set D, using Prim's algorithm;
Generate a linear representation $L(D)$ of D;

FOR $i = 1$; **UNTIL** $i = \|D\| - K$ **DO**
    FOR $j = i + K$; **UNTIL** $j = \|D\|$ **DO**
        IF (elements of $L(D)[i, j]$ forms a subtree of T_D **AND**
           left- and right-boudanry edges of $L(D)[i, j]$ are larger than edges
of $T_{L(D)[i,j]}$)
           THEN output $L(D)[i, j]$ as a cluster.

Checking if a subset of vertices of a tree forms a subtree of the tree can be done in linear time of the number of vertices [12]. So this algorithm takes $O(\|D\|^3)$ time to locate all clusters. Improved computing time may be possible using advanced data structures, but the current running time is adequate for our applications. **Theorem 4 implies that the ClusterIdentification procedure finds all the clusters (defined by our definition) in the data set and finds clusters only.**

To make the discussion on clusters complete, we have the following result regarding the structure among all clusters in a data set. The proof is straightforward.

Corollary: For any data set D and its two clusters A and B, if $A \cap B \neq \emptyset$, then $A \subseteq B$ or $B \subseteq A$. $\square$

4   Applications

For a given set of upstream regions of genes, possibly collected through finding genes having correlated expression profiles, our procedure finds the conserved short fragments, say with k bases, as follows. First, it breaks the genomic sequences into k-mers (if k is not known we will go through all k's within some range provided by the user), denoted as a set S. For two k-mers $A = a_1...a_k, B = b_1...b_k \in S$, we define their distance $\rho(A, B) = \sum_1^k \sigma_i M(a_i, b_i)$, where $M(x, y) = 0$ if $x = y$ otherwise 1. Initially, all σ_i is set to $1/K$, where K is the number of sequences containing at least one of the k-mers A or B. Then we apply the **ClusterIdentification** Procedure to identify all clusters, using ρ as the distance measure. As we discussed earlier, this procedure identifies subsets of S that satisfy the necessary condition of a cluster (for a particular distance measure) while the sufficient condition is problem specific. For the binding site identification problem, the problem-specific conditions should include the following: (1) the position-specific information content [13] of the gapless multiple-sequence alignment, among all the sequence fragments represented by a cluster, should be relatively high; (2) elements of an identified cluster should not be among long, simple repeats; and (3) the data density within a cluster should be relatively higher than the one of the overall background.

To incorporate the information-content constraint into our binding-site identification procedure, we will do the following. After a cluster is identified using the procedure of **ClusterIdentification**, we will measure the position-specific information content. If the overall information content is lower than some threshold, we will discard this cluster for further consideration. Otherwise, we will modify our original distance measure $\rho()$ by the calculated information content as follows. For each position i, we use its information content as σ_i in the next iteration of applying **ClusterIdentification**; and set $M(a_i, b_i) = 2 - (p_i(a_i) + p_i(b_i)) + |p_i(a_i) - p_i(b_i)|$, where $p_i(x)$ represents the frequency of letter x among all letters in position i. We will iterate this procedure for a fixed number of iterations. Our observation has been that for a real binding-site cluster, the procedure converges quickly. Otherwise it diverges.

To "guarantee" some level of uniformity within an identified cluster, we examine the edge-distance distribution of the subtree (of the MST) representing the cluster. If data points are approximately "uniformly" distributed, the edge-distance distribution should be generally unimodular; a multi-modular distribution typically indicates that there are regions within the cluster, which have different density levels. Our procedure discards clusters with non-unimodular

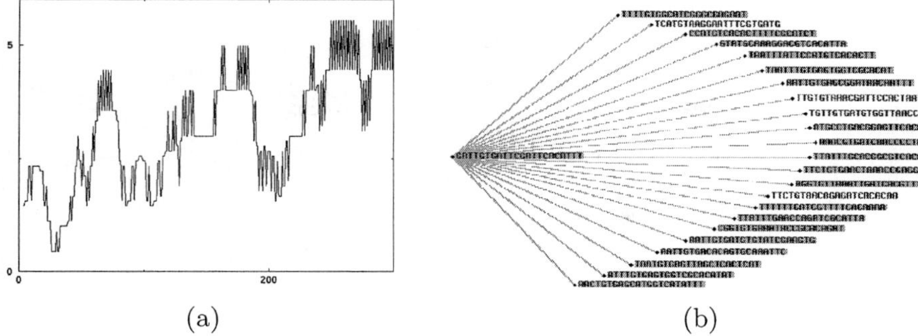

Figure 3: Identification of CRP binding sites. (a) Edge-distance plot. The deepest valley in the plot corresponds precisely to the subtree of (b). (b) The (sub)tree structure of identified CRP binding sites. The shaded fragments represent known CRP sites. The edge-distance is color-coded, ranging from green to yellow, with green being the shortest.

distributions from further consideration.

Our implemented program has a number of parameters, e.g., the minimum cluster size. The values of these parameters are selected, using a simple search program to find an "optimal" performance on a set of genomic sequences with known regulatory binding sites. We have applied this program to a number of binding-site identification problems. We present three case studies here. The computing time for each case took a few minutes on a Unix workstation.

CRP binding sites: This is a widely used testing set from *E. Coli* for validating a method of binding-site identification [1]. The test set consists of 18 sequences (each of length 105 bps) with 23 experimentally verified CRP binding sites (22-mers). Our iterative procedure stopped after two iterations. The only cluster identified is the one shown in Figure 3. The cluster consists of 24 fragments, of which 20 are known CRP sites (out of 23) and the remaining four may or may not be true CRP sites. This result is at least as good as results reported previously by other approaches [1,14].

Yeast binding site: The second application example is for the nucleosome complex proteins promoter binding sites in yeast [15]. There are 8 regulatory sequences, each containing 1000 bp. By using 9-mers, our method identified several clusters (see Fig. 4 (a)). The most populated cluster is TTAC-CACCG, which also connects several other clusters with similar motifs on the MST (see Fig. 4 (b)). The cluster TTACCACCG and its connected clusters have high information content and they appear in all 8 regulatory sequences. It turned out that TTACCACCG is an experimentally verified motif [15].

Human binding sites: We have applied our method to a set of human

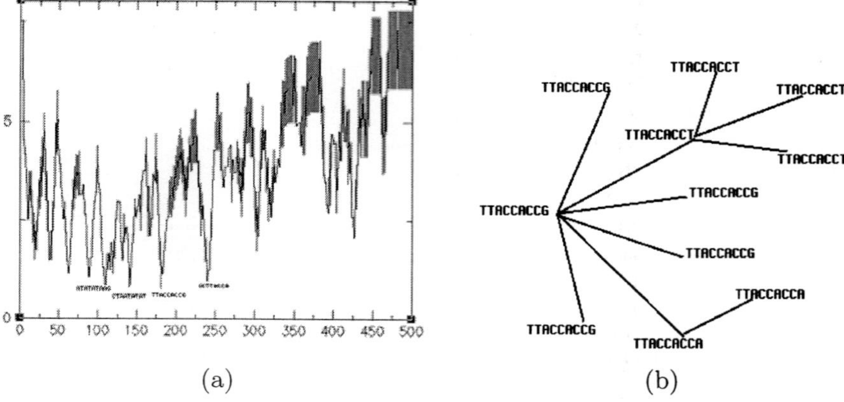

Figure 4: Binding site identification for the yeast data set. (a) Edge-distance plot. (b) The subtree structure of an identified binding site TTACCACCG.

genomic sequences containing binding sites, collected in [16]. The data set has 113 regulatory sequences containing regulatory regions. Each sequence is 300 bp long, with 250 bp upstream and 50 bp downstream of the transcriptional start site. It is known that the TATA box is one of the binding sites. It is also believed that other protein-binding sites may exist in the data set, although no experimental data is available to support this yet. One of the best identification methods for binding sites [17] identified the 6-mer TATAAA as the only identifiable binding motif consensus. Our method has easily identified this motif (see Fig. 5 (a)). In addition, it has identified a number of other possible binding sites, including GCAGCC as shown in Fig. 5 (b). The GCAGCC motif with at most one mis-match appears in 96 regulatory sequences, even more frequently than the TATAAA motif, where appears in 66 regulatory sequences with at most one mis-match. The information content of the GCAGCC motif is also higher:

```
TATAAA: 0.819569 0.793079 0.861977 0.869478 0.802716 0.821651

GCAGCC: 0.755810 0.798584 0.692777 0.906340 0.823345 0.849254
```

where the information content at each position is normalized to the range of 0 to 1. Based on this, we believe that our method is more sensitive in detecting conserved sites.

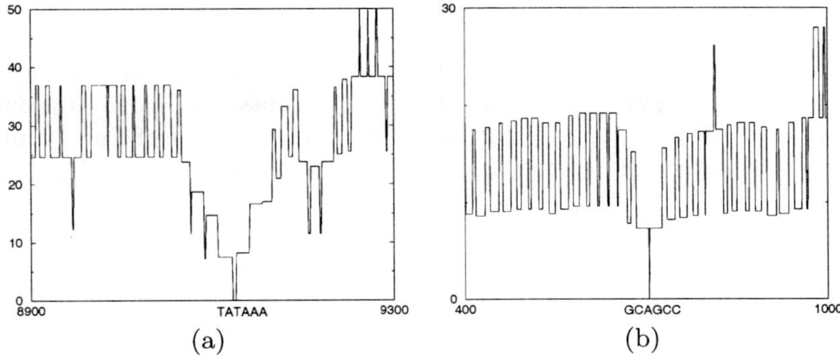

Figure 5: Edge-distance plot for binding site identification for the human data set. (a) Around the cluster containing the TATAAA binding site. (b) Around the cluster containing the GCAGCC binding site.

5 Discussion

Based on a formal definition of *clusters*, we have developed a rigorous framework for identifying and extracting clusters from a noisy background. As we believe that many data analysis problems can be formulated as a cluster identification problem or its variant, we expect that this framework will find many interesting applications. The linear representation of a data set allows us to "directly" visualize the cluster structures even for high dimensional data sets. Having such a visualization capability should clearly improve our confidence in our clustering results, since we can "see" the clusters.

Based on the examples we have tested so far, our method clearly shows some advantages over the existing methods. The first advantage is that our method can take advantage of the fact that a regulatory sequence contains the same motif at the different locations, and it will increase the population of the cluster containing the binding site. Another advantage of our method is its sensitivity. Our method is based on a combinatorial approach, which can identify all clusters of possible binding sites. Existing methods generally use sampling techniques, which are likely to miss some binding sites that do not have very strong patterns, as shown in the example of human binding site.

A method is currently being developed for assessing the statistical significance of each identified cluster based on the "depth" and the "width" of the valley representing the cluster in the edge-distance plot (unpublished results).

Acknowledgements

The authors thank Dr. Gary Stormo for providing us the CRP binding site data. The work was supported by ORNL LDRD funding and by the Office of Biological and Environmental Research, U.S. Department of Energy, under Contract DE-AC05-00OR22725, managed by UT-Battelle, LLC.

1. G. D. Stormo and G. W. Hartzell 3rd. *Proc. Natl. Acad. Sci. USA*, 86:1183–1187, 1989.
2. Y. V. Kondrakhin, et al. *Comput. Appl. Biosci.*, 11:477–488, 1995.
3. R. Staden. *Comput. Appl. Biosci.*, 89:293–298, 1989.
4. G. Z. Hertz and G. D. Stormo. *Bioinformatics*, 15:563–577, 1999.
5. T. L. Bailey and M. Gribskov. *J. Comput. Biol.*, 5:211–221, 1998.
6. J. C. Gower and G. J. S. Ross. *Applied Statistics*, 18:54–64, 1969.
7. R. O. Duda and P. E. Hart. *Pattern Classification and Scene Analysis*. Wiley-Interscience, New York, 1973.
8. R. C. Prim. *Bell System Technical Journal*, 36:1389–1401, 1957.
9. T. H. Cormen, C. E. Leiserson, and R. L. Rivet. *Introduction to Algorithms*. MIT Press, Cambridge, 1989.
10. Y. Xu, V. Olman, and D. Xu. *Bioinformatics*, 18:536-545, 2002.
11. Y. Xu, V. Olman, and D. Xu. *Proceedings of the 12th GIW*, pages 24–33. Universal Academy Press, Tokyo, 2001.
12. A. V. Aho, J. E. Hopcroft, and J. D. Ullman. *The Design and Analysis of Computer Algorithms*. Addison-Wesley, Reading, MA, 1974.
13. T. D. Schneider, G. D. Stormo, L. Gold, and A. Ehrenfeucht. *J. Mol. Biol.*, 188:415–431, 1986.
14. P. A. Pezner and S. Sze. *ISMB*, 8:269–278, 2000.
15. P. Pavlidis, et al. *Pac. Symp. on Biocomp.*, 2001:151–63, 2001.
16. J. W. Fickett and A. G. Hatzigeorgiou. *Genome Res.*, 7:861–878, 1997.
17. M. Q. Zhang. *Genome Res.*, 8:319–326, 1998.

INTRASPLICING - ANALYSIS OF LONG INTRON SEQUENCES

S. OTT*[†] Y. TAMADA*[‡] H. BANNAI[†] K. NAKAI[†] S. MIYANO[†]

[†]*Human Genome Center, Institute of Medical Science, The University of Tokyo*
4-6-1 Shirokanedai, Minato-ku, Tokyo 108-8639, Japan
{ott, bannai, knakai, miyano}@ims.u-tokyo.ac.jp

[‡]*Department of Mathematical Sciences, Tokai University*
1117 Kitakaname, Hiratsuka-shi, Kanagawa, 259-1292, Japan
tamada@ims.u-tokyo.ac.jp

We propose a new model for the splicing of long introns, which we call *intrasplicing*. The basic idea of this model is that the splicing of long introns may be facilitated by the splicing of inner parts of the intron prior to the splicing of the long intron itself. Since long introns have up to about 100,000 bases, this model seems to be a likely explanation of their splicing. To investigate the possibility of this model, we develop a new computational method for the analysis of DNA sequences with respect to splicing. We analyze the genomic sequence of four species with our method and derive several results indicating that intrasplicing may be an appropriate model for the splicing of at least part of the long intron sequences.

1 Introduction

Nuclear splicing is known to play an essential role in the expression of genetic information of eukaryotes. The molecular components responsible for carrying out the splicing process, which removes the introns from the pre-mRNA sequence, are becoming more and more elucidated and a large amount of research is spent on alternative splicing, aberrant splicing and on splicing inhibitors as well as splicing enhancers[2,3,4]. However, while the splicing reaction can be easily imagined to happen on short introns of a few hundred bases, it is poorly understood how long introns can be correctly recognized by the splicing machinery. Such introns can contain up to about 100,000 bases (and even more for some species[10]) while the known splice signals are limited to an area of about 50 bases around the beginning and the end of the intron. Therefore, it seems rather unlikely that such introns are removed from the pre-mRNA in a single reaction, because this would require the beginning and the end of such introns to get spatially very close.

Recursive splicing has been proposed as one explanation for the splicing

---

*These authors contributed equally to this work. Please address correspondence to these authors.

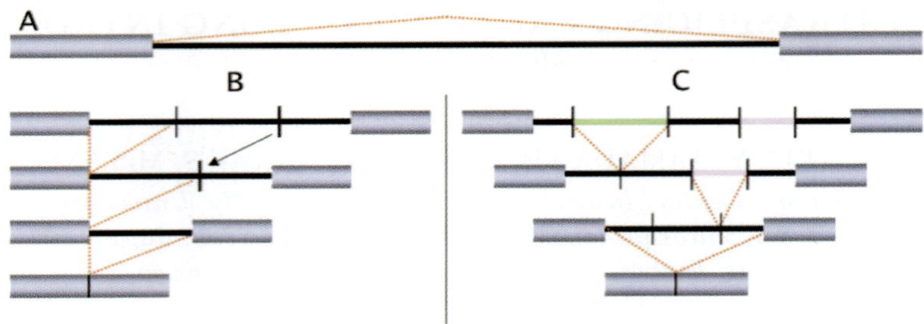

Figure 1. Three models of long intron splicing: A) splicing in one step, B) recursive splicing, C) intrasplicing.

of long introns[6]. In this model, splice sites of consecutive splicing reactions are generated by preceding splicing reactions. Recursive splicing has been shown to occur in one of the long introns of *Drosophila*[6], but it might be only one of several ways of splicing a long intron. Therefore, we propose a more general model for the splicing of long introns, which we call *intrasplicing*. In this model the long intron is shortened by several splicing reactions occurring within the long intron without the use of the long intron splice sites itself. As soon as the remainder of the long intron becomes sufficiently short, it is cut out of the pre-mRNA by a single splicing reaction. This model is less restrictive, because splicing reactions do not have to yield new splice sites and the process as a whole can take place in many different ways. We use the term *intraintron* to denote (putative) introns within introns.

The aim of this work is to investigate the possibility that intrasplicing is an accurate model for the splicing of long introns in a computational way (see Methods). We make use of an intron prediction program as well as the genomic sequence of human (*H. sapiens*), mouse (*M. musculus*), fly (*D. melanogaster*), and zebrafish (*D. rerio*).

We derive several results indicating that intrasplicing may be an explanation or a partial explanation for the splicing of long introns.

2 Methods

2.1 Outline of our Approach

In order to evaluate the possibility of intrasplicing, we have developed a computer program, which can be outlined as follows. First, our program learns an intron predictor for short introns (at most 400 bases) using two thirds of the currently available mRNA data of the particular species. Then it uses the remaining third of the data to evaluate the probability that a prediction is true with respect to its score. Using the intron predictor for short introns and the probability function, our program recursively searches the highest scoring intraintron candidate within a long intron and cuts it out, until the remainder of the long intron becomes sufficiently short.

Step 1:   Learn a short intron predictor from mRNA data.
Step 2:   Evaluate the intron predictor.
Step 3:   Use the intron predictor to compute a likely way for intrasplicing for all long introns.

By doing so for all long introns longer than 10,000 bases, the program computes several statistics concerning the intraintron candidates used. This cutting procedure is then conducted for several different sequences such as sequences derived from Markov Chains, and statistics are computed for these sequences in the same way. If intrasplicing is an appropriate model for long intron splicing, we expect to find many high scoring intraintron candidates in long intron sequences. It is also expected that intrasplicing statistics between intronic sequences and sequences which are close to exons, will show significant differences. We explain the three main steps in more detail.

2.2   Intron Predictor

We have implemented an intron prediction program, which is similar to the one used in Lim and Burge[8]. This predictor includes inhomogeneous first-order Markov Models for the splice sites, a weight matrix model for the branch point, a length score, and an intron composition score. All introns longer than a certain threshold are removed from the data during training of this predictor. We only describe differences to Lim and Burge's method here.

The intron predictor of Lim and Burge is designed for predicting very short introns of length at most 134. For our purpose, we also need to predict introns longer than this, because introns of length 200 or 300 can easily be thought to be spliced in a single reaction. In order to get higher prediction accuracies for introns of length up to 400, we changed the

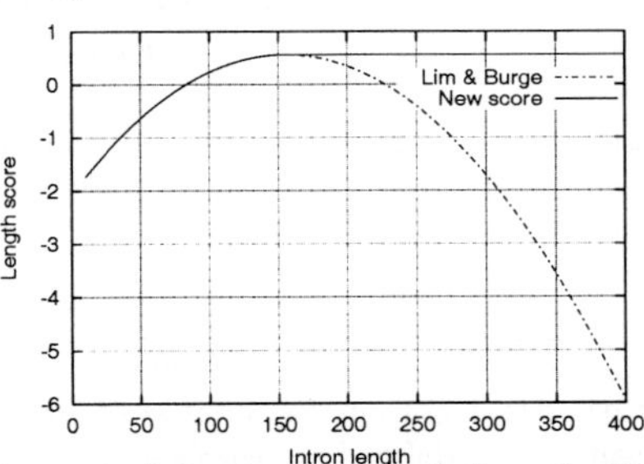

Figure 2. Length score graph [length=300].

length score as shown in Figure 2.

The intron composition score used in Lim and Burge[8] is defined as the sum of the pentamer scores contained in the intron candidate. Scoring in this way yields a length dependent score, since longer candidate regions get a higher score if they are composed of high-scoring pentamers. To remove this length-dependency, we used the average of the pentamer scores instead and evaluated the optimal weight for this score.

Furthermore, in order to improve the prediction accuracy, we conduct a clustering of splice sites in the training procedure of our intron predictor and learn one inhomogeneous first-order Markov Model for each cluster. We use this set of Markov Models by applying all models and choosing the highest score. We use five clusters for the 5' site and two clusters for the 3' site, because this configuration performed best in preliminary computations. The branch point score was not calculated for species other than human, because it has been reported to contribute very weakly to intron prediction accuracy[8].

Figure 3 shows short intron prediction accuracies with respect to the maximal length of predicted introns[a].

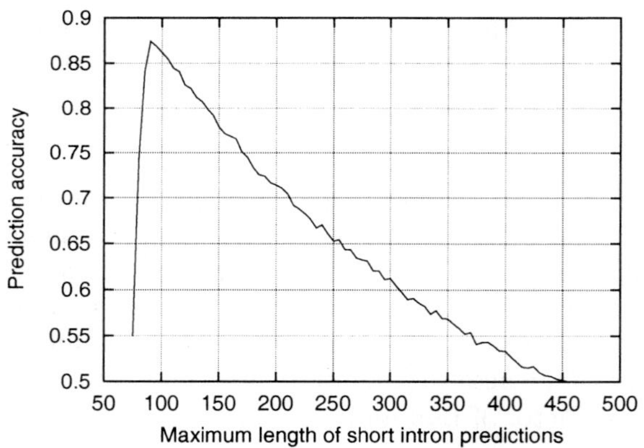

Figure 3. Short intron prediction accuracy. The accuracy is measured as sensitivity plus specificity divided by two. Therefore, 0.5 can be achieved in a trivial way and corresponds to a powerless predictor.[b]

---

[a]We used human mRNA sequences listed in RefSeq[9] for this computation.

[b]With TP denoting the number of true positives and FP, TN, FN correspondingly, the sensitivity is defined as $\frac{TP}{TP+FN}$, and the specificity is defined as $\frac{TP}{TP+FP}$. Therefore, a predictor selecting all candidates as positives has score 0.5.

2.3 Evaluating the Intron Predictor

After training the intron predictor on two thirds of the mRNA data set, we use the remainder of the data set to evaluate the predictor. We compute the specificity of predictions with respect to the score as shown in Figure 4. That means for each score s, we compute the ratio of the number of real introns scoring above s and the total number of predictions scoring above s. This ratio can be viewed as the probability that a predicted intron candidate of score s is in fact an intron. For very high scores, the function shows a minor instability, because such scores are very rare.

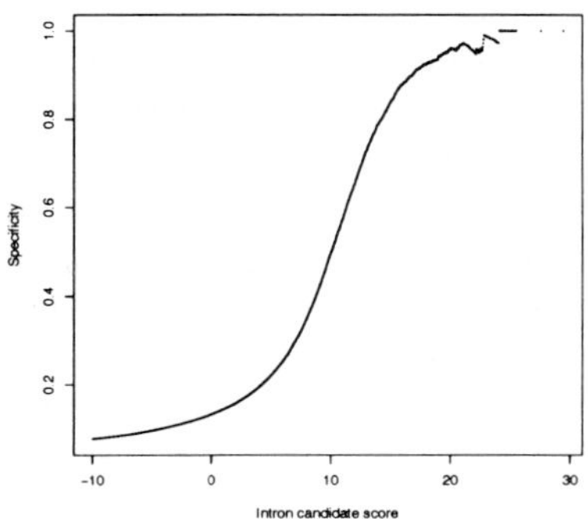

Figure 4. The probability function for short intron predictions.

2.4 Simulated Intrasplicing

The intrasplicing procedure recursively searches for the highest scoring intraintron candidate within the sequence of a long intron and cuts out the region of this candidate until the remainder of the long intron becomes short enough. The reason for selecting only the highest scoring candidate at a time is that non-overlapping candidates are not affected by the cutting and can still be selected in a later step. Since overlapping candidates cannot be selected later, it would be an interesting approach to try different possibilities, but a computation of all possibilities is not feasible. Therefore, choosing the candidate with the highest likelihood is a straightforward way around this problem.

The main parameters of this procedure are the degree to which the long intron is shortened, and the maximum length of short intron predictions. While predicting longer introns promises an easier intrasplicing, increasing the maximum length would limit the quality of intrasplicing predictions, since the accuracy of short intron predictions would decrease, as seen in Figure 3.

We also generated several different sequences to do a comparison of the results. For each long intron, we generated sequences of the same length to exclude the influence of sequence lengths. The sequences were generated from Markov Chain models of exon sequences, intron sequences, gene sequences, and intergenic sequences. Markov Chains are widely used to model DNA sequences[5]. In most computations, we used fifth-order Markov Chains, but we also evaluated the influence of the order of the Markov Chains (see Results).

Furthermore, we concatenated the sequences of all exons of a species and chose random substrings of this sequence of length equal to the particular long intron. Finally, we made use of a 0th-order Markov model of the whole available genomic data, i.e. using the base content of the genome as the only information for sequence generation.

In order to analyze the interior sequence of long introns only, influence of the 5' and 3' splice sites of the long intron had to be excluded. Therefore, the first 8 characters and the last 20 characters of each sequence were not used by our intrasplicing procedure. These values correspond to the length of the intron part of the splice sites as used in Lim and Burge[8].

We used data of the Ensembl database[7] for all of the computations[b]. The Markov Chain models of intergenic DNA were calculated only for human using GenBank[1] data (December 2001). Most of the computations were also performed using mRNA sequences from GenBank, which are listed in the RefSeq[9] database, but the results did not show significant differences to the results derived from Ensembl data. Only the Ensembl results are presented here.

The minimum length for long introns was set to 10,000 bases for all computations concerning a single intron. We excluded long introns with more than 0.5% of non-**acgt**-characters or more than 20 non-**acgt**-characters within a part of 100 characters from the analysis. Furthermore, to avoid the influence of characters other than **a,c,g,t** in the sequence data, we replaced each such character in a random way by one of the nucleotide characters it represents.

The number of long introns meeting the above criteria was 6468 for human, 330 for mouse, 71 for fly and 31 for zebrafish. For fly and zebrafish, no long intron was excluded by our criteria, but for mouse the number of ruled out long introns was very high. Therefore, 3% of non-**acgt**-characters were allowed and the 100-bases-criteria was skipped for mouse in some computations (see Results) yielding a set of 1052 long introns.

The computation was conducted using a PC cluster with 64 Pentium4 CPUs with 2 GHz for about two weeks.

3 Results

3.1 Analysis of Human Sequences

We applied the intrasplicing procedure to all known human long introns meeting the criteria described in Methods. Figure 5 shows the distribution of prob-

---

[b]*H. sapiens* release 4.28, *M. musculus* release 4.1, *D. melanogaster* release 4.3, *D. rerio* release 4.06.

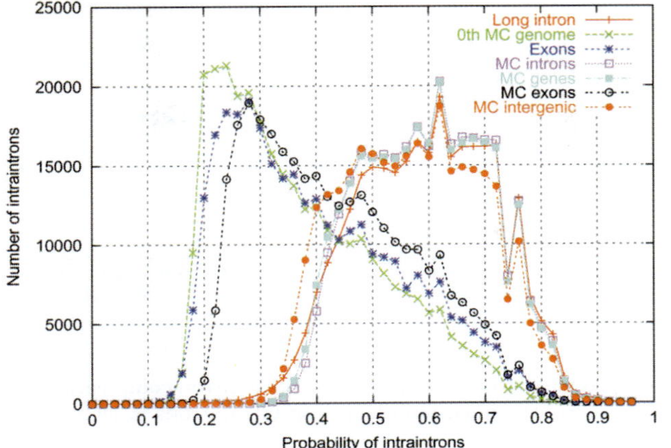

Figure 5. Comparison of probability distributions of various sequences, at least 60% of long introns are removed by simulated intrasplicing.

abilities of putative intraintrons for human. Each line represents one type of sequence. "MC" abbreviates Markov Chain. The sequences clearly fall in two distinct groups. One group consists of long intron sequences, the Markov Chain models for introns, genes, and intergenic sequences. The other group contains sequences from both real and Markov modeled exons, and the 0th-order Markov Chain of the genome. Therefore, long introns as well as the Markov Models for introns and genes show a clearly different behavior than the sequences in the second group. Since the major part of genes is constituted by introns, it is not surprising, that the Markov Models for introns and genes fall in the same group. Interestingly, the Markov Model for intergenic sequences also falls in the same group, though it shows differences to the other three members of the first group (orange line). The reason may be, that the splicing signal is weak enough to allow many sufficiently strong splice sites to occur in non-coding regions. This would imply that the splicing process can be stable against the insertion of intergenic sequences into introns.

Since the 0th-order Markov Model contains the nucleotide frequencies as the only information, it can be considered as a baseline for comparison with the other sequences. Surprisingly, exons as well as their Markov Model achieve only a very slight distinction from this baseline.

Figure 6 shows the average probability of intraintrons as a function of the parameter for stopping the intrasplicing procedure. This parameter specifies the percentage of the length of the original intron which is to be left. Therefore 0.7 indicates, that 30% of the intron have to be removed, while in the strictest

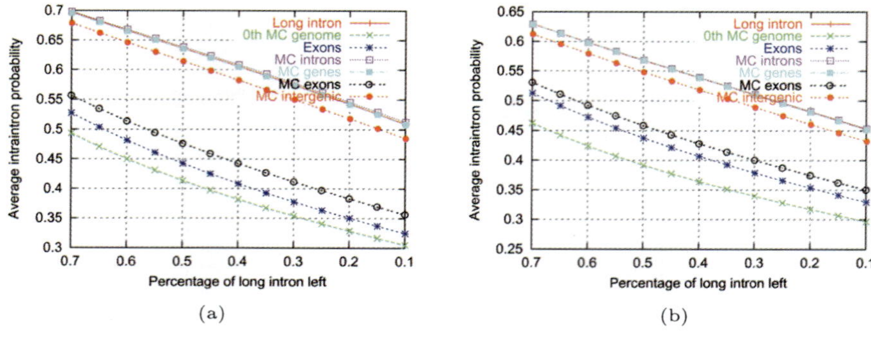

Figure 6. Average probability of intraintrons, (a) with intron composition score, (b) without intron composition score, the maximum length of intraintrons is set to 250.

case 90% are to be removed. The left graph shows the same clustering as we observed in Figure 5. Since the intron prediction program we used includes an intron composition score, we asked whether this might be the explanation for our observation. But as shown in the right graph, the general trends do not change, even if the intron composition score is not used.

Interestingly, while the Markov Model for exons achieves significantly higher intraintron probabilities than the sequences consisting of real exons, the Markov Model for introns does not separate from the intron sequences. This indicates that a fifth-order Markov Chain is a very good model for introns.

We also observe that the Markov Model for intergenic DNA shows a small difference to the intron sequences, if compared to the gap between the two groups. This could be explained, if introns have evolved from intergenic DNA and have further developed since then.

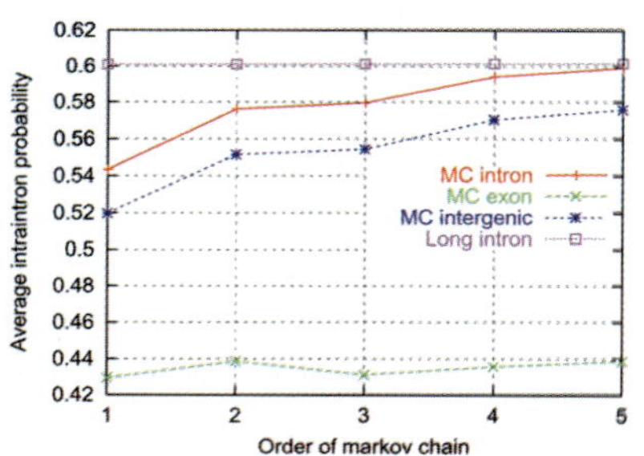

Figure 7. Influence of the order of the Markov Chain.

The influence of the order of the Markov Chain is shown in Figure 7, the value for long introns is shown for comparison. The maximum length of intraintrons is set to 250, and the long introns are removed to at least 60%. The 0th-order models all have low intraintron probabilities, but the 0th-order model for exonic sequences separates from intergenic and intronic sequences. The values for higher order Markov Chains of introns and intergenic DNA are converging towards the value for long introns, but the values for Markov Chains of exonic sequences stay on the same level.

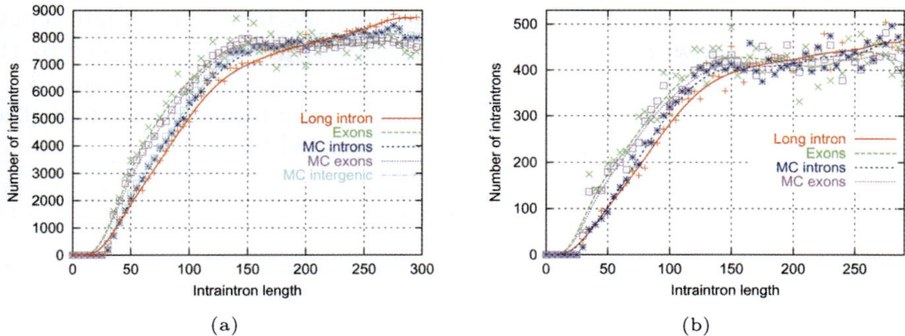

(a)   (b)

Figure 8. Length distributions of intraintrons, data points and smooth functions, (a) human, (b) mouse. At least 75% of long introns are removed by simulated intrasplicing.

In Figure 8, we show the distribution of lengths of intraintrons. We observe that except for long introns all sequences show a length distribution of the same shape as the length score we used for intraintron predictions (see Figure 2). But the long intron sequences show a different trend favoring longer intraintrons. That means that the long introns show a distinction to all other sequences used. This is an indication of a structure of intraintrons in long introns. It would be very interesting to see whether this trend continues for a longer range of the x-axis, but in lack of a powerful predictor of introns longer than 300 or 400 bases, our sight is limited to this range. Therefore, we suggest that this statistic shows a meaningful trend, which could be seen more clearly if such a predictor would be at hand.

For mouse, this observation cannot be made as clearly, but the amount of complete long intron sequences for mouse is still limited: using the criteria described in Methods, only 330 long introns were selected.

3.2 Comparison between Species

We also applied the intrasplicing procedure to the long intron data sets of mouse, fly and zebrafish described in Methods. Figure 9 shows the average intraintron probability of four species as a function of the upper bound for intraintron prediction. The curve for zebrafish shows a different degree of variation, probably due to the low amount of long intron

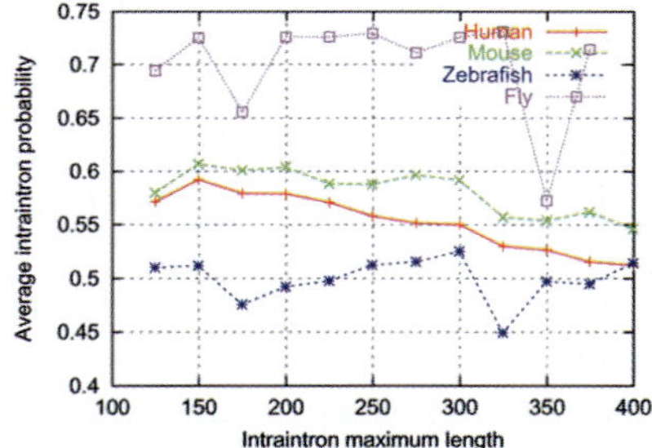

Figure 9. Species comparison of average intraintron probabilities, at least 75% of long introns are removed by simulated intrasplicing.

sequence data for zebrafish.[c] The overall tendency observed for human and mouse is an increase in intraintron probability, when the upper bound is increased from 125 to 150 bases, but further increase of the maximum length does not yield higher intraintron probabilities, which can be explained by the rapid drop of prediction accuracy for longer introns as shown in Figure 3.

It is interesting that fly shows higher intraintron probabilities than other species, because recursive splicing has been reported for fly[6].

3.3 Intrasplicing Simulation on Human Genes

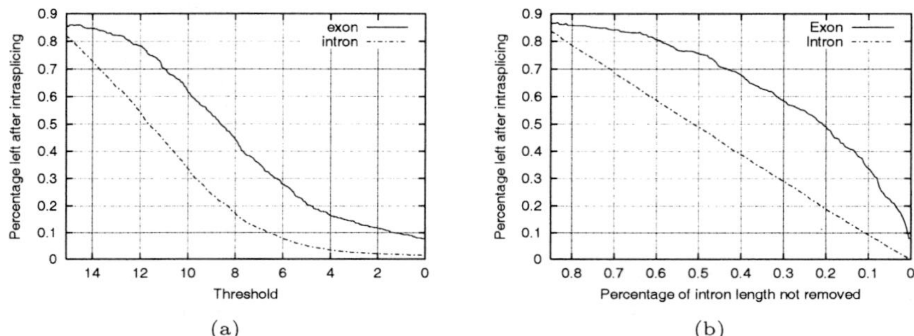

Figure 10. Comparison of percentage left after intrasplicing simulation.

To evaluate the plausibility of the intrasplicing model as a model for splicing in general, we applied it to whole genes, i.e., instead of trying to remove large parts of long introns, we used complete genes as an input to our intrasplicing procedure and examined to which degree exons and introns are removed for different threshold values. Figure 10 (a) shows the results when applying our intrasplicing procedure to each gene sequence until all putative intraintrons (maximum length 175) scoring above a threshold are removed from the sequence. Figure 10 (b) shows the percentage of intron/exon sequences remaining, as a function of the percentage of the remaining intron sequence.

Our results show that exons are highly stable under intrasplicing. Thus, our model may suggest that simply everything (including introns and intergenic regions) except exons can be removed through intrasplicing.

---

[c]For this computation the dataset of 1052 mouse long introns was used. A computation on the set of 330 long introns with less non-**acgt**-characters did not yield a significantly different result.

4 Discussion

We have introduced a new model for the splicing of long introns and developed a new computational method in order to analyze whether this model might be appropriate or not. The results show a large difference in the average probability of putative intraintrons between two groups of sequences. Also the shape of the probability distributions is distinctive between both groups. This raises the possibility that splicing of intraintrons precedes and facilitates the splicing of the long intron itself, and that long introns contain a structure of (possibly nested) intraintrons.

Though intergenic DNA and long introns fall into the same group, their average intraintron probabilities and the length distribution of intraintrons show differences. This could be explained, if long introns have evolved from intergenic sequences. The observed intraintron length distribution for intraintrons shows a preference for longer intraintrons. A possible explanation is that longer intraintrons may facilitate a faster intrasplicing process.

The strength of our analysis method is limited by some different factors. First, the current intron prediction programs are powerful only for very short introns, and also these predictors are far from perfect. But in fact, if the intrasplicing model is the true splicing model for at least some introns, the model itself would explain the weakness of intron predictors: if intraintrons are (correctly) predicted as introns by an intron predictor, these predictions would be nevertheless classified as wrong predictions, because there is no data about intraintrons. This mistake would happen increasingly often for longer introns, since they are more likely to harbour intraintrons, which could explain the steep falling of the prediction accuracy (Figure 3).

Predictors making use of the context of introns within genes cannot be used for the analysis of intrasplicing. Therefore, this work shows that stronger intron prediction programs, which do not make use of the context information, would be very useful. This may become reality in the future, as the knowledge of splicing enhancers and splicing inhibitors grows.

We also note that if there are different ways for the splicing of long introns, the significance of our results may be limited by the usage of the whole set of long introns. Also the availability of long intron sequences is still limited for species other than human.

Another approach to improve the results of this work might be to try different ways of selecting intraintrons out of the long intron. For example, it can be thought that some intraintrons are already spliced before the long intron is completely transcribed. In this case, a procedure starting at the 5' region and progressing to the 3' region might yield stronger evidence.

In this work, we presented a computational approach to the known problem of long intron splicing. The intrasplicing model we proposed seems to be a likely explanation for long intron splicing and is worth to be examined. Since *in vitro* experiments on intermediate splicing products of long introns are presently difficult to conduct, there is the possibility that long introns contain a structure of intraintrons, though there are no experimental results supporting this hypothesis so far. The attractive model we proposed should therefore be pursued more extensively.

Acknowledgments

This research was supported in part by Grant-in-Aid for Encouragement of Young Scientists and Grant-in-Aid for Scientific Research on Priority Areas 'Genome Information Science' from the Ministry of Education, Culture, Sports, Science and Technology of Japan. We would like to thank the anonymous referees for valuable comments on this research.

References

1. D.A. Benson, I. Karsch-Mizrachi, D.J. Lipman, J. Ostell, B.A. Rapp, D.L. Wheeler. GenBank. *Nucleic Acids Research*, 30(1):17–20, 2002.
2. B. J. Blencowe. Exonic splicing enhancers: mechanism of action, diversity and role in human genetic diseases. *TIBS*, 106–110, March 2000.
3. L. Cartegni, S.L. Chew, and A.R. Krainer. Listening to silence and understanding nonsense: exonic mutations that affect splicing. *Nature reviews*, 3:285–298, 2002.
4. J.F. Cáceres and A.R. Kornblihtt. Alternative splicing: multiple control mechanisms and involvement in human disease. *TRENDS in Genetics*, 18(4):186–193, 2002.
5. R. Durbin, S.R. Eddy, A. Krogh, G. Mitchison. Biological Sequence Analysis. Cambridge University Press, 1998.
6. A.R. Hatton, V. Subramaniam, A.J. Lopez. Generation of alternative Ultrabithorax isoforms and stepwise removal of a large intron by resplicing at exon-exon junctions. *Molecular Cell*, 2(6):787–797, 1998.
7. T. Hubbard, et al. The Ensembl genome database project. *Nucleic Acids Research*, 30(1):38–41, 2002.
8. L.P. Lim and C.B. Burge. A computational analysis of sequence features involved in recognition of short introns. *PNAS*, 98:11193–11198, 2001.
9. K.D. Pruitt, D.R. Maglott. RefSeq and LocusLink: NCBI gene-centered resources. *Nucleic Acids Research*, 29(1):137–140, 2001.
10. D. Tollervey and J.F. Caceres. DNA processing marches on. *Cell*, 103(5):703–709, 2000.

TRAJECTORY CLUSTERING: A NON-PARAMETRIC METHOD FOR GROUPING GENE EXPRESSION TIME COURSES, WITH APPLICATIONS TO MAMMARY DEVELOPMENT

T.L. PHANG, M.C. NEVILLE, M. RUDOLPH, L. HUNTER

University of Colorado School of Medicine,
Denver, Colorado 80262, USA.
tzu.phang@uchsc.edu

Trajectory clustering is a novel and statistically well-founded method for clustering time series data from gene expression arrays. Trajectory clustering uses non-parametric statistics and is hence not sensitive to the particular distributions underlying gene expression data. Each cluster is clearly defined in terms of direction of change of expression for successive time points (its 'trajectory'), and therefore has easily appreciated biological meaning. Applying the method to a dataset from mouse mammary gland development, we demonstrate that it produces different clusters than Hierarchical, K-means, and Jackknife clustering methods, even when those methods are applied to differences between successive time points. Compared to all of the other methods, trajectory clustering was better able to match a manual clustering by a domain expert, and was better able to cluster groups of genes with known related functions.

1 Introduction

Clustering is one of the most widely used approaches for analysis of genome-wide expression data. All clustering methods make assumptions about the nature of the items clustered and the definition of "similarity" among those items. For example, the popular K-means clustering method assumes that items to be clustered can be described by values drawn from K univariate Normal distributions. When the assumptions underlying a clustering method are violated, that method is unlikely to produce clusterings that reflect the true underlying groupings of the data. In addition to their distributional assumptions, K-means and other widely used clustering methods are not specifically able to take into account the relationship among adjacent points in a time series.

In this paper, we introduce "trajectory clustering," a non-parametric method of clustering gene expression data from time course experiments. No assumptions about the distributional nature of gene expression levels are required, nor are uniformly spaced time points or any assumptions about the behavior of expression between time points. Furthermore, the method itself involves no free parameters (such as the K in K-means) which must be estimated separately. The trajectories used in our clustering method are defined by the direction of change between adjacent time points in a series. The direction of change can take on one of three possible values: increasing, decreasing or flat. For a time series containing N time points, there are N-1 changes, and 3^{N-1} possible trajectories.

We apply trajectory clustering to a dataset from mouse mammary gland development, and show that the trajectory clusters correspond better to a manually derived expert clustering, and group genes with known biological function more accurately than two other popular clustering methods, Hierarchical and K-means. Data was acquired by Affymetrix oligonucleotide-based microarray data showing secretory activation in the mouse mammary gland. Secretory activation is a unidirectional process that takes place with high temporal coherence during the physiological transition from pregnancy to lactation (Neville et al, 2002). Many of the biochemical events in this process have been studied extensively for more than three decades (Wilde et al. 1986; Mellenberger and Bauman, 1974; Kuhn, 1968), and it is clear that many of the changes are transcriptionally regulated (Rosen et al. 1999). However, the molecular mechanisms that regulate and coordinate these changes *in vivo* are not well understood. Because the process is complex, the most efficient way to approach the problem is to begin with a global analysis of gene expression prior to, during and subsequent to secretory activation.

2 Methods

2.1 Acquisition of time course data from mice

Five time points were collected between day 12 of pregnancy and day 9 of lactation, with 4 replicates at each data point. Four FVB mice were sacrificed for each of the time points investigated (P12, P17, Lac1, Lac2, and Lac9). Both fourth mammary glands were removed from each animal and the imbedded lymph nodes excised. The mammary tissue was stored in RNA*later* stabilization buffer (Qiagen, Valencia, CA) at -20 °C according to protocol. Total RNA was isolated and purified from each sample following the Qiagen RNA extraction/clean-up protocol. Using a spectrophotometer and the RNA 6000 Nano Assay (Agilent Technologies, Palo Alto, CA), purity, concentration and integrity of the total RNA was verified. If the samples qualified, the RNA was amplified, labeled, and fragmented following the 2002 protocol for eukaryotic target preparation (Affymetrix, Santa Clara, CA). The labeled and fragmented cRNA products of the Affymetrix protocol were again verified for sample integrity and concentration using RNA 6000 Nano Assay. Accepted samples were hybridized to Affymetrix Mu74Av2 microarray chips. Raw data were gathered from scanned array chips using Affymetrix Microarray Suite version 5.0. All animal procedures were approved by the Institutional Animal Care and Use Committee of the University of Colorado Health Sciences Center.

2.2 Computational analysis

We describe our computational approach in three phases. First, we describe how we selected portions of the raw data for further analysis. Then, we describe the details of the trajectory clustering algorithm itself. Finally, we describe the processes by which we evaluated the method and compared it to other approaches. All original algorithms were implemented in Matlab v6.1.R12 (MathWorks Inc); others were either implemented in Matlab or were from the GeneSpring (Silicon Genetics, Inc.) expression array analysis toolkit.

2.2.1 Identifying genes with significant changes during the time course

Many mammary genes are not related to secretory activation, and therefore most genes' expression will not change significantly over the course of this experiment. Before analyzing the genes putatively related to secretory activation, we applied three computational methods to filter out genes which did not vary significantly during the course of the experiment.

The first filter (Hogg & Craig, 1978, p175) identifies genes with at least moderate variance over the entire experiment; genes which do not vary at all are not likely to be related to secretory activation. Since we can assume that most genes are not related to activation, then the gene with the median variance is a reasonable model of null variation, that is, the variation due to factors other than secretory activation. We calculate the variance s^2 for each gene. The null hypothesis is that these variances represent random and Normally distributed noise. We can then compute the statistic $W=(N-1)s^2/\mathrm{median}(s^2)$ where N is the number of observations of the gene, which is approximately chi-square distributed with N-1 degree of freedom. We calculate a p value for rejecting the null hypothesis that the gene did not vary, and perform the False Discovery Rate (FDR) multiple testing correction, (Benjamini et al, 1995) setting the false discovery rate to be 10%. This results in a list of genes with significantly greater variation than the median variation gene, with at most 10% of that list including genes having true variation less than or equal to median variation.

Our second filter uses Affymetrix's mRNA detection call to exclude all genes with an Absolute Call of "Absent" in all experiments. The third filter is used to test the consistency of the gene across replicates of a particular time point. Genes whose within-replicate coefficient of variation was greater than 0.03 were removed. These preprocessing steps screen out genes with low variance, low mRNA levels, and inconsistent expression measurements.

The final preclustering step is to apply the non-parametric Kruskal-Wallis statistic to select genes whose expression levels are significantly different between at least two time points. Kruskal-Wallis is the non-parametric equivalent of an

ANOVA test. We then again perform the False Discovery Rate test for multiple comparison correction for these genes, setting FDR to 0.015. The initial filtering steps greatly reduce the number of genes tested, and hence reduce the penalty in statistical power incurred by this correction.

2.2.2 The Trajectory Clustering algorithm

Trajectories are defined to be a sequence of length T-1, where T is the number of time points, and each element of the sequence is either I (increase), D (decrease), or F (flat). For example, all the genes whose expression decreased at each point in a four measurement series would be assigned to cluster DDD. Given a list of genes which varied significantly across at least two time points, the goal of the clustering algorithm is to assign each of these genes to a particular trajectory. Because the Kruskal-Wallis test requires at least one significant difference, no gene should ever be assigned to the sequence of all flat (FFFF).

When the Kruskal-Wallis test identifies a significant difference between an adjacent pair of points, the assignment of the gene to the I or D trajectory is trivially based on the sign of the difference. However, it is possible that a significant difference is found between expression levels that are not adjacent, yet none of the adjacent pairs themselves are found to be significantly different. For example, the expression of a gene at time point 3 may be significantly greater than at time point 1, yet there may not be a significant difference between the expression levels at time points 1 and 2, nor between the expression levels at time points 2 and 3. The challenge in this situation is to determine whether to assign this gene to the II trajectory, the FI trajectory or the IF trajectory.

We will first present a solution to this problem in the three point case, and then generalize it to N points. Let T_{ij} represent the interval between time points i and j. The nontrivial case arises when T_{13} is significant, but not T_{12} or T_{23}. It is possible that T_{12} and T_{23} contribute equally to the difference in T_{13}, or the difference might be heavily weighted toward T_{12} or T_{23}. We discriminate between these possibilities as follows. For each of these genes, we sort the expression levels and assign a rank to each measurement, then calculate the mean rank difference $C_{i,j,k,l} = |R_i - R_j| - |R_k - R_l|$ for the transitions from i to j and k to l, where the Rs are the average rank of the expression level at a particular time point (over all replicates). By assumption, the difference between points i and l is statistically significant, but the differences between i and j and between k and l are not. The C value is a non-parametric measure of the relative contribution of the two transitions to the difference between the first and last point. A large positive C value means the first transition made more of a contribution to the total difference than the second transition, and therefore that we should assign that gene to the (I or D)F trajectory depending on the sign of the overall difference. A large negative value implies an F(I or D)

```
for i = number of time points down to 3
            row = 1; col = i;
            for 1:length of diag of Matrix(i-1)
                  if Hrow,col == 1
                        if Crow,col-1,row+1,col >> 0
                              Hrow,col-1 = 1
                              if Hrow+1,col != 1;
                                    Hrow+1,col = 0;
                              end
                        elseif Crow,col-1,row+1,col << 0
                              if Hrow,col-1 != 1
                                    Hrow,col-1  = 0
                              end
                              Hrow+1,col = 1;
                        elseif Crow,col-1,row+1,col  ~= 0
                              Hrow,col-1 = 1 and Hrow+1,col = 1;
                        end
                  end
                  row = row + 1;
                  col = col + 1;
            end
      end
```

Figure 1 Pseudocode for iterative clustering of more than 3 time points. diag(i) is the ith diagonal of the matrix counting from the main diagonal; it is equivalent to Matlab's diag function.

trajectory, and a value near zero means that the relative contributions are similar and the trajectory should be either II or DD. In the three time point example above, we are interested in $C_{1,2,2,3}$.

All that remains is to determine where the cutoff should be for being "near" zero. Since there are many orderings in which both mean ranks are the same, many $C_{i,j,k,l}$s will be exactly zero. However, we might also reasonably treat small differences in average rank by assuming that the relative contributions are similar. We used an ad hoc, percentile-based cutoff C*, calling the top C* percentile of the negative differences, and the bottom C* percentile of the positive differences to be near zero. In this dataset, the clustering is not very sensitive to the particular choice of C*. We chose C* to allow average rank differences of <= 1 to count as near zero. Any C* between 26 and 31 achieved this for most of the clusters, so we selected C*=30 for the analysis below. If the distribution of C scores were known, a statistically sound cutoff could be specified, but that is currently an open problem.

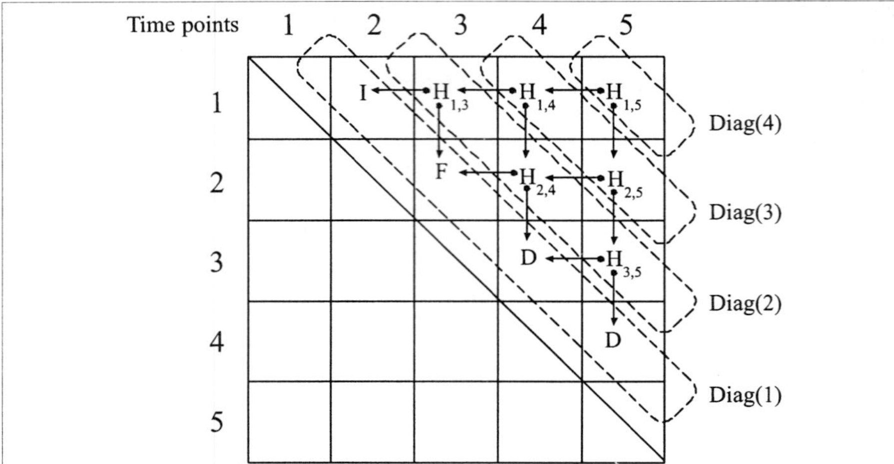

Figure 2 Recursive expansion of the three point case, starting with the final time point (see pseudocode in Figure 1) and executing along successive diagonals. Diag(x) represents the values in x's diagonal, shown by the dashed-boxes. The last diagonal is translated to the a trajectory by sign.

The above approach can be generalized to more than three time points. We initialize an upper diagonal matrix H with bits set to one based on the significance tests, i.e. $H_{ij} = 1$ if the null hypothesis was rejected for T_{ij}, and 0 otherwise. We work three points at a time, starting with the final time point and executing along successive diagonals. Additional bits are set to 1 based on the following process. At the end of the process, the last diagonal of the H matrix can be straightforwardly translated to a particular trajectory, substituting F for 0s and either I or D for 1s, depending on the sign of the difference. Figures 1 (pseudocode) and 2 describe the process in detail.

2.3 Manual clustering

One of the authors [Neville] has long experience in secretory activation, and performed an ad hoc, semi-manual clutering of the genes, using tools available in GeneSpring (Silicon Genetics, www.sigenetics.com) and her extensive knowledge of the biology of secretory activation. The ad hoc method required direct user interaction and took many hours to complete, and relied on a variety of unsupported assumptions. The procedure was as follows. Beginning with the same list of genes that varied significantly over at least one pair of time points, each pair of adjacent time points was tested for differences using the Mann-Whitney U test (which is equivalent to the Kruskal-Wallis test when there are only two conditions) with a critical level of P< 0.05. These were assigned to the I or D trajectory using the fold change filter with the fold difference set at 1.1. All genes that did not fit this criterion were initially assigned to the F trajectory. These cutoffs were set to coincide with intuition for a subset of the important genes which we examined manually. The one interval sets were combined into 81 preliminary classes reflecting the patterns of expression over the four intervals in the dataset.

Classes containing two or more flat sets in a row (e.g. DDFF, IFFD, IFFF, FFFF) were further examined to determine whether there was a statistically significant change over two or more consecutive intervals. For each FF class an initial sort was made into genes that changed or remained flat over two intervals using the Mann-Whitney test as describe above. For FFF or FFFF classes the genes were initially examined in sets of FF classes. Those that showed no significant change were examined by eye over 4 or 5 time points. A significant change was seen in about 200 genes in classes containing the FF, FFF and FFFF patterns, about half the genes in FF and FFF trajectories and all the genes in the FFFF trajectories as predicted.

As with the automated method, the hardest problem is to apportion the changes in these genes over the two, three, or four intervals over which a significant change was noted. To do this each of the intervening adjacent pairs was tested using a Welsh t-test with a critical value of 0.20. Again, this particular value was selected because it agreed with the expert's intuitions about the assignments; also note that the Normality assumptions underlying this test are not valid for this data. Adjacent pairs which were different under this test we assigned to either I or D, all others were left as F.

2.4 Other clustering techniques

For comparison to more established techniques, hierarchical and K-mean clustering were used to cluster our time series data. In hierarchical clustering (Eisen MB et al. 1998), two types of similarity metric were calculated; Euclidean distance and jackknife correlation (Heyer LJ, et al. 1999). A complete-linkage hierarchical clustering was used for the purpose of computing a dendrogram that represent all elements into a single tree. We used Matlab's "cluster" function to draw a horizontal line on the dendrogram tree, and produced the user defined number of clusters. We divided the hierarchical tree into 20 and 9 clusters, to obtain results comparable to the trajectory method (see results). In addition, K means clustering (Tavazole, S. et al, 1999) is designed to partition the data into K groups by minimizing the within-group sum-of-squares. We used the K means algorithm in GeneSpring with Pearson correlation to partition the genes into 20 and 9 clusters.

358

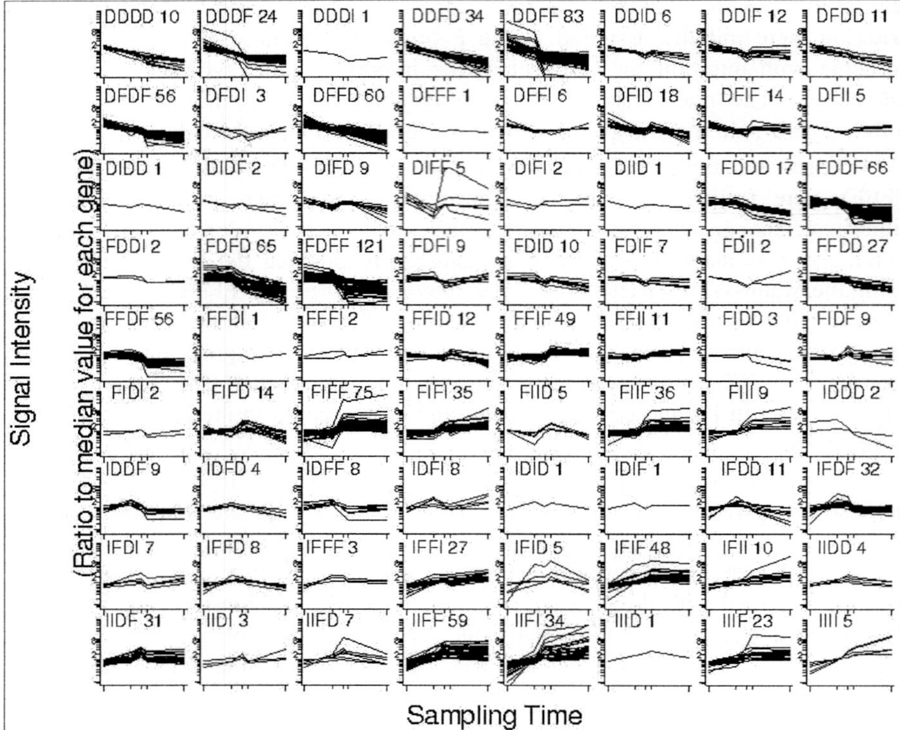

Figure 3 Clusters produced by automatic trajectory clustering for secretory activation in the mammary gland. Four replicates at each of five time points, Pregnancy days 12 and 17 and lactation days 1, 2 and 9, are represenedt in each plot. Intensities were normalized to the median for each gene and plotted on a log scale.

3   Results

3.1 Significantly varying genes and their trajectory clusters

The preprocessing and Kruskal-Wallis test resulted in 1358 genes with at least one significant difference (with FDR=0.015, so approximately 20 of these are likely false positives). There are 4 intervals between the points and therefore the potential to generate 3^4=81 trajectories (actually 80, since FFFF cannot be occupied with this method). Figure 3 shows the 72 populated clusters, identifying the trajectory and the number of genes in the cluster; 20 of these were occupied by 20 or more genes.

These 20 large clusters contained 975 (or 72%) of the total genes. We therefore used 20 as the target for K means and hierarchical clustering.

Number of interval	Number of manual Cluster	Other cluster	Number of other Cluster	Percent of other Cluster to manual Cluster	Discrimination index
4	73	TC	72	55.1	0.80
4	20	TC	20	61.0	0.75
2	9	TC	9	77.4	0.56
4	20	KM	20	41.0	0.78
2	9	KM	9	28.6	0.59
4	20	JK	20	39.4	0.59
2	9	JK	9	30.0	0.34
Using mean difference values					
4	20	KM	20	34.6	0.47
2	9	KM	9	32.9	0.42
4	20	JK	20	26.9	0.57
2	9	JK	9	43.9	0.48

Table 1 Clustering quality measures (see text). TC is trajectory clustering, KM is K-means and JK is hierarchical jackknife.

3.2 Comparison with manual trajectory clustering

Manual clustering resulted in 73 clusters which in general overlapped the results of the automated algorithm, particularly in the twenty large clusters. Table 1 shows the degree of overlap between each of the clustering methods and the manual method. In mouse mammary gland secretory activation studies, the two middle intervals (P17-Lac1-Lac2) showed the greatest number of significant changes, we therefore also combined clusters that had the same trajectories for these two central transitions, creating 9 clusters. In this latter analysis 77.4% of the genes mapped to the same cluster in the automatic and manual methods (Table 1). Furthermore, inspection suggested that genes that.differed between the automatic and manual methods diverged only slightly in assigned trajectories, and then only when the difference between two time points was small relative to the variance of the time points.

3.3 Comparison with other clustering methods

The same mapping approach was used to compare 20 and 9 clusters from K-Means and hierarchical jackknife methods to the manual clusterings. We matched each K-means and hierarchical cluster to the manual cluster that had the most genes in common, without allowing multiple matches. As seen in Table 1, 41% and 28.6% of the genes in the 20 and 9 K-Means clusters mapped on to the most closely corresponding manual cluster. Similarly, 39.4% and 30% of the 20 and 9 hierarchical Jackknife clusters were mapped on to the manual clusters. To ensure a fair comparison, we performed another set of K-Means and hierarchical Jackknife clusters using differences between adjacent time points rather than the raw values. The results show that 34.6% and 32.9% of 20 and 9 K-Means clusters, and 26.9% and 43.9% of hierarchical Jackknife clusters mapped on to the closest manual cluster.

3.4 Separation of functional groups by trajectory clustering

Although trajectory clustering produces a statistically well-founded grouping that is much more similar to the ad hoc expert grouping than traditional methods, it is unclear how well the manual clustering represents biological reality. To determine whether trajectory clustering could separate and/or identify biologically relevant genes we examined the genes associated with six functional classes known to be important in secretory activation (milk proteins, energy metabolism, fatty acid synthesis, cholesterol synthesis, adipocyte-specific and fatty acid degradation), and measured the purity of each cluster with respect to these classes. A discrimination index was calculated as follows. For each cluster, consider each pair of functional groups A and B. If the cluster contains only genes of one or the other functional group, the cluster gets a discrimination score of 1. If the cluster contains both groups, it gets a score of $1 - \{(G_A + G_B) / (Tot_A + Tot_B)\}$, where G_A and G_B are the number of genes of each functional type in the cluster and Tot_A and Tot_B are the total number of genes in that functional group. The discrimination score for the cluster is the mean score over all pairs of functions. Table 1 shows the discrimination scores. Both automatic and manual trajectory clustering gave a discrimination index of 0.80 using the full number of clusters derived here. All other methods gave lower discrimination scores, although the K-means method for 20 clusters gave an index of 0.78 which is quite close.

We also examined each of the genes with known function specifically in the trajectory clustering. With one exception milk genes clustered into related groups all beginning with I, indicating that they all increase significantly at the end of pregnancy. All continued to increase in subsequent intervals falling into 4 related groups with slightly different patterns of expression. The one gene that does not increase during pregnancy (clustered to FFIF 75) is PTHrP, a gene encoding a protein hormone involved in calcium regulation (Neville et al, 2002), which could be deleterious in the pregnant animal and may, therefore be differentially regulated. The genes of energy metabolism, with two exceptions, fell into downward going clusters, suggesting a relative fall in ATP generation in this tissue, which must devote most of its energy to synthetic reactions and transport. Of greatest interest of the genes for fatty acid and cholesterol synthesis; both groups cluster predominantly to FIFF 75 and are, therefore, turned off during pregnancy, turning on immediately after birth of the pups, a point at which our histological studies show that secretion is activated (McManaman, J. and Neville, M.C., unpublished). These genes are likely, therefore, to be coordinately regulated and indeed many in both classes are up-regulated by the transcription factor SREBP-1 (Horton, et al. 2002). Interestingly SREBP-1 itself falls into the cluster FFIF 75. Genes that mediate fatty acid degradation, in general by the β-oxidation pathway were distributed among four clusters showing different butrelated patterns of decrease. The mammary gland contains several different tissue compartments whose proportion changes with secretory development. We know that the milk protein genes reflect the epithelial compartment and assume that the changes in the metabolic pathways illustrated also represent this compartment. In order to evaluate changes in another tissue compartment, the adipose compartment which is quite prominent in the mammary gland, we examined the expression pattern of 5 adipose specific genes. These genes cluster into 5 distinct clusters, which have in common a D in first interval, the interval that reflects late pregnancy. We know from other studies (McManaman and Neville, unpublished) that a the relative proportion of adipose tissue declines steeply during pregnancy, with much small changes during lactation as reflected in the variation in the last three time intervals. We also examined two categories of genes known to decrease during secretory activation, adipocyte specific genes and genes that are involved in fatty acid degradation. Both segregated into predominantly down-going clusters.

4 Discussion

Trajectory clustering is a non-parametric clustering method using only the direction of change between subsequent time points to group genes in time course study. Clustering using various ad hoc schemes to conform as closely as possible to expert intuition gave quite similar results to trajectory clustering, and rather different than the other methods. More importantly, trajectory clustering also showed the ability to discriminate among genes in relevant functional categories better than the alternative methods.

Trajectory clustering has a natural interpretation, unlike the other methods studied. In this case, where each interval studied represents a well-characterized and different process, linking gene change directions to each of these intervals facilitates interpretation. For example, two important synthetic processes, milk protein synthesis, and lipid and sterol synthesis show distinct temporal activation. Thus the synthesis of all but one of the proteins we have classified as milk proteins are turned on in late pregnancy; many of these (see group IFIF 48) do not change between late pregnancy and the first day after birth, increasing sharply on the second day of lactation. Many of these molecules including β-casein, and whey acidic protein are known to be regulated by stat5, a mediator of prolactin signaling, whose mRNA and phosphorylation change little over the period of parturition (Rosen, et al. 1999). On the other hand the genes that regulate lipid synthesis are known to be regulated by the transcription factor, SREBP-1. A large proportion of these genes does not increase during pregnancy but are activated on the first day of lactation. SREBP-1 shows a similar pattern of activation, suggesting that its activation and up-regulation are needed to turn on lipid and cholesterol synthesis. A number of other transcription factors are found in these two clusters, JunD1, Pou 11 and TCFL4 (MCX) are located in IFIF whereas NFAT and Sox13 are found in FIFF. These genes are candidates for further investigation.

The spread of some classes of functionally related genes across similar clusters (e.g. FIFF and FIIF) suggests that collapsing some distinctions may be of biological value. Certainly as the number of time points increases, the number of trajectories increases exponentially, and therefore some cluster combining is probably warranted. Since each trajectory has a well defined relationship to all the others, we expect in future work to be able to identify a well-founded method for combining trajectory clusters as defined here.

Acknowledgements

Anna Baron and Sonia Leach provided valuable advice. This work was supported by NIH R37HD 19547 & P01Hd38129 to MCN and NIH R24 AA13162-01 to LH.

References

Benjimini, Y. and Hochberg, Y. (1995) J. R. Stat. Soc. Ser. B57, 289-300.

Eisen, M.B., Spellman, P.T., Brown, P.O., and Botstein, D. (1998) Proc. Natl. Acad. Sci. USA 95, 14863-14868.

Heyer, L.J., Krugliak, S., and Shibu, Y. (1999) Genome Res., 9:1106-1115.

Hogg, R. and Craig, A. (1978). Introduction to Mathematical Statistics. Macmillan Publishing Company.

Horton, J.D., Goldstein, J.L. and Brown, M.S. (2002) J.Clin.Invest. 109:1125-1131.

Hunter, L., Taylor, R.C., Leach, S.M., and Simon, R. (2001) Bioinformatics. 17 Supplement 1, S115-S122.

Kuhn, N.J. (1968) Biochem.J. 106, 743-748.

Mellenberger, R.W. and Bauman, D.E. (1974) Biochem.J, 138, 373-379.

Neville, M.C., McFadden, T.B. and Forsyth, I. (2002) J.Mammary Gland Biol.Lact, 7, 49-66.

Rosen, J.M., Wysolmerski, S.L. and Hadsell, D. (1999) Annu. Rev. Nutr. 19, 407-436.

Tavazole, S., et al, (1999) Nature Genetics, 22, 281-285.

Wilde, C.J., Henderson, A.J. and Knight, C.H. (1986) J.Reprod.Fert, 76, 289-298.

ALGORITHMS FOR MULTIPLE GENOME REARRANGEMENT BY SIGNED REVERSALS

SHIQUAN WU, XUN GU

Center of Bioinformatics and Biological Statistics
Iowa State University, Ames, IA 50011, USA
squwu@cs.iastate.edu, xgu@iastate.edu

We discuss a multiple genome rearrangement problem by signed reversals: Given a collection of genomes, we generate them in the minimum number of signed reversals. It is NP-hard and equivalent to finding an optimal Steiner tree to connect the genomes by reversal paths. We design two algorithms to find the optimal Steiner nodes of the problem: Neighbor-perturbing algorithm and branch-and-bound algorithm. The first one is a polynomial running time approximation algorithm. It searches for the optimal Steiner nodes by perturbing initial Steiner nodes nearby their neighborhoods and improving them better and better until convergence. The second one is an exact exponential running time algorithm for a median problem. It finds the optimal Steiner node by checking all candidates that satisfy the necessary conditions for optimal Steiner nodes. We implement the algorithms into two programs respectively and show by experimental examples that they are more efficient than other similar ones, such as GRAPPA, BPAnalysis, and MGR, etc.

1　Introduction

Phylogenetic trees show the evolutionary relationship of species. They are inferred based on DNA or amino acid sequences. Three major methodologies are widely used for the purpose: Parsimony method, maximum likelihood method, and distance-based method [10,11]. Genome projects have generated enormous types of informative genome data for phylogeny reconstructions [9]. One interesting and challenging problem is about comparative genome-wide gene orders for multiple genomes and phylogeny inference. Most earlier discussions focus on edit distances defined by local mutations such as insertion, deletion, substitution [10,11], and furthermore recombination [17]. Recently, distance was widely discussed based on the orders of genes. Breakpoint analysis first investigated such kind of distances [14,15]. However, the biological meaning of breakpoint distance is not very clear. To overcome this, reversal distance was then introduced. It is defined as the minimum number of signed/unsigned reversals needed to account the difference of the gene orders of two genomes [14,15]. Sorting by reversal is an important problem in comparative genomics. It transforms one genome into another by signed or unsigned reversals [2,3,4]. Sorting by unsigned reversals is NP-hard [7,8], while sorting by signed reversals is polynomial-time solvable [12,13]. Indeed, the signed reversal distance between any two signed permutations can be computed by linear running time algorithms[1].

A multiple genome rearrangement problem discusses how to reconstruct optimal distance-based phylogenetic trees for a collection of genomes[15,16,18]. The problem is NP-hard[8]. Therefore, the practical purpose turns out to find good approximation solutions. Various approximation algorithms/programs (such as GRAPPA, BPAnalysis, MGR, etc.) are developed for solving different versions of multiple genome rearrangement problems[5,6,15,16,18].

In this paper, we discuss a multiple genome rearrangement problem by signed reversals[18]: Given a collection of genomes, we generate them in the minimum number of signed reversals. It is NP-hard and equivalent to finding an optimal Steiner tree to connect the genomes by reversal paths. We focus on designing efficient algorithms to find the optimal Steiner nodes for the genomes.

The rest of the paper consists of five parts. In Section 2, we introduce the mathematical model of our multiple genome rearrangement problem. In Section 3, we review some related previous algorithms on the problem. In Section 4, we design an approximation algorithm and an exact algorithm for the problem. In Section 5, we compare our algorithms/programs with other similar ones (such as GRAPPA, BPAnalysis, and MGR, etc.) and show that ours are more efficient than those. And finally, we discuss some possible further work and appliocations in Section 6. We obtain the following results.

(1) Design a neighbor-perturbing algorithm. It is a polynomial running time approximation algorithm and searches for the optimal Steiner nodes for all given genomes by perturbing initial Steiner nodes nearby their neighborhoods and improving them better and better until convergence.

(2) Find the necessary conditions for optimal Steiner nodes.

(3) Based on the necessary conditions, we design a branch-and-bound algorithm. It is an exact exponential running time algorithm for a median problem and finds the optimal Steiner node by checking all possible candidates that satisfy the necessary conditions for optimal Steiner nodes.

(4) Two programs are implemented from the algorithms, respectively. Experimental examples show that our algorithms/programs are more efficient than other similar ones, such as GRAPPA, BPAnalysis, and MGR, etc.

2　Problem and model

First of all, we introduce the notations and set up the mathematical model of the multiple genome rearrangement problem by signed reversals[18].

Definition (1) Let X denote an alphabet (e.g., the set of genes) and $|X| = n$. Assume $p = (p_1 p_2 \cdots p_{i-1} \; \underline{p_i p_{i+1} \cdots p_j} p_{j+1} \cdots p_n)$ is a signed permutation (i.e., genome) on X. Each p_i stands for a gene and its sign represents its strand of DNA. A positive (or negative) sign means that the gene is in the forward (or backward) strand. A signed reversal on a segment $[i, j]$ of p is defined as the mutation $r(p; i, j) = (p_1 p_2 \cdots p_{i-1} \overline{-p_j \cdots -p_{i+1} - p_i} p_{j+1} \cdots p_n)$, where the segment $[i, j]$ changes both its orientation and signs (i.e., strand).

(2) Define $N(p) = \{q | q = r(p; i, j)$ for all $1 \le i \le j \le n\}$, which is the collection of all permutations that can be obtained from p by one signed reversal. $N(p)$ is called a reversal neighborhood (or sphere) of p. Let P be a collection of permutations. A reversal neighborhood (or sphere) of P is defined as $N(P) = \cup_{p \in P} N(p)$. Define $N_1(P) = N(P)$, $N_2(P) = N(N_1(P))$, and $N_k(P) = N(N_{k-1}(P))$, the $k-$neighborhood (or $k-$sphere) of P.

MGRBSR Problem (Multiple Genome Rearrangement By Signed Reversal) Given a collection of permutations $G = \{g_1, g_2, \cdots, g_m\}$, we generate G from some $p \in G$ in the minimum number of signed reversals, i.e., to find a collection of signed permutations $t_k (1 \le k \le s)$ on X such that (1) any g_j is obtained from p by a series $t_{k_1}, t_{k_2}, \cdots, t_{k_j}$, where each $t_{k_{i+1}}$ is obtained from t_{k_i} by one signed reversal, i.e., $t_{k_{i+1}} \in N(t_{k_i})$, and (2) s is minimized. Denote $d(p, G) = s$.

Various kinds of multiple genome rearrangement problems are widely discussed. The problems are NP-hard[8] and equivalent to finding optimal Steiner trees for G in the space of permutations [6,10,14,15,16,18]. We here discuss the **MGRBSR Problem** involving only pure signed reversals.

3   Previous related work

Many algorithms are designed to solve various kinds of multiple genome rearrangement problems. Most of them are approximation algorithms.

Breakpoint analysis   The breakpoint distance between two genomes is defined as the number of consecutive pairs of genes that are adjacent in one genome but not in the other. Breakpoint analysis is one of the earliest methods for genome rearrangement problem based on gene orders. There are several efficient approximation algorithms and programs for the problems, such as GRAPPA and BPAnalysis [5,15]. However, breakpoint analysis gives many approximation solutions that do not have a clear biological meaning. It could not find a more biologically accurate rearrangement (reversal) distance[6].

Grid search   Sankoff et al [16] discussed a multiple genome rearrangement problem for reversals and transpositions. For three genomes, a local optimal solution was searched upon a grid consisting of a series of reversal paths (see Fig.1a). A general problem with m genomes is recursively approximated by a series of groups of three genomes.

Nearest path search   Wu and Gu [18] discussed a multiple genome rearrangement problem for pure signed reversals. An approximation solution was obtained by a nearest path search algorithm upon a simple grid (see Fig.1b). For three genomes, suppose g_1 and g_2 are the pair with the greatest reversal distance. The algorithm at first finds a shortest reversal path P_1 from g_1 to g_2, then improves P_1 to another better reversal path P_2 (where P_2 is in some neighborhood of P_1 and closer to g_3 than P_1), and repeatedly improves a series of reversal paths to get a closest reversal path P_k (in some neighborhood of P_{k-1}) to g_3. Next, it constructs a grid by P_k and g_3. Finally, a local optimal solution is obtained by searching on the grid. This simplifies Sankoff's *Grid search* algorithm [16] and is shown to be efficient by experimental examples.

Greedy split   Bourque and Pevzner [6] designed a multiple genome rearrangement algorithm based on recursively greedy splitting. For the given genomes $g_1, g_2, \cdots, g_m$, the algorithm at first connects the nearest pair, say g_1 and g_2, by a shortest reversal path. Suppose $g_1, g_2, \cdots, g_{m-1}$ have been connected by some reversal paths. Then the algorithm finds a nearest point (the split site) on all these paths and connects g_m to the split site (see Fig.1c). The algorithm can be applied to both unichromosomal and multichromosomal genomes.

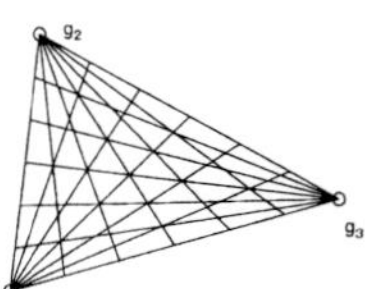

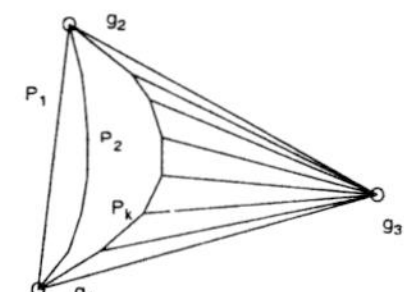

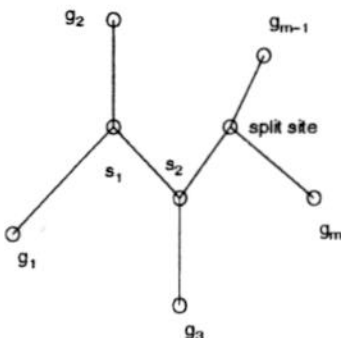

Figure 1: (a) Grid search   (b) Nearest path search   (c) Greedy split

4   Approximation algorithms

In this section, we design two algorithms: Neighbor-perturbing algorithm and branch-and-bound algorithm. Multiple genome rearrangement problems by signed reversals can be efficiently solved by applying these two algorithms.

4.1   Neighbor-perturbing algorithm

Any minimum spanning tree is a 2-approximation solution of the optimal Steiner tree (to prove this, add a multiple edge for each edge of the optimal

Steiner tree T^* and compare a spanning tree T_0 with these edges. Then we have an inequality on the lengths of the trees: $L(T_{MinSpanningTree}) \leq L(T_0) \leq 2L(T^*))$. Any minimum spanning tree can be regarded as a trivial Steiner tree with each given g_i as a trivial Steiner node. Within $N(g_i)$, we can find some Steiner node better than g_i. For a median problem, we at first choose g_2 as an initial Steiner node s_0. Then find a Steiner node $s_1 \in N(s_0)$ better than s_0, and next find $s_2 \in N(s_1)$ better than s_1, and so on. The Steiner nodes are improved better and better. This is the idea of our neighbor-perturbing algorithm. Neighbor-perturbing algorithm is to perturb an initial Steiner node within its neighborhood and improve it better and better until convergence.

The neighbor-perturbing algorithm consists of two steps: Initialization and iteration. For a median problem, in the initialization step, we rearrange the genomes so that the path $g_1 g_2 g_3$ forms a minimum spanning tree. g_2 is then chosen as the initial Steiner node, i.e., $s_0 = g_2$. We start perturbing from g_2 so that the iteration converges fast. In the iteration step, we perturb each s_i and find a better $s_{i+1} \in N(s_i)$ (Fig.2b). The Steiner nodes are improved better and better from $s_0(= g_2)$ to $s_1, s_2, \cdots, s_i, \cdots$, until s_i finally converges at s (Fig.2a).

Algorithm   **Neighbor-perturbing (Median problem)**

Input   Permutations: $G = \{g_1, g_2, g_3\}$.

Output   Optimal Steiner node.

Step 1   Initialization:

(1.1)   Rearrange each g_i: $d(g_1, g_3) = max\{d(x,y)|x,y \in G\}$.

(1.2)   Initial Steiner node $s_0 = g_2$.

Step 2   Iteration:

(2.1)   Suppose s_i have been defined. Check each $x \in N(s_i)$:
Denote $d^*(x, G) = d(x, g_1) + d(x, g_2) + d(x, g_3)$.
If x is better than s_i, i.e., $d^*(x, G) < d^*(s_i, G)$,
then define $s_{i+1} = x$.

(2.2)   Repeat (2.1) until convergence.

For the general multiple genome rearrangement problem, in the initialization step, we choose a minimum spanning tree T as an initial Steiner tree. We rearrange the genomes so that g_1 and g_m are the two longest leaves in T. There are at most $m - 2$ Steiner nodes. If we perturb g_1 and g_m to get two Steiner nodes, it may take a longer time for the convergences. So we choose $g_2, g_3, \cdots, g_{m-1}$ as the initial Steiner nodes so as to get a better approximation solution and a fast convergence. In the iteration step, we perturb and improve each Steiner node s by checking all $x \in N(s)$. If $S \cup \{x\} - \{s\}$ generates a better Steiner tree than S does, then replace s by x (Fig.2c). The Steiner nodes/tree are improved from the initial ones better and better until convergence.

Algorithm	**Neighbor-perturbing (General case,** see Fig.2c)
Input	Permutations: $G = \{g_1, g_2, \cdots, g_m\}$.
Output	Optimal Steiner nodes S.
Step 1	Initialization:
(1.1)	Define the reversal distance graph for G (vertex set) by all pairwise signed reversal distances $d(x, y)(x, y \in G)$.
(1.2)	Find a minimum spanning tree and its two longest leaves x_0 and y_0. Choose the initial $S = G - \{x_0, y_0\}$.
Step 2	Iteration:
(2.1)	For each $s \in S$, check all $x \in N(s)$: If the optimal Steiner tree generated by $S \cup \{x\} - \{s\}$ is better than that generated by S, then replace s by x.
(2.2)	Repeat (2.1) until convergence.

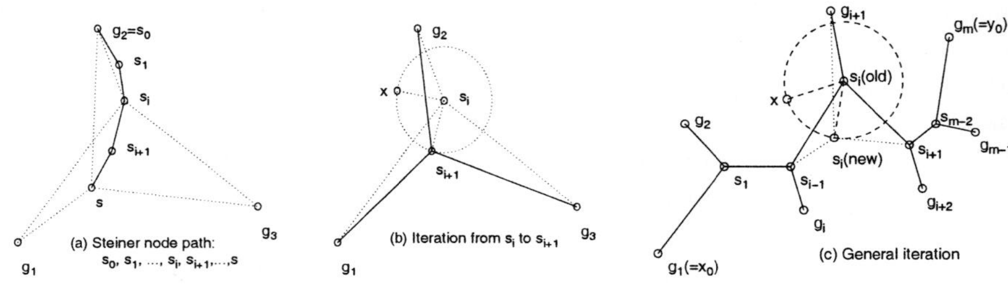

Figure 2: (a) Initial Steiner node $s_0 = g_2$. Recursively iterate/update the Steiner nodes: $s_0, s_1, \ldots, s_i, \cdots$ until convergence (at s). (b) Process of iteration from s_i to s_{i+1}: Suppose s_i have been generated. Check each $x \in N(s_i)$, if x is better than s_i, then update $s_{i+1} = x$. If no better x is found, then return s_i (optimal). (c) General iteration: Assume g_1 and g_m are the two longest leaves in the minimum spanning tree. Then $g_2, g_3, \cdots, g_{m-1}$ are chosen as the initial Steiner nodes. Each s_i is generated from $g_{i+1}(i = 2, 3, \cdots, m - 1)$ by a series of iterations. In each iteration, for Steiner node s_i, check all $x \in N(s_i)$, if $S \cup \{x\} - \{s_i\}$ generates a better Steiner tree, then replace s_i by x, i.e., s_i is updated by a new s_i. The three edges $s_{i-1}s_i$, $g_{i+1}s_i$, $s_{i+1}s_i$ are replaced by three new $s_{i-1}s_i$, $g_{i+1}s_i$, $s_{i+1}s_i$, respectively. The old Steiner tree is updated by a new one. The iterations are repeated until convergence.

Theorem 1 *The Neighbor-perturbing algorithm is a 2-approximation algorithm and has a running time $O(m^3 n^4 log m)$.*

Proof The algorithm is obvious 2-approximation. $O(m^3 n^4 log m)$ is from all steps: $O(mn)$ for all $d(g_i, g_j)$ and $O(m^2 n^3 log m)$ for minimum spanning trees.

4.2 Branch-and-bound algorithm

In this part, we design an exact algorithm, branch-and-bound algorithm, to find the optimal Steiner node for a median problem. Assume $G = \{g_1, g_2, g_3\}$. We rearrange the genomes so that the path $g_1g_2g_3$ forms a minimum spanning tree. Denote $d^*(x, G) = d(x, g_1) + d(x, g_2) + d(x, g_3)$. For any optimal Steiner s, the correspoding optimal Steiner tree must at first connect s to some $s_1 \in N(g_2)$, and then connect s_1 to g_2 (see Fig.3a). $g_2s_1g_1$ and $g_2s_1g_3$ are two optimal reversal paths from g_2 to g_1 and g_3, respectively. They share a common part g_2s_1 $(s_1 \in N(g_2))$. We can see that $d(g_2, s_1) + d(s_1, g_i) = d(g_2, g_i)(i = 1, 3)$ and $d^*(s_1, G) < d^*(g_2, G)$, i.e., s_1 is a Steiner node better than g_2.

Generally, any optimal Steiner tree connects s and g_2 through a series of Steiner nodes $s_0(= g_2), s_1, s_2, \cdots, s_j, \cdots$ such that $s_j \in N(s_{j-1})(j \geq 0)$ (see Fig.3b). $s_0s_1s_2 \cdots s_j \cdots g_1$ and $s_0s_1s_2 \cdots s_j \cdots g_3$ are two optimal reversal paths from g_2 to g_1 and g_3, respectively. They share a common part $s_0s_1s_2 \cdots s_j$. Each s_q is better than s_{q-1} $(q \geq 1)$. We state this as a theorem.

Theorem 2 (Necessary conditions for an optimal Steiner node) *Let $G = \{g_1, g_2, g_3\}$ (genomes). If s is an optimal Steiner node, then*

$$g(g_2, s) + d(s, g_i) = g(g_2, g_i) \ (i = 1, 3). \tag{1}$$

Moreover, there exist a series of Steiner nodes $s_0 = g_2, s_1, s_2, \cdots, s_j, \cdots, s_p = s$ such that $g(g_2, s_j) + d(s_j, g_i) = g(g_2, g_i)$ $(i = 1, 3; j \geq 0)$ and $d^*(s, G) < d^*(s_j, G) < d^*(s_{j-1}, G) < d^*(s_0, G), (p > j \geq 2)$.

Proof By the previous statements and Fig.3.

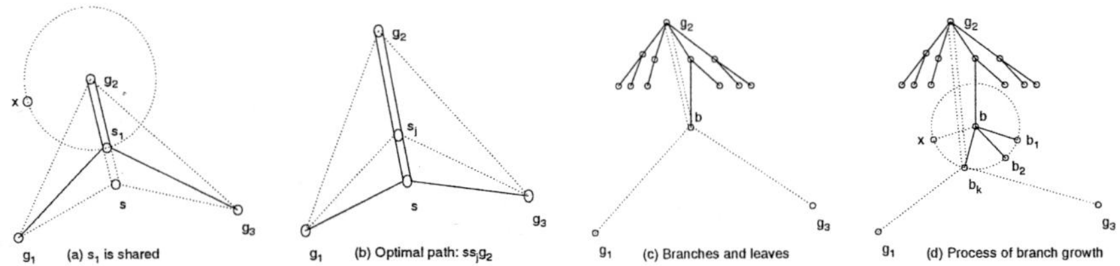

Figure 3: (a) Two optimal reversal paths $g_2s_1g_1$ and $g_2s_1g_3$ share a common part g_2s_1. $d(g_2, s_1) + d(s_1, g_i) = d(g_2, g_i)(i = 1, 3)$ and $d^*(s_1, G) < d^*(g_2, G)$. (b) s is an optimal Steiner node. $g_2 \cdots s_j \cdots g_1$ and $g_2 \cdots s_j \cdots g_3$ are two optimal reversal paths from g_2 to g_1 and to g_3, respectively. $g(g_2, s_j) + d(s_j, g_i) = g(g_2, g_i)(i = 1, 3; j \geq 0)$ and $d^*(s, G) < d^*(s_j, G) < d^*(s_{j-1}, G) < d^*(s_0, G)$ for each j. (c) Generation relations are shown by the branches/leaves, each leaf is obtained from one branch (candidate) of last generation. (d) Process of growing branches: For each leaf b, check all $x \in N(b)$, if x satisfies Eq. (1), then take x as a new branch/leaf. $b_1, b_2, \cdots, b_k$ are from b. Update s by x if $d^*(x, G) < d^*(s, G)$.

Theorem 2 gives us the idea of our branch-and-bound algorithm. Any s satisfying Eq. (1) is a candidate of the optimal Steiner node. By $s_j \in N(s_{j-1})(j \geq 0)$ and Eq. (1), we can recursively search for the optimal Steiner node from all possible candidates starting from g_2. At first, we check all $x \in N(g_2)$. If

$s_1 \in N(g_2)$ satisfies Eq. (1), then it is one of the first generation candidates of g_2. Next, we find all second generation candidates s_2 from each first generation candidate s_1 by $s_2 \in N(s_1)$ and Eq. (1). We repeatedly find all candidates for all generations until convergence. The optimal Steiner node can be found by $min_j d^*(s_j, G)$ over all candidates s_j. The relations between all candidates of different generations can be expressed as a tree (see Fig.3c-d). Suppose at some iteration step we have found some candidates of some generations (see Fig.3c). Let B be the set of all candidates (denoted by branches) and L the set of all the current generations (denoted by leaves). For each leaf $b \in L$, check all $x \in N(b) - B$, if x satisfies Eq. (1), then x is a new candidate and it is then put into both the branch set B and the leaf set L. If x is better than s, i.e., $d^*(x, G) < d^*(s, G)$, then update the optimal Steiner node s by x. New leaves (new generation candidates) $b_1, b_2, \cdots, b_k$ are obtained from b (see Fig.3d). b is removed from L and becomes an internal node in next iteration. We repeat the process until convergence, i.e., no new branch can be found.

Branch-and-bound algorithm consists of two steps: Initialization and iteration. It starts at g_2, grows branches and update the optimal s (see Fig.3d).

Algorithm	**Branch-and-bound** (Median problem, see Fig.3)
Input	Permutations: $G = \{g_1, g_2, g_3\}$.
Output	Optimal Steiner node s.
Step 1	Initialization:
(1.1)	Rearrange each g_i: $d(g_1, g_3) = max\{d(x, y)\|x, y \in G\}$.
(1.2)	Initial $B = \{g_2\}$, $L = \{g_2\}$, $s = g_2$.
Step 2	Iteration: (see Fig.3d)
(2.1)	Grow branches and update the optimal Steiner node:
	For each $b \in L$, check every $x \in N(b) - B$:
	If $g(g_2, x) + d(x, g_i) = g(g_2, g_i)(i = 1, 3)$, then
	x is a new candidate: $B = B \cup \{x\}$, $L = L \cup \{x\}$.
	If $d^*(x, G) < d^*(s, G)$, then update $s = x$.
	Remove b from L after all $x \in N(b) - B$ is checked.
(2.2)	Repeat (2.1) until convergence.

Theorem 3 *The branch-and-bound algorithm finds the optimal Steiner node in a running time $O(n^3 e_n^{[r_0/2]})$. Where $r_0 = min\{d(x, y)\|x, y \in G\}$ and e_n deponds on n (e.g., $e_n \leq 3$ in the examples in next section).*

Proof By Theorem 2 for the optimum and Theorem 3[18] for the running time.

5 Comparisons and examples

We show that our algorithms are more efficient than other similar ones. Theoretically, our neighbor-perturbing algorithm is more efficient than the greedy-split algorithm [6] and other similar algorithms. The greedy-split algorithm greedily connects all given genomes one after another. Usually, a greedy algorithm may generate an approximation solution far away from the optimal one in the worse case. However, our neighbor-perturbing algorithm starts from a minimum spanning tree, which is a 2-approximation solution. The neighbor-perturbing algorithm improves each Steiner node again and again and has more chances to get the optimal one. This is its advantage over other algorithms.

The two algorithms are implemented into two programs, GenomeNP and GenomeBnB, respectively, which take a reversal distance program "signed.c" as a subroutine written by Hannenhalli (http:// www-hto.usc.edu/software). We show by experimental examples that our algorithms/programs are more efficient than other similar ones, such as GRAPPA, BPAnalysis, MGR, etc.

Example 1 $G = \{p, q_1, q_2\}$ (the genomes of human, sea urchin,fruit fly) [6,16].

$$p = \quad 26\ 13\ 17\ 12\ -24\ 15\ 18\ 32\ -2\ -16\ -3\ -33\ 4\ -28\ 7\ 5\ 1\ 10\ 19\ 25\ 22\ 11$$
$$29\ 14\ 20\ -21\ -8\ 6\ 30\ -23\ 9\ 27\ 31$$

$$q_1 = \quad 26\ 4\ 25\ 22\ 5\ 1\ -28\ 19\ 11\ 29\ 20\ -21\ 6\ 9\ 27\ 8\ 30\ 23\ -24\ 16\ 14\ -2\ 32$$
$$3\ -31\ 15\ -7\ 33\ 10\ 13\ 17\ 12\ 18$$

$$q_2 = \quad 26\ 14\ 17\ -28\ -6\ -21\ 23\ -30\ 22\ 11\ 20\ 9\ -8\ -29\ -16\ -25\ -2\ -19\ -10\ -1$$
$$-7\ -5\ -13\ -4\ 33\ 3\ -32\ -18\ -15\ 24\ -12\ 27\ 31$$

By GenomeNP, we obtain an optimal Steiner node:

$$s = \quad 26\ 13\ 17\ 12\ -24\ 15\ 18\ 32\ -2\ -3\ -33\ 4\ 14\ 20\ -21\ -8\ 6\ -7\ 28\ -29\ -11$$
$$-22\ -25\ -19\ -10\ -1\ -5\ -16\ 30\ -23\ 9\ 27\ 31$$

Therefore, $d(s,p) = 4, d(s,q_1) = 20, d(s,q_2) = 15$, and $d(p, \{q_1, q_2\}) \leq 39$. GRAPPA and BPAnalysis obtained $d(p, \{q_1, q_2\}) = 43$, **MGR** also obtained $d(p, \{q_1, q_2\}) \leq 39$ [5,6,15]. However, can 39 be improved better? **MGR** did not answer. Since $d(s,p) + d(s,q_1) \geq d(p,q_1) = 24, d(s,p) + d(s,q_2) \geq d(p,q_2) = 19$, $d(s,q_1) + d(s,q_2) \geq d(q_1,q_2) = 32$, so $d(p, \{q_1, q_2\}) \geq [(24 + 19 + 32)/2] = 38$. GenomeBnB checks all possible candidates and obtains $d(p, \{q_1, q_2\}) = 39$.

Example 2 $G = \{A, B, \cdots, K\}$ are the following the genomes of: human, asterina pectinifera, paracentrotus lividus, drosophila yakuba, artemia franciscana, albinaria coerulea, cepaea nemoralis, katharina tunicata, lumbricus terrestris, ascaris suum, onchocerca volvulus, respectively [6].

$$A = \quad 1\ -32\ 17\ 2\ 23\ 12\ 3\ 20\ 6\ 30\ 7\ 8\ 21\ 31\ 24\ 9\ -10\ -18\ 11\ 33\ -28$$
$$19\ 14\ 34\ 13\ 25\ 4\ 22\ -29\ 26\ 5\ 35\ -15\ -27\ -16\ -36$$

$$B = \quad 1\ 30\ 7\ 2\ 23\ 12\ 3\ -32\ 6\ 8\ 21\ 31\ 9\ -10\ 11\ 19\ 14\ 18\ 33\ -13\ -5$$
$$-22\ -4\ -25\ -20\ -36\ 17\ -26\ 34\ -16\ -35\ 15\ -24\ -27\ 29\ -28$$

$$C = \quad 1\ 30\ 7\ 2\ 23\ 12\ 3\ -32\ 6\ 8\ 21\ 31\ 9\ -10\ 11\ 19\ 14\ 18\ 33\ 28\ -29$$
$$27\ 24\ -15\ 35\ 16\ -34\ 26\ -17\ 36\ 20\ 25\ 4\ 22\ 5\ 13$$

372

$$D = \quad 1\ 25\ 2\ 23\ 17\ 12\ 3\ 20\ 6\ 15\ 30\ 27\ 31\ 18\ \text{-}19\ \text{-}9\ \text{-}21\ \text{-}8\ \text{-}7\ 33$$
$$\text{-}28\ 10\ 11\ 32\ \text{-}4\ \text{-}24\ \text{-}13\ \text{-}34\ \text{-}14\ 22\ \text{-}29\ 26\ 5\ 35\ \text{-}16\ \text{-}36$$
$$E = \quad 1\ 25\ 2\ 23\ 17\ 12\ 3\ 20\ 6\ 15\ 30\ 27\ 31\ 18\ \text{-}19\ \text{-}9\ \text{-}21\ \text{-}8\ \text{-}7\ 33$$
$$\text{-}28\ 10\ 11\ 32\ \text{-}4\ \text{-}24\ \text{-}13\ \text{-}34\ \text{-}14\ 26\ 5\ 35\ \text{-}22\ \text{-}29\ \text{-}16\ \text{-}36$$
$$F = \quad 1\ 34\ 13\ 24\ 28\ 15\ 10\ 9\ 4\ 7\ 11\ 17\ 16\ 19\ 2\ 36\ 35\ 20\ 21\ \text{-}29\ \text{-}25$$
$$\text{-}27\ \text{-}12\ \text{-}30\ \text{-}18\ \text{-}14\ \text{-}26\ \text{-}6\ \text{-}32\ 31\ 8\ \text{-}33\ \text{-}3\ 22\ 5\ 23$$
$$G = \quad 1\ 34\ 13\ 24\ 15\ 10\ 28\ 9\ 4\ 7\ 11\ 17\ 16\ 19\ 2\ 36\ 35\ 20\ 21\ \text{-}29\ \text{-}25$$
$$\text{-}27\ \text{-}12\ \text{-}30\ \text{-}18\ \text{-}14\ 26\ \text{-}6\ \text{-}32\ \text{-}33\ \text{-}3\ 31\ 8\ 22\ 5\ 23$$
$$H = \quad 1\ 17\ 2\ 12\ \text{-}19\ \text{-}9\ \text{-}21\ \text{-}8\ \text{-}7\ 33\ \text{-}32\ \text{-}11\ \text{-}10\ 28\ \text{-}4\ \text{-}25\ \text{-}24\ \text{-}13$$
$$\text{-}34\ \text{-}14\ \text{-}26\ \text{-}16\ \text{-}36\ \text{-}35\ \text{-}29\ \text{-}20\ \text{-}18\ 3\ 23\ 15\ 30\ 27\ 22\ 6\ 31\ 5$$
$$I = \quad 1\ 27\ 2\ 17\ 36\ 20\ 3\ 29\ 10\ 11\ 35\ 12\ 30\ 21\ 9\ 19\ 18\ 28\ 33\ 7\ 8\ 16$$
$$26\ 14\ 34\ 13\ 24\ 15\ 32\ 25\ 4\ 22\ 23\ 6\ 31\ 5$$
$$J = \quad 1\ 16\ 26\ 17\ 20\ 2\ 21\ 13\ 6\ 9\ 15\ 28\ 34\ 10\ 7\ 35\ 18\ 14\ 32\ 27\ 36\ 4$$
$$12\ 23\ 25\ 31\ 5\ 22\ 30\ 29\ 19\ 11\ 24\ 3\ 33\ 8$$
$$K = \quad 1\ 35\ 10\ 30\ 29\ 11\ 24\ 3\ 23\ 15\ 25\ 27\ 26\ 7\ 14\ 36\ 4\ 19\ 12\ 22\ 20\ 2$$
$$21\ 13\ 6\ 16\ 32\ 28\ 17\ 34\ 9\ 18\ 31\ 5\ 33\ 8$$

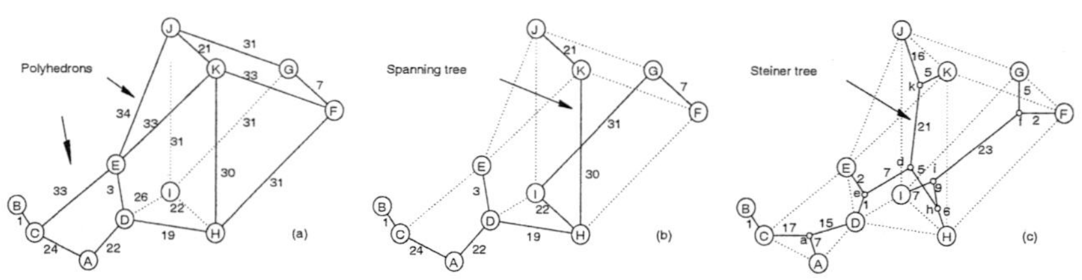

Figure 4: (a) The distance graph for the 11 genomes. (b) The minimum spanning tree of the genomes. (c) Optimal Steiner tree. The total distance 149 < 150 (obtained by **MGR**).

We at first construct the signed reversal distance graph for the genomes from all pairwise signed reversal distances (Fig.4a), then find a minimum spanning tree T of the graph (Fig.4b), and finally generate the following optimal Steiner nodes a, d, e, f, h, i and k by perturbing T by GenomeNP (Fig.4c).

$$a = \quad 1\ \text{-}32\ 17\ \text{-}11\ 10\ \text{-}9\ \text{-}27\ 18\ 33\ \text{-}28\ 19\ 14\ 34\ 13\ 25\ 4\ 22\ \text{-}29\ 26\ 5\ 35$$
$$30\ 7\ 8\ 21\ 31\ 24\ 2\ 23\ 12\ 3\ 20\ 6\ \text{-}15\ \text{-}16\ \text{-}36$$
$$d = \quad 1\ 25\ 31\ 5\ \text{-}35\ \text{-}29\ \text{-}22\ \text{-}12\ \text{-}17\ \text{-}23\ \text{-}3\ 14\ 34\ 13\ 24\ 15\ 30\ 27\ 2\ 20\ 6$$
$$4\ \text{-}32\ \text{-}11\ \text{-}10\ 28\ \text{-}33\ 7\ 8\ 21\ 9\ 19\ \text{-}18\ \text{-}26\ \text{-}16\ \text{-}36$$
$$e = \quad 1\ 25\ 2\ 23\ 17\ 12\ 3\ 20\ 6\ 15\ 30\ 27\ 31\ 18\ \text{-}19\ \text{-}9\ \text{-}21\ \text{-}8\ \text{-}7\ 33\ \text{-}28\ 10$$
$$11\ 32\ \text{-}4\ \text{-}24\ \text{-}13\ \text{-}34\ \text{-}14\ 22\ 29\ 26\ 5\ 35\ \text{-}16\ \text{-}36$$

$$f = \quad 1\ 34\ 13\ 24\ 15\ 10\ 9\ 4\ 7\ 11\ 17\ 16\ 19\ 2\ 36\ 35\ 20\ 21\ \text{-}29\ \text{-}25\ \text{-}27\ \text{-}12$$
$$\text{-}30\ \text{-}18\ \text{-}14\ \text{-}26\ \text{-}6\ \text{-}32\ 31\ 8\ \text{-}33\ \text{-}28\ \text{-}3\ 22\ 5\ 23$$
$$h = \quad 1\ 17\ 12\ \text{-}19\ \text{-}9\ \text{-}21\ \text{-}8\ \text{-}7\ 33\ \text{-}28\ 10\ 11\ 32\ \text{-}4\ \text{-}25\ \text{-}23\ \text{-}3\ 18\ 22\ 29$$
$$35\ 36\ 16\ 26\ 14\ 34\ 13\ 24\ 15\ 30\ 27\ 2\ 20\ 6\ 31\ 5$$
$$i = \quad 1\ 27\ 2\ 16\ 26\ 14\ 34\ 13\ 24\ 15\ 10\ 11\ 35\ 12\ \text{-}18\ \text{-}19\ \text{-}9\ \text{-}21\ \text{-}8\ \text{-}7\ \text{-}33\ \text{-}28$$
$$30\ 17\ 36\ 20\ \text{-}4\ \text{-}25\ \text{-}23\ \text{-}3\ 22\ 29\ 32\ 6\ 31\ 5$$
$$k = \quad 1\ 35\ 18\ 27\ 32\ \text{-}9\ \text{-}34\ \text{-}17\ \text{-}28\ 10\ 30\ 29\ 11\ 24\ 3\ 23\ \text{-}7\ \text{-}26\ \text{-}16\ \text{-}6\ \text{-}13$$
$$\text{-}21\ \text{-}2\ \text{-}20\ \text{-}22\ \text{-}12\ \text{-}19\ \text{-}4\ \text{-}36\ \text{-}14\ 15\ 25\ 31\ 5\ 33\ 8$$

GenomeNP perturbs A, E, F, I, K, D, H into a, e, f, i, k, d, h, respectively. The minimum spanning tree (see Fig.4b, length: 180) is perturbed into the optimal Steiner tree (Fig.4c) with a length 149, which is better than 150 obtained by **MGR**[6]. Our algorithms/programs are better than **MGR** and others.

6 Discussion

We have designed two algorithms (Neighbor-perturbing algorithm and branch-and-bound algorithm) and two programs (GenomeNP and GenomeBnB) to find the optimal Steiner nodes. The experimental examples show that the algorithms/programs are more efficient than other similar ones. Our discussions are based on signed reversals. However, we can extend the discussions to other mutations, such as insertion, deletion, substitution, reversal, transposition, translocation, fusion, fission, etc. The algorithms have great potential applications in a wide range of genomes including multichromosome genomes.

In the iteration steps of the algorithms, we always check all nodes in $N(x)$. We can get better approximation solutions by checking $N_k(x)$ for some greater k. By randomly checking only some nodes in $N(x)$, or $N_k(x)$, we can also expect to get some satisfactory approximation solutions. Similarly, if we randomly check only some of the candidates, the branch-and-bound algorithm then becomes a nice approximation algorithm.

Acknowledgment This work is supported by the NIH grant RO1 GM62118 to X.G. and Wu is also is supported in part by NSF of China (19771025).

References

1. Bader, D.A., Moret, B.M.E. and Yan, M., (2001) *A linear-time algorithm for computing inversion distances between signed permutations with an experimental study.* Lecture Notes in Computer Science, 2125: 365-376.
2. Bafna,V. and Pevzner, P. (1994) *Genome rearrangements and sorting by reversals.* In Proc. 34th IEEE Symp. of the Foundations of Computer Science, 148-157. IEEE Computer Society Press.

3. Bafna, V. and Pevzner, P. (1995) *Sorting permutations by transpositions.* Proceedings of the 6th Annual Symposium on Discrete Algorithms, pages 614-623. ACM Press, January 1995.

4. Bafna, V. and Pevzner, P. (1996) *Genome rearrangements and sorting by reversals.* SIAM Journal on Computing, 25(2):272-289.

5. Blanchette, M., Bourque, G., and Sankoff, D. (1997) *Breakpoint phylogenies.* In *Genome Information Workshop (GIW 1997)*, (eds. Mivano S. and Takagi, T.), pp.25-34. University Academy Press, Tokyo.

6. Bourque, G. and Pevzner, P. (2002) *Genome-Scale Evolution: Reconstructing Gene Orders in the Ancestral Species.* Genome Res.12:26-36.

7. Caprara, A. (1997) *Sorting by Reversals is Difficult.* Proceedings of the First Annual International Conference on Computational Molecular Biology (RECOMB'97), ACM Press.

8. Caprara, A. (1999) *Formulations and hardness of multiple sorting by reversals.* Proceedings of the Third Annual International Conference on Computational Molecular Biology (RECOMB'99), ACM Press.

9. Gu, X., (2000) *A simple evolutionary model for genome phylogeny inference based on gene content.* Comparative genomics (ed. Sankoff and Nadeau), pp.515-524. Kluwer Academic Publishers.

10. Gusfield,D., Algorithms on Strings, Trees, and Sequences: Computer Science and Computational Biology. Cambridge University Press, 1997.

11. Hillis, D., Motitz, C., Mable, B., Molecular systematics. Sinauer Association, Inc. Massachusetts USA, 1996.

12. Hannenhalli, S. and Pevzner, P. (1995a) *Transforming men into mice (polynomial algorithm for genomic distance problems.* Proc. IEEE Symp. of the Foundations of Computer Science.

13. Hannenhalli, S. and Pevzner, P. (1995b) *Transforming cabbage into turnip (polynomial algorithm for sorting signed permutations by reversals).* Proc. of 27th Ann. ACM Symp. on Theory of Comput., 178-189.

14. Sankoff, D. (1999) *Genome rearrangement with gene families.* Bioinfomatics, 15:909-917.

15. Sankoff, D. and Blanchette, M. (1998). *Multiple genome rearrangement and breakpoint phologeny.* J. Comp. Biol. 5: 555-570.

16. Sankoff, D., Sudaram, G. and Kececioglu, J. (1996) *Steiner points in the space of genome rearrangements.* International Journal of the Foundations of Computer Science, 7:1-9.

17. Wu, S. and Gu, X., (2001) *A greedy algorithm for optimal recombination.* Lecture Notes on Computer Science, 2108:86-90.

18. Wu, S. and Gu, X., (2002) *Multiple Genome Rearrangement By Reversals.* Pacific Symposium on Biocomputing 7:259-270.

DIGIT: A NOVEL GENE FINDING PROGRAM BY COMBINING GENE-FINDERS

T. YADA, T. TAKAGI

Institute of Medical Science, University of Tokyo
4-6-1, Shirokanedai, Minato-ku, Tokyo 108-8639, Japan

Y. TOTOKI, Y. SAKAKI

Genomic Sciences Center, RIKEN
1-7-22 Suehiro-cho, Tsurumi-ku, Yokohama, Kanagawa 230-0045, Japan

Y. TAKAEDA

Mitsubishi Research Institute, Inc.
2-3-6, Otemachi, Chiyoda-ku, Tokyo 100-8141, Japan

We have developed a general purpose algorithm which finds genes by combining plural existing gene-finders. The algorithm has been implemented into a novel gene-finder named DIGIT. An outline of the algorithm is as follows. First, existing gene-finders are applied to an uncharacterized genomic sequence (input sequence). Next, DIGIT produces all possible exons from the results of gene-finders, and assigns them their exon types, reading frames and exon scores. Finally, DIGIT searches a set of exons whose additive score is maximized under their reading frame constraints. Bayesian procedure and a hidden Markov model are used to infer exon scores and search the exon set, respectively. We have designed DIGIT so as to combine the results of FGENESH, GENSCAN and HMMgene, and have assessed its prediction accuracy by using recently compiled benchmark data sets. For all data sets, DIGIT successfully discarded many false-positive exons predicted by individual gene-finders and yielded remarkable improvements in sensitivity and specificity at the gene level compared with the best gene level accuracies achieved by any single gene-finder.

1 Introduction

Draft sequences corresponding to approximately 90% of the human genome have been produced [1], and interest in exhaustive gene finding in the genome has increased enormously. Over the last decade, many gene finding programs (gene-finders) for the human genome have been developed, but none of them is entirely reliable [2].

Gene-finders may be categorized into two classes: empirical gene-finders and *ab initio* gene-finders. Empirical gene-finders, which are also called 'sequence similarity-based gene-finders', detect genes by aligning known cDNA and protein sequences onto uncharacterized genomic sequences [a]. The remarkable feature of empirical gene-finders is their high specificity, that is, genes exist with high probability in genomic regions which these gene-finders detect. However, empirical gene-finders can detect only a limited number of genes (low sensitivity) because it is extremely difficult experimentally to collect mRNAs of all genes. On the other hand, *ab initio* gene-finders do not utilize sequence similarity and rely on intrinsic gene measures such as coding potential and splice signals. The remarkable feature of *ab initio* gene-finders is their high sensitivity, that is, these gene-finders are capable of detecting almost all genes. However, *ab initio* gene-finders also predict many false-positive genes (low specificity) because known gene measures are insufficient to distinguish true positives from false positives.

Several attempts to complete the human gene catalogue have been launched since the production of the draft sequences. Among them, Ensembl (`http://www.ensembl.org/`) is the most popular project. Ensembl adopts a conservative gene annotation protocol which mainly uses empirical gene-finders. Therefore, the gene catalogue compiled by Ensembl contains only a small number of false-positive genes but misses a large number of unknown genes.

Ab initio gene-finders are essential for the completion of the human gene catalogue. Hence, an improvement in their specificity has become an important problem. To address this problem, hybrid gene-finders which combine coding potential and splice signals (*ab initio* approach) with similarity to known gene sequences (empirical approach) have been recently developed [3,4]. On average, hybrid gene-finders enable us to improve the specificity of *ab initio* gene-finders. However, their specificity drops to levels comparable with that of *ab initio* gene-finders when remote homologous genes or no homologous genes are available.

Ab initio gene-finders are immensely powerful tools to find genes when there are no known homologues. It is for these cases where specificity should be improved. The sequencing project of human chromosome 21 addressed this problem by using plural *ab initio* gene-finders, namely, only genome regions which were simultaneously detected by more than one of them were chosen as exons [5]. Although this heuristic approach considerably improves the specificity of *ab initio* gene-finders, the following problems have newly arisen. When we assemble such exons into a gene, the frame consistencies between adjoining exons are not always ensured. Moreover, in the case when all exon scores given by plural gene-finders are low, it is questionable as to

---

[a] Gene-finders based on genome-genome comparisons may also be categorized into this class.

Table 1. Patterns of exons predicted by two gene-finders, where X and Y are the 5' and the 3' ends of actual exons, respectively. For example, 'Case 1' is the case where gene-finders 1 and 2 predict the same genomic region as an exon and this region corresponds exactly to an actual exon.

Case	Exons predicted by Gene-finder 1 ($x_1 < y_1$)		Exons predicted by Gene-finder 2 ($x_2 < y_2$)	
	5' end (x_1)	3'end (y_1)	5' end (x_2)	3' end (y_2)
1	$x_1 = X$	$y_1 = Y$	$x_2 = X$	$y_2 = Y$
2	$x_1 = X$	$y_1 = Y$	$x_2 = X$	$y_2 \neq Y$
3	$x_1 = X$	$y_1 = Y$	$x_2 \neq X$	$y_2 = Y$
4	$x_1 = X$	$y_1 \neq Y$	$x_2 = X$	$y_2 = Y$
5	$x_1 \neq X$	$y_1 = Y$	$x_2 = X$	$y_2 = Y$
[1]6	$x_1 = X$	$X < y_1 < Y$	$X < x_2 < Y$	$y_2 = Y$
[2]7	$X < x_1 < Y$	$y_1 = Y$	$x_2 = X$	$X < y_2 < Y$
8	$x_1 = X$	$y_1 > Y$	$x_2 < X$	$y_2 = Y$
9	$x_1 < X$	$y_1 = Y$	$x_2 = X$	$y_2 > Y$
10	$x_1 = X$	$X < y_1 < Y$	$x_2 < X$	$y_2 = Y$
	$x_1 = X$	$y_1 > Y$	$X < x_2 < Y$	$y_2 = Y$
11	$x_1 < X$	$y_1 = Y$	$x_2 = X$	$X < y_2 < Y$
	$X < x_1 < Y$	$y_1 = Y$	$x_2 = X$	$y_2 > Y$
12	$x_1 = X$	$y_1 = Y$	$x_2 \neq X$	$y_2 \neq Y$
13	$x_1 \neq X$	$y_1 \neq Y$	$x_2 = X$	$y_2 = Y$

[1] where $y_1 > x_2$.   [2] where $x_1 < y_2$.

whether we should choose it or not. Independently of the above sequencing project, Murakami & Takagi have reported that different *ab initio* gene-finders will often correctly predict different exons, suggesting that they could complement one another, yielding better predictions [6]. However, they have not proposed an algorithm which address the above problems.

We present here a general purpose algorithm which finds genes by combining plural existing gene-finders. The algorithm addresses the two problems stated above by applying the frameworks of hidden Markov model (HMM) [7] and Bayesian procedure [8], namely, an HMM is used to ensure the frame consistency between exons within a gene, and Bayesian procedure is used to take into account the exon scores given by the gene-finders. A remarkable feature of the algorithm is that it can combine most gene-finders in a systematic manner. The algorithm has been implemented into a novel gene-finder named DIGIT (Digit Integrates Gene Identification Tools). Currently, DIGIT has been designed so as to combine plural *ab initio* gene-finders. Our extensive testing has clearly shown that DIGIT can accurately identify genes with a low rate of false-positives. In this paper, we report the prediction accuracy of DIGIT as well as presenting the detailed algorithm behind DIGIT.

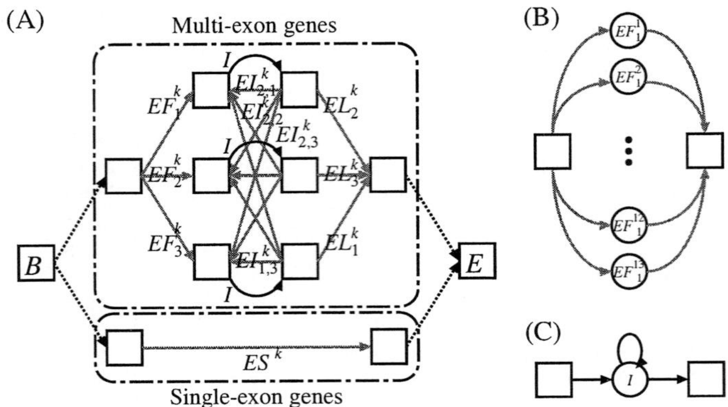

Figure 1. (A) the basic architecture of the HMM which DIGIT employs. EF_i^k indicates a first coding exon which ends at the i-th position of a codon. $EI_{i,j}^k$ indicates an internal coding exon which starts at at the i-th position of a codon and ends at the j-th position of a codon. EL_i^k indicates a last coding exon which starts the i-th position of a codon. ES^k and I indicate single exon genes and introns, respectively. Squares are null states. B and E are the begin and the end states, respectively. This topology includes frame constraints within genes. For example, a downstream exon which adjoins with the first exon ending at the first base of codon (EF_1^k) through an intron always starts at the second base of codon ($EI_{2,j}^k$ or EL_2^k). (B) a detailed architecture corresponding to EF_1^k in the left figure, where k indicates a pattern of exons predicted by gene-finders. There are thirteen possible patterns in the case of combining two gene-finders. Each of them are listed in Table 1. Note that these states are able to emit patterns of the predicted exons. (C) a detailed architecture corresponding to I in the left figure. This state is able to emit four bases i.e. A, C, G and T.

2　Methods

We describe below the computer algorithm which is employed by DIGIT. Currently, DIGIT combines two kinds of gene-finders. However, we can easily extend the algorithm so as to combine more than two kinds of gene-finders. In such case, although the number of model parameters which we should estimate increases exponentially, we can apply the same technique described in 'Parameter estimation' in order to reduce the number.

2.1　HMM architecture

Since DIGIT employs an HMM whose architecture includes frame constraints in gene structure (Figure 1 A), the parsing of genome sequences by DIGIT

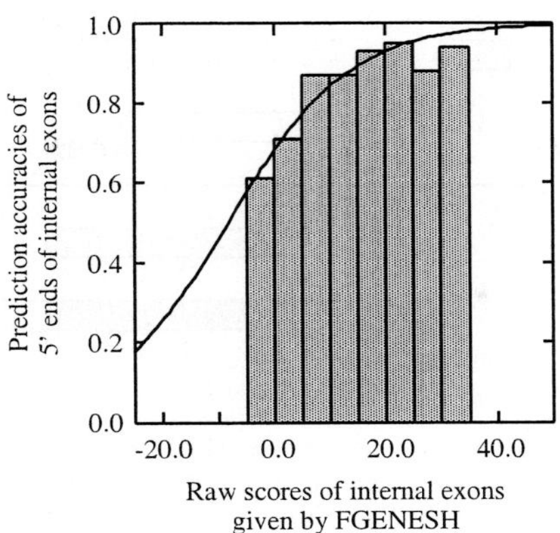

Figure 2. The logistic function which transforms raw exon scores given by gene-finders into probabilities. The solid curve is a logistic function which enables us to transform raw exon scores given by FGENESH into the prediction accuracies of 5' ends of internal coding exons. This function is adjusted to the correlation, shown here by the bar graph, between the exon scores and the prediction accuracies observed in the training data.

necessarily results in finding of frame consistent genes. This feature seems to be very similar to that of Genie [9]; however, DIGIT employs a more complicated architecture. The state transitions corresponding to each exon type in Figure 1 A are divided into thirteen sets of state transitions and states in fact (Figure 1 B), each of which corresponds to a pattern of exons predicted by the gene-finders. In other words, these states emit patterns of the predicted exons. According to the positional relationship between both ends of the predicted exons and the actual exons, the patterns can be classified into thirteen cases (Table 1). Note that the number of cases exponentially increases as the number of gene-finders increases. That is, the number of model parameters which we should estimate increases exponentially in such cases. As for introns, states which emit four bases and possess in/out- and self-transitions are prepared (Figure 1 C).

In addition, several state transitions, i.e. from the begin state to $EI_{i,j}^k$ and EL_i^k, and from those to the end state, are defined in order to find 'partial' genes. For example, transitions from the begin state to $EI_{i,j}^k$ enables us to find partial genes starting in internal coding exons. These transitions are not shown in Figure 1 for reason of simplicity.

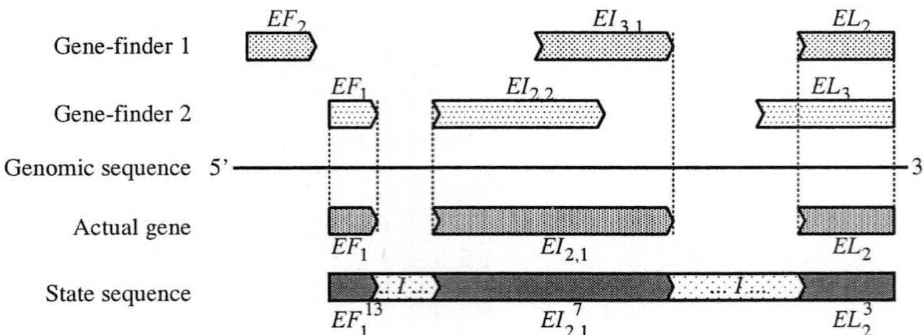

Figure 3. Identification of state sequences in the HMM shown in Figure 1. When we apply gene-finders to genomic sequences of the training data, we obtain sets of predicted exons with their types and frames. Since the training data contains information concerning the types and frames of actual exons, we can easily identify state sequences in the HMM by refering to Figure 1 and Table 1. For example, the actual exon EF_1 in this figure corresponds to 'Case 13' in Table 1 because gene-finder 2 exactly predicted this exon but gene-finder 1 missed.

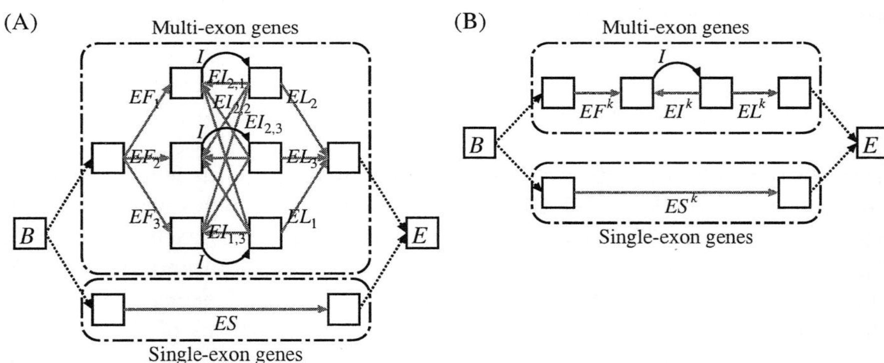

Figure 4. Two types of HMMs introduced in order to reduce the number of parameters which need to be estimated. (A) the HMM where information for the pattern of predicted exons is removed from the HMM shown in Figure 1. (B) the HMM where frame constraints are removed from the HMM shown in Figure 1.

2.2 Parameter estimation

Emission probabilities of exon states must reflect probabilities that a genomic region $[X \ldots Y]$ is an actual exon for each case in Table 1. We assume that such probabilities can be calculated by $\alpha\beta$, where α and β are probabilities that X and Y are actual 5' and 3' ends of exons, respectively. Although α and β should be calculated from exon scores given by the gene-finders, these scores are usually given as log-odds measure. Thus, we have to begin with the transformation of raw exon scores into probabilities. First, we plot graphs representing correlations between the exon scores given by each gene-finder and the prediction accuracies of 5' and 3' ends of exons based on training data (Figure 2). Next, we adjust logistic functions to the plots by using a non-linear least square method, namely, the Marquardt method [10]. These logistic functions enable us to transform raw exon scores given by gene-finders into prediction accuracies of 5' and 3' ends of exons. Finally, we calculate probabilities that a genomic region $[X \ldots Y]$ is an actual exon for each case in Table 1. Bayesian procedure tells us that these probabilities can be calculated in the following manner. For example, the probability for 'Case 1' is given by

$$\frac{\alpha_1 \alpha_2}{\alpha_1 \alpha_2 + (1 - \alpha_1)(1 - \alpha_2)} \times \frac{\beta_1 \beta_2}{\beta_1 \beta_2 + (1 - \beta_1)(1 - \beta_2)}$$

where α_1 and β_1 are the prediction accuracies of 5' and 3' ends of exons by 'Gene-Finder 1', respectively. α_2 and β_2 are the prediction accuracies of 5' and 3' ends of exons by 'Gene-Finder 2', respectively. The first term corresponds to α, and the second term corresponds to β. $\alpha_1 \alpha_2$ indicates the probability that X is the 5' end of an actual exon. $\beta_1 \beta_2$ indicates the probability that Y is the 3' end of an actual exon. $(1 - \alpha_1)(1 - \alpha_2)$ indicates the probability that X is not the 5' end of an actual exon. $(1 - \beta_1)(1 - \beta_2)$ indicates the probability that Y is not the 3' end of an actual exon. We assume that the background probability is given by setting $\alpha_1 = \alpha_2 = \beta_1 = \beta_2 = 0.5$ in the equation above. We can calculate the probabilities for other cases in a similar way and assign them to the emission probabilities of exon states.

Emission and transition probabilities related to introns (Figure 1 C) are set so as not to contribute to the Viterbi score [7]. That is, emission probabilities of all bases are 0.25, and the corresponding background probabilities are also 0.25. All transition probabilities are 1.0.

Transition probabilities to exon states must reflect the frequency for each case in Table 1 as observed in the training data. A straightforward way for the estimation of transition probabilities is to count the number of degrees for each case observed in the training data. However, since the HMM shown in Figure 1 contains a large number of transition parameters to exon states, an overfitting problem might occur. Thus, we adopted the following strategy in order to reduce the number of parameters which need to be estimated. First, we apply the gene-finders to genomic sequences of the training data and identify state sequences of the HMM for the genomic sequences (Figure 3). Next,

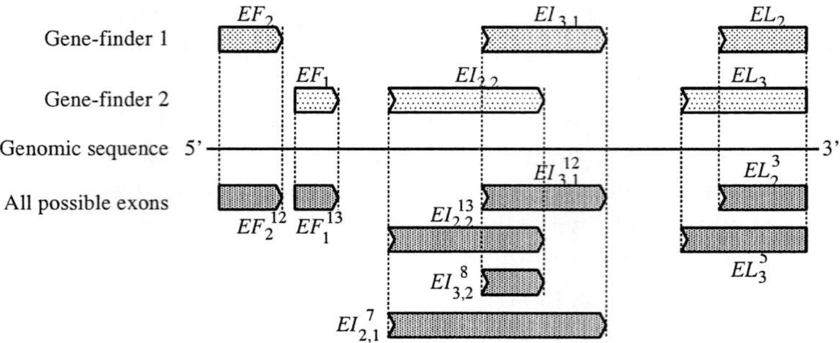

Figure 5. All possible exons produced from the prediction results of gene-finders and the identification of their state labels. This figure shows the case when exons which overlap each other have the same reading frame, i.e. $EI_{3,1}$ and $EI_{2,2}$ have the same frame. If $EI_{3,1}$ and $EI_{2,2}$ have different reading frames, we do not generate possible exons $EI^8_{3,2}$ and $EI^7_{2,1}$ because their frames can not be determined. When two overlapping exons have the same reading frame but their exon types are different, we do not generate exons for their intersection and union regions.

we prepare two types of HMMs (Figure 4); one is an HMM where information for the pattern of predicted exons is removed from the HMM shown in Figure 1, and the other is an HMM where frame constraints are removed from the HMM shown in Figure 1. For each HMM, we count the number of degrees corresponding to the transition probabilities based on the state sequences of the training data, add simple pseudocounts [7] to these numbers, and then calculate the frequencies from these numbers. Finally, we obtain the transition probabilities for the exon states in Figure 1 by assuming that these probabilities can be calculated by multiplying the corresponding parameters of the two HMMs. For example, a transition probability for EF_i^k can be calculated by multiplying the transition probability for EF_i by that of EF^k. This procedure enables us to reduce the number of transition probabilities which need to be estimated. For example, although the HMM contains 39 (3 × 13) transition probabilities for EF_i^k, we only have to estimate 16 (3 + 13) probabilities by applying the above procedure.

2.3 Prediction algorithm

Since gene-finders tend to perform better if provided with a genomic region containing only a single gene and its immediate neighborhood [2], the prediction algorithm should begin with the extraction of such genomic regions. First, we apply an *ab initio* gene-finder FGENESH [11] to an uncharacterized genomic sequence and extract genomic regions as candidate gene regions, each of which includes a single predicted gene and the surrounding regions (up to 1,000 bps on each side). Our experiments preliminary to this study have shown that FGENESH is capable of detecting gene boundaries with high reliability (data not shown). Second, we apply gene-finders to the candidate gene regions. This second analysis often leads to different results from the first analysis [2]. Third, we generate all possible exons and identify their state labels in the HMM shown in Figure 1 (Figure 5). Fourth, we calculate emission probabilities for all possible exons based on the Bayesian procedure explained in 'Parameter estimation'. The state labels are used to calculate these probabilities. Last, we parse the candidate gene regions by using the HMM shown in Figure 1. For the parsing algorithm, semi-global search algorithm [7], which aligns an entire HMM to partial genomic sequences, is used. The local search algorithm enables us to find plural genes within a candidate gene region. The search threshold is set so as to maximize the average value of sensitivity and specificity of gene level in the training data (see below).

3   Data

In order to evaluate the prediction accuracy of DIGIT, we used three various data sets each of which has different characteristics. The first one is the data set, HMR195, compiled by Rogic *et al.* [12]. They attempted to create a data set which did not have many overlaps with the training sets used for the gene-finders. Then, they selected only sequences entered in GenBank after Aug., 1997. Since all sequences in HMR195 are relatively short and each of them contains exactly one complete gene, it can be said that HMR195 is a typical benchmark set for gene-finders. HMR195 consists of 195 human/murine DNA sequences, 43 of them contain single-exon genes, and 152 of them contain multi-exon genes. The average sequence length is 7,096 bps, the proportion of coding sequence is 14%, of intronic sequence is 46% and of intergenic sequence is 40%. Note that these sequences are highly gene dense. The second data set, Gen178, was compiled by Guigò *et al.* [2]. They attempted to overcome the lack of well-annotated large genomic sequences, by preparing a set of well-annotated DNA sequences each of which contains exactly one complete gene and embedding them in simulated intergenic DNA. Gen178 is a more realistic benchmark set for gene-finders, that is, all sequences are fully long, some of them contain several genes including partial ones, and some of them do not contain any genes. Gen178 consists of 43 semiartificial genomic sequences and contains 178 human genes, 40 are single-exon genes, and 138 are multi-exon

Table 2. Exon- and gene-level prediction accuracies of DIGIT and three *ab initio* gene-finders on three data sets. Statistics on annotation are also summarized, i.e. numbers of actual exons and actual genes in each data set. Exon level sensitivity (Sn) is the percent of annotated exons predicted correctly. Exon level specificity (Sp) is the percent of predicted exons which are correct. Gene level Sn is the percent of annotated genes predicted correctly. Gene level Sp is the percent of predicted genes which are correct.

Data	Program	Exon level			Gene level		
		#	Sn	Sp	#	Sn	Sp
HMR195	Annotated	948			195		
	DIGIT	899	0.795	0.838	188	0.507	0.526
	FGENESH	1006	0.819	0.772	222	0.476	0.418
	GENSCAN	1011	0.773	0.725	221	0.364	0.321
	HMMgene	1181	0.754	0.605	299	0.446	0.290
Gen178	Annotated	900			178		
	DIGIT	911	0.786	0.777	166	0.449	0.481
	FGENESH	1055	0.771	0.657	187	0.376	0.358
	GENSCAN	1332	0.665	0.449	222	0.185	0.148
	HMMgene	1711	0.696	0.366	362	0.258	0.127
Chr22	Annotated	3660			522		
	DIGIT	4513	0.695	0.563	654	0.123	0.098
	FGENESH	5734	0.708	0.452	866	0.115	0.069
	GENSCAN	6588	0.707	0.393	803	0.067	0.044
	HMMgene	6840	0.629	0.336	1508	0.090	0.031

genes. The average sequence length is 177,160 bps, the proportion of coding sequence is 2.3%, of intronic sequence is 6.3% and of intergenic sequence is 91.4%. Note that these statistics are highly similar to ones observed in real genomic sequences. The third data set is the finished human chromosome 22 (Chr 22) sequences (May 19, 2000, version) [13] with the annotation provided by the Sanger Institute (March 6, 2001, version). It is one of the few well-annotated large genomic sequence sets for human. Chr22 consists of 12 contiguous segments covering 33.4 million bps. It consists of 504 multi-exon genes and 18 single-exon genes. The proportion of coding sequence is 1.7%, of intronic sequence is 32.3% and of intergenic sequence is 66.0%. Note that some uncharacterized genes may still remain in these sequences.

4   Results

In this study, we have constructed DIGIT so as to combine plural *ab initio* gene-finders. More explicitly, DIGIT has been designed so as to combine the first coding exons predicted by FGENESH and HMMgene [14], the internal and last coding exons predicted by FGENESH and GENSCAN [15], and single exon genes predicted by FGENESH and HMMgene. Among *ab initio* gene-finders, these programs are highly reliable and widely used. The selection and the combination of gene-finders was determined on the basis of our experiments preliminary to this study. We have evaluated the prediction accuracy of nearly twenty *ab initio* gene-fingers by using various benchmark data. Our experiments showed that (1) FGENESH had the best prediction accuracy on average, (2) the second best was GENSCAN, and (3) HMMgene was good at predicting coding regions including start codons compared with other gene-finders.

We performed three tests to evaluate the prediction accuracy of DIGIT. The first test was a 10-fold cross validation test using the HMR195 data set. The second and third tests were open tests using the Gen178 and Chr22 data sets, respectively. In the open tests, the model parameters of DIGIT were optimized by using the HMR195 data set. In all tests, RepeatMasker was used to mask the interspersed repeats (replace interspersed repeats with Ns) in the sequence data in order to reduce the number of false positive exons detected by *ab initio* gene-finders. All the *ab initio* gene-finders including DIGIT were applied to the masked sequences.

The prediction accuracy of DIGIT with three *ab initio* gene-finders on the three data sets is summarized in Table 2. For all data sets, although the exon-level sensitivity of DIGIT was roughly comparable to that of the other gene-finders, the exon-level specificity of DIGIT was remarkably higher than that of the other gene-finders. Remarkably, DIGIT outperformed the other gene-finders for sensitivity and specificity of gene level in all data sets. Note that the numbers of exons and genes detected by DIGIT decreased drastically in comparison with the other gene-finders. However, these numbers are closer to the actual numbers of the annotated exons and genes.

5   Discussion

Table 2 clearly shows that DIGIT significantly improves exon-level specificity in comparison with the other *ab initio* gene-finders. This indicates that DIGIT successfully discards many false positive exons predicted by the gene-finders. In such cases, although actual exons are also frequently discarded, DIGIT prevents the sensitivity from dropping by generating all possible exons and applying Bayesian procedure to the inference of exon scores. That is, in addition to candidate exons detected by the other gene-finders, DIGIT produces intersection and union regions of overlapping exons as novel candidate exons, and then, Bayesian procedure tells us the likelihood of all possible exons on the basis of positional relationships between these exons and their scores given

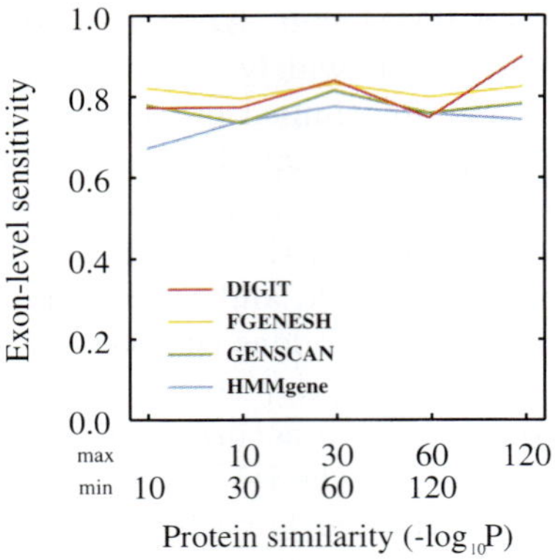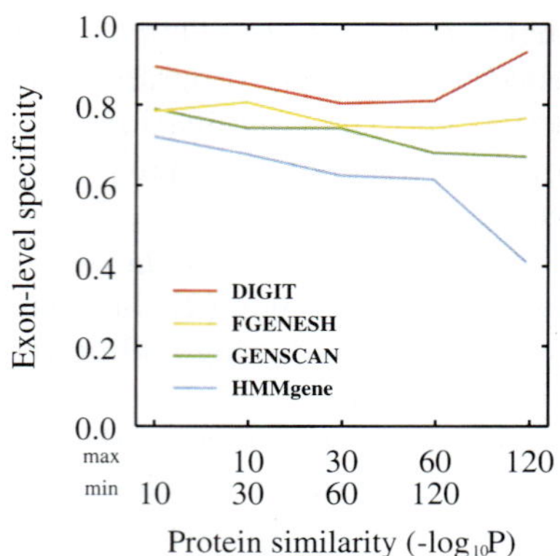

Figure 6. Exon-level accuracies of DIGIT and three *ab initio* gene-finders on the HMR195 data set as a function of protein similarity. Left, exon-level sensitivity. Right, exon-level specificity. These measures were calculated for subsets of the HMR195 data set and grouped according to the level of BLASTX similarity between HMR195 entries and the Swiss-Prot database http://www.expasy.ch/sprot/sprot-top.html. The definitions of the subsets and number of genes per subset were as follows: $P > 10^{-10}$ (29); $10^{-10} > P > 10^{-30}$ (48); $10^{-30} > P > 10^{-60}$ (45); $10^{-60} > P > 10^{-120}$ (36); and $10^{-120} > P$ (37).

by the other gene-finders. This significant improvement of exon-level specificity naturally leads DIGIT to remarkable improvements in sensitivity and specificity at the gene level as compared with the best gene-level accuracies achieved by any single *ab initio* gene-finder.

One remarkable feature of the algorithm which DIGIT employs is that it can combine most gene-finders in a systematic manner. Although we introduced here the algorithm as it combines two gene-finders, we can easily extend it so as to combine more than two gene-finders (see 'Methods'). Moreover, the algorithm can combine any gene-finder which gives predicted exons their types, reading frames and scores. Therefore, it enables us to combine not only *ab initio* gene-finders but also empirical gene-finders. When we combined an *ab initio* gene-finder and an empirical gene-finder by using the algorithm, the exon-level sensitivity was significantly improved instead of the exon-level specificity (data not shown). This is due to differences in genes which *ab initio* and empirical gene-finders can detect. On the other hand, the main limitation of the algorithm is that it cannot predict exons whose boundaries and regions are not detected by gene-finders. We can however reduce this limitation by extending DIGIT so as to combine more than two gene-finders.

We strongly believe that DIGIT is an essential program for exhaustive gene finding in the human genome. Figure 6 shows the exon-level accuracy of DIGIT on the HMR195 data set as a function of protein similarity. The prediction accuracy of DIGIT remains high (the sensitivity is comparable to that of other gene-finders, and the specificity is higher than that of other gene-finders) even if only remote homologous genes are available ($P > 10^{-10}$). For $P > 10^{-10}$, it is known that the prediction accuracy of empirical gene-finders drops remarkably [2], and that the prediction accuracy of hybrid gene-finders drops to levels comparable with that of *ab initio* gene-finders [3]. This strongly suggests that DIGIT is needed to help complete the human gene catalogue, since DIGIT can find genes with high accuracy which empirical and hybrid gene-finders almost never find. DIGIT is made available upon request to the authors.

Acknowledgments

We thank Kiyoshi Asai and Kyosuke Nagasaka for their helpful discussions and valuable advice; Todd Taylor for critically reading the manuscript. This work was partly supported by a Grant-in-Aid for Scientific Research on Priority Areas (C) *"Genome Information!Science"* from The Ministry of Education, Culture, Sports, Science and Technology of Japan.

References

1. International Human Genome Sequencing Consortium, *Nature* **409**, 860 (2001).
2. R. Guigò *et al.*, *Genome Res.* **10**, 1631 (2000).
3. R.-F. Yeh *et al.*, *Genome Res.* **11**, 803, (2001).
4. I. Korf *et al.*, *Bioinformatics* **17**, S140, (2001).
5. The Chromosome 21 Mapping and Sequencing Consortium, *Nature* **405**, 311, (2000).
6. K. Murakami and T. Takagi. *Bioinformatics* **14**, 665, (1998).
7. R. Durbin *et al.*, *Biological sequence analysis: Probabilistic models of proteins and nucleic acids*, (Cambridge University Press, Cambridge, 1998).
8. A. Gelman *et al.*, *Bayesian Data Analysis*, (Champman & Hall/CRC, Boca Raton, FL, 1995).
9. D. Kulp *et al.*, *Proc. of the ISMB* **4**, 134, (1996).
10. W.H. Press *et al.*, *Numerical Recipes in C: The art of scientific computing*, (William H. Press, Cambridge, MA, 1993).
11. A.A. Salamov and V.V. Solovyev, *Genome Res.* **10**10, 516, (2000).
12. S. Rogic *et al.*, *Genome Res.* **11**, 817, (2001).
13. I. Dunham *et al.*, *Nature* **402** 489, (1999).
14. A. Krogh, *Proc. of the ISMB* **5**, 179, (1997).
15. C. Burge and S. Karlin, *J. Mol. Biol.* **268**, 78, (1997).

LINKING BIOLOGICAL LANGUAGE, INFORMATION AND KNOWLEDGE

L. HIRSCHMAN

The MITRE Corporation, 202 Burlington Road,
Bedford, MA 01730, USA

C. FRIEDMAN

Columbia University, 622 West 168 St, VC-5,
New York, NY 10032,USA

R. MCENTIRE

GlaxoSmithKline Pharmaceuticals, 709 Swedeland Road
King of Prussia, PA 19406-0939, USA

C. WU

Georgetown University Medical Center, Box 571414 , 3900 Reservoir Road, NW
Washington, DC 20057-1414, USA

Information access is a major challenge for biologists today. Results are pouring in from microarray experiments, more model organisms are being sequenced and results are being used to expedite drug discovery. There is a growing demand to combine information from different sources and across multiple disciplines, such as clinical medicine, pharmacology, and molecular biology. The volume of literature is increasing exponentially, making it almost impossible for biologists to keep up with current research or to find the particular pieces of information that they need. This makes linkage among existing biological information resources a critical problem.

Information resides in the biological literature. It resides in biological databases that distill information on specialized topics. Such databases include genomic databases (Genbank[a]), model organism databases (FlyBase[b], Mouse[c], Yeast[d]) and protein databases (e.g., SWISS-PROT[e], PIR[f]).   Linking the literature and the databases are nomenclatures and ontologies that provide standardized references to biological entities and topics – e.g., gene names and symbols, protein names, and standard terminology for diseases and symptoms or biological function.

---

[a] http://www.ncbi.nlm.nih.gov/Genbank/
[b] http://flybase.bio.indiana.edu/
[c] http://www.informatics.jax.org/
[d] http://genome-www.stanford.edu/Saccharomyces/
[e] http://kr.expasy.org/sprot/
[f] http://pir.georgetown.edu/

This session focuses on techniques for managing the linkages among the literature, the databases, and existing terminologies (e.g., UMLS[1]) and ontologies such as the Gene Ontology.[2] All of these sources mediate the information through natural language. The literature consists of free text, with accompanying figures and tables. The databases are typically a mixture of structured information (for example, DNA, RNA and protein sequences and structures), pointers to the underlying primary source in the literature, and text annotations describing function, form, location, organism, and other relevant information. These annotations are expressed in a controlled vocabulary, or, for more complex information, in short phrases or even short paragraphs of text. Thus typical biological databases contain significant amounts of content expressed in natural language. The nomenclatures and ontologies also rely on natural language – they contain entries expressed as terms (a word or phrase) with an accompanying unique identifier.

There is an urgent need for tools to maintain these linkages. New knowledge is being added very rapidly. From a linguistic point of view, this means that the language of biology is changing. For example, the Mouse Genome issues a weekly report on "Nomenclature Events." For the week of 8/25/01, there were 166 such events, reporting name additions or name withdrawals for the Mouse Genome database[g]. Tools are needed that can recognize mentions of new entities or new relations in text, to capture these new pieces of information for inclusion in ontologies and entry into databases.

Since genes and proteins are often named by their function, they can have lengthy names, which are then abbreviated. Abbreviations are a major source of ambiguity, especially when searching across multiple subdisciplines. Two papers in this session address the issue of identifying abbreviations and their expansions when they appear in parenthetical expressions. The paper by Liu and Friedman uses collocations to determine the appropriate expansion; they report a precision of 96.3% and an estimated recall of 88.5% for 380,000 parenthetical expressions automatically extracted from MEDLINE abstracts. The paper by Schwartz and Hearst addresses the same problem. Their approach identifies candidate "long forms" in the neighborhood of the short form and defines some simple rules to map from long form to short form. This method was evaluated on a small corpus of 1000 abstracts and 871 abbreviation/expansion pairs, with a reported recall of 83% and precision of 96%.

There are many databases (over 280 according to recent estimates), and also multiple nomenclatures and ontologies. To provide a uniform view across information sources, it would be useful to "translate" between different nomenclatures and/or ontologies. This is explored in the paper by Sarkar *et al.* who

---

[g] (ftp://ftp.informatics.jax.org/pub/informatics/reports/index.html#statistics

describe several methods for mapping from GO terminology to UMLS. They report a precision of 89% and recall of 90% by normalizing text strings and removing stop words from both terminologies before matching.

Other tools can assist by identifying mentions of relevant biological entities in running text. The names of these entities serve as indices into the article, to characterize the subject of the article. Extraction of more detailed information, such as interaction information, depends critically on the ability to identify the underlying participants in an interaction relation. Two papers in this session describe two different approaches to identifying biological entities. The paper by Hanisch *et al.* describes the process of assembling and correcting a large-scale lexical resource to identify gene and protein names in text; their method analyzes complex names into token classes, which can then be selectively matched against the lexicon. The paper by Narayanaswamy *et al.* describe a linguistically motivated approach to capturing names of key biological entities, including genes, proteins, and chemical names. Their approach uses features, such as case, or suffix to define semantic classes for individual words. These are then assembled into complex terms, which can be labeled with the appropriate class.

Finally, the paper by Glenisson *et al.* makes use of the multiple information sources including terminology lists from databases, reference pointers, and journal articles, to generate clusters of related proteins. The paper presents several evaluation methods of the clusters, including both measures of internal cluster cohesion and an external method based on comparison to a gold standard.

These papers provide significant contributions to linking the literature, databases and ontologies. Progress in this area is accelerating, but evaluation remains a major stumbling block. There are still no standard measures or test sets that can be used to compare the efficacy of one approach to another. Each research group is still forced to spend a significant amount of time creating training and test corpora and defining appropriate evaluation measures. The associated special session will focus on the creation of common resources and challenge evaluations to facilitate cross-system comparison and accelerate progress in this important area.

References

1. Lindberg DA, Humphries BL, McCray AT. The Unified Medical Language System. Methods Info Med. 32(4):281-291 (1993)
2. Gene Ontology Consortium. Gene Ontology: tool for the unification of biology. Nature Genetics 25: 25-29 (2000); also see http://www.geneontology.org/ for further information.

EVALUATION OF THE VECTOR SPACE REPRESENTATION IN TEXT-BASED GENE CLUSTERING

P. GLENISSON, P. ANTAL, J. MATHYS, Y. MOREAU, B. DE MOOR

Department of Electrical Engineering, ESAT-SISTA,
Kasteelpark Arenberg 10,
B-3001 Leuven, Belgium

Thanks to its increasing availability, electronic literature can now be a major source of information when developing complex statistical models where data is scarce or contains much noise. This raises the question of how to deeply integrate information from domain literature with experimental data. Evaluating what kind of statistical text representations can integrate literature knowledge in clustering still remains an unsufficiently explored topic. In this work we discuss how the bag-of-words representation can be used successfully to represent genetic annotation and free-text information coming from different databases. We demonstrate the effect of various weighting schemes and information sources in a functional clustering setup. As a quantitative evaluation, we contrast for different parameter settings the functional groupings obtained from text with those obtained from expert assessments and link each of the results to a biological discussion.

1 Introduction

More and more, a successful understanding of complex genetic mechanisms (such as regulation, functional understanding,...) critically depends on the interaction between statistical analysis and various knowledge sources, such as annotations databases, specialized literature, and curated cross-links between them (Baxevanis[1]). Despite these efforts, the current interaction between the experimental (data) analysis and text-based information requires extensive user intervention. Gene expression experiments, which measure large-scale genetic activity under a variety of biological conditions, are excellent examples of environments that rely strongly on this interaction. Indeed as (1) the cost of data collection is high, (2) measurements are often noisy or unreliable, and (3) established relationships in the transcriptome are fragmentary at best, a deeper integration between data and text-based information will benefit the knowledge discovery process.

The present strategies for knowledge-based expression data analysis rely on the premise that statistical data analysis and biological knowledge can complement each other by *linking* two independently constructed sources that contain conceptually related records (Masys[2] and Vidal[3]).

In yeast for example, interpreting cluster patterns involves the consultation of curated functional databases such as the Saccharomyces Genome

Database[a] (SGD), which offers concise functional annotations and a variety of cross-references to other repositories. For more elaborate information, researchers can resort to MEDLINE, an online bibliographic source of citations and abstracts in biomedical research dating from 1966 till present. While the use of a controlled and curated index, like MeSH[b], is already common in automatically associating gene functions (see for example Jenssen[4], Masys[5], Kankar[6]), we tested additionally the use of free-text as a potentially more informative, and in the future possibly more dominant, information source (see also Stapley[7], Stephens[8], Renner[9], Iliopoulos[10], Raychaudhuri[11]).

In this work, we explore how representations borrowed from the field of information retrieval can be adopted for clustering genes based on their associated literature. We encode text-based information from various sources in a typical bag-of-words representation following the vector space model, a work horse in information retrieval research. We investigate the effect of *pooling* and *expanding* these sources, together with the question of which type of representation is more appropriate. To evaluate the biological usefulness of literature clustering, we formulate a clustering problem with gene sets from *Saccharomyces cerevisiae* for which the functional associations are well-established and biologically distinct. The reason not to start immediately from expression-based gene clusters is that these data-based clusters are often biologically complex and cannot provide a gold standard to interpret and quantify the correspondence between various data mining methods. Additionally, we seek to identify some inherent biases of the vector-space model by testing and quantifying its performance on a fairly simple biological problem. To compare different versions of the representation with respect to clustering performance, we use both external and internal scores for cluster validation (see Section 2). The aim of these evaluations is to establish a powerful statistical text representation as a foundation for knowledge-based gene expression clustering.

2   Methods

2.1   Compilation of Information Sources

We collect and compile (as of September 2001) several sources for textual annotations of the genes. Firstly we retrieve the gene descriptions from the Saccharomyces Genome Database (SGD)[c]. Secondly, we use SWISS-PROT (SP)[d], a curated protein sequence database. We pool the SGD and SP information

---

[a]http://genome-www.stanford.edu/Saccharomyces/
[b]http://www.nlm.nih.gov/mesh/meshhome.html
[c]http://genome-www.stanford.edu/Saccharomyces/
[d]http://www.expasy.org/sprot/

into a local database we denote by YeastCard (YC). It serves as an extended textual resource for yeast genes. Finally, as a source for more detailed information, we use a collection of 493,923 yeast-related MEDLINE abstracts dated between January 1982 and November 2000. They were selected by retaining those abstracts coming from a list of 59 journals that was composed according to both impact factor[e] and relevance. The aim of this trimming is to retain a more domain-specific subset of abstracts, which is still diverse enough to hold essential genetic information. We evaluate how these sources influence text-based gene clustering and, more specifically, we investigate how the expansion of the SGD and YeastCard annotations with MEDLINE abstract information (see Section 2.3) affects clustering performance.

2.2 Text Representation

The representation called the vector space model encodes a document in a k-dimensional space where each component v_{ij} represents the weight of term t_j in document d_i. The grammatical structure of the text is neglected and therefore it is also referred to as a bag-of-words representation. As a basic index for each document in the collection, we construct a vocabulary consisting of 26,420 (possibly multi-word) terms extracted from the Gene Ontology[f] *Term* field. The Porter stemmer is used to canonize the words. Based on the *Term* field in GO and *Synonym* field in SWISS-PROT, we process candidate phrases and replace known synonyms. In this work we used the following common used indexing schemes (Baeza-Yates[12] and Korfhage[13]):

- $v_{ij}^{\text{bool}} = 1$ if $t_j \in d_i$, 0 otherwise

- $v_{ij}^{\text{freq}} = \frac{f_{ij}}{\max_{\forall j}(f_{ij})}$, where f_{ij} is the number of occurrences of t_j in d_i

- $v_{ij}^{\text{tf.idf}} = f_{ij} \log(\frac{N}{n_i})$, where N is the total number of documents and n_i is the number of documents containing term i in the collection

Additionally, we define another type of index called the reference representation (see Shatkay[14]). When a document contains references to other documents in the same or another repository, we can encode this as follows:

- $v_{ij}^{\text{ref}} = 1$ if annotation i contains a reference to document j, 0 otherwise.

---

[e] http://jcrweb.com/

[f] http://www.geneontology.org

2.3   Relevance and Similarity

We express similarity between pairs of documents d_i and d_j, or between a text document d_i and a query document d_j, by the cosine of the angle between the corresponding normalized vector representations:

$$\text{sim}(d_i, d_j) = \cos(d_i, d_j).$$

The underlying hypothesis states that high similarity equals strong relevance. Further, the method termed *pseudo-relevance feedback* is geared towards *expanding* a query document with the n most similar documents in a collection and aims at refining the search or clustering process by a recalculation of the term weights (Yates[12]). We denote the annotations A expanded with n documents from collection C by $A\text{-}C_n$. A related application of pseudo-relevance feedback in combination with the reference representation can be found in Shatkay[14].

2.4   Cluster Algorithm

As divisive clustering algorithm we used the K-medoids algorithm (Rousseeuw[15]), which minimizes the objective function

$$\sum_{k=1}^{K} \sum_{j \in C_k} d(x_j, m_k)$$

over multiple partitionings $C = \{C_1, ..., C_K\}$ with $\{m_1, ..., m_K\}$ the corresponding representative points (called medoids) of each cluster. The parameter K denotes the number of clusters and is fixed in advance. One advantage of this algorithm over centroid-based methods, such as K-means, is that each medoid constitutes a robust representative data point for each cluster.

2.5   Cluster Quality

To measure the performance and quality of the clustering we define three scores: the silhouette coefficient, the performance of the clustering as a classifier, and the Rand index. The first two are termed *internal* scores since they rely on statistical properties of the clustered data, the last one is called *external* because it involves a comparison with a known, external labeling.

Silhouette Coefficient

As a first internal score for cluster quality we use the silhouette coefficient per cluster $S_k = \frac{1}{n_k} \sum_{i=1}^{n_k} s_{ik}$ and the overall silhouette coefficient $S = \frac{1}{n} \sum_k \sum_{i=1}^{n_k} s_{ik}$

with n_k is the size of cluster k, n the number of objects, and

$$s_{ik} = \frac{b(i) - a(i)}{\max(a(i), b(i))},$$

where $a(i)$ is the average dissimilarity of member i to all other members of its cluster and $b(i)$ the dissimilarity of member i to the nearest member of the nearest cluster. It is a metric-independent measure designed to describe the ratio between cluster coherence and separation and to assist in choosing which clustering is preferable according to the data (Rousseeuw[15]).

k-NN Learnability

For the second measure of internal cluster quality, we look upon the problem as being semi-supervised. Using the clustering result as a labeling for all the points, we assess the performance of a given classifier on a class (or cluster) in a cross-validated leave-one-out setup. Following Pavlidis[16], we use a k-NN classifier jointly with the $(1 - \cos(\cdot, \cdot))$ distance measure to compute a misclassification score for each class. The statistical significance of this score m, is expressed by a p-value derived from a binomial $\mathcal{B}(m, p_{\text{misclass}})$ with p_{misclass} the prior chance of misclassification, which can be computed analytically in case of a k-NN classifier (details can be found in Pavlidis[16]).

Rand Index

As an external measure for cluster validity we use the adjusted Rand index[17]. Given a set of n points, an external partition $P = \{P_1, ..., P_k\}$, and a clustering $C = \{C_1, ..., C_l\}$, define a as the number of pairs of points that co-occur in a group in the partitioning P as well as in the clustering C, d the number of pairs of points that are in different groups in P as well as in C, and b and c as the number of pairs of points that co-occur in a group in P, but not in C or vice-versa. The Rand index is then defined by

$$R = \frac{a + d}{a + b + c + d}.$$

The correction for random partitioning is $R_{\text{adj}} = \frac{R - E(R)}{\max(R) - E(R)}$, where a hypergeometric baseline distribution is used to compute the expected values. In a comparative study[17], the adjusted Rand index is recommended as the external measure of choice.

3   Results

3.1   Construction of Test Set

We construct a set of genes for which the functional associations are well-established. From the MIPS catalogue[1], we select three biologically distinct functional groups consisting of 116 genes in total. For all genes we select their corresponding SGD and YC annotations (see Section 2.1) and proceed with the 105 genes that have entries in both databases[g]. The first group holds 63 genes that encode lysosomal proteins. The second group consists of 30 genes involved in translational control and the third contains 23 genes related to amino acid transport.

3.2   Cluster Performance

Following the strategies outlined in Section 2, all gene annotations are represented by various indices and subsequently *expanded* with the 20 best matching MEDLINE abstracts. More specifically, we perform the expansion by re-indexing the enriched annotations, again following various indexing schemes. Table 1 summarizes the impact of these settings on cluster performance, expressed by means of the Rand index R_{adj}. Firstly we discuss the effect of information source, afterwards follows the results on the indexing schemes.

Table 1: R_{adj} scores for clustering the three groups using various representations. Note that some results are duplicated along the blocks to facilitate discussion.

	Representation	Weight	R_{adj}	$\sharp(t_i)$
Source	SP Keywords	*bool*	0.1767	3
	SGD	*tf*	0.4050	
	YeastCard	*tf · idf*	0.4617	
Index	SGD	*bool*	0.3386	8
	SGD	*tf*	0.4050	
	YeastCard	*bool*	0.3323	
	YeastCard	*tf*	0.4028	
	YeastCard	*tf · idf*	0.4617	26
	$YC\text{-}ML_{20}$	*bool*	0.3726	
	$YC\text{-}ML_{20}$	*tf*	0.2953	
	$YC\text{-}ML_{20}$	*tf · idf*	0.7344	396
	$YC\text{-}ML_{20}$	*ref*	0.2354	20
Expansion	$SGD\text{-}ML_{20}$	*tf · idf*	0.5920	
	$YC\text{-}ML_{20}$	*tf · idf*	0.7344	

Effect of Indexing Scheme   In the second block of Table 1 we write the performance of the boolean (*bool*), frequency (*tf*) and *tf · idf* index on typical

---

[g] ftp.esat.kuleuven.ac.be/sista/glenisson/reports/webSuppl_TR02_121/yeastcardTable.htm

free-text entries in annotation databases, and on a set of our top 20 retrieved MEDLINE abstracts. For very brief keyword-based descriptions (less than 8 words) the boolean representation is found to be the best one. If all fields from SGD are used, tf (0.4050) improves on *bool* (0.338). For the YC database, typically containing 30 to 50 terms per entry, $tf \cdot idf$ (0.4617) outperforms *bool* (0.3323) and tf (0.4028) slightly.

In the expansion step, we collect and re-index the 20 best matching MEDLINE abstracts for each gene. This operation provides a profile for each gene with the number of terms ranging typically between 200 and 400. Among the indexing options for this set of abstracts, $tf \cdot idf$ (0.7344) scores considerably higher than *bool* (0.3726) and tf (0.2953), even after stopword removal.

Basing ourselves on the same 20 top-scoring abstracts we also evaluate the performance of the reference representation ref, which characterizes a gene in document space instead of term space. It has an R_{adj} value of 0.2354, indicating that it is a less descriptive representation. This can be explained by the fact that ref is probably more dependent on the retrieval of highly relevant abstracts (see also Shatkay[14]).

Effect of Information Source In the first block of Table 1, we see that for the gene groups considered, the *keywords* field in SWISS-PROT does not provide sufficient information for an acceptable clustering result (0.1767). For instance, the SWISS-PROT keyword list only provides an average of 2 to 3 meaningful keywords for 86 out of 105 genes. The remaining genes are described with no or irrelevant keywords such as *hypothetical protein*, which will not allow for correct classification. Using the GO entries and especially the description line of SGD improves the results and raises the Rand score to 0.4050. Only two genes have no meaningful representation, YKL002w and YLR309c, whereas the others are now described by 7 to 8 biologically relevant terms. When resorting to our pooled information source YC (see Section 2.1), we obtain a score of 0.4617, misclassifying 21 out of 105 genes. Although the clustering itself is not dramatically influenced by the expansion with YC, for most of the genes, the textual representation is greatly improved (e.g., the weights of specific terms are increased and additional specific terms are incorporated). For instance, Table 2 shows the text profiles of the medoids of the vacuolar cluster for various representations.

In the clustering based on SWISS-PROT keywords, the vacuolar cluster itself is not found. Instead, the algorithm identifies a cluster of ATP-binding proteins that contains the vacuolar ATPases but also a number of ATP-binding proteins involved in translational control. The SGD representation ensures the grouping of vacuolar proteins solely based on one relevant term, *vacuolar*. Both

Table 2: Text profiles of the medoids for group1 (only 25 top-scoring terms are shown).

SP keywords	SGD	YC	YC-ML$_{20}$
ATP (0.45)	vacuolar (0.38)	vacuolar (0.54)	vacuolar (0.54)
ATP bind (0.45)	vps41 (0.38)	ATPas (0.4)	vacuol (0.45)
bind (0.45)		vacuolar membran (0.32)	snare (0.36)
		vma13 (0.21)	vacuolar membran (0.18)
		subunit (0.2)	T snare (0.17)
		associ (0.17)	syntaxin (0.16)
		organel (0.16)	vacuolar assembli (0.12)
		vacuolar acidif (0.16)	Golgi (0.1)
		acidif (0.15)	carboxypeptidas (0.1)
		sector (0.14)	vam3 (0.09)
		hydrogen (0.13)	pep12 (0.09)
		membran (0.1)	V snare (0.08)

Table 3: Text profiles of gene YPL029w based on the SGD and YeastCard representations.

SGD	YC
ATP (0.27)	helicas (0.57)
ATP depend helicas (0.27)	mitochondri (0.36)
depend (0.27)	ATP depend helicas (0.29)
helicas (0.53)	suv3 (0.29)
RNA (0.27)	ATP (0.23)
RNA helicas (0.27)	depend (0.2)
suv3 (0.27)	RNA (0.2)
	RNA helicas (0.19)
	post (0.19)
	ATP depend RNA helicas (0.18)
	elem (0.16)
	translat (0.13)
	control (0.13)
	interact (0.11)
	transcript (0.09)

Table 4: Text profiles of gene YLL048c and YPL149w based on the YeastCard representation and the corresponding expansion to MEDLINE (only the top-scoring terms are shown).

| YLL048c | | YPL149w | |
YC	YC-ML$_{20}$	YC	YC-ML$_{20}$
bile (0.68)	bile (0.92)	autophagi (0.89)	autophagi (0.87)
transport (0.46)	bile acid transport (0.28)	apg5 (0.43)	apg5 (0.17)
bile acid transport (0.25)	bile acid (0.22)		conjug (0.15)
ybt1 (0.25)	hepatocyt (0.06)		apg1 (0.13)
ATP (0.20)	transport (0.06)		cAMP (0.13)
abc (0.15)	abc (0.05)		starvat (0.11)
ATP bind (0.14)	ATP (0.05)		kinas (0.11)
integr membran (0.14)	ATPas (0.03)		phosphati-dylinositol (0.08)
integr (0.13)	apic (0.03)		vacuol (0.08)
membran (0.11)	vesicl (0.03)		apoptosis (0.08)
acid (0.1)	cotransport (0.03)		hepatocyt (0.07)
similar (0.1)	sister (0.03)		antagonist (0.06)
depend (0.09)	voltag (0.03)		ubiquitin (0.06)
bind (0.07)	glycoprotein (0.02)		apg12 (0.06)
famili (0.03)	triphosph (0.02)		amino-peptidas (0.04)

the YC representation and the MEDLINE expansion of the YC annotation result in a large cluster containing most of the vacuolar proteins. The text profiles of the corresponding medoids confirm the success of the MEDLINE expansion and the feasibility of our approach to identify relevant terms that characterize individual genes or groups of genes. For the other two groups a similar improvement is observed.

Table 3 shows two examples of text profiles of individual genes that were misclassified when the SGD representation was used whereas the YC representation assigned the genes to the correct cluster. For the RNA helicase, YPL029w, terms like *mitochondri* and *translat* are added to the text profile. The clustering of YBR024w in the group of translation-related proteins is based on terms such as *mitochondri, inner,* and *membrane.*

Expansion to MEDLINE improves the text profiles of almost all of the genes and even the clustering of a few genes such as YLL048c, a lysosomal bile transporter, and the genes that encode autophagy-related proteins. In the clustering based on the YC representation, YLL048c was wrongly assigned to the group of amino acid transporters. However, the expansion strongly decreased the weight of the term *transport* and introduced the term *ATPase* in the text profile, resulting in a correct classification of the gene. For the autophagy-related genes, retrieval of the term *vacuol* ensures correct grouping after MEDLINE expansion as shown in Table 4. However, some of the genes are incorrectly clustered no matter what representation or weighting scheme was used. For instance, Group 1 and Group 3 include several proteins that regulate transcription, a process that is closely related to translation and shares many of its keywords. The proteins YLR025w (Group 1), YLR375w (Group 3) and YDL048c (Group 3) are therefore persistently misclassified into Group 2. One gene, YLR309c (Group 1), is consistently assigned to the wrong cluster because it lacks proper annotation. The only terms that characterize YLR309c are vague, aspecific words such as *gene product* and the name of the gene *imh1.* This information is insufficient for successful expansion with MEDLINE. A manual search via the PUBMED engine did not reveal much information on *imh1* (YLR309c) either.

3.3 *Cluster Quality*

Because of the absence of a gold standard or prior knowledge in regular clustering problems, internal measures of quality are used to evaluate a cluster result (Jain[17]). They are based on various statistical properties of the grouped data and provide clues to choose between different parameterizations of a single algorithm (such as the number of clusters) or even between various cluster algorithms. Here we use two measures, the silhouette coefficient and a k Nearest Neighbour (k-NN) learnability index to study the influence of the text representation on a standard clustering procedure. Our concern is that, although high Rand scores may be encouraging, they do not provide any level of confidence in the result: it might be that clusters or groups lie very close to each other or that clusters exhibit a high spread. Therefore, we compute for each clustering their score over the major text representations. From Table 5 we see that the silhouette score does not contain any indications towards the optimal representation (i.e., the one with the highest R_{adj}). The more local 10-NN misclassification rate performs better, indicating that individual cluster structures should be examined more carefully. We expect that groups that are easy (and therefore do not need an elaborate representation to be learned) will end up in clusters having low misclassification rates over all the representations. Harder groups will behave inversely.

In Figure 1 we plot the misclassification rate against the silhouette coefficient to look for possible discrepancies between the two scores in this problem. We show the results for 10-NN. From Figure 1 (left) we estimate the group of translational control as the hardest to learn from text, since it has the highest misclassification rate, even with the MEDLINE expanded representation. Additionally, there exists a discrepancy between the silhouette and learnability score. For the amino acid group there exists great variation in the silhouette value, while the misclassification rate stays below 0.1. This indicates that the shape and constitution of the cluster changes over the representation without changing its relative position with respect to the other clusters.

The quality of the cluster is highly affected by the presence of distant genes: genes that have a poor or biased description (and hence representation) will end up far away from the cluster center (i.e., the medoid). We illustrate this in Figure 1 (right), where we plot the growing of the silhouette score (from right to left on the x-axis) while increasingly dropping members beyond a given distance. Flat regions indicate the absence of members in that distance region and sudden changes in silhouette scores show the detrimental effect of those, more distant, genes on the scores. Since a biologist is not always interested in clustering *all* the genes *per se*, this information can be utilized to prune genes from the clustering process or to check the information given by that gene.

Table 5: Various cluster quality scores for the three major text representations.

Representation	Silhouette	10-NN p-value (miscl. rate)	R_{adj}
SGD	0.220	$4.634^{-4}(0.2286)$	0.4050
YC	0.1576	$1.2^{-3}(0.2095)$	0.4617
$YC\text{-}ML_{20}$	0.2192	$10^{-9}(0.1143)$	0.7344

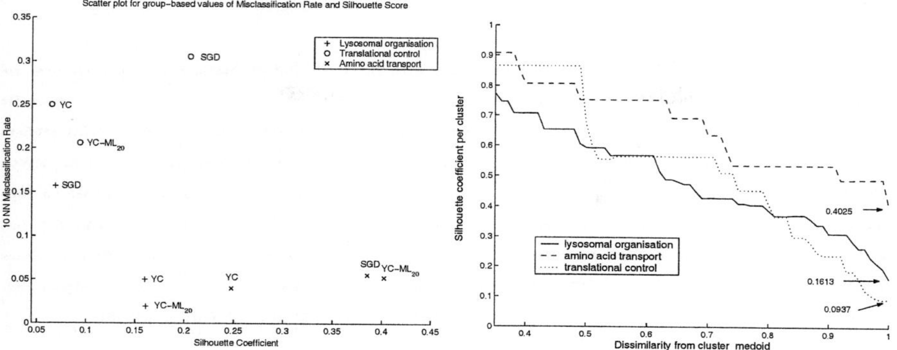

Figure 1: Correspondence between spatial cluster information as captured by the silhouette coefficient and learnability in Nearest Neighbour sense (left) and effect of distant members in each cluster on its silhouette score for (right).

4   Conclusion

Our aim was to investigate the potential of the vector-based representation for functional and text-based gene clustering.We looked into which bag-of-words representation was optimal for what type of information source. We expanded various gene annotations with abstracts that were closest for the cosine measure. Since similarity ranking scores are often hard to threshold and provide a poor quantification for relevance, we retained the top 20 matching entries. This approach considerably improved clustering results because of the inclusion of important terms not present in the annotation databases or because of a relative weight change of already included terms.

Next to a biological evaluation, we computed two complementary internal cluster quality measures to examine some statistical properties of the text representations. The k-NN learnability score gave useful clues on how difficult a class or cluster was to learn. The outcome matched our biological expectations, indicating that our recommended representation is usable in an unsupervised learning task. The silhouette profiles gave more insight into the nature of the clustered annotations and were used to prune or check the information of genes distant from a cluster's medoid.

Finally, the ultimate goal of our approach is to use key elements of the shallow-statistical approach as extra background information in the clustering of expression data.

Acknowledgments

Patrick Glenisson and Peter Antal are research assistants with the KUL. Janick Mathys is a post-doctoral researcher with the KUL. Yves Moreau is a post-doctoral researcher with the FWO Vlaanderen. Dr. Bart De Moor is a full professor at the KUL, Belgium. Research supported by Research Council KUL: GOA-Mefisto 666, IDO (IOTA Oncology, Genetic networks), several PhD/postdoc/fellow grants; Flemish Government: FWO: PhD/postdoc grants, projects G.0115.01 (microarrays/oncology), G.0413.03 (inference in bioi), G.0388.03 (microarrays for clinical use), G.0229.03 (ontologies in bioi), research communities (ICCoS, ANMMM); AWI: Bil. Int. Collaboration Hungary/ Poland; IWT: PhD Grants, STWW-Genprom (gene promotor prediction), GBOU-McKnow (Knowledge management algorithms), GBOU-SQUAD (quorum sensing), GBOU-ANA (biosensors); Belgian Federal Government: DWTC (IUAP IV-02 (1996-2001) and IUAP V-22 (2002-2006); EU: CAGE; ERNSI;

References

1. A.D. Baxevanis. The molecular biology database collection: 2002 update. *Nucleic Acids Research*, 30:1–12, 2002.
2. D.R. Masys. Linking microarray data to the literature. *Nature Genetics*, 28:9–10, 2001.
3. M. Vidal. A biological atlas of functional maps. *Cell*, 104:333–339, 2001.
4. T.K. Jenssen, A. Laegreid, J. Komorowski, and E. Hovig. A literature network of human genes for high-throughput analysis of gene expression. *Nature Genetics*, 28:21–28, 2001.
5. D.R. Masys, J.B. Welsh, J.L. Fink, M. Gribskov, I. Klacansky, and J. Corbeil. Use of keyword hierarchies to interpret gene expression patterns. *Bioinformatics*, 17:319–326, 2001.
6. P. Kankar, S. Adak, A. Sarkar, K. Murari, and G. Sharma. Medmesh summarizer: text mining for gene clusters. In *Proceedings of the Second SIAM International Conference on Data Mining*, 2002.
7. B.J. Stapley and G. Benoit. Biobibliometrics: Information retrieval and visualization from co-occurences of gene names in medline abstracts. In *Proceedings of the Fifth Annual Pacific Symposium on Biocomputing (PSB 2000)*, 2000.
8. M. Stephens, M. Palakal, S. Mukhopadhyay, R. Raje, and J. Mostafa. Detecting gene relations from medline abstracts. In *Proceedings of the Sixth Annual Pacific Symposium on Biocomputing (PSB 2001)*, 2001.
9. A. Renner and A. Aszodi. High-throughput functional annotation of novel gene products using document clustering. In *Proceedings of the Sixth Annual Pacific Symposium on Biocomputing (PSB 2000)*, 2000.
10. I. Iliopoulos, A.J. Enright, and C.A. Ouzounis. Textquest: document clustering of medline abstracts for concept discovery in molecular biology. In *Proceedings of the Sixth Annual Pacific Symposium on Biocomputing (PSB 2001)*, 2001.
11. S. Raychaudhuri, J.T. Chang, P.D. Sutphin, and R.B. Altman. Associating genes with gene ontology codes using a maximum entropy analysis of biomedical literature. *Genome Research*, 12:203–214, 2002.
12. R. Baeza-Yates and B. Ribeiro-Neto. *Modern Information Retrieval*. ACM Press, 1999.
13. R. Korfhage. *Information Storage and Retrieval*. New York: Wiley Computer Pub., 1997.
14. H. Shatkay, S. Edwards, W.J. Wilbur, and M. Boguski. Genes, themes and microarrays: Using information retrieval for large-scale gene analysis. In *Proceedings of the Eighth International Conference on Intelligent Systems for Molecular Biology, Menlo Park, CA, USA*, pages 317–328. AAAI, 2000.
15. L. Kaufman and P. Rousseeuw. *Finding groups in data*. Wiley-Interscience, 1990.
16. P. Pavlidis, D.P. Lewis, and W.S. Noble. Exploring gene expression data with class scores. In *Proceedings of the Seventh Annual Pacific Symposium on Biocomputing (PSB 2002)*, 2002.
17. A. Jain and R. Dubes. *Algorithms for clustering data*. Prentice Hall, 1988.

PLAYING BIOLOGY'S NAME GAME: IDENTIFYING PROTEIN NAMES IN SCIENTIFIC TEXT

DANIEL HANISCH[a], JULIANE FLUCK[a], HEINZ-THEODOR MEVISSEN[a]

Fraunhofer Institute for Algorithms and Scientific Computing (SCAI)
Schloss Birlinghoven, D-53754 Sankt Augustin, Germany
{Daniel.Hanisch, Juliane.Fluck, Heinz-Theodor.Mevissen}@scai.fhg.de

RALF ZIMMER

Institut für Informatik, Ludwig-Maximilians-Universität München
Theresienstraße 39, D-80333 München, Germany
Ralf.Zimmer@bio.informatik.uni-muenchen.de

A growing body of work is devoted to the extraction of protein or gene interaction information from the scientific literature. Yet, the basis for most extraction algorithms, i.e. the specific and sensitive recognition of protein and gene names and their numerous synonyms, has not been adequately addressed. Here we describe the construction of a comprehensive general purpose name dictionary and an accompanying automatic curation procedure based on a simple token model of protein names. We designed an efficient search algorithm to analyze all abstracts in MEDLINE in a reasonable amount of time on standard computers. The parameters of our method are optimized using machine learning techniques. Used in conjunction, these ingredients lead to good search performance. A supplementary web page is available at http://cartan.gmd.de/ProMiner/.

1 Introduction

Biological articles provide a wealth of information on genes and proteins and their interaction under different experimental conditions. To make this amount of data manageable to biological experts and to utilize these data in conjunction with bioinformatics methods (e.g. for contextual analysis of DNA microarray data) it is desirable to condense the free text information into a machine-readable, well-defined form. One example is the generation of biological interaction networks from scientific abstracts [1,2,3]. A requirement for all these approaches is an accurate, sensitive and efficient recognition of the entities under consideration, i.e. proteins and genes, in the free text. In our opinion, however, this building block of higher level information extraction did not receive sufficient attention.

A fundamental problem of gene name search in biological articles is the quite frequent deviation of authors from a recommended gene nomenclature or the absence of such a standard in some cases. Consequently, each gene might

---

[a] All authors contributed equally to this paper.

have several synonymous aliases and functionally unrelated genes might bear the same name [4,5]. For protein names the situation is even more complicated, as synonyms often consist of several words and permutations of these tokens may occur.

Methods for identifying gene and protein names in free text can be divided into methods utilizing a name dictionary and methods relying on other means. Methods which do not use a dictionary [6,7,8] can identify potential proteins in texts which are previously not contained in standard dictionaries and can thus be used to compose such dictionaries semi-automatically. In contrast, they face the problem to unify different aliases of found entities. This makes them hard to use as a building block for higher level analysis. A more straightforward method is to utilize a database of protein and gene names. Krauthammer et al. [9] treat the search problem as an alignment of a protein name against the database of scientific abstracts. Their approach is character-based and utilizes the BLAST algorithm. While the character-oriented approach has the advantage of finding slightly modified forms of words (e.g. plural forms), it faces the problem of how to detect semantically significant mismatches of characters (e.g. modifications in gene names). The success of all dictionary-based approaches obviously depends on the quality of the dictionary. Ono et al. [1] use a manually constructed dictionary which contains only a problem-specific subset of proteins. Jenssen et al. [3] employ a dictionary which only contains gene symbols and short gene names extracted from the HUGO database [10]. Thereby they circumvent the problems associated with longer synonyms, but face a sensitivity penalty.

In this paper we will show that simple text search of gene names leads to poor sensitivity, whereas naïve search of protein synonyms incurs a loss in specificity. Consequently, we build a large curated dictionary of protein and gene names and a corresponding token-based search algorithm to achieve our goal. In the following section we will discuss the underlying model for protein and gene names. Based on this, we describe the automated generation and curation of our synonym dictionary. This dictionary is used in the search algorithm which is presented in section 4. In section 5 we optimize the parameters of our method and validate our findings. The paper closes with a discussion of possible further enhancements of our method.

2　Model for protein and gene names

A crucial characteristic of protein names is that they are often composed of more than one word (or token). The order of these words is only semantically significant up to a certain limit, i.e. permutations of tokens may occur (cf. Ta-

Name	Description	Examples
Modifier	Semantic-modifying tokens	receptor, inhibitor
Non-descriptive	Annotating tokens	fragment, precursor
Specifier	Numbers and Greek letters	1,VI, alpha, gamma
Common	Common English words	and, was, killer
Delimiter	Separator tokens	( ) , . ;
Standard	Standard tokens	TNF, BMP, IL

Table 1: Definition of token classes with differing semantic significance.

ble 2 Examples 6 and 7). Moreover, we face the problem that general-purpose dictionaries of protein names must be automatically composed (e.g. from protein databases) as only the number of human proteins in the SWISSPROT and TREMBL databases is approximately 40.000. However, some tokens included in those databases are only rarely used in free text (cf. Table 2 Example 1, 2). An important observation to overcome these problems when identifying synonyms in free text is that words can be partitioned according to their semantic significance into *token classes*.

2.1 Definition of token classes

To assess the different significance of tokens, we extract all words from the dictionary with frequency of occurrence greater than one hundred. On the basis of these data, we manually define token classes which influence curation of the dictionary and the match procedure in various ways. The class of *non-descriptive tokens* contains words which often occur in databases but are rarely used in free text (cf. Table 2 Example 1, 'precursor') or have no influence on the significance of the match (Example 6, 'type'). In contrast, the class of *modifier tokens* contains words which are crucial for the correct recognition of the underlying entity. Names 3 and 5 of Table 2 clearly describe different proteins. The difference is expressed through the modifier token 'receptor'. Along the same lines, the class of *specifier tokens* which is comprised of Arabic and Roman numbers and Greek letters is usually used to discriminate among the members of a protein family (cf. examples 3 and 4). *Delimiter tokens* are used to gain specificity in the matching procedure. Name boundaries in the free text which are unknown *a-priori* can be detected more easily with the help of delimiters. Usually, the capitalization of protein names is insignificant. However, some gene identifiers are exceptions to this rule. For example, the gene names 'KILLER' or 'WAS' will be detected erroneously in free text when the search is case-insensitive. Tokens of this type can be assigned to the class of *common words* by comparison to a standard English dictionary. During

1.	Interleukin	-	1	beta	precursor
2.	INTERLEUKIN	1	-	beta	PROTEIN
3.	INTERLEUKIN	1		beta	
4.	Interleukin	2		beta	
5.	Interleukin	1	RECEPTOR	BETA	
6.	Collagen	type	XIII	alpha	1
7.	Alpha	1	type	XIII	collagen

Table 2: Examples of protein names tagged with token classes. Names 1-3 refer to the same entity, but differ in spelling and *non-descriptive tokens*. Examples 4 and 5 represent distinct entities differing in *modifier-* and *specifier tokens*. Examples 6 and 7 are synonyms demonstrating the presence of permutations of tokens.

the matching procedure, synonyms composed of only one common word are considered case-sensitive. All tokens not explicitly classified are termed *standard tokens*. This class also includes gene identifiers as they cannot be easily assigned to a separate class. The current definition of token classes is available through the supplementary web page (http://cartan.gmd.de/ProMiner).

3   The protein and gene name dictionary

The quality of the used dictionary is essential for the success of the matching procedure. Indeed, if a perfect general purpose dictionary of names in all occurring spelling variants were available, the matching procedure would be trivial. Unfortunately, it is highly unlikely that such a dictionary will be available in the near future. As a manual definition of a general purpose dictionary is infeasible, we focus on automatic generation of a dictionary and subsequent sensible curation and expansion steps. In this paper, we tested our approach on the basis of human genes and proteins as this field is especially relevant for clinical and pharmaceutical research.

3.1   Automatic generation of the dictionary

We extracted gene symbols, alias names, and full names for all human genes from the HUGO Nomenclature database[10] and created an entry in the dictionary (called an *object*) for each official gene symbol and added the corresponding names available in the OMIM database[11]. Furthermore, we extracted all synonyms of human proteins of the SWISSPROT and TREMBL databases[12] and matched these to HUGO entries.

#	Original syn.	Curated syn.	Status	Reason
1a	IL1	IL1	keep	—
1b		IL 1	add	separation
1c		Interleukin 1	add	acronym expansion
2	Transcription Factor	—	remove	expert curation list
3	Phosphohexokinase	—	ambiguous list	occurs in several objects
4	EPO	—	ambiguous list	occurs in several objects
5	BETA SUBUNIT	—	remove	regular expression match
6	fragment	—	remove	regular expression match

Table 3: Examples of curation of the dictionary. Synonyms may be modified, added or removed during the curation procedure for various reasons (cf. section 3.2).

3.2 Curation of the dictionary

To resolve ambiguities and to remove nonsensical names from the dictionary, a curation procedure consisting of an expansion and a pruning phase was implemented. Table 3 shows several curated synonyms which serve as examples in the following explanations of the curation procedure.

In the expansion phase, further synonyms for existing records were generated. This was achieved by separating alphanumeric tokens into numbers and words (cf. Example 1b), expanding known unambiguous acronyms within synonyms to broaden the scope of our dictionary (cf. Example 1c), and collecting a list of curation items which are maintained by biological experts to add or remove synonyms from the dictionary (cf. Example 2).

In the pruning phase, redundancies, ambiguities and irrelevant synonyms were removed from the dictionary. First, each token in the dictionary was tagged with its corresponding token class generating a string of token class identifiers for each synonym. Each string was then matched against a set of regular expressions representing patterns for unspecific synonyms (e.g. only *non-descriptive tokens*, only *specifier tokens* etc.). If a regular expression matched, the corresponding synonym was pruned (cf. Example 5,6). After an additional step of pruning manually defined superfluous synonyms, only one synonym of several alternatives differing only in capitalization was retained, as the matching procedure is basically case-insensitive (cf. section 4). Finally, a list of ambiguous names found in the curated dictionary was extracted and these names were stored in a separate dictionary with reference to their original records. Generally, ambiguities may arise for several reasons. One problem is the parallel invention of gene and protein names by several biologists (cf. Example 4). Moreover, reorganization of names for protein classes leads to inconsistent names. Besides these inherent name ambiguities, we encounter ambiguous names because databases do not only store name aliases but sometimes also protein class identifiers (cf. Example 3). Obviously, these identifiers apply to several entities described in our dictionary. In effect, the ambiguity list can be used to identify such entries and move them to our manual curation list based on their frequency of occurrence.

Using this curation procedure, we arrived at a name dictionary of reasonably high quality which can be used as the input for our matching procedure described in the next section. This dictionary consists of approximately 38.200 entries with 151.700 synonyms after curation.

4 Efficient detection of names in scientific abstracts

The MEDLINE database [13] contains approximately 11 million abstracts and is rapidly growing. As our ultimate goal is the detection of protein and gene names in all abstracts of the database, the running time of the detection algorithm is an important concern. Consequently, our search procedure should take time linear in the number of tokens of the parsed text. The basic idea is to sweep over the abstract, processing one token at a time and keep a set of candidate solutions and two associated scoring measures for the present position. One scoring measure, the *boundary score* s_β controls the end of the extension of a candidate match and is increased on a token mismatch (cf. Algorithm 1, line 13). If this score rises above a threshold, i.e. if a certain number of mismatches has occurred, the candidate is pruned from the candidate set and checked for reporting. Then, the second score measure, the *acceptance score* s_α, determines whether the candidate is reported as a match. The term s_α is a linear combination of token class specific match- and mismatch terms. A *match term* is defined as the *percentage of matched tokens* of the respective token class. A *mismatch term* counts for each token class the *number of tokens additionally found in the text* and, thus, mismatched in the candidate synonym. With appropriate weighting, the mismatch term allows to disregard false substring matches. To illustrate this, consider the following example:

synonym	Interleukin	1		precursor
candidate match	Interleukin	1	receptor	

As only the *non-descriptive token* "precursor" is unmatched in the candidate, a nearly maximal match score would be computed (if *non-descriptive tokens* receive a small weight). However, the presence of the semantically significant *modifier token* "receptor" leads to a substantial mismatch term for this token class (if weights are set appropriately).

Variable	Description
S	Set of all synonyms to search
T	Set of all tokens
$C = \{c_1, c_2, \ldots, c_n\} \subseteq S$	Current set of candidates
$\tau(c), c \in C$	Unmatched tokens of candidate c
$\sigma(t), t \in T$	Set of synonyms containing token t
$\#\text{token}(c), c \in C$	Number of tokens of candidate c

```
  for each token t_j ∈ T read from the abstract do
2     for each synonym s ∈ σ(t_j) do
          if (s ∉ C) then
4             C = C ∪ s;
      end
6     for each candidate c ∈ C do
          if (t_j ∈ τ(c)) then do
8             Update match terms of s_α(c);
              τ(c) = τ(c)\t_j;
10        end
          else do
12            Update mismatch terms of s_α(c);
              s_β(c) += 1 / #token(c);
14        end
          if (s_β(c) > boundaryThreshold or
16            mismatchedDelimiter found) then do
              C = C\c;
18            if (s_α(c) > acceptanceThreshold) then
                  report c;
20        end
      end
22 end
```

Algorithm 1: Linear algorithm for detection of protein names in free text.

Delimiter tokens play a special role in the algorithm. If the current token t_j is a delimiter but is unmatched in the current candidate (cf. line 16 in Algorithm 1), the candidate match will not be extended further but checked for reporting immediately (cf. line 18). The rationale is that a delimiter signifies a break in the text possibly followed by another protein name which should not be mixed up with previous candidates. To further extend the sensitivity of the search, we maintain a list of synonymous token. Most importantly, Arabic and Roman numbers are treated equivalently as both variants frequently occur in abstracts. In general, the algorithm ignores the order of the synonym tokens. However, permuted matches of some synonyms (e.g. EC numbers) are meaningless. Therefore, we extended the search algorithm to respect the order of tokens when this is necessary. This complicates the operations on τ (lines 7 and 9), but the concept is applicable analogously.

The parameters of our search procedure, namely the match- and mismatch-weights, can be used to control the fuzziness of the search as demonstrated in the above example. A method to optimize these parameters using a set of training instances is discussed in the next section.

5 Parameter optimization and validation

To examine the performance of our name identification procedure, we need a standard-of-truth of annotated abstracts. As creation of a substantial data-set is extremely work-intensive, we based our benchmark set on the TRANSPATH database [14] on regulatory interactions (Version 2.3). We extracted all human proteins with SWISSPROT annotations and all corresponding MEDLINE identifiers and re-examined all abstracts for further occurrences. Additionally, we discarded abstracts if no text was available or a protein was described for the first time and, thus, was not associated with any name, at all. Our resulting benchmark set consists of 611 associations (141 objects in 470 abstracts). The subset of the curated dictionary relevant to this benchmark is available through the supplementary web page (http://cartan.gmd.de/ProMiner).

As the acceptance score s_α is a linear function of the token class specific scoring terms, we can use *robust linear programming (RLP)* [15] to compute a set of sensible weights. This supervised machine learning technique uses a set of positive samples, i.e. correctly identified protein names, and a set of negative examples, i.e. erroneous matches. The RLP procedure computes a separating hyperplane in the vector space of scoring contributions, thus determining an optimized scoring function. We calibrated the match and mismatch weighting parameters for *delimiter-*, *specifier-*, *modifier-* and *standard-tokens*. To this end, we generated a number of training instances for the RLP algorithm by setting weighting parameters to unity and acceptance and mismatch thresholds to lenient values and classified the resulting matches according to our standard-of-truth. The RLP parameter optimization consistently led to plausible weightings which penalize mismatch terms of *modifier* and *number tokens* and reward matching terms of the other token classes to various extents.

To assess the performance of our method, we use a 5-fold cross-validation procedure, i.e. the set of abstracts is divided into five parts. Four parts are used for optimization of parameters and the remaining one determines matching performance. Figure 1 depicts the averaged results of our matching procedure individually optimized for different dictionaries. Additionally, results for a

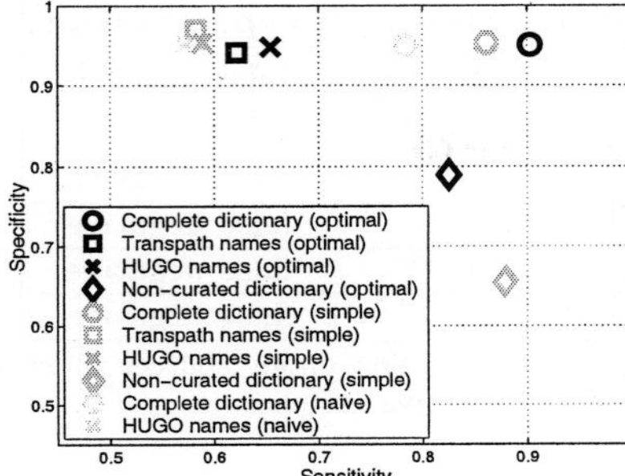

Dictionary	Method	Spec	Sens
Compl. Dict.	opt.	0.95	0.90
Transpath	opt.	0.94	0.62
HUGO	opt.	0.94	0.65
Non-cur. Dict.	opt.	0.78	0.82
Compl. Dict.	simple	0.95	0.86
Transpath	simple	0.97	0.58
HUGO	simple	0.95	0.58
Non-cur. Dict.	simple	0.65	0.87
Compl. Dict.	naïve	0.94	0.78
HUGO	naïve	0.95	0.57

Figure 1: Specificity (true positives/(true positives + false negatives)) and sensitivity (true positives/(true positives + false positives)) of matching procedure using different dictionaries of synonyms and parameters for the matching procedure. These concepts are closely related to the alternative measures of *precision* and *recall*. *Complete dictionary* and *non-curated dictionary* refer to the comprehensive dictionary in its fully curated and non-curated form, respectively. For *Transpath names* and *HUGO names* only the names defined in the respective databases are considered. In addition, three different modes of matching are employed. *Optimal* denotes cross-validated parameter optimization, *simple* refers to a simpler choice of parameters (non-descriptive tokens ignored, permutations allowed, no additional mismatches) and *naïve* to performance of naïve string matching.

simpler choice of weight parameters are shown.

Clearly, the optimized version in conjunction with the curated dictionary performs best. The reason is twofold. On one hand, there is a significant gain in sensitivity when switching from either HUGO or TRANSPATH names to the complete dictionary. On the other hand, our matching procedure adequately addresses the characteristics of protein names, again resulting in increased sensitivity and near constant specificity. In contrast, the use of the non-curated version of the dictionary leads to a large specificity penalty.

Table 4 summarizes major sources of errors in our protein name identification algorithm. One source of error for false positive matches are unspecific dictionary entries (e.g. glutamate receptor). Another source are semantically ambiguous names (e.g. HEK protein and HEK cells) which could not be discovered in the dictionary curation process. Imperfect parameter settings are a cause of further false positive matches. This shortcoming could possibly be remedied by a larger training set or an extension of token classes. Main reasons for failures in name identification are missing synonyms. This includes renaming of synonyms not reflected in our constituent databases (e.g. lymphotoxin to lymphotoxin 1), use of delimiters within synonyms (FGF.6 for FGF 6), or spelling variants of individual authors (e.g humEAA1 for EAA1 or pRB2 for RB2). In some cases, synonyms were written as one token (e.g. 'IL1beta' in-

Description of error	No. abstracts	No. objects	Type of error
ambiguous synonyms	7	3	false positive
unspecific synonyms	14	4	false positive
not in dictionary	25	15	false negative
plural form	2	2	false negative
tokenization problem	4	4	false negative
linguistics required	10	10	false negative

Table 4: Major sources of detection errors. Upper part contains major sources of false positive matches, lower part summarizes reasons for false negative matches.

stead of 'IL 1 beta') or plural forms were not recognized. In other cases, more linguistic information is required to recognize the protein name.

6 Discussion and future work

Analysis of today's large-scale experiments requires knowledge of relationships among a large number of genes and proteins. The successful identification of these objects including all known synonyms requires a comprehensive, curated, general-purpose dictionary. As the manual generation of such a protein name dictionary is infeasible, we presented a semi-automated method to arrive at a dictionary of reasonable high quality. The procedure is based on the definition of token classes for name components of different semantic significance and the specification of acronym and manual curation lists. Through improvement and adaptation of these lists to user needs, the quality of the resulting dictionary can be enhanced. An important point is that the invested knowledge is reused on integration or update of the underlying synonym databases. For example, the identification of further *modifier tokens* could enhance specificity especially when considering members of protein families. Manual addition or subtraction of correct synonyms is also essential, e.g. to increase specificity for certain objects (cf. Table 4) or to take into account lists of unspecific annotations (e.g. in the automatically annotated TREMBL database). To this end, ambiguities identified in the curation process can be inspected and selectively added to the manual curation lists. As our name dictionary is based on the most relevant reference databases, it is easily possible to link other datatypes to the generated results. For example, co-occurrence networks can be visualized in conjunction with data from gene expression experiments.

The inclusion of protein names to enhance sensitivity complicates the design of the search algorithm as longer phrases, permutations of tokens, and spelling variants must be taken into account (cf. Figure 1, "Non-curated dictionary" and "Complete dictionary (naïve)"). The presented algorithm was designed for high efficiency while respecting the defined characteristics of protein names. To this end, we approximated protein phrase boundaries by a mismatch threshold and the use of delimiter tokens. As demonstrated in Figure 1, the method is both specific and sensitive. The biological use of the presented work is currently evaluated in the context of the "Leitprojekt Osteoarthrose" [16], a research project dedicated to elucidate the pathomechanisms of the degenerative joint disease osteoarthritis. In that context, name search is employed to generate co-occurrence networks of proteins associated with the disease under consideration. Examples of networks can be found on the supplementary web page (http://cartan.gmd.de/ProMiner/).

As future work, we would like to enhance the performance of our method further. The incorporation of lexical rules (e.g. plural forms) during token matching could improve sensitivity. Moreover, recent work on disambiguating the semantic context of found names [17] might be used to decrease the number of false positive matches. In conjunction with methods that work independent of predefined dictionaries, the name dictionary could be extended further in a semi-automated fashion.

In addition, we plan to utilize ontological information to control ambiguous synonyms and abstract our search results in a hierarchical manner. A good candidate for this enhancement is GO, the gene ontology database [18], as it provides a direct mapping to SWISSPROT which is a constituent database of our name dictionary. In general, we feel that there is an urgent need for publicly available, annotated, domain-specific corpora for more accurate estimation of matching parameters and, even more important, for standardized comparison of different algorithms for name identification on a large scale.

We plan to develop a publicly available web-interface to the matching procedure itself, when further work on information extraction algorithms has been incorporated. Used in conjunction, these methods promise to make the huge body of biological literature accessible to bioinformatics methods and biological experts alike.

Acknowledgments This work was in part supported by Aventis Pharma, Frankfurt. We thank the members of the "Leitprojekt Osteoarthrose" for discussion on co-occurrence matrices and networks. We are indebted to Alexander Zien for his implementation of the RLP algorithm. We thank Biobase Inc. for providing us with a TRANSPATH license.

1. T. Ono, H. Hishigaki, A. Tanigami, and T. Takagi. Automated extraction of information on protein-protein interactions from the biological literature. *Bioinformatics*, 17:155, 2001.
2. J. Thomas, D. Milward, C. Ouzounis, S. Pulmann, and M. Carroll. Automatic extraction of protein interactions from scientific abstracs. *Pacific*

Symposium on Biocomputing, 5:514, 2000.

3. T.K. Jenssen, A. Lagreid, J. Komorowski, and E. Hovig. A literature network of human genes for high-throughput analysis of gene expression. *Nature Genetics*, 28:21, 2001.

4. J. Pustejovsky, J. Castao, B. Cochran, M. Kotecki, M. Morrell, and A. Rumshisky. Extraction and disambiguation of acronym-meaning pairs in medline. *Medinfo*, 2001.

5. H. Pearson. Biology's name game. *Nature, Features*, 2001.

6. K. Fukada, A. Tamura, T. Tsunoda, and T. Takagi. Toward information extraction: identifying protein names from biological papers. *Pacific Symposium on Biocomputing*, page 701, 1998.

7. D. Proux, F. Rechenmann, L. Julliard, V. Pillet, and B. Jacq. Detecting gene symbols and names in biological texts: a first step toward pertinent information extraction. *Genome Informatics Workshop*, pages 72–80, 1998.

8. N. Collier, C. No, and J. Tsujii. Extracting the names of genes and gene products with a hidden markov model. In *Proc. COLING 2000*, pages 201–207, 2000.

9. M. Krauthammer, A. Rzhetsky, P. Morozov, and C. Friedmann. Using blast for identifying gene and protein names in journal articles. *Gene*, 259:245, 2000.

10. H.M Wain, M. Lush, F. Ducluzeau, and S. Povey. Genew: the human nomenclature database. *Nucleic Acids Research*, 30:169, 2002.

11. Online Mendelian Inheritance in Man, OMIM (TM), 2000.

12. A. Bairoch and R. Apweiler. The swiss-prot protein sequence database and its supplement trembl in 2000. *Nucleic Acids Research*, 28:45, 2000.

13. Pubmed - national library of medicine. http://www.ncbi.nlm.nih.gov/entrez/, 2000.

14. F. Schacherer, C. Choi, U. Gtze, M. Krull, S. Pistor, and E. Wingender. The transpath signal transduction database: a knowledge base on signal transduction networks. *Bioinformatics*, 17:1, 2001.

15. K.P. Bennett and O.L. Mangasarian. Robust linear programming discrimination of two linearly separable sets. *Optimization Methods and Software*, 1:23–34, 1992.

16. Leitprojekt Osteoarthrose. http://www.leitprojekt-oa.de, 2000.

17. Vasileios Hatzivassiloglou, Pablo A. Duboue, and Andrey Rzhetsky. Disambiguating proteins, genes, and RNA in text: a machine learning approach. *Bioinformatics*, 17(90001):97S–106, 2001.

18. The Gene Ontology Consortium. Creating the gene ontology resource: design and implementation. *Genome Research*, 11:1425, 2001.

MINING TERMINOLOGICAL KNOWLEDGE IN LARGE BIOMEDICAL CORPORA

HONGFANG LIU, CAROL FRIEDMAN

Department of Medical Informatics, Columbia University
New York, NY 10032

Terminological knowledge of the biomedical domain is important for natural language processing (NLP) and information retrieval (IR) applications, and a number of terminological knowledge sources, such as LocusLink, GeneBank, and the UMLS, already exist. However, because of the tremendous amount of research activity in the field, new terms and symbols are continually being created, many of which are published in the literature, but are not available in any of the other resources. Therefore, effective mining of the literature for new terminology is critical for furthering NLP and IR applications. Abbreviations are widely used in the biomedical domain, and the understanding of abbreviations requires a terminological knowledge base that consists of abbreviations with their associated senses. In previous work, several methods have been developed for automatic construction of abbreviation knowledge bases from parenthetical expressions. However, these methods pair abbreviations and their expansions based on manually crafted patterns or rules. In this paper, we propose an automatic method, which is not based on patterns or rules but is based on the use of collocations, to extract a set of related terms from parenthetical expressions including abbreviations associated with their expansions and other types of related terms such as synonyms, or hyponyms etc. Our method is based on the observation that terms associated with parenthetical expressions i) are usually related, and ii) are often collocations because they tend to co-occur more often than expected by chance. Our method was applied to the collection of MEDLINE abstracts. The method and the results were evaluated using two collections: Berman's handcrafted abbreviation list and the LocusLink collection.

1   Introduction

In recent years, there has been a growing interest in automatic methods that construct and manage terminology resources for natural language processing (NLP) applications using large online collections of documents [1-4]. Although it is feasible to construct and manage terminologies manually in very limited domains, automatic or at least semi-automatic methods are required for applications that apply to domains with rich genres of context [5;6]. The domain we consider for this study is biomedical literature. We seek to explore the use of parenthetical expressions in text for automatic acquisition of terminological knowledge for NLP applications that are applied to biomedical text. Specifically, we consider two tasks: i) to find expansions for abbreviations, and ii) to find other types of semantically related terms.

It has been known that the wide use of abbreviations in the biomedical domain affects NLP applications, such as information retrieval systems or information extraction systems [6-8]. In order for NLP applications to process and interpret abbreviations that are not defined in documents, the associated terminology should include abbreviations together with their corresponding expansions in that domain. In addition, terminologies are more useful for NLP purposes if they not only list single-word or multi-word terms but also provide semantic knowledge, such as semantic categories and various other kinds of semantic relations (such as hypernymy/hyponymy, or synonymy) [9].

Observing that terms associated with parenthetical expressions are usually related terms, and expansions together with their abbreviations are often collocations (i.e., they tend to co-occur more often than expected by chance), we propose an automatic method for acquiring terminological knowledge from parenthetical expressions. In the following, we first introduce background material and related work for this study. We then describe our method in detail. The acquisition of terminological knowledge from the 2002 version of the MEDLINE abstracts, the evaluation of the method, and the results are presented next. We then discuss the results. Finally, we point out future directions of this work and conclude.

2   Background and Related Work

In this study, the uses of parentheses that we consider are parenthetical expressions "B̲ (A)" where there is a space separating B̲ from $\mathbf{A}^1$. A is the complete text string inside the parentheses (called the inner-text) and B̲ is the text string to the left of the parentheses (called the outer-text) within a certain window size. In the biomedical literature domain, parenthetical expressions are popularly used to define abbreviations as in "estrogen receptor (**ER**)" or as in "GABA (**gamma-aminobutyric acid**)". They are also used to specify semantic relations such as synonymy as in "natural toxin (i.e., **aflatoxin**)" or hypernymy as in "an inactive H-Ras protein (**RasN17**)". Parenthetical expressions can also be citations as in "here by using a recently developed ultrasensitive HPLC technique (**Sakhi et al. J. Chromatogr. A 828:451-460, 1998**)" or measures as in "CGRP failed to inhibit glucose-stimulated (**16.7 mM**)", etc.

There have been studies that report on the automatic construction of abbreviation terminologies using parenthetical expressions. Hisamitsu and Niwa [10] identified technical terms using parenthetical expressions that were statistically significant and then applied a set of simple rules to identify whether the text string inside parentheses was an abbreviation for a phrase that was at the left side of parentheses. Oh et al.[7] proposed a statistics-based model for constructing technical terminology by selecting similar phrase pairs "A (B)" including abbreviation pairs or translation pairs from parentheses, where a phrase pair was considered as an abbreviation pair if a half of the uppercase letters in A appeared in B sequentially. The method of Yoshida et al. [11] first identified terms representing biological substances using the PROPER system, and then extracted abbreviations using parenthetical expressions from terms identified by the PROPER system. Pustejovsky et al. [12] developed a system called ACROMED that applied a restricted pattern for

---

[1] Uses of parentheses without a space between B̲ and A usually imply that A and B̲ exist as a whole and they have no semantic relations with each other, for example, uses of parentheses in chemical names, e.g., Ca(OH)2.

identifying expansions for abbreviations from parentheses. Yu et al. [13] used a set of patterns to extract expansions for abbreviations from parenthetical expressions in full articles. Almost all above studies reported a precision of around 97% when matching abbreviations to their expansions.

However, all above studies used manually crafted patterns or rules for identification of expansions for abbreviations. Manually crafted patterns or rules are limited and often incomplete. For example, all above studies consider that alphabetic letters in an abbreviation occur in the corresponding expansions. Those methods may miss an abbreviation pair (*1H MRS, proton magnetic resonance spectroscopy*), or incorrectly identify the expansion as *magnetic resonance spectroscopy* from the sentence "*we used proton magnetic resonance spectroscopy (1H MRS) here*". In this study, we propose an automatic method to associate abbreviations with expansions for the purpose of automatic acquisition of terminological knowledge. Our method does not require patterns or rules, and is based on collocations. In addition to pairs of abbreviations and expansions, our method extracts other types of related terms from parenthetical expressions. The following provides background information about collocations as well as background information for resources used in this study.

Collocations considered in this paper are cohesive lexical clusters according to Smadja [14], where a collocation is a set of words such that the presence of one or several words of the set often implies or suggests the rest of the collocation. Parenthetical expressions, especially those used for defining abbreviations, are collocations. For example, given a parenthetical expression "*congestive heart failure (CHF)* ", the presence of words *CHF*, *heart*, and *failure* implies the presence of *congestive*. There are several methods to select collocations from text including simple frequency-based methods (such as eliminating all collocations with a frequency of less than a threshold) as well as complicated methods such as hypothesis-testing methods or mutual information methods [2;14]. Our method of selecting collocations is a complex frequency-based method.

We used the MEDLINE free-text collection for the experiment. MEDLINE [15] is the NLM bibliographic database that contains over 11 million references to journal articles in life sciences with a concentration in biomedicine. Each entry contains a unique MEDLINE identifier and citation information for the corresponding journal article, and often an abstract.

Our method, which is described in the Methods section, is based on word normalization that utilizes the SPECIALIST Lexicon, which is a UMLS Knowledge Source developed by the National Library of Medicine (NLM) as part of the Unified Medical Language System project [16]. The SPECIALIST Lexicon is a general English lexicon that includes a comprehensive set of biomedical terms. The lexical entry for each word or term provides syntactic information such as part of speech information (e.g., *acid* is a noun), and morphological information that maps textual variants to base forms (e.g., *discharging* and *discharged* to *discharge*). Additionally, the SPECIALIST Lexicon includes an abbreviation table which includes 11,051 pairs.

Two collections of abbreviations associated with their expansions were used as a gold standard. The first collection was a list of 12,098 pathology-related abbreviations that were manually collected by Berman [17]. The second collection consists of pairs of gene symbols together with their definitions. The list was extracted from LocusLink, which was developed by the NCBI of the NLM [18]. We excluded pairs where definitions contain the corresponding symbol as a sub-string with boundaries (e.g., *KIAA0042* vs *KIAA0042 gene product*). Additionally, there were 11,516 pairs that belong to RIKEN cDNA genes (i.e., *5730583K22Rik* vs *RIKEN cDNA 5730583K22*) that were excluded in this study in order to avoid biases in measures because they were formed using a simple pattern but contained almost 20% of the total entries. The resulting collection consisted of 42,875 unique pairs. After normalization the total collection obtained from the two sets amounted to 49,536 unique pairs.

3   Methods

The method contains three steps. The first step, COLLECT, collects parenthetical expressions from a large collection of text and filters out certain expressions (e.g., citations and measures) since they are not helpful for acquiring terminological knowledge. The second step, DETECT, uses the results of the first step to derive a set of pair-wise terms. In the third step, SEPARATE, we assess the set of pair-wise terms and separate them into two sets: a set of (abbreviation, expansion) pairs and a set of other types of related terms such as synonyms and hyponyms, etc.

3.1   Collect

We collect all uses of parenthetical expressions "<u>B</u> (A)" from sentences in a large collection of text (we use a heuristic to determine sentence boundaries and the window size for extraction is generally twice the length of A). Additionally, we collect all uses of parenthetical expressions within nested parentheses and all uses of square brackets because authors use them to avoid nested parentheses. After each successful extraction, the text string inside the parentheses is deleted and the resulting string is used for subsequent extraction. For example, we extract four uses of parenthetical expressions (i.e., *the substitution of asparagine for threonine at position 1405* **(T1405N)**, *... by a functional polymorphism* **(the substitution of asparagine for threonine at position 1405)**, *...the current produced by the Na(+)-Ca(2+) exchanger* **(I(NCX))**, and *the release of* endothelin **(ET)**)   from the following two sentences a) and b).

a.   ... by a functional polymorphism (the substitution of asparagine for threonine at position 1405 [T1405N]) in ....

b.   ...the current produced by the Na(+)-Ca(2+) exchanger (I(NCX)) working in ... the possible autocrine role played by the release of endothelin (ET) in ...

Parenthetical expressions where the inner-texts occur only once in the corpus are filtered out since most likely they occur together by chance. In addition, we filter out certain expressions such as citations or measures since they are useless for the automatic acquisition of terminological knowledge. The filtering process is achieved using several heuristics that are consistent with patterns for citations and measures. For example, *here by using a recently developed ultrasensitive HPLC technique* (*Sakhi et al. J. Chromatogr. A 828:451-460, 1998*) and *CGRP failed to inhibit glucose-stimulated* (*16.7 mM*) are filtered out since the former one is a citation and the latter is a measure.

3.2   Detect

After we have a collection of parenthetical expressions, we need to detect collocations from all outer-text strings of expressions that share the same inner-text. This step contains several components: a normalization module, a collocation generator, and a collocation selector.   Table 1 shows an example of the overall process, where the input to the process consists of all unique outer-text strings (e.g., *treatment of community acquired pneumonia*, ..., *hospitalized community acquired pneumonia*) that correspond to the same inner-text (i.e., *CAP*) and their frequency. The output of the overall process is a set of pair-wise terms (e.g. CAP, community acquired pneumonia)

The purpose of normalization is to unify textual variants since they usually represent the same sense in terminologies. The normalization module changes an outer-text string into lower case, removes all non-letter characters and a small set of stop words. It then normalizes each word in the text string by transforming it to the base form in the SPECIALIST Lexicon, if applicable, and represents the normalized words as an array. For example, the output for an outer-text string for CAP, *Patients with community acquired pneumonia*, is an array (*patient, community, acquired, pneumonia*).   Note that for simplicity, we use the last base form listed in the SPECIALIST Lexicon if there are multiple entries for the corresponding word (e.g., *acquired* has two base forms: *acquire* when it occurs as a past particle, and *acquired* when it occurs as an adjective); otherwise, a syntax parser would be needed in order to choose the most likely one.

The collocation generator generates candidate collocations associated with frequency information. For an array that contains l words, we generate l potential collocations $\{pc_j^l,$ j ranges from 1 to $l\}$, where pc_j^l is formed by concatenating the last $l-j+1$ words in the array. For example, the above outer-text string for *CAP* after normalization generates four potential collocations (i.e., "*pneumonia*", "*acquired pneumonia*", "*community acquired pneumonia*", "*patient community acquired pneumonia*"). The number of occurrences for each potential collocation is then counted. For the sake of saving computational resources, we eliminate all potential collocations that occur only once since these words most likely occur together by chance (note that for parenthetical expressions that are not used for capturing abbreviations, this statement may not be true). The last module selects a set of collocations based on frequency. The collocation selector selects a set of collocations based on frequency information. There are two main processing phases

Step	Examples	FREQ
Outer-texts for CAP	treatment of community acquired pneumonia	14
	Patients with community acquired pneumonia	10
	patients with community acquired pneumonia	3
	group of patients with community acquired pneumonia	2
	blood culture of community- acquired pneumonia	2
	hospitalized community acquired pneumonia	1
	including pneumonia	1
Normalization	(treatment, community, acquired, pneumonia) (14)	14
	(patient, community, acquired, pneumonia) (13)	13
	(group, patient, community, acquired, pneumonia) (2)	2
	(blood, culture, community, acquired, pneumonia) (2)	2
	(hospitalize, community, acquired, pneumonia) (1)	1
	(include, pneumonia)	1
Potential Collocations	pneumonia	33
	acquired pneumonia	32
	include pneumonia	1
	community acquired pneumonia	32
	treatment community acquired pneumonia	14
	patient community acquired pneumonia	15
	culture community acquired pneumonia	2
	hospitalize community acquired pneumonia	1*
	blood culture community acquired pneumonia	2
	group patient community acquired pneumonia	2
After Eliminating	pneumonia	33
	acquired pneumonia	32
	community acquired pneumonia	32
After Subsuming	community acquired pneumonia	32
Final Collocation	(CAP, community acquired pneumonia)	32

Table 1: An example illustrating the process of detecting collocations from all outer-texts that share the same inner-text CAP.

for the selection: eliminating, and subsuming, as shown in Figure 1. Both phases contain a loop associated with the number of words (LEN, ranges from 2 to the maximum number of words in a potential collocation, i.e., MAX) in a potential collocation. Let pc be a collocation with LEN words, pc' be a collocation formed from the last LEN-1 words in pc, and PC(pc) be the set of potential collocations formed by adding a prefix word to pc. The elimination of potential collocations is achieved using two formulas a and b, which are described as following:

Formula a) states that if the number of elements in PC(pc) is larger than a certain threshold t_0, we consider these prefix words to occur with pc by chance, and therefore all elements in PC(pc) are eliminated from the final set. For example in Table 1, when LEN = 3, pc = "*community acquired pneumonia*", t_0 =3, all potential collocations containing pc and having more than three words are eliminated from the final set.

```
Selecting:
FOR LEN = 2 to MAX {
       FOR each pc that is a potential collocation with LEN words {
          Let pc' be the potential collocation formed by the last LEN-1 words of pc
```
$$IF(a).\ \ \left| PC(pc) \right| > t_0\ \{$$
```
          Delete all potential collocations from PC where the last LEN words are the  same as pc
          }
```
$$IF(b).\ \ \frac{freq(pc)}{freq(pc')} < t_1\ \{$$
```
             Delete pc and all potential collocations from PC where the last LEN words     are the
          same as pc
             }
          }
}
Subsuming:
FOR LEN = 2 to MAX {
       FOR each pc that is a potential collocation with LEN words {
          Let sc be the summation of the frequency of all collocations formed by adding one more
       word to the left of pc
```
$$IF(c).\ \ \frac{sc}{freq(pc)} > t_2\ \{$$
```
          Delete pc from PC
          }
       }
}
RETURN PC
```

Figure 1: The process of selecting collocations.

Formula b) states that if the ratio of the frequency of pc compared to the frequency of pc' is less than a certain threshold t_1, we consider it to be relatively less frequent, and eliminate it together with potential collocations that have it as the postfix string. For example in Table 1, when LEN = 2, pc = *"include pneumonia"*, and $t_1 = 0.1$, the string *"include pneumonia"* is eliminated since the ratio of the frequency of *"include pneumonia"* to the frequency of *"pneumonia"*, i.e., 1/35, is less than 0.1.

Let sc be the summation of the frequency of all collocations in PC(pc). The subsuming process is achieved using the formula c in Figure 1, i.e., if the ratio of sc to the frequency of pc is larger than certain threshold t_2, we consider pc to be subsumed by elements in PC(pc), and delete it from the final set. For example in Table 1, when $t_2=0.9$, *"pneumonia"* is subsumed by *"acquired pneumonia"* and *"acquired pneumonia"* is then subsequently subsumed by *"community acquired pneumonia"*. Note that thresholds can be set based on experience or learned from a gold standard. Pairs generated by pairing the inner-text string with each collocation in the final set (e.g., (**CAP**, *community acquired pneumonia*)) become the output of the DETECT phase.

3.3 Separate

After collocations are detected and a set of pair-wise terms are generated, we then separate them into two sets so that relations that are not abbreviations are treated separately: i) abbreviations associated with expansions, and ii) other semantically related terms, according to lengths of two items in a pair and the existence of capitalized letters since usually, an abbreviation is short and contains capitalized letters, and an expansion is much longer than the corresponding abbreviation. Note that the semantic relations for some pairs can be determined by checking the use of simple patterns in the original text. For example, if we assume NP represents a noun phrase, the patterns "a NP (NP)" or "NP (a NP)" such as, "*indomethacin (a cyclooxygenase inhibitor)*", usually implies a hypernymy relation while the pattern "NP (i.e., NP)" such as "*the congenital neutropenia (i.e., Kostmann's syndrome)* " implies a synonymy relation.

4   Experiments and Results

Our method was applied to the 2002 version of MEDLINE. The threshold values for t_0, t_1, and t_2 (3, 0.25, 0.9 respectively) were estimated using the abbreviation list in the SPECIALIST Lexicon. In the following, we described the evaluation of our method and assessment of the results.

4.1 Evaluation

We first evaluated the soundness of ignoring potential collocations with frequency equal to one during the selection process by counting the number of pairs in the gold standard set (i.e., the combination of Locus-link collection and Berman's abbreviation list) that occurred only once in MEDLINE and the number of pairs that occurred more than once after normalization using the SPECIALIST Lexicon. We then evaluated the selection process by computing the percentage of pairs with the expansions that were correctly identified to pairs that occurred more than once in MEDLINE.

We performed a manual analysis of the performance of the SEPARATE step using 50 randomly chosen pairs from each set (i.e., a set of abbreviations associated with expansions, and a set of other types of pairs). We then computed the precision based on the manual analysis.

The assessment of the results concentrated on the acquired abbreviation knowledge base. We did not assess other types of pairs since it required a detailed investigation of the original context as well as expert knowledge.

The acquired abbreviation knowledge base was assessed through several measures: frequency distribution, ambiguity (i.e., the number of unique expansions for the same abbreviation), the coverage of the abbreviation knowledge base, the relation of ambiguity to the length of abbreviations, the percentage of the number of abbreviations to the number of abbreviations that were ambiguous, and the percentage of abbreviations that did/did not contain digits, which were ambiguous

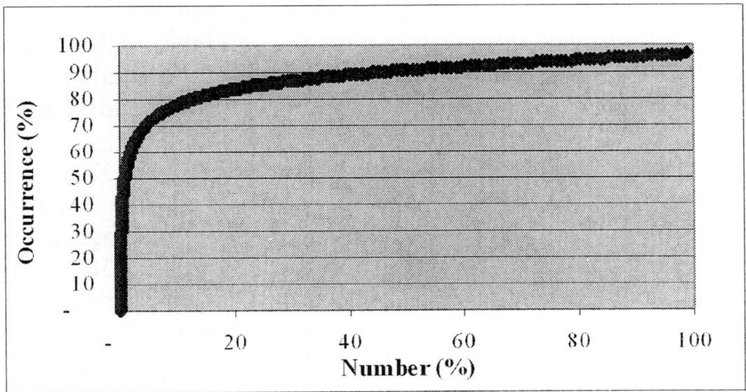

Figure 2: The most frequent pairs in relation with their occurrences. The X-Axis denotes the percentage of the number of the most frequent pairs, and the Y-Axis denotes the percentage of the number of their occurrences.

Len	With Digits		Without Digits	
	# Pairs (% Amb)	AVG	# Pairs (% Amb)	AVG
1	2,039 (37.0)	3.96	240 (56.3)	48.11
2	5,676 (28.3)	1.95	2,571 (56.3)	19.21
3	6,094 (26.1)	1.74	12,016 (66.9)	7.68
4	3,433 (19.7)	1.50	27,540 (36.0)	2.01
>=5	7,802 (8)	1.22	32,657 (17.1)	1.44

Table 2: The ambiguity study results with respect to the existence of digits in abbreviations and the number of letters in the abbreviations.

(all measures were computed by transforming pairs to lower-case except frequency distribution).

4.2   Results

We collected 16,068,562 uses of parenthetical expressions. After filtering out certain uses, such as citations and measures as well as uses where the inner-text occurred only once in the current version of MEDLINE, there were 6,626,790 uses remaining for generating pair-wise terms.

Among 49,536 unique normalized pairs in the gold standard, 5,232 occurred in MEDLINE using parenthetical expressions, and 4,809 (91.9%) occurred more than once in MEDLINE. Expansions associated with 96.3% of the pairs were detected correctly, which suggests that the recall of the method was around 88.5% (i.e., the product of 91.9% and 96.3%) if abbreviations were defined using parenthetical expressions and they were presented in the gold standard.

We collected 381,126 unique pairs, where 308,339 were incorporated into the abbreviation knowledge base, and 72,787 were considered as other types of pairwise terms. Two of the 50 pairs from the abbreviation knowledge base that we manually checked were not abbreviation pairs (i.e., and two of the 50 pairs from the set of other types of pair-wise terms were actually abbreviations (i.e., (**NESP, darbepoetin alfa,** *novel erythropoiesis stimulate protein*), and (**GnRH-A,** *GnRH agonist*).

Figure 2 shows the frequency distribution where the X axis denotes the percentage of the number of the most frequent pairs, and the Y axis denotes the percentage of the number of their occurrences. The coverage of the acquired abbreviation knowledge base differed in the two collections: 38.3% for Berman's abbreviation list, and 3% for Locuslink collection. The ambiguity of an abbreviation was related to the existence of digits in the abbreviation as well as to the number of alphabetic letters in the string. Table 2 shows the result, where LEN is the number of letters, and AVG is the average number of expansions.

5   Discussion

We have presented an automatic method for the purpose of automatic acquisition of terminological knowledge for NLP applications using parenthetical expressions in large corpora. A novel aspect of the method is consideration of parenthetical expressions as collocations. Utilizing a collocation selector based on frequency, expansions for abbreviations are automatically recognized. The method has an advantage that it does not require manually crafted patterns or rules for the recognition of expansions. Many abbreviations (i.e., symbols) for biological substances do not follow rules or patterns, and including them in terminologies for NLP applications is important. We believe our method has a higher sensitivity for acquisition of pairs of abbreviations and expansions that occur frequently and should be useful as a complement to pattern-based methods. However, we were unable to. compare our method to other methods because of the absence of a common gold standard set

Our method is not suitable for recognizing expansions that occur only once in the corpora. However, since the purpose of our method is to automatically acquire terminological knowledge, the exclusion of expansions of abbreviations that occur only once from terminological knowledge has almost no impact on NLP applications. Newly defined pairs of abbreviations and expansions will most likely be captured using the most current version of the corpus based on the intuition that if they are accepted by the community, they will be repeatedly referred to and defined in literature; otherwise, they are likely to occur infrequently.

The frequency distribution of pairs of abbreviations and expansions follows the 20/80 rule, i.e., 20% of the pairs contribute to 80% of the total occurrences.

From the result of the coverage study, we found that abbreviations suggested or collected by humans may not be used in the literature. For example, around 10% of expansions from both collections existed in our abbreviation set obtained from MEDLINE but were associated with different abbreviations (1,361 out of 12,098 for Berman's abbreviation collection, and 4,376 out of 42,875 for LocusLink collection). The low coverage of LocusLink was because there are many pairs associated with digits or letters specifying members, sub-families, or types. For example, there are 50 members in the gene family of *ATP-binding cassette* where the symbol for *ATP-binding cassette* is *ABC* (which is not listed in LocusLink), such that each has been assigned a symbol by attaching a sub-family symbol (from *A* to *G*) and a member number (e.g., *ABCA1*, *ABCB1*, and *ABCG8* etc). The resulting abbreviation knowledge base contains 429 occurrences of the pair (*ABC*, *ATP-binding cassette*), but does not contain symbols of most members in the family, signifying that researchers did not use the same conventions as the LocusLink curators.

The assessment of the ambiguity study shows that abbreviations are highly ambiguous, which is consistent with our previous study [19]. For example, the abbreviation *CAP* represents not only *community-acquired pneumonia*, but also dozens of others, including the following protein names: *catabolic activator protein, cystine aminopeptidase, cellulose acetate phthalate, cyclase-associated protein, cementum attachment protein, calcium-activated protease, capsid-associated protein* etc. We noticed that abbreviations containing digits are much less ambiguous than those without digits. The ambiguity of an abbreviation depends on the number of letters it contains: ones with fewer characters are more ambiguous. In order to allow NLP applications to process and interpret ambiguous abbreviations that are not defined in documents, a disambiguation method is needed. In several previous studies [19], we developed disambiguation methods with a precision of 97% for disambiguating abbreviations given a set of known expansions. The abbreviation knowledge base acquired here can be used as the set of expansions for our disambiguation methods.

Although our method also extracts a set of related terms, we did not focus on this feature in the current paper. A further investigation will be performed in order to assign appropriate semantic relations and semantic categories to those pairs. Future work will also involve use of multiple biomedical databases and use of contextual clues other than parentheses to discover semantic collocations associated with terminology.

6   Conclusion

We proposed and evaluated an automatic method for automatic acquisition of terminological knowledge based on the observation that terms associated with parenthetical expressions are collocations. We identified expansions for abbreviations using frequency information associated with the collocations without the requirement of manually crafted patterns or rules. The method had a precision 96.3% and a recall of around 88.5% for abbreviations that were defined using parenthetical expressions and were presented in the gold standard. We acquired 381,126 unique pairs from the 2002 version of MEDLINE abstracts, where 308,339 were incorporated into the abbreviation knowledge base, and 72,787 were considered as other types of pair-wise terms. Abbreviations are highly ambiguous and the ambiguity is related to the number of alphabetic letters in the abbreviation and the existence of digits: ones with fewer characters are more ambiguous, and ones with digits are less ambiguous. In the future, we plan on expanding this method in order to obtain a more comprehensive terminological knowledge using other corpora such as full articles and online databases.

Acknowledgment

This study was supported in part by grants LM06274 from the NLM and ILS-9817434 from the NSF.

Reference List

1. Lauriston A. Ph.D Dissertation *Automatic Term Recognition: performance of Linguistic and Statistical Techniques.* 1996 UMIST, Manchester.
2. Justeson J, Katz S. *Technical terminology: some linguistic properties and an algorithm for indentification in text.* Nat Lang Eng 1995; 1(1):9-27.
3. Frantzi K, Ananiadou S. *The C-value/NC-value domain indendent method for multi-word term extraction.* Journal of Natural Language Processing 1999; 6(3):145-180.
4. Georgantopoulos B, Piperidis S *Automatic Acquisition of Terminological Resources for Information Extraction Applications.* Proc. of New Information Technologies 1998
5. Friedman C, Hripcsak G, DuMouchel W, Johnson SB, Clayton PD. *Natural language processing in an operational clinical information system.* Nat Lang Eng 1995; 1(1): 83-108.
6. Friedman C *A Broad Coverage Natural Language Processing System.* Proc AMIA Symp 2000: 270-274.
7. Aronson A *Effective Mapping of Biomedical Text to the UMLS Metathesaurus: The MetaMap Program.* Proc. AMIA Symp 2001: 17-21.
8. Nadkarni P, Chen R, Brandt C. *UMLS concept Indexing for Production Databases: A Feasibility Study.* J Am Med Inf Assoc 2001; 8(1): 80-91.
9. Finkelstein-Landau M, Morin E *Extracting Semantic Relationships between Terms: Supervised vs Unsupervised Methods.* Proc. International Workshop on Ontological Engineering on the Global Information Infrastructure 1999:71-80
10. Hisamitsu T, Niwa Y *Extraction of useful terms from parenthetical expressions by using simple rules and statistical measures.* The First Workshop on Computational Terminology Computerm'98 1998: 36-42.
11. Yoshida M, Fukuda K, Takagi T. *PNAD-CSS: a workbench for construction a protein name abbreviation dictionary.* Bioinformatics 2000; 16(2):169-175.
12. Pustejovsky J, Castano J, Cochran B, Kotechi M, Morrell M, Rumshisky A *Extraction and Disambiguation of Acronym-Meaning Pairs in Medline.* Medinfo 2001
13. Yu H, Hripcsak G, Friedman C. *Mapping abbreviations to full forms in electronic articles.* JAMIA 2002: 9(3):262-272.
14. Smadja F. *Retrieving Collocations from Text: Xtract.* Computational Lingustics 1993;19:143-176.
15. MEDLINE. http://www.nlm.nih.gov . 2001.
16. NIH. Unified Medical Language System. NIH. 2000.
17. Berman J. A classification for Medical Abbreviations. 2002. http://www.pathinfo.com/jjb/berman1j.htm.
18. Maglott D, Katz K, Sicotte H, Pruitt K. *NCBI's LocusLink and RefSeq.* Nucleic Acids Research 2000; 28:126-128.
19. Liu H, Johnson SB, Friedman C. Automatic Resolution of Ambiguous Terms Based on Machine Learning and Conceptual Relations in the UMLS. J Am Med Inf Assoc 2002: 9(6): 621-636

A BIOLOGICAL NAMED ENTITY RECOGNIZER

MEENAKSHI NARAYANASWAMY AND K. E. RAVIKUMAR
AU-KBC Research Centre
Chennai 600044 INDIA

K. VIJAY-SHANKER
Department of Computer and Information Sciences,
University of Delaware
Newark, DE 19716, USA

In this paper we describe a new named entity extraction system. Our system is based on a manually developed set of rules that rely heavily upon some crucial lexical information, linguistic constraints of English, and contextual information. This system achieves state of art results in the protein name detection task, which is what many of the current name extraction systems do. We discuss the need for detection of chemical names and show that we not only obtain a high degree of success in recognizing chemicals but that this task can help improve the precision of protein name detection as well. We use context and surrounding words for categorization of named entities and find the results obtained are encouraging.

1   Introduction

Research in biology in the past decade has generated a large volume of biological data that is available only in the literature, i.e. in the textual form, which are stored in databases such as MEDLINE. The quality and amount of continuously updated knowledge makes this an extremely useful form of information that needs to be trapped for research and development. There is a need to develop automated intelligent text analysis tools in order to extract useful information from the literature.

As a first step towards information extraction on various interactions between biological entities, a system must identify biological enities, e.g. gene, protein, chemical, cell and organism names. Handling of unknown words, long compounding of word sequences are some of the difficulties in name recognition in this domain. A few biological name entity recognition tools (e.g., Collier et al.[2], Friedman et al.[3] Fukuda et al.[4]) exist which were developed as an integral part of different information extraction applications. Much of these efforts are geared to extract protein and gene names. Some of the tools use statistical or machine learning approaches in identifying names from biological texts.

428

1.1 Main Features of Our System

Our approach is a symbolic one and based on a set of manually developed rules. These rules exploit surface clues and simple linguistic and domain knowledge in identifying the relevant terms in the biomedical papers. While our approach is not based on machine learning or statistical methods, we wish to point out that the results we obtain are due to the type of features we extract from text and that these can just as easily be exploited in a statistical or machine learning setting.

There are advantages and disadvantages of using either a manually designed system or one based on statistical or machine learning. A statistical or (supervised) machine learning approach requires large amount of text where the named entities are marked. One of the chief reasons for not having used a supervised machine learning approach is because of a lack of an annotated corpus. Creating an error-free annotated corpus will take extensive time and effort. Besides, we have noticed that there doesn't appear to be an exact consensus on exactly what constitutes a name (even when we limit it to proteins/genes). Additionally, as more information extraction work proceeds, the need to extract named entities of other types might arise (for instance, we discuss below why it might be useful to extract chemical names as well). So until these matters are better settled, we decided not to take a supervised learning approach (to avoid reannotations and retraining of models).

In many respects, our approach is inspired by the design of PROPER (PROtein Proper-noun Extracting Rules), also based on a manually designed set of rules (Fukuda et al.[4]). We believe our work extends the term recognition capability of this system, and other systems, in two significant ways. First, we believe that we improve upon the precision and recall (standard measures for evaluating name/term recognition systems). Second, while most systems tend to focus on recognizing protein/gene names alone, we recognize other types of names/terms as well. In particular, we also recognize chemical names. While the work in the GENIA project (see e.g., Collier et al.[2]) is also meant to recognize terms beyond names and proteins, they do not identify chemical names. There are two major reasons why we felt it was important to capture chemical names. First, biomedical papers often contain plenty of chemical names. These chemical names do share various features that are used to identify protein names. Hence, we believe name detection systems can wrongly identify chemical terms as protein names. Our approach then is to identify the chemical names and classify them as such and thereby improve the precision of the (protein) name recognition. Second, most of the current work on information extraction from biomedical papers is geared towards extracting protein-protein interactions. Hence most of the biological named entity recognizers are geared towards identifying protein and gene names only. In biological interactions, the entities involved are not just proteins and genes but chemicals too. From information perspective, which might be important in aiding drug discovery process it is very crucial to identify chemicals in the text. We are only aware of work by Wilbur et.al.[5]

for extracting chemical names automatically. Existing tools mostly use a dictionory or ontology lookup to identify chemical and drug names.

In our name recognition module we intend to capture the following biological terms and classifying them under their respective category:

* **Protein/gene** – Terms that correspond to names of proteins or genes. Like most other systems, but unlike Hatzivassiloglou et al.[6], we do not distinguish between protein and gene names.

* **Protein/gene parts** - Terms corresponding to protein/gene parts. Examples of these include MH2 domain, and lysine residue. Protein name detection tools such as KeX (Fukuda[4]) which is based on PROPER do not attempt to distinguish protein names from terms of this category.

* **Chemical** – Examples include Indomethacin, N-methylformamide, and suberoylanilide hydroxamic acid.

* **Chemical parts** – Terms, like methyl groups, that correspond to parts of chemicals.

* **Source** - Terms that represent source terms including cells, cell parts and organisms.

* **General** (biological) – Terms that can truly be classified as belonging to more than one class or related to the above class but can not be classified as belonging to one of them. A protein-chemical complex, such as hematoporphyrin-LDL complex, would belong to the latter case.

2. Design of the System

The text (Medline abstracts, in our current experiment) are split into sentences, tokenized and then part of speech tagged using Brill's tagger (Brill[1]). In this section and in Section 4 we describe the main modules of our system. These modules are applied in the order they are described below.

2.1 Identifying Abbreviations

Abbreviations are widely used in medical literature. Hence identifying them and their expanded form is a necessary task in biological name entity recognition. Furthermore in our setting we have to associate with each occurrence of an abbreviation its classification (protein/gene, protein/gene parts, chemical, chemical parts and source).

Our system works on the premise that typically the first occurrence of an abbreviation occurs in a pattern in which the original term is followed by the abbreviation within the parenthesis. E.g., *testosterone repressed prostrate message 2 (TRPM-2)*

We have currently implemented a simple algorithm to identify such pairs. It only covers cases where the acronym appears in parenthesis and is to the right of its definition. It works by matching as many characters of the acronym as possible. It currently identifies 94.2% of such pairs in our test set but however has an accuracy of 87.4% only. We are working on improving this component and incorporating ideas from Taghva et. al[7].

Once a pair, say for example *CBP/p300 associated factor (CAF)*, is identified we attempt to classify it (as described later) based on the expanded form *CBP/p300 associated factor*. This type information is associated with the abbreviation so that remaining occurences (within the same abstract/paper) of the abbreviation can be classified appropriately. Note that while future occurrences of the abbreviation *(CAF)* can be detected, it would be difficult not be possible to figure out its classification just based on its surface form.

2.2 Core Terms and Functional Terms

Like in KeX that incorporates PROPER, the basic idea involves identifying two types of terms: core terms (c-terms) and functional terms (f-terms). This categorization typically applies to individual words. C-terms have surface features (such as capital letters, numerals, and special symbols) that are used by most name recognition systems. On the other hand, the set of functional terms are specific to the name recognition in biomedical domain and play a key role in the classification of the extracted terms into our five categories as described below. Perhaps since the latter issue is not considered in KeX, despite making a distinction between c-terms and f-terms, the algorithm does not appear to make distinction between the two types of terms.

2.2.1 C-term Recognition

Like noted above, various surface features of words such as the use of capital letters are used as important clues in most name recognition systems. Following Fukuda et al.[4], we designate words with such surface features as c-terms. However, while these features are useful in identifying names, they clearly do not provide any clue for the appropriate categorization of the identified names. Hence we call the words that contain these features as *general* c-terms (where recall our use of the word general refers to the situation where appropriate categorization is not possible). Names that include general c-terms might however get the appropriate category when they combine with other words (such as f-terms) that provide the necessary information. In addition to the general c-terms we also have two other types of c-terms: *protein* c-terms and *chemical* c-terms. These c-terms contain some information that can be associated with protein/gene names and chemical names respectively.

The recognition of chemical and protein c-terms is obviously not based on clues for names in general. For instance, the module for extracting chemical c-terms includes recognition of chemical root forms. We base this on IUPAC conventions followed in naming chemicals. We also exploit various morphological features and suffixes used extensively in naming chemicals.

For example, consider "We have developed a class of **HDAC** inhibitors, such as ***suberoylanilide hydroxamic acid (SAHA)***, that were initially identified based on their ability to induce differentiation of cultured murine erythroleukemia cells." *Suberoylanilide hydroxamic acid (SAHA)* is a chemical, which inhibits HDAC an enzyme. The suffix –ic followed by acid helps in identifying the two words as chemical c-terms.

In contrast, in the sentence, "Polar organic solvents such as *methanol* or *N-methylformamide* inactivate lipases.", methanol and N-methylformamide are both chemical names. These are first identified as chemical c-terms because they contain the chemical root forms methyl and meth.

KeX identified *N-methylformamide* as protein (it is meant only to recognize protein names) as it has a capital letter. Likewise occurences of SAHA lowers its precision as it misidentified *SAHA* as a protein name. In contrast due to the chemical c-term detection together with the treatment of abbreviations helps us overcome these problems.

We extract protein c-terms purely on the basis of suffixes such as *ase*. Table 1 shows some of the main features for recognizing the remaining c-terms. These features clearly apply to the names of any of our categories.

Table 1 – Word features to capture the c-terms

S.No	Word Feature	Example
1	Caps only	CBP
2	More than one Cap	hCG
3	letters and Digits	TRPM2, H4, p53
4	Single Cap	Aspirin, Asp1
5	Terms having special symbols (/, -, etc)	IL-7 etc

Among the core terms that were selected, some of them are eliminated if they contain only numerals and special characters. Also, special names for experimental techniques and units are eliminated.

2.2.2 Extraction of Functional Terms

Functional words are not only helpful for locating biological terms, they are also very useful for purposes of categorizing the terms properly.

Table 2- Description of functional terms

CLASS	FUNCTIONAL WORDS	DESCRIPTION	EXAMPLE
Protein/Gene	Receptor, protein, factor, gene etc	Protein/Gene/RNA	CREB binding protein
Protein/Gene parts	motif, domain, promoter, etc	Parts of Gene/Protein/RNA	MH2 domain
Chemicals	Steroid, drugs, etc.	Lipid, steroid, organic, inorganic compounds, and carbohydrates	Tertbutyldimethyl silyoxyandrost-4-ene steroid (9)
Chemical parts	Radical, ions, groups etc.	Chemical radicals, inorganic and organic ions etc	acetyl groups, methyl groups etc
Source terms	cells, cell lines, phage	Cell, Organisms, cell parts etc	MCF-10F cells
General terms	Mutants, molecules	Can belong to any of the above category,	asf1mutants Corby mutant

The functional words were similar to the feature terms used by Fukuda et al.[4] However our name recognition tool differs from KeX in dealing with the functional words. The Kex module does not distinguish between the feature terms and the core terms. However we not only distinguish the functional words from the core terms but also classify them under the categories mentioned in the Table 2.

2.3 Concatenation and Extension Rules

So far, we have only identified individual words. Names that might need to be extracted can be several words long. We now consider how such names are extracted. These are done by applying a few concatenation and extension rules.

The first rule we consider says that two f-terms that are next to each other can be combined into one. The category assigned to the new term is determined as follows. If one of the f-terms is of general category and the other is not of general category then the latter's category is adopted. For example, we identify *protein complex* as being of category protein while the f-term complex is of general category.

On the other hand, when both f-terms being concatenated are not general then the category of the one on the right is adopted. This rule is based on the characteristics of the noun phrases in English whose parse structure is typically right branching and whose head is typically rightmost. An example of the application of this rule characterizes *protein fragment* as protein part since the two individual f-terms are of protein and protein part category respectively.

Concatenation of a c-term and f-term is similar. With few exceptions, the f-term is to the right of the c-term and usually determines the type. In *SIR3 Protein fragment*, as argued above, *protein fragment* is a f-term phrase of type protein part. SIR3 is of course a general c-term. Together they form a protein part. There are a few cases where the f-term is of general category in which case the c-term's category provides the necessary information. One such example is *acetyltransferase family*. *Acetyltransferase* is a protein c-term and provides the category information for the composite name.

Concatenation of two c-terms is similar. *H4 acetyltransferase*, and *acetyltransferase Sas2* are two examples of concatenation of c-terms. The right c-term provides the category information in the first case. In the second example, the right c-term has a general category whereas the left category is more specific and gives the required information.

We now present two examples where a pair of terms being combined are not of general category. The first involves a chemical c-term which combined with a term to its right is no longer of chemical category. (On the other hand, we have rarely noticed this with protein categories.) *Xanthine* is a chemical core because of the chemical root *Xanth*. However, the concatenated phrase *Xanthine oxidase* is designated a protein category c-term phrase because of the nature of the c-term to its right. In *Ras guanine nucleotide exchange factor*, every combination from left to right of two terms leads to a different category until the final assignment of protein category is done.

To conclude this section, we present a few other rules to extend terms beyond individual c-terms. We connect two non-adjacent terms as long as every word in between them is a noun, an adjective or a numeral. This allows us to extract *CREB binding protein* and *MOZ leukemia gene* where the leftmost and rightmost words are already marked (as c-term and f-term) but the one in the middle is not. Also extensions to the left upto a determiner is allowed subject to some conditions and extensions to include greek letters are considered. Finally we have a few rules to drop annotated phrases. For example, a f-term that is not extended to left or right is dropped. Thereby the single word protein by itself is not considered a name.

3. Preliminary Evaluations

To evaluate our system, we collected 55 Medline abstracts, hand annotated them and associated the categories with each marked name. These abstracts were obtained by searching for *acetylates, acetylated* and *acetylation*. This choice was made so that we could have a good proportion of protein, protein part as well as chemical names.

Of the 620 names according to the hand annotation, 593 terms were correctly identified. Thus, for the pure name *detection* task (i.e., without considering the assignment of categories), we get precision, recall and F-measure of 90.39%, 95.64% and 92.94% respectively. These numbers however mask a certain problem that become clear on examination of the following table.

Table 3 – Disambiguation of biological names (preliminary version)

S. No	Total	Terms Disambiguated	Precision	Recall	F-mes
Protein/ Gene	302	104	93.69%	34.44%	50.37%
Protein/ Gene parts	99	43	95.56%	43.33%	59.62%
Chemical	158	116	92.06%	73.42%	81.69%
Chemical parts	13	8	100.00%	61.54%	76.19%

Source terms	36	29	96.66%	80.56%	87.88%
Protein + Protein Parts	401	147	94.23%	40.05%	56.21%

While the precision is still high, the recall falls down significantly. The reason is that for a vast number of cases none of the above categories could be assigned. Of all the terms detected, we can see from Table 3 that only 200 (sum of the second column entries except for the last row – the last row is just the sum of the first two rows) were assigned categories. That is, the category information could not be discerned after the application of the concatenation and extension rules and the term remains categorized as general. In order to compare the performance with other protein name detection system, we should consider the protein and protein parts together (see last row). Since such protein name detectors would consider all the names they detected as protein names, we can also treat all the names detected which our system has not classified as chemical, chemical part or source as protein name. That is, all names that were left in the general category are also considered as proteins (and protein parts). Such an evaluation leads to a F-measure of 86.54 (precision=78.61 and recall=96.26). Notice that the recall has risen sharply. This is because of the 336 terms that were in the general category, 187 protein terms and 52 protein part terms. On the other hand the number of chemical terms was 38. By this strategy, these 38 (although of the 158 chemical terms in the test set) are mistagged as proteins. This causes the precision to fall. Another reason for the reduction in precision is that almost all false positives fell into the general category. And in the situation we were able to identify the category information, the number of false positives is small.

4. PostProcessing Rules

While our system has high precision in disambiguating the terms into the respective classes its recall is low. For many identified names the correct category could not be inferred from the surface string (e.g., SP-A, TAF (II) 30, CD40). We apply additional rules to categorize those names that are in this category with the hope of increasing the recall. These postprocessing rules are essentially of two types. The first considers the surface strings and applies heuristic rules which we were not ready to apply at an earlier stage but are probably more reliable now that we know there is no other clue to categorize it otherwise. The other type that we believe to be novel in the name detection setting is the use of adjoining context for purposes of disambiguation. Hence, given space constraints we will only discuss these.

It is well-known in computational linguistics that words surrounding a phrase (in our case, a detected name) often help in disambiguation. This idea also underlies the work reported by Hatzivassiloglou et al.[6] and Liu et. al[8] who use it for disambiguating ambiguous expressions. Our situation is different in that there is only possible category (from our list) but is unknown at this stage of processing.

We have come up with a list of help-words (h-terms) that are like f-terms in that they provide clue about the category but unlike f-terms are not considered part of the name. A few examples of h-terms are expression, homolog, and recombinant, which are all associated with protein. Although there is much scope of extending this idea, currently we have a small list of h-terms for proteins only. If we observe a *name* and *h-term* adjacent to each other or the pattern *h-term of name* and the name has been classified as general (ior unknown) then we assign the category associated with the h-term to the detected name. Thus, SP-A, TAF (II) 30, and CD40 each get the category protein/gene because they appeared in the context *SP-A expression*, *homolog of TAF (II) 30*, and *expression of CD40* respectively. Despite a preliminary version of such disambiguation rules, as noted below we get very encouraging results. We plan to investigate the use of context in the disambiguation process more thoroughly in the future. We now show our results in a format similar to Table 3. However these results now reflect the application of our postprocessing rules.

Table 4 - Disambiguation of biological names (after post processing)

S. No	Total	Terms Disambiguated	Precision	Recall	F-mes
Protein/ Gene	302	189	96.43%	62.58%	75.86%
Protein/ Gene parts	99	88	97.78%	88.89%	92.75%
Chemical	158	136	93.15%	86.08%	90.86%
Chemical parts	13	11	100.00%	84.62%	91.67%
Source terms	36	35	97.22%	97.22%	97.22%

Only protein terms	401	277	96.85%	69.08%	80.64%

The number of terms which are now in the general or unknown category drops sharply. 102 terms are protein terms after the application of the postprocessing rules as compared to 187 before. However, the biggest change occurs in the protein parts category where we now only have 7 as compared to 52 before. Those in the chemical category drop to 18 from 38 before.

As discussed earlier, we finally categorize all the remaining terms in the general (or unknown) category as proteins or protein parts. Thus in comparison to the results (86.54 for F-measure, 78.61% for precision and 96.26% for recall) obtained before the postprocessing results were applied, we obtain as the final performance results on the Protein Name recognition task as 91.90% for F-measure with 87.92% for precision and 96.26% for recall.

5 Comparison with KeX

The only (protein) name detection system we are able to compare with is KeX which incorporates PROPER, perhaps considered a standard and among the better-known of the existing systems. We could do this because this software is freely available on the internet. Since we are not able to run the other name detection systems on our test set (and neither do we know what their test set was) we are unable to compare our results with them. As the following discussion suggests the nature of test set can make a substantial difference in the results and we didn't choose to compare with the results cited in papers as well.

Our system substantially outperforms KeX on the protein name recognition task when we applied it to our test set. A fairly large proportion of the difference in precision can be attributed to the presence of chemical and source names. The results reported in Fukuda et al.[4] were obtained on a set of Medline abstracts based on SH2 and SH3 domains and perhaps these abstracts didn't have the same proportion of chemical names. And perhaps our test corpus, which was based on search for keywords related to acetylation, might have a higher proportion of chemical names than a randomly chosen Medline abstract.

These results do show two critical points: even within Medline abstracts, results of KeX has varied considerably and also that considering chemical name detection can improve precision of protein name detection. Of course, one of our main motivations for chemical name detection is to aid extraction of information involving chemical compounds and drugs.

We are currently working on improving our acronym-definition detection component; generalize and systematize our approach embodied in the postprocessing rules for assigning categories to the name detected when no surface clue in the name exists; and increasing the categories of named entities we recognize.

References

E. Brill, "Some Advances in Transformation-Based Part of Speech Tagging" in *Proceedings of the Twelfth National Conference on Artificial Intelligence (AAAI-'94)*, Washington 1994

N. Collier, C. Nobata, and J. Tsujii, "Extracting the names of genes and gene products with a hidden Markov model" in *Proceedings of the 18th International Conference on Computational linguistics (COLING 2000)*, Saarbr¨ucken, 2000.

C. Friedman, P. Kra, H. Yu, M. Krauthammer, and A. Rzhetsky, "GENIES: a natural language processing system for the extraction of molecular pathways from journal articles" *Bioinformatics.* **17 Suppl 1**(2001)

K. Fukuda, T. Tsunoda, A. Tamura, and T. Takagi, "Information extraction: Identifying protein names from biological papers" in *Proceedings of the Pacific Symposium on Biocomputing '98 (PSB'98)*, Hawaii, January 1998.

W.J. Wilbur, G.F. Hazard, G. Divita, J.G. Mork, A.R.Aronson, and A.C. Browne, "Analysis of biomedical text for chemical names: a comparison of three methods" in *Proc. AMIA Symp 1999*, Washington, 1999.

V. Hatzivassiloglou, P.A. Duboue, and A. Rzhetsky, "Disambiguating proteins, genes, and RNA in text: a machine learning approach" *Bioinformatics.***17 Suppl 1**(2001)

K. Taghva and J. Gilbreth, "Recognizing acronyms and their definitions", IJDAR. **1**. 1999.

H. Liu, Y.A. Lussier, and C. Friedman, "Disambiguating ambiguous biomedical terms in biomedical narrative text: an unsupervised method" *J Biomed Inform.* **34 (4)** (2001)

LINKING BIOMEDICAL LANGUAGE INFORMATION AND KNOWLEDGE RESOURCES: GO AND UMLS

I.N. SARKAR[*,1], M.N. CANTOR[*,1], R. GELMAN[1], F. HARTEL[2] AND Y.A. LUSSIER[§,1,3]

1- *Department of Medical Informatics, Columbia University College of Physicians and Surgeons, New York, NY 10032 USA*

2- *Center for Bioinformatics, National Cancer Institute, National Institutes of Health, Bethesda, MD 20892 USA*

3- *Department of Medicine, Columbia University College of Physicians and Surgeons, New York, NY 10032 USA*

Integration of various informatics terminologies will be an essential activity towards supporting the advancement of both the biomedical and clinical sciences. The GO consortium has developed an impressive collection of biomedical terms specific to genes and proteins in a variety of organisms. The UMLS is a composite collection of various medical terminologies, pioneered by the National Library of Medicine. In the present study, we examine a variety of techniques for mapping terms from one terminology (GO) to another (UMLS), and describe their respective performances for a small, curated data set attained from the National Cancer Institute, which had precision values ranging from 30% (100% recall) to 95% (74% recall). Based on each technique's performance, we comment on how each can be used to enrich an existing terminology (UMLS) in future studies and how linking biological terminologies to UMLS differs from linking medical terminologies.

1 Introduction

1.1 Need to Link Structured Information Resources

As is often the case in an advancing science, there are divisions between various people working towards a common goal. This is no different in the information sciences – whether in medical or bio- informatics[1,2,3]. The quick advancement of technology, both computational and laboratory, has enabled the recent integration of medical and bio- informatics[1,2]. To this end, we have explored how various methods can be used to help integrate the essential terminologies of these diverse, yet, similar fields.

The respective terminologies that serve the medical and biological sciences communities are, by themselves, of great importance to each individual field. Links between the two fields, however, are growing, as medicine increasingly incorporates basic biological science advances into clinical practice, and biologists or bioinformaticians validate their experiments using real patient data. These growing

---

* These authors contributed equally to the work
§ Corresponding author

interactions necessitate a standardized method for communicating results between the fields.

Among the medical terminologies, the Unified Medical Language System (UMLS)[4] sets the standard for breadth of inclusion. The included terminologies, however, are generally focused on clinical medicine, so representation of more basic biological terms is often lacking. There is a need for representation of gene and gene products, such as those from the Human Genome Project (HGP), in the UMLS. There has been initial exploration into strategies to map genomic knowledge into the UMLS[5]. Aside from mapping specific terms from projects such as the HGP, however, there has been minimal documented speculation about the inclusion of entire biological terminologies, such as the Gene Ontology (GO)[6,7].

Before terminologies such as GO are included within the UMLS, there needs to be an exploration into methodologies that may be utilized for the augmentation process. Analyses of the use of various text mining and information extraction techniques need to be performed. In the present study, we provide a preliminary analysis using a variety of applicable methodologies.

1.2 Not Really a New Art, but Still the Same Problems

Mapping of various medical terminologies to the UMLS has been studied extensively.[7,8,9,10,11,12] The UMLS 'Lexical Tools', provided by the National Library of Medicine (NLM)[13], offer some assistance with mapping of terms, but with varying results. However, the attempted methods have demonstrated limited success, generally able to map 13 - 60% of terms.[9,10]

Part of the complexity lies in the variety of ways that a single concept may be represented. As disparate systems often use the same information resources, it is imperative that redundancy be kept to a minimum. However, this lends itself to a great challenge when trying to augment composite terminologies, such as the UMLS, with highly structured terminologies such as GO.

Additionally, different terminologies may represent the same concept in very different ways. There may be future endeavors that will incorporate intelligent Natural Language Processing (NLP) techniques that will enable such disparities to be resolved[12]. Indeed, there has been some investigation of using the UMLS as a resource for language processing systems.[15,16]

However, using various non-NLP mapping techniques it should be possible to identify the correct concept classification of a significant portion of the non-UMLS terms. To date, there have been no published results of mapping GO terms to the UMLS. There have been some recent efforts in annotating genes from biomedical literature using the Gene Ontology.[17] Admittedly, GO was designed for the annotation of genes and gene products. For this reason, most of the recent focus has been on the annotation of gene and gene products in the biomedical literature and not in mapping terminologies. However, there has been some previous investigation into how other biomedical resources such as OMIM and GENBANK can be mapped to the medical information structures.[18] This previous work demonstrates that there is a desire and a need for the integration of the various biomedical (clinical and basic science) terminologies.

2 Materials

2.1 The Unified Medical Language System

We used the 2001 version of the UMLS, created and maintained by the National Library of Medicine. The NLM began constructing the UMLS in 1986, to facilitate not only the retrieval and integration of information from diverse biomedical sources but also the linkage of disparate information systems.[4] The 2001 version of the UMLS consists of about 800,000 unique concepts (797,359) from over 60 diverse terminologies.[20] Each individual concept is represented in the UMLS by a 'Concept Unique Identifier' (CUI). There may be multiple text string variants (UMLS terms) affiliated with each CUI. The variants are identified by 'String Unique Identifiers' (SUIs). In the 2001 UMLS there are 1,728,075 SUIs.

The principal component of the UMLS used in the present study is the Metathesaurus. The UMLS Metathesaurus links "terminology and concepts from a range of vocabularies and classifications" through the UMLS Semantic Network, which provides a "consistent high level of categorization" to the individual concepts.[4] The UMLS, while generally focused on medical terminologies, does contain some gene and gene product names. These are mostly a result of the Medical Subject Headings initiative for indexing biomedical articles.[21,22]

2.2 Gene Ontology

The Gene Ontology consortium is focused on developing structured, coded, vocabularies for molecular function, biological processes, and cellular components that can be used across species.[6] Significantly, the vocabularies are independent of the associations between specific gene products and GO terms, leading to flexibility and precision in the use of the framework. The focus of the present study involved the May 2001 version of the GO. Each GO concept is represented by one unique text string also known as GO term. We refer to each GO concept by their identifying number (GOID). 2.3 Gold Standard

The National Cancer Institute (NCI) provided us with several files containing mappings between a subset of GO (NCI-GO version May 2001) and the UMLS to their internal metathesaurus. From these files, we derived a subset of 332 distinct GOIDs that had been mapped to CUIs from the UMLS.

For the present study, we treated these distinct 332 GOID-CUI pairs as our gold standard (GS). We note that the GS contained 314 distinct CUIs, indicating that some CUIs mapped to more than one GOID. Since mapping methods make extensive use of the UMLS terms related to the CUIs, it is noteworthy to mention that 6,113 SUIs are associated to these 314 individual CUIs.

2.3 Applications and Scripts

All the applications and scripts to implement the methods discussed in this paper were written in C, C++, Java, and Perl. Additionally, Lexical Tools, the MMTx tool obtained from the National Library of Medicine, and applications associated with the National Center for Biotechnology Information's BLAST were also used. Most applications were run on a Sun machine running the SunOS 5.8 operating system. MMTx was run on a Personal Computer running Microsoft Windows 2000.

3 Methods

3.1 Approaches to Mapping Terminologies

The final goal of this project is to accurately capture and map the largest possible subset of the GOIDs to the UMLS CUIs. In order to find the most effective method for doing so, we evaluated four different coupling approaches to matching the two vocabularies. Because we did not use any machine learning methods a training set was not required. For all the methods implemented, we used the textual, flat file representations of both GO and UMLS knowledge bases.

Exact String Matching. The first and simplest method was exact string matching, which finds the lexical matches between UMLS terms and GO terms. This was implemented through a join of MySQL tables containing the set of terms from both the GO and the UMLS.

Norm. Norm represents one of two methods utilized from the lexical tools available from the UMLS.[13] As its name implies, *norm* converts text strings into a normalized form, removing punctuation, capitalization, stop words, and genitive markers. Following the normalization process the remaining words are sorted in alphabetical order. After processing each GO term with *norm,* we matched each distinct GO term against the normalized form of UMLS terms, as represented by a normalized English table that was obtained from the NLM (MRXNS_ENG).

MMTx. MMTtx is an implementation of MetaMap, a highly configurable program used to map concepts found in biomedical texts to the UMLS Metathesaurus.[22] After normalizing and parsing text into noun phrases, MetaMap uses UMLS knowledge sources to generate both synonyms and derivational variants. The program then retrieves the set of all candidate strings contained in the Metathesaurus, evaluates each one against the input text, and produces a result list of candidates and their UMLS CUIs, ordered by mapping strength (scored on a scale from 0-1000). The highest-scoring combinations of the candidates are presented as an additional list called 'Meta Mappings' that also scores the results.

We used MMTx to perform two types of analyses, which we term 'Loose' and 'Strict'. Loose MMTx analysis consists of all distinct GOID-CUI pairs provided by the 'Meta-Mappings' regardless of their scores. For 'Strict' analysis, we determined a threshold score (T) and neglected all mappings below the selected threshold. We varied the thresholds from T=600 to T=1000, in increments of 100, to provide five different sets GOUI-CUI pairs.

BLAST-based Matching. Based on the work of Krauthammer, et al., the Basic Local Alignment Search Tool (BLAST)[24] can be used to compare text characters to each other that have been converted into nucleotide sequences.[25].BLAST allows for approximate string matching, with the respective flat files serving as reference databases. BLAST provides efficient identification of sequences that have a high probability of correspondence with the query sequence. We transcribed the text strings of both the GO and the UMLS terms into nucleotide sequences by substituting each character with a specified nucleotide combination, as outlined by Krauthammer, et al[25]. After converting the transcribed nucleotide sequences into FASTA format, we created a BLAST database using the *formatdb* program of the transcribed UMLS terms. Using the *blastn* application, each transcribed GO term was used as a query sequence to search the entire UMLS database for the most similar match.

3.2 Implementation

Every mapping methodology was used to map a subset of GO terms provided by the NCI to the entire UMLS. While GO has only one term per concept, UMLS supports multiple variant terms for every concept. This results in a potential total of 574 million pairs (332 GO terms * 1,728,075 SUIs) that need to be searched. Since we are only concerned with mapping the *concepts* (as opposed to *terms*) of the respective terminologies, this number is further reduced since we consider only the possible *distinct* GOID-CUI pairs stemming out of the mapping methods (this is a smaller subset since many UMLS terms can be represented by a single CUI). The resulting concept combinational space consists of 265 million individual GOID-CUI pairs (332*797,359).

444

3.3 Evaluation

The GS, consisting of 332 GO unique identifier/UMLS Concept Unique Identifier pairs, was examined for each method and analyzed in the following manner: relevant pairs ('True Positive'; TP) were pairs found by the coupling method that were also in the GS; non-relevant ('False Positive'; FP) matches were those that were not found in the GS; relevant, but *not* retrieved ('False Negative'; FN) were in the original GS but not matched by the coupling method. Using the GS GOID-CUI pairs as our gold standard, we performed an evaluation of each of the methodologies implemented.

We measured the efficacy of each of the methods using precision and recall. Recall was calculated as the ratio of the number of *distinct* GOID-CUI pairs that were identified by the mapping method that *match* GOID-CUI pairs in the GS, divided by the total number of pairs in the GS, TP/(TP+FN). Precision was measured as the ratio of the number of *distinct* GOID-CUI pairs returned by the mapping method that *match* GOID-CUI pairs in the Gold Standard, divided by the total number of putative GOID-CUI pairs found by the mapping method, TP/(TP+FP).

4 Results and Discussion

Table 1 and figure 1 summarize the analysis of our results using each of the four described coupling methodologies.

Within the context of the gold standard, our evaluation shows that each of the four coupling methods provides mapping with varying degrees of success. The simpler coupling approaches, exact string match and *norm*, fared well in terms of precision, but suffered in recall. Exact match, the simplest of the methods tested, had a recall of only 65% but had a high level of precision (94%). Because of what constitutes an exact match, this high level of precision is to be expected. On the other hand, *Norm* yielded both a high recall (90%) and precision (89%). The 4-5% rate of non-relevant matches (FP) for these relatively simple matching methods is reflective of the variance, and hence the difficulty in finding all the match pairs, that exists in the UMLS of a single concept's terms. In an ideal world, if everyone used the same terms for the same concepts, then exact string algorithms would yield 100% precision.

The more complex methods, BLAST and Loose-MMTx, both suffered in precision (36% and 62% respectively) as a result of how their general algorithms work and how the results were tabulated. That is, both of these methods attempt to make a classification for almost every term that it is given resulting in as many FP as FN counts (Loose-MMTx recall = 82%; BLAST recall = 36%). Imposing the strictest threshold (T=1000), Strict-MMTx achieved a precision of 95%. However, recall dropped to 74%. We explored changing the threshold to various values, and report the values at threshold equal to 900 and 800 graphically in figure 1. There was a consistent trade-off between precision and recall as we changed the threshold value. As there was no significant effect for thresholds tested below 800, these values are not shown in the graph.

Table 1. Analysis of GO and UMLS Mappings according to the methodologies (Count of Distinct GOI-CUI Pairs)

	Exact String	Norm	MMTx Loose	MMTx T=1000	BLAST
Relevant GS Matches (TP)	216	299	272	247	121
Non-Relevant GS Matches (FP)	13	38	170	12	211
Not Retrieved GS Matches (FN)	116	33	60	85	211

BLAST's low precision may be due in part to how the BLAST algorithm functions and highlights some of the issues of the heuristic method that BLAST implements. BLAST is designed to try to determine the best alignment between two nucleotide sequences, a 'query' and 'subject' sequence, based on contiguous word lengths as defined by the algorithm. Certain penalties are incurred for the incorporation of a gap (or space) that may be inserted into a sequence to help optimize and alignment. In the present study, no gaps were inserted in any sequences. Possible lowering of the incurred penalty for inserting a gap may have yielded a higher recall value.

It is important to note that the alignment that is sought by BLAST is for the longest sequence with the least amount of dissimilarity. This can be problematic in implementations where slight variants of words result in the transcribed sequences being different enough to not make them look similar. Also it is important to point out that BLAST is an algorithm of *similarity*, not *homology*, which can make the heuristic prone to misclassifications. For example, BLAST addresses the question: "Which item(s) in the database are *most similar* to my query sequence?" This is opposed to addressing the question: "Which item(s) in the database *is the same as* my query sequence?"

As a result, for every sequence an attempt is made by the algorithm to find a match. This produces not only a low recall value, but (by definition) results in a high number of irrelevant matches (64%). Similar issues have been addressed in relation to attempting to use BLAST as a primary classification tool for nomenclature of traditional biological sequences.[26]

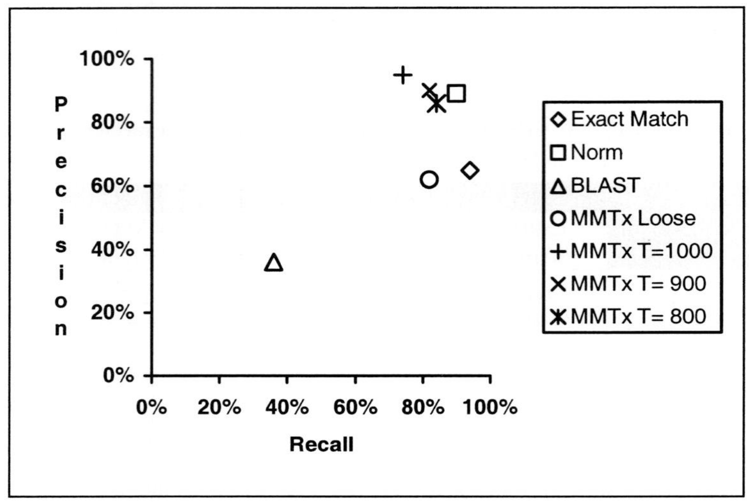

Figure 1: Precision versus Recall according to the mapping method

As the gold standard file used for the present study consisted of a heavily curated data set that is maintained by the National Cancer Institute, our results may have been biased, since it represented a subset of the UMLS, particular to only one domain of medicine (oncology). As a result, specific UMLS concepts that may not be contained in the NCI metathesaurus may not have been mapped to a GOID. Additionally, because of the millions of SUIs that a single GO term could map to, it is unreasonable to expect manual curation of all combinations. This may contribute to an inflated number of calculated False Positive values. To this end, we qualitatively explored some of the False Positives reported by the mapping methods.

Two experts reviewed some of the distinct GOID-CUI pairs that were classified as False Positive. They unveiled that some GOID-CUI pairings labeled as FP, because they did not exist in our GS, appeared to be relevant. For *exact string match* and *norm*, 13 and 8 of their respective GOID-CUI FP matches appeared correctly mapped and thus misclassified as FPs. The recalculated precision of the two methods would reach 100% and 91%, respectively. Similarly, some of the irrelevant matches returned by both of MMTx and BLAST also appeared misclassified, but their large sets of FP counts precluded quantitative analyses. This issue is reflective of the complex task of mapping terms to a composite terminology like the UMLS, which may have multiple synonymous terms and contains millions of distinct terms.

This last point emphasizes the complexity of the issue that is at the core of ontologies for use in a practical setting. In order to address the non-homogeneous manner in which professionals choose to represent the same concept a number of approaches have been implemented. The Systematized Nomenclature of Medicine (SNOMED) is an example of a terminology that has taken this factor into account and allows concepts to be defined in terms of other concepts using formal predicate logic[27]. This formalism allows for computable evaluations of incomplete 'identity' matches in which one of the following relationships could be assigned: 'putative ancestor', 'putative descendant' or 'putative sibling'.

These results demonstrate the difficulty and applicability of existing coupling methods for mapping even a very select set of terms between terminologies. The difficulty of mapping arises, in part, because the UMLS is a composite terminology. That is, it is made of not just a single resource, but multiple resources. To this end, UMLS inherits all the terms from all the various terminologies that it includes. For this reason, NLM has developed a number of tools, such as *norm* and MMTx to map novel terms to the existing unique concept identifiers. To date, the tools from NLM have been solely used to append additional medical terminologies. While some gene and gene product terms may have been incorporated from existing medical terminologies there has been no attempt, to our knowledge, for adding strictly basic biological terms.

Appending basic biological terminologies to the UMLS is a different art than adding terms from medical terminologies. Processing text strings with tools such as MMTx or *norm,* for example, de-emphasizes numeric values, which are less important in clinical medicine (though there are exceptions, such as 'Diabetes Mellitus Type 2') than in biology, where such differences are extremely important. Basic biological terms often consist of subtle textual, yet significant definitional differences, such as the difference between 'Transcription Factor I', 'Transcription Factor II', or 'Transcription Factor III'. Using CUIs or GOIDs provides a method for overcoming these types of ambiguities, since appropriate concept mapping is essential for proper integration of the respective terminologies.

5 Caveats and Implications for Future Work

It is important to emphasize that the gold standard used in the present study was only a relatively small, domain-specific subset based on work that was not specifically aimed at mapping GO terms to the UMLS. Since creating a gold standard is a resource intensive endeavor, we conveniently used relevant work from the NCI. However, as future work will inevitably include GO in the UMLS, it will be imperative to be aware of such mappings, especially in the case of certain concept terms that may be described very differently by either clinicians or biologists. These will likely involve either manual curation or sophisticated Natural Language Processing (NLP) tools to create mappings.

Nonetheless, this work demonstrates the feasibility of mapping a significant portion of GO to existing UMLS concepts. It is conceivable that using a combination of the methods described in the present study, as well as others that may exist, will enable a complete mapping of these terminologies as well as any future terminology from other related domains. We are conducting additional work that will address the insertion of concepts in the UMLS schema with algorithms allowing the mapping of terminologies with additional relationships beyond the "identity" relationship, such as the "ancestor", "descendant" and "sibling" relationships.

Beyond examining additional methods for simple mapping terms, integration of these techniques into a voting scheme may also yield more favorable results by incorporating the strengths of each of the methods used. While the present study primarily focused on the ability of each of the methods to function independently, we believe that by linking the methods together in a simple voting scheme may result in higher levels of precision and recall for the overall mapping process.

6 Conclusions

This work lays the foundation for mapping biological and medical terminologies, which are essential for information systems. The task of knowledge mapping will be essential for the future integration of the various biomedical domains, such as nursing, public health, etc. The primary challenge that these future mapping face, as is faced with GO, is that clinical informaticians may not necessarily design a significant portion of these non-medical terminologies, leading to potential ambiguities, redundancies, or incompatibilities Using proven text mining and information extraction techniques it will be possible to create a mapping of large sections of existing knowledge resources to each other.

Acknowledgments

INS and MNC are funded by National Library of Medicine Medical Informatics Training Grant LM07079-09. Additional financial support for this work was from LM06274 of the National Library of Medicine. Special thanks goes to Sherri De Coronado of the National Cancer Institute for her assistance and providing us with their UMLS and GO mappings. Thanks also goes to David Figurski and his laboratory for their encouragement and insightful discussions, without which this paper would not be possible.

References

1. Altman RB & Klein TE. Challenges for Biomedical Informatics and Pharmacogenomics. Annual Review of Pharmacology & Toxicology. 42:113-133. (2002)
2. Nakamura RM. Technology That Will Initiate Future Revolutionary Changes in Healthcare and the Clinical Laboratory. J Clin Lab Anal. 13:49-52. (1999)
3. Shortliffe EH & Perrault LE (eds). Medical Informatics: Computer Applications in Health Care and Biomedicine. (Springer, New York, 2001)
4. Lindberg DA, Humphries BL, McCray AT. The Unified Medical Language System. Methods Inf Med. 32(4):281-291. (1993)
5. Yu H, Friedman C, Rhzetsky A, Kra P. Representing Genomic Knowledge in the UMLS Semantic Network. 1999 Proc Annu Symp Am Med Inf Assoc:181-186. (1999)
6. Gene Ontology Consortium. Creating the Gene Ontology Resource: Design and Implementation. Gen Res. 11(8):1425-1433. (2001)
7. Bodenreider O & McCray AT. The Lexical Properties of the Gene Ontology. Proceedings of the American Medical Informatics Association 2002 Annual Symposium.
8. Cimino JJ & Johnson SB. From ICD-9 to MeSH Using the UMLS: A How-to Guide. 1994 Proc Annu Sump Am Med Inf Assoc: 730-734.
9. Zeng, Q & Cimino JJ. Mapping Medical Vocabularies to the Unified Medical Language System. 1996 Proc Annu Symp Am Med Inf Assoc: 105-109. (1996)
10. Tuttle MS, Suarez-Munist ON, Olsen NE, et al. Merging Terminologies. 1995 MEDINFO. 8(Pt 1):162-166. (1995)
11. Tuttle MS, Cole WG, Sheretz, DD, Nelson SJ. Navigating to Knowledge. Methods Inf Med. 34(1-2):214-231. (1995)
12. Tuttle MS, Sherertz DD, Erlbaum MS, et al. Adding Your Terms and Relationships to the UMLS Metathesaurus. 1991 Proc Annu Symp Comput Appl Med Care:219-223. (1991)
13. National Library of Medicine. UMLS Lexical Tools. Application and Documentation available at http://umlsks.nlm.nih.gov.
14. Lussier YA, Shagina L, Friedman C. Automating SNOMED Coding Using Medical Language Understanding: A Feasibility Study. 2001 Proc Annu Symp Am Med Inf Assoc: 418-422. (2001)
15. McCray AT, Bodenreider O, Malley JD, Browne AC. Evaluating UMLS Strings for Natural Language Processing. 2001 Proc Annu Symp Am Med Inf Assoc: 448-452. (2001)
16. Friedman C, Liu H, Shagina L, et al. Evaluating UMLS as a Source of Lexical Knowledge for Medical Language Processing. 2001 Proc Annu Symp Am Med Inf Assoc: 189-193. (2001)

17. Raychaudhuri S, Chang JT, Sutphin PD, Altman RB. Associating Genes with Gene Ontology Codes Using a Maximum Entropy Analysis of Biomedical Literature. Gen Res. 12:203-214. (2002)
18. Sperzel WD, Abarbanel RM, Nelson SJ, et al. Biomedical Database Inter-connectivity: An Experiment Linking MIM, GENBANK, and META-1 via MEDLINE. 1991 Proc Annu Symp Comput Appl Med Care:190-193. (1991)
19. National Library of Medicine. Unified Medical Language System. 12th Ed. January 2001.
20. Tuttle MS, Olsen NE, Campbell KE, et al. Formal Properties of the Metathesaurus. 1994 Proc Annu Symp Comput Appl Med Care:500-504. (1994)
21. Coletti MH & Bleich HL. Medical Subject Headings Used to Search the Biomedical Literature. J Am Med Inf Assoc. 8(4):317-323. (2001)
22. Lipscomb CE. Medical Subject Headings (MeSH). Bull Med Libr Assoc. 88(3):265-266. (2000)
23. Aronson AR. Effective Mapping of Biomedical Text to the UMLS. 2001 Proc Annu Symp Am Med Inf Assoc: 17-21. (2001)
24. Altschul SF, Gish W, Miller W, et al. Basic Local Alignment Search Tool. J Mol Biol. 215(3):403-410. (1990)
25. Krauthammer M, Rzhetsky A, Morosov P, Friedman C. Using BLAST for Identifying Gene and Protein Names in Journal Articles. Gene. 259(1-2):245-252. (2000)
26. Sarkar IN, Thornton J, Planet PJ, et al. An Automated Phylogenetic Key for Classifying Homeoboxes. Mol Phylogenet Evol: 24:388-399. (2002)
27. Spackman KA & Campbell KE. Compositional Concept Representation using SNOMED: Towards Further Convergence of Clinical Terminologies. 1998 Proc Annu Symp Am Med Inf Assoc: 875-879. (1998)

A SIMPLE ALGORITHM FOR IDENTIFYING ABBREVIATION DEFINITIONS IN BIOMEDICAL TEXT

ARIEL S. SCHWARTZ

Computer Science Division
University of California, Berkeley
Berkeley, CA 94720
sariel@cs.berkeley.edu

MARTI A. HEARST

SIMS
University of California, Berkeley
Berkeley, CA 94720
hearst@sims.berkeley.edu

Abstract

The volume of biomedical text is growing at a fast rate, creating challenges for humans and computer systems alike. One of these challenges arises from the frequent use of novel abbreviations in these texts, thus requiring that biomedical lexical ontologies be continually updated. In this paper we show that the problem of identifying abbreviations' definitions can be solved with a much simpler algorithm than that proposed by other research efforts. The algorithm achieves 96% precision and 82% recall on a standard test collection, which is at least as good as existing approaches. It also achieves 95% precision and 82% recall on another, larger test set. A notable advantage of the algorithm is that, unlike other approaches, it does not require any training data.

1 Introduction

There has been an increased interest recently in techniques to automatically extract information from biomedical text, and particularly from MEDLINE abstracts.[3, 4, 7, 15] The size and growth rate of biomedical literature creates new challenges for researchers who need to keep up to date. One specific issue is the high rate at which new abbreviations are introduced in biomedical texts. Existing databases, ontologies, and dictionaries must be continually updated with new abbreviations and their definitions. In an attempt to help resolve the problem, new techniques have been introduced to automatically extract abbreviations and their definitions from MEDLINE abstracts.

In this paper we propose a new, simple, fast algorithm for extraction of abbreviations from biomedical text. The scope of the task addressed here is the same as the one described in Pustejovsky et al.:[14] identify <"short form", "long form"> pairs where there exists a mapping (of any kind) from characters in the short form to characters in the long form.[a]

---

[a] Throughout the paper we use the terms "short form" and "long form" interchangeably with "abbreviation" and "definition". We also use the term "short form" to indicate both abbreviations and acronyms, conflating these as have previous authors.

Many abbreviations in biomedical text follow a predictable pattern, in which the first letter of each word in the long form corresponds to one letter in the short form, as in *methyl methanesulfonate sulfate (MMS)*. However, there are many cases in which the correct match between the short form and long form requires words in the long form to be skipped, or matching of internal letters in long form words, as in *Gcn5-related N-acetyltransferase (GNAT)*. In this paper, we describe a very simple, fast algorithm for this problem that achieves both high recall and high precision.

2   Related Work

Pustejovsky et al.[13, 14] present a solution for identifying abbreviations based on hand-built regular expressions and syntactic information to identify boundaries of noun phrases. When a noun phrase is found to precede a short form enclosed in parentheses, each of the characters within the short form is matched in the long form. A score is assigned that corresponds to the number of non-stopwords in the long form divided by the number of characters in the short form. If the result is below a threshold of 1.5, then the match is accepted. This algorithm achieved 72% recall and 98% on "the gold standard," a small, publicly available evaluation corpus that this group created, working better than a similar algorithm that does not take syntax into account.[b]

Pustejovsky et al.[13] also summarize some drawbacks of other earlier pattern-based approaches, noting that the results of Taghva et al.[17] look good (98% precision and 93% recall on a different test set), but do not account for abbreviations whose letters may correspond to a character internal to a definition word, a common occurrence in biomedical text. They also find that the Acrophile algorithm of Larkey et al.[8] does not perform well on the gold standard.

Chang et al.[5] present an algorithm that uses linear regression on a pre-selected set of features, achieving 80% precision at a recall level of 83%, and 95% precision at 75% recall on the same evaluation collection (this increases to 82% recall and 99% precision on a corrected version).[c] Their algorithm uses dynamic programming to find potential alignments between short and long form, and uses the results of this to compute feature vectors for correctly identified definitions. They then use binary logistic regression to train a classifier on 1000 candidate pairs.

Yeates et al.[19] examine acronyms in technical text. They address a more difficult problem than some other groups in that their test set includes instances that do not have distinct orthographic markers such as parentheses to indicate the

---

[b] There are some errors in the gold standard. The results reported by Pustejovsky et al.[13] are on a variation of the gold standard with some corrections, but the actual corrections made are not reported in the paper. Unfortunately, the corrections needed on the standard are not standardized.
[c] Personal communication, H. Schuetze.

proximity of a definition to an abbreviation (they report that only two thirds of the examples take this form). Their algorithm creates a code that indicates the distance of the definition words from the corresponding characters in the acronym, and uses compression to learn the associations. They compile a large test collection consisting of 1080 definitions; training on two thirds and testing on the remainder, reporting the results on a precision/recall curve.

Park and Byrd[12] present a rule-based algorithm for extraction of abbreviation definitions from general text. The algorithm creates rules on the fly that model how the short form can be translated into the long form. They create a set of five translation rules, a set of five rules for determining candidate long forms based on their length, and a set of six heuristics for determining which definition to choose if there are many potential candidates. These are: syntactic cues, rule priority, distance between definition and abbreviation, capitalization criteria, number of words in the definition, and number of stopwords in the definition. Rule priority is based on how often the rule has been applied in the past. They evaluate their algorithm on 177 abbreviations taken from engineering texts, achieving 98% precision and 95% recall. No mention is made of the size and nature of the training set, or whether it was distinct from the test set.

Yu et al.[21] present another rule-based algorithm for mapping abbreviations to their full forms in biomedical text. Their algorithm is similar to that of Park and Byrd. For a given short form, the algorithm extracts all the candidate long forms that start with the same character as the short form. The algorithm then tries to match the candidate long forms to the short form starting from the shortest long form, by iteratively applying 5 pattern-matching rules. The rules include heuristics such as prioritizing matching the first character of a word, allowing the use of internal letters only if the first letter of a word was matched, and so on. The algorithm was evaluated on a small collection of biomedical text containing 62 matching pairs, achieving 95% precision and 70% recall on average.

Adar[2] presents an algorithm that generates a set of paths through the window of text adjacent to an abbreviation (starting from the leftmost character), and scores these paths to find the most likely definition. Scoring rules used include "for every abbreviation character that occurs at the start of a definition word, add 1", and "A bonus point is awarded for definitions that are immediately adjacent to the parenthesis". After processing a large set of abbreviation-definition pairs, the results are clustered in order to identify spelling variants among the definitions. N-gram clustering is coupled with lookup into the MeSH hierarchy to further improve the clusters. Performance on a smaller subset of the gold standard yielded 85% recall and 94% precision; the author notes that 2 definitions identified by his algorithm should have been marked correct in the standard, resulting in a precision of 95%.[d]

---

[d] Results verified through personal communication with the author.

The work described in this paper arose because the authors found difficulties making the Park and Byrd algorithm work well on biomedical text. The rules it produces are very specific to the format of candidate abbreviations, and so many abbreviations were being represented by patterns that had not yet been encountered by the algorithm, and thus rule priority was not often applicable.

The approach closest to the one we present here is the algorithm of Yoshida et al.[20] Their algorithm assumes that the definition or the abbreviation occurs adjacent to parentheses, but their paper does not state how the length of candidate definitions is determined. Their algorithm scans words from the end of the abbreviation and candidate definition to the beginning, trying at each iteration to find a match for the substring of the abbreviation in the definition. The algorithm assumes that in order for a character from the abbreviation to be represented in the interior of a word in the definition, there must be a match of some other character from the abbreviation on the first letter of that word. In addition, characters that match in the interior of the word must either be adjacent to one another following that initial letter, or adjacent to one another following a syllable boundary. Each iteration of the algorithm requires a check to see if a subsequence can be properly formed according to these rules. They test this algorithm on a very large collection (they had an independent assessor evaluate more that 15,000 categorizations), achieving 97.5% precision and 95.5% recall.

Another important processing issue for abbreviations is disambiguation of multiple senses of the same short form. Pustejovsky et al.[13] describe an algorithm that yields abbreviation sense disambiguation accuracies of 98%, and Pakhomov[9] achieves accuracies of 89% on clinical records.

Yet another issue is normalization of different spellings of the same abbreviation. It is difficult to define what it means for two biomedical terms to refer to the same concept; Cohen et al.[6] provide one set of rules.

3   Methods and Implementation

3.1   Identifying Short Form and Long Form Candidates

The process of extracting abbreviations and their definitions from medical text is composed of two main tasks. The first is the extraction of <short-form, long-form> pair candidates from the text. The second task is identifying the correct long form from among the candidates in the sentence that surrounds the short form. Most approaches, including the one presented here, use a similar method for finding candidate pairs. Abbreviation candidates are determined by adjacency to parentheses.

The two cases are:

    (i) long form '(' short form ')'

    (ii) short form '(' long form ')'

In practice, most <short form, long form> pairs conform to pattern (i). Whenever the expression inside the parentheses includes more than two words, pattern (ii) is assumed, and a short form is searched for just before the left parenthesis (word boundaries are indicated by spaces). Short forms are considered valid candidates only if they consist of at most two words, their length is between two to ten characters, at least one of these characters is a letter, and the first character is alphanumeric. For simplicity, pattern (i) is assumed in the discussion below.

The next step is to identify candidates for the long form. The long form candidate must appear in the same sentence as the short form, and as in Park and Byrd[12], it should have no more than $\min(|A| + 5, |A| * 2)$ words, where $|A|$ is the number of characters in the short form.

Although the algorithm of Park and Byrd allows for an offset between the short and long forms, we consider only long forms that are adjacent to the short form. For a given short form, a long form candidate is composed of contiguous words from the original text that include the word just before the short form.

3.2   Algorithm for Identifying Correct Long Forms

When the previous steps are completed there is a list of long form candidate words for the short form, and the task is to choose the right subset of words. Figure 1 presents the code that performs this task. The main idea is: starting from the end of both the short form and the long form, move right to left, trying find the shortest long form that matches the short form. Every character in the short form must match a character in the long form, and the matched characters in the long form must be in the same order as the characters in the short form. Any character in the long form can match a character in the short form, with one exception: the match of the character at the beginning of the short form must match a character in the initial position of the first (leftmost) word in the long form (this initial position can be the first letter of a word that is connected to other words by hyphens and other non-alphanumeric characters).

The implementation in Figure 1 uses two indices, *lIndex* for the long form, and *sIndex* for the short form. The two indices are initialized to point to the end of their respective strings. For each character *sIndex* points to, *lIndex* is decremented until a matching character is found. If *lIndex* reaches the beginning of the long form candidate list before *sIndex* does, the algorithm returns *null* (no match found).

456

```java
/**
   Method findBestLongForm takes as input a short-form and a long-
   form candidate (a list of words) and returns the best long-form
   that matches the short-form, or null if no match is found.
**/
public String findBestLongForm(String shortForm, String longForm) {
   int sIndex;      // The index on the short form
   int lIndex;      // The index on the long form
   char currChar;   // The current character to match

   sIndex = shortForm.length() - 1;   // Set sIndex at the end of the
                                       // short form
   lIndex = longForm.length() - 1;    // Set lIndex at the end of the
                                       // long form
   for ( ; sIndex >= 0; sIndex--) {   // Scan the short form starting
                                       // from end to start
      // Store the next character to match. Ignore case
      currChar = Character.toLowerCase(shortForm.charAt(sIndex));
      // ignore non alphanumeric characters
      if (!Character.isLetterOrDigit(currChar))
        continue;
      // Decrease lIndex while current character in the long form
      // does not match the current character in the short form.
      // If the current character is the first character in the
      // short form, decrement lIndex until a matching character
      // is found at the beginning of a word in the long form.
      while (
        ((lIndex >= 0) &&
        (Character.toLowerCase(longForm.charAt(lIndex)) != currChar))
        ||
        ((sIndex == 0) && (lIndex > 0) &&
        (Character.isLetterOrDigit(longForm.charAt(lIndex - 1)))))
           lIndex--;
      // If no match was found in the long form for the current
      // character, return null (no match).
      if (lIndex < 0)
        return null;
      // A match was found for the current character. Move to the
      // next character in the long form.
      lIndex--;
   }
   // Find the beginning of the first word (in case the first
   // character matches the beginning of a hyphenated word).
   lIndex = longForm.lastIndexOf(" ", lIndex) + 1;
   // Return the best long form, the substring of the original
   // long form, starting from lIndex up to the end of the original
   // long form.
   return longForm.substring(lIndex);
}
```

Figure 1 – Java Code for Finding the Best Long Form for a Given Short Form

Otherwise, each time a matching character is found, *sIndex* and *lIndex* are decremented. When *sIndex* is at the initial (leftmost) character of the short form, a match is considered only if it occurs at the beginning of a word in the long form. This is accomplished by decrementing *lIndex* until it reaches a non-alphanumerical character or reaches the beginning of the long form. Only then is the character it is pointing to checked for a match against the character *sIndex* points to (this allows for matches in the beginning of the long form just before hyphens). *lIndex* is then decremented until it reaches a space, or the beginning of the long form (whichever comes first), in order to include all the words that are connected, usually by hyphens, to the leftmost matched word in the long form. Finally, the algorithm returns the substring of the original long form, starting from *lIndex* up to the end of the original long form.

To increase precision, the algorithm discards long forms that are shorter than the short form, or that include the short form as one of the words in the long form.[e]

To illustrate the algorithm, consider the following pair *<HSF, Heat shock transcription factor>*. The algorithm starts by setting *sIndex* to point to the end of the short form (*HSF*), and *lIndex* to point to the end of the long form (*factor*). It then decrements *lIndex* until a match is found (*factor*). *sIndex* is decremented by one (*HSF*). *lIndex* is decremented until a match is found (*transcription*). *sIndex* is decremented again (*HSF*). Since *sIndex* now points to the beginning of the short form, the next match should be found at a beginning of a word in the long form. Therefore, *lIndex* is decremented until a valid match is found (*Heat*). Note that another match was skipped (*shock*) because it was not in the beginning of a word. Also note, that although the algorithm did not match the second character correctly (*transcription* instead of *shock*) it still found the right long form.

To illustrate when the algorithm might fail, consider the following example. *<TTF-1, Thyroid transcription factor 1>*. In this case the algorithm finds the following wrong match *<TTF-1, Thyroid transcription factor 1>*. Our experiment results show that this kind of error is very rare.

The algorithm is based on the observation that it is very rare for the first character of the short form to match an internal letter of the long form. By adding the constraint that the first character of the short form matches the beginning of a word in the long form, together with the limitation on the length of the long form, the precision is increased by removing most of the false positives, without significantly reducing the recall. By contrast, adding additional constraints, as is done by most other algorithms, does not seem to help much in terms of precision, but can severely reduce the recall. To illustrate this point consider the results of Yu et al.[21], which is a similar algorithm to ours, but has additional constraints. While the precision of both algorithms is very similar, the recall of our algorithm is higher.

---

[e] This part of the algorithm is omitted from Figure 1.

458

4   Evaluation and Results

To evaluate the algorithm, 1000 MEDLINE abstracts were randomly selected from the results of a query on the term "yeast". These were then hand tagged, producing a list of 954 correct <short form, long form> pairs. The algorithm was also tested against a publicly available tagged corpus, the Medstract Gold Standard Evaluation Corpus,[22] which includes 168 <short form, long form> pairs.

On a corrected version of the gold standard, the algorithm identified 143 pairs. Out of these, 137 pairs were correct, resulting in a recall of 82% at precision of 96%. For comparison, the algorithm described in Chang et al.[5] achieved 83% recall at 80% precision, and that of Pustejovsky et al.[14] achieved 72% recall at 98% precision.[f]

Analysis of the 6 incorrect pairs reveals that in actuality, 2 of them are correct, but were overlooked by the creators of the gold standard.[g] The other 4 pairs are counted as incorrect, since they only partially matched the correct long form. For example, the algorithm found the pair <Pol II, polymerase II> instead of <Pol II, RNA polymerase II>. Allowing for reasonable partial matches, as was done in Chang et al.[5] and considering the 2 missing pairs as correct, the precision is increased to 99%, and the recall to 84%.

The algorithm missed 31 pairs: 9 (38%) pairs have skipped characters in the short form (e.g. <CNS1, cyclophilin seven suppressor>), 7 (23%) do not have any pattern match between the short form and long form (e.g. <5-HT, serotonin>), 5 (16%) have an out of order match (e.g. <ATN, anterior thalamus>), for 3 (10%) pairs the long form includes an additional words to the left of the match resulting in a partial match (e.g. <Pol I, RNA polymerase I>), 2 pairs have a short definition inside the parenthesis, 2 pairs have the long form inside parenthesis and the short form inside nested parenthesis, 1 pair has a comma in the long form, 1 pair has no parenthesis, and for 1 pair the algorithm found a wrong partial match (see the example at the end of section 3).

---

[f] It is important to note that because of the errors in the gold standard these results cannot be accurately compared. Each of these evaluations used its own interpretation of the standard, fixing different parts of it. In our evaluation, we followed the guidelines of Pustejovsky et al.[14] (received through personal communication), since they have developed and maintained the standard. We did not include here the results of Adar[2], since he used a subset of the original gold-standard with only 144 pairs, which did not include most of the pairs missed by the other algorithms.

[g] The two missing pair are <l'sc, lethal of scute>, and <cAMP, 3',5' cyclic adenosine monophosphate>. The second long form was only partially extracted by the algorithm (without the leading numbers).

On the larger test collection, the algorithms identified 827 pairs. Out of these, 785 pairs were correct, resulting in a recall of 82% at precision of 95%.[h] Analysis of the 42 incorrect pairs reveals that 17 are completely incorrect, 12 pairs have a matched long form that is a superset of the correct long form (this happens when the correct mapping includes unused characters, out of order mappings, first-character matches to internal letters, or when the first word of the correct long form is connected by hyphens to preceding words), 11 are partial matches where the extracted long form is a subset of the correct long form (this happens when the algorithm is able to match all the characters of the short form to a subset of the long form, and when the correct match does not start from the first word of the long form), and for 2 pairs the extracted long form includes left brackets that are not part of the correct long form.

Out of the 169 missed pairs, 70 (41%) have unused characters in the short form, 23 (14%) have an out of order match, 12 (7%) have first-character matches to internal letters, 12 (7%) have nested parentheses, 11 (7%) have some kind of transformation involved in their mapping (like 2D -> two-dimensional), 11 (7%) have partial matches (see above), 5 (3%) have a non-continuous long forms, 4 (2%) involve multiple concurrent definitions, 4 (2%) have a short form of only one character, and the rest of the pairs, 19 (11%), have miscellaneous issues.

5   An Alternative Algorithm

When we began to investigate the problem of abbreviation definition identification, we devised a much more complex algorithm than that presented here. This algorithm uses the representation of Park and Byrd[12] in combination with a variation on the decision lists algorithm, as applied by Yarowsky[18] to the lexical ambiguity resolution task. The algorithm makes use of training data to rank features that are combinations of matching rule transformations.

Space restrictions prevent detailed description of that algorithm (the interested reader should refer to Schwartz and Hearst[16] for a complete description of the algorithm). However, we found that it performed mildly better than our simple algorithm on both training sets, achieving for the gold standard 97% precision and 82% recall, which is a reduction in error of 17% over the simpler algorithm. For the larger test collection, it achieves 96% precision and 82% recall, which is an error reduction of 22% over the simpler algorithm.

---

[h] The dataset was originally annotated by a graduate student in computational and biosciences. We furthered verified the data by comparing any questionable pairs against other occurrences of the same abbreviation in other abstracts, using the web site provided by Chang et al.[5] A pair extracted by the algorithm is considered correct only if it exactly matches a pair labeled in the dataset.

Because the simple algorithm is so much easier to implement and requires no training data, we recommend its use, in combination with checking the entire dataset for redundancy in definitions in order to further reduce the error rates.

6   Conclusions

In this paper we introduced a new algorithm for extracting abbreviations and their definitions from biomedical text. Although the algorithm is extremely simple, it is highly effective, and is less specific – and therefore less potentially brittle – than other approaches that use carefully crafted rules. Although we are staunch advocates of machine learning approaches for problems in computational linguistics, it seems that in the case of this particular problem, simpler is better. One can argue that the problem may vary across collections or languages, and so machine learning can help in these cases, but our experience with a machine learning approach to sentence boundary determination[11] suggests that most practitioners do not want to bother with labeling training data for relatively simple tasks.

Another advantage of the simplicity of the algorithm is its fast running time performance. The task of extracting the definition of an abbreviation, is a common pre-processing step of larger multi-layered text-mining tasks.[1, 10] Therefore, it is essential that this step be as efficient as possible. Since our algorithm needs to consider only one possible long form per short form, it is much faster than the alternative algorithms that first extract many possible long forms and then pick the best of them. To provide a rough comparison, using an IBM T21 laptop with a single CPU (800 MHz, 256 Mb RAM) running MS-Windows 2000, it takes our algorithm about 1 second to process 1000 abstracts, while the algorithm in Chang et al.[5] using a 5 processor Sun Enterprise E3500 server, processed only 25.5 abstracts per second. While our algorithm is clearly I/O bound (running time depends almost entirely on the time it takes to read the files from disk, and write the results back to the disk), the algorithm of Chang et al. seem to be heavily CPU bound.

The algorithm performs better or the same as the best results of other work, with the possible exception of that of Yoshida et al.[20] However, the main advantage of the proposed algorithm over the alternatives is its simplicity, and transparency. It was implemented with 260 lines of Java code and requires no training data to run. The Yoshida et al. algorithm is more complex, in that it requires a module for recognizing syllable boundaries, and it performs a substring check at each iteration of the loop.

Analysis of the errors produced indicates that further improvement of the algorithm requires the use of syntactic information, as suggested in Pustejovsky et al.[13] Shallow parsing of the text as a preprocessing step might help correct some of the errors inherent in the algorithm, by helping to identify the noun phrases near the abbreviations. In addition, combining evidence from more than one MEDLINE abstract at a time, as was done in Adar[2], might also prove to be beneficial for increasing both precision and recall. Finally, the algorithm currently only considers candidate definitions when the abbreviation is enclosed in parentheses (and vice versa); finding all possible pairs is a more difficult problem and requires additional study.

Acknowledgements

We thank Barbara Engelhardt for producing the marked-up data for the large test collection, and Jeff Chang and Hinrich Schuetze, Youngja Park and Roy Byrd, Jose Castaña and James Pustejovsky, and Eytan Adar for providing details about their data. This research was supported in part by a gift from Genentech, and in part by a grant from ARDA.

References

1. L. A. Adamic, D. Wilkinson, B. A. Huberman, and E. Adar, "A Literature Based Method for Identifying Gene-Disease Connections" *Proceedings of the 2002 IEEE Computer Society Bioinformatics Conference*, Stanford, CA, August 2002.
2. E. Adar, "S-RAD: A Simple and Robust Abbreviation Dictionary" *HP Laboratories Technical Report*, September 2002.
3. M.A. Andrade et al. "Automatic annotation for biological sequences by extraction of keywords from MEDLINE abstracts. Development of a prototype system." *ISMB 1997*. **5**: 25-32.
4. C. Blasche et al. "Automatic extraction of biological information from scientific text: protein-preotein interactions" *ISMB 1999*. **7**: 60-67.
5. J.T. Chang, H Schütze, and R.B. Altman, "Creating an Online Dictionary of Abbreviations from MEDLINE" *JAMIA*, to appear.
6. B. Cohen, A. E. Dolbey, G. K. Acquaah-Mensah, and L. Hunter, "Contrast and variability in gene names" *Proceedings of the ACL Workshop on Natural Language Processing in the Biomedical Domain*, Philadelphia, PA, July 2002.
7. M. Craven, and J Kumlien, "Constructing Biological knowledge Bases by Extracting information from Text Sources" *Proceedings of the 7th International Conference on Intelligent Systems for Molecular Biology*, 77-86, Heidelberg, Germany. AAAI Press.
8. L.S. Larkey, P. Ogilvie, M.A. Price, and B. Tamilio, "Acrophile: an automated acronym extractor and server" *Proceedings of the ACM Fifth International Conference on Digital Libraries*, DL '00, Dallas TX, May 2000.

9. S. V. Pakhomov, "Semi-supervised Maximum Entropy-based Approach to Acronym and Abbreviation Normalization in Medical Texts" *Proceedings of ACL 2002*, Philadelphia, PA, July 2002.

10. M. Palakal, M. Stephens, S. Mukhopadhyay, R. Raje, and S. Rhodes, "A Multi-level Text Mining Method to Extract Biological Relationships" *Proceedings of the 2002 IEEE Computer Society Bioinformatics Conference*, Stanford, CA, August 2002.

11. D. Palmer, and M. Hearst, "Adaptive Multilingual Sentence Boundary Disambiguation" *Computational Linguistics*, 23 (2), 241-267, June 1997.

12. Y. Park, and R.J. Byrd, "Hybrid Text Mining for Finding Abbreviations and Their Definitions" *Proceedings of the 2001 Conference on Empirical Methods in Natural Language Processing*, Pittsburgh, PA, June 2001: 126-133

13. J. Pustejovsky, J. Castaño, B. Cochran, M. Kotecki, M. Morrell, and A. Rumshisky, "Extraction and Disambiguation of Acronym-Meaning Pairs in Medline" unpublished manuscript, 2001.

14. J. Pustejovsky et al. "Automation Extraction of Acronym-Meaning Pairs from Medline Databases" *Medinfo* 2001;10 (Pt 1): 371-375.

15. T.C. Rindflesch et al. "Extracting Molecular Binding Relationships from Biomedical Text" *Proceeding of the ANLP-NAACL 2000*, pages 188-195 Association for Computational Linguistics, Seattle, WA, 2000.

16. A. Schwartz, and M. Hearst, "A Rule Based Algorithm for Identifying Abbreviation Definitions in Biomedical Text Using Decision Lists" *University of California, Berkeley, Technical Report*, to appear.

17. K. Taghva, and J. Gilbreth, "Recognizing Acronyms and their Definitions" *IJDAR 1* (4), 191-198, 1999.

18. D. Yarowsky, "Decision Lists for Lexical Ambiguity Resolution: Application to Accent Restoration in Spanish and French" *Proceedings of the ACL, 1994*: 88-95

19. S. Yeates, D. Bainbridge, and I. H. Witten, "Using compression to identify acronyms in text" *in Data Compression Conference*, 2000.

20. M. Yoshida, K. Fukuda, and T. Takagi, "PNAD-CSS: a workbench for constructing a protein name abbreviation dictionary" *Bioinformatics*, 16 (2), 2000.

21. H. Yu, G. Hripcsak, and C. Friedman, "Mapping abbreviations to full forms in biomedical articles" *J Am Med Inform Assoc 2002*; 9(3): 262-272.

22. http://www.medstract.org/gold-standard.html

HUMAN GENOME VARIATION: HAPLOTYPES, LINKAGE DISEQUILIBRIUM, AND POPULATIONS

FRANCISCO M. DE LA VEGA

Applied Biosystems, 850 Lincoln Centre Dr., Foster City, CA 94404, USA
E-mail: delavefm@appliedbiosystems.com

KENNETH K. KIDD

Department of Genetics, Yale University School of Medicine
333 Cedar Street, New Haven, CT 06520, USA
E-mail: kidd@biomed.med.yale.edu

ISAAC S. KOHANE

Children's Hospital Informatics Program & Harvard Medical School
320 Longwood Avenue Boston, MA 02115, USA
E-mail: Isaac_Kohane@Harvard.edu

The working draft of a reference sequence of the human genome is nearing completion and is providing a basis for studies in a variety of domains. Computational challenges exist in all of these domains because of the massive amounts of data, the multiple often complex relationships among the types of relevant data, and the need to make the data accessible to researchers approaching the data from different perspectives. One aspect of genomic data of broad relevance is the variation in the DNA sequence among the billions of separate copies existing in the several billion living humans. Some of that variation is "abnormal" and the basis for inherited diseases. However, most of the variation is normal and simply makes each of us unique. Yet this common, normal variation is of great biomedical relevance because it can alter disease susceptibility, physiologic reactions to drugs, and response to environmental stimulus. The variation is also relevant to anthropology and understanding human evolution. In fact, several aspects of DNA sequence variation are consequences of recent human evolution: the amount of variation, the distribution of variation among human populations, and the organization of variation along the DNA sequence. This last issue has become increasingly interesting as millions of single nucleotide polymorphisms (SNPs) have been identified and mapped. With multiple SNPs mapped to every small segment of DNA the focus has shifted from the individual SNP to considering groups of SNPs as haplotypes (haploid genotypes) with the common finding that for n SNPs in a small segment of DNA there are usually far fewer than the 2^n haplotypes expected by chance. This non-randomness, commonly referred to as linkage disequilibrium (LD), is adding an additional level of complexity to genetic databases and analytic programs.

In previous years this session of the Pacific Symposium on Biocomputing has dealt with many of the issues related to DNA sequence variation. Last year the focus was relating genotype to phenotype and the handling of the data generated by high-throughput genotyping technologies. This year the focus has shifted to haplotypes and their computational challenges. Nine accepted manuscripts comprise this year's original work presented at the conference.

The recent empirical findings about the patterns of LD across samples of the genome have generated enormous interest, since the extent and intensity of this parameter relates to the statistical power that a set of SNP markers can have in association studies. In particular, the observation of large "blocks" of strong LD and low haplotype diversity [1,2] predicts that, at least in those areas, power would be adequate and that a highly dense SNP map could be redundant if the aim is to detect the common haplotypes. The contribution of Avi-Itzhak *et al.* to this volume is to present a simple algorithm to select minimum subsets of SNPs that can "tag" all the common haplotypes of a block. Utilizing Shannon Entropy as the metric of the haplotype diversity within the block, the authors find that the algorithm can efficiently reduce the number of SNPs required for genotyping, the extent to which depends on the population studied. Is the very concept of haplotype block what is studied in the manuscript of Koivisto *et al.* Here, the authors describe a new method for finding haplotype blocks based on the use of the minimum description length principle, paying special attention to the robustness of the block boundary predictions. The authors compare the accuracy of this new methodology with other published methods.

Many of the theoretical frameworks developed for studying the patterns of genetic variation in the Human genome and their application for mapping complex disease assume that in a given experiment, the researcher can obtain accurate data about the variants present in a set of DNA samples. However, as is clear for those who have spent some time as experimentalists, genotyping error is inherent in routine laboratory work. In their contribution, Gordon and colleagues describe what the interaction between genotyping error and LD is, and how this affects statistical power and the required sample size for case-control studies. Their report suggest that high LD can mitigate the reduction in power induced by genotyping errors, and that marker selection for genetic studies needs to take this additional factor into account. Comeron *et al.,* deal with the requirements necessary to detect quantitative trait loci for complex disease from a theoretical point of view. Using results from coalescent simulations, the authors suggest that the use of a highly dense SNP map does not necessarily results in increased power, unless recombination is high in the region under study. Now that empirical data on LD is becoming available, the comparison between the theoretical predictions the authors' present and the experimental data, would test our understanding on the origin of genomic variation. On the other hand, Rannala and Reeve outline a new Bayesian method to jointly estimate the position of a disease mutation and its age utilizing LD. The work of these authors is a valuable addition to the analytical tools needed for analysis of the impending volumes of genotyping data coming from the many large scale projects that are underway.

Novel methods to visualize and analyze large amounts of genotyping data that are robust and can be applied in the clinic are urgently needed. Tsalenko *et al.* present a number of such methodologies with the aim of selecting subsets of SNPs that can predict disease state and can be used for patient classification. By using data available in the public domain from a previous study, the authors demonstrate the utility of the proposed methods. In another contribution to the conference, Lancaster *et al.* address the issue of the availability of tools for analysis of genetic studies, by describing a software framework for large scale analysis of Human population data that undoubtedly would be extremely useful for those in the community just getting started with efforts of such magnitude. Finally, Stryke *et al.* remind us that these large genetic studies are in many cases product of multidisciplinary collaborations, where easy and efficient access to the results of the data analysis is crucial. These authors present their implementation for handling the data analysis flow of one of such projects.

As the papers in this session illustrate, the biocomputing aspects of human genome variation are numerous and diverse. There is ongoing need for innovation and implementation of new approaches to help individual researchers manage their own data and to make these complex data readily accessible to the broad scientific community.

Acknowledgements

We would like to acknowledge the generous help of the anonymous reviewers that supported the selection process for this session, as well as the panelists that joined us to discuss the challenges in this field.

References

1. Daly, M.J., Rioux, J.D., Schaffner, S.F., Hudson, T.J. and Lander, E.S. *Nat Genet* 29:229-232 (2001).
2. Gabriel, S.B., Schaffner, S.F., Nguyen, H., Moore, J.M., Roy, J., Blumenstiel, B., Higgins, J., DeFelice, M., Lochner, A., Faggart, M., Liu-Cordero, S.N., Rotimi, C., Adeyemo, A., Cooper, R., Ward, R., Lander, E.S., Daly, M.J. and Altshuler, D. *Science* 296:2225-2229 (2002).

SELECTION OF MINIMUM SUBSETS OF SINGLE NUCLEOTIDE POLYMORPHISMS TO CAPTURE HAPLOTYPE BLOCK DIVERSITY

HADAR I. AVI-ITZHAK[1,2], XIAOPING SU[1], FRANCISCO M. DE LA VEGA[1]

[1]Applied Biosystems, Bioinformatics R&D, 850 Lincoln Center Drive, Foster City CA 94404, and [2]Imagenix Corporation, P.O. Box 735, Los Altos CA 94023

We present a simple numerical algorithm to select the minimal subset of SNPs required to capture the diversity of haplotype blocks or other genetic loci. This algorithm can be used to quickly select the minimum SNP subset with no loss of haplotype information. In addition, the method can be used in a more aggressive mode to further reduce the original SNP set, with minimal loss of information. We demonstrate the algorithm performance with data from over 11,000 SNPs with average spacing of 6 to 11 Kb, across all the genes of chromosomes 6, 21, and 22, genotyped on DNA samples of 45 unrelated African-Americans and 45 Caucasians from the Coriell Human Diversity Collection. With no loss of information, we reduced the number of SNPs required to capture the haplotype block diversity by 25% for the African-American and 36% for the Caucasian populations. With a maximum loss of 10% of haplotype distribution information, the SNP reduction was 38% and 49% respectively for the two populations. All computations were performed in less than 1 minute for the entire dataset used.

1   Introduction

Single nucleotide polymorphisms (SNPs) are the most abundant form of genetic variation. These single nucleotide changes are found approximately every 500 base pairs (bp) in the human genome. Almost all SNPs are bi-allelic (i.e., only two different alleles exist), and typically one allele is present in the majority of the chromosomes of a population, and the alternative variant (i.e., the minor allele) is present with less frequency.

SNPs are promising tools for mapping susceptibility mutations that contribute to complex diseases. Although most SNPs are neutral (i.e., do not affect phenotype), they can be used as surrogate markers for positional cloning of genetic loci, because of the allelic association, known as linkage disequilibrium (LD), that can be shared by groups of adjacent SNPs. LD is eroded by gene conversion [3] and recombination [12], and the amount of LD depends on the age of the mutations and on the demographic history of the population. The extent of LD across a genomic region dictates the density of SNP markers necessary to ensure association between a marker and the causative allele sought.

Early attempts to model the extent of LD on theoretical grounds predicted very short regions of LD, extending only a few kilobases (Kb) [13]. However, empirical surveys reported average LD distances between 5 Kb and 60 Kb, with the upper range extending up to hundreds of Kb [2,4,14]. More recently, studies have reported a discontinuous structure in the patterns of LD across specific genomic regions, where long stretches of strong LD are punctuated by recombination hot-spots [4,8,9]. These LD "blocks" show little evidence of historical recombination. These results suggest that a reduced set of contiguous chromosomal segments, or haplotypes, exist in specific populations. For example, for a block spanning tens of Kbs for which 10 SNPs exist, instead of the 2^{10} theoretically possible haplotypes, it has been found that 95% of the haplotype diversity is made up of only 4 to 6 so-called common haplotypes [8].

It is noteworthy that these LD block patterns change depending on the population sampled because of historical differences; for example, populations that have experienced bottlenecks (e.g., Caucasians) show longer LD blocks and less evidence of historical recombination events than other populations [8]. The haplotype diversity in a given population is typically constant in a given block irrespective of the number of SNPs sampled [8]; therefore typing an arbitrarily large number of SNPs within a LD block is unnecessary. Selecting the minimum subset of SNPs within LD blocks, or any other discrete genetic loci, that enable discrimination of the common haplotypes present in a block without loss of information is the optimum approach.

Most experimental methods for typing the specific SNP alleles present in a DNA sample produce unphased genotypes (i.e., the alleles detected cannot be assigned to either the maternal or the paternal chromosome). Although cumbersome methods exist to directly determine haplotypes, algorithms are widely used to infer the haplotypes from genotypes using maximum-likelihood or Bayesian principles. Family relationships of the DNA donors, if available, can be used to increase haplotype inference accuracy. Even in absence of family information, the Expectation-Maximization (EM) algorithm introduced by Excoffier and Slatkin [7] is quite accurate in most realistic situations, especially in regions of low diversity [16]. The analysis of haplotype distributions has been reported to provide more power for finding associations in genetic studies undertaken to find susceptibility mutations in case-control populations [6].

2   The Algorithm for Determining the Minimum Set

Given a block containing N SNPs and M haplotypes, define P, a probability vector of length M, where P_i is the relative frequency of the ith haplotype. Also define A, a haplotype/SNP allele state matrix of N columns and M rows, where A_{ij}, (the ith row of the jth column of this matrix) indicates the allele state ('1' or '2') of the jth SNP for the ith haplotype. The algorithm will eliminate as many columns of A, while preserving as much of the information in P as possible. For the purpose of quantifying the information in P, we used the information-theoretic quantity known as Shannon Entropy [15] as the measure of haplotype diversity information within each LD block [11]:

$$H = -\sum_{i=1}^{M} P_i \log_2(P_i)$$

However, it is quite useful in many cases to use the algorithm in *lossless* mode, in which case it is irrelevant which information measure is used for the haplotype distribution. In addition, the algorithm can use any other measure of information preferred by the user.

2.1 Lossless Mode

The algorithm consists of two sequentially performed phases. Phase I is *elimination by columns*. Phase II is *elimination by rows*. Initially, we will define these operations in *lossless* mode only:

Phase I: **Any column that is identical to another column, or is the exact opposite of another column, can be eliminated.**

A column that is identical to another column represents a SNP that behaves identically to another SNP for all tested samples. Thus, under the assumption that we have enough DNA samples to infer the major haplotypes, this redundant SNP will not provide any useful information. Similarly, a column that is the exact opposite of another column is a SNP whose behavior can always be predicted from the behavior of another SNP simply by inverting it; therefore this SNP will not provide new information. Note that merely selecting a set of basis vectors from the matrix A would often miss elimination of columns that are an exact opposite of other columns. The 2x2 identity matrix illustrates this.

Assume that after phase I, the N columns of matrix A have been reduced to N' unique columns where $N' \leq N$.

Phase II: **Any column whose elimination does not reduce the number of unique rows should be eliminated.**

Each row represents the allelic states of the SNPs for a specific haplotype. Removing a "useful" SNP would eliminate the ability to detect at least one haplotype. In such a case, two or more haplotypes would register the same allelic states at the remaining SNPs, thereby reducing the number of unique rows. Therefore, if the elimination of a column does not reduce the number of unique rows, it can be omitted.

Note that phase II actually subsumes phase I, in the sense that if phase I were skipped, phase II would eliminate the SNPs that phase I would have eliminated. However, without the "pruning" that takes place in phase I, phase II can quickly become computationally unmanageable, especially in *lossy* mode.

The following example illustrates the method. Table 1 shows four SNPs and the four haplotypes they participate on a hypothetical LD block.

Table 1. Haplotype/SNP Allele State Matrix

	SNP_1	SNP_2	SNP_3	SNP_4
Haplotype 1	1	1	1	2
Haplotype 2	2	2	1	1
Haplotype 3	2	2	2	1
Haplotype 4	1	2	2	2

The fourth column is the exact opposite of the first column. This implies that either SNP_4 or SNP_1 is redundant. If SNP_4 is removed from the SNP set, no information is lost. When SNP_1 registers allele "1", the state of SNP_4 is known as allele "2", and conversely, when SNP_1 registers allele "2", the state of SNP_4 is known as allele "1". Removing SNP_4 leaves the matrix in Table 2.

Table 2. Haplotype/SNP Allele State Matrix after Phase I

	SNP_1	SNP_2	SNP_3
Haplotype 1	1	1	1
Haplotype 2	2	2	1
Haplotype 3	2	2	2
Haplotype 4	1	2	2

The three columns are unique (including accounting for opposites), thus phase I is complete. $N = 4$ has been reduced to $N' = 3$, as phase II is entered.

Table 3 depicts the three remaining matrices, following the removal of SNP_1, SNP_2, or SNP_3, respectively. The first and the third matrices only have three unique rows, whereas the second matrix has four unique rows. Thus, if the haplotype list is exhaustive, we can eliminate SNP_2 with no loss of haplotype detection.

Table 3. Three Possible Haplotype/SNP Allele State Matrices

	SNP_2	SNP_3		SNP_1	SNP_3		SNP_1	SNP_2
Haplotype 1	1	1		1	1		1	1
Haplotype 2	2	1		2	1		2	2
Haplotype 3	2	2		2	2		2	2
Haplotype 4	2	2		1	2		1	2

We have shown that the set $\{SNP_1, SNP_3\}$ provides the same haplotype detection ability as the full set $\{SNP_1, SNP_2, SNP_3, SNP_4\}$. In the above example, each phase caused the elimination of exactly one SNP. However, in general, each phase could result in the elimination of multiple SNPs, or possibly none.

2.2 Lossy Mode

An alternative version of the algorithm can be used to further reduce the retained SNP set while minimizing the loss of haplotype detection. Phase I will remain unchanged, however phase II will now select the optimal SNPs to eliminate by performing an exhaustive search.

Phase II: **For the** $\binom{N'}{k}$ **possible selections of k SNPs, compute the entropy** H **for the resulting** P**. Choose the selection with the highest** H **as the best selection.**

When only k out of N' SNPs are selected, $N' - k$ columns are eliminated. The resulting matrix (with k columns) could have fewer unique rows than the full matrix (with N' columns). When a row is repeated more than once, it implies that several "minor" haplotypes will now be measured as a single "major" haplotype. This is because with fewer SNPs, we have lost the ability to make the finer distinction between them. The relative frequency (probability) of this "major" haplotype is equal to the sum of the frequencies of the "minor" haplotypes. Thus, when elimination of columns results in repeating rows, the repeating rows can be combined into a single row, and their respective probabilities summed to form a new probability. The vector P will be shorter and have larger numbers. This will always reduce the value of the entropy, H. The combination with the smallest reduction of entropy is deemed the optimal selection. Obviously, if all the rows are unique after elimination of $N' - k$ columns, the entropy is not reduced, and k SNPs can be used with no loss of information, as in the *lossless* case.

3   Implementation and Test Data

The algorithm was implemented in MATLAB v6.1 (The MathWorks Inc., Natick, MA, USA) without further optimization. The computations were completed on a 700MHz PC for all the 2,874 blocks in our test dataset in less than 1 minute. To validate and assess the utility of the algorithm on a realistic dataset, we used genotyping data from 11,160 SNPs distributed in a gene-centric fashion across chromosomes 6, 21, and 22 (see Table 7 below). The SNPs were scored with 5'nuclease assays with TaqMan®-MGB probes from Applied Biosystems' Assays-on-Demand™ SNP Genotyping Products (Foster City, CA, USA)[5]. The samples typed included 45 African-American and 45 Caucasian DNAs from the Coriell Human Diversity Collection (Coriell Institute for Medical Research, Camden, NJ, USA). LD blocks and haplotypes were computed independently for each population using methods described elsewhere [1,8]. Only blocks of 3 or more SNPs were considered in this study. Therefore, 4,864 SNPs were used for the African-American population and 7,347 SNPs were used for the Caucasian population, which is known, in general, to have more and longer LD blocks.

4   Experimental Results

To exemplify the usefulness of the algorithm, we applied it in *lossy* mode to an LD block that we discovered using the Caucasian population panel, in chromosome 6, overlapping the human gene *TTK* (RefSeq ID NM_003318, Celera ID hCG401205). The block consists of 17 SNPs, and the EM algorithm inferred 8 haplotypes, including two major haplotypes: haplotype 2 and haplotype 7 with frequencies of approximately 43% and 33% respectively. The remaining 24% of the diversity is spread among the remaining 6 haplotypes. Table 4 below summarizes the allelic states of the 17 SNPs, as well as the respective probability, for each of the 8 haplotypes.

Table 4. Original Haplotype/SNP Allele State Matrix

Haplotype Number	P	1	2	3	4	5	6	7	8	9	10	11	12	13	14	15	16	17
1	0.1136	1	1	1	1	1	2	1	2	1	2	1	1	2	2	2	1	2
2	0.4318	1	1	1	1	1	2	1	2	1	2	1	1	2	2	2	2	2
3	0.0114	1	1	1	2	2	1	2	1	2	1	2	2	1	1	1	1	1
4	0.0454	1	2	2	1	1	2	1	2	1	2	1	1	2	2	2	1	1
5	0.0454	2	1	1	1	1	2	1	2	1	2	1	1	2	2	2	1	2
6	0.0118	2	2	2	2	2	1	2	1	2	1	2	1	1	1	1	1	1
7	0.3292	2	2	2	2	2	1	2	1	2	1	2	2	1	1	1	1	1
8	0.0114	2	2	2	2	2	1	2	1	2	2	1	2	1	1	1	1	1

After running phase I of the algorithm, the number of SNPs is reduced to 7, with the remaining SNP set {SNP1, SNP_2, SNP_4, SNP_{10}, SNP_{12}, SNP_{16}, SNP_{17}}. All the haplotype information is preserved, including their distribution, as shown in Table 5 below. The entropy of the original distribution of haplotypes is $H(P) = 2.0351$ bits.

Table 5. Haplotype/SNP Allele State Matrix After Phase I

Haplotype Number	P	SNP_1	SNP_2	SNP_4	SNP_{10}	SNP_{12}	SNP_{16}	SNP_{17}
1	0.1136	1	1	1	2	1	1	2
2	0.4318	1	1	1	2	1	2	2
3	0.0114	1	1	2	1	2	1	1
4	0.0454	1	2	1	2	1	1	1
5	0.0454	2	1	1	2	1	1	2
6	0.0118	2	2	2	1	1	1	1

Table 5. Haplotype/SNP Allele State Matrix After Phase I (continued)

7	0.3292	2	2	2	1	2	1	1
8	0.0114	2	2	2	2	2	1	1

After running phase II of the algorithm in *lossless* mode, it is apparent that the number of SNPs in the set can be reduced to 5 without any loss of information (Table 6). Reducing the set to 4 SNPs causes haplotype 8, the rarest haplotype, to merge into haplotype 7 because we have lost the ability of distinguish between the two. As a result 3.5% of the entropy is lost in the resultant haplotype distribution. The least amount of entropy that can be lost by reducing the SNP set to 3 is 9.2% when haplotypes 3 and 4 are merged, and haplotypes 6, 7, and 8 are merged.. Interestingly, the optimal single SNP is SNP_{16}. With SNP_{16} , we can detect only "haplotype 2" or "other". Haplotype 2 is the most common haplotype, with 43.2% of the frequency. Therefore it seems intuitively obvious that if we were only allowed to choose a single SNP, SNP_{16} would be our most useful choice.

Table 6. Lossy Min. SNP Set Example

No. of SNPs (k)	No. of Combinations $\binom{7}{k}$	Optimal Set of k SNPs	Haplotype Distribution Resulting from the Optimal SNP Set	Resulting Entropy (H) (bits)
7	1	{ SNP_1, SNP_2, SNP_4, SNP_{10}, SNP_{12}, SNP_{16}, SNP_{17}}	(0.114, 0.432, 0.011, 0.045, 0.045, 0.012, 0.329, 0.011)	2.0351
6	7	{ SNP_1, SNP_4, SNP_{10}, SNP_{12}, SNP_{16}, SNP_{17}}	(0.114, 0.432, 0.011, 0.045, 0.045, 0.012, 0.329, 0.011)	2.0351
5	21	{ SNP_1, SNP_{10}, SNP_{12}, SNP_{16}, SNP_{17}}	(0.114, 0.432, 0.011, 0.045, 0.045, 0.012, 0.329, 0.011)	2.0351
4	35	{ SNP_1, SNP_{12}, SNP_{16}, SNP_{17}}	(0.114, 0.432, 0.011, 0.045, 0.045, 0.012, 0.341)	1.9631
3	35	{ SNP_1, SNP_{16}, SNP_{17}}	(0.114, 0.432, 0.057, 0.045, 0.352)	1.8475
2	21	{ SNP_{12}, SNP_{16}}	(0.216, 0.432, 0.352)	1.5311
1	7	{SNP_{16} }	(0.5682 , 0.4318)	0.9865

Table 7 below summarizes the results after applying the algorithm to the full dataset of haplotype blocks detected for chromosomes 6, 21, and 22. The African-American population panel is denoted by 'A' and the Caucasian population panel is denoted by 'C'. The results show that the mean number of SNPs per block is reduced by on average 36% for the Caucasian population, whereas the reduction for the African-American panel is 25%, using the *lossless* mode of the algorithm. Using the *lossy* mode and accepting <10% information loss, the algorithm accomplishes an additional 17% reduction in the average number of SNPs per block for the African-American panel, versus 19% for the Caucasian samples.

Table 7. Experimental Results Summary

Chr.	Pop.	Total No. of SNPs	Mean Intragenic Spacing Between SNPs (bp)	No. of Haplotype Blocks	Mean Block Size (bp)	Mean No. of SNPs per Block	Mean No. of Min. SNP per block	
							Lossless	<10% Loss
6	A	2,504	10,840	646	23,000	3.88	2.94	2.44
	C	4,009	10,630	883	34,000	4.54	2.86	2.33
21	A	955	7,382	242	14,933	3.95	2.92	2.39
	C	1,555	7,031	336	21,032	4.63	2.88	2.32
22	A	1,405	6,035	350	13,714	4.01	2.99	2.47
	C	1,783	7,760	417	17,505	4.28	2.81	2.27

Figure 1 graphically illustrates the relationship between the original number of SNPs in an LD block (horizontal axis) and the minimum number of SNPs required to genotype the LD block with no loss of information (vertical axis). The thickness of the 'x' corresponds to the number of different blocks found in chromosome 6 with the same properties. This figure shows that irrespective of the original number of SNPs per LD block (up to 18), the maximum number of minimum SNPs levels off at about 6 for the Caucasian population, and is not too different for the African-American panel which has a few outlier blocks with 7 and 9 minimum SNPs.

Figure 1. SNP per block vs. minimum informative SNP subset

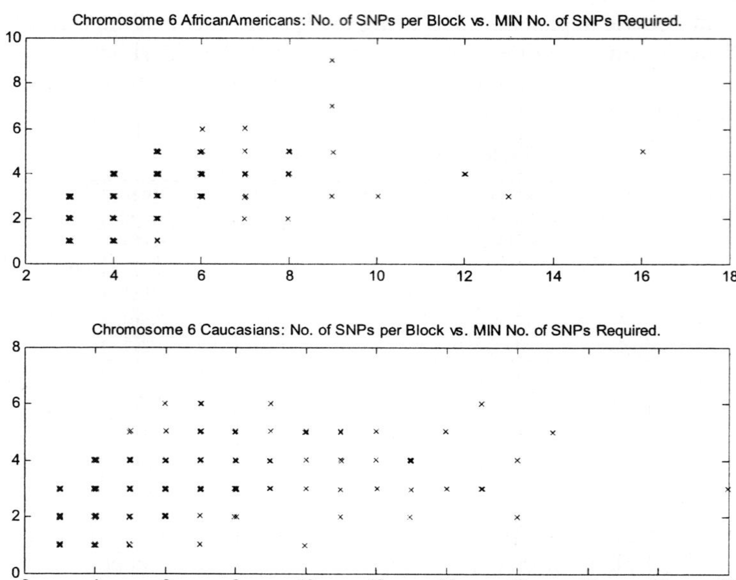

5   Discussion

Frequently, when SNPs are initially selected for typing, not much is known about the existence or location of LD blocks, nor about the number and relative frequencies of haplotypes within the blocks. It is therefore typical to "over-sample" the chromosomal region, (i.e., select SNPs as densely as one's budget permits). However, using large numbers of SNPs on a study adds significant cost; therefore the algorithm introduced in this paper can be useful for reducing the set of SNPs to the minimum number required for adequate coverage with no loss of haplotype information. Furthermore, the method can be used to eliminate additional SNPs while minimizing loss of haplotype information.

Previous work attempting to find the minimum SNP sub-set [10,11] have generally focused on complete genes or randomly selected loci, as opposed to LD blocks. In such cases, the number of haplotypes was expected to be higher, and more importantly, the amount of information in the haplotype distribution was expected to be much greater. As a result, these studies were challenged by numerical complexity, and solutions were optimized for relatively small regions and specific local conditions. Our method computes the global optimum, whether in *lossless* or *lossy* mode, making it more broadly applicable to a variety of purposes.

The method described by Judson *et al* [11] is essentially equivalent to phase II of the *lossy* version of our method. However, it is limited to $k \leq 11$. This is expected, because without the efficient pruning of SNPs performed by phase I of our method, the exponential nature of phase II can require practically infinite execution time. For example, the largest block we encountered consisted of 22 SNPs. With Judson's method, comparing sub-sets of 22 SNPs requires examining over almost 4.2 million combinations. Even using only half of the SNPs, sub-sets 1 to 11, would require examining over 2.4 million combinations, after which it would not be clear whether the solution is optimal, since it could represent a local optimum. Our method uses phase I to quickly reduce the 22 SNPs to a subset of 4 SNPs. As a result, phase II can find the global optimum (3 SNPs in lossless mode or 2 SNPs with less than 10% loss) by examining only 15.

The method described in the on-line supplement to the paper by Johnson et al. [10] appears to strive to compromise the maximization of the information detected by the SNP set with other considerations (e.g., maximization of the individual SNPs' properties). However, there is little explanation of the method details. One example asserts that any haplotype matrix with full rank cannot be pruned. However, we show in Table 1 a haplotype matrix of full rank can be pruned with no loss of information. The on-line supplement also provides executable programs, but the maximum subset size is limited to $k \leq 5$, which would lead to suboptimal results, because we have found the global optimum to be greater than 5 in some blocks.

Previous reports have suggested that the Caucasian population has longer LD blocks than the African-American population [8]. As Table 7 shows, the Caucasian population initially yielded more blocks, and more SNPs per block. However, after "compression" of the SNP sets of each block by our algorithm into the minimum required to represent the information (with no loss), the set size is almost the same for the two populations. Further compression with our *lossy* method, produced a very similar reduction for both populations. We believe that this reflects the arbitrary nature of the criteria to define LD blocks, which was applied uniformly to both populations.

It is important to note that the premise of the algorithm, as well as any other that attempts to reduce the size of the SNP set, is that the DNA sample size for each population is large enough so that the inferred haplotypes adequately represent reality. The accuracy of the computational haplotype inference is dependent on sample size, among other factors. There is always a risk that a SNP whose behavior is identical to another SNP (and thus deemed worthless in terms of new information) for the sample size used, could differentiate an additional haplotype inferred in a larger sample size. Although this risk is low for common haplotypes, it is possible that a rare haplotype can harbor the causative mutation sought and be present in higher frequency in affected individuals. Another factor that must be considered is the possibility of experimental errors that can eliminate data points and thus render suboptimal the minimum SNP subset. Therefore, it will always be necessary to supplement the minimum SNP subset with additional SNPs to enhance robustness. An optimal procedure for choosing additional SNPs that increase robustness, as well as the ability to bias the selection of the minimum SNP subset as a function of cost or availability of specific SNP assays, is currently under investigation.

6   Conclusions

We have described a simple algorithm that can select a minimum subset of SNPs without loss of haplotype information, or an even smaller subset with some acceptable loss of information. In a practical example encompassing 3 human chromosomes, we were able to reduce the SNP set by 25% for the African American population and by 36% for the Caucasian population with no loss of haplotype distribution information. It is clear that for an arbitrarily large genomic segment with numerous SNPs and haplotypes, the number of computations required by our algorithm would grow exponentially even after our phase I heuristic, and become unpractical. This situation would require different approaches to reduce the combinatorial complexity of the problem. However, for most practical situations our algorithm shall suffice and produce optimal results in a reasonable time, even allowing for the real-time calculation of minimum SNP subsets for haplotype blocks.

7   Acknowledgments

We are indebted to Charles Scafe, Heinz Hemken, Yu Wang, Marion Webster, Lewis Wogan, Lily Xu, Xiaoqing You, and Janet Ziegle for their role in the selection of SNPs, design of assays, genotyping of DNA samples, and management of the experimental results used in this work. Thanks are also due to Eugene Spier, Sorin Istrail, and Andrew Clark for many helpful discussions, and Mignon Fogarty for assistance with the manuscript.

References

1. Abecasis, G.R., Cherny, S.S., Cookson, W.O. and Cardon, L.R. *Nat Genet* 30:97-101 (2002).
2. Abecasis, G.R., Noguchi, E., Heinzmann, A., Traherne, J.A., Bhattacharyya, S., Leaves, N.I., Anderson, G.G., Zhang, Y., Lench, N.J., Carey, A., Cardon, L.R., Moffatt, M.F. and Cookson, W.O. *Am J Hum Genet* 68:191-197 (2001).
3. Ardlie, K., Liu-Cordero, S.N., Eberle, M.A., Daly, M., Barrett, J., Winchester, E., Lander, E.S. and Kruglyak, L. *Am J Hum Genet* 69:582-589 (2001).
4. Daly, M.J., Rioux, J.D., Schaffner, S.F., Hudson, T.J. and Lander, E.S. *Nat Genet* 29:229-232 (2001).
5. De La Vega, F.M., Dailey, D., Ziegle, J., Williams, J., Madden, D. and Gilbert, D.A. *Biotechniques* Suppl:48-50, 52, 54 (2002).
6. Drysdale, C.M., McGraw, D.W., Stack, C.B., Stephens, J.C., Judson, R.S., Nandabalan, K., Arnold, K., Ruano, G. and Liggett, S.B. *Proc Natl Acad Sci U S A* 97:10483-10488 (2000).
7. Excoffier, L. and Slatkin, M. *Mol Biol Evol* 12:921-927 (1995).
8. Gabriel, S.B., Schaffner, S.F., Nguyen, H., Moore, J.M., Roy, J., Blumenstiel, B., Higgins, J., DeFelice, M., Lochner, A., Faggart, M., Liu-Cordero, S.N., Rotimi, C., Adeyemo, A., Cooper, R., Ward, R., Lander, E.S., Daly, M.J. and Altshuler, D. *Science* 296:2225-2229 (2002).
9. Jeffreys, A.J., Kauppi, L. and Neumann, R. *Nat Genet* 29:217-222 (2001).
10. Johnson, G.C., Esposito, L., Barratt, B.J., Smith, A.N., Heward, J., Di Genova, G., Ueda, H., Cordell, H.J., Eaves, I.A., Dudbridge, F., Twells, R.C., Payne, F., Hughes, W., Nutland, S., Stevens, H., Carr, P., Tuomilehto-Wolf, E., Tuomilehto, J., Gough, S.C., Clayton, D.G. and Todd, J.A. *Nat Genet* 29:233-237 (2001).
11. Judson, R., Salisbury, B., Schneider, J., Windemuth, A. and Stephens, J.C. *Pharmacogenomics* 3:379 (2002).
12. Ke, X., Tapper, W. and Collins, A. *Bioinformatics* 17:581-586 (2001).
13. Kruglyak, L. *Nat Genet* 22:139-144 (1999).
14. Reich, D.E., Cargill, M., Bolk, S., Ireland, J., Sabeti, P.C., Richter, D.J., Lavery, T., Kouyoumjian, R., Farhadian, S.F., Ward, R. and Lander, E.S. *Nature* 411:199-204 (2001).
15. Shannon, C.E. *Bell System Technical J.* 27:379-423 (1948).
16. Xu, C.F., Lewis, K., Cantone, K.L., Khan, P., Donnelly, C., White, N., Crocker, N., Boyd, P.R., Zaykin, D.V. and Purvis, I.J. *Hum Genet* 110:148-156 (2002).

ON THE POWER TO DETECT SNP/PHENOTYPE ASSOCIATION IN CANDIDATE QUANTITATIVE TRAIT LOCI GENOMIC REGIONS: A SIMULATION STUDY

JOSEP M. COMERON[1], MARTIN KREITMAN

Department of Ecology and Evolution, University of Chicago,
1101 East 57th Street, Chicago, IL 60637, USA
[1]Current address: Department of Biological Sciences, University of Iowa,
212 Biology Bldg. (BB), Iowa City, IA 52242, USA

FRANCISCO M. DE LA VEGA

Applied Biosystems, 850 Lincoln Centre Drive,
Foster City, CA 94404, USA

We use coalescent methods to investigate the ability of linked neutral 'markers" to reveal in simulated population samples the presence of one or more single nucleotide polymorphisms that is contributing to a trait having a complex genetic basis (QTN: quantitative trait nucleotide). Realistic mutation and recombination rates in our simulations allow us to generate SNP data appropriate for analyzing human variation across short chromosomal intervals corresponding to approximately 100 kilobases. We investigate the performance of both single marker and multiple-marker (haplotype) data for several *ad hoc* procedures. Our results with single SNP markers indicate that (1) the density of SNP markers need not be much higher than 10% in order to achieve near-maximal detection of a QTN; (2) a higher density of markers does not improve much on the ability to localize a QTN within an interval unless the recombination rate is high. Haplotype-based tests were investigated for the case in which more than one QTN is present in the studied interval. Larger sample sizes improve both the probability of detecting the haplotype with the largest number of QTNs, as well as the ability to infer correct haplotypes from genotypic data. Testing a series of short haplotypes across a longer interval can also be beneficial. The rate of false positives (*i.e.,* when the most significant haplotype does not contain the greatest number of QTNs in the sample) can be very high when the contribution of individual QTNs to a trait is small. The elimination of low-frequency haplotypes does not substantially reduce the probability of detecting the haplotype with the largest number of QTNs but it can reduce the rate of false positives.

1 Introduction

Many diseases and traits are influenced by combinations of mutations acting at more than one site in the genome. The genes underlying these traits are generally referred to as Quantitative Trait Loci (QTL). The ultimate goal of association studies is to detect the presence of a causative mutation, deemed for simplicity a QTN (Quantitative Trait Nucleotide), by testing whether or not there is a difference in the frequency of individual mutations or haplotypes (*i.e.,* linked markers) associated with differences in phenotype. Using a population-based approach, we have studied and compared the power to detect marker/haplotype(s)-trait associations using different statistical methods, data analysis approaches and experimental strategies. Following Long and Langley[1], our study uses coalescent theory to simulate selectively neutral SNP in a sample of chromosomes. This methodology uses realistic genetic and population parameters (neutral mutation rate, recombination rate and evolutionary effective population size) but assumes a somewhat simplified population structure and demographic history compared to those likely for human populations. The samples it generates are reasonable approximations of observed human variation with respect to density and number of SNPs, the frequency spectrum of these SNPs, and magnitude of linkage disequilibrium between them. Thus, despite simplifications in generating simulated samples, these *in silico* data should be useful for studying statistical properties of association measures applicable to human data. In all cases, we use phenotype distributions compatible with the subtle genetic and environmental effects expected for most QTLs.

2   General Methods

2.1   *Simulations of neutral genealogies*

The coalescent theory[2,3] provides an efficient framework for generating neutral population samples. These simulations assume a random mating population at equilibrium, neutrality and an infinite-sites model. Coalescent theory always looks backward in time, with lineages joining up ("coalescing") as a genealogy extends back into the past. Mutations are "sprinkled" along the lineages according to a Poisson process in proportion to branch lengths of the lineages, producing a sample of SNPs that obey the expected equilibrium distribution and frequency spectrum for selectively neutral mutations.

In our simulated samples we choose one SNP to be contributing to a phenotypic trait, and call it the QTN. That is to say, an individual carrying the derived mutation at this SNP site is given a phenotypic score that deviates from the mean by a certain amount. The magnitude of this deviation is described in the "Phenotypic Distribution" section below. One consequence of this scheme is that the frequency distribution of QTNs follows a neutral frequency spectrum, and will therefore be skewed towards low frequency variants.

The selectively neutral SNPs in the sample, excluding the QTN, are taken to be "markers". Then, by virtue of when each one occurred in the history of the genealogy relative to that of the QTN, which lineages they occurred on relative to the QTN, and how much recombination occurred between the marker and the QTN, each marker (and each haplotype) will have a certain informative value in predicting the presence of the QTN. The statistical properties of the associations between markers and QTN are what we investigate. Many of our analyses of QTN - SNP associations use haploid data. Similar results are expected to obtain for diploids when there is no dominance. In our analyses of QTN – haplotype associations where haplotypes are reconstructed (inferred) from genotypic data, we form genotypes by randomly pairing chromosomes produced by the simulations.

Two properties of neutral samples are worth mentioning with implications for our study: 1) population genetics theory predicts that the expected mean frequency of a new mutation in a sample decreases with the number of sequences (n), and 2) the number of haplotypes—ordered combinations (phase) of genotypic variants (e.g., SNPs), which may or may not be closely linked or inherited together—constituted by a number of SNPs (S) in a sample is an increasing function of S, the number of sequences (n), and the recombination rate, but the number of expected haplotypes is *always* much smaller than 2^S.

2.2 *Mutation and recombination for human populations*

The ratio of polymorphisms or SNPs : recombination events has a strong influence on the power to detect SNP/phenotype associations. We used realistic values of mutation and recombination rates for human populations to assure the relevance and applicability of the simulation studies to complex traits in our species. The expected number of polymorphisms (S) per physical distance can be estimated from published studies of nuclear sequence variation[4]. These studies suggest an average of $\approx 1 \times 10^{-4}$ mutations or differences per site when comparing two randomly chosen sequences (popularly referred to as the nucleotide diversity per site). Nucleotide diversity may not be constant across all regions of the human genome, but a large fraction of the genome is expected to have densities of SNP near this average value. The number of SNPs in a sample, in contrast, is not constant, but is an increasing, nonlinear function of the sample size[5] (n); for instance, in a study of a 10 kb region, an average $S\approx 60$ and ≈ 36 is expected when the number of sequences is 200 and 20, respectively. The recombination rate in the humans varies across the genome; in regions of normal recombination 1cM (1% recombinants/generation) corresponds approximately to 1 Mb of DNA. The evolutionary relevant recombination rate between adjacent sites (4 *Ne c*) in humans might be on the order of ≈ 0.0001-0.001 in regions of low and high recombination, respectively. These are the range of values we investigate.

2.3 *Phenotype distribution*

A core element in this kind of study is the choice of an appropriate phenotype distribution congruent with QTLs. We used as phenotype distribution Y a modification of the distribution proposed in [1],

$$Y_i = z(1-\pi)^{1/2} + 1.96 \, Q_i(\pi)^{1/2}$$

where $Q_i=1$, 0 represents presence and absence of the QTN in chromosome i, respectively, π is the proportion of phenotype variation attributable to the QTN, and z is a random normal deviate (mean=0, variance =1). Unlike the somewhat more elaborate formula for generating phenotypic scores given in [1] ours has no dependency on allele frequency[6,7]. Thus low- and high-frequency QTNs have the same average individual contribution to phenotype.

Most QTLs in humans might be compatible with our simulated scheme with π = 10-25%. Therefore, unless indicated, we have used for most of the subsequent analyses a conservative $\pi = 0.1$. The statistical power estimated will be conservative and, more importantly, the methodological approaches and experimental strategies shown to be adequate will be qualitatively accurate when the genetic contribution is stronger. At a practical level, for instance establishing the optimum density of markers in a study, our scheme might be easily modified based on external information about the phenotype distribution and the underlying genetics.

2.4 Statistical methods

Several authors have proposed the F-statistic ANOVA test (in particular a Model II ANOVA for two groups) to study association between phenotype and SNP variation[8,9]. We first compared this approach to nonparametric tests, namely the Mann-Whitney (MW) U-test and the Kruskall-Wallis (KW) H-test. In all cases, the significance of the estimated statistic is obtained by comparison to an empirically derived null distribution of this statistic in samples in which the phenotypic scores have been randomly permuted among individual sequences with the same number of markers. This assures that both multiple tests and the non-independence of tightly linked markers are taken into account. Our results reveal unambiguously that the two nonparametric tests (MW and KW) have greater power than the F-statistic for single QTNs with contribution to the phenotype (π) of 50% or lower. Therefore, we have used KW, which may be applied to both marker-based and haplotype-based association studies, in our analyses.

3   Results

3.1 Effect of single causative mutations (QTNs)

3.1.1   Optimal density of markers in the light of plausible recombination rates for humans

Clearly, increasing the density of marker SNPs within a candidate genomic region will increase the likelihood of including the QTN itself. On the other hand, increasing the density of SNPs studied will increase the number of tests, hence reducing the probability of detecting a statistically significant association of any one marker with a QTN. This is especially true when the QTN has a subtle phenotypic effect unless the sample size is very large. Also, the low levels of effective recombination evident in human population genetic data further suggest that studies that exhaustively include every SNP in a region of interest may not increase dramatically the power compared to those studies analyzing only a fraction of the total variability.

To investigate these possible tradeoffs, we studied a case equivalent to analyzing a human genomic region of 100 kb and a sample of n=200, with average level of polymorphism (i.e., 600 SNPs), and for a range of recombination rates typical of the human genome. SNPs used in the analysis were randomly distributed across the region and only one of the 600 SNPs is the QTN. The results reveal that beyond 10%, increasing the density of studied SNPs will only slightly increase the overall power to detect genetic association between one SNP and the phenotype. That is, for realistic recombination rates observed in the human genome, 5-10 % density almost gives the maximum power, and greater power is achieved by increasing sample size than by increasing the density of markers. The optimal density mostly depends on the number of samples and the recombination environment of the candidate genomic region. Such a density is the one that might indicate that the genomic region under study likely includes a QTN although it does not necessary allow the precise specification of its location. Higher density of SNPs (such as complete resequencing) will increase the probability of localizing the QTN (although see below), especially in regions of high recombination.

3.1.2   Power *vs.* location

We investigated the average distance between the QTN and the SNP with the greatest (significant) association with the phenotype. Again, our simulation design assumes a genomic region with a total of 600 SNPs and the analysis of a varying percentage (density) of these SNPs (1 – 40%). In all cases under scrutiny the average distance between the 'significant' SNP and the QTN is considerable (e.g., always greater than 50 SNPs apart). Increasing the density of studied SNPs helps to locate the QTN with more accuracy, but this is mostly noticeable in regions of high recombination. Again, densities higher than 10-15% will not substantially increase the localization of the QTN, even in regions of high recombination, unless the sample size is very large.

Studies in regions of high recombination in the human genome will give overall reduced power to detect association between *any* SNP and phenotype variability compared to regions with low recombination. But when a significant association SNP/phenotype is detected, there is a higher probability that the detected SNP might be the QTN or close to it than when the region has low rates of recombination. This result follows because tighter linkage will enhance the probability that SNPs with significant association may be more physically distant from the QTN. Moreover, the lower the recombination rate, the higher the probability of finding multiple SNPs with similarly high statistical significance, also due to tight linkage.

3.2 Effect of multiple causative mutations (QTNs)

So far, we have investigated statistical properties of markers associated with one and only one QTN, and with the ability to detect the QTN in a population-based sample. However, common disease may be influenced by combinations of causative mutations at a candidate locus or gene (multiple QTNs). In these cases, common sense dictates that association studies should focus on haplotype-based tests, as they can better capture the presence of these combinations of mutations. As indicated, the number of haplotypes depends on the number of sequences, the number of SNPs, the recombination rate, and population structure. We investigate the probability of detecting haplotypes with the largest number of QTNs in a sample (i.e, the most extreme haplotype). Note that because most SNPs segregate at low frequency, the number of QTNs in this most extreme haplotype usually will not represent the totality of QTNs present in the sample that might be influencing the phenotype, only a detectable subset. To carry out this study, we designated several SNPs in a region as QTNs, each contributing independently and equally to phenotype.

In diploid organisms the ability to detect association between individual haplotypes and phenotype will be influenced by our ability to discern the haplotype structure from heterozygous individuals. Further, in individuals with heterozygous QTNs, the phenotype is most likely to be less conspicuous (we assume additivity). As a first approach, we compare the probability of detecting the most extreme haplotype in three cases: 1) in a haploid case, 2) in a diploid case where haplotypes are known with certainty, and 3) in a diploid cases where no effort is made to discern the actual haplotypes, and hence haplotypes are constructed randomly from the genetic information. The results show that increasing the sample size causes (as expected) an overall increase in the probability of detecting a significant association for the haplotype with the most QTNs. The results also reveal that the difference between knowing or not the actual haplotype becomes critically important as sample size increases. In other words, in order to have a high probability of detecting association, a large sample is required and diploid genotypes need to be resolved as haplotypes (either by inference or by experiment).

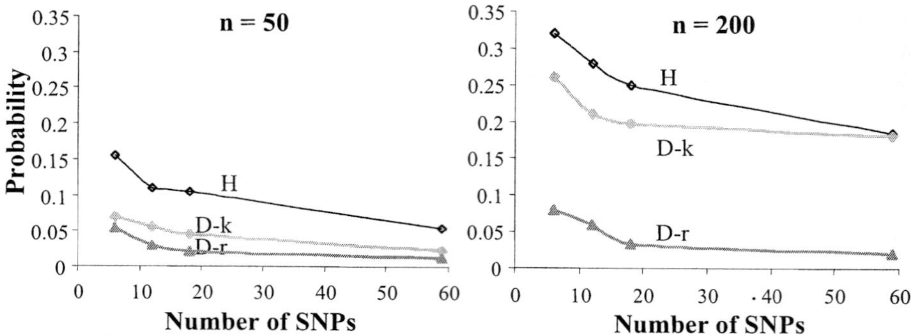

Figure 1. Probability of detecting the haplotype with the largest number of QTNs (i.e, the most extreme haplotype) as a function of the number of SNPs under study. 3 QTNs are segregating in the sample, each with a contribution to phenotype (π) of 25%. Results are shown for the case of no recombination when n= 50 and 200. Three cases are compared: Haploid (H), Diploid with known (D-k) haplotypes, and Diploid with randomly constructed (D-r) haplotypes.

3.2.1 Inference of haplotypes from diploid populations

We investigated a widely used method proposed by A. Clark[10]. Clark's method uses a parsimonious approach to infer the minimum number of haplotypes in a sample: it utilizes information from homozygous or single-site heterozygous individuals to sequentially resolve multiply-heterozygous genotypes into haplotypes. Figure 2 shows two extreme cases. Clark's method always performs well when the number of SNPs is small.

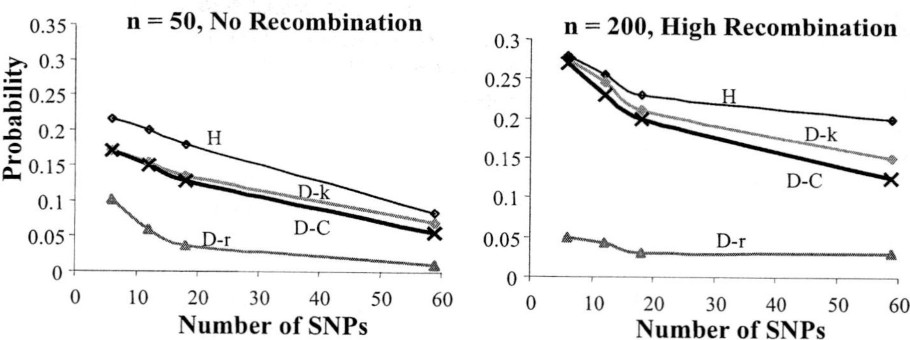

Figure 2. Probability of detecting the haplotype with the largest number of QTNs as a function of the number of SNPs under study. 3 QTNs are segregating in the sample, each with a contribution to phenotype (π) of 25%. Results are shown for four cases: Haploid (H), Diploid with known (D-k) haplotypes, Diploid with randomly constructed (D-r) haplotypes, and Diploid after applying Clark's method[10] (D-C) to infer haplotypes.

As expected, recombination produces unresolved or wrongly inferred haplotypes, a problem that is enhanced by increasing the number of SNPs. The 'sequential' problem (i.e., the fact that a different order in the sequential solution of haplotypes might cause different solutions) also increases with the number of SNPs. Large sample sizes increase the power to detect associations and also the probability of observing homozygous individuals, required in Clark's method. But large sample sizes will also increase the probability of observing individuals with two or more SNPs present only once in the sample, causing unresolved haplotypes (but see below the small practical consequences of this problem).

Another method, proposed by Stephens, Smith and Donnelly[11] applies a Markov-chain Monte-Carlo (MCMC) algorithm with population genetics assumptions. The results (data not shown) indicate that for the conditions studied in Figure 2, it performs similar or a little worse than Clark's method when recombination occurs and the number of SNPs is high.

3.2.2 Number of SNPs used to define haplotypes: Partial haplotypes

We investigated the effect of haplotype 'size' --the number of adjacent SNPs used to define a haplotype-- on the probability of detecting significant association. *A priori*, we expect that the use of more SNPs will increase the likelihood of haplotypes including more QTNs. Also, the use haplotypes based a small number of markers (henceforth called 'small' haplotypes) will increase substantially the number of tests performed with the consequent reduction of statistical power. We studied a scenario equivalent to a situation in humans in which all QTNs are restricted to 1 kb ($\approx$ 6 SNPs) but SNP markers are dispersed across a 10 kb interval (59 SNPs).

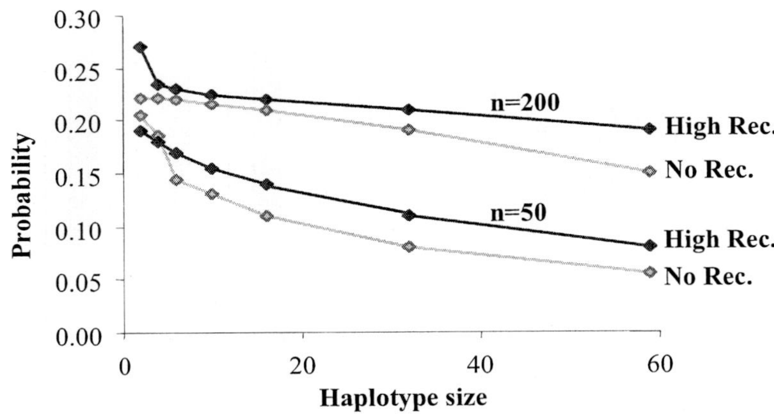

Figure 3. Relationship between the number of adjacent SNPs used to construct the haplotypes and the probability of detecting the haplotype with the largest number of QTNs. 5 QTNs are segregating in the sample, each with a contribution to phenotype (π) of 25%.

The use of haplotypes constructed by a small number of SNPs increases the probability of detecting the haplotype with the highest number of QTNs. Interestingly, the number of adjacent SNPs used to construct the haplotype with the highest probability of detecting the most extreme haplotype is smaller than the actual number of QTNs, in agreement with the idea that it is highly unlikely to observe a haplotype with all QTNs. Note also that small haplotypes are those more accurately predicted by most algorithms.

The use of haplotypes constructed only by a subset of adjacent SNPs (partial haplotypes) can be put to good advantage in surveys of longer regions by utilizing a 'sliding window' analysis across these regions (or a 5'- vs central vs 3' study). A sliding window approach might better localize the region encompassing QTNs since it would give a quantitative idea of the signal observed in the 'background'. For instance, in a study of 200 sequences with five adjacent QTNs, partial haplotypes give 17% significant detection (false positive) when they are located 50 SNPs apart from the QTN positions when recombination is high, while this percentage jumps to 23% around the QTN positions. Overall, the higher the recombination rate or the more distant the regions, the higher the chance of detecting differences between regions, and hence of localizing the region with more QTNs

3.2.3   Probability of false positives

We studied the probability that the haplotype showing the strongest significant association with phenotype variability is not the one with the highest number of QTNs.

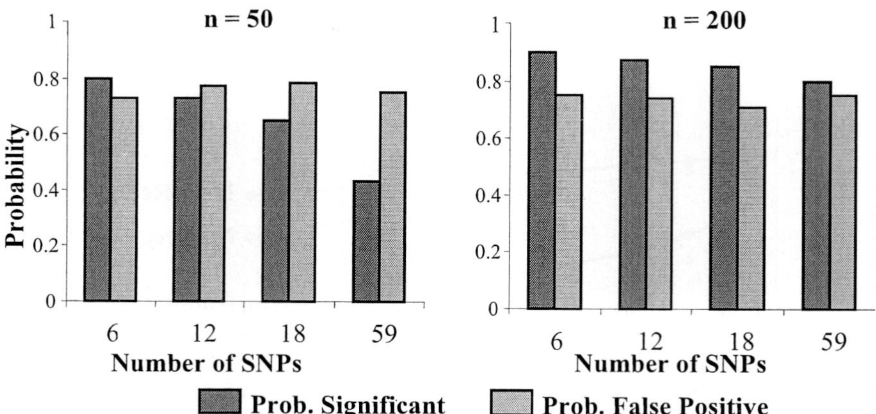

Figure 4. Probability of detecting a significant association between any haplotype and phenotype variability and the probability that the detected haplotype is not the one with the highest number of QTNs in the sample (false positives). 5 QTNs are segregating in the sample, each with a contribution to phenotype (π) of 25%.

As shown in Figure 4, the probability of false positives is very high for QTNs with small contribution to phenotype. Increasing the sample size improves the probability of detecting significant associations between haplotype and phenotype. But the probability of detecting the haplotype with more information about the causative mutations increases less rapidly.

3.2.4 Haplotype frequency

As discussed above, the highest probability of detecting significant association between SNP (or haplotype) and phenotype is attained when QTNs are at intermediate frequency. This is also true when we take into consideration the frequency of false positives, since higher QTN frequencies in the sample does increase the probability of detecting a significant association, but it does not alter the probability of false positive (data not shown). Hence, the use of a sample with high-frequency QTNs will also increase the reliability of the results. We have studied whether the elimination of haplotypes at low frequency in the sample also reduces the probability of false positives.

Indeed, as suspected the elimination of haplotypes at low frequency substantially reduces the probability of false positives. Eliminating low frequency haplotypes has a practical advantage as well: haplotypes at low frequency are also more difficult to infer from diploid data.

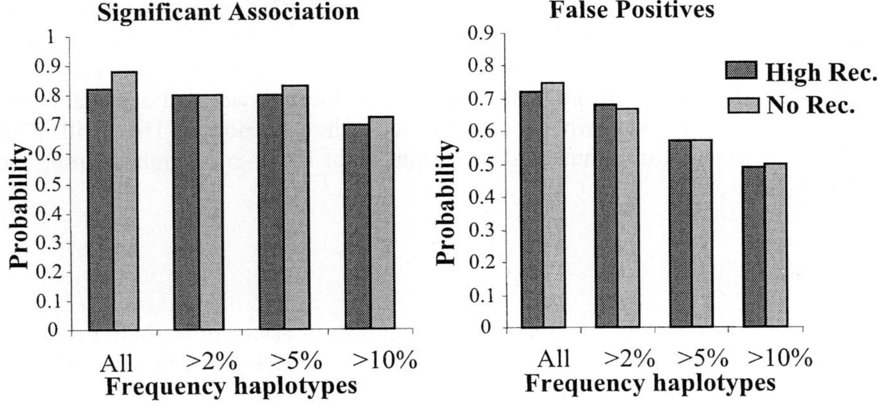

Figure 5. Probability of detecting significant association between haplotype presence and phenotype variability and of false positives (among those cases with the strongest significant association) when the haplotypes used in the analysis are those at frequency higher than 2%, 5% or 10% in the sample. n = 200.

4   Conclusions

Overall, the study shows that the probability of detecting a significant association between nucleotide and phenotype variability is low for most conditions suitable to most QTLs ($\pi < 50\%$). The low, albeit variable, rate of recombination present in the human genome also contributes to a very high percentage of false positives, a percentage that decreases with recombination. On the other hand, empirical investigation of linkage disequilibrium in the human genome suggests a strong haplotype structure, possibly caused by recombination cold- and hot-spots[12]. If true, the study of only a small percentage of all SNPs present in a genomic region gives almost the maximum power to detect association, and greater power is achieved by increasing sample size than by increasing the density of markers. The use of nonparametric tests, the study of extreme phenotypes, and the analysis of common haplotypes based on a small number of adjacent SNPs are all methodologies that increase the chance of detecting and locating a QTN.

The recent molecular evolutionary history of humans almost certainly includes intense selection on many sites across the genome, and many QTNs may be selected mutations. Such a scenario, taken together with relatively low recombination rates, implies a high probability of false positives, a problem that might be exacerbated by population expansion. Overall, recent selection and/or population expansion makes the problem of discerning a QTN among all mutations or SNPs across a small genomic region (i.e., the same exon or gene) a more difficult task. As shown here, some analytical and experimental approaches can improve the chance of being successful. Nothing, however, is likely to overcome the need for very large population sample sizes in order to achieve reasonable power and acceptably low levels of false positives when scanning the whole genome. This will place additional incentive for commercial development of lower-cost, higher-throughput SNP assays.

References

1. A. D. Long, and C. H. Langley, "The power of association studies to detect the contribution of candidate genetic loci to variation in complex traits" *Genome Res.* **9,** 720 (1999)
2. J. F. C. Kingman, "The coalescent" *Stoch. Proc. Appl.* **13,** 235 (1982)
3. R. R. Hudson, "Gene genealogies and the coalescent process" *Oxf. Surv. Evol. Biol.* **7,** 1 (1990)
4. M. Przeworski, R. R. Hudson, and A. Di Rienzo, "Adjusting the focus on human variation" *Trends Genet.* **16,** 296 (2000)
5. G. A. Watterson, "On the number of segregating sites in genetical models without recombination" *Theor. Popul. Biol.* **7,** 256 (1975)

6. D. E. Reich, and E. S. Lander, "On the allele spectrum of human disease" *Trends Genet.* **17,** 502 (2001)
7. J. K. Pritchard, "Are rare variants responsible for susceptibility to complex disease?" *Am. J. Hum. Genet.* **69,** 124 (2001)
8. G. A. Churchill, and R. W. Doerge, "Empirical threshold values for quantitative trait mapping" *Genetics* **138,** 963 (1994)
9. A. D. Long, R. F. Lyman, C. H. Langley, and T. F. Mackay, "Two sites in the Delta gene region contribute to naturally occurring variation in bristle number in Drosophila melanogaster" *Genetics* **149,** 999 (1998)
10. A. G. Clark, "Inference of haplotypes from PCR-amplified samples of diploid populations" *Mol. Biol. Evol.* **7,** 111 (1990)
11. M. Stephens, N. J. Smith, and P. Donnelly, "A new statistical method for haplotype reconstruction from population data" *Am. J. Hum. Genet.* **68,** 978 (2001)
12. S. B. Gabriel, S. F. Schaffner, H. Nguyen, J. M. Moore, J. Roy, *et al.*, "The Structure of Haplotype Blocks in the Human Genome" *Science* **296,** 2225 (2002)

ERRORS AND LINKAGE DISEQUILIBRIUM INTERACT MULTIPLICATIVELY WHEN COMPUTING SAMPLE SIZES FOR GENETIC CASE-CONTROL ASSOCIATION STUDIES

D. GORDON[1], M. A. LEVENSTIEN[1], S. J. FINCH[2], AND J. OTT[1]

[1]*Laboratory of Statistical Genetics, Rockefeller University*
1230 York Avenue, New York, NY 10021-6399
[2]*Department of Applied Mathematics and Statistics, State University of New York at Stony Brook, Stony Brook, NY 11794*

Single nucleotide polymorphisms (SNP) may be used in case-control designs to test for association between a SNP marker and a disease. Such designs may assume that the genotype data are reported without error. Our goal is quantifying the effects that errors have on sample size for case-control studies with haplotypes formed by a disease locus and a SNP marker locus in the presence of linkage disequilibrium (LD). We consider the effects of a recently published error model on 2×3 chi-square analysis. We study the joint relation of LD and errors with sample size for three specific genetic disease models and two settings each of marker allele frequencies (total of 6 studies). Minimal sample size necessary for fixed asymptotic power is estimated as a 4th degree polynomial in the variables S (error) and D' (LD measure) via a backward step-wise regression.

We find that increased error rates lower power. In all studies, we observe that LD and errors interact in a non-linear fashion. In particular, regression analyses shows that several higher order interaction terms have coefficients significantly different from 0 in each study, with fraction of variance explained greater than 0.9999. Finally, the increase in sample size necessary to maintain constant asymptotic power and level of significance as a function of S is smallest when $D' = 1$ (perfect LD). The increase grows monotonically as D' decreases to 0.5 for all studies.

1 Introduction

Single nucleotide polymorphisms (SNPs) may be used in case-control designs to test for genetic association between marker and disease. Such designs usually assume that genotype data are reported without error. In statistical genetics, errors in genotyping or phenotyping (incorrectly assigning a case to be a control, or vice versa) can significantly affect linkage and genetic association studies. A number of authors have studied such effects[1-10]. Sobel et al.[11] summarize results to date. Major findings are that errors lead to inflation in genetic map distances, an increase in type I error or a decrease in power for statistical methods designed for gene localization, and biased estimates of parameters such as the recombination fraction among loci and the amount of linkage disequilibrium (LD) between two loci.

For case-control studies of genetic association, researchers[12,13] have found that, for a particular error model (not presented here), errors lead to a loss in power to detect association between a disease and a locus. However, to our knowledge, there has been no quantitative assessment of the relation between errors and LD in genetic case-control association studies for multiple disease models, although other authors[6,14-17] have developed methods that allow for errors in genetic linkage and/or association analyses.

The purpose of this work is therefore a quantitative assessment, in terms of increased sample size, of error rates in genetic case-control association studies. The data we consider is haplotype data for cases and controls from a SNP marker locus that is in LD with a disease locus. The SNP marker is observed, and the disease locus is unobserved. The test statistic considered is the standard χ^2 on 2×3 tables. We compute asymptotic power analytically by means of a non-centrality parameter. Errors affect the power of such statistics by deviating genotype frequencies in cases and controls away from their true values. Furthermore, determining sample size for fixed power level is equivalent to determining power for a fixed sample size, and it is this first question that we study in this work.

For three particular genetic disease models and two different settings of SNP marker allele frequencies (a total of 6 studies), we compute genotype frequencies for cases and controls in the presence of errors, and compute the sample size necessary to maintain constant asymptotic power and level of significance for different values of the error model parameters. Finally, we perform model fitting by regressing the minimal sample size necessary to maintain constant power on a 4th degree polynomial in the variables S (error parameter) and D' (LD parameter).

2 Materials and Methods

2.1 Notation

The following notation is used through the remainder of this work:

Count parameters:
N_A = number of cases
N_U = number of controls

Frequency parameters:
p_1 = allele frequency of SNP marker *1* allele
p_2 = allele frequency of SNP marker *2* allele = $1 - p_1$
p_d = allele frequency of disease locus *d* allele
p_+ = allele frequency of disease wild-type allele = $1 - p_d$
p_{Aij} = frequency of SNP marker genotype *ij* in the case population ($ij \in \{11, 12, 22\}$)
p_{Uij} = frequency of SNP marker genotype *ij* in the control population ($ij \in \{11, 12, 22\}$)
Disequilibrium parameters:
D = disequilibrium (non-standardized as defined in Hartl and Clark[18]) [Note: max ($-p_1 p_+$, $-p_2 p_d$) $\leq D \leq$ min ($p_1 p_d$, $p_2 p_+$)]
$D_{\max}$ = min ($p_1 p_d$, $p_2 p_+$) (we assume in this work that disequilibrium is positive)

D' = proportion of total disequilibrium (or standardized disequilibrium [19])

$\quad = D/\ D_{max}$

Penetrances:

$f_0 = \Pr(\text{affected} \mid ++ \text{ at disease locus})$

$f_1 = \Pr(\text{affected} \mid +d \text{ at disease locus})$

$f_2 = \Pr(\text{affected} \mid dd \text{ at disease locus})$

Conditional probabilities:

$p_{A11} = \Pr(11 \text{ genotype at SNP locus} \mid \text{affected})$
$p_{A22} = \Pr(22 \text{ genotype at SNP locus} \mid \text{affected})$
$p_{U11} = \Pr(11 \text{ genotype at SNP locus} \mid \text{unaffected})$
$p_{U22} = \Pr(22 \text{ genotype at SNP locus} \mid \text{unaffected})$

Prevalence and other parameters:

$$\phi = \text{disease prevalence} = (1 - p_d)^2 f_0 + 2(p_d)(1 - p_d)f_1 + p_d^{\ 2} f_2$$

(Note: We assume Hardy-Weinberg equilibrium (HWE) at the disease locus; no such assumption is made for the marker locus)

h_{ij} = haplotype frequency of i allele at disease locus (i = + or d) and j allele at marker locus ($j = 1$ or 2) (see Methods)

Error model parameters:

$\varepsilon_1 = \Pr(\text{true heterozygote incorrectly coded as a homozygote})$,

$\varepsilon_2 = \Pr(\text{true heterozygote has one allele misread})$,

$\varepsilon_3 = \Pr(\text{jointly misreading both alleles of a genotype})$,

$\varepsilon_4 = \Pr(\text{falsely adding an allele to a true homozygote})$,

$\varepsilon_5 = \Pr(\text{pre-gel error})$.

Sobel et al. [11] describe these parameters more completely. It should be noted that, for a di-allelic locus, the parameter $\varepsilon_2 = 0$, since it is not possible for one heterozygote to be incorrectly read as another heterozygote for a di-allelic locus.

When considering the χ^2 statistic on 2 × 3 tables, the sample size determination for fixed asymptotic power and significance level is completely determined by the non-centrality parameter λ, which is a function of the genotype frequencies in the case and control populations and the ratio of cases to controls. In section 2.2, we demonstrate how to compute genotype frequencies in each population as a function of the genetic model parameters (penetrance values, disease allele frequency), an LD parameter and the SNP marker allele frequency. In section 2.3, we present an error model and compute precisely how genotype frequencies determined in section 2.2 are altered for general settings of the error model parameters

2.2   *Computation of genotype frequencies*

We assume that we know the following six parameter values: the penetrance values f_0, f_1, f_2, the SNP marker allele frequency p_1, the disease allele frequency p_d, and the standardized disequilibrium D'. Using the definition of conditional probability, we calculate all such values Pr(ab at SNP marker locus | affection status)[20,21]. For example, we have the following case genotype frequency expressions:

$$p_{A11} = \Pr(11 \mid \text{affected}) = [1/(\phi)]\{(h_{+1})^2 f_0 + 2(h_{+1})(h_{d1})f_1 + (h_{d1})^2 f_2\},$$

$$p_{A12} = \Pr(12 \mid \text{affected}) = [2/(\phi)]\{(h_{+1})(h_{+2})f_0 + (h_{+1}h_{d2} + h_{d1}h_{+2})f_1$$
$$+ (h_{d1})(h_{d2})f_2\},$$

$$p_{A22} = \Pr(22 \mid \text{affected}) = [1/\phi]\{(h_{+2})^2 f_0 + 2(h_{+2})(h_{d2})f_1 + (h_{d2})^2 f_2\}.$$

To compute the corresponding genotype frequencies for controls, replace ϕ by $1-\phi$ and each f_i by $1 - f_i$ in each expression. The haplotype frequencies are functions of the parameters p_1, p_2, p_+, p_d, and D'. Using the notation defined above, we have:

$$h_{+1} = p_+ p_1 + D'D_{\max},$$

$$h_{+2} = p_+ p_2 - D'D_{\max},$$

$$h_{d1} = p_d p_1 - D'D_{\max},$$

$$h_{d2} = p_d p_2 + D'D_{\max}.$$

To obtain the genotype frequency expressions as functions of LD, substitute the haplotype relations above in the genotype frequency expressions.

2.3   *Error model*

Recently, Sobel, Papp, and Lange[11] proposed a model to describe how errors affect genotypes, in terms of the probabilities Pr(observed genotype is ab | true genotype is cd) (where $\{ab, cd\} \in \{11, 12, 22\}$). We call these probabilities *error penetrances*. While their model generalizes to a marker locus with any number of alleles, we present in table 1 the error penetrances for a di-allelic locus.

Table 1 – Error penetrances for a SNP marker locus using the Sobel-Papp-Lange error model

Observed Genotype	True Genotype		
	11	12	22
11	$1-(\varepsilon_3+\varepsilon_4+\varepsilon_5)$	$(\varepsilon_1+\varepsilon_5)/2$	$\varepsilon_3+\varepsilon_5/2$
12	$\varepsilon_4+\varepsilon_5/2$	$1-(\varepsilon_1+\varepsilon_5)$	$\varepsilon_4+\varepsilon_5/2$
22	$\varepsilon_3+\varepsilon_5/2$	$(\varepsilon_1+\varepsilon_5)/2$	$1-(\varepsilon_3+\varepsilon_4+\varepsilon_5)$

Using table 1, we compute the observed genotypes for either cases or controls when errors are present. If table 1 is thought of as a 3×3 matrix M, we can compute the vector of observed case genotype frequencies in the presence of errors, $A=(p_{A11},p_{A12},p_{A22})^T$, (here, T is the transpose operator) by performing the matrix multiplication $M\times A$. For example,

$$p^*_{A11}=[1-(\varepsilon_3+\varepsilon_4+\varepsilon_5)]p_{A11}+[(\varepsilon_1+\varepsilon_5)/2]p_{A12}+[\varepsilon_3+\varepsilon_5/2]p_{A22}.$$

Note that the observed genotype frequencies are a function of both the error rates and the LD parameter. While the Sobel-Papp-Lange error model assumes 5 parameters, in order for us to present 3-dimensional plots of the interaction between LD and errors, we must reduce it to a single parameter. Therefore, we use fixed multiples of the settings: $\varepsilon_1=0.0125,\varepsilon_2=0,\varepsilon_3=0.005,\varepsilon_4=0.01,\varepsilon_5=0.0025$ from 0 up to 6 (increments of 0.5) from this point forward. Sobel et al. give these settings as the default settings for their error model parameters when considering a di-allelic locus [11]. The notation S represents the sum $k\sum_{i=1}^{5}\varepsilon_i$, where k =0.0, 0.5, 1,0, ..., 6.0.

2.4   Non-centrality parameter

Using the notation above and a general result proved by Mitra[22], Gordon et al.[23] found that that the non-centrality parameter λ for the test of genotype frequency differences among cases and controls is given by:

$$\lambda=N_A N_U\left[\frac{(p^*_{A11}-p^*_{U11})^2}{N_A p^*_{A11}+N_U p^*_{U11}}+\frac{(p^*_{A12}-p^*_{U12})^2}{N_A p^*_{A12}+N_U p^*_{U12}}+\frac{(p^*_{A22}-p^*_{U22})^2}{N_A p^*_{A22}+N_U p^*_{U22}}\right]. \quad (1)$$

This formula provides us with the sample size for a fixed value of the non-centrality parameter. Assuming a fixed power and significance level, the non-centrality is known. It is then possible to solve equation (1) for sample sizes. We compute this solution for all genetic models presented in the next section.

2.5 Genetic models

Here we present values for the parameters in section 2.2. Each set of genetic model parameters (penetrances + disease allele frequency) comes from a genetic disease model in which the disease prevalence is 0.03 and the disease allele frequency is 0.2. In all studies, the non-centrality parameter is set to 15.4408, which corresponds to a fixed asymptotic power of 0.95 at the 0.05 level of significance for a χ^2 distribution with 2 degrees of freedom. Also, the LD parameter D' is varied between 0.5 and 1.0 in increments of 0.05. Finally, the SNP marker 1-allele frequency p_1 is set at both 0.2 and 0.5 in all studies. The genetic model parameter values are:

(Dominant model)
$$f_0 = 0.004, f_1 = 0.07, f_2 = 0.07, p_d = 0.2$$
(Additive model)
$$f_0 = 0.014, f_1 = 0.028, f_2 = 0.042, p_d = 0.2$$
(Multiplicative model)
$$f_0 = 0.011, f_1 = 0.028, f_2 = 0.071, p_d = 0.2$$

2.6 Regression analysis

As a further means of describing the quantitative relationship among sample size, LD, and errors, we perform a backward step-wise regression analysis. For each setting of error parameter S and the LD parameter D', the value of the dependent variable is the sample size necessary for asymptotic power 0.95 at level of significance 0.05. The general form of the fitted regression equation (i.e., the upper model) is:

$$\hat{Y} = \hat{\beta}_0 + \sum_{\substack{i=0 \\ i+j\leq 4}}^{4} \sum_{j=0}^{4} \hat{\beta}_{i,j} S^i D'^j ,$$

where $\hat{Y}$ is the fitted sample size (in terms of case individuals) corresponding to a given setting of S and D', and the terms $\hat{\beta}_{i,j}$ are the parameters of the regression (*regression coefficients*) that minimize the sum of squares of differences between the fitted values for settings of S and D' (using equation 1) and the observed values for the same settings. The regression coefficients are determined using the S-PLUS 6.0 software (see Electronic Database Information).

3   Results

We have three main results. Our first is that, for the genetic models considered in section 2.5, there is multiplicative interaction between the error parameter S and the standardized LD D'. This interaction is documented graphically in figures 1 and 2 and quantitatively in our regression analysis results (Table 2).

Table 2 – Regression coefficients for all genetic model studies and SNP allele frequency settings

Exponent pair (i,j) for term $S^i D'^j$	Genetic Model[a]/SNP allele frequency					
	Dom/0.5	Dom/0.2	Mult/0.5	Mult/0.2	Add/0.5	Add/0.2
(0,0)(intercept)	6476	1837	17826	4906	46617	12518
(1,0)	7889	2753	21147	8932	54787	21280
(0,1)	-25030	-7134	-68367	-18727	-179104	-48223
(2,0)	5030	0	17624	1931	48206	7670
(0,2)	39822	11466	108940	30051	285705	77505
(1,1)	-23081	-8256	-61853	-25499	-160530	-61320
(0,3)	-29568	-8605	-81041	-22564	-212776	-58192
(3,0)	6946	3022	3924	6915	9685	12864
(2,1)	-10598	0	-33696	-3566	-93658	-17031
(1,2)	24739	8797	66397	26550	172977	64664
(0,4)	8449	2482	23195	6520	· 60972	16805
(2,2)	6365	0	16655	2312	47213	10279
(3,1)	-7629	-2834	0	-7629	0	-12526
(1,3)	-9269	-3209	-24737	-9625	-64782	-23796

[a]*(Dom = Dominant, Mult = Multiplicative, Add = Additive)*

Figures 1 and 2 present the minimal sample size necessary to maintain constant asymptotic power of 0.95 at the 0.05 significance level for our dominant model with SNP 1-allele frequency of 0.5 and our additive model with SNP 1-allele frequency of 0.2, respectively. The sample size, as indicated above, is a function of S and D'.

We comment that in table 2, the non-zero coefficients, when tested (using the t-test) for being non-zero, are all significant at the 0.001 level (data not shown). The observations that several interaction terms in table 2 are significantly non-zero and that the fraction of variance (multiple R^2 value) for each regression is at least 0.9999 (data not shown) indicate that, for these error models, sample size is well fit by a high degree polynomial in the variables S and D', and hence there is significant interaction between these two variables in explaining the sample size increase.

Our second result is that the general trend of sample size increase as a function of the two variables S and D' is robust to genetic model specification for the models we consider here. This result may be observed quantitatively by noting that, for each monomial term in table 2, the sign of the regression coefficient for the non-zero coefficients is the same across genetic models and SNP allele frequency specifications, and may be observed graphically by studying figures 1 and 2. We comment the shape of the surfaces in figures 1 and 2 is identical to the shape of the surfaces for those figures determined by all other genetic model and SNP allele frequency specifications (data not shown).

The third result is that, for all values of S, sample size increase as a function of S is smallest when $D' = 1$, and is largest when $D' = 0.5$ (table 2; figures 1 and 2). This result suggests that high levels of LD, in addition to increasing power for genetic case-control studies, may have the additional benefit of mitigating the effects of errors in data in the sense of requiring the smallest possible increase in sample size for a given error setting.

4 Summary and Discussion

In this work, we have demonstrated that it is possible to compute analytically sample size requirements for genetic case-control studies in the presence of errors. In sections 2.2-2.5, we have described how these computations are done for the test of genotypic association using the 2×3 contingency table. Further, we have shown that, for our genetic model, error model, and LD parameter settings, sample size is accurately predicted by a polynomial of high degree in the variables S and D'. From the viewpoint of marker selection, we have documented that high levels of LD have the smallest cost, in terms of increased sample size, for a given setting of error parameters. This result should be reassuring to researchers who are planning association studies and who are concerned about errors in their data.

This work generalizes to an analytic method for sample-size calculations in the presence of errors when the observed data are haplotypes or multi-locus genotypes. One only needs to specify multi-locus error models. Perhaps the simplest and most reasonable model is one in which errors in individual marker loci are independent of errors in other marker loci. Also, this work is not restricted to just di-allelic loci; it can also be extended to markers loci with any number of alleles. The analytic price is that one has to specify multiple LD parameters and multiple allele or haplotype frequency parameters for the marker loci.

We have considered the question of interaction between errors and LD over a larger set of values for the genetic model parameters specified in section 2.2; our observation is that the interaction between S and D' is robust to genetic model specifications. That is, the shape of figures 1 and 2 is repeated for every set of genetic model parameters considered (data not shown).

An important question for this work regards parameter estimation. We are currently working on methods to determine genotyping error rates. Also, LD parameters can be estimated using inter-marker LD patterns. With traits for which the genetic model parameters are difficult to estimate, one can specify genetic model-free parameters[23] rather than the genetic model-based parameters we have specified in this work.

Software performing these calculations will be available from our website http://linkage.rockefeller.edu/pawe/ by January 2003. The program is called PAWE (<u>P</u>ower of <u>A</u>ssociation Tests <u>W</u>ith <u>E</u>rrors).

Acknowledgments
The authors gratefully acknowledge grants K01-HG00055 and MH59492 from the the National Institutes of Health.

Electronic Database Information
S-PLUS 6.0 Academic Site Edition Release 2. Copyright 1988-2001 Insightful Corp.

References

1. Douglas, J.A., Boehnke, M. & Lange, K. A multipoint method for detecting genotyping errors and mutations in sibling-pair linkage data. Am J Hum Genet **66**, 1287-97 (2000).
2. Shields, D.C., Collins, A., Buetow, K.H. & Morton, N.E. Error filtration, interference, and the human linkage map. Proc Natl Acad Sci **88**, 6501-5 (1991).
3. Buetow, K.H. Influence of aberrant observations on high-resolution linkage analysis outcomes. Am J Hum Genet **49**, 985-94 (1991).
4. Terwilliger, J.D., Weeks, D.E. & Ott, J. Laboratory errors in the reading of marker alleles cause massive reductions in lod score and lead to gross overestimates of the recombination fraction. Am J Hum Genet **47**, A201 (1990).
5. Gordon, D., Matise, T.C., Heath, S.C. & Ott, J. Power loss for multiallelic transmission/disequilibrium test when errors introduced: GAW11 simulated data. Genet Epidemiol **17 Suppl 1**, S587-92 (1999).
6. Gordon, D., Heath, S.C., Liu, X. & Ott, J. A transmission/disequilibrium test that allows for genotyping errors in the analysis of single-nucleotide polymorphism data. Am J Hum Genet **69**, 371-80 (2001).
7. Goldstein, D.R., Zhao, H. & Speed, T.P. The effects of genotyping errors and interference on estimation of genetic distance. Hum Hered **47**, 86-100 (1997).
8. Cherny, S.S., Abecasis, G.R., Cookson, W.O., Sham, P. & Cardon, L.R. The effect of genotype and pedigree error on linkage analysis: analysis of three asthma genome scans. Genet Epidemiol **2001**, S117-22 (2001).
9. Abecasis, G.R., Cherny, S.S. & Cardon, L.R. The impact of genotyping error on family-based analysis of quantitative traits. Eur J Hum Genet **9**, 130-4 (2001).
10. Akey, J.M., Zhang, K., Xiong, M., Doris, P. & Jin, L. The effect that genotyping errors have on the robustness of common linkage-disequilibrium measures. Am J Hum Genet **68**, 1447-56 (2001).

11. Sobel, E., Papp, J.C. & Lange, K. Detection and integration of genotyping errors in statistical genetics. Am J Hum Genet **70**, 496-508 (2002).

12. Bross, I. Misclassification in 2 x 2 tables. Biometrics **10**, 478-486 (1954).

13. Gordon, D. & Ott, J. Assessment and management of single nucleotide polymorphism genotype errors in genetic association analysis. Pac Symp Biocomput, 18-29 (2001).

14. Goring, H.H. & Terwilliger, J.D. Linkage analysis in the presence of errors II: marker-locus genotyping errors modeled with hypercomplex recombination fractions. Am J Hum Genet **66**, 1107-1118 (2000).

15. Goring, H.H. & Terwilliger, J.D. Linkage analysis in the presence of errors I: complex-valued recombination fractions and complex phenotypes. Am J Hum Genet **66**, 1095-106 (2000).

16. Goring, H.H. & Terwilliger, J.D. Linkage analysis in the presence of errors III: marker loci and their map as nuisance parameters. Am J Hum Genet **66**, 1298-309 (2000).

17. Goring, H.H. & Terwilliger, J.D. Linkage analysis in the presence of errors IV: joint pseudomarker analysis of linkage and/or linkage disequilibrium on a mixture of pedigrees and singletnos when the mode of inheritance cannot be accurately specified. Am J Hum Genet **66**, 1310-27 (2000).

18. Hartl, D.L. & Clark, A.G. Principles of population genetics, (Sinauer Associates, Sunderland, 1989).

19. Lewontin, R.C. The interaction of selection and linkage. I. General considerations; heterotic models. Genetics **49**, 49-67 (1964).

20. Risch, N. A general model for disease-marker association. Ann Hum Genet **47**, 245-52 (1983).

21. Sham, P. Statistics in Human Genetics, (J. Wiley and Sons, Inc., New York, 1998).

22. Mitra, S.K. On the limiting power function of the frequency chi-square test. Ann Math Stat **29**, 1221-1233 (1958).

23. Gordon, D., Finch, S.J., Nothnagel, M. & Ott, J. Power and sample size calculations for case-control genetic association tests when errors are present: application to single nucleotide polymorphisms. Hum Hered **(in press)** (2002).

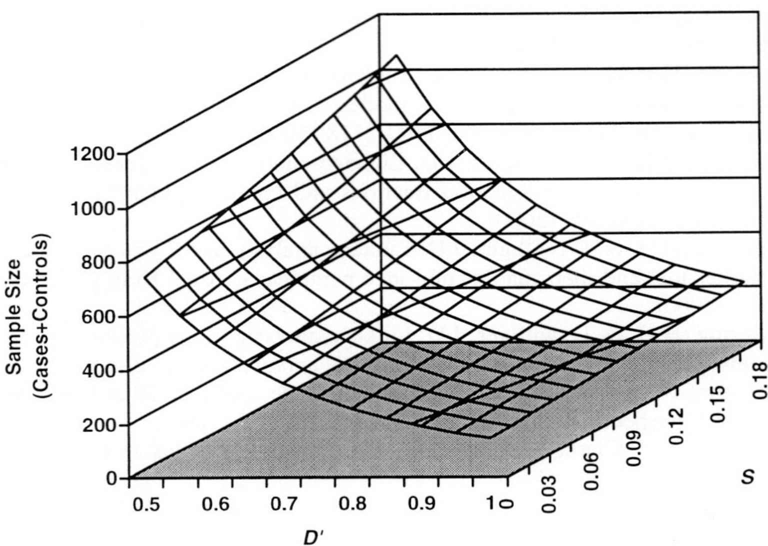

Minimum Sample Size: 146, for D'=1, S=0
Maximum Sample Size: 1056, for D'=0.5, S=0.18

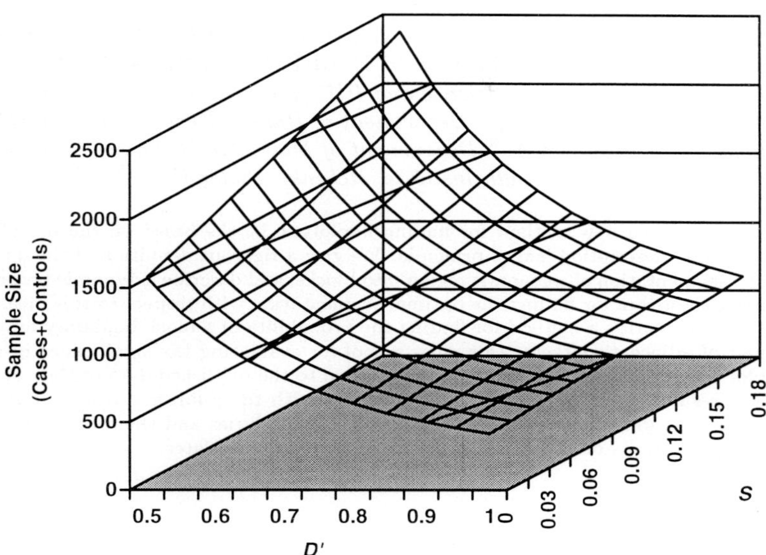

Minimum Sample Size: 416, for D'=1, S=0
Maximum Sample Size: 2376, for D'=0.5, S=0.18

AN MDL METHOD FOR FINDING HAPLOTYPE BLOCKS AND FOR ESTIMATING THE STRENGTH OF HAPLOTYPE BLOCK BOUNDARIES

M. KOIVISTO,[1] M. PEROLA,[2,3] T. VARILO,[3] W. HENNAH,[3] J. EKELUND,[3] M. LUKK,[1] L. PELTONEN,[2,3] E. UKKONEN,[1] H. MANNILA [1]

[1] *Department of Computer Science and HIIT Basic Research Unit, University of Helsinki, Helsinki, Finland;* [2] *Department of Human Genetics, UCLA School of Medicine, Los Angeles, CA;* [3] *Department of Molecular Medicine, National Public Health Institute, Helsinki, Finland.*

We describe a new method for finding haplotype blocks based on the use of the minimum description length principle. We give a rigorous definition of the quality of a segmentation of a genomic region into blocks, and describe a dynamic programming algorithm for finding the optimal segmentation with respect to this measure. We also describe a method for finding the probability of a block boundary for each pair of adjacent markers: this gives a tool for evaluating the significance of each block boundary. We have applied the method to the published data of Daly *et al.*[1] The results are in relatively good agreement with the published results, but also show clear differences in the predicted block boundaries and their strengths. We also give results on the block structure in population isolates.

1   Introduction

Haplotype blocks [1,2,3,4] form fascinating small-scale structure in the human genome. While several studies have confirmed that some type of block structures exist, the recent data about haplotype blocks in the human genome have left multiple uncertainties concerning block boundaries and their variation. The published algorithms by Daly *et al.*,[1] Patil *et al.*,[2] and Zhang *et al.*[3] have used segmentation algorithms with fairly ad hoc criteria for block quality. Also, the existing methods produce a segmentation without any clear indication on how strong or weak the different block boundaries are.

We describe a new method for finding haplotype blocks based on the use of the minimum description length principle. We give a rigorous definition of the quality of a segmentation of a genomic region into blocks, and describe a dynamic programming algorithm for finding the optimal segmentation with respect to this measure. We also describe a method for finding the probability of a block boundary for each pair of adjacent markers: this gives a tool for evaluating the significance of each block boundary.

Intuitively, a haplotype block can be considered to be a sequence of markers such that for those markers most of the haplotypes in the population fall into a small number of classes; each class consists of identical or almost identical haplotypes. This notion can be formalized by considering the problem of describing the haplotypes in a succinct way. This approach is an instance of the minimum description length principle (MDL) by Rissanen [5,6] widely used in statistics, machine learning, and data mining (see, e.g., Li and Vitanyi [7] or Hansen and Yu [8]). Similar ideas have also been applied to partitioning homogeneous DNA domains.[9]

We have applied the method to the published data of Daly *et al.* [1] The results are in relatively good agreement with the published results, but also show clear differences in the predicted block boundaries and their strengths. The method has also been applied to samples from isolated populations.

The rest of this paper is organized as follows. In Section 2 we describe the MDL principle and the encoding of the data. Section 3 gives the dynamic programming algorithm, and Section 4 shows how dynamic programming can be used to compute the probabilities of block boundaries. Section 5 gives an overview of the empirical results, and Section 6 is a short conclusion.

2   MDL principle and coding of haplotype data

Let D be an $n \times p$ matrix of n observations over p markers. We refer to the jth allele of observation i by D_{ij}. For simplicity, we first assume that $D_{ij} \in \{0, 1\}$.

A marker interval $[a, b] = \{a, a + 1, \ldots, b\}$ is defined by two marker indices $a, b \in \{1, \ldots, p\}$. A *segmentation* is a set of non-overlapping non-empty marker intervals. A segmentation is *full* if the union of the intervals is $[1, p]$. The data matrix limited to interval $[a, b]$ is denoted by $D(a, b)$ and the values of the ith observation are denoted by $D(i, a, b)$.

The minimum description length principle by Rissanen [5,6] considers the description of the data using two parts: the model B and the description of the data D given the model. The description length for the data and the model is

$$L(B, D) = L(B) + L(D \mid B),$$

where $L(B)$ is the length of the description of the model and $L(D \mid B)$ is the length of the description of the data, when the data is described using the model B.

The minimum description length principle states that the desired descriptions of the data are ones having the minimum length $L(B, D)$ of the total description. For a good survey of the (sometimes intricate) connections between MDL, Bayesian statistics, and machine learning, see Li and Vitanyi.[7] The MDL principle has successfully been used in various applications.[10,11,12,8]

The haplotype data set D can be described by telling first of all how many blocks there are and where the blocks start and end. For each block, we have to specify how many typical haplotypes (*class center*) there are and what they are. For each observation and each block, we tell from which typical haplotype the observation comes from.

More formally, a block model B consists of the following components.

1. A segmentation S, i.e., the start and end markers s_h and e_h for each block $(h = 1, \ldots, \ell)$.[a] Implicitly, the segmentation specifies the number of blocks ℓ.

2. For each block h, the class centers $\theta_h = (\theta_{hc})$ $(c = 1, \ldots, k_h)$ specifying the coordinates θ_{hcj} for each marker $j = s_h, \ldots, e_h$. Implicitly, each θ_h also specifies the number of centers k_h.

The coordinates θ_{hcj} are real numbers, encoding the probability of seeing 1 in marker j of an observation stemming from class center c of block h. So, strictly speaking, a class center is not a typical haplotype, but a mean vector of the haplotypes associated with the class.

Given such a block model $B = ((s_h, e_h), (\theta_{hcj}))$, the data can be encoded as follows. For each observation $i = 1, \ldots, n$, and for each block $h = 1, \ldots, \ell$, we first have to tell which of the k_h class centers does the observation $D(i, s_h, e_h)$ belong to; let this center be c. This takes $\log k_h$ bits per observation. Then we have to describe $D(i, s_h, e_h)$ using the center coordinates θ_{hcj}, for $j = s_h, \ldots, e_h$. This is done by assuming independence of marker values given the class center. Thus the probability is

$$P(D(i, s_h, e_h) \mid (\theta_{hcj})) = \prod_{j=s_h}^{e_h} \theta_{hcj}^{D_{ij}} (1 - \theta_{hcj})^{1-D_{ij}}. \tag{1}$$

Using the relation of coding lengths and probabilities we get a code of length $-\log P(D(i, s_h, e_h))$ for the data D, given the segmentation model B.

Assuming the segmentation in the block model B is full, it can be coded using $\ell \log p$ bits for encoding the block boundaries, and $\ell \log n$ bits for the number of centers in the block, and $\alpha k_h(e_h - s_h + 1)$ bits for coding the centers, where α is the number of bits needed for the coding of a real number. Theoretical arguments [5,6,8] indicate that the appropriate accuracy is obtained by choosing $\alpha = (\log n)/2$.

Thus the length of the description of the block model is

$$L(B) = \ell \log p + \ell \log n + \sum_{h=1}^{\ell} k_h \alpha(e_h - s_h + 1).$$

---

[a] Of course, $s_1 = 1$ and $e_\ell = p$. In some of our models we allow parts of the data to be uncoded, so $s_{h+1} = e_h + 1$ does not necessarily hold.

The length of the description of the data is

$$L(D \mid B) = \sum_{h=1}^{\ell} \sum_{i=1}^{n} \left[\log k_h - \log P(D(i, s_h, e_h)) \right].$$

Thus the goal of the segmentation procedure is to find a block model B such that the overall coding length $L(B, D) = L(B) + L(D \mid B)$ is minimized.

The description method is easily extended to handle missing or unknown data values. If the values D_{ij} are interpreted as a degree of certainty that the correct value is 1, the expression in Eq. 1 can be used directly. For instance, one can assign the value 0.5 for each missing allele in the data. To obtain a proper probability model a normalizing factor should be included in Eq. 1. However, the factor behaves as an irrelevant constant and therefore can be ignored.

3 Dynamic programming algorithm

We use a dynamic programming algorithm to compute an optimal block structure, and then estimate the probabilities of each block boundary.

The MDL cost function is, as defined above, a function of the whole segmentation. However, it is straightforward to see that it can be decomposed into the blocks of the segmentation. Given a marker interval $[a, b]$, let $\hat{k}$ is the optimum number of centers, and let $\hat{\theta}_{ij}$ be the corresponding center coordinates associated with jth allele of ith observation, such that the cost

$$
\begin{aligned}
f(a, b) \quad = \quad & \log p + \log n + n \log \hat{k} + \frac{1}{2}\hat{k}(b - a + 1) \log n \\
& + \sum_{i=1}^{n} \sum_{j=a}^{b} \left[- D_{ij} \log \hat{\theta}_{ij} - (1 - D_{ij}) \log(1 - \hat{\theta}_{ij}) \right] \quad (2)
\end{aligned}
$$

is minimized. Then for the MDL optimal block model B_{mdl} we have

$$L(B_{mdl}, D) = \min_{S} \sum_{[a,b] \in S} f(a, b),$$

where S runs through all full segmentations on $[1, p]$. Thus the minimum description length of haplotype data can be defined as the sum of costs of coding of individual blocks.

Denote by $F(b)$ the cost of the optimal segmentation of the haplotypes from marker 1 to marker b. We have the typical dynamic programming equation

$$F(b) = \min_{1 \leq a \leq b} (F(a - 1) + f(a, b));$$

additionally $F(0)$ is defined to be 0. Namely, the coding from marker 1 to marker b is either produced by coding all the markers in one block (with cost $F(0)+f(1,b)$), or by coding for some a from marker 1 to marker $a-1$ optimally (cost $F(a-1)$) and then coding from marker a to marker b in one block (cost $f(a,b)$). Given the costs $f(a,b)$, the computation can be done in $O(p^2)$ time.

The cost $f(a,b)$ is computed by using k-means clustering on the data set $D(a,b)$. The number of cluster centers is varied from 1 to 10, and for each number we produce 5 different clusterings. For each clustering, the coding cost of $D(a,b)$ is computed, and as the cost $f(a,b)$ we select the smallest cost.[b] The computation of $f(a,b)$ takes time $O(n(b-a+1))$ for a fixed number of iterations in the k-means algorithm. Thus the total amount of time needed for computing the costs $f(a,b)$ is $O(np^3)$.

In many cases it is interesting to see how optimal segmentations behave when some (bad) markers are allowed to be ignored in the data. We call such ignored markers *gaps* between haplotype blocks. A natural extension of the problem of finding the optimum segmentation is to find a segmentation that gives the shortest description length and includes at most u gaps. Denoting by $F(b,u)$ the cost of optimal segmentation from marker 1 to marker b using at most u gaps, we have

$$F(b,u) = \min(F(b-1,u-1), \min_{1 \le a \le b}(F(a-1,u) + f(a,b))).$$

Namely, if a gap is used at the bth marker, then the prefix segmentation of $[1, b-1]$ is allowed to contain at most $u-1$ gaps. Otherwise, a block $[a,b]$ is introduced and the maximum allowed number of gaps from marker 1 to marker $a-1$ is still u. The computation of $F(b,u)$ can be arranged to take $O(np^3)$ time.

4 Computing the probability of a block boundary

We next consider the probability that there is a block boundary between markers j and $j+1$. Denote by $\mathcal{S}_{j,j+1}$ the set of all full segmentations having a boundary between markers j and $j+1$. Then we are interested in the probability of any segmentation from $\mathcal{S}_{j,j+1}$, given the data D:

$$P(\mathcal{S}_{j,j+1} \mid D) = \sum_{S \in \mathcal{S}_{j,j+1}} P(S \mid D).$$

---

[b]The problem of finding the best cluster centers is NP-hard; thus the approach does not guarantee that the shortest description for the single block from a to b is found.

Denoting by $\mathcal{S}[1,p]$ the set of all full segmentations on $[1,p]$, this can be written

$$P(\mathcal{S}_{j,j+1} \mid D) = \frac{\sum_{S \in \mathcal{S}_{j,j+1}} P(S,D)}{\sum_{S' \in \mathcal{S}[1,p]} P(S',D)}. \tag{3}$$

The probabilities $P(S,D)$ come naturally from our description method. For any segmentation S and data set D we define

$$P(S,D) = Z^{-1} 2^{-\sum_{[a,b] \in S} f(a,b)},$$

where $f(a,b)$ are the minimum description lengths for the corresponding blocks as described in Eq. 2, and Z is a normalization constant that does not depend on S and D. Note that the normalization constant cancels out when substituted into Eq. 3.

Define

$$q(a,b) = 2^{-f(a,b)},$$

and for any interval $[j,j']$,

$$Q(j,j') = \sum_{S \in \mathcal{S}[j,j']} \prod_{[a,b] \in S} q(a,b),$$

where $\mathcal{S}[j,j']$ denotes the set of all full segmentations on marker interval $[j,j']$. Then, since $\mathcal{S}_{j,j+1}$ is equal to the Cartesian product $\mathcal{S}[1,j] \times \mathcal{S}[j+1,p]$, we have

$$P(\mathcal{S}_{j,j+1} \mid D) = \frac{Q(1,j)Q(j+1,p)}{Q(1,p)}.$$

(For a similar development, see Durbin $et\ al.,$[13] Eq. 3.14, and also Liu and Lawrence.[14])

Again, dynamic programming can be applied. The equations are

$$Q(1,b) = \sum_{1 \leq a \leq b} Q(1,a-1)q(a,b)$$

and

$$Q(a,p) = \sum_{a \leq b \leq p} q(a,b)Q(b+1,p).$$

(Here, of course, we define $Q(1,0) = Q(p+1,p) = 1$.) Thus the probabilities $P(\mathcal{S}_{j,j+1} \mid D)$ can be computed for all j in time $O(p^2)$.

Table 1: Haplotype classes of the block 2 found by the MDL method for the data of Daly *et al.* The block consists of the markers 15–24. There are two classes of haplotypes. For each class the size (the number of associated haplotypes), and the center coordinates and the most commonly occuring haplotype is shown. The right-most columns show the number of haplotypes that differ from the most commonly occuring haplotype at 0, 1, 2, or $\geq$ 3 markers.

Class	Size	Center and the most frequent haplotype										Frequencies			
												0	1	2	≥ 3
1	190	.16	.02	96	.64	.06	.15	.05	.02	.94	.02				
		0	0	1	1	0	0	0	0	1	0	115	44	20	11
2	68	.91	.88	.18	.20	.82	.20	.82	.92	.15	.91				
		1	1	0	0	1	0	1	1	0	1	34	18	4	12

5 Empirical results

We have tested the above methods on synthetic (generated) and real data. The results on synthetic data show that the method finds the block structure that have been used to generate the data, and that the method is quite robust against noise (data not shown).

Figure 1 shows the results when the MDL method is applied to the data of Daly *et al.*[1] The segmentations reported by Daly *et al.*[1] and those produced by the MDL method are overall in quite good agreement. There are, however, some differences. For example, the 3 block boundaries around marker 40 reported in [1] are clearly weaker than some of the others, as shown by the log odds curve, and also in the plot of the optimal segmentations with a varying number of gap markers.

An interesting difference can also been found at the block around marker 20. Since Daly *et al.*[1] count exact matches they report 3 distinctive common haplotypes. Our method, however, suggests 2 haplotype classes, for it allows variability within a class, see Table 1.

Figure 2 displays the results when random noise is added to the data of Daly *et al.* The block structure as well as the probabilities for block boundaries show quite good stability. As expected, when the marker order is randomly permuted, the block structure disappears. Note that some information already shown in Figure 1 is repeated, but using the physical location of the markers yields additional insights into the nature of the blocks.

We also applied the method to SNP data on samples from three subpopulations from Finland, which are representative for the settlement history of Finland, inhabited by two periods of immigration 4000 and 2000 years ago.[15] The early settlement sample ($n = 32$ chromosomes) consisted of descendants of the early settlers ($\approx$100 generations) on the coastal areas of Finland, the late

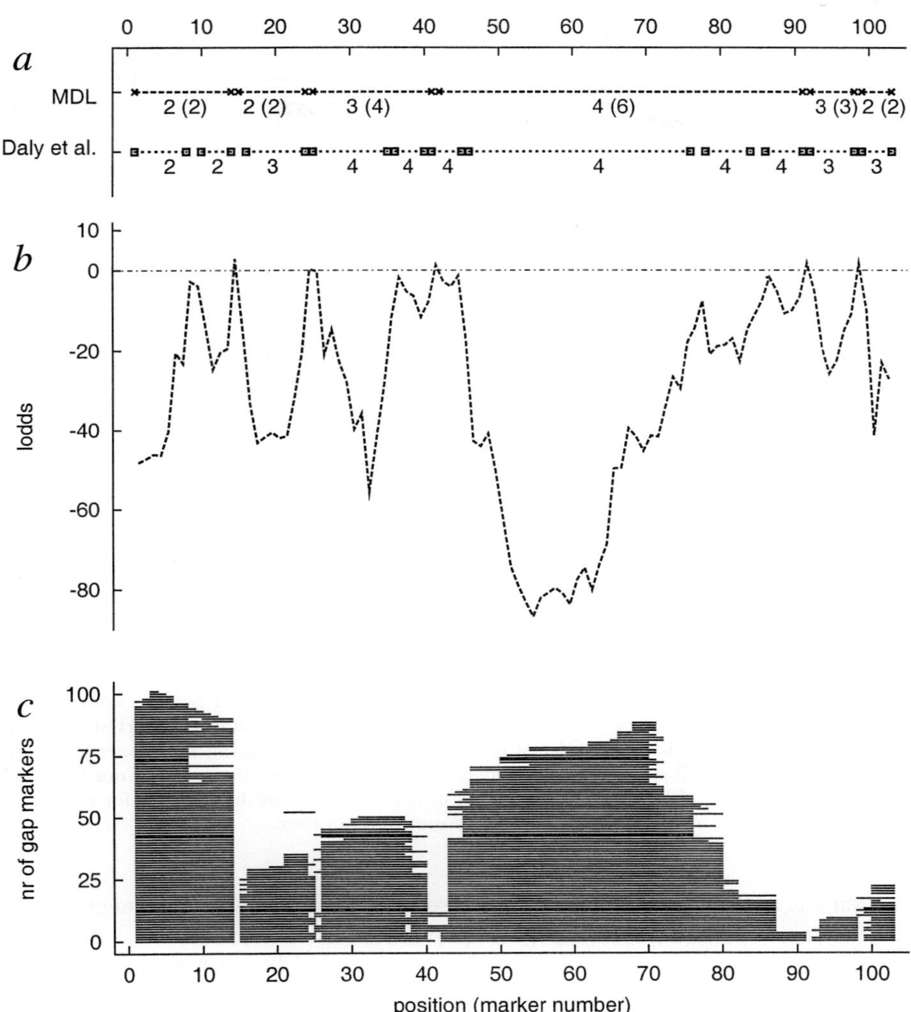

Figure 1: Haplotype block structure in the data of Daly *et al.* The x-axis is the number of the marker. (a) The optimal block structure produced by the MDL scoring function compared against the block boundaries reported by Daly *et al.* The numbers associated with the MDL blocks give the number of haplotype classes that suffice to cover at least 85 percent of the block and, in parenthesis, the total number of the classes. The numbers associated with the Daly *et al.* blocks give the number of haplotypes in the block that suffice to cover at least 90 percent of the block. (b) The log odds of the probability of block boundaries for each pair of adjacent markers. (c) The optimal segmentation when k markers are allowed to be left outside the blocks, for varying k.

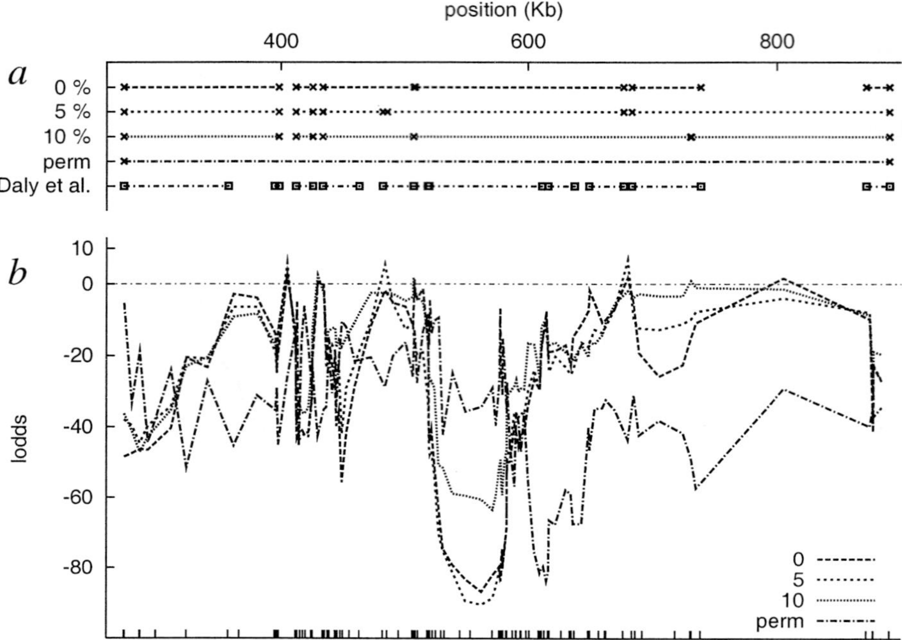

Figure 2: Haplotype block structure in the data of Daly *et al.* when noise is added. The x-axis is the physical location of the marker. (a) The block boundaries reported by Daly *et al.* and the optimal block structures produced by the MDL scoring function when 0, 5, and 10 percent of random noise was added to the data and when the order of markers was randomly permuted. (b) The log odds of the probability of block boundaries for each pair of adjacent markers after adding noise and permuting the order.

settlement sample ($n = 108$ chr.) representing a younger (15–20 generations) population that has gone through a population bottleneck in the 1500s and the third regional subisolate ($n = 108$ chr.), founded by 40 families some 300 years ago, followed by a major expansion.

A total of 45 SNPs were genotyped over 1Mb area on chromosome 1q. Figure 3 shows the results. Five haplotype blocks were identified in three populations. The identified blocks varied in size from 12kb to 361kb. As expected, the haplotype blocks do not differ in different subpopulations of Finland, most probably reflecting the limited set of original founder chromosomes shared by all analyzed populations.

The log odds curve for estimating the probability of boundaries shows that generally the boundaries are stronger for larger values of n; this is also visible in experiments done by sampling from the data of Daly *et al.* (results not

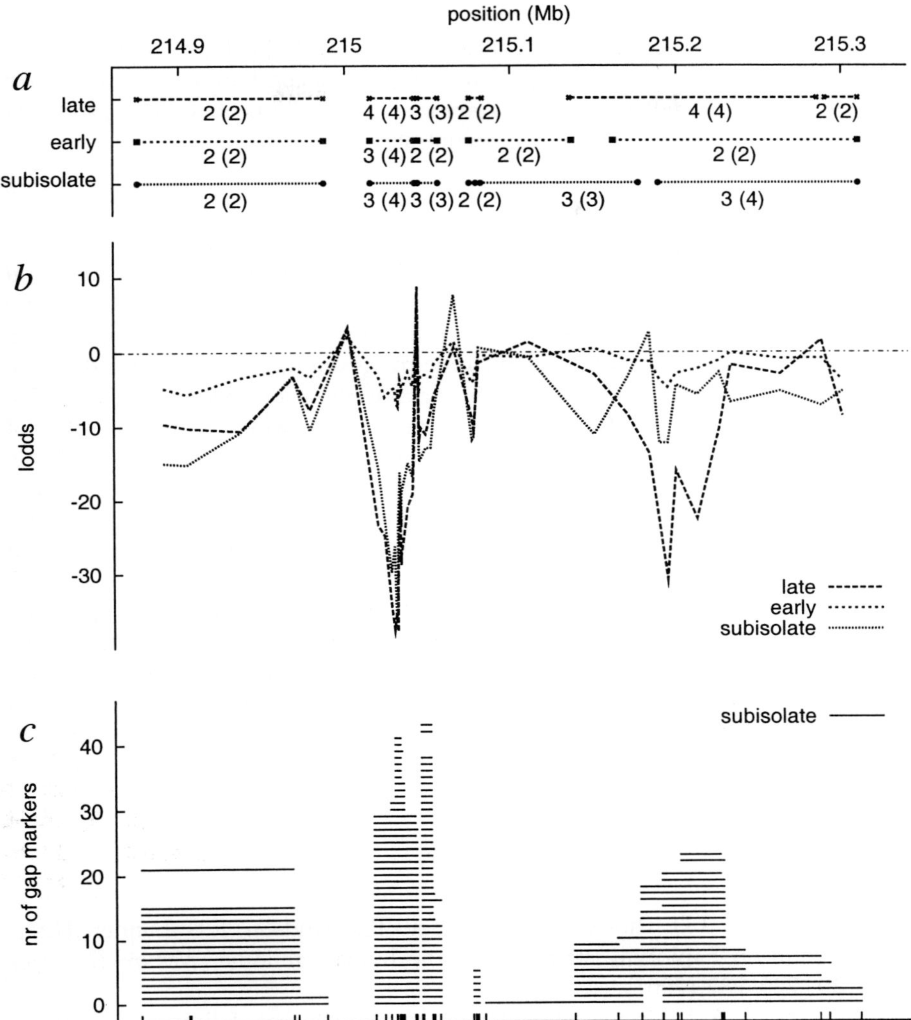

Figure 3: Haplotype block structure in the data from three subpopulations in Finland. Sample sizes: late settlement, $n = 108$; early settlement, $n = 32$; subisolate, $n = 108$. The x-axis is the physical location of the marker. (a) The optimal block structure produced by the MDL scoring function in the three subpopulations. The numbers refer to the number of haplotype classes covering at least 85 percent of haplotypes and, in parenthesis, the full block. (b) The log odds of the probability of block boundaries for each pair of adjacent markers. (c) The subisolate: the optimal segmentation when k markers are allowed to be left outside the blocks, for varying k.

shown).

The significance of block boundaries was also tested using bootstrap methods by investigating whether the optimal segmentation had a block boundary in resampled data sets. The results are similar to the probabilities produced by the probabilistic approach (results not shown).

6 Concluding remarks

We have described a method for defining and finding haplotype blocks based on the use of the minimum description length principle. The clearer the haplotype block structure is, the shorter a description can be given to the data. The coding cost function is such that dynamic programming can be applied to the problem, yielding an $O(np^3)$ algorithm for n observations over p markers. The method can also be used to search for segmentations where a given number of markers are allowed to be left as gaps. We also showed how the MDL principle can be used to obtain probabilities for block boundaries for all pairs of adjacent markers, giving a clear way of evaluating the significance of block boundaries. Experiments on synthetic and real data show that the method produces useful results.

Our method can be extended in many directions. The clustering approach and k-means algorithm could be replaced by closely related but directly probabilistic mixture models and the usual expectation maximization algorithm, respectively. This would also help in modifying the method to handle microsatellite markers. A challenging open problem is to develop an efficient method that is able to discover block structure using genotypic data with unknown phase.

References

1. M.J. Daly, J.D. Rioux, S.F. Schaffner, T.J. Hudson, E.S. Lander, High-resolution haplotype structure in the human genome, *Nature Genetics* **29**, 229–232 (2001).
2. N. Patil, A.J. Berno, D.A. Hinds *et al.*, Blocks of limited haplotype diversity revealed by high-resolution scanning of human chromosome 21, *Science* **294**, 1719–1723 (2001).
3. K. Zhang, M. Deng, T. Chen, M.S. Watermanm, F. Sun, A dynamic programming algorithm for haplotype block partitioning, *PNAS* **99**, 7335–7339 (2002).
4. S.B. Gabriel, S.F. Schaffner, H. Nguyen *et al.*, The structure of haplotype blocks in the human genome, *Science* **296**, 2225–2229 (2002).

5. J. Rissanen, Modeling by shortest data description, *Automatica* **14**, 465–471 (1978).

6. J. Rissanen, Stochastic Complexity, *Journal of the Royal Statistical Society* B **49**, 223–239 (1987).

7. M. Li, P. Vitanyi, *An Introduction to Kolmogorov Complexity and its Applications* (Springer-Verlag, New York, 1997).

8. M.H. Hansen, B. Yu, Model Selection and the Principle of Minimum Description Length, *Journal of the American Statistical Association* **96**, 746–774 (2001).

9. W. Li, New stopping criteria for segmenting DNA sequencies, *Physical Review Letters* **86**, 5815–5818 (2001).

10. J.R. Quinlan, R.L. Rivest, Inferring decision trees using the Minimum Description Length principle, *Information and Computation* **80**, 227–248 (1989).

11. P. Kilpeläinen, H. Mannila, E. Ukkonen, MDL learning of unions of simple pattern languages from positive examples, In *Proceedings of the Second European Conference on Computational Learning Theory (Euro-COLT)*, ed. Paul Vitanyi (Springer-Verlag, Berlin, 1995).

12. P. Domingos, The role of Occam's razor in knowledge discovery, *Data Mining and Knowledge Discovery* **3**, 1–19 (1999).

13. R. Durbin, S.R. Eddy, A. Krogh, G. Mitchison, *Biological Sequence Analysis: Probabilistic Models of Proteins and Nucleic Acids* (Cambridge University Press, Cambridge, 1998).

14. J.S. Liu, C.E. Lawrence, Bayesian inference on biopolymer models, *Bioinformatics* **15**, 38–52 (1999).

15. T. Paunio, J. Ekelund, T. Varilo *et al.*, Genome-wide scan in a nationwide study sample of schizophrenia families in Finland reveals susceptibility loci on chromosomes 2q and 5q, *Human Molecular Genetics* **10**, 3037–3048 (2001).

PYPOP: A SOFTWARE FRAMEWORK FOR POPULATION GENOMICS: ANALYZING LARGE-SCALE MULTI-LOCUS GENOTYPE DATA

ALEX LANCASTER, MARK P. NELSON, DIOGO MEYER AND GLENYS THOMSON

Department of Integrative Biology, University of California, Berkeley,
3060 Valley Life Sciences, Berkeley, CA, 94720, USA
E-mail: alexl@socrates.berkeley.edu

RICHARD M. SINGLE

Department of Biometry, University of Vermont,
Hills Science Building, Burlington, VT, 05405, USA

Software to analyze multi-locus genotype data for entire populations is useful for estimating haplotype frequencies, deviation from Hardy-Weinberg equilibrium and patterns of linkage disequilibrium. These statistical results are important to both those interested in human genome variation and disease predisposition as well as evolutionary genetics. As part of the 13th International Histocompatibility and Immunogenetics Working Group (IHWG), we have developed a software framework (PyPop). The primary novelty of this package is that it allows integration of statistics *across* large numbers of data-sets by heavily utilizing the XML file format and the R statistical package to view graphical output, while retaining the ability to inter-operate with existing software. Largely developed to address human population data, it can, however, be used for population based data for any organism. We tested our software on the data from the 13th IHWG which involved data sets from at least 50 laboratories each of up to 1000 individuals with 9 MHC loci (both class I and class II) and found that it scales to large numbers of data sets well.

1 Introduction

Several major factors account for variation in the human genome: *mutation, random genetic drift, migration* (or *gene flow*) and *natural selection*. Understanding of the relative roles of these evolutionary forces is important for the study of both complex and Mendelian diseases, since they can affect our ability to identify and localize disease predisposing variants and our power to recognize underlying functional mechanisms through which predisposing genes can become relatively common in a population.

In particular, genetic systems which are highly polymorphic can implicate natural selection as an important factor in maintaining variation. Genetic systems such as the Human Leukocyte Antigen (HLA) region (the Major Histocompatibility Complex [MHC] for humans) are highly polymorphic. Six classical class I and II loci each contain up to 399 alleles [11].

Several basic population genetics statistics from multi-locus genotype data can help us understand these patterns of variation, and their implications for disease studies and evolutionary genetics. These statistics include, but are not limited to estimating haplotype frequencies, identifying deviation from Hardy-Weinberg equilibrium and locating patterns of linkage disequilibrium in a given population.

In population studies several implementations of programs and routines to calculate these basic statistics exist, but are mainly oriented towards analyzing statistics on a population-by-population basis. The ability to cross-correlate these statistical features across many population data-sets will enable the identification of features in the genetic data that can further our understanding of the functional and disease predisposing role of specific alleles, and conversely allow us to rule out others.

Currently, packages for analyzing population data already exist such as `Arlequin` [13], `PHYLIP` [1] and `Genepop` [3]. In general, however, they are not oriented towards large-scale cross-population data analyses. Analogous to the tools being developed for sequence analysis and search, we seek a framework in which basic statistical data from population genetic analyses can be housed, interrogated and visualized in such a way that important features of interest to both the biomedical investigator and the evolutionary biologist can be highlighted.

We also did not want to reinvent the wheel, so, where possible we can inter-operate with existing population genetic analysis packages (either as part of the framework or through file formats). In this way, our software, *PyPop* (*Py*thon for *Pop*ulation Genetics) can be viewed as an integrating framework which draws on the strengths of existing tools in the community.

1.1 *IHWG: International Histocompatibility Working Group*

The primary motivation for developing this project was our role as the 'Biostatistics Core', part of the International Histocompatibility Working Group (IHWG (`http://www.ihwg.org/`)). The IHWG collected population data on the HLA region and largely focused on the HLA classical class I (A, B, and C) (1.8 Mb) and class II (DR, DQ, and DP) (1.2 Mb) genes, which flank the class III region on chromosome 6 (ch. 6p21.31).

As part of the Anthropology and Human Diversity component of the IHWG, data from populations from upwards of 50 laboratories was made available via a database housed at the Fred Hutchison Cancer Research Institute (the 'Database Core'). The IHWG provided standardized typing reagents

516

to each lab involved in the component, which resulted in high resolution genotype data for each population.

The molecular characterization of these alleles thus allowed us to use both the allele frequency (the raw allele 'calls') and the underlying sequence information (the 'calls' can be converted into sequences if desired) as input to the analysis framework. PyPop does not distinguish between allele calls or sequence data and can transparently handle both.

1.2 Population genetic tests and statistics

The particular cross-population statistics we wished to address included: (1) conformity to Hardy-Weinberg expectations, (2) tests for balancing selection; (3) haplotype distributions and patterns of linkage disequilibrium among populations; and (4) other tests such as worldwide patterns of genetic differentiation.

Hardy-Weinberg Hardy-Weinberg equilibrium essentially states that unless there are counteracting forces, the frequencies of alleles will not change in a population and the expected genotype frequencies each generation are determined by the allele frequencies, and are termed Hardy-Weinberg proportions (HWP). In the context of multi-population analyses we can use deviation from HWP to determine whether this results from: (1) typing error (the first possibility investigated); (2) an 'admixed', or merged population; (3) operation of natural selection; or (4) inbreeding.

Ewens-Watterson test of neutrality This is a test with the null hypothesis of neutral evolution and determines the probability that the observed homozygosity under HWP, for a given sample size and observed number of alleles, is more extreme than the expected homozygosity under random neutral mutations and genetic drift (neutrality). This test can tell us whether selection, either directional (observed homozygosity > expected homozygosity) or balancing (observed homozygosity < expected homozygosity) is in operation on a particular locus across populations.

Haplotype estimation and linkage disequilibrium Linkage disequilibrium (LD) describes the non-random association of alleles at different genetic loci. Through estimating haplotype frequencies, it is possible to estimate LD in a population, the presence of significant LD can be due to history for very closely linked genes and can also indicate the operation of selection.

Other statistics Other individual-level population genetic statistics can also be calculated, such as F_{st}, which describes the apportionment of genetic diversity within subpopulations, relative to a larger population, allowing us the estimate the amount of admixture in a population.

1.3   Requirements

The data sets we wished to analyze were highly heterogeneous. Datasets varied considerably in the number of loci typed (from two to nine), number of individuals sampled (from less than 50 to 1000), and number of alleles at each locus (from 5 to 179). Given that we had such highly diverse data sets, we nevertheless wished to generate analyses that could be integrated in a systematic way, this led to a set of requirements for the software framework:

- **modular** each analysis (e.g. Hardy Weinberg) can be run stand-alone or as part of a battery of tests

- **configurable** analyses can be switched on or off as required by the user in a simple configuration file

- **a filter** all output from the analyses should be available as input to other programs

- **standardized output** output should be generated in the open standard format XML

- **integrating** platform should allow simple integration of modules written in other languages (e.g., C) and/or third party software

1.4   Integrating with existing software

Where possible, we would like to integrate with existing software. `Arlequin` [13] and `PHYLIP` [1] are packages that deal with some aspect of population genetic statistics. In addition, we had some existing in-house software, such as `emhaplofreq`, a program for haplotype estimation for the highly polymorphic HLA loci, and `gthwe`, a program for implementing Guo and Thompson's Hardy-Weinberg exact test [4].

In using these programs, however, we realized a need for a tool that could integrate features of existing software, and where needed, implement missing features that would realize our goals of doing large-scale population genetic data analysis. Specific limitations that we wished to address were:

1. **Modularity** Programs should allow different options to be set for analysis and produce content that is easily parsable, not a monolithic output set of statistics (such as unstructured plain ASCII text or HTML).

2. **Batch-ability** It should be simple to set up an entire job in an unattended 'batch mode', involving the creation of (or modification of) a configuration file with a text editor, followed by the invocation of a script.

Software that relies on a 'captive user interface', for a population to be analyzed and requires user interaction such as mouse clicks and menus, makes it difficult to analyze hundreds of data sets.

3. **Scalability** It should be straightforward, for example, to gather a single statistic for several populations and display it in a table without a laborious manual search across many files. This relates partly to the previous point: often existing software was oriented towards smaller and less heterogeneous datasets. More typical in evolutionary genetics studies is integrating results across several populations at one or two loci, rather than many populations with a large number of loci and high variability in the number of individuals.

4. **Open-source** We plan to release our software under an open source license [9]. Software that is open-source allows others to extend and re-use components, allows interoperation via an open and published interfaces, and can reduce duplication of effort within the community. Some existing software that was not open-source required reverse engineering of their file formats and run-time behaviour in order to be able to communicate and write interfaces to them.

5. **Cross-platform** Software should be cross platform, and not be tied to proprietary features. In particular it should be be available to run high-performance UNIX platforms such as GNU/Linux or Solaris as well as Windows and Macintosh platforms.

2   Method

2.1   Overall design

To integrate the data analyses for multiple populations, the analysis pipeline for PyPop has been designed in two major parts, or phases, the first is the basic population genetic analyses, and the second integrates the results of these analyses across multiple populations. The overall data and work flow is shown in Figure 1.

2.2   Implementation

We decided to implement the project in the object-oriented scripting language Python [10]. It is an interpreted language allowing for rapid prototyping of modules and has a convenient standard library of functions. Through use of the

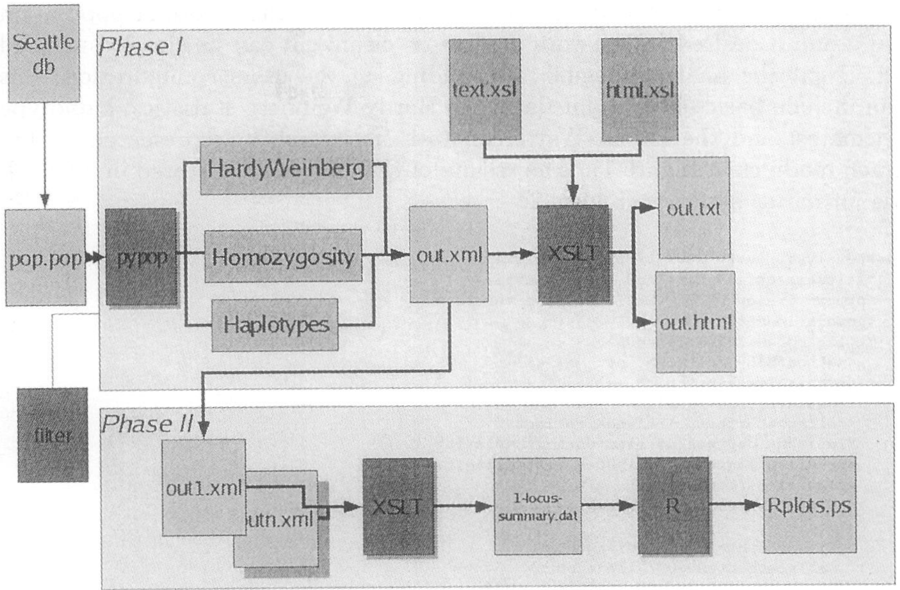

Figure 1. Work flow

the Simple Wrapper Interface Generator (SWIG [12]), it is also straightforward to 'wrap' existing code in C, C++ or Java and call it from Python.

In line with our philosophy of integrating with 'best of breed' open-source components, we leveraged the work of the Numeric Python [8] project, which provides efficient data structures for holding large arrays of data, and R, the open-source implementation [6] of the programming language S. As our XSLT parser, we chose the fast C-based `libxslt` [7] from the GNOME project [2].

2.3 Data flow

In phase I, before analysis begins, the multi-locus genotype data for each population (stored in a text file with the `.pop` extension), is passed through a filter module for data cleaning. Next, the basic population genetic analyses are run for each population file from the database. The filter module is a set of rules that traps any allele name which does not have a close match in the database of HLA alleles [5] maintained by Steve Marsh of the Anthony Nolan Trust in the UK. Although at present this is HLA-specific, it has been designed in such a way that it is simple to write a filter for other types of population data, and if desired (if the data is already 'clean'), it can be simply switched off. Next the analysis begins. Depending on the users configuration, this can include basic allele count data; the Hardy-Weinberg statistics; haplotype estimates; and the Ewens-Watterson test of neutrality (represented by the green modules in Figure 1). The results of the analyses are stored in an XML file format as shown in Figure 2.

```
<dataanalysis date="2002-07-09-05-01-18">
   <filename>ukimid_nireland.pop</filename>
   <pypop-version>DEVEL_VERSION</pypop-version>
   <populationdata>
     <labcode>UKIMID</labcode>
     <method>SSOP</method>
     <ethnic>Irish</ethnic>
     <contin>Europe</contin>
     <collect>Northern Ireland</collect>
     <latit>54 degrees 40 minutes north</latit>
     <longit>6 degrees 45 minutes west</longit>
     <complex>3</complex>
     <popname>NIreland</popname>
     <totals>
       <indivcount>1000</indivcount>
       <allelecount>2000</allelecount>
       <locuscount>9</locuscount>
       <lociWithDataCount>4</lociWithDataCount>
     </totals>
   </populationdata>
   <locus name="A">
     <allelecounts>
       <untypedindividuals>0</untypedindividuals>
       <indivcount>1000</indivcount>
       <allelecount>2000</allelecount>
       <distinctalleles>26</distinctalleles>
       <allele name="0101">
         <frequency>0.20200 </frequency><count>404</count>
       </allele>
...
```

Figure 2. Extract from sample XML output file

XML was chosen as the output format because: 1) it can be read as input by other programs and; 2) it is readily transformable into human-readable form (e.g., a text file) or web-form (e.g., HTML) via XSLT (eXtensible Stylesheet Language for Transformations) [14].

In phase II, the results of the data analyses of the individual data files are integrated, and the benefits of using XML as the storage and exchange format are realized. There are two major benefits: one relevant for displaying results of each individual run, the second, far more powerful benefit, for aggregating the data for cross-population meta-analyses and transforming it as input for third party packages.

First, since many of the analyses (notably estimating the significance of all pairwise linkage disequilibrium) can take a considerable amount of time, especially if the population consists of many individuals and is highly polymorphic, the *analysis generation* stage is decoupled from the *analysis presentation* stage. This enables tweaking of the 'human-readable' text output of the individual files of the presentation without completely re-running the analyses. A sample text output is shown in Figure 3. The same XML content can also be used to generate an HTML version or web version of the same data set.

```
Performed on the 'ukimid_nireland.pop' file at: 2002-07-09-05-01-18

Population Summary
==================

          Lab code: UKIMID
     Typing method: SSOP
         Ethnicity: Irish
         Continent: Europe
   Collection site: Northern Ireland
          Latitude: 54 degrees 40 minutes north
         Longitude: 6 degrees 45 minutes west
   Population Name: NIreland
   [...]


1.1. Allele Counts [A]

----------------------

Untyped individuals: 0
Sample Size (n): 1000
Allele Count (2n): 2000
Distinct alleles (k): 26

Counts ordered by frequency   | Counts ordered by name
Name       Frequency (Count)  | Name       Frequency (Count)
0201       0.27400    548     | 0101       0.20200    404
0101       0.20200    404     | 0201       0.27400    548
```

Figure 3. Extract from sample plain text output generated from XML data

Second, the results of Phase I are used to investigate patterns of variation across populations and within populations. Examples include comparisons of evidence of selection, using the homozygosity test of neutrality for a given locus for all populations (e.g., DPA1 for all populations), or comparing the sets of loci in significant linkage disequilibrium across populations. This is implemented using XSLT to transform all the desired sets of output XML files into tables of data that are read by the statistical package R.

A sample output of this process is shown in Figure 4. This is an output plot from R and depicts for each of the loci analyzed as part of the IHWG workshop, the proportion of populations in which the linkage disequilibrium, as measured by the W_n statistic, exceed 0.6.

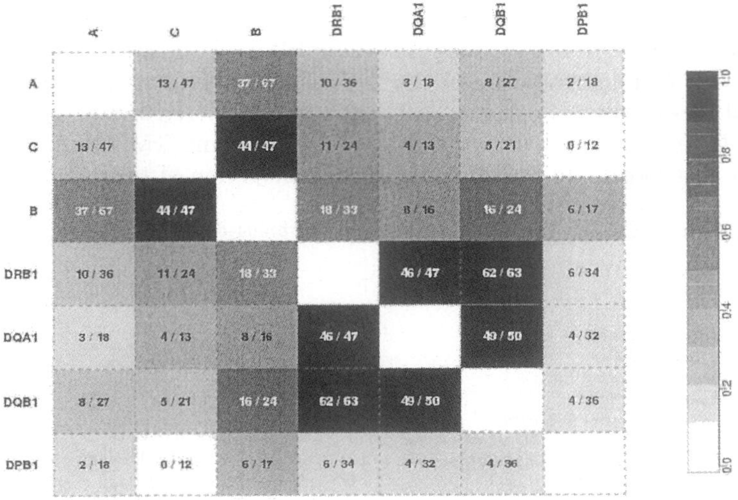

Figure 4. Sample graphical output from meta-analysis: proportions of populations with linkage disequilibrium measure Wn greater than 0.6

The XML output (both the individual population-level data files, and the aggregated multi-population data) can be transformed (via XSLT) into formats suitable for input to other programs. Currently we have a prototype module for generating input for **PHYLIP** and a prototype module to generate **Arlequin .arp** files.

3 Results and discussion

3.1 Analyzing IHWG workshop data

The **PyPop** framework was used to analyze the full set of IHWG data from both the 12th and 13th workshops. With 119 separate data-sets (some of which individually required up to 9 days to complete the basic per-population statistics), it was straightforward to set up a batch program to generate the individual output analyses. For each file, an individual XML file was generated. From these individual files, using XSLT stylesheets as described above in phase II of PyPop, it was straightforward to generate the input data files for the statistical package, R. The R code was set up to generate overall graphs for many population genetic parameters of interest. In particular, the analysis allows 'slicing' of the data along a number of axes. As an example, we can view Hardy Weinberg deviation, for all geographic regions at a given locus, or view all loci for a given region. As another example, we can view the number of populations that have data for a given region (Figure 5), or number of populations for which data was provided for a given locus (Figure 6).

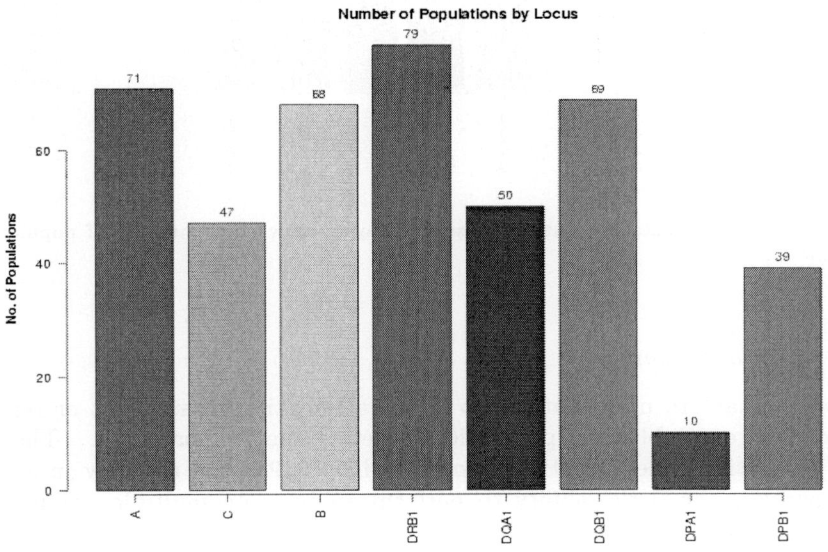

Figure 5. Sample metadata output: viewing number of populations by locus

This 'meta-analysis' code is modular and can cope with non-HLA data, and thus can potentially be useful for other large-scale population genetic analyses. The significance is that the complete pipeline, from analyzing the individual data files, to the generation of the overall 'meta' output can be completely automated from the command-line. Further, the flexibility of the XML format, allows us to easily extract and output data for future data analyses without requiring regeneration of the basic population genetics statistics.

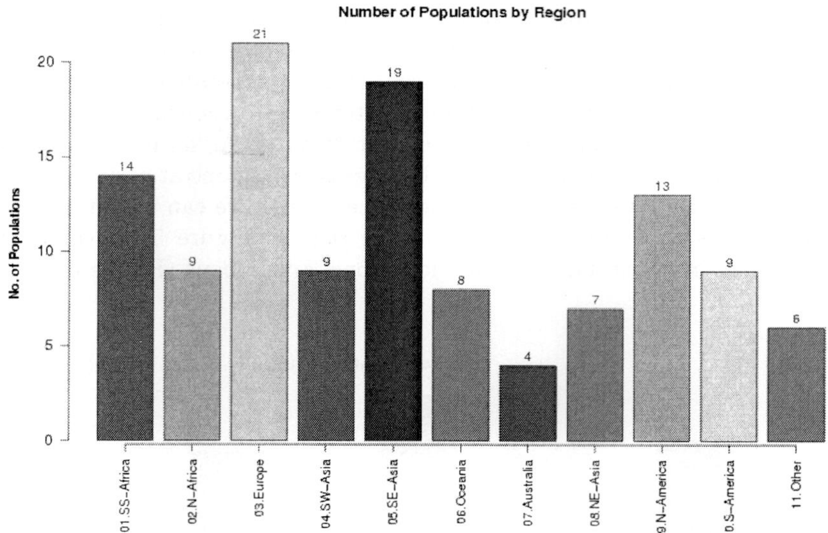

Figure 6. Sample metadata output, a second way to view data: number of populations within each region

3.2 Role and significance

It's important to note that PyPop is not attempting to supplant or replace other, more established, population genetics analysis packages. The basic population genetic statistics computed by PyPop are not new in and of themselves, nor is the ability to perform analyses on multiple populations (Arlequin can currently do this). Indeed, PyPop can use existing packages to calculate them. However the approach of integrating information in highly heterogeneous datasets on a large scale is new, and not currently available in the evolutionary genetics community. In addition to tests unavailable in existing projects, the uniqueness of the present project is that it is intended to be a high-throughput system that enables population genetics to join the realm of genomics.

3.3 Future directions

In the future we plan to continue development on the modules that can inter-operate with PHYLIP and Arlequin. We also plan to take the Ewens-Watterson test of neutrality down to the amino-acid level (by considering each amino acid site as a genetic locus), necessitating the translation of the allele 'calls' into sequence data (where possible). This will result in a useful module for those wishing to analyze allele data at the sequence level. Longer-term possibilities include developing a graphical front-end, possibly web-based, integration with data mining tools such as clustering analysis, and currently under discussion is integration with the NCBI's new dbMHC database [15]. We also plan to release PyPop under an open source license. Details will be made available at the Thomson lab website: `http://allele5.biol.berkeley.edu/`

Acknowledgements

This work has benefited from the support of NIH grant AI49213 (13th IHW). Thanks to Steve Mack.

References

1. J. Felsenstein, 'PHYLIP Phylogeny Inference Package (Version 3.2) `http://evolution.genetics.washington.edu/phylip.html`' *Cladistics* **5**:164-166 (1989).
2. GNOME Project `http://www.gnome.org/`.
3. Genepop `ftp://ftp.cefe.cnrs-mop.fr/genepop`.
4. SW Guo and EA Thompson, 'Performing the exact test of Hardy-Weinberg proportion for multiple alleles' *Biometrics* **48**:361-72 (1992).
5. IMGT/HLA Database `http://www.ebi.ac.uk/imgt/hla/`.
6. Ross Ihaka and Robert Gentleman, 'R: A Language for Data Analysis and Graphics `http://www.r-project.org/`' *Journal of Computational and Graphical Statistics* **5**(3):299-314 (1996).
7. libxslt, XSLT C library for GNOME `http://xmlsoft.org/`.
8. Numeric Python `http://numpy.sf.net/`.
9. Open Source Initiative `http://www.opensource.org/`.
10. Python `http://www.python.org/`.
11. D A Rhodes and J Trowsdale, 'Genetics and molecular genetics of the MHC' *Rev. Immunogenetics* **1**(1):21-31 (1999).
12. Simple Wrapper Interface Generator `http://www.swig.org/`.
13. S. Schneider, D. Roessli, and L. Excoffier, 'Arlequin: A software for population genetics data analysis. `http://lgb.unige.ch/arlequin/`', Genetics and Biometry Lab, Dept. of Anthropology, University of Geneva (2000).
14. eXtensible Stylesheet Language `http://www.w3.org/Style/XSL/`.
15. dbMHC `http://www.ncbi.nlm.nih.gov/IEB/Research/GVWG/MHC/`.

JOINT BAYESIAN ESTIMATION OF MUTATION LOCATION AND AGE USING LINKAGE DISEQUILIBRIUM

B. RANNALA, J.P. REEVE

Department of Medical Genetics, University of Alberta, Edmonton
Alberta T6G2H7, Canada

Associations between disease and marker alleles on chromosomes in populations can arise as a consequence of historical forces such as mutation, selection and genetic drift, and is referred to as "linkage disequilibrium" (LD). LD can be used to estimate the map position of a disease mutation relative to a set of linked markers, as well as to estimate other parameters of interest, such as mutation age. Parametric methods for estimating the location of a disease mutation using marker linkage disequilibrium in a sample of normal and affected individuals require a detailed knowledge of population demography, and in particular require users to specify the postulated age of a mutation and past population growth rates. A new Bayesian method is presented for jointly estimating the position of a disease mutation and its age. The method is illustrated using haplotype data for the cystic fibrosis $\Delta F508$ mutation in europe and the DTD mutation in Finland. It is shown that, for these datasets, the posterior probability distribution of disease mutation location is insensitive to the population growth rate when the model is averaged over possible mutation ages using a prior probability distribution for the mutation age based on the population frequency of the disease mutation. Fewer assumptions are therefore needed for parametric LD mapping.

1 Introduction

The term "linkage disequilibrium" (LD) describes a population distribution of alleles (at two or more loci) among chromosomes that is not independent [1]. In other words, alleles at different loci co-occur on chromosomes at a frequency that does not equal the product of the marginal frequencies of the alleles. One mechanism by which LD commonly arises is mutation. A variant allele at a locus arises by mutation on a particular chromosome (bearing specific alleles at other linked loci). Initially, the new allele is found exclusively on this chromosomal background, but over time the association of the new allele with alleles present at other loci on the ancestral chromosome breaks down due to recombination and mutation. Under certain conditions, the association among alleles at the linked loci may disappear and alleles will occur together on chromosomes in proportion to the product of the marginal frequency of each allele, a situation referred to as "linkage equilibrium."

The extent of linkage disequilibrium (LD) among alleles in a population is determined by many factors including the map distances among loci, the rates of mutation at the loci, natural selection, and genetic drift (population demographic history, etc). Recently, human geneticists have begun to exploit linkage disequilibrium to map disease mutations [2], to estimate the ages of known mutations [3], and to reconstruct ancient demographic events [4].

1.1 Linkage disequilibrium gene mapping

One of the most important practical applications of linkage disequilibrium studies in humans is to map the positions of mutations that cause disease [2]. Linkage mapping methods for finding mutation locations using patterns of marker-disease co-segregation on pedigrees have limited resolution, usually less than 1 cM (roughly 1 Mb). To identify a disease mutation by positional cloning greater resolution is needed. LD mapping methods can have much greater resolution than linkage analysis because the methods exploit recombination events occurring in the extended genealogy relating a random sample of individuals from a population. The population genealogy will typically involve thousands of meioses versus a few hundred, at most, for linkage analysis using even very large pedigrees.

1.2 Estimating mutation ages

Several recent papers have proposed methods for estimating the age of a mutation with a known location using information from variation at linked genetic markers and the population frequency of the mutation [3]. These methods typically assume that the location of the mutation is known relative to a set of linked markers. Although it has been suggested that it might be possible to jointly estimate mutation ages and locations [5] the complexity of the analysis is such that this has so far not been achieved, except in simplified cases which are generally unrealistic for human populations [6].

1.3 Joint estimation of mutation age and location

In this paper, we explore the use of LD to map genes in the face of an unknown disease allele age and uncertain past population growth rates using Bayesian Markov chain Monte Carlo (MCMC) methods. Previous methods for LD mapping have assumed that one or more of the mutation age, location, or population growth rate parameters are known [7,8,9,10,11] or have made unrealistic assumptions about population demography [12,13,6]. Here, we show that by averaging over the possible age of a disease mutation, using a prior age distribution based on the present population frequency, estimates of mutation location may be obtained that are quite insensitive to the population growth rate; this is encouraging since the past growth rates are usually poorly known and a less parameterized model is therefore desirable.

2 Theory

Following standard terminology [1] we define a genetic locus to be a specific physical position in the nucleotide sequence of a chromosome. An allele is defined to be a variant of a locus with one or more nucleotides altered by point mutation, nucleotide insertion or deletion, etc. Define $\mathbf{X}$ to be a matrix of haplotypes for a specified set of marker loci (i.e., phase-determined alleles for the markers) for a sample of chromosomes bearing a disease mutation of unknown location. Let $\mathbf{Z}$ be a matrix of the marker haplotypes from a random (ethnically matched) sample of normal individuals. Define $\mathbf{p}$ to be a matrix of the (unknown) marker allele frequencies in the population of normal chromosomes. Let τ be the unobserved ancestral genealogy underlying the sample of disease chromosomes, $\mathbf{Y}_{-0}$ be a matrix of the ancestral haplotypes in the genealogy, $\mathbf{Y}_0$ be the (unknown) ancestral haplotype on which the disease mutation first arose, and t_0 be the (unknown) age of the mutation. Define θ to be the position (in Morgans) of the mutation relative to marker locus 1, Θ to be a vector of genetic parameters, such as the map distances among marker loci, etc, and Λ to be a vector of the demographic parameters, including the fraction of the population of disease chromosomes sampled, f, and the population growth rate, r (assuming exponential growth). In this paper, we consider haplotype data but our method also can be used with genotype data under simple models of inheritance [14]

2.1 Likelihood and prior distributions of parameters

The method we present is an extension of the Bayesian LD mapping method of Rannala and Reeve [11,14]. Details of the likelihood and the priors for parameters other than t_0 can be found in the earlier papers. Here, we focus on the addition of a prior for t_0 and the steps involved in integrating over this prior. The likelihood of the sampled disease haplotypes and the (unobserved) ancestral haplotypes is

$$f(\mathbf{X}, \mathbf{Y}_{-0} | \theta, \Theta, \tau, \mathbf{p}, \mathbf{Y}_0, t_0).$$

The prior probability density that we use for t_0 is proportional to the likelihood of the observed sample frequency, i, of the disease allele. From Slatkin and Rannala [15] this is

$$f(t_0 | i, r, N, q) \propto \left\{ \frac{1 - e^{-rt_0}}{1 - e^{-rt_0} + e^{-rt_0}(2Nqr/i)} \right\}^{i-1} \frac{1}{1 - i/(2Nqr) + e^{rt_0}/(2r)},$$

where we have substituted $f = i/(Nq)$, where i is the number of disease chromosomes in the sample, N is the population size and q is the relative population frequency of disease chromosomes (e.g., estimated based on disease incidence and mode of inheritance). The above equation is sometimes multiplied by an additional term, e^{-rt_0} to take account of the decreased influx of mutations at time t_0 in the past due to the smaller past population size under a model of exponential population growth [16] but this addition has little effect on estimates of disease location for the examples we present below.

2.2 *Posterior distribution of parameters*

The joint posterior density of the parameters is given by

$$
\begin{aligned}
&f(\theta, \mathbf{Y}, \tau, \mathbf{p}, t_0 | \mathbf{X}, \mathbf{Z}, \Theta, N, q, r, i) = \\
&\frac{f(\mathbf{X}, \mathbf{Y}_{-0} | \theta, \Theta, \tau, \mathbf{p}, \mathbf{Y}_0, t_0) f(\tau | N, q, r) f(\mathbf{Y}_0 | \mathbf{p}) f(\mathbf{Z} | \mathbf{p}) f(\theta) f(\mathbf{p}) f(t_0 | i, N, q, r)}{f(\mathbf{X}, \mathbf{Z} | \Theta, N, q, r, i)},
\end{aligned}
$$

where the priors for the variables other than t_0 are as given in Rannala and Reeve [11]. We used a uniform prior for the position of the disease mutation, θ, although a prior based on an annotated human genome sequence and a mutation database could also have been used [11].

Markov chain Monte Carlo (MCMC) methods were used to generate the joint posterior density of the parameters in the above equation based on a Metropolis-Hastings algorithm. The basic principle of MCMC analysis is to simulate observations from a Markov chain with a stationary distribution that is the joint posterior probability density of the parameters. The joint and marginal posterior densities can be estimated by running this chain on a computer until it converges and then sampling parameter values from the chain at equal intervals. The above method has been implemented in version 2.2 of the LD mapping program DMLE+ available for downloading from dmle.org.

3 Examples

To illustrate the method, we apply it to two data sets for which haplotypes are available and a disease mutation has been cloned. This will allow us to directly test the accuracy of the method using relevant empirical data. The first data set that we examine is for a common mutation causing cystic fibrosis (CF) in europeans, the $\Delta F508$ mutation; the second data set we examine is for a founder mutation in Finland that causes diastrophic dysplasia (DTD). All analyses were carried out using DMLE+ version 2.2.

530

3.1 *Cystic fibrosis $\Delta F508$ mutation in Europe*

The $\Delta F508$ mutation is the most common cause of CF in european populations, accounting for roughly 70 percent of CF mutations. The data set we analyze was originally used to map the CF gene [17]. In total, 63 chromosomes from this data set carry the $\Delta F508$ mutation. We excluded one of these chromosomes from our analysis as it appears to belong to a very different haplogroup and possibly represents a recurrent mutation. The chromosomes were typed for 23 biallelic markers (RFLPs) that span 1.8 Mb, with the mutation located 880 kb from marker 1. The two closest markers to the mutation are at 869.8 and 889.8 kb, respectively. Haplotype phase was inferred via linkage analysis of relatives of probands. Approximately 6 percent of the markers in the disease chromosomes are missing data; we integrate over the missing markers using data augmentation in our algorithm [14].

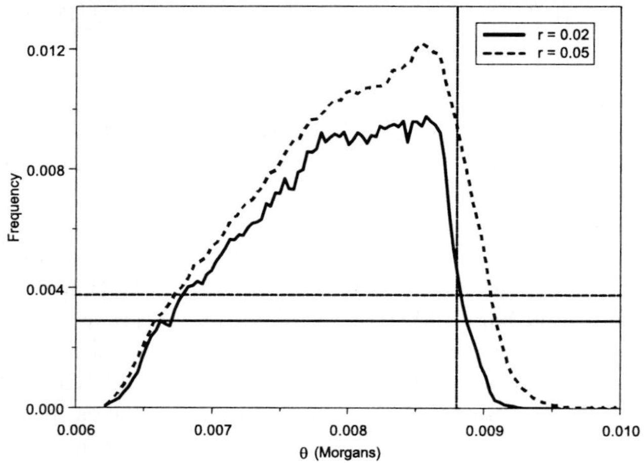

Figure 1: The posterior probability density of the position, θ, of the CF mutation $\Delta F508$ relative to marker 1. A total of 23 markers were typed for 62 disease chromosomes carrying the $\Delta F508$ mutation. Separate analyses were carried out using two different population growth rates $r = 0.05$ (dashed line) and $r = 0.02$ (solid line). The 95 percent credible set of values for each posterior density is indicated by the (dashed and solid) horizontal lines at the bottom of the figure. The true position of the mutation (assuming 1 cM = 1 Mb) is indicated by the vertical line. The posterior density is little affected by the growth rate. The program is simultaneously estimating the age of the mutation, t_0.

The results of the analysis are shown in Figures 1 and 2. Using either a growth rate of $r = 0.02$ or $r = 0.05$ ($r = 0.05$ is often assumed to be the recent growth rate for european populations) the 95 percent credible set for θ, the

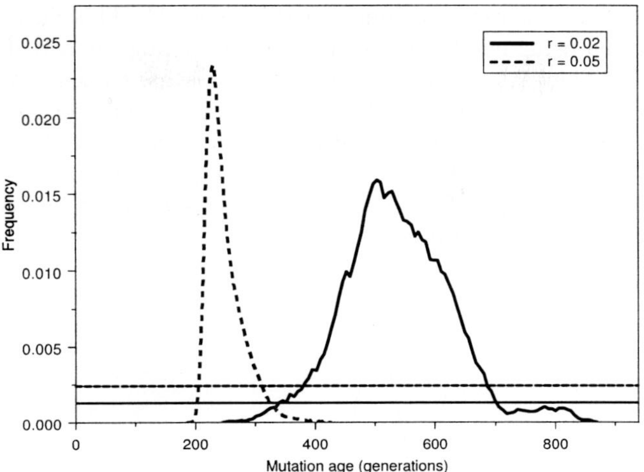

Figure 2: The posterior probability density of the age, t_0, of the CF mutation $\Delta F508$. A total of 23 markers were typed for 62 disease chromosomes carrying the $\Delta F508$ mutation. Separate analyses were carried out using two different population growth rates $r = 0.05$ (dashed line) and $r = 0.02$ (solid line). The 95 percent credible set of values for each posterior density is indicated by the (dashed and solid) horizontal lines at the bottom of the figure. The posterior density is little affected by the growth rate. The program is simultaneously estimating the position of the mutation, θ.

position of the mutation relative to marker 1, ranges from roughly .65 cM to either 0.9 cM (for $r = 0.02$) or 0.925 cM (for $r = 0.05$) as shown in Figure 1. In both cases, the true position of the mutation (0.88 cM assuming 1 cM = 1 Mb) is bracketed by the 95 percent credible set. The growth rate parameters have a much larger effect on the posterior density of the allele age parameter, t_0 as shown in Figure 2. In that case, the 95 percent credible set is (200,310), for $r = 0.05$, and is (350,700) for $r = 0.02$ (in units of generations). This indicates that estimates of allele ages are highly sensitive to inferred population growth rates but estimates of mutation location are much less sensitive to the population growth rate if one integrates over the mutation age.

3.2 Diastrophic dysplasia mutation in Finland

DTD is an autosomal recessive disease with an unusually high frequency (roughly 2 percent are carriers) in the Finnish population, probably due to a founder event that occurred about 2,000 years ago (100 generations). We used 5 markers (2 RFLPs and 3 microsatellites) from an unpublished data set provided by

532

J. Hästbacka that was used to clone the DTD gene [18]. This data set was previously analyzed using our DMLE+ program and fixing the age of t_0 to be 100 generations [11]. The data set is comprised of 148 disease chromosomes typed for the 5 markers and 126 controls. There are no missing marker genotypes. The five markers span 20 kb with the mutation located roughly 86 kb centromeric to marker 1. Haplotype phase was inferred via linkage analysis of relatives of probands. A growth rate of $r = 0.085$ was used based on records of the demographic expansion in Finland [11].

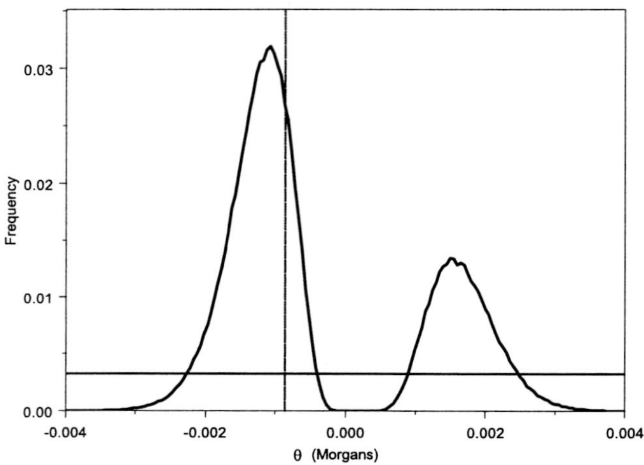

Figure 3: The posterior probability density of the position, θ, of the DTD mutation relative to marker 1. A total of 5 markers were typed for 148 disease chromosomes carrying the DTD mutation. Analyses were carried out using a population growth rate of $r = 0.085$. The 95 percent credible set of values is indicated by the horizontal line at the bottom of the figure. The true position of the mutation (assuming 1 cM = 1 Mb) is indicated by the vertical line. the program is simultaneously estimating the age of the mutation, t_0.

The results of the analysis are shown in Figures 3 and 4. Integrating over the allele age has little effect on the posterior density of θ in this case. Figure 3 shows the posterior density of θ. This density is nearly identical to the posterior density that is obtained by assuming $t_0 = 100$ (see e.g., Figure 4 of Rannala and Reeve [11]). As in the earlier analysis, the 95 percent credible set for the location of the DTD mutation includes the true location (assuming 1 cM = 1 Mb). Figure 4 shows the posterior density of the mutation age. This density has a mode of roughly 80 generations with a 95 percent credible set of ages ranging from 65 generations to 105 generations; this range includes the postulated time of the founding of the Finnish population roughly 100

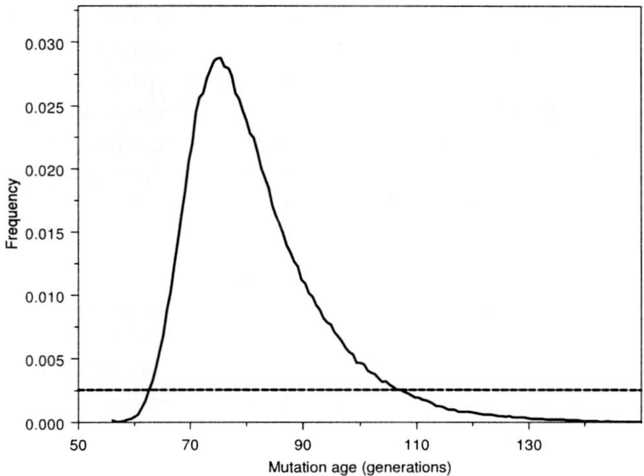

Figure 4: The posterior probability density of the age, t_0, of the DTD mutation . A total of 5 markers were typed for 148 disease chromosomes carrying the DTD mutation. Analyses were carried out using a population growth rate of $r = 0.085$. The 95 percent credible set of values for the posterior density is indicated by the horizontal line at the bottom of the figure. The program is simultaneously estimating the position of the mutation, θ.

generations ago supporting the idea that the mutation was introduced at the founding event. Thus, the analysis of the DTD data set supports the idea that it is possible to jointly estimate mutation location and age as both the estimates obtained in this case appear quite reasonable.

4　Discussion

In this paper, we have explored the feasibility of jointly estimating the age of a disease mutation and its location using multilocus marker haplotypes for a sample of chromosomes bearing a disease mutation and a sample of normal chromosomes. We have used MCMC methods, extending our previously developed LD mapping program DMLE+. In the case of the two data sets analyzed, one for the DTD mutation in Finland and another for the CF $\Delta F508$ mutation, the method performs well, allowing joint estimates of the two parameters and leading to reasonable results for mutation ages and successfully locating the disease mutation with 95 percent probability, One new finding that is particularly encouraging is that the posterior density of the mutation location appears insensitive to the (unknown) population growth rate when the method integrates over possible allele ages. Thus, fewer assumptions may be needed for parametric LD-based disease mutation mapping than was previously thought. The estimate of mutation age, on the other hand, appears quite sensitive to the assumed growth rate and should therefore be interpreted with greater caution in future, perhaps instead considering values for the estimated mutation age based on a range of plausible population growth rates.

Acknowledgments

This research was supported by grants to B. Rannala from the Alberta Heritage Foundation for Medical Research, the CIHR (MOP 44064), the Peter Lougheed Foundation (CIHR-PLS 47851), and the NIH (R01-HG01988). J. Reeve was supported by a Killam Memorial Postdoctoral Fellowship.

References

1. J.H. Gillespie *Population Genetics: A Concise Guide* (John Hopkins Press, Baltimore, 1998).
2. E.S. Lander and D. Botstein, *Cold Spring Harb. Symp. Quant. Biol.* **51**, 49 (1986).
3. M. Slatkin and B. Rannala, *Annu. Rev. Genomics Hum. Genet.* **1**, 225 (2000).
4. S.A. Tishkoff, et al., *Science* **271**, 1380 (1996).
5. S-W. Guo and M. Xiong, *Hum. Hered.* **47**, 315 (1997).
6. A.P. Morris, J.C. Whittaker and D.J. Balding, *Am. J. Hum. Genet.* **70**, 686 (2002).
7. J. Hastbacka et al., *Nat. Genet.* **2**, 204-211 (1992).
8. N.L. Kaplan, W.G. Hill and B.S. Weir, *Am. J. Hum. Genet.* **56**, 18 (1995).
9. B. Rannala and M. Slatkin, *Am. J. Hum. Genet.* **62**, 459 (1998).
10. J. Graham and E.A. Thompson, *Am. J. Hum. Genet.* **63**, 1517 (1998).
11. B. Rannala and J.P. Reeve, *Am. J. Hum. Genet.* **69**, 159 (2001).
12. S.K. Service et. al., *Am. J. Hum. Genet.* **64**, 1728 (1999).
13. A.P. Morris, J.C. Whittaker and D.J. Balding, *Am. J. Hum. Genet.* **67**, 155 (2000).
14. J.P. Reeve and B. Rannala, *Bioinformatics* **18**, 894 (2002).
15. M. Slatkin and B. Rannala, *Am. J. Hum. Genet.* **60**, 447 (1997).
16. M. Slatkin and B. Rannala, *Genetics* **147**, 1855 (1997).
17. B. Kerem et al., *Science* **245**, 1073 (1989).
18. J. Hästbacka et al., *Cell* **78**, 1073 (1994).

SNP ANALYSIS AND PRESENTATION IN THE PHARMACOGENETICS OF MEMBRANE TRANSPORTERS PROJECT

DOUG STRYKE, CONRAD C. HUANG, MICHIKO KAWAMOTO,
SUSAN J. JOHNS, ELAINE J. CARLSON, JOSEPH A. DEYOUNG,
MAYA K. LEABMAN, IRA HERSKOWITZ,
KATHLEEN M. GIACOMINI, THOMAS E. FERRIN
Departments of Pharmaceutical Chemistry,
Biopharmaceutical Sciences, and Biochemistry & Biophysics
University of California San Francisco, 513 Parnassus Avenue,
San Francisco, CA 94143, USA

The multidisciplinary UCSF Pharmacogenetics of Membrane Transporters project seeks to systematically identify sequence variants in transporters and to determine the functional significance of these variants through evaluation of relevant cellular and clinical phenotypes. The project is structured around four interacting cores: genomics, cellular phenotyping, clinical phenotyping, and bioinformatics. The bioinformatics core is responsible for collecting, storing, and analyzing the information obtained by the other cores and for presenting the results, in particular, for the genomic data. Most of this process is automated using locally developed software written in Python, an open source language well suited for rapid, modular development that meets requirements that are themselves constantly evolving. Here we present the details of transforming ABI trace file data into useful information for project investigators and a description of the types of data analysis and display that we have developed.

1 Introduction

1.1 Motivation

Membrane transporters, a major determinant of pharmacokinetics, are of great pharmacological importance. By controlling the amount of drugs within the body, they determine whether drug levels are sufficient for therapeutic effect. Transporters play a second important role in pharmacology in that about 30% of the most commonly used prescription drugs target transporters.

The UCSF Pharmacogenetics of Membrane Transporters Project (PMT) includes investigators from diverse disciplines who are conducting a series of integrated studies to elucidate the pharmacogenetics of membrane transport proteins. To accomplish this, the investigators are systematically identifying sequence variants in transporters and determining the functional significance of these variants through evaluation of relevant cellular and clinical phenotypes. The goal of the PMT is to understand the genetic basis for variation in drug response for drugs that interact with membrane transporters. As a part of the PMT project, the

536

goal of the bioinformatics core is to provide support to the genomic, cellular phenotyping, and clinical phenotyping cores.

1.2 Overview

Our multidisciplinary project is structured around four interacting cores: genomics, cellular phenotyping, clinical phenotyping, and bioinformatics. The genomics core identifies sequence variants in the targeted transporters from a collection of DNA samples. The cellular phenotyping core tests the functional significance of the sequence variants in cell-based assays. The clinical phenotyping core tests hypotheses about the functional significance of variation in membrane transporters in clinical drug response. The bioinformatics core develops and maintains a database of the information obtained by the other cores and by individual project investigators. A particularly important additional role of the bioinformatics core is to analyze the genomic data, *e.g.*, determine haplotypes, identify evolutionarily conserved and nonconserved positions in proteins, and calculate population-genetic parameters for mutation rate. The project's web site is http://pharmacogenetics.ucsf.edu/.

In the first year, the genomics core screened 367 amplicons, comprising 24 genes. The sample set consisted of 247 ethnically identified DNA samples from the Coriell Institute[1]. The bioinformatics core collected, stored, and analyzed these data and presented the results in various formats. Most of this process was automated using software written by the bioinformatics core in Python, an open source programming language[2]. This dynamically typed, object oriented language is well suited for rapid development of software tools in a dynamic, evolving environment. Our software falls roughly into two categories: analysis and presentation. The analysis software we call RefMap and SnpMap; the presentation software, SnpWeb.

In the bioinformatics core, one person began working part-time on this project in late 2000. In early 2001, two more people were assigned part-time to the project, bringing the full-time equivalent staff to one-and-one-third. All development is done on clustered HP Alpha servers running Tru64 Unix and TruCluster Server V5.

2 Analysis

2.1 Data Flow

The process begins when the project investigators select a transporter for study (Figure 1). The bioinformatics core retrieves the relevant GenBank entry from NCBI, preferably a curated RefSeq entry[3]. A Python program extracts gene coding sequence (CDS) and chromosome position from the GenBank entry. Using

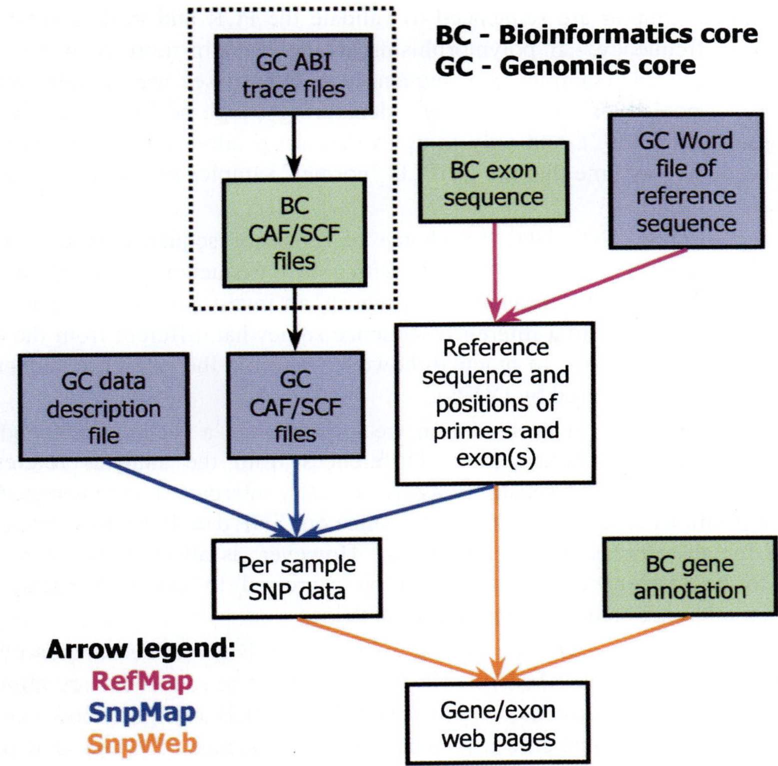

Figure 1 Work flow schematic (dashed box represents planned addition of PolyPhred for base calling).

BLAST[4], the CDS is queried against the high-throughput genomic sequences to locate exons. If necessary, the nonredundant nucleotide database is also searched. The exon boundaries are then verified by looking for splice junctions.

Whenever multiple genomic locations for the same exon are found, these regions are checked via alignments to determine the nature of the event. All the areas are reported; however, the area that best fits the organization of the rest of the gene is recommended. In the case of an extremely short first exon, we combine the exon with its respective 5' UTR region to aid in its location. Once confirmed, the exon sequences, the genomic sequences, the CDS, and the annotations are forwarded to the genomics core.

The genomics core designs PCR primers to amplify each exon and a minimum of 35 flanking 3' and 5' intronic bases. The amplicons are typically 250-500 bases long. Small, closely spaced exons are combined; larger exons are sequenced using multiple, overlapping amplicons.

Six random samples are sequenced to validate the PCR and to determine the polymorphism frequency. If polymorphisms are detected in more than 3 of 12 chromosomes, all 247 samples are sequenced. Otherwise, the sample set is screened in pools of three using denaturing high-performance liquid chromatography (DHPLC), and only samples that are positive for polymorphisms are sequenced. We assume that the DHPLC "normal" samples match the reference sequence.

When presenting SNP data, the choice of reference sequence is somewhat arbitrary. For example, an 'A' to 'G' SNP with a 40% frequency is equivalent to a 'G' to 'A' SNP with a 60% frequency. In theory, an investigator could choose to present SNP data relative to a reference sequence somewhat different from the one used by the genomics core. In practice, however, altering the reference sequence changes the genotype inferred by DHPLC.

To permit flexibility in presentation, we implemented a second, independent reference sequence for presentation. Differences from the analysis reference sequence in the form of base substitutions are correctly interpreted and presented so long as the positions changed contain SNPs, since our SNP data file format contains genotypes for all samples at those positions. However, as all other positions are assumed to be reference, care still must be exercised to avoid changing the presentation reference sequence at non-SNP positions.

The genomics core sequences the samples using BigDye Terminator cycle sequencing and an ABI PRISM 3700 DNA Analyzer. The sequences are aligned and edited, and heterozygous bases are assigned IUPAC-IUB ambiguity codes using Sequencher[5]. The genomics core trims each sample sequence of primer and poor quality base calls, both of which would confound subsequent analysis. From Sequencher, aligned contigs are exported as a Common Assembly Format[6] (CAF) file and all sample sequence data as Standard Chromatogram Format[7] (SCF) files.

The plain text CAF file contains all of the predominant sample base calls aligned into one or more contigs, each with its own consensus sequence. Each consensus sequence entry also contains a list associating each sample sequence with start and stop positions in two coordinate systems: the sample's own and that of the consensus sequence. The genomics core adds a copy of the amplicon reference sequence to each contig, thereby aligning it to the consensus as well. This creates a two-step alignment from each sample sequence to the reference sequence via the contig consensus sequences. The Sequencher-scored heterozygous base calls are contained in the binary SCF files. Optimally, each sample is sequenced in both sense (forward) and antisense (reverse) directions.

In addition to the CAF file's inherent data structure, we have defined a naming scheme for contigs and samples that allows most polymorphisms to be interpreted using automated software analyses. Heterozygous insertions and deletions (indels), however, throw chromatograms out of phase, making their interpretation beyond the indel difficult. The genomics core places heterozygous indel sequences into separate CAF file contigs, one contig for each unique event. They encode the number of bases involved and the SNP types in the contig name. For example, "EXON_7_HET_2_DEL" indicates a two-base heterozygous deletion in exon seven. To further simplify analysis, the genomics core removes the phase-shifted base calls beginning immediately beyond the last indel base in each direction.

Assuming the availability of data from both forward and reverse directions, the above indel protocol provides for automated analysis of up to two heterozygous indels in a single sample by separating the forward and reverse reads into distinct contigs. This proved sufficiently robust for all but one SNP in the first set of 367 amplicons.

More complicated samples require manual intervention. The genomics core gathers these sequences into a contig named, appropriately, "nonstandard." They also describe the indel events in a text file already included with each amplicon. The nonstandard contig signals the need to build a data file by hand to augment the automatically derived data.

The genomics core transmits data to the bioinformatics core in amplicon bundles. Each bundle includes the reference sequences for the amplicon and primers, a CAF file, and the SCF and ABI trace files. Also included is a text file describing information not available in the sequence data. For example, a "no SNPs" flag is used to distinguish an amplicon without trace files for which DHPLC screening found no SNPs from one that is simply missing data.

The bioinformatics core program RefMap begins with the amplicon reference sequence, received as a Microsoft Word document and formatted to delineate exon and PCR primer boundaries. RefMap transforms this file into an HTML file by running it through a program called mswordview[8]. The HTML format preserves the relevant formatting contained in the Microsoft Word document yet is easily parsed. RefMap confirms the exon sequence and adjusts the exon boundaries using the exon sequence originally identified by the bioinformatics core. The amplicon sequence and the primer and exon locations are written to a text file in a modified FASTA format.

Next, SnpMap parses the CAF file. The Sequencher-scored heterozygous base calls contained in the SCF files are substituted for the predominant base calls obtained from the ABI 3700. Using CAF file contig alignment information, sample sequences are aligned relative to the reference sequence. Sample alleles are collected by contig for each reference position. Each multi-base insertion is collated into the appropriate reference position and the downstream positions adjusted for the offset created by the inserted bases. The per-contig SNP sites are then collated into a single file-wide set.

SnpMap subjects the data to quality checks at each step. As stated, almost all aspects of the CAF file follow a defined naming convention. Nonconforming items are flagged. The original reference amplicon is compared against the reference sequence contained in each CAF file contig. Multiple sample reads for a single ID of the same direction are compared for consistency, as are the forward and reverse reads for each sample.

Reference positions lacking polymorphisms are then dropped. The remaining polymorphic sites are written to a tab-delimited file containing sample number, read direction, reference position, genotype, and the source of the base calls, *i.e.*, observed by direct sequencing or inferred via DHPLC. Subsequent analyses are based upon this file.

Not every amplicon successfully passes this process the first time. SnpMap classifies problems into two categories: warnings and errors. Warnings typically apply to individual samples, which can be dropped from the amplicon pending correction. Errors, on the other hand, halt the analysis. Descriptive messages are written to log files, which are linked to and summarized by a color-coded web page that provides at-a-glance status of all PMT amplicons.

2.2 Products

The data are analyzed over four scopes: by SNP site, by gene, by gene family, and over the entire gene set. Each SNP is categorized by type, including exonic vs. intronic, coding vs. noncoding, cytoplasmic vs. extracellular vs. transmembrane, and substitution vs. insertion vs. deletion. SNP statistics, including the chi-square probability that the difference between the observed allele distributions and the predicted Hardy-Weinberg equilibrium could be due to chance alone[9], are calculated overall and by ethnic group.

Gene analyses include molecular genetics statistics over categories such as transitions, transversions, and specific variations within CpG islands. We compare our SNPs to those reported in dbSNP[10]. Haplotypes are estimated using PHASE[11], a program for reconstructing haplotypes (Figure 2). Haplotype-based genetic diversity is calculated by treating each gene haplotype as an allele, and counting the number of differences for all pairs of sample haplotypes[12].

If at least two mammalian homologs of the human transporter can be found, we use multiple sequence alignments to determine evolutionarily conserved amino acid positions. Transmembrane domains and SNP locations are also indicated.

We calculate population genetics statistics, such as nucleotide polymorphism, nucleotide diversity, and the Tajima test of the neutral mutation hypothesis[13] for each gene. These statistics are also computed for gene families or across the entire gene set.

Results are currently written out as tab-delimited files. Some of these files are reformatted into Extensible Markup Language[14] (XML), conforming to schemas defined by the Pharmacogenetics and Pharmacogenomics Knowledge Base[15]

541

(PharmGKB), and then transmitted to the PharmGKB. As part of the Pharmacogenetics Research Network[16], our resource disseminates data publicly via the PharmGKB. The results also become the input for the suite of programs that we call SnpWeb, which presents these results in a scientifically useful manner to the PMT investigators.

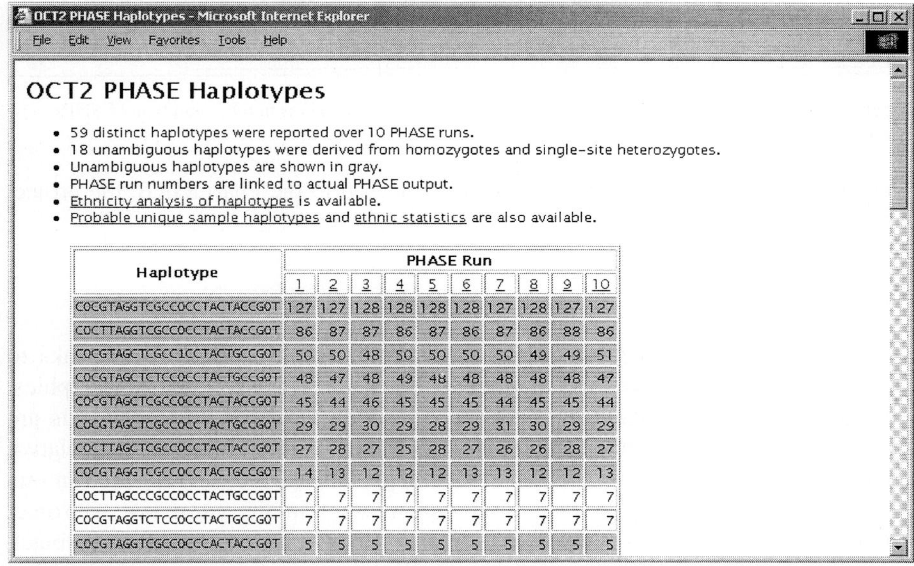

OCT2 PHASE Haplotypes

- 59 distinct haplotypes were reported over 10 PHASE runs.
- 18 unambiguous haplotypes were derived from homozygotes and single-site heterozygotes.
- Unambiguous haplotypes are shown in gray.
- PHASE run numbers are linked to actual PHASE output.
- Ethnicity analysis of haplotypes is available.
- Probable unique sample haplotypes and ethnic statistics are also available.

Haplotype	PHASE Run									
	1	2	3	4	5	6	7	8	9	10
COCGTAGGTCGCCOCCTACTACCGOT	127	127	128	128	128	128	127	128	127	127
COCTTAGGTCGCCOCCTACTACCGOT	86	87	87	86	87	86	87	86	88	86
COCGTAGCTCGCC1CCTACTGCCGOT	50	50	48	50	50	50	50	49	49	51
COCGTAGCTCTCCOCCTACTGCCGOT	48	47	48	49	48	48	48	48	48	47
COCGTAGCTCGCCOCCTACTACCGOT	45	44	46	45	45	45	44	45	45	44
COCGTAGCTCGCCOCCTACTGCCGOT	29	29	30	29	28	29	31	30	29	29
COCTTAGCTCGCCOCCTACTACCGOT	27	28	27	25	28	27	26	26	28	27
COCGTAGGTCGCCOCCTACTGCCGOT	14	13	12	12	12	13	13	12	12	13
COCTTAGCCCGCCOCCTACTGCCGOT	7	7	7	7	7	7	7	7	7	7
COCGTAGGTCTCCOCCTACTGCCGOT	7	7	7	7	7	7	7	7	7	7
COCGTAGGTCGCCOCCCACTACCGOT	5	5	5	5	5	5	5	5	5	5

Figure 2 Partial PHASE haplotype estimation for OCT2.

3 Presentation

3.1 Overview

SnpWeb presents the analysis results in two formats. Tab-delimited files are made available to project investigators and researchers for further analysis. These files are easily imported into desktop spreadsheet or database programs. Basic SNP statistics, per sample data, and summary population genetics statistics are made available this way.

The primary output format that SnpWeb generates is World Wide Web content, including HTML, graphics, and plots. The advantages of web pages include cross-platform availability, familiarity to scientists, and availability from multiple

542

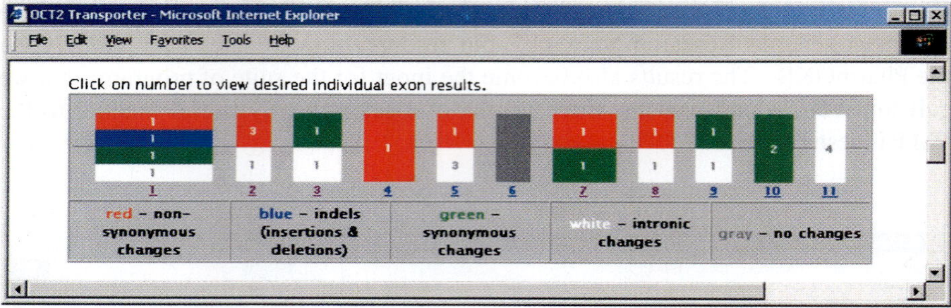

Figure 3 Exon bar showing number and relative size of exons, as well as number and type of SNPs found for the gene OCT2.

geographic locations. Until final results are released to the PharmGKB, the project web site is kept on a password-protected intranet server.

3.2 Web Pages

The main page for each gene displays background information, off-site links to targeted NCBI data, on-site links to additional presentation pages, and two graphics. The first graphic is the exon bar (Figure 3). Along the bar's horizontal axis are drawn various shaped rectangles, one for each exon, sized to reflect the relative sequence length of the exons. The exon rectangles themselves are depicted in one or more of four colors, for each SNP type present in the exon (*i.e.*, intronic, synonymous, indels, nonsynonymous), or gray to indicate no variants. A black exon indicates a lack of analysis results. Within each color region, a number

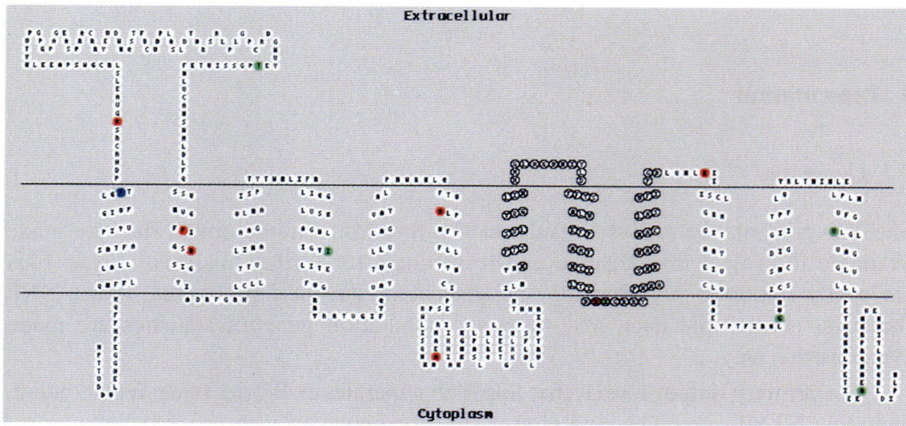

Figure 4 TOPO2 transmembrane prediction image showing OCT2 coding SNPs and location of exon 7.

indicates how many variants of that type were found. The exons are numbered with hyperlinks, which lead to the amplicon web pages.

The second gene page graphic is generated by TOPO2[17], which uses transmembrane predictions to create a secondary-structure graphic of the amino acid locations relative to the cell membrane (Figure 4). We obtain transmembrane predictions from SwissProt annotations and from published papers. Coding SNP locations are indicated. A TOPO2 image is also generated for each amplicon, which additionally highlights exon boundaries.

OCT2, Exon 1

Polymorphisms and Allele Frequencies

Exon	SNP #	CDS Pos	Exon Pos	Nucleotide Change	Amino Acid Position	Amino Acid Change	Total Freq	AA Freq	CA Freq	AS Freq	ME Freq	PA Freq
							n=494 o=492 i=0	n=200 o=200 i=0	n=200 o=198 i=0	n=60 o=60 i=0	n=20 o=20 i=0	n=14 o=14 i=0
1	1		(-47)	C -> T	–	–	0.002 (0.975) n=492	0.000 (n/a) n=200	0.000 (n/a) n=198	0.017 (0.926) n=60	0.000 (n/a) n=20	0.000 (n/a) n=14
1	2	134	134	T -> +A	45	Phe -> INS	0.002 (0.975) n=492	0.000 (n/a) n=200	0.005 (0.960) n=198	0.000 (n/a) n=60	0.000 (n/a) n=20	0.000 (n/a) n=14
1	3	160	160	C -> T	54	Pro -> Ser	0.002 (0.975) n=492	0.005 (0.960) n=200	0.000 (n/a) n=198	0.000 (n/a) n=60	0.000 (n/a) n=20	0.000 (n/a) n=14
1	4	390	390	G -> T	130	syn	0.280 (0.036) n=492	0.205 (0.270) n=200	0.394 (0.126) n=198	0.200 (0.819) n=60	0.250 (0.292) n=20	0.143 (0.659) n=14

Figure 5 SNP statistics by ethnic group and overall for OCT2 exon 1.

We present the basic SNP statistics on another set of web pages, one for each amplicon, in tabular form (Figure 5). Three different position numbers are presented: the CDS position, the exon relative position, and the amino acid position. The amino acid change, if any, is shown. The number of chromosomes constituting the nominal sample set (n), the number observed by sequencing (o), and those inferred by DHPLC (i) are listed by ethnic group and overall. Frequency figures are broken down by ethnic group, as are Hardy-Weinberg probability figures. High frequency SNPs and presumed out-of-equilibrium distributions are highlighted in red. The full amplicon sequence is presented in color to indicate primer and exon locations (not shown). SNPs are mapped onto the amplicon in red and keyed to the table.

Per-sample diplotype data are presented in tabular format (not shown). One table uses colored squares to indicate whether each sample has data at each SNP position and, if so, whether it is homozygous, heterozygous, or reference. Reference samples are further differentiated as to whether they were observed from sequencing or inferred by DHPLC screening. A second table presents similar data in a predominantly textual format.

We present a summary of population genetics statistics in tabular form. This summary provides statistics by ethnicity for various combinations of SNP characteristics, such as coding vs. noncoding, synonymous vs. nonsynonymous, conserved vs. unconserved, and transmembrane vs. loop. A web-based plotter allows researchers to plot the level of nucleotide diversity (θ), average heterozygosity (π), and Tajima's D statistic (Figure 6). Plots can be generated for any combination of genes or gene families over all samples or by ethnic group and for various sequence regions.

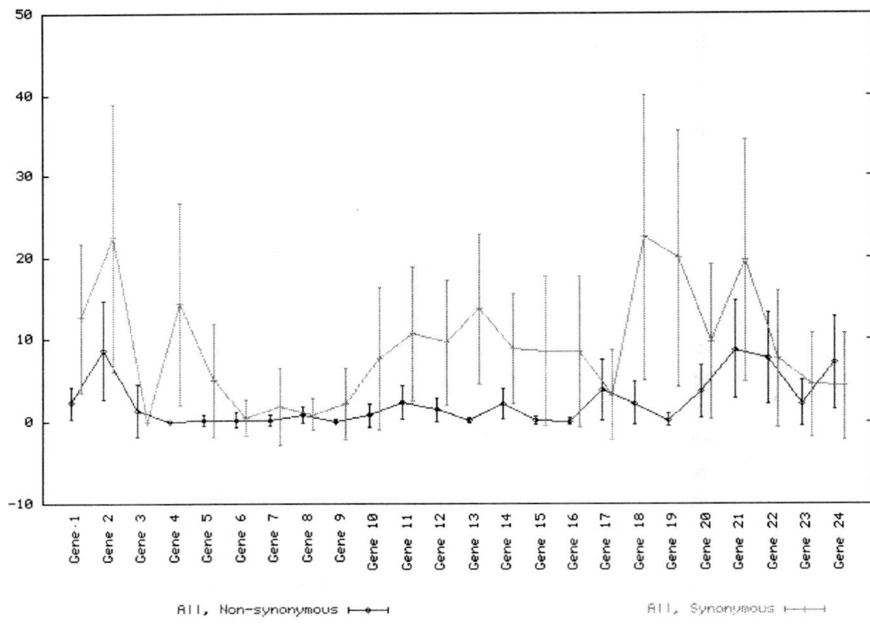

Figure 6 Average heterozygosity (π) for synonymous and non-synonymous SNPs for all genes.

Alignments to other mammalian species are displayed in a color-coded multiple alignment (Figure 7). Gray shading indicates conserved regions. Orange bars above the alignments highlight transmembrane domains. The positions of the coding SNPs are indicated on the alignment, making it readily apparent whether a nonsynonymous SNP affects an evolutionarily conserved or nonconserved position.

In an identical manner, transporter gene families are also presented in graphical multiple alignments.

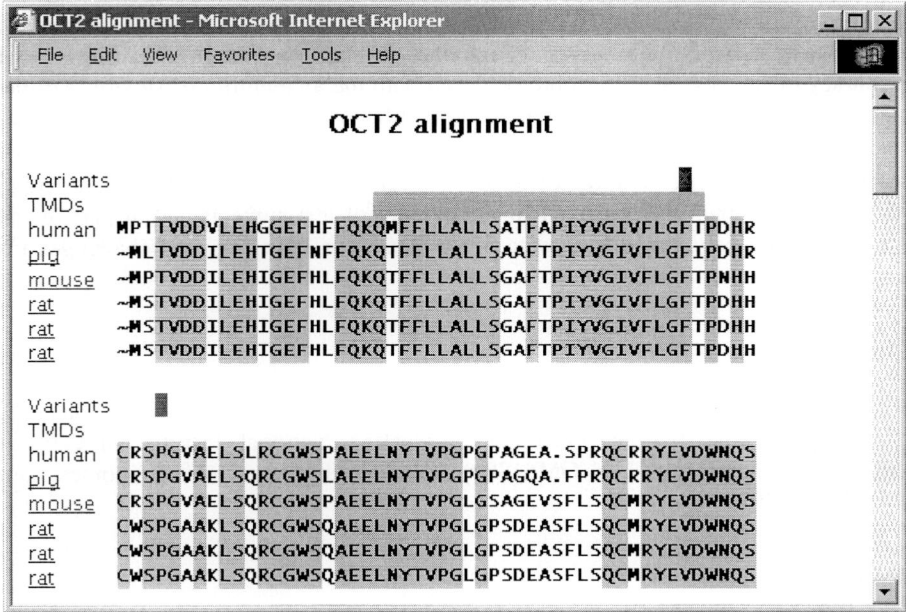

Figure 7 Consensus alignment with mammalian homologs showing conserved locations, predicted transmembrane domains, and SNP locations for a portion of OCT2.

4 Conclusion

4.1 Current Status

We currently store data and analysis results as files, using the file system to keep everything organized. The first year files number over 500,000 and consume approximately 30 gigabytes of disk space. The web pages, though created by Python scripts, are static. Programmatically, the gene is the unit of analysis, and the average time to rerun the analysis for a gene is 11 minutes. Software maintenance has been kept manageable by distributing the programming code over more than 30 Python modules. Evolving requirements for analysis have encouraged some modules to grow overly complex, however. This fact and the continued growth of the data set have led us to consider using a relational database to store intermediate results. Storing intermediate results in a database would facilitate modular programming, speed some analysis updates, and greatly enhance our ability to mine the data.

Another planned change is to use the PolyPhred[18,19] suite of programs from the University of Washington for primary base calling and SNP detection (Figure 1). Preliminary testing with version 3 of the program has largely confirmed the accuracy of our SNP detection protocol, but with the availability of version 4 of the program we opted to put off implementation in favor of further testing.

To date, the bioinformatics core has interacted primarily with the genomics core and, to a lesser extent, the cellular phenotyping core. The clinical phenotyping core has been primarily engaged in recruitment of test subjects. That part of the project should begin producing data in the next year, bringing new challenges for analysis and presentation.

Acknowledgments

The UCSF Pharmacogenetics of Membrane Transporters (PMT) Project is sponsored by the National Institutes of Health's National Institute of General Medical Sciences (grant U01 GM61390). Support for this project also comes from NIH P41-RR01081.

References

1. Coriell Cell Repository, http://locus.umdnj.edu/nigms/
2. http://www.python.org/
3. K.D. Pruitt and D.R. Maglott, "RefSeq and LocusLink: NCBI gene-centered resources" *Nucleic Acids Research* **1**, 137 (2001)
4. S.F. Altschul, W. Gish, W. Miller, E.W. Myers and D.J. Lipman, "Basic local alignment search tool" *Journal of Molecular Biology* **215**, 403 (1990)
5. http://www.genecodes.com/
6. S. Dear, R. Durbin, L. Hillier, G. Marth, J. Thierry-Mieg, and R. Mott, "Sequence assembly with CAFTOOLS" *Genome Research* **3**, 260 (1998)
7. S. Dear and R. Staden, "A standard file format for data from DNA sequencing instruments" *DNA Sequence* **3**, 107 (1992)
8. http://www.wvware.com/
9. D.L. Hartl and A.G. Clark, *Principles of Population Genetics, Third Edition*, 80-82 (1997)
10. http://www.ncbi.nlm.nih.gov/SNP/
11. M. Stephens, N. Smith, and P. Donnelly, "A new statistical method for haplotype reconstruction from population data" *American Journal of Human Genetics* **68**, 978 (2001)

12. J.C. Stephens, J.A. Schneider, D.A. Tanguay, J. Choi, T. Acharya, S.E. Stanley, R. Jiang, C.J. Messer, A. Chew, J. Han, J. Duan, J.L. Carr, M.S. Lee, B. Koshy, A.M. Kurnar, G. Zhang, W.R. Newell, A. Windemuth, C. Xu, T.S. Kalbfleisch, S.L. Shaner, K. Arnold, V. Schulz, C.M. Drysdale, K. Nandabalan, R.S. Judson, G. Ruano, and G.F. Vovis, "Haplotype variation and linkage disequilibrium in 313 human genes" *Science* **293**, 489 (2001)
13. F. Tajima, "Statistical method for testing the neutral mutation hypothesis by DNA polymorphism" *Genetics* **123**, 585 (1989)
14. http://www.w3.org/XML/
15. M. Hewett, D. E. Oliver, D. L. Rubin, K. L. Easton, J. M. Stuart, R. B. Altman and T. E. Klein, "PharmGKB: the Pharmacogenetics Knowledge Base" *Nucleic Acids Res.* **30**, 163 (2002)
16. http://www.nigms.nih.gov/pharmacogenetics
17. S.J. Johns and R.C. Speth, "TOPO, Transmembrane protein display software" http://www.sacs.ucsf.edu/TOPO/topo.html
18. B. Ewing, L. Hillier, M.C. Wendl and P. Green, "Base-calling of automated sequencer traces using Phred. I. Accuracy assessment" *Genome Research* **8**, 186 (1998)
19. B. Ewing and P. Green, "Base-calling of automated sequencer traces using Phred. II. Error probabilities" *Genome Research* **8**, 186 (1998)

METHODS FOR ANALYSIS AND VISUALIZATION OF SNP GENOTYPE DATA FOR COMPLEX DISEASES

ANYA TSALENKO[1], AMIR BEN-DOR[1], NANCY COX[2] AND ZOHAR YAKHINI[1]

[1] *Agilent Laboratories, 3500 Deer Creek Road,Palo Alto CA, 94304.*
E-mail: anya_tsalenko@agilent.com
amir_ben-dor@agilent.com
zohar_yakhini@agilent.com

[2] *Department of Human Genetics, University of Chicago, 920 East 58th Str.,*
Chicago, IL 60637.
E-mail: ncox@genetics.bsd.uchicago.edu

SNP markers are becoming central for studying genetic determinants of complex diseases. Large SNP data collected in such studies call for the development of specialized analysis tools. We present methods for selecting sets of SNPs that can be associated to sample properties in case/control studies. We also describe how scoring and selection can be statistically tested. This is done at the single locus as well as at the set level.

1 Introduction

Much of human DNA sequence variation is due to *single nucleotide polymorphisms* (SNPs), which are single base pair positions in genomic DNA at which different sequence alternatives (*alleles*) exist in normal individuals in some population(s). They are distinguished from rare sequence variations by a requirement for the least abundant allele to have a frequency above 1% in the population. The density of SNPs differ between different genomic regions and different populations, but the overall frequency, for the global human population, is estimated to be around 1 in every 200-600 base pairs. The importance of SNPs in various areas of clinical medicine is gaining increasing attention [14]. Association studies, using polymorphic markers (such as SNPs), in genome-wide scans have been advocated as the most efficient way of identifying genetic regions or genes implicated in common complex diseases and traits [15]. The collection of SNP variants/allelles that an individual possesses in a number of key genes (termed his/her *genotype* over this set of loci) are assumed to play an important role in conferring drug response variability. Therefore, *pharmacogenomics* (and other) association studies are expected to reveal sets of SNPs that separate phenotypically distinct classes of samples according to their genotype signatures.

The sequencing of most of the human genome, the improved understanding of this sequence, mostly of its coding part, and the fast development of parallel measurement platforms such as microarrays are three important recent advances in molecular biology. The combination of these advances drive an increasing interest and activity in measuring gene expression profiles of different cell types and disease stages or types as well as in understanding the role of human sequence variation in influencing disease and treatment susceptibility. Gene expression profiling data are accumulating at a fast rate and the association between profile properties and clinical attributes is being explored [13,5,10,17]. Such studies reveal sets of genes that separate phenotypically distinct classes of samples according to their expression signatures. The study of naturally occurring DNA sequence variations and the relationship between genetic variants and clinically meaningful phenotypes precedes the interest in expression profiling by many years. The recent developments mentioned above, however, bring them together and allow for the exploitation of common characteristics and for the study of joint properties.

In this paper we describe statistical methods, visualization tools and algorithmic approaches to questions that arise in pursuing correlations between SNPs as well as *sets* of SNPs and sample properties. Some of the methods draw on the common characteristics of expression data [2,4,3] and genotyping data in case/control studies.

A pioneering effort to positionally clone a gene that affects susceptibility to type 2 diabetes in Mexican Americans is reported by Horikawa *et al* [12]. The authors show that certain combinations of polymorphisms in the gene encoding calpain-10 are associated with the risk of type 2 diabetes. By indicating a role for a calpain protease, these findings propose a fundamentally new hypothesis for diabetes research.

This study exemplifies the long process and the various stages involved in finding genetic determinants of any clinically meaningful condition. Significant evidence for linkage of type 2 diabetes to the distal long arm of human chromosome 2 was reported in 1996 by Hanis *et al* [9] and the locus was designated *NIDDM1*. The implicated region was large, with the 1-lod support interval, which is expected to contain the responsible gene in 80 to 90% of the cases, extending over 12 cM. In later studies this region was narrowed down to 7 cM, corresponding, in this case, to a relatively short span of 1.7 million base pairs, rather than the expected 7 million [a]. Finding the causative gene(s) in 1.7 Mb of sequence is a difficult task: this particular interval contains at least

---

[a]the average ratio of physical distance to genetic distance across the human genome is approximately 1 million base pairs per cM

7 known genes and 15 ESTs; none of these are obvious candidates. Horikawa *et al.* [12] chose to screen polymorphisms in the region for association with diabetes, relying on linkage disequilibrium (LD). Further investigation of the results of the said screening led to genotyping 63 SNPs in a larger set of about 100 diabetic cases and controls. The authors applied simulation based statistical tests to the results and implicated *CAPN10* as associated with increased risk of diabetes. We demonstrate our methods on data that further extends this study, finding interactions between genes in different chromosomes and identifying a specific set of SNPs and a specific genotype profile for these that helps explain evidence for linkage in the *CAPN10* region.

The main contribution of the current work is in providing methods for selecting and for statistically benchmarking sets of SNPs (as opposed to single SNPs) that jointly associate with a property of interest. An approach to identifying sets of SNPs, associated with disease, is described by Ott *et al* in [11]. In this pioneering work SNPs were ordered by a score that combined allele association, Hardy-Weinberg equilibrium and evidence for genotyping errors. Contributions from the highest scoring SNPs were combined to form a single genome-wide test. The statistical significance of the scores is assessed by simulations. Our approach differs in the way we score individual SNPs and sets of SNPs and in the way we model and compute the related statistics. We also apply less greedy selection methods. These typically perform better when the features are highly dependent.

The current paper and [11] share an emphasis on rigorous and critical statistical assessment of conclusions based on the data. As genotyping and association studies explore new frontiers it is important that methods remain grounded in sound statistics.

The paper is organized as follows. We start by describing the running example data, in Section 2. In Section 3 we describe an information theory driven method for scoring SNPs for their relevance to a partition of the set of samples. Assigning statistical meaning to this score is also discussed and an application introduced in Section 4. Data visualization is demonstrated in Section 5. Finding SNP *set* association, including a statistical test, is the topic of Section 6. We conclude with methods applicable for quantitative traits, in Section 7, and a discussion in Section 8.

2   Data

We analyzed 216 SNPs typed in Mexican-Americans (from Starr County, Texas) with type 2 diabetes [12]. Of these, 88 SNPs were on chromosome 2 in the *NIDDM1* region, 63 SNPs were on chromosome 15 in the CYP19 region and 65 SNPs were on chromosome 7. In previous studies these regions had shown at least nominally significant evidence for linkage [9], moreover it was shown that there is statistical interaction between genes on chromosomes 2 and 15 [7]. The study consisted of 170 families, 330 possible affected sib-pairs. One patient from each affected sib-pair was selected into a set of representatives for the analysis, but not all representatives from all families were typed for all SNPs in the study. 108 families were typed for SNPs on chromosomes 2 and 15, one member from each of 96 families was typed on chromosome 7. The overlap between families typed on chromosomes 2, 15 and 7 is 57 families. A random sample of 112 individuals from Starr County, Texas not diagnosed with diabetes at the time of the study was also typed for most of the SNPs.

3 Scoring SNP for relevance

Consider a partition of the samples into disjoint classes $C = \{C_1, ..., C_n\}$ (for example affected/unaffected individuals). Denote the number of samples in each class by $d_1, d_2, ...d_n$ respectively, and the total number of samples by D. We want to score each SNP locus l according to its relevance to the partition C. One way of mathematically defining locus relevance (with respect to C), is by the *mutual information score*. Let G denote the partition of the tissues induced by the genotypes at l. The *mutual information* of the partitions G and C is defined as the difference between the measure-theoretic entropy of the partition C and the conditional entropy of C conditioned on G:

$$s = H(C) - H(C|G), \tag{1}$$

where H is the entropy [6], i.e. $H(C) = -\Sigma_{i=1}^{n} d_i/D \cdot log(d_i/D)$. This score measures the amount of information the genotype at the locus under consideration gives about membership in each one of the sets $C_1, C_2, ...C_n$. Figure 1 shows an example of loci with high and low mutual information scores.

Together with the mutual information score we compute the corresponding significance level (p-value) in the following way. Consider a random assignment of the samples to n groups of the appropriate sizes $d_1, d_2, ..., d_n$. We call such assignments *admissible*. For a locus l, let S be a random variable obtained by computing (1) for a partition uniformly drawn over the set of all admissible assignments and fixed partition G defined by the genotypes at l. Then given a score s we define:

$$p\text{-value}(l, s) = Prob(S \geq s). \tag{2}$$

Note that this way of scoring SNPs, and of computing their p-values, allows to compare relevance of SNPs on the same scale, since it takes into account possible missing genotype data. In the example of Figure 1, the p-value for locus 1 is 0.0002, the p-value for locus 2 is 1. Whenever no confusion arises we will denote p-value(l, s) by p-value(s).

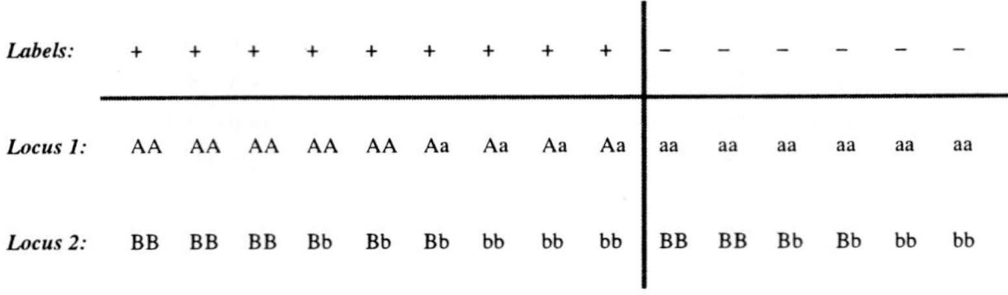

Figure 1. Illustration of the mutual information score for two classes: one with 9 individuals labels by '+', the other one with 6 individuals labelled by '-'. At locus 1, all people in the first class have genotypes AA or Aa, and all people in the second class have genotypes aa. This locus is informative and has the score of 0.97 (p-value=0.0002). At locus 2, there is no difference between genotype frequencies in different classes, the mutual information score for this locus is 0 (p-value=1).

For modest size data sets like the one we are considering, these p-values can be computed exactly, exhaustively counting over all possible admissible label assignments. For data sets with many samples, for micro-satellite loci or for haplotype data these p-values can be estimated by simulations.

In the diabetes data, we considered several ways to partition the samples. We compared all patients with all controls, and we compared subgroups of patients. One way of partitioning the patients was generated using linkage signal. Patients were partitioned into 3 groups: patients from families that showed evidence for linkage (NPL score > 0.6) in the *NIDDM1* region, patients from families with evidence against linkage (NPL score < -0.6) and patients from families with no linkage signal (-0.6 ≤ NPL score ≤ 0.6). The first group has 38 patients, the second group has 16 patients and the third group has 54 patients. Analogous partitions were considered using NPL scores on chromosomes 15 and on chromosome 7. We will use the 'linkage' partition L generated by NPL scores on chromosome 2 in the examples below. SNPs with high scores for L may help explain evidence for linkage in this region and point to genes related to diabetes susceptibility.

Results for the highest scoring chromosome 2 SNPs and the partition L are shown in Table 1. We identified 10 informative SNPs on chromosome 2 with p-values less than 0.05. Interestingly, 7 of these are tightly linked

Table 1. Top scoring SNPs from chromosome 2. The nucleotide indicates the location of the SNP relative to the A of the ATG of the initiator Met of the *CAPN10*.

Chromosome number	SNP number	Nucleotide (bp)	Score	p-value
2	30	11098	0.15	0.0058
2	48	15111	0.16	0.0062
2	19	7917	0.14	0.0082
2	59	6018	0.17	0.0099
2	56	5415	0.13	0.0124
2	28	23527	0.12	0.0154
2	65	23527	0.11	0.0269
2	33	(65kb)	0.11	0.0403
2	31	(60kb)	0.10	0.0414
2	43	4852	0.08	0.0451
2	27	(100kb)	0.06	0.0619
2	18	(330kb)	0.08	0.0657
2	36	41509	0.08	0.0807
2	49	41943	0.08	0.0806
2	51	41959	0.07	0.1038

polymorphisms in the calpain-10 gene; others are linked SNPs from the region between *GPR35* and *ATSV* on chromosome 2.

4 Overabundance analysis

To estimate the overall significance of a set of SNPs with respect to a given partition we compare the observed number of SNPs with score $\geq s$ for each score level s with the expected number of SNPs with scores $\geq s$ for random admissible labellings of the samples. The higher the gap is between the observed number of significant SNPs (high mutual information score) and the expected number of significant SNPs, the more significant the sample partition is.

At a given score level s, let $p = p\text{-value}(s)$. Suppose that in the data we observe $N(s)$ SNPs with score $\geq s$. In a data set with N SNPs we expect to see $E(s) = pN$ SNPs with a score better than s. Moreover, the number of SNPs with score $\geq s$ we observe for uniformly and independently drawn labellings is a random variable $n(s)$, with $n(s) \sim Binom(N,p)$. The surprise rate at s is defined as

$$\sigma(s) = Prob(n(s) \geq N(s)) = \sum_{k=N(s)}^{N} \binom{N}{k} p^k (1-p)^k.$$

Finally, the maximum surprise score for the partition is

$$\Omega = \max_{s}(-\log(\sigma(s))).$$

Figure 2 shows results of overabundance analysis for chromosome 2 SNPs and the partition into diabetes patients and random sample.

Note that the exact numerical value of the partition score Ω should not be taken literally, as it relies on strong statistical assumption (namely, that the different loci are not linked and thus can be treated as independent). Nevertheless, this score is useful when we want to compare two partitions of the samples (to select which is more supported by the SNP data). Another application for the partition score is class discovery [3].

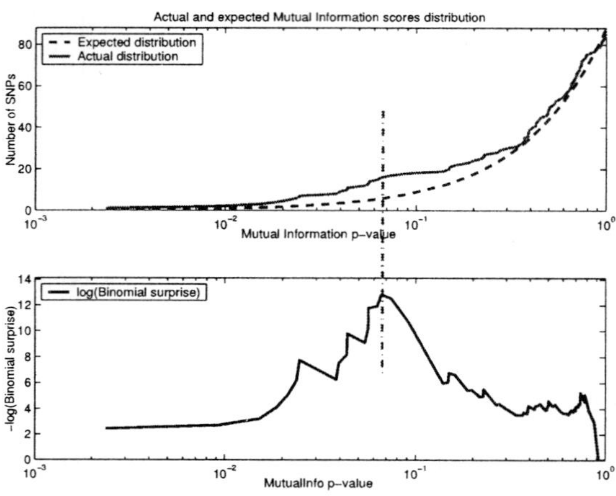

Figure 2. Overabundance analysis of SNPs of chromosome 2 with respect to the affected/random sample partition. In the upper part of the figure we plot the distribution of SNPs p-values with respect to this partition (green curve) vs. the expected distribution of p-values for random admissible partition (blue curve). The gap between the two curves show that there is an overabundance of significant SNPs. The red line (at p-value $=0.0659$) correspond to the max-surprise score 12.84 (depicted at the lower half of the figure).

5 Visualization of the data

In this Section we present a few figures that allow us to visualize the SNP data with respect to different sample partitions, and different SNP ordering methods [b]. As a result we can highlight different aspects of the data.

---

[b]These figures were generated by SNPTool, part of BioTools, a software package developed at Agilent Labs, for internal research and scientific collaborations.

Figure 3 shows data for SNPs from chromosome 2 with highest mutual information scores. Note that this Figure shows that the top SNPs 2_59, 2_48, 2_30, 2_19, 2_19, 2_65 are very similar. Indeed these SNPs are polymorphisms (see Table 1) from the *CAPN10* gene in strong linkage disequilibrium.

Figure 4 shows the same data, but now each row is sorted by genotype with each class. This way of plotting illustrates well the mutual information score. The first 5 SNPs got high scores because individuals from 'not-linked' group do not have homozygous genotypes for the rare allele at these SNPs, plotted in yellow.

SNPs could also be sorted by their chromosomal location, which together with their scores may provide additional insight to the interesting genomic regions that may be related to disease susceptibility.

6   Selection of SNP subsets for classification

Assume we are given a sample partition C (e.g., diabetics/random samples or 'linked'/'not-linked' families). For complex, multi-genic diseases like diabetes, it is not necessarily the case that a single SNP would suffice to explain the genetic origin of the disease. In this Section we describe an approach to select a set of SNPs that *together* contains a strong *genotype signature* of the sample class. To rigorously define the SNP-subset search problem, we need to choose an appropriate optimization criteria, and search for a set of SNPs that maximize it. In this paper we describe a simple approach that is based on the *classification accuracy* of the SNP subset. Intuitively, a good set of SNPs, A, is such that knowing the genotypes over A for an unknown sample will allow us to make a good guess about the class membership of this sample. We first show how to construct a classifier (Naive Bayesian Classifier) using a SNP set A. We shortly describe how the classification accuracy of such a classifier can be estimated (using LOOCV approach). We then describe simple search heuristics to select the set of SNPs, and evaluate the success rate of these methods with respect to the 'linked'/'not-linked' partition. Finally, we conclude the section with a comparison of the best classification accuracy achieved, and a classification accuracy for a random admissible assignment of classes.

6.1   *The naive Bayesian classifier*

Consider a set of SNPs A. One simple way to construct a classifier is to consider the naive Bayesian classifier based on the probabilistic approach to this problem [8]. For each SNP, we compute the probability of a given label, given genotypes of the training set of samples. Then these one-SNP classifiers are combined together to predict the labels of test samples.

For a sample x with unknown label, in the case of two classes (e.g. 'linked'/'not linked') labelled by '+' and '-', applying Bayes rule to the set of SNPs:

$$\log \frac{P(+|x)}{P(-|x)} = \log \frac{P(+)}{P(-)} + \log \frac{P(x|+)}{P(x|-)} = \log \frac{P(+)}{P(-)} + \sum_{a \in A} \log \frac{P(x_a|+)}{P(x_a|-)}$$

$$= \log \frac{P(+)}{P(-)} + \sum_{a \in A} \left(\log \frac{P(x_a|+)}{P(x_a|-)} - \log \frac{P(+)}{P(-)} \right), \tag{3}$$

where x_a is the genotype of sample x at locus a. In the above formula we assumed independence of SNP loci in the set A given the partition of the samples. For positive $\log \frac{P(+|x)}{P(-|x)}$, we predict that the label of x should be '+', otherwise the label is '-'.

The accuracy of the classifier can be measured by the number c of correct predictions it makes for the test samples, and we can use it to find the 'best' subset of SNPs. Training and test sets of samples can be defined using leave one out cross-validation technique (LOOCV). On each step of the LOOCV algorithm we 'hide' one sample and construct a classifier using the remaining samples. Then this classifier is used to predict the label of the 'hidden' sample, and the procedure is repeated for every samples in the data. LOOCV method can be modified to hide more than one sample at a time and can be applied to more than 2 classes.

6.2 SNP subset selection

As it is not computationally feasible to exhaustively try all possible SNP subsets we describe here a few simple efficient methods to select SNP subsets.

One approach to find the SNP subset is to order SNPs by their scores, e.g. by mutual information score, and consider classifiers using k top scoring SNPs A_k for each $k = 1, 2, ..., N$. Then, we choose the subset A_{k_0} for which the classifier with k_0 SNPs makes the biggest number of correct predictions $c_{k_0} = max_{1 \leq k \leq N} c_k$. In the Mexican-American diabetes data set this approach did not work well, because genotypes at many SNPs are not independent since many SNPs in this data set are very close to each other and are in strong linkage disequilibrium. The number of correct predictions for 'linked'/'not linked' classes was close to the prior probability of making a correct prediction 0.7 (Figure 5). This approach is very computationally efficient, since the calculation time is linear with the number of SNPs, and will probably work best in the data sets with independent SNPs.

Another approach is to select the subset using forward (backward) sequential search [16,1]. On the first step of the forward sequential search, we select a SNP a_1 out of the whole set of SNPs A such that the corresponding classifier makes the biggest number of correct predictions. We set $A_1 = \{a_1\}$. On each step $k = 2, ..., N$, we find a SNP a_k such that the classifier corresponding to the set $A_k = A_{k-1} \cup \{a_k\}$ makes the biggest number of correct predictions among the classifiers with subsets $A_{k-1} \cup \{a\}, a \in A \backslash A_{k-1}$. The 'best' subset is defined by k_0 for which the classifier with the set A_{k_0} makes the biggest number of correct predictions. Backward sequential search works in the reverse direction, i.e. we start with the set of all SNPs, and on each step of the algorithm remove a SNP such that the classifier build using remaining subset of SNPs makes the biggest number of correct predictions.

Using backward sequential search we identified a subset of 11 SNPs from chromosome 2 with combined genotypes that predict the 'linkage status' with 87% accuracy (Figure 5). Similarly, a set of 11 SNPs from chromosomes 2 and 2 SNPs from chromosome 15 was found that predicts the 'linkage status' with 90% accuracy. Another interesting result was found for chromosome 7 and 15 SNPs. Out of 57 families typed for SNPs on both chromosomes, 22 showed evidence for linkage on chromosome 7 (NPL > 0.6), 17 showed evidence against linkage (NPL < -0.6). A set of 7 SNPs from chromosomes 7 and 15 was found that predicts 'linkage status' on chromosome 7 correctly in 38 out of 39 samples (Figure 6).

Note that the number of steps in forward/backward sequential searches may become quite big for data sets with many SNPs. To save on the computational time, we can limit the searches to top scoring SNPs only.

6.3 Statistical significance

We estimated the significance of LOOCV results by simulations, i.e. we simulated random admissible labels of the samples and ran LOOCV with the corresponding set selection method for each labelling. Then we compared the observed probability of the maximal number of correct predictions for random labels with the number of correct predictions for the original labels.

In the current example, 100 random admissible labellings of the samples were simulated and we ran LOOCV with backward sequential search algorithm for these labels. The probability of finding a subset of SNPs on chromosome 2 that better predicts set membership than the original subset is 0.04. Also observe that prediction quality of 'linkage' status is consistently better than the average prediction quality for random labellings (Figure 5).

7 Quantitative traits

In many studies, clinical quantitative information is available for the samples. This information may help define more homogeneous subgroups of patients for which SNP association with disease susceptibility will be easier to detect. Similar to (1), we can score each SNP l, and each quantitative measurement q, in the following way. We find a threshold t such that the mutual information score of l with respect to the partition of samples into samples with $q \geq t$ and samples with $q < t$ is maximal. Analogous to (2) we can compute the p-value for this score, counting over all possible assignments of samples to groups. This computation can be effectively and efficiently carried out using dynamic programming methods, as described in the supplementary information (http://dogbert.cs.technion.ac.il).

8 Discussion

We presented methods for analyzing and visualizing SNP case/control data. In particular we described processes for selecting subsets of loci that jointly correlate with sample properties. The selection process an be compared to a null-model by means of simulations. In future work we will further address this crucial statistical testing. We also briefly discussed similar methods that apply to quantitative traits. We are planning to test these and other means of scoring quantitative trait predictors on biological data. More scientific activity in this space will drive the emergence of appropriate methodology.

Acknowledgments

Supported by DK-55889, DK-47486, DK-20595 and Agilent Laboratories. Authors would like to thank Professor Graeme Bell for useful discussions and SNP genotype data on chromosomes 2 and 15 and Professor Pat Concannon for SNP genotype data on chromosome 7.

References

1. Aha D and Bankert R. A Comparative Evaluation of Sequential Feature Selection Algorithms. *Artificial Intelligence and Statistics, D. Fisher and J. H. Lenz, New York* (1996).

2. Ben-Dor A, Bruhn L, Friedman N, Nachman I, Schummer M and Yakhini Z. Tissue classification with gene expression profiles. Journal of Computational Biology **7**, 3-4:559-83 (2000).

3. Ben-Dor A, Friedman N and Yakhini Z. Class Discovery in Gene Expression Data. *Recomb* (2002).

4. Ben-Dor A, Friedman N and Yakhini Z. Scoring genes for relevence. *Agilent technical report,*
http://www.labs.agilent.com/resources/techreports.html (2001).

5. Bittner M *et al.* Molecular classification of cutaneous malignant melanoma by gene expression profiling. Nature **406**, 6795:536-40 (2000).

6. Cover T and Thomas J. Elements of Information Theory. *Jon Wiley Sons Inc* (1993).

7. Cox N, Frigge M *et al.* Loci on chromosomes 2 (*NIDDM1*) and 15 interact to increase susceptibility to diabetes in Mexican Americans. Nature Gen **21**, 213-215 (1999).

8. Duda R and Hart P. Pattern Classification and Scene Analysis. *New York, Jorn Wiley and Sons*(1973).

9. Hanis C *et al.* A genome-wide search for human non-insulin-dependent (type 2) diabetes genes reveales a major susceptibility locus on chromosome 2. Nature Gen **13**, 161-166 (1996).

10. Hedenfalkl I *et al.* Gene expression profiles in hereditary breast cancer. N Engl J Med **344**, 8:539-48 (2001).

11. Hoh J, Wille A and Ott J. Trimming, Weighting, and Grouping SNPs in Human Case-Control Association Studies. Genome Research **11**, 2115-2119 (2001).

12. Horikawa Y *et al.* Genetic variation in the gene encoding calpain-10 is associated with type 2 diabetes mellitus. Nature Gen **26**, 2:163-75 (2000).

13. Golub T, Slonim D *et al.* Molecular classification of cancer: class discovery and class prediction by gene expression monitoring. Science **286**, 5439:531-7 (1999).

14. Risch N, Merikangas K. The future of genetic studies of complex human diseases. Science **273**, 1516-1517 (1996).

15. Risch N. Linkage strategies for genetically complex traits. III. The effect of marker polymorphism on analysis of affected relative pairs. American Journal of Human Genetics **46**, 242-253 (1990).

16. Pudil P, Novovicova J and J. Kittler. Floating search methods in feature selection.Pattern Recognigion Letters **15**, 1119-1125 (1994).

17. Shipp M *et al.* Diffuse large B-cell lymphoma outcome prediction by gene-expression profiling and supervised machine learning. Nature Med **8**, 1:68-74 (2002).

560

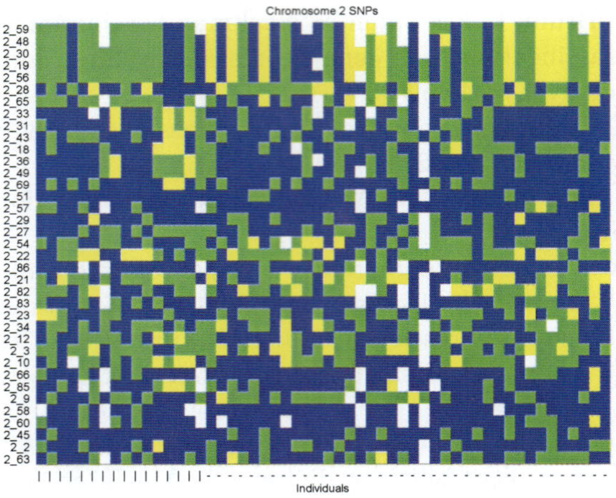

Figure 3. Graphical representation of the highest scoring SNPs from chromosome 2 and 'linked'/'not-linked' partition. Each column represents all genotypes for a given person; each row **represents all genotypes for a given SNP**. Blue corresponds to homozygous genotype for **common** allele, **yellow corresponds to homozygous genotype for rare allele and green corresponds to heterozygous genotype**. White corresponds to missing data. Loci are ordered with **respect** to **mutual information** score. Columns marked by '|' on the x-axis correspond to patients from 'not linked' group, columns marked by '-' on the x-axis correspond to patients from 'linked' group.

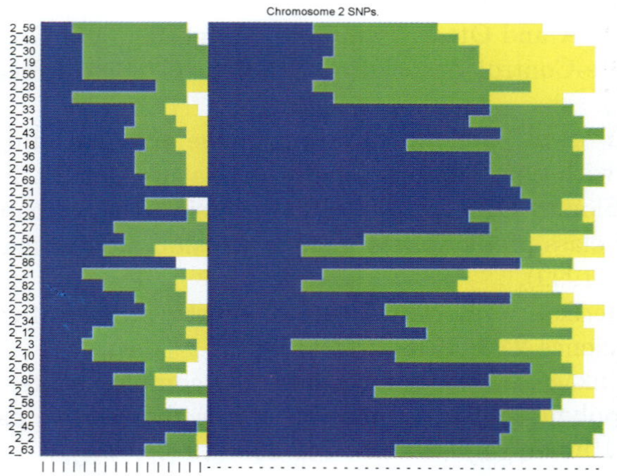

Figure 4. Same data as in Figure 3, now data in each row is sorted by genotypes within each group. This plot helps to visually assess mutual information score, since it is clear that the top 5 SNPs got high scores because individuals in 'not-linked' group do not have homozygous genotypes for the rare alleles at these SNPs. Note that columns no longer correspond to a particular individual.

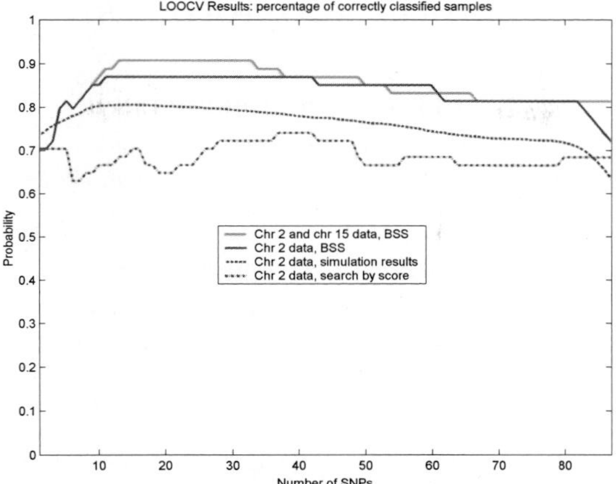

Figure 5. LOOCV results for SNPs on chromosomes 2 and 15. 'Linkage' status was correctly predicted for 91% of patients, using subset of SNPs identified by backward sequential selection method. Using SNPs from chromosome 2 only 'linkage' status was correctly predicted for 87% of patients.

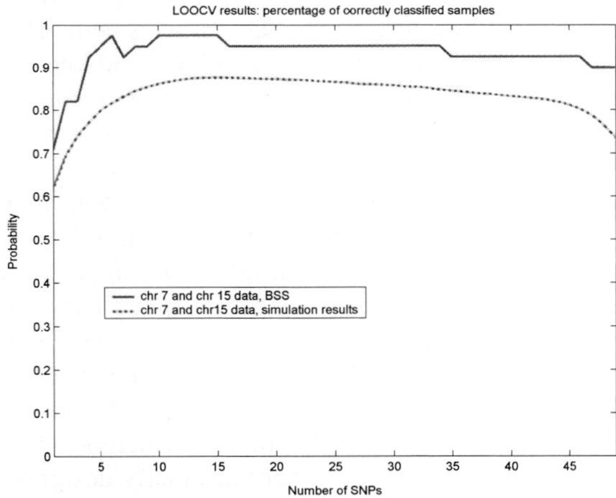

Figure 6. LOOCV results for SNPs on chromosomes 7 and 15. 'Linkage' status on chromosome 7 was correctly predicted for 98% of patients using subset of SNPs identified by backward sequential selection method.

BIOMEDICAL ONTOLOGIES

OLIVIER BODENREIDER

U.S. National Library of Medicine
8600 Rockville Pike, MS 43, Bethesda, Maryland, 20894, USA
E-mail: olivier@nlm.nih.gov

JOYCE A. MITCHELL

U.S. National Library of Medicine and
University of Missouri, Department of Health Management & Informatics,
Columbia, Missouri, 65211, USA
E-mail: MitchellJo@health.missouri.edu

ALEXA T. MCCRAY

U.S. National Library of Medicine
8600 Rockville Pike, MS 52, Bethesda, Maryland, 20894, USA
E-mail: mccray@nlm.nih.gov

Biomedical ontologies provide an organizational framework of the concepts involved in biological entities and processes in a system of hierarchical and associative relations that allows reasoning about biomedical knowledge. In contrast, biomedical terminologies promote a standard way of naming these concepts. Differences among various kinds of terminological systems can be briefly summarized as follows. *Controlled vocabularies* define a set of terms to be used for a given purpose (e.g., indexing the literature, annotating gene functions). *Thesauri* organize the terms in a system of relations designed to help navigate among terms as needed, for example, in information retrieval tasks. *Ontologies*, on the other hand, aim at representing what exists independently of any specific use; they also typically follow general theories (e.g., mereology) and carefully distinguish between the various kinds of relations among things that exist. Thesauri are often limited to tasks such as information retrieval, whereas ontologies support reasoning. Both can be shared, but ontologies lend themselves to reuse, sometimes in widely differing applications from the ones for which they were originally designed. Although more than sixty terminological systems exist in the biomedical domain, few actually qualify as an ontology. Interestingly, the most recent systems tend to be ontologies, developed either from the top down (e.g., GALEN[1]) or from reengineering the knowledge present in older systems (e.g., SNOMED-CT[2]).

---

[1] http://www.opengalen.org/

[2] http://www.snomed.org/

Biological knowledge is evolving so rapidly that it is difficult for most scientists to assimilate and integrate the new information with their existing knowledge. One advantage of ontologies over terminological systems is to support reasoning. The formal structure and rules of inference provided by logic may be coupled with the properties of the relations among things in an ontology in order to draw inferences. The uses of bioinformatics ontologies include natural language processing, knowledge discovery, and supporting interoperability among the many knowledge resources now available. Bridging between the terminological resources of the UMLS® (Unified Medical Language System®)[3] and the biotechnology information resources is another important issue. Increasingly, natural language processing techniques are applied to massive biomedical corpora such as the MEDLINE® bibliographic database[4] in order to extract information and discover knowledge. In these tasks, while terminology is needed for identifying the concepts in the text, ontologies help identify the relationships among concepts suggested by syntactic and discourse structures.

Ontologies are not tied to any kind of particular formalism for their representation, nor are they concerned, in principle, with issues of computer tractability. In practice, however, some representations such as frames (used, for example, in Protégé[5]) or description logics (e.g., DAML+OIL[6]) represent a trade-off between expressivity (what can be represented) and tractability (what can be inferred), needed if the ontology is to be used in computational tasks. Both representations are expressed in or can be translated to some flavor of first-order logic. Recent biomedical ontologies such as GALEN and SNOMED-CT are based on description logics; others such as the Foundational Model of Anatomy[7] are frame-based.

Although the number of ontologies available for biomedicine has not yet reached that of terminological systems, it is expected that applications relying on domain knowledge will have to deal with multiple ontologies, either because no single ontology offers a broad enough coverage, or because the task is to interoperate between applications using different ontologies. For example, a repository of interconnected ontologies represented in a standard formalism is what is envisioned as a possible infrastructure for the Semantic Web[8]. Different approaches can be used to reconcile the knowledge from distinct ontologies. The

---

[3] http://umlsinfo.nlm.nih.gov/

[4] http://www.nlm.nih.gov/

[5] http://protege.stanford.edu/index.html

[6] http://www.daml.org/

[7] http://sig.biostr.washington.edu/projects/da/

[8] http://www.semanticweb.org/

classical approach consists of developing methods and tools for aligning or merging several ontologies. Proposed more recently, ontology negotiation would allow applications operating on multiple ontologies to cooperate in performing a task by using similarities and differences in the ontologies in order to establish communication among them.

The availability of domain ontologies capturing the knowledge of specific subdomains of biomedicine (e.g., molecular functions, subcellular localization) is important to many applications. Conversely, by providing a framework for these domain ontologies to hook to, upper level ontologies represent an important, yet less popular, aspect of ontology development. The theories represented in upper level ontologies are general theories such as the theory of parts and wholes, the theory of dependence, and the theory of boundaries. Although not sufficient in itself for representing the knowledge of a domain, an upper level ontology provides the basis for making explicit the difference between, for example, substances and processes. IEEE's Standard Upper Ontology[9] (SUO) working group is developing a standard for specifying "a structure and a set of general concepts upon which domain ontologies (e.g., medical, financial, engineering, etc.) could be constructed". Many general properties will be defined at this upper level and these can be inherited by the domain ontologies hooked underneath them.

Biological knowledge is evolving from structural genomics towards functional genomics. The tremendous amount of DNA sequence information that is now available provides the foundation for studying how the genome of an organism is functioning, and microarray technologies provide detailed information on the mRNA, protein, and metabolic components of organisms. This knowledge allows researchers to discover new metabolic pathways, to model metabolic and regulatory networks in living organisms, and ultimately to understand the pathogenesis of diseases. In this context, in the perspective of acquiring knowledge from the literature or from various and often heterogeneous databases, it is fundamental that not only biologic knowledge, but also medical knowledge be accurately represented and the appropriate linkages be made between these domains. Existing and yet to be developed biomedical ontologies play a critical role in this effort.

---

[9] http://suo.ieee.org/

FUNCTIONAL DISCRIMINATION OF GENE EXPRESSION PATTERNS IN TERMS OF THE GENE ONTOLOGY

LIVIU BADEA[1]

AI Lab, National Institute for Research and Development in Informatics
8-10 Averescu Blvd., Bucharest, Romania, badea@ici.ro

The ever-growing amount of experimental data in molecular biology and genetics requires its automated analysis, by employing sophisticated knowledge discovery tools. We use an Inductive Logic Programming (ILP) learner to induce functional discrimination rules between genes studied using microarrays and found to be differentially expressed in three recently discovered subtypes of adenocarcinoma of the lung. The discrimination rules involve functional annotations from the Proteome HumanPSD database in terms of the Gene Ontology, whose hierarchical structure is essential for this task. While most of the lower levels of gene expression data (pre)processing have been automated, our work can be seen as a step toward automating the higher level functional analysis of the data. We view our application not just as a prototypical example of applying more sophisticated machine learning techniques to the functional analysis of genes, but also as an incentive for developing increasingly more sophisticated functional annotations and ontologies, that can be automatically processed by such learning algorithms.

1 Introduction and motivation

The success of the various whole genome sequencing projects (including the Human Genome Project) has paved the way towards *'functional genomics'*, an undertaking whose main goal is to uncover the functions of the genes and their protein products.

Although painstaking experimental work has revealed essential details on the functions of *individual* genes and proteins, many of their most important effects depend on the orchestration of the activities of entire pathways, comprising numerous genes and proteins.

The major experimental breakthrough that allowed the measurement and analysis of the expression patterns of (tens of) thousands of genes simultaneously is the technology of microarrays and oligonucleotide arrays [1]. It is currently technically feasible to measure the expression levels of the entire set of genes of a living cell and compare such gene expression profiles (patterns) obtained under different experimental conditions. For example, different snapshots of the gene expression patterns in a developing organism can be used to study its development at a global genetic level. In other experiments, different samples of normal and diseased cells can be compared to reveal the genetic abnormalities causing the malady.

Many complex diseases, such as carcinomas, cannot be simply attributed to a single gene or a very small number of genes. They typically show complex disruptions of the gene expression patterns of normal cells. And very frequently, distinguishing between the different subgroups of the disease is essential for its

---

[1] This work was partially supported by the European Community projects SILK and ILPnet2.

566

correct treatment, each of the subgroups requiring a different therapy. Unfortunately, currently applied tests are not always capable to distinguish reliably between the various subtypes. The analysis of microarray gene expression data for various tissue samples has enabled researchers to determine gene expression profiles characteristic of the disease subtypes. The groups of genes involved in these genetic profiles are rather large and a deeper understanding of the functional distinction between the disease subtypes might help not only to select highly accurate 'genetic signatures' of the various subtypes, but hopefully also to select potential targets for design drugs.

Most current approaches to microarray data analysis use (supervised or unsupervised) clustering algorithms to deal with the numerical expression data [2]. The functional interpretation of the resulting clusters is however difficult in certain cases, although a large number of *functional annotations* for the genes involved is already available in public databases. This is probably due to the fact that most functional annotations are in free text and thus of little use to an automated analysis program. Despite this, significant efforts are under way to develop a unifying language[2] in terms of which to describe functions of genes, proteins, biological processes and cellular locations. The *Gene Ontology* (GO) [3] is a large, constantly growing hierarchy of molecular biology concepts, which can be used to annotate genes and proteins with functional information. Such annotations are currently available, e.g. in the Proteome HumanPSD database [4].

In this paper, we use Inductive Logic Programming (ILP) [9] to *induce discrimination rules between two recently discovered subtypes of adenocarcinoma* (AC) [5] *in terms of functional knowledge from the GO*. In other words, we would like to explain the differences between two AC subtypes in terms of the functions of the genes that are differentially expressed in these subtypes. As already hinted above, this should enable a deeper understanding of the mechanisms of the two AC subtypes.

2   Microarray data analysis of adenocarcinoma of the lung

Different experimental approaches were developed to study the gene expression profiles of entire genomes under precise conditions. *Microarrays* and *oligonucleotide chips* [1] allow the simultaneous measurement of the concentration of virtually every transcript in a cell, leading to a holistic understanding of cell physiology. In this paper, we are concentrating on the gene expression profiles produced in the study of Garber et al. of adenocarcinoma (AC) of the lung [5]. We have chosen this particular dataset, because unlike other similar diseases (like leukemias [6] or lymphomas [7] – which are more uniform, and thus much easier to analyse), carcinomas of the lung are quite heterogeneous and sometimes hard to distinguish histologically (under the light microscope). Adenocarcinomas (AC)

---

[2] In the current state of the art, it is more of a *vocabulary* rather than a full-fledged knowledge representation language.

comprise about 30% of all cases, but they are still heterogeneous: Garber et al. have distinguished at least three AC subgroups (AC1-3) with a large difference in survival rate between patients from group AC1 and those from group AC3 (in favour of AC1). In fact, the three subtypes of AC were determined by clustering microarray gene expression profiles (and not by looking at survival rates).

Unlike many diseases which show a rather localized disruption of the genetic profile of normal cells, carcinomas are much more heterogeneous and display more complex disruptions of the gene expression programs of the cells. Tracing these modifications back to the molecular level thus represents a challenging task.

Garber et al. have analysed the expression levels of 23,100 cDNA clones (corresponding to 17,108 unique genes, as defined by Unigene) in 73 different tissue samples (41 AC, 16 SCC, 5 LCLC, 5 SCLC, 5 normal and 1 fetal[3]). After an initial elimination of the low-quality measurements, Garber et al. have selected 918 cDNA clones (representing 835 unique genes) whose expression varied widely among the tissue samples[4] while being most similar within samples originating from the same patient. The resulting dataset taking the form of a 918×73 (gene×sample) matrix was subsequently median centered. Average linkage *hierarchical clustering* of the *samples* was performed using M. Eisen's CLUSTER and TREEVIEW programs [2].

Although hierarchical clustering is an unsupervised method that groups genes exclusively according to the degree of similarity in the patterns of gene expression, it almost perfectly distinguished the known types of lung carcinomas (AC, SCC, LCLC and SCLC) from each other as well as from the normal tissue samples. However, AC did not result as homogeneous cluster, but as 3 distinct sub-clusters: AC1, AC2 and AC3.

Garber et al. have searched for individual genes that were differentially expressed in the 3 AC subgroups and obtained a list of 149 genes (see the supplementary information to [5]). Now, although the distinction between the three AC subgroups may be due to many genes, it is unlikely that a list of 149 genes is the shortest, most comprehensive explanation of the distinction between the AC subgroups. Therefore, Garber et al. tried to subjectively select a subset (of the 149 differentially expressed genes) that would best explain the distinctions between the 3 AC subgroups (Fig.5. of [5]). This selection process consisted in the expert looking at each of the 149 genes and deciding whether it is interesting (as an explanation) or not. While this can be done in this way for one experiment and 149 genes, we argue that processing the ever-growing flood of available microarray data will have to be assisted in a more automatic manner, especially because the functional annotations available are becoming increasingly more sophisticated and hard to manage by a

---

[3] In this paper, we concentrate on AC – the most heterogeneous, and thus the most interesting group.

[4] A gene whose expression level is constant across all tissue samples is obviously not directly involved in the phenomenon under study.

person's working memory. For example, Garber et al. used functional annotations in the form of text descriptions, which can be easily obtained from various molecular biology and genetics databases (such as NCBI Entrez, Expasy/SwissProt etc.).

Obviously, such text descriptions cannot be used by an automated system. However, we are fortunate that more sophisticated annotations are already available, for example in the Proteome HumanPSD database [4]. In the following we show how such functional annotations can be used to *automatically induce discrimination rules* between group AC3 (with a significantly lower survival rate) and the other AC subgroups (AC1 and AC2, showing a much better prognosis).

This should help in selecting a functional explanation of the differences between the AC subgroups in a semi-automatic manner (and thereby speed up microarray data analysis for larger datasets).

3   Functional discrimination of genes using ILP

The HumanPSD Proteome database contains, among others, *Gene Ontology* annotations of a large number of human genes (we could find in Proteome 459 of the 918 cDNA clones of Garber et al.). Functional annotations are however of little use without some sort of background knowledge that relates the various concepts involved.

The *Gene Ontology* (GO) [3] is a hierarchy[7] of concepts used in molecular biology and genetics, which can be employed as *background knowledge* for an inductive learner. (The ontology is organized according to *'Molecular Function'*, *'Biological Process'* and *'Cellular Component'*, for which we have annotations in the Proteome database.)

Inductive Logic Programming (ILP, or Relational Learning) [9] deals with learning logic programs from positive and negative examples with respect to some existing *background knowledge*. ILP genererealizes other machine learning approaches not just by dealing with more complex hypotheses (first order logic programs), but also by taking into account a given background knowledge.

In the following, we show how an ILP learner can be used to induce functional discrimination rules between the genes that correlate well with the AC3-AC1,2 distinction and those that do not. The discrimination rules will involve GO concepts (either GO annotations from the Proteome database, or their generalisations in the GO hierarchy, used as background knowledge).

3.1   Setting up the learning problem: positive and negative examples

We have used the dataset from [5] restricted to the samples classified in AC.

---

[7] In GO, there are two main types of relationships between concepts: *'isa'* (inheritance) and *'part-of'*.

The selection of positive examples (genes correlated with the AC3-AC1,2 distinction) was performed using a nonparametric t-test with a (very strict) P value cutoff of 0.0005, while all genes with $P \geq P_{neg} = 0.3$ were considered negative examples. Finally, only the examples with a counterpart in the HumanPSD database (thus having a GO annotation) were kept. (We thus selected 39 positive and 87 negative examples.)

3.2 The background knowledge

The GO annotations of the examples selected above were used as background knowledge.[8] For instance, the GO annotation of example *'CDKN2A'* (cyclin-dependent kinase inhibitor 2A) is:

    'Molecular Function'('CDKN2A','cyclin-dependent protein kinase inhibitor').
    'Molecular Function'('CDKN2A','tumor suppressor').
    'Biological Process'('CDKN2A','cell cycle checkpoint').
    'Biological Process'('CDKN2A','regulation of CDK activity').
    'Biological Process'('CDKN2A','oncogenesis').
    'Biological Process'('CDKN2A','cell cycle arrest').
    'Cellular Component'('CDKN2A','nucleus').

As many of the annotations are quite specific, they may be unable to cover more than one example. To allow non-trivial generalisations, we added as background knowledge an encoding of the GO hierarchy in the form of Prolog rules of the form

    go ('nucleic acid binding', X) : − go('DNA binding', X).
    go('DNA binding', X) : − go('DNA replication factor', X).

For example, the first rule states that annotation X is in the GO category *'nucleic acid binding'* if it is in category *'DNA binding'*. Of course, the leafs of the GO hierarchy verify a fixpoint clause:

    go(X, X).

3.3 The hypotheses language

Using the ILP learner Progol [8], we looked for hypotheses with (an arbitrary number of) *function* literals in the body (with the second argument instantiated to a constant − in this case a GO annotation), such as:

    target(Gene) : − function(Gene, 'calcium binding'), function(Gene, 'protein binding').

where

    function(Gene, GOterm) : − ('Molecular Function'(Gene,X)   ;

---

[8] Unfortunately, the Proteome database contains numerous mistakes, ranging from typos to inconsistencies with the current version of the GO. All of these mistakes were corrected by hand.

```
'Biological Process'(Gene,X)    ;
'Cellular Component'(Gene,X)    ),  go(GOterm, X).
```

3.4   Obtaining all "best" discriminating hypotheses

For each positive example, Progol's heuristic search selects *only one* "best" clause[9] discriminating this positive example from the negative examples, although there might be several clauses with the same covering and the same heuristic estimate. For instance, the positive example *target('S100P')* admits the following alternative "best" discriminating clauses:

```
target(G) : − function(G,'calcium binding'), function(G,'protein binding').
target(G) : − function(G,'calcium binding'), function(G,'cell-cell signaling').
target(G) : − function(G,'calcium binding'), function(G,'cell communication').
target(G) : − function(G,'cell-cell signaling'), function(G,'ligand binding or carrier').
target(G) : − function(G,'protein binding'), function(G,'cell-cell signaling').
target(G) : − function(G,'protein binding'), function(G,'cell communication').
```

Returning just one, as Progol does, may not be enough in our application, where
- the number of negative examples may be too small to invalidate the wrong alternatives (candidate hypotheses), and/or
- alternative viewpoints on the distinction between classes are highly desirable.

Since Progol conducts a *complete* admissible search, it was relatively straightforward to extract *all* alternative hypotheses (having the same covering and the same heuristic estimate as the hypothesis selected by Progol). More precisely, we return for each seed example a list of alternative hypotheses, each with the set of positive examples covered (in general, containing also other examples besides the seed example). For example:

Alternative hypotheses	Positive examples covered
target(G) : − function(G,'substrate-bound cell migration')	VEGFC
target(G) : − function(G,'lymph gland development')	VEGFC
target(G) : − function(G,'cell migration')	VEGFC

3.5   Results

Table 1 (on the last page) summarizes the results obtained. Each line represents a discriminating hypothesis and the positive examples covered by it (none of the hypotheses cover negative examples). Lines in the same group represent alternative hypotheses (for example, group 18 contains the alternative hypotheses from the previous section, covering VEGFC). Note that we do not return only the *most*

---

[9] Best according to a covering-based heuristic.

general alternative hypotheses (such as 'transcription' from group 3), since the available example set is small (as well as inherently incomplete) and a general hypothesis like 'transcription' may be too general to be informative. (Generalization is also controlled via the negative examples.) Nor do we keep just the *most specific* alternative hypotheses, since these may be too specific. It is ultimately up to the molecular biologist to choose the most significant hypothesis from the set of alternative ones. (The GO hierarchy is "shallow", thereby making this task easy.)

We found functional discrimination rules for virtually all the genes discussed in Garber et al. [5] in relation to AC [11] and all of them were found to be relevant by an expert molecular biologist.[12] (The descriptions are biologically informative, although a bit too general – probably due to the quality of the initial annotations). The most significant ones are:

- the *'cell cycle control'* genes, especially the cyclin-dependent kinase inhibitor CDKN2A (p16) and the polo-like serine/threonine kinase (PLK),[13] which are both up-regulated in AC3.
- the *'cystein-type endopeptidase'* cathepsin L (CTSL), involved in (extracellular) proteolysis, which is also up-regulated in AC3 with respect to AC1,2.
- the *'signal transduction'* and *'growth factor'* Dickkopf (DKK1), which is known to contribute to neoplastic processes (it may play a key role in the transition from an epithelial to a mesenchymal phenotype - which is significant since AC involve epithelial cells).
- the *'transporter'*, *'membrane'* solute carrier protein SLC7A5, known to be closely linked to cellular activation and division (the transcripts of the SLC7A5 gene are rapidly induced and degraded, which is unusual for an integral plasma membrane protein and resembles more closely the kinetic seen for protooncogenes and lymphokines in T cells. It is thought to be up-regulated to support the high protein synthesis for cell growth and activation.)
- in the *'tumor antigen'* category, the carcinoembryonic antigen-related cell adhesion molecule CEACAM1 is involved in tumor angiogenesis. (CEACAM1 is down-regulated in AC3 with respect to AC1,2.)

Note that the discrimination rules induced strike a reasonable balance between generality and specificity: they are specific enough to be informative, but also general enough to explain occasional groups of genes that can be distinguished (from the negative examples) by certain *common* functions.

---

[11] An exception was the thyroid transcription factor TITF1, which did not have a Proteome annotation.

[12] The evaluation of the resulting hypotheses has to be done by human analysis rather than more traditional methods employed in machine learning, mostly due to the small number of genes under study and since we are aiming at automating the last (result interpretation) phase of gene expression analysis, whose validation requires biological experiments.

[13] which is not mentioned in [5] but shows significant correlation with the AC3-AC1,2 distinction.

There was one important functional distinction that could not be discovered due to the incompleteness of the Proteome GO annotation, namely *'angiogenesis'* (involving for example the vascular endothelial growth factor C, VEGFC, which nevertheless was distinguished as being related to *'substrate-bound cell migration'*).

Note that none of the discriminatory descriptions induced by Progol contained general concepts (e.g. 'oncogenesis'), which are common of all the AC subgroups. Discriminatory descriptions are thus informative: they tell us what distinguishes the genes expressed differentially in AC3 and AC1,2 from the others.

The P_{neg} = 0.3 t-test margin for negative examples may be somewhat arbitrary. Too few negative examples would produce too general and thus useless discriminators. On the other hand, too many negative examples combined with a very coarse grained ontology like GO, may produce very specific discriminators or even none at all.

Still, our experiments showed a surprising *robustness* with respect to the choice of the P_{neg} margin: the discriminators for P_{neg} = 0.3 and 0.1 respectively were very similar (the latter are not shown due to space limitations).

Also, the fact that almost all alternative hypotheses (having the same covering and heuristic estimate as the 'best' hypothesis chosen by Progol) *covered the same set of genes* increases our confidence that the distinctions made by the discriminators reflect true functional distinctions rather than contingencies due to a biased initial functional annotation of the genes in the Proteome database. Indeed, for P_{neg} = 0.1 and the seed PLK, Progol learns 3 alternative hypotheses covering different groups of genes:

protein serine/threonine kinase, *biological_process*	CIT,PLK
cell cycle control, enzyme	BUB1B,DUSP4,PLK
cell cycle control, *molecular_function*	BUB1B,CDKN2A,DUSP4,PLK

However 2 of the 3 hypotheses (marked in *italics*) involve two very general GO concepts (*'biological_process'* and *'molecular_function'*) and the distinctions they make are most probably annotation artifacts – in this case some negative examples (which would otherwise be covered) simply happen to lack annotations for *'biological_process'* and *'molecular_function'*.

The uniformity of the coverings of the other groups of alternative definitions strongly suggests that the distinctions are true functional distinctions and not annotation artifacts (they were also confirmed by an expert).

4   Towards more complex annotations

Most current approaches to functional analysis of experimental data in genetics assign genes to one of a number of typically disjoint functional categories (which can be drawn from the GO, but without using the GO hierarchy). However, while such approaches are useful for obtaining aggregate overviews of whole genomes, they are less appropriate for more fine-grained functional discrimination of groups of genes, for which a predefined set of functional categories may turn out to be too coarse grained.

The hierarchical structure of GO, as well as the presence of negative examples proved to be essential in our application. It is encouraging that the discriminations obtained are biologically sensible – this heavily relies on the GO and the HumanPSD annotations. But this also automatically prompts the question of whether more sophisticated knowledge representation formalisms, such as Description Logics (DL) [13] might allow even more precise functional distinctions to be made.

The discriminatory hypotheses induced typically involve (combinations of) very specific GO properties of the target genes, in order to avoid covering any negative examples. However, the learning algorithm tries not only to avoid negative examples, but also to produce "short" hypotheses. In settings like ours involving complex disruptions of gene networks, we expect to be able to obtain more general descriptions of such disrupted groups of genes (in terms of concepts placed higher in the GO hierarchy). This was the case for example with the '*cell cycle control*' genes, but this *generalization* capability of GO was limited to a certain extent by the *fixed* hierarchy.

A DL may allow an "on-the-fly" construction of concepts, rather than relying on a fixed hierarchy. Thus, we wouldn't need to explicitly record in the ontology all generalizations of existing concepts. For example, the current GO contains not just specific concepts like '*cyclin-dependent protein kinase inhibitor*' or '*transmembrane receptor protein tyrosine kinase activator*', but also their generalization '*kinase regulator*'. On the other hand, a DL may take advantage of the intrinsic composite nature of the concepts above and represent them as $\exists inhibits.CDK$ and $\exists activates.TRPTK$. Their generalization need not be explicitly represented, since it can be computed by taking the *least general generalization* ("least common subsumer" in DL terminology) $\exists regulates.kinase$ of the two concepts above.

Thus, we think there are two types of possible extensions of the GO that would enhance its utility in functional analyses:

- allowing a more sophisticated knowledge representation language (like DL) for describing GO concepts (this would generalize the hierarchical structure of GO and allow more refined concept generalizations)
- integrating GO with metabolic, regulatory and cell signalling pathway databases (which would allow more precise causal reasoning – for example, determining possible primary causes for complex genetic disruption profiles).

Already the most basic background knowledge on functional annotations, which involves *hierarchies of concepts* (as in GO, where the main type of relational information is in the form of *inheritance* relationships), is not directly treatable by propositional learners like C4.5 [10], but could be dealt with our approach.

However, as the degree of sophistication of functional annotations will increase in the near future, it is important to know whether our approach will still be applicable in such more complex settings. (C4.5 will definitely be out of the question, due to its inherent difficulty of constructing hypotheses involving

relational knowledge.) We argue that our approach is indeed applicable in such extended settings, even in the presence of arbitrarily more sophisticated relational annotations, such as:
- *'part-of'* relationships (already present in the Gene Ontology)
- causal relationships between genes/proteins such as *activates(G1,G2)* or *inhibits(G1,G2)* (such types of knowledge exist to a certain degree[14] in genetic regulatory and metabolic pathway databases, such as KEGG).

For example, hypotheses involving *'part-of'*, such as[15]

```
target(Gene)  :–  'Biological Process'(Gene, 'cell cycle control'),
                  'Cellular Component'(Gene, X),
                  active(G1),                              %gene G1 is active
                  'Cellular Component'(G1, X1), part_of(X, X1).
```

can only be induced using a true relational learner.

5   Discussion

The need for large-scale functional analysis of genomic and proteomic data requires increasingly more sophisticated knowledge discovery tools, able to take advantage of complex functional annotations as well as existing background knowledge on those. The functional discrimination of genes using the GO hierarchy seems a natural exploitation of the knowledge available in GO, but, as far as we know, hasn't been tried before. ([11] also uses the GO, but for inducing signatures of temporal gene expression profiles rather than functional discrimination.)

We view our application not just as a prototypical example of applying more sophisticated machine learning techniques to gene expression analysis, but also as an incentive for developing increasingly more sophisticated functional annotations and ontologies, that can be automatically processed by such learning algorithms.

With the increasing efforts towards more complete functional annotations of genes, proteins and their pathways, the ability to learn complex *relational* descriptions will become even more important, making sophisticated learning techniques, such as ILP, indispensable.

Several data- and knowledge bases of biological networks and pathways, designed with the aim of allowing automated processing, are currently under development. (As opposed to existing databases which were designed mainly for human users.) We argue that a tight feedback should exist between the annotation effort, the experimental conditions [14], but also the knowledge discovery tools to be used on these annotations.

---

[14] This knowledge is currently very fragmentary and also not well integrated with GO, so we couldn't use it yet in our experiments.

[15] In English: "a gene is a target if it is involved in cell cycle control and it is localized in a sub-component of a cellular component in which another gene is active".

In settings like ours, most of the lower levels of data preprocessing and clustering have been automated. With the ever-growing amount of such expression data available, their high level functional analysis seems to be the major bottleneck, which can be appropriately addressed using more sophisticated machine learning techniques, like ILP, able to deal with complex background knowledge.

The specificities and complexities of functional genomics may require modifying existing learning algorithms, as we did in the case of constructing all alternative "best" hypotheses covering a given seed example. This modification turned out to be crucial in our application (together with the completeness of the hypotheses search).[16]

References

1. Basset D.E., M.B. Eisen, M.S. Boguski. Gene expression informatics – it's all in your mine, Nature Genetics supplement, Vol. 21, Jan. 1999, 51-55.
2. Eisen M.B., P.T. Spellman, P.O. Brown, D. Botstein. Cluster analysis and display of genome-wide expression patterns, Proc.Natl. Acad. Sci., Vol.95, 14863-14868, Dec.1998.
3. Gene Ontology Consortium. Gene Ontology: tool for the unification of biology. Nature Genet. 25:25-29, 2000.
4. http://www.incyte.com/proteome.
5. Garber M.E. et al. Diversity of gene expression in adenocarcinoma of the lung. Proc. Natl. Acad. Sci., Vol. 98, 13784-13789, November 20, 2001.
6. Golub T.R. et al. Molecular classification of cancer: class discovery and class prediction by gene expression monitoring, Science, Vol. 286, 15 October 1999, 531-537.
7. Alizadeh A.A. et al. Distinct types of diffuse large B-cell lymphoma identified by gene expression profiling, Nature, Vol. 403, 503-511, Feb. 2000.
8. Muggleton S. Inverse entailment and Progol. New Gen. Computing J., 13:245-286, 1995.
9. S.H. Nienhuys-Cheng, R. de Wolf. Foundations of Inductive Logic Programming, Vol. 1228 of Lecture Notes in Artificial Intelligence. Springer-Verlag, 1997.
10. Quinlan, J.R. C4.5: Programs for Machine Learning, Morgan Kaufmann, 1993.
11. T.R. Hvidsten, J. Komorowski, A.K. Sandvik, and A. Lægreid. Predicting Gene Function from Gene Expressions and Ontologies, PSB 6:299-310 (2001).
12. Liviu Badea, S.H. Nienhuys-Cheng. A Refinement Operator for Description Logics. Proc. Intl.Conf. on Inductive Logic Programming ILP-2000, LNAI 1866, 40-59, Springer.
13. F. Baader, D. McGuinness, D. Nardi, P. P. Schneider. The Description Logic Handbook, Cambridge University Press, 2003.

---

[16] Indeed, without it, the learner may return some spurious discriminators without any means of discerning these from the "real" ones. Such spurious discriminators were in fact obtained when we tried a propositional decision tree learner (C4.5rules) on a preprocessed example set taking into account the inheritance relationships in the Gene Ontology. The inappropriateness of C4.5 in this case may also be due to the fact that we are not really dealing with a classification problem, but more with a descriptive induction problem. (C4.5rules also produces rules activated by the *absence* of annotations. However, due to the inherent incompleteness of the GO annotations, we need to restrict ourselves to positive information only and this seems difficult to do using C4.5.)

14. Microarray Gene Expression Data (MGED) Society Ontology Working Group
http://www.cbil.upenn.edu/Ontology/MGED_ontology.html.

Discriminatory hypotheses	Positive examples covered
calcium binding, protein binding	S100P
calcium binding, cell-cell signaling	S100P
calcium binding, cell communication	S100P
cell-cell signaling, ligand binding or carrier	S100P
protein binding, cell-cell signaling	S100P
protein binding, cell communication	S100P
integral plasma membrane proteoglycan	ICAM1,PTK7
transcription, from Pol II promoter	IRX5,TRIM29
transcription, DNA-dependent	IRX5,TRIM29
transcription	IRX5,TRIM29
tumor antigen	CEACAM1,STEAP
cell surface antigen	CEACAM1,STEAP
furin	PACE
cell cycle control	BUB1B,CDKN2A,DUSP4,PLK
6-phosphofructokinase	PFKP
glucose metabolism	PFKP
hexose metabolism	PFKP
monosaccharide metabolism	PFKP
carbohydrate metabolism	PFKP
transporter, membrane	ABCC2,AQP8,SLC2A1,SLC7A5,STEAP
RHO small monomeric GTPase	ARHE
actin cytoskeleton reorganization	ARHE
peripheral plasma membrane protein	ARHE
small monomeric GTPase	ARHE
protein kinase cascade	CIT,DUSP4,STAT4
monooxygenase	CYP24
cysteine-type peptidase	CTSH,CTSL
lysosomal cysteine-type endopeptidase	CTSH,CTSL
cysteine-type endopeptidase	CTSH,CTSL
transcription co-repressor	ID3
transcription regulation, from Pol II promoter	ID3,STAT4
transcription regulation	ID3,STAT4
respiration	SFTPC
membrane, ligand	ICAM4,SCYD1
plasma membrane, ligand	ICAM4,SCYD1
RNA helicase	DBY
adenosinetriphosphatase	DBY
helicase	DBY
signal transduction, growth factor	DKK1
substrate-bound cell migration	VEGFC
lymph gland development	VEGFC
cell migration	VEGFC
excretion	UNC13
basement membrane	LAD1

Table 1. Hypotheses discriminating genes differentially expressed in classes AC3 and AC1,2.

TOWARDS A BROAD-COVERAGE BIOMEDICAL ONTOLOGY BASED ON DESCRIPTION LOGICS

U. HAHN [a]   S. SCHULZ [a,b]

[a] ⟨CꟼF⟩ *Text Knowledge Engineering Lab, Freiburg University, D-79085 Freiburg, Germany*
`http://www.coling.uni-freiburg.de`

[b] *Department of Medical Informatics, University Hospital, D-79104 Freiburg, Germany*

We describe an ontology engineering methodology by which conceptual knowledge is extracted from an informal medical thesaurus (UMLS) and automatically converted into a formal description logics system (LOOM). Our approach consists of four steps: concept definitions are automatically generated from the UMLS, integrity checking of taxonomic and partonomic hierarchies is performed by LOOM's terminological classifier, cycles and inconsistencies are eliminated, as well as incremental refinement of the evolving knowledge base is performed by a domain expert. We report on experiments with a very large knowledge base composed of 164,000 concepts and 76,000 relations.

1 Introduction

Unlike many other disciplines, medicine has a long standing tradition in structuring its domain knowledge, e.g, disease taxonomies, medical procedures, anatomical terms, in a wide variety of medical terminologies, thesauri and classification systems. These efforts are typically restricted to the provision of broader and narrower terms, related terms or synonymous terms. This is most evident in the UMLS, the *Unified Medical Language System*[1], an umbrella system which covers more than 60 medical thesauri and classifications (e.g., MeSH, ICD, SNOMED, Digital Anatomist).

From a conceptual perspective, the UMLS can be divided into two major parts. The UMLS *Semantic Network (SN)*, on the one hand, forms the upper ontology and consists of 134 semantic types linked by 54 types of semantic relations, which makes a total of 7,473 edges, The UMLS *Metathesaurus*, on the other hand, contains 776,940 concepts (2002 version), each of which is assigned to one or more UMLS SN types. These concepts are linked by the semantic relations also supplied by the UMLS SN. In total, 10,147,419 semantic links between these Metathesaurus concepts exist, most of them directly taken from the sources, some added by the UMLS developers. The vast majority of these links introduce thesaurus-like broader/narrower term relationships.

Both, the UMLS SN and the Metathesaurus, form a huge semantic network. Its semantics is shallow and entirely intuitive, which is due to the fact that their usage was primarily intended for humans in order to support health-related knowledge management. Given the size, the evolutionary diversity and inherent heterogeneity of the UMLS, there is no surprise that the lack of a formal semantic foundation leads to inconsistencies, circular definitions, etc. [2,3]. This may not cause utterly severe problems when humans are in the loop and its use is limited to disease or procedure encoding, accountancy or document retrieval tasks. However, anticipating its use for more knowledge-intensive applications such as medical decision making or the understanding of medical narratives those shortcomings might lead to an impasse.

As a consequence, formal models for dealing with medical knowledge have been proposed such as conceptual graphs, semantic networks or description logics [4,5,6,7]. Not surprisingly, there is a price to be paid for more expressiveness and formal rigor, *viz.* increasing modeling efforts and, hence, increasing maintenance costs [8]. Operational systems making full use of such rigid approaches, especially those which employ high-end knowledge representation languages, are usually restricted to rather small subdomains. The most comprehensive of these sources we know of is the GRAIL-encoded GALEN knowledge base which covers up to 9,800 concepts [6]. The limited coverage then hampers their routine usage, an issue which is always highly rewarded in the medical informatics community.

Almost all of the knowledge bases developed on the basis of formal representation languages have been designed from scratch – without making systematic use of the large body of knowledge contained in widely spread medical terminologies. Hence, it would be an intriguing approach to join the massive *coverage* offered by informal medical terminologies with the high level of *expressiveness* and *reasoning capabilities* supported by rigid knowledge representation systems in order to develop formally solid medical knowledge bases on a larger scale. This idea has already been fostered by Pisanelli et al. [9] who extracted knowledge from the UMLS SN and from parts of the Metathesaurus, and merged them with logic-based top-level ontologies from various sources. Another example is the re-engineering of SNOMED[10] from a multi-axial coding system into a formally founded ontology[11,12].

Unfortunately, the efforts made so far are entirely focused on generalization-based reasoning along taxonomies and lack a reasonable coverage of partonomies. As the latter form a crucial part of medical knowledge, modeling efforts not only have to be directed at generalization hierarchies and taxonomic reasoning but must incorporate part-whole knowledge and partonomic reasoning, as well. We here describe such an integrated knowledge engineering methodology. The resulting medical ontology will form the domain knowledge backbone of MEDSYNDIKATE, a system for the automatic acquisition of knowledge from medical finding reports [13].

2 Reasoning Along Part-Whole Hierarchies

Medical knowledge is typically organized around generalization hierarchies – on which taxonomic reasoning along *is-a* relations is based – and part-whole hierarchies – allowing partonomic reasoning along *part-of* or *has-part* relations. Unlike generalization-based reasoning in concept taxonomies, no fully conclusive mechanism exists up to now for reasoning along part-whole hierarchies (for a survey of different approaches, cf. Artale et al.[14]).

Within the description logics (DL) paradigm several extensions to representation languages have been proposed which provide special constructors for part-whole reasoning[6,15]. This seems a reasonable way to proceed as long as the transitivity property of a relation can be assumed, in general. In the medical[16], as well as in commonsense domains [17,18], however, this view has been invalidated. Obviously, various exceptions exist such that the transitivity of *part-of* relations cannot be granted, in general. Hence, both the expression of regular transitive use, as well as exception handling for nontransitive *part-of* relations have to be taken into consideration. Another challenge comes with the propagation of properties across part-whole hierarchies, often referred to as 'inheritance across transitive roles' (e.g., *inflammation-of* ∘ *part-of* → *inflammation-of*) [19]. Especially with biomedical knowledge this reasoning pattern cannot be generalized, since it faces lots of exceptions, too.

Motivated by previous approaches [20,21], we formalized a model of partonomic reasoning [16] that accounts for the above considerations and, also, does not exceed the expressiveness of the well-understood, parsimonious concept language $\mathcal{ALC}$ [22].[a] Our proposal is centered around a particular data structure for partonomic reasoning, so-called *SEP triplets* (cf. Figure 1). They define a characteristic pattern of *is-a* hierarchies which support the emulation of inferences typical of transitive *part-of* relations. In this formalism, the relation *anatomical-part-of* describes the partitive relation between physical parts of an organism.

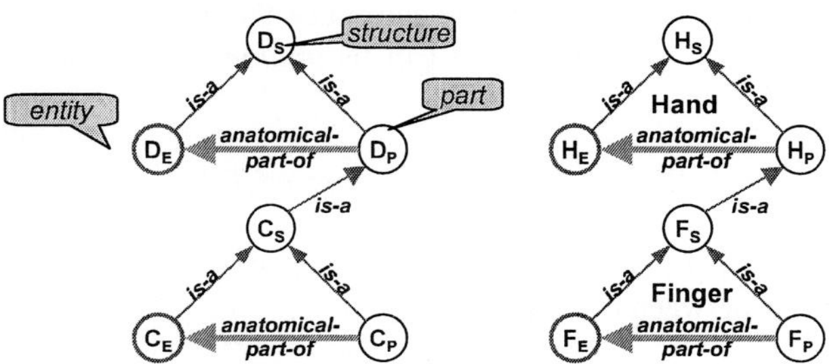

Figure 1: SEP Triplets: Partitive Relations within Taxonomies

A triplet consists, first of all, of a composite *Structure* concept, the so-called *S-node* (e.g., *Hand-Structure*, H_S). Each *Structure* concept subsumes directly an anatomical *Entity* concept, on the one hand, and a common subsumer of anything that

---

[a]$\mathcal{ALC}$ allows for the construction of concept hierarchies, where '⊑' denotes subsumption and '≐' definitional equivalence. Existential (∃) and universal (∀) quantification, negation (¬), disjunction (⊓) and conjunction (⊔) are also supported. Role filler constraints (e.g., typing by C) are linked to the relation name R by a dot, $\exists R.C$.

580

is a *Part* of that entity concept, on the other hand. These two concepts are called *E-node* and *P-node*, e.g., *Hand-Entity* (H_E) and *Hand-Part* (H_P), respectively. Whereas E-nodes denote the anatomical concepts proper to be modelled in our domain, S-nodes and P-nodes constitute representational artifacts required for the formal reconstruction of the systematic patterns and exceptions underlying partonomic reasoning. More precisely, a P-node is the common subsumer of those concepts that have their role *anatomical-part-of* filled by the corresponding E-node concept, as an existential condition. For example, *Hand-Part* subsumes those concepts all instances of which have a *Hand-Entity* as a necessary whole. As an additional constraint, E-nodes and P-nodes can be modelled as being mutually disjoint. This is a reasonable assumption for most concepts denoting singleton objects, where parts and wholes cannot be of the same type (a red blood cell cannot be part of yet another red blood cell). On the contrary, masses and collections can have parts and wholes of the same type, e.g., a tissue can be part of another tissue.

For the reconstruction of the *anatomical-part-of* relation by taxonomic reasoning, we assume C_E and D_E to denote E-nodes, C_S and D_S to denote the S-nodes that subsume C_E and D_E, respectively, and C_P and D_P to denote the P-nodes related to C_E and D_E, respectively, via the role *anatomical-part-of* (cf. Figure 1). These conventions can be captured by the following terminological expressions:

$$C_E \sqsubseteq C_S \sqsubseteq D_P \sqsubseteq D_S \tag{1}$$

$$D_E \sqsubseteq D_S \tag{2}$$

The P-node is defined as follows (we here introduce the disjointness constraint between D_E and D_P, i.e., no instance of D can be *anatomical-part-of* any other instance of D):

$$D_P \doteq D_S \sqcap \neg D_E \sqcap \exists anatomical\text{-}part\text{-}of.D_E \tag{3}$$

Since C_E is subsumed by D_P (according to (1)), we infer that the relation *anatomical-part-of* holds between C_E and D_E, too.[b]

$$C_E \sqsubseteq \exists anatomical\text{-}part\text{-}of.D_E \tag{4}$$

The encoding of concept hierarchies in terms of SEP triplets allows the knowledge engineer to switch the transitivity property of part-whole relations off and on, dependent on whether the E-node or the S-node, respectively, is addressed as the target concept for a conceptual relation. In the first case, the propagation of roles

---

[b]An extension of this encoding scheme which allows additional reasoning about *has-part* in a similar way, is proposed in[23], though it has not been considered in the knowledge base described in this paper.

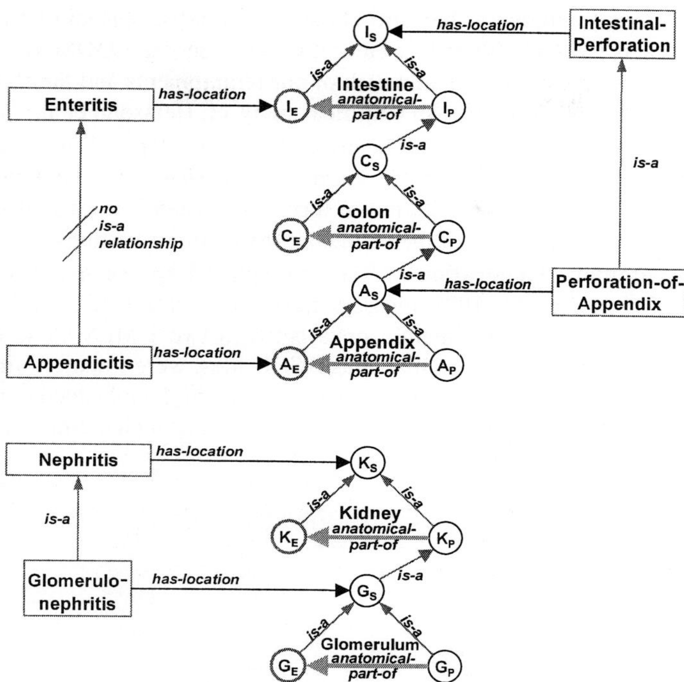

Figure 2: Enabling/Disabling Role Propagation in a SEP-Encoded Partonomy

across part-whole hierarchies is disabled, in the second case it is enabled. As an example (cf. Figure 2), *Enteritis* is defined as *has-location Intestine$_E$*, i.e., the range of the relation *has-location* is restricted to the E-node of *Intestine*. This precludes, e.g., the classification of *Appendicitis* as *Enteritis* though the *Appendix* is related to the *Intestine* via an *anatomical-part-of* relation. In the 'switch-on' mode, however, *Glomerulonephritis* (*has-location Glomerulum$_S$*) is classified as *Nephritis* (*has-location Kidney$_S$*), with *Glomerulum* being an *anatomical-part-of* the *Kidney*. In the same way, *Perforation-of-Appendix* is classified as *Intestinal-Perforation* (cf. Hahn et al.[16] for an in-depth analysis of these phenomena).

3 Ontology Engineering Workflow

Our goal is to extract conceptual knowledge from two major subdomains of the UMLS, *viz.* anatomy and pathology, in order to construct a formally sound knowledge base based on ALC-type description logics[22]. Such a research program puts considerable pressure on well-engineered terminological reasoning systems, since we

582

require their inference engine to perform classification-based reasoning reliably on very large data sets ($\gg$ 100,000 items). Hence, we consider the LOOM system[24], at the system level, the most convenient match of our requirements and the state of the art in terminological reasoning (for an empirical study, cf. Heinsohn et al.[25]).

The ontology engineering task will be divided into four steps: (1) automatic generation of terminological expressions, (2) automatic consistency checking by a terminological classifier, (3) manual restitution of formal consistency in case of inconsistencies, and, (4) manual curation of the formal representation structures.

Step 1: Automatic Generation of Terminological Expressions. Sources for concepts and relations were the 1999 release of the UMLS SN and the UMLS metathesaurus. Figure 3 exhibits the semantic links between two UMLS CUIs (concept unique identifier).[c] These tables, available as ASCII files, were imported into a Microsoft Access relational database and manipulated using SQL embedded in the VBA programming language. For each CUI in the *mrrel* subset its alphanumeric code was substituted by the English preferred term.

CUI1	REL	CUI2	RELA	X	Y
C0005847	CHD	C0014261	part_of	MSH99	MSH99
C0005847	CHD	C0014261		CSP98	CSP98
C0005847	CHD	C0025962	isa	MSH99	MSH99
C0005847	CHD	C0026844	part_of	MSH99	MSH99
C0005847	CHD	C0026844		CSP98	CSP98
C0005847	CHD	C0034052		SNMI98	SNMI98
C0005847	CHD	C0035330	isa	MSH99	MSH99
C0005847	CHD	C0042366	part_of	MSH99	MSH99
C0005847	CHD	C0042367	part_of	MSH99	MSH99
C0005847	CHD	C0042367		SNM2	SNM2
C0005847	CHD	C0042449	isa	MSH99	MSH99

Figure 3: Semantic Relations in the UMLS Metathesaurus

After manual remodeling of the top-level concepts of the UMLS SN we extracted, from a total of 85,899 metathesaurus concepts, 38,059 anatomy and 50,087 pathology concepts. The criterion for the inclusion into one of these sets is the assignment to predefined UMLS SN types. Also, 2,247 concepts appeared in both sets, anatomy and pathology. Since we wanted to keep the two subdomains strictly disjoint, we duplicated these overlapping concepts, and prefixed all concepts by *ana-* or *pat-* according to their proper subdomain. Indeed, these hybrid concepts mirror multiple meanings. For instance, *tumor* has the meaning of a malignant disease, on the one hand, and of an anatomical structure, on the other hand.

---

[c]As a convention in UMLS, any two CUIs must be connected by at least a shallow relation (in Figure 3, CHilD relations in the column REL are assumed between CUIs). Shallow relations may be refined in the column RELA, if a thesaurus is available which contains more specific information. Some CUIs are linked either by *part-of* or *is-a*. In any case, the source thesaurus for the relations and the CUIs involved is specified in the columns X and Y (e.g., MeSH 1999, SNOMED International 1998).

As target structures for the anatomy domain we chose SEP triplets. Only UMLS-supplied *part-of* (including *conceptual-part-of*), *has-part* (including *has-conceptual-part*) and *is-a* relation attributes were considered for the construction of taxonomic and partonomic hierarchies. Hence, for each anatomy concept one SEP triplet was created. For the pathology domain, we treated *CHD* (*child*) and *RN* (*narrower* relation) from the UMLS as indicating taxonomic (*is-a*) links. No part-whole relations were considered, since there were only two occurences of the relation attribute *part-of* between pathology concepts in the whole metathesaurus. Furthermore, for all anatomy concepts contained in the definitional statements of pathology concepts the S-node is the default concept to which they are linked, thus enabling the propagation of roles across the part-whole hierarchy.

As a fundamental assumption, all roles generated in this process were considered as being existentially quantified. This means that any relation r (*part-of, has-location*, etc.) which holds between two concepts, A and B, is mapped to a corresponding role $\tilde{r}.B$ which is a necessary condition in the definition of the concept A. All conceptual constraints for a concept definition are mapped to a conjunction of constraints.

Another basic assumption concerns the strict antisymmetry of *is-a* and *part-of*. As a consequence, we did not process reflexive relations such as "C1 *broader-than* C1" ($n = 9{,}546$ cases of reflexive hierarchical relations were found).

Furthermore, type-to-type relations from the UMLS SN (e.g., *Anatomical Structure location-of Disease or Syndrome*) were not considered, since they indicate "possible" relations that cannot be adequately expressed in description logics.

Step 2: Automatic Consistency Checking by the LOOM **Classifier.** The import of UMLS anatomy concepts resulted in 38,059 *deftriplet* expressions for anatomical concepts and 50,087 *defconcept* expressions for pathological concepts. Each *deftriplet* was expanded into three *defconcept* (S-, E-, and P-nodes), and two *defrelation* (*anatomical-part-of-x, inv-anatomical-part-of-x*) expressions, summing up to 114,177 concepts. This yielded (together with the concepts from the UMLS SN) a total of 240,764 definitory LOOM expressions.

From 38,059 anatomy triplets, 1,219 *deftriplet* statements contained a *:has-part* clause followed by a list of a variable number of triplets, with more than one argument in 823 cases (3.3, on average). 4,043 *deftriplet* statements contained a *:part-of* clause, only in 332 cases followed by more than one argument (1.1, on average). We then submitted the resulting knowledge base to the terminological classifier, the inference engine which computes subsumption relations, and checked for terminological cycles and consistency. In the anatomy subdomain, one terminological cycle and 2,328 inconsistent concepts were found, in the pathology subdomain 355 terminological cycles though not a single inconsistent concept were determined.

Step 3: Manual Restitution of Consistency. The inconsistencies in the anatomy part of the knowledge base identified by the classifier could all be traced back to the simultaneous linkage of two triplets by both *is-a* and *part-of* links, an encoding that raises a conflict due to the disjointness required for corresponding P- and E-nodes (cf. expression (3)). In most of these cases the parents involved belonged to a class of concepts that obviously cannot be appropriately modeled as SEP triplets, e.g., *Subdivision-Of-Ascending-Aorta* or *Organ-Part*. The meaning of each of these concepts almost paraphrases that of a P-node, so that the violation of the SEP-internal disjointness condition could be accounted for by substituting the triplets involved with simple LOOM concepts, by matching them with already existing P-nodes, or by disabling *is-a* or *part-of* links. B In the pathology part of the knowledge base, we expected a large number of terminological cycles to occur, simply as a consequence of interpreting the extremely weak *narrower term* and *child* relations in terms of taxonomic subsumption (*is-a*). Bearing in mind the size of the knowledge base, we consider 355 cycles quite a low number (Bodenreider[3] found a total of 1,920 cycles in the complete 2001 metathesaurus release.). Those cycles were primarily due to very similar concepts, e.g., *Arteriosclerosis vs. Atherosclerosis, Amaurosis vs. Blindness*, and residual categories ("other", "NOS" = *not otherwise specified*). These concepts were directly inherited from the source terminologies and are notoriously difficult to interpret out of their definitional context, e.g., *Other-Malignant-Neoplasm-of-Skin vs. Malignant-Neoplasm-of-Skin-NOS*. In many cases the decision which relations could be maintained and which ones had to be eliminated was taken arbitrarily, since for biomedical terminology often no consensus can be reached on the exact meaning of terms. As the result of the analysis we obtained a negative list which consisted of 630 concept pairs. In a subsequent extraction cycle we incorporated this list in the automated construction of the LOOM concept definitions and, with these new constraints added, a fully consistent knowledge base was generated, finally.

Step 4: Manual Curation of the Knowledge Base. To set up this high-volume knowledge base including the aforementioned working steps required three months of work for a single person. The fourth step – when performed for the whole knowledge base – is very time-consuming and requires broad and in-depth medical expertise. Random samples from both subdomains were analyzed by the second author, a domain expert. The data we here supply refer to the analysis of two random samples of each 100 anatomy and 100 pathology concepts. This took one person about a single month. From the experience we gained so far, the following workflow can be derived:

- *Checking the correctness of the taxonomic and partonomic hierarchies.* Taxonomic and partonomic links are manually added or removed.

  Results: In the *anatomy* sample, only 76 in 100 concepts could be unequivocally classified as belonging to 'canonical' anatomy. (The remainder, e.g., *ana-Phalanx-of-Supernumerary-Digit-of-Hand*, referring to pathological anatomy, was immediately excluded from analysis.) Besides the assignment to the UMLS semantic types, only 27 (direct) taxonomic links were found. 83 UMLS relations (mostly *child* or *narrower* relations) were manually upgraded to taxonomic links. 12 (direct) *part-of* and 19 *has-part* relations were found. Four *part-of* relations and one *has-part* relation had to be removed, since we considered them as implausible. 51 UMLS relations (mostly *child* or *narrower* relations) were manually upgraded to *part-of* relations, and 94 UMLS relations (mostly *parent* or *broader* relations) were upgraded to *has-part* relations. After this workup and upgrade of shallow UMLS relations to semantically more specific relations, the sample was checked for completeness again. As a result, 14 *is-a* and 37 *part-of* relations were still considered as missing.

In the *pathology* sample, the assignment to the pathology subdomain was considered plausible for 99 of 100 concepts. A total of 15 false *is-a* relations was identified in 12 concept definitions. 24 *is-a* relations were found to be missing.

- *Check of the* :has-part *arguments assuming 'real anatomy'.* In the UMLS sources *part-of* and *has-part* relations are considered as symmetric. According to our transformation rules, the attachment of a role *has-anatomical-part* to an E-node B_E, with its range restricted to A_E, implies the existence of a concept A for the definition of a concept B. On the other hand, the classification of A_E as subsumee of the P-node B_P, the latter being defined via the role *anatomical-part-of* restricted to B_E, implies the existence of B_E given the existence of A_E.

These constraints do not always conform to 'real' anatomy, i.e., anatomical entities that may exhibit pathological modifications. Figure 4 (left) sketches a concept A that is necessarily *anatomical-part-of* a concept B, but whose existence is not required for the definition of B. This is typical of the results of surgical interventions, e.g., a large intestine without an appendix, or an oral cavity without teeth, etc.

Results: All 112 *has-part* relations obtained by the automatic import and the manual workup of our sample were checked. The analysis revealed that more than half of them (62) should be eliminated in order not to obviate a coherent classification of pathologically modified anatomical objects. For instance, maintaining *has-anatomical-part.Thumb* as an existential restriction in the definition of *Hand* would disallow to classify as *Hand* all those anatomical entities that have no thumb due to congenital or acquired abnormalities. As another example, most instances of *Ileum* do not contain a *Meckel's Diverticulum*, whereas all instances of *Meckel's Diverticulum* are necessarily *anatomical-part-of Ileum*. Many surgical interventions that remove anatomical structures (appendix, gallbladder, etc.), produce similar patterns.

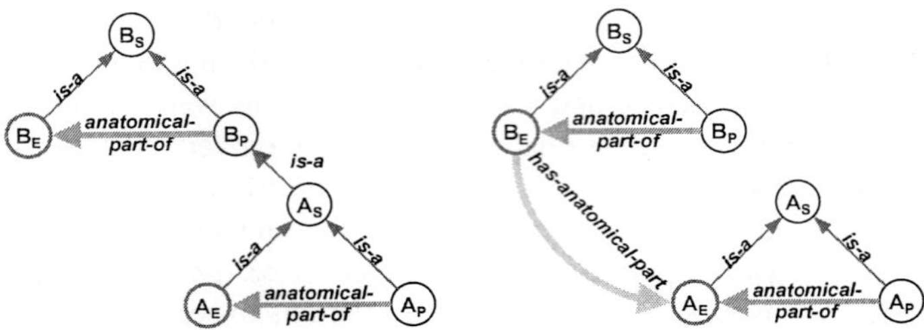

Figure 4: Patterns for Part-Whole Reasoning Using SEP triplets

In formalism we propose, this corresponds to a single taxonomic link between an S-node and a P-node (cf. Figure 4, left part). The contrary is also possible (cf. Figure 4, right part): the definition of A_E does not imply that the role *anatomical-part-of* be filled by B_E, but B_E does imply that the inverse role, *has-anatomical-part*, be filled by A_E. As an example, a *Lymph-node* necessarily contains *Lymph-follicles*, but there exist *Lymph-follicles* that are not part of a *Lymph-node*. This pattern is typical of the mereological relation between macroscopic (countable) objects, such as organs, and multiple uniform microscopic objects.

- *Analysis of the sibling relations.* In UMLS, the *SIB* relation links concepts that share the same parent in a taxonomic or partonomic hierarchy.

  Results: We found on the average 6.8 siblings per concept in the anatomy domain, 8.8 in the pathology domain. So far, the analysis of sibling relations has been performed only for the anatomy domain. From a total of 521 sibling relations, 9 were identified as *is-a*, 14 as *part-of*, and 17 as *has-part*, whereas 404 referred to topologically disconnected concepts.

- *Completion and modification of anatomy–pathology relations.* Surprisingly, only very few pathology concepts contained an explicit reference to a corresponding anatomy concept. Therefore, these relations have to be added by a domain expert. In each case, a decision must be made whether the E-node or the S-node has to be addressed as the target concept for modification such that the propagation of roles across part-whole hierarchies is disabled or enabled, respectively (cf. Figure 2).

  Results: In the sample, we found 522 anatomy-pathology relations, from which 358 (i.e., 69%!) were judged incorrect by the domain expert. In 36 cases an adequate anatomy-pathology relation was missing. All 164 *has-location* roles were analyzed as to whether they had to be filled by an S-node or an E-node of an anatomical triplet. In 153 cases, the S-node (which allows propagation across the part-whole hierarchy) was considered to be adequate, in 11 cases the E-node was preferred. The analysis of the 100 pathology concepts revealed that only 17 had to be linked with an anatomy concept. In 15 cases, the default linkage to the S-node was considered as correct, in one case the linkage to the E-node was preferred, in another case the linkage was considered as false.

The high number of implausible constraints points to the lightweight semantics of *has-location* links in the UMLS sources. While we interpreted them as a conjunction for the mapping procedure, a disjunctive meaning seems to prevail implicitly in many definitions of top-level concepts such as *Tuberculosis*. In this example, *all* anatomical concepts that can be affected by this disease are linked by *has-location*. All these constraints (e.g., *has-location Urinary-Tract*) are inherited to subconcepts such as *Tuberculosis-of-Bronchus*. A thorough analysis of the top-level pathology concepts is necessary, and conjunctions of constraints will have to be substituted by disjunctions where necessary.

In conclusion, our study shows that it is relatively straightforward to restitute consistency of the UMLS knowledge base, but it is nearly impossible to reach a high degree of both adequacy and completeness due to the huge amount of manual work required. Restituting adequacy should, however, not be considered primarily as eliminating obvious categorization 'errors' from the UMLS sources, but rather as making choices between alternative conceptualizations of medical terms whose meaning differs slightly due to the heterogeneity of the knowledge sources. Another aspect is the need for the curation of concept definitions. These may have become obsolete as a consequence of imposing particularly rigid axiomatic assumptions on them. For instance, the conjunctive reading of defining attributes is not true in all cases, and, thus, necessarily requires individual manual specification. Finally, from a technical perspective, we found the implications of using the terminological classifier of utmost importance and of outstanding heuristic value. Hence, the knowledge refinement cycles are truly semi-automatic, fed by medical expertise on the side of the human knowledge engineer, but also driven by the reasoning system which makes explicit the consequences of (im)proper concept definitions.

References

1. A. McCray and S. Nelson. The representation of meaning in the UMLS. *Methods of Information in Medicine*, 34(1/2):193–201, 1995.
2. J. Cimino. Distributed cognition and knowledge-based controlled medical terminologies. *Artificial Intelligence in Medicine*, 12(1):153–168, 1998.
3. O. Bodenreider. Circular hierarchical relationships in the UMLS: Etiology, diagnosis, treatment, complications and prevention. In *Proc. AMIA 2001*, pages 57–61, 2001.

4. J. Cimino, P. Clayton, G. Hripsack, and S. Johnson. Knowledge-based approaches to the maintenance of a large controlled medical terminology. *JAMIA*, 1(1):35–50, 1994.

5. E. Mays, R. Weida, R. Dionne, M. Laker, B. White, C. Liang, and F. Oles. Scalable and expressive medical terminologies. In *Proceedings AMIA'96*, pages 259–263, 1996.

6. A. Rector, S. Bechhofer, C. Goble, I. Horrocks, W. Nowlan, and W. Solomon. The GRAIL concept modelling language for medical terminology. *Artificial Intelligence in Medicine*, 9:139–171, 1997.

7. F. Volot, M. Joubert, and M. Fieschi. Review of biomedical knowledge and data representation with Conceptual Graphs. *Methods of Inf. in Medicine*, 37(1):86–96, 1998.

8. A. Rector. Clinical terminology: Why is it so hard? *Meth. Inf. Med.*, 38:147–157, 1999.

9. D. Pisanelli, A. Gangemi, and G. Steve. An ontological analysis of the UMLS metathesaurus. In *Proceedings AMIA'98*, pages 810–814, 1998.

10. R. Côté. *SNOMED International.* College of American Pathologists, 1993.

11. K. Spackman and K. Campbell. Compositional concept representation using SNOMED: Towards further convergence of clinical terminologies. In *Proceedings AMIA'98*, pages 740–744, 1998.

12. K. A. Spackman. Normal forms for description logic expression of clinical concepts in SNOMED RT. In *Proc. AMIA 2001*, pages 627–631, 2001.

13. U. Hahn, M. Romacker, and S. Schulz. Creating knowledge repositories from biomedical reports: The MEDSYNDIKATE text mining system. In *Proceedings PSB 2002*, pages 338–349, 2002.

14. A. Artale, E. Franconi, N. Guarino, and L. Pazzi. Part-whole relations in object-centered systems: An overview. *Data & Knowledge Engineering*, 20(3):347–383, 1996.

15. I. Horrocks and U. Sattler. A description logic with transitive and inverse roles and role hierarchies. *Journal of Logic and Computation*, 9(3):385–410, 1999.

16. U. Hahn, S. Schulz, and M. Romacker. Part-whole reasoning: A case study in medical ontology engineering. *IEEE Intelligent Systems & Applications*, 14(5):59–67, 1999.

17. D. Cruse. On the transitivity of the part-whole relation. *J. Ling.*, 15:29–38, 1979.

18. M. Winston, R. Chaffin, and D. Herrmann. A taxonomy of part-whole relationships. *Cognitive Science*, 11:417–444, 1987.

19. A. Rector. Analysis of propagation along transitive roles: Formalisation of the GALEN experience with medical ontologies. In *Proceedings DL 2002*, 2002.

20. J. Schmolze and W. Mark. The NIKL experience. *Comput. Intell.*, 7(1):48–69, 1991.

21. E. Schulz, C Price, and P. Brown. Symbolic anatomic knowledge representation in the Read Codes Version 3: Structure and application. *JAMIA*, 4(1):38–48, 1997.

22. M. Schmidt-Schauß and G. Smolka. Attributive concept descriptions with complements. *Artificial Intelligence*, 48(1):1–26, 1991.

23. S. Schulz and U. Hahn. Necessary parts and wholes in bio-ontologies. In *Proceedings KR 2002*, pages 387–394, 2002.

24. R. MacGregor and R. Bates. The LOOM knowledge representation language. Technical Report RS-87-188, Information Sciences Institute, Univ. of Southern California, 1987.

25. J. Heinsohn, D. Kudenko, B. Nebel, and H.-J. Profitlich. An empirical analysis of terminological representation systems. *Artificial Intelligence*, 68(2):367–397, 1994.

EVALUATION OF ONTOLOGY MERGING TOOLS
IN BIOINFORMATICS

P. LAMBRIX, A. EDBERG

Department of Computer and Information Science
Linköpings universitet, 581 83 Linköping, Sweden

Ontologies are being used nowadays in many areas, including bioinformatics. One
of the issues in ontology research is the aligning and merging of ontologies. Tools
have been developed for ontology merging, but they have not been evaluated for
their use in bioinformatics. In this paper we evaluate two of the most well-known
ontology merging tools with a bioinformatics perspective. As test ontologies we
have used Gene Ontology and Signal-Ontology.

1 Introduction

Ontologies define the basic terms and relations comprising the vocabulary of
a topic area, as well as the rules for combining terms and relations to define
extensions to the vocabulary. They are being used nowadays in many areas,
including bioinformatics, for several reasons. They are used for communication
between people and organizations by providing a common terminology over a
domain. They provide the basis for interoperability between systems. They can
be used for content explicitation for information sources and serve as an index
to a repository of information. They can also be used as a basis for integration
of information sources and as a query model for information sources. To assist
users in developing and maintaining ontologies a number of tools have been
developed and comparative studies of ontological engineering tools have been
performed. In [DSWKB99] Ontolingua, WebOnto, ProtégéWin, OntoSaurus,
ODE and KADS22 were compared. The authors used ontologies concerning
academia and university studies for testing. In [HP02] we used Gene Ontology to compare the basic functionality of the ontology tools Protégé-2000,
Chimaera, DAGEdit and OilEd.

Within the bioinformatics area there are a number of bio-ontologies, each
with their own focus, that cover different aspects in molecular biology such
as molecular function and cell signaling. Many of these ontologies contain
overlapping information. For instance, a protein can be involved in both cell
signaling and other biological processes. In applications using ontologies it is
therefore of interest to be able to use multiple ontologies. However, to obtain
the best results, we need to know how the ontologies overlap and align them or
merge them into a new ontology. Another reason for merging ontologies is that

it allows for the creation of ontologies that can later be composed into larger ontologies. Also, companies may want to use de facto standard ontologies and merge them with company-specific ontologies. There exist a number of tools for merging ontologies that have been used outside the bioinformatics area. In this paper we evaluate how well these tools work for merging bio-ontologies. We have chosen to evaluate Protégé-2000 with PROMPT and Chimaera, which are, currently, two of the main ontology merging tools. We describe these tools in section 3. We looked at a number of bio-ontologies and decided to use two ontologies for testing: Gene Ontology, which has become a de facto standard and Signal-Ontology. The overlap between these ontologies gives interesting test cases. In section 2 we briefly describe these ontologies. In section 4 we discuss the evaluation criteria, the evaluation method and the set-up of the evaluation. Section 5 discusses, because of space limitations, a high-level view of the results of the evaluation. More details can be found in [Edb02].

2   Ontologies

2.1   Gene Ontology

The Gene Ontology (GO) Consortium is a joint project with as goal to produce a structured, precisely defined, common and dynamic controlled vocabulary that describes the roles of genes and proteins in all organisms [GO]. Currently, there are three independent ontologies publicly available over the Internet: biological process, molecular function and cellular component. The biological process ontology deals with biological objectives to which the gene or gene product contribute. A process is accomplished via one or more ordered assemblies of molecular functions. The molecular function ontology deals with the biochemical activities of a gene product. It only describes what is done without specifying where or when the event takes place. The cellular component ontology describes the places where a gene product can be active. The GO ontologies are becoming a de facto standard and many different bio-databases are today annotated with GO terms. The ontologies grow continuously. The terms in GO are arranged as nodes in a directed acyclic graph, where multiple inheritance is allowed.

2.2   Signal-Ontology

The purpose of the Signal-Ontology (SO) [SO] project is to extract common features of cell signaling in the model organisms, try to understand what cell signaling is and how cell signaling systems can be modeled. SO is a publicly available controlled vocabulary of the cell signaling system. It is based on the knowledge of the Cell Signaling Networks databank [TNK98] and treats complex knowledge of living cells such as pathways, networks and causal relationships among molecules [TT00]. SO consists of four different conceptual classes: signal motif, reaction, signal family and cellular function. In the signal motif class the units of the signaling network that are conserved through species are defined. In the reaction class the biochemical properties of individual molecular reactions in the signaling pathways are defined. The signal family class provides a functional classification of signal molecules and the cellular function class classifies biological functions controlled by the cell signaling system. SO is composed of signal modules that are minimal units of signal processing common to the model species. Each module models a concept extracted from biological information and constitutes a unit in the signal transduction. The ontology consists of a flow diagram of signal transduction and a conceptual hierarchy of biochemical attributes of signaling molecules.

3 Tools for merging ontologies

3.1 Protégé-2000 with PROMPT

Protégé-2000 is software for creating, editing and browsing ontologies developed by Stanford Medical Informatics. The design and development of Protégé-2000 has been driven by two goals: to be compatible with other systems for knowledge representation and to be an easy to use and configurable tool for knowledge extraction. Protégé-2000 is available as free software and should be installed locally. It also has a number of plug-ins, among others PROMPT, which is an algorithm for merging and aligning ontologies [PROMPT, NM00].

When merging two ontologies PROMPT creates a list of suggested operations. An operation can, for instance, be to merge two terms or to copy a term to the new ontology. The user can then perform an operation by choosing one of the suggestions or by specifying an operation directly. PROMPT performs the chosen operation and additional changes that follow from that operation. The list of suggestions is then updated and a list of conflicts and possible solutions to these conflicts is created. This is repeated until the new ontology is ready. PROMPT was previously called SMART and a high-level description of the algorithm is given in [NM99].

3.2 Chimaera

Chimaera is developed by the Knowledge Systems Laboratory at Stanford University to provide assistance to users for browsing, editing, merging and diagnosing of ontologies. It was built on top of the Ontolingua Distributed Collaborative Ontology Environment. The initial goal was to develop a tool that could give substantial assistance for the task of merging knowledge bases produced by different users for different purposes with different assumptions and different vocabulary. Later the goals of supporting testing and diagnosing ontologies arose as well. The user interacts with Chimaera through a browser such as Netscape or Microsoft Internet Explorer [MFRW00].

The two main tasks when merging two ontologies in Chimaera are to merge two semantically identical terms from different ontologies so that they are referred to by the same name in the resulting ontology, and to identify terms that should be related via the is-a, disjointness or instance relationships and provide support for introducing those relationships. Chimaera also supports the identification of the locations for editing and performing the edits. Today, Chimaera has support for merging taxonomies of terms and for merging attributes. The support for merging relations, functions, facets, individuals and axioms will be developed later. To assist the user Chimaera generates name resolution lists that suggest terms that are candidates to be merged or to have taxonomic relationships not yet included in the merged ontology. Chimaera also generates a taxonomy resolution list where it suggests taxonomy areas that are candidates for reorganization. On the basis of these lists the user decides what should be done.

4   Evaluation Methods

In this section we present the evaluation of the ontology merging tools Protégé-2000 with PROMPT and Chimaera. We only discuss the systems in terms of their support for merging ontologies in some detail. We have also studied the systems in a more general study of ontology tools and refer to [HP02] for detailed results concerning the other functionality of these systems.

4.1   Methods

The evaluation of the tools has been divided into two parts. In the first part we evaluated the systems based on a number of predefined criteria. The evaluation was done through a literature study and tests using GO and SO. The criteria that were evaluated are presented in section 4.3.

In the second part of the evaluation the user interface of the systems was evaluated. For this we used an empirical approach. In this technique end users test the systems. This approach is usually time-consuming and requires the use of test persons, but has the advantage that the problems that are discovered are real problems. The use of this method requires the development of good instructions for the test persons. In our evaluation eight test persons participated, four with a computer science background and four with a biology background. The biologists were used to work with computers. They had no experience of working with ontologies. As evaluation method we used the

REAL method [Löw93]. This method was chosen for its ease of use, because it requires few test persons and it usually gives good results. Before the actual evaluation the test persons were given information on ontologies, what they are, how they can be represented, what they are used for etc. Then the test persons familiarized themselves with the tools using a tutorial and by doing another evaluation [HP02]. This took about 1.5 to 2.5 hours per person. The actual evaluation for each system was divided into two parts. During the first part the test persons were given a number of tasks to perform. They had a manual on paper and the system help (when available) for support. They were asked to think loud and an evaluator took notes during the process. Afterwards, the test persons were asked to fill in a questionnaire. The questionnaire consisted of questions with coded values. The test persons were also asked to write comments on their grading. In total this took about 1.5 to 2.5 hours per person.

4.2 Description of test ontologies

In our tests two chosen 'cases' from GO and SO were used. Each case consists of one part of SO and one part of GO. Each case was chosen in such a way that there was an overlap between the GO part and the SO part. The first case, *behavior*, contains approximately 60 terms from GO and approximately 10 terms from SO. The second case, *immune defense*, contains approximately 70 terms from GO and approximately 15 terms from SO. A complete listing of the terms that are included can be found in [Edb02]. We used more terms from GO than from SO because the granularity of GO is higher than the granularity of SO for these topics. The terms in SO and GO include concepts, attributes and instances. The concepts are arranged in a hierarchy. Concepts may have attributes. The attributes are not hierarchically organized. Each concept may also have instances.

4.3 Evaluation criteria

Our criteria for the first part of the evaluation are partly based on previous work on evaluating ontology tools [DSWKB99] and ontology exchange languages for bioinformatics [McE00]. We investigated the availability of the tool. We tested the systems during a period of time and checked their stability. Further, we looked at the representation language that is used by the tools. For the merging we evaluated the tools with respect to the functionality they provide. This deals with issues such as what kind of ontologies can be merged and whether consistency checking is performed. We also looked at the assistance that is given to the user while merging. Both tools generate suggestions for merging. We looked at how the suggestions are generated and

594

evaluated their quality measuring precision and recall using our test ontologies. The precision measures how many of the suggestions are relevant, while recall measures how many of the total number of the relevant suggestions the systems actually proposed. Further, we also measured the time it took to merge the test ontologies using the tools.

The user interface was evaluated using the REAL approach. The aspects studied in this evaluation are Relevance, Efficiency, Attitude and Learnability. Relevance measures how well a user's needs are satisfied by the tool. Efficiency measures how fast users can perform their tasks using the tool. The subjective feelings towards the tools are measured by attitude. Finally, learnability measures how easy or difficult it is to learn the tool for initial use as well as how much a user can remember the workings of the tool.

5 Evaluation Results and Discussion

5.1 General

Protégé-2000 with PROMPT is available as free Java software and is installed locally. Chimaera is available over the Internet and requires a relatively fast connection to be able to work efficiently. With high use of the network even simple operations can take a long time to perform.

Protégé-2000 with PROMPT is not entirely stable. During about 35 hours (excluding user interface evaluation time) of work the Protégé-2000 program had to be terminated in an abnormal way five times. Together with PROMPT this happened about half of the times we worked with merging. During about 10 hours (excluding user interface evaluation time) of work with Chimaera, the system has been stable. On one occasion the server at Stanford was not accessible.

Both Protégé-2000 with PROMPT and Chimaera allow import and export of ontologies in different formats. Protégé-2000 also allows user-programmed plug-ins to import and export to other formats. The internal representations in both systems are based on the Open Knowledge Base Connectivity (OKBC), which is an application programming interface for frame-based knowledge representation systems [CFFKR98]. The internal language of Protégé-2000 is less expressive than the language of Chimaera. Both are frame based and are expressive enough to represent most of the current bio-ontologies, including our test ontologies.

5.2 Merging

Protégé-2000 with PROMPT allows for merging and copying concepts, attributes and instances from two ontologies to a new ontology. PROMPT generates suggestions for merge operations on concept and attribute levels together

Table 1: Quality of suggestions

Tool	Case	Suggestions	Correct	Missing	Recall	Precision
PROMPT	B	3	3	2	0.6	1
Chimaera	B	16	4	1	0.8	0.25
PROMPT	ID	4	4	5	0.44	1
Chimaera	ID	10	4	5	0.44	0.4

with an explanation of why these suggestions were made. When the user decides to follow the suggestion, also a number of other changes (logically derived) are made automatically and new suggestions may be derived too. PROMPT also identifies possible conflicts that could occur as a result of the merging and proposes possible solutions. Concepts in the original ontologies that are not merged need to be copied into the new ontology.

The generation of suggestions is based on similarities in the concept and attribute names, but in [NM99] the authors suggest that the similarity measure is not hard-coded and that users may define their own similarity measures. For the tests, however, we used the system as it is provided in the distribution. Extensions to the basic algorithm are being made, such as Anchor-PROMPT [NM01] where paths between merged concepts in the two ontologies are investigated to find new suggestions.

Chimaera provides about seventy commands in the user interface, thereby providing a taxonomy and slot editor. The applicable commands at each point in time are made available by the interface. Some of these commands are related to ontology merging such as 'merge classes' and 'move class x to become a subclass of class y'. There are also commands related to diagnosis that, among others, check for incompleteness, cycles and value-type mismatches.

As a help to the user, Chimaera generates a list of concepts and attributes that are candidates for merging. The generation of this list is based on similarities in names, definitions, acronyms, name extensions, names that appear as suffixes in other names etc. Also a list with suggestions for areas that may need restructuring is generated based on heuristics. For instance, one heuristic looks for concepts that have direct sub-concepts in both ontologies and suggest these as restructuring areas [MFRW00]. A difference between PROMPT and Chimaera is that Chimaera suggests concepts and places *where* actions may need to be taken, while PROMPT also suggests *what* actions should be taken.

In table 1 we show the results regarding precision and recall for the cases behavior (B) and immune defense (ID), respectively. The 'suggestions' column informs about the number of suggestions that were made. The 'correct'

Table 2: Time (in minutes) for merging

Tool	Case	Merging based on suggestions	Additional time	Total
PROMPT	B			15
Chimaera	B	8	0	8
PROMPT	ID			26
Chimaera	ID	10	5	15

column shows how many of the suggestions were relevant while the 'missing' column shows how many of the relevant suggestions were missing. In total there were 5 possible cases for merging in behavior and 9 possible cases in immune defense. PROMPT got perfect precision in both test cases, while Chimaera's precision was below 50%. However, PROMPT gave much fewer suggestions than Chimaera. Chimaera's recall was higher than PROMPT's recall, but both performed relatively poorly on immune defense. Examples of concepts to merge that were missing were: 'inflammatory response' and 'inflammation', 'antigen processing and presentation' and 'antigen presentation', and 'complement activation' and 'complement signaling'. The last pair would have been found if we compiled a synonym list based on the synomyms given in GO and SO. The two other pairs might have been found with less restrict string matching.

In table 2 we show the results regarding the time it took to merge the test ontologies for the cases behavior (B) and immune defense (ID), respectively. For Chimaera we calculated the time to merge the ontologies based on the suggestions and the additional time it took to merge the concepts and attributes that were not suggested by the program. For PROMPT we only calculated the complete time as much of the work was spent on copying the concepts and attributes that did not need to be merged. It is clear that merging was much faster with Chimaera than with PROMPT. This may be a key factor when we want to merge large ontologies.

5.3 User interface

The results from the questionnaire are found in tables 3 and 4. In this section we summarize these results as well as the observations reported from the comments the test persons gave during the evaluation.

Relevance The test persons felt it was better to use PROMPT than Chimaera for the merging of ontologies. This depended, among others, on the fact that it takes relatively long time to get a response for some operations in Chimaera.

Table 3: Evaluation of graphical user interface for Protégé-2000 with PROMPT

Question	Mean	Min	Max
Relevance			
To use PROMPT for merging ontologies feels (bad=1, good=10)	8.25	3	10
Efficiency			
To merge ontologies with PROMPT feels (difficult=1, easy=10)	5.50	3	8
To choose wanted operation feels (difficult=1, easy=10)	6.63	3	10
The overview over the ontologies you wish to merge feels (bad=1, good=10)	8.50	5	10
The overview over the terms, attributes and instances you merge feels (bad=1, good=10)	7.13	4	9
Attitude			
To merge ontologies with PROMPT seems (boring=1, fun=10)	5.88	1	10
To decide whether two terms, attributes or instances should be merged or not seems (difficult=1, easy=10)	7.63	6	9
The naming of the operations is (unclear=1, clear=10)	8	4	10
The naming feels (inconsistent=1, consistent=10)	8.50	6	10
Learnability			
To learn how to merge ontologies seems (difficult=1, easy=10)	6.25	4	9
The available help feels (insufficient=1, sufficient=10)	5.63	2	8
To learn how to load the ontologies you wish to merge feels (difficult=1, easy=10)	8.25	5	10
To learn how to merge terms, attributes and instances feels (difficult=1, easy=10)	7	4	10
To learn how to copy the terms, attributes and instances that you do not wish to merge, feels (difficult=1, easy=10)	7	4	10

Efficiency The test persons found that it was better to use PROMPT than Chimaera for specific operations and the fact that some operations with Chimaera took a long time was one of the reasons for the lower score. However, they also thought that merging ontologies required too much work with PROMPT. In particular, the need to remember and copy all the concepts, attributes and instances that were not merged was considered awkward. The overview over the ontologies in the tool was considered to be better in PROMPT than in Chimaera. In particular, the fact that the original ontologies were kept and represented in different colors in PROMPT was considered good. In Chimaera it was difficult to see where in the hierarchy a concept was situated. The choice of operations to perform was easier in PROMPT than in Chimaera. This may be because Chimaera allows for many more operations than PROMPT,

598

Table 4: Evaluation of graphical user interface for Chimaera

Question	Mean	Min	Max
Relevance			
To use Chimaera for merging ontologies feels (bad=1, good=10)	6.25	2	10
Efficiency			
To merge ontologies with Chimaera feels (difficult=1, easy=10)	5.25	2	9
To choose wanted operation feels (difficult=1, easy=10)	6	1	10
The overview over the ontologies you wish to merge feels (bad=1, good=10)	5.25	2	9
The overview over the terms and attributes you merge feels (bad=1, good=10)	5.63	2	8
Attitude			
To merge ontologies with Chimaera seems (boring=1, fun=10)	3.88	1	8
To decide whether two terms or attributes should be merged or not seems (difficult=1, easy=10)	5.75	5	9
The naming of the operations is (unclear=1, clear=10)	5.75	1	10
The naming feels (inconsistent=1, consistent=10)	7.63	6	10
Learnability			
To learn how to merge ontologies seems (difficult=1, easy=10)	6.63	3	10
The available help feels (insufficient=1, sufficient=10)	6.50	5	10
To learn how to load the ontologies you wish to merge feels (difficult=1, easy=10)	6.63	2	10
To learn how to merge terms, attributes and instances feels (difficult=1, easy=10)	7	2	10

which may make it harder to find and choose the right operation.

Attitude The test persons found it was more fun to use PROMPT than Chimaera. Chimaera's graphical user interface was considered to be boring and unclear. Also, it was felt that the names of the operations were more self-explaining in PROMPT than in Chimaera.

Learnability To learn how to merge ontologies felt equally hard for both systems. The available help for Chimaera was considered better than the help for PROMPT. The hardest to learn was how to copy the concepts, attributes and instances that are not merged in PROMPT.

5.4   Method critique

Although we have some experience with ontologies, we are not the common end users of ontology merging tools. This may have an influence on the results of the first part of the evaluation. However, we have tried to minimize this influence by studying other evaluations, learning to work with the tools and clearly defining the criteria before the actual evaluation.

The test persons in the user interface evaluation were novices in using ontologies and ontology merging. Therefore, there may have been an influence on the way the test persons performed the given tasks. In particular, there may have been a larger influence for the results of Chimaera than for PROMPT as the choice of operations in Chimaera is significantly larger.

We were also concerned about the fact that doing a task in one tool may make it easier to perform a similar task in the next tool. Therefore, to minimize the effects on the results of having done a similar task before we varied the order in which the different test persons evaluated the two tools.

As we wanted to investigate whether a test person's knowledge of biology would influence the results in the user interface evaluation, we took test persons that work daily with biology and persons that have only high school knowledge of biology. However, there was no significant difference between the results of these two groups.

6   Conclusion

In this paper we have evaluated Protégé-2000 with PROMPT and Chimaera for their use as ontology merging tools in bioinformatics. We used two cases based on Gene Ontology and Signal-Ontology as test cases. Both systems are expressive enough to model most of the current bio-ontologies. The user interface of Protégé-2000 with PROMPT was considered better than Chimaera. It gave a better overview over the ontologies and it was easier to work with. Chimaera, however, provides more functionality and better help. It is also much faster to merge ontologies with Chimaera. This may be an important factor when merging large ontologies. The quality of the suggestions could be improved for both systems and work is ongoing by the developers. We actually implemented our own algorithm for merging using synonym lists and less restricted string matching and got better results for precision and recall [Edb02]. Another issue that may be taken into consideration are the results of [HP02] where it is shown that Protégé-2000 with PROMPT has characteristics that make it appropriate to use in the early phases of ontology creation and Chimaera is a good choice for later phases such as analysis and maintenance. As the tools allow for some common import languages, using both tools may actually be an option.

References

[CFFKR98] Chaudhri, V., Farquhar, A., Fikes, R., Karp, P., Rice, J., 'OKBC: A Programmatic Foundation for Knowledge Base Interoperability', *Proceedings of*

600

the Fifth National Conference on Artificial Intelligence, pp 600-607, Madison, WI, USA, 1998.

[Chimaera] http://www.ksl.stanford.edu/software/chimaera/

[DSWKB99] Duineveld, A.J., Stoter, R., Weiden, M.R., Kenepa, B., Benjamins, V.R., 'Wondertools? A comparative study of ontological engineering tools', *Proceedings of the 12th Workshop on Knowledge Acquisition, Modeling and Management*, pp 4.6.1-4.6.20, Banff, Canada, 1999.

[Edb02] Edberg, A., *Ontology merging in bioinformatics*, M.Sc. thesis LiTH-IDA-Ex-02/62, Department of computer and information science, Linköpings universitet, 2002. In Swedish.

[GO] The Gene Ontology Consortium, 'Gene Ontology: tool for the unification of biology', *Nature Genetics*, 25:25-29, 2000. http://www.geneontology.org/

[HP02] Habbouche, M., Pérez, M., *Evaluation of ontology tools for bioinformatics*, B.Sc. thesis LiTH-IDA-Ex-Ing-02/10, Department of computer and information science, Linköpings universitet, 2002. In Swedish.

[Löw93] Löwgren, J., *Human-Computer Interaction. What every system developer should know.* Studentlitteratur, Lund, 1993.

[NM99] Noy, N.F., Musen, M., 'SMART: Automated Support for Ontology Merging and Alignment', *Proceedings of the 12th Workshop on Knowledge Acquisition, Modeling and Management*, Banff, Canada, 1999.

[NM00] Noy, N.F., Musen, M., 'PROMPT: Algorithm and Tool for Automated Ontology Merging and Alignment', *Proceedings of Seventeenth National Conference on Artificial Intelligence*, pp 450-455, Austin, TX, USA, 2000.

[NM01] Noy, N.F., Musen, M., 'Anchor-PROMPT: Using Non-Local Context for Semantic Matching', *Proceedings of the IJCAI01 Workshop on Ontologies and Information Sharing*, 2001.

[McE00] McEntire, R., Karp, P., Abernethy, N., Benton, D., Helt, G., DeJongh, M., Kent, R., Kosky, A., Lewis, S., Hodnett, D., Neumann, E., Olken, F., Pathak, D., Tarczy-Hornoch, P., Toldo, L., Topaloglou, T., 'An Evaluation of Ontology Exchange Languages for Bioinformatics', *Proceedings of the Eight International Conference on Intelligent Systems for Molecular Biology*, pp 239-250, San Diego, California, 2000.

[MFRW00] McGuinness D., Fikes R., Rice J., Wilder S., 'An Environment for Merging and Testing Large Ontologies', *Proceedings of the Seventh International Conference on Principles of Knowledge Representation and Reasoning*, pp 483-493, Breckenridge, CO, USA, 2000.

[PROMPT] Protégé (http://protege.stanford.edu/index.html) with PROMPT (http://protege.stanford.edu/plugins/prompt/prompt.html)

[SO] Signal-Ontology, http://ontology.ims.u-tokyo.ac.jp/signalontology/

[TNK98] Takai-Igarashi, T., Nadaoka, Y., Kaminuma, T., 'A Database for Cell Signaling Networks', *Journal of Computational Biology* 5(4):747-754, 1998. CSNDB: http://geo.nihs.go.jp/csndb/

[TT00] Takai-Igarashi, T., Takagi, T., 'Cell Signaling Ontology', *Third Annual Bio-Ontologies Meeting*, La Jolla, CA, USA, 2000.

SEMANTIC SIMILARITY MEASURES AS TOOLS FOR EXPLORING THE GENE ONTOLOGY

P.W.LORD[a], R.D. STEVENS, A. BRASS AND C.A.GOBLE

Department of Computer Science
University of Manchester
Oxford Road
Manchester
M13 9PL
UK

`p.lord@russet.org.uk`
`robert.stevens@cs.man.ac.uk`
`abrass@man.ac.uk`
`carole@cs.man.ac.uk`

Abstract

Many bioinformatics resources hold data in the form of sequences. Often this sequence data is associated with a large amount of annotation. In many cases this data has been hard to model, and has been represented as scientific natural language, which is not readily computationally amenable. The development of the Gene Ontology provides us with a more accessible representation of some of this data. However it is not clear how this data can best be searched, or queried. Recently we have adapted information content based measures for use with the Gene Ontology (GO). In this paper we present detailed investigation of the properties of these measures, and examine various properties of GO, which may have implications for its future design.

1 Introduction

Historically bioinformatics has largely grown out of efforts to deal with the increasingly large amount of data produced by molecular biology in the form of protein or DNA sequences. This sort of data can be modelled straightforwardly as a list of characters, and then searched, stored, and manipulated computationally.

During the development of the many repositories that store this data, a large amount of "annotation" has been associated with these sequences. This ranges from semi-structured data, such as species information, to unstructured free text descriptions. Often there is a large amount of annotation. Although,

---

[a]To whom correspondence should be addressed

for example, SWISS-PROT is often described as a protein sequence database, it could also be considered to be a protein annotation database.

This has served the community well in the past, when the annotation was meant for humans to read. However it causes difficulties when trying to analyse the annotation computationally for the purpose, for example, of summarising many different SWISS-PROT entries comprising a protein family.[1] While the text is accessible by computer applications, it is not easy to interpret computationally.

It is partly because of these difficulties that there has been growing interest in ontologies within bioinformatics.[2] They provide a mechanism for capturing a community's view of a domain in a shareable form, that is accessible by humans and also computationally amenable. An ontology provides a set of vocabulary terms that label domain concepts. These terms should have definitions and be placed within a structure of relationships, the most important being the "is-a" relationship between *parent* and *child* and the "part-of" relationship between *part* and *whole*.[3,4]

The Gene Ontology (GO) is one of the most important ontologies within the bioinformatics community.[5] It is specifically intended for the purpose of extending free text annotation commonly found, with ontological annotation. As the name suggests it is limited to annotation of gene products. It comprises three orthogonal taxonomies or "aspects", that hold terms describing the *molecular function, biological process,* and *cellular component* for a gene product. GO is rapidly growing having over 11 000 terms (as of April 2002). Additionally new ontologies covering other regions of biology are being developed.[b]

GO represents terms within a Directed Acyclic Graph (DAG) consisting of a number of terms, represented as nodes within the graph, connected by relationships, represented as edges. Terms can have multiple parents, as well as multiple children along the "is-a' relationships ("photoreceptor" and "transmembrane receptor" are children of "receptor"), together with part-of relations that describe, for instance, that "mitochondrial membrane" is part of "mitochondrion".

The terms held within this structure are used to annotate database entries.[c] For example, the SWISS-PROT protein, OPSR_HUMAN, has the molecular function annotation of "red-sensitive opsin", (GO:0015061). By providing a standard vocabulary across many biological resources such as SWISS-PROT and InterPro, this shared understanding should enable querying across these databases. One obvious way to query these databases would be to ask

---

[b]http://www.geneontology.org/doc/gobo.html
[c]http://www.geneontology.org/goa

for proteins which are *semantically similar* to a query protein.

In a previous paper[6] we adapted an existing measure for semantic similarity for use with GO. This measure was based on the *information content*, which uses the notion that the less frequently used terms are more informative. We tested this measure by analysing semantic similarity, and correlating it with sequence similarity, showing that, as would be expected, the more closely similar two sequences are, the more similar their ontological annotation is. We also demonstrated how this measure could be used as the basis for a simple search tool, operating over the ontological annotation.

In this paper we extend our analysis to two different methods for measuring semantic similarity, again validating the measures against sequence similarity. We also use these measures to investigate the annotation from different aspects of GO. We discuss the implications that these results have for future development of a search tool, and speculate on the implications this may have for future development of the Gene Ontology.

2 Semantic Similarity Measures

All of the measurements used here are based on the information content of each term. This is defined as the number of times each term, or any child term, occurs in the corpus. This is expressed as a probability. Although there are several available corpora we have limited our analysis to SWISS-PROT-Human (see Section 3). It is possible to interpret "child term" in a number of ways, by considering links of all semantic type, or just is-a inks. In this paper all links are used, as the distribution of link types across the different aspects, differ widely (molecular function, 6207 is-a's to 35 part-of's, cellular component, 542 to 619, biological process, 5697 to 989).

In Figure 1 these probabilities are shown diagrammatically for a small section of GO. From the definition, we can guarantee that the information content of each node increases monotonically toward the root node, which will have an information content of 1. As the three aspects of GO are disconnected subgraphs, this also holds true if we ignore the top level node ("Gene Ontology", (GO:0003683)), and take, for example, "molecular function", (GO:0003674) as our root node instead.

Given these probabilities, there are several measures of semantic similarity.[7,8,9] All three of these measures use the information content of the shared parents of the two terms, as defined in Equation (1), where $S(c1, c2)$ is the set of parental concepts shared by both $c1$ and $c2$. As GO allows multiple parents for each concept, two terms can share parents by multiple paths. We take the minimum $p(c)$, where there is more than one shared parent. We call this p_{ms}

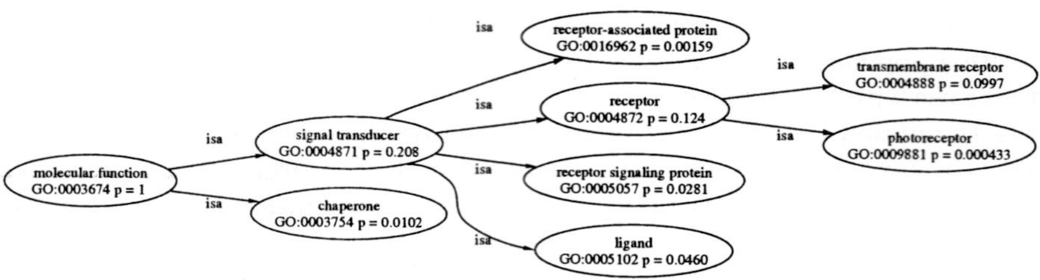

Figure 1: Probabilities in the Gene Ontology. Each node is annotated with its GO accession and the probability of this term occurring in the SWISS-PROT-Human database. This figure was produced from GO, using the graphviz tools (http://www.graphviz.org).

for *probability of the mimimum subsumer.*

$$p_{ms}(c1, c2) = \min_{c \in S(c1,c2)} \{p(c)\} \tag{1}$$

The first of the three measures shown in Equation (2), is after Resnik, [7] and uses only the information content of the shared parents. As p_{ms} can, in general, vary between 0 and 1, this measure varies between infinity (for very similar concepts) to 0. In practise, for terms actually present in the corpus, the maximum value of this measure is defined by $-\ln(1/t) = \ln(t)$ where t is the number of occurrences of any term in the corpus.

$$\mathrm{sim}(c1, c2) = -\ln p_{ms}(c1, c2) \tag{2}$$

The next measure, after Lin, [9] uses both the information content of the shared parents, and that of the query terms. In this case, as $p_{ms} \geq p(c1)$ and $p_{ms} \geq p(c2)$, this value varies between 1 (for similar concepts) and 0.

$$\mathrm{sim}(c1, c2) = \frac{2 \times [\ln p_{ms}(c1, c2)]}{\ln p(c1) + \ln p(c2)} \tag{3}$$

The final measure, after Jiang, [8] shown in Equation (4), is of semantic distance, which is the inverse of similarity. It uses all the same terms as Equation (3), but not in the same order. [8] As with Equation (2) this can give arbitrarily large values although in practice has a maximum value of $2\ln(t)$.

$$\mathrm{dist}(c1, c2) = -2\ln p_{ms}(c1, c2) - (\ln p(c1) + \ln p(c2)) \tag{4}$$

For our purposes we are most interested in the semantic similarity between proteins, rather than GO terms *per se*, so we needed to combine these measures when a protein was annotated with several terms. In previous work, based on WordNet,[10] a similar problem was found, as individual words have more than a single sense.[11] In this case the maximum similarity between the word senses was taken, as generally only a single word sense is used at a time. With GO annotated gene products this is not the case. A gene product will generally have all of the roles attributed to it by the annotators at the same time. We have therefore taken the average similarity between all terms. For this paper only those terms with evidence codes of "Traceable Author Statement"[d] have been used. With this dataset, most proteins have been annotated with a single term from each aspect, so values are rarely combined in this way.

3   Implementation

All results shown are from analysis performed on the April 2002 release of GO database available from `http://www.godatabase.org/dev`. The perl API available from the same source was used as an interface to this database, running over a MySQL RDBMS. The work was limited to those associations between GO terms, and SWISS-PROT proteins. In this paper SWISS-PROT-Human refers to those proteins in SWISS-PROT for which GO annotations were available, which, at the time of writing was limited to approximately 7 000 human proteins. Only those associations with "Traceable Author Statement" tags were used. The semantic similarity measures were implemented using a perl library developed for this work. All software is available on request.

BLAST searches were performed using local copies of the NCBI BLAST program over the complete SWISS-PROT protein database, available from the NCBI FTP site (`http://www.ncbi.nlm.nih.gov/BLAST/`). An "expect" value of 100 was used for all searches. Self matches, usually the best match for any protein, were excluded from the analysis. Searches were launched using the "bioperl" API (`http://www.bioperl.org`).

Results shown in Figure 2(a) and similar figures was analysed from the raw data, by taking "slices" down the X axis (ln[bit score]), and calculating the average values at each point. Scripts were written in perl, and results displayed using gnuplot (`http://www.gnuplot.info`).

Correlation is calculated as shown in Equation (5), where x_i and y_i are the semantic similarity between two proteins, over different aspects of GO, for all possible pairs of proteins in the SWISS-PROT-Human dataset.

$$\mathrm{corr}(x, y) = \frac{\sum (x_i - \bar{x})(y_i - \bar{y})}{\sqrt{\sum (x_i - \bar{x})^2 \sum (y_i - \bar{y})^2}} \tag{5}$$

---

[d]see `http://www.geneontology.org/doc/GO.Evidence.html`

4 Similarity Measures Over Different Aspects

In previous work,[6] we investigated the semantic similarity measure described in Equation (2). In order to validate that this measure was producing appropriate results we compared them to sequence similarity. Results of this comparison are shown in Figure 2(a). Combined with a correlation coefficient measure it appears that sequence similarity is strongly correlated with semantic similarity based on the "molecular function" aspect of GO. This fits with the biological expectations. The sequence of a protein determines its molecular function, but does not necessarily relate to the biological process that it is involved in, or its cellular localisation.

We therefore extended this analysis to the other sequence similarity measures given in Section 2. The results of this analysis are shown in Figure 2, and Table 1.

Aspect	Resnik	Lin	Jiang
Molecular Function	0.577	0.541	-0.483
Biological Process	0.280	0.303	-0.312
Cellular Component	0.368	0.452	-0.414

Table 1: Correlation co-efficients between BLAST bit scores, and semantic similarity. Data was calculated as for Figure 2. Correlation co-efficients were calculated for each data set.

For all three measures the correlation coefficients show that sequence similarity is most tightly correlated (or in the case of the distance measure inversely correlated) with the Molecular Function aspect of GO, followed by the Cellular Component aspect, and finally the Biological Process aspect. Of the three, the measure after Resnik, shows the strongest correlation with sequence similarity. Interestingly this measure also provides the weakest correlation against the biological process aspect. This suggests that the Resnik measure may be the most discriminatory. However there is no *a priori* reason to suspect that relationships will be linear, which may affect the correlation co-efficients. By inspection of Figure 2, it appears that the Resnik measure is the more linear.

5 Correlating Aspects

As discussed previously, GO is split into three different aspects. In terms of the GO DAG, although these three aspects are collected under a single top level term, "Gene Ontology", (GO:0003673), they are entirely orthogonal, being disconnected subgraphs. This part of the design of GO is justified, because these aspects "are all attributes of genes[...]. Each of these may be assigned

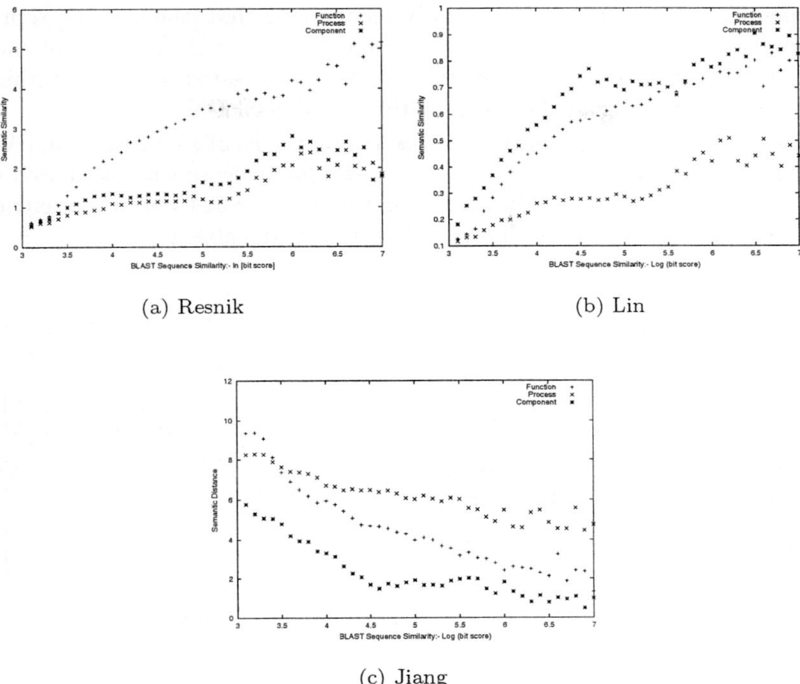

(a) Resnik

(b) Lin

(c) Jiang

Figure 2: Comparing sequence and semantic similarity. BLAST searches were performed for each SWISS-PROT-Human protein, and all matches analysed for semantic similarity with the search protein. Intervals were taken along the x-axis, $ln[bitscore]$, and values averaged, see Section 3 for details.

independently".[5] Furthermore "simply recognizing that [the aspects] represent independent attributes is by itself clarifying". Although in terms of the GO DAG these aspects are independant, we were interested in whether this was also true of the usage of the terms within SWISS-PROT.

To test this, we performed pairwise comparisons of all proteins in SWISS-PROT-Human. For each pair a semantic similarity score was calculated using each of the three aspects. The individual pairs of scores were then extracted, and compared. Results are shown in Figure 3 and Table 1.

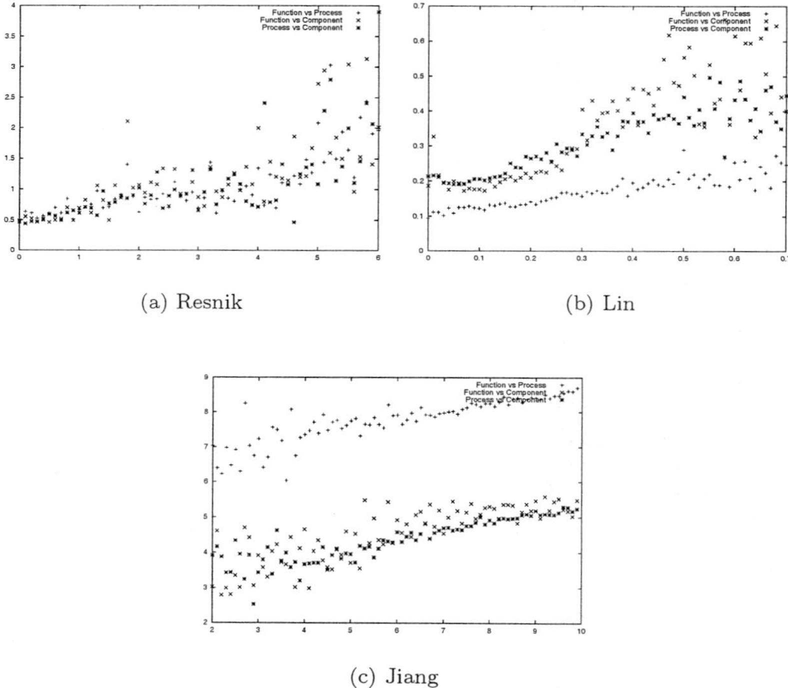

(a) Resnik

(b) Lin

(c) Jiang

Figure 3: Comparing semantic similarity over aspects of GO. Pairwise comparisons of semantic similarity over all three aspects of GO, and for all proteins in SWISS-PROT-Human were performed. Results were split into pairs, and averaged as in Figure 2.

It is clear that for all the measures there is a significant, but weak correlation between all three of aspects. For the two similarity measurements the ordering of the correlation is conserved (molecular function to cellular com-

Aspects	Resnik	Lin	Jiang
Molecular Function - Cellular Component	0.290	0.318	0.0877
Molecular Function - Biological Process	0.219	0.244	0.269
Biological Process - Cellular Component	0.202	0.175	0.166

Table 2: Correlation coefficients for semantic similarity scores over different aspects of GO. Data was collected as for Figure 3, and correlation coefficients calculated for each data set.

ponent, molecular function to biological process, biological process to cellular component). This order is different for the distance measure after Jiang (Table 2). We are unclear whether the unexpectedly low correlation observed with this measure for the molecular function versus cellular component aspects is either because the Jiang measure is one of distance rather than similarity, or because of its behaviour over small ontologies (the cellular component ontology is about 1/5 the size of the other two aspects).

It therefore appears that while the use of the three aspects of GO is not, in fact, independent, the correlation between their use is quite weak.

6 Discussion

One of the obvious uses for these semantic similarity measures is in the development of a "semantic search" tool. A previous study compared these measures, in a different context, and found that the measure after Jiang, gave the best results. [12]

The results presented in Section 4, suggest that all three of the measures show a strong correlation between sequence similarity and molecular function semantic similiarity. The Resnik measure shows the highest correlation, as well as having the lowest correlation for the other two aspects, so it may be the most discriminatory. Further, it provides us with more information. Results are bounded between 0 and $\ln(t)$, where t is the number of terms in the corpus, while the Lin measure is bounded between 0 and 1. A large numerical value therefore indicates a large corpus. The numerical value also reveals information about the usage within corpus of the part of the ontology queried. The score from comparing a term with itself depends on where in the ontology the term is, with less frequently occurring terms having higher scores.

The Lin measure hides this information, as term compared to itself will always score 1. However, it has a significant advantage. One difficulty with using these measures in a search is that many protein pairs share identical scores (see Table 3) which hinders ranking. As the Resnik measure depends solely on the information content of the shared parents, there are only as many discrete scores as there are ontology terms. By using the information content of the query terms the Lin measure increases the number of discrete scores at least quadratically with the ontology size. The example search shown in Table 3 demonstrates this point. The Resnik measure ranking places a protein annotated as "Androgen Receptor", (GO:0004882), and "RNA polymerase II Transcription Factor", (GO:0003702) equally, because the former term is a child of the latter. The Lin measure, however, ranks all proteins annotated with "RNA polymerase II Transcription Factor", (GO:0003702) first, as it is capable of differentiating between a term and one of its children.

Swissprot ID	Description	Similarity	GO Term
Resnick			
ANDR_HUMAN	Androgen receptor (Dihydrotestosterone receptor)	3.412	"Androgen receptor", (GO:0004882)
AP1_HUMAN	Transcription factor AP-1	3.412	"RNA polymerase II transcription factor", (GO:0003702)
ATF4_HUMAN	Cyclic-AMP-dependent transcription factor	3.412	"RNA polymerase II transcription factor", (GO:0003702)
Lin			
AP1_HUMAN	Transcription factor AP-1	1	"RNA polymerase II transcription factor", (GO:0003702)
ATF4_HUMAN	Cyclic-AMP-dependent transcription factor	1	"RNA polymerase II transcription factor", (GO:0003702)
BTF3_HUMAN	Transcription factor	1	"RNA polymerase II transcription factor", (GO:0003702)
Jiang			
ENL_HUMAN	ENL protein.	0.634	"RNA polymerase II transcription factor", (GO:0003702)
AF4_HUMAN	AF-4 protein (Proto-oncogene AF4)	0.934	"Transcription factor", (GO:0003700)
AIRE_HUMAN	Autoimmune regulator (APECED protein)	0.934	"Transcription Factor", (GO:0003700)

Table 3: Tables shows results from a search against SWISS-PROT-Human, with the HXA1_HUMAN protein, against the molecular function aspect of GO, using the three different similarity measures for ranking. The top three results are shown for each measure. Following ranking by semantic similarity, proteins were sorted alphabetically.

The Jiang measure, which is the only distance measure, has the weakest correlation between the molecular function similarity and sequence similarity. However, as with the Lin measure, it combines information content from the shared parent, and the query terms.

Further investigation is required to determine which of the three measures is most appropriate for use within a search tool. It also seems likely that the relative advantages and disadvantages will change as GO increases in size and usage. The information hidden by the Lin measure will be less relevant if we know that both ontology and corpus are large, while the advantages of ranking for the Lin measure will increase with the size of the ontology. As well as further theoretical studies, we are developing a web delivered tool which should allow practical experimentation and user feedback.

We have also shown that the results from different aspects are only weakly correlated. *A priori* it is unclear whether users would prefer to perform semantic similarity searches over the GO as a whole, or over the different aspects independently. The data presented here suggests that as the aspects are largely independent combining results from the different aspects would be of little value, unless the user is looking for identically annotated proteins.

Although the aspects are largely independent, there is a correlation between them. This may have implications for the design of GO. Currently the aspects are completely disconnected subgraphs, which reflects the notion that these attributes are independent, when, in reality, they are not. Yet there is no formal linkage between, for example, the concept of "taste", (GO:0007607), which is a biological process term, and the concept of a "taste receptor", (GO:0008527) which is a molecular function term. As new ontologies emerge, covering further areas of biology, this problem may become more acute. Moving GO to a more expressive description logic based representation may be a good way to achieve this.[13]

In summary we have investigated several different measures of semantic similarity, all using information content as their basis. None of the three measures stand out as having a clear advantage over the others, although each has strengths and weaknesses. We have also investigated the behaviour of the different aspects of GO, and shown that they are largely independent, so that it will clearly be profitable to provide searches over different aspects of GO, rather than combining results. We believe that this work paves the way toward the development of a semantic similarity search tool, which will be a valuable additional tool in the armoury of the researcher.

Acknowledgements

This work was funded under the ESPRC/BBSRC Bioinformatics Programme (Grant number: BIF/10507)

1. P.W.Lord, J.R.Reich, A.Mitchell, R.D.Stevens, T.K.Attwood, and C.A.Goble. PRECIS: An Automated Pipeline for Producing Concise Reports About Proteins. In *IEEE International Symposium on Bioinformatics and Biomedical engineering*, pages 59–64. IEEE press, 2001.

2. R. Stevens, C.A. Goble, and S. Bechhofer. Ontology-based Knowledge Representation for Bioinformatics. *Briefings in Bioinformatics*, 1(4):398–416, 2000.

3. M. Winston, R. Chaffin, and D. Herrmann. A Taxonomy of Part-Whole Relations. *Cognitive Science*, 11:417–444, 1987.

4. J.J. Odell. *Six Different Kinds of Aggregation*, pages 139–149. Cambridge University Press, 1998.

5. The Gene Ontology Consortium. Creating the gene ontology resource: design and implementation. *Genome Res*, 11(8):1425–33, August 2001.

6. P. W. Lord, R.D. Stevens, A. Brass, and C. A. Goble. Investigating semantic similarity measures across the Gene Ontology: the relationship between sequence and annotation. *Bioinformatics*, 2002. Submitted.

7. P. Resnik. Using information content to evaluate semantic similarity in a taxonomy. In *IJCAI*, pages 448–453, 1995.

8. J. J. Jiang and D. W. Conrath. Semantic similarity based on corpus statistics and lexical taxonomy. In *Proceedings of International Conference on Research in Computational Linguistics*, Taiwan, 1998. ROCLING X.

9. D. Lin. An information-theoretic definition of similarity. In *Proc. 15th International Conf. on Machine Learning*, pages 296–304. Morgan Kaufmann, San Francisco, CA, 1998.

10. C. Fellbaum, editor. *WordNet:- An electronic lexical database*. MIT Press, Cambridge, Massachusetts, 1998.

11. P. Resnik. Semantic similarity in a taxonomy: An information-based measure and its application to problems of ambiguity in natural language. *Journal of Artificial Intelligence Research*, 11:95–130, 1999.

12. A. Budanitsky and G.Hirst. Semantic distance in WordNet: An experimental, application-oriented evaluation of five measures. In *Workshop on WordNet and Other Lexical Resources, Second meeting of the North American Chapter of the Association for Computational Linguistics*, Pittsburgh, 2001.

13. C. J. Wroe, R. D. Stevens, C. A. Goble, and M. Ashburner. An evolutionary methodology to migrate the Gene Ontology to a description logic environment using DAML+OIL. In *Pacific Symposium of Biocomputing.*, 2003.

LINKING MOLECULAR IMAGING TERMINOLOGY TO THE GENE ONTOLOGY (GO)

P.K. TULIPANO,[1] W.S. MILLAR,[2] J.J. CIMINO[1]

Department of Medical Informatics[1], Department of Radiology[2]
Columbia University, New York, NY 10032, USA
Email: tulipano@dmi.columbia.edu, wsm8@columbia.edu, cimino@dmi.columbia.edu

The rapidly developing domain of molecular imaging represents the merging of current advances in the fields of molecular biology and imaging research. Despite this merger, an information gap continues to exist between the scientists who discover new gene products and the imaging scientists who can exploit this information. The Gene Ontology (GO) Consortium seeks to provide a set of structured terminologies for the conceptual annotation of gene product function, process and location in databases. However, no such structured set of concept-oriented terminology exists for the molecular imaging domain. Since the purpose of GO is to capture the information about the role of gene products, we propose that the mapping of GO's established ontological concepts to a molecular imaging terminology will provide the necessary bridge to fill the information gap between the two fields. We have extracted terms and definitions from an already published molecular imaging glossary as well as molecular imaging research articles, and developed molecular imaging concepts. We then mapped our molecular imaging concepts to the existing gene ontology concepts as a method to comprehensively represent molecular imaging.

1 Introduction

Advances in medical imaging such as improved image resolution, new imaging agents, and experimental micro imaging devices, have stimulated interest in the *in vivo* assessment of molecular interactions and pathways.[1] These advancements coupled with the latest genomic discoveries have led to the rapid development of the molecular imaging domain. The integration of molecular sciences with imaging research has created a new cross-domain environment for molecular imaging scientists. Molecular imaging scientists are challenged with "learning the language" of the basic molecular sciences. In addition, they must keep up with the vast amount of information generated in the maturing fields of molecular imaging and molecular biology.

We have developed a controlled terminology of molecular imaging as the foundation for the integration of this cross-domain knowledge. The terminology will provide a comprehensive representation of the concepts in molecular imaging. We propose that linking our terminology to the concepts organized in the Gene Ontology will facilitate communications among the disciplines of molecular imaging and molecular biology.

614

2 Background

Molecular imaging is defined as "the *in vivo* characterization and measurement of biological processes at the cellular and molecular level through the use of imaging devices".[2] While conventional imaging captures the phenotypic changes at the gross anatomic level that result from molecular processes, molecular imaging attempts to detect detailed information about the underlying molecular and cellular processes themselves. The key elements necessary for molecular imaging are: 1) highly specific imaging probes with high affinity for their targets and acceptable biological delivery, 2) identification of suitable targets, 3) appropriate amplification strategies, and 4) sensitive and fast imaging systems with high resolution.[3,4] The goals of molecular imaging research are the exploration of these key elements and the development of new agents, strategies and imaging techniques for *in vivo* imaging.[4] For example, EgadME is one of a new class of chemicals dubbed "smart contrast agents",[5] so called because it is activated solely in the presence of the gene encoding beta-galactosidase. Researchers captured the MR signal produced by EgadME's interaction with cells expressing the beta-galactosidase gene using conventional *in vivo* MR imaging.

Traditional biological techniques have already provided detailed molecular *in vitro* diagnostic information.[6] Molecular imaging research can draw upon these *in vitro* techniques to provide *in vivo* diagnostic information. For example, many of the detectable molecular imaging parameters, such as cell surface receptors and enzymatic activity, should be identical to those found *in vitro*. Information acquired about molecular function and pathways will aid in the determination of specific molecular targets as well as the development of novel imaging agents. The clinical success of a contrast agent or molecular probe, however, may be related to additional in vivo factors, such as biocompatibility and directional transport to the target molecule.[6]

The description, classification, and organization of biological objects has become increasingly important, particularly in bioinformatics.[7-9] The structuring of biological information can be accomplished through the use of ontological methods. The Gene Ontology (GO) Consortium was established in 1998 to develop shared, structured terminologies for molecular characteristics across three model organism databases: SGD, the *Saccharomyces* Genome database, FlyBase, the *Drosophila* genome database, and MGD/GXD, the Mouse Genome Informatics databases.[10] The GO project became involved in the development of a database resource that allows access to datasets that utilize a standardized terminology for genes and gene product.[10] The GO polyhierarchy consists of three ontologies: 1) molecular function, 2) biological process, and 3) cellular component. These three ontologies were chosen because they are common to all living organisms and are basic to annotations of information about genes and gene products.[11] The GO project has expanded considerably since its inception to include other databases such as WormBase and Rat Genome Database (RGD).[12]

Currently, an information gap exists between the molecular scientists who discover new gene products and the imaging scientists who can exploit the gene product functions into new noninvasive imaging methods.[4] A structured set of terminologies for the conceptual annotation of gene product function, process and location in databases would be useful in this regard; however, no such structured set of concept-oriented terminology exists for the molecular imaging domain.

The need for concept-oriented terminologies has been recognized and addressed by many medical informatics researchers.[13] The development of a controlled terminology is valuable in specifying and organizing concepts important in the domain. In addition, a controlled terminology provides the framework on which the development of informatics tools can evolve. Informatics tools such as automatic information retrieval systems, indexed image retrieval databases, and decision support systems have been developed and rely on existing terminologies such as the Unified Medical Language Systems (UMLS) and the Medical Entities Dictionary (MED).[14-17]

There are well established informatics standards for the development of a controlled terminology.[18,19] The Desiderata is a set of standards necessary for the development of a standard, reusable multipurpose terminology.[19] Some of the requirements of the Desiderata include domain content coverage, concept orientation, nonsemantic concept identifiers, polyhierarchy, multiple granularities, and multiple consistent views. A terminology must be able to provide appropriate coverage of the domain's concepts (*domain coverage*). The concept is the unit of representation and must have a single, coherent meaning within the terminology (*concept orientation*). The concepts must be represented by meaningless, unique identifiers that are free of hierarchical meaning (*nonsemantic identifiers*). Such identifiers allow for multiple classifications and rearrangement of concepts within a hierarchy. A terminology should have a hierarchical arrangement that allows assignment of concepts in one or more areas of the hierarchy (*polyhierarchy*). In addition, to provide for multiple user functionality, a terminology must provide different levels of granularity and must maintain a consistent view throughout its hierarchy (*multiple granularities and consistent views*).

A number of formal representations exist for the modeling of controlled terminologies. One particular model is frame-based that includes a directed acyclic graph (DAG) as its hierarchy structure.[20,21] In a DAG hierarchy, concept nodes are children of one or more parent nodes. The hierarchical relationships from child to parent are of 'is a' type. In a frame-based model, each concept node can also be viewed as a frame with named slots. Slots may have values associated with them. Each concept node can also be viewed as having nonhierarchical relationships to other nodes through named slots or semantic links. The MED, currently in place at Columbia University, is an example of a frame-based model that closely adheres to the guidelines enumerated in the Desiderata. The development of our controlled terminology is based on this representation.

We propose that the establishment of a controlled terminology of molecular imaging and its linkage to the GO's ontological concepts will provide the necessary bridge to fill the information gap between domains as well as facilitate communications and knowledge sharing between domains.

3   Methods

A literature review of the molecular imaging domain was performed in order to extract concepts and terms. Thirty molecular imaging papers were selected and manually reviewed; topics ranged from a broad molecular imaging overview[3] to more specific applications.[22,23] In addition, a molecular imaging glossary was retrieved and used as the starting point for collection of molecular imaging terms.[2]

The terms defined in the molecular imaging glossary were extracted and initially divided into four general classifications. Several iterations were required to classify, and then reclassify, the terms into appropriate classes. Terms from other articles were subsequently added. The final iteration involved the dissection of terms into general classifications, placement of terms into a hierarchy, and the assignment of concept node attributes.

A frame-based representation model was developed with a directed acyclic graph (DAG) as its hierarchical structure, similar to the representation found in the MED.[20] Each concept in the terminology was assigned a unique identifier and a unique name. Each of the concepts was assigned named attributes that may or may not have values. The top-level node concept and its four descendants are listed and defined in Table 1.

Table 1. Top Level Concepts Names and Definitions

Concept Name	Definition
Molecular Imaging Entity	A broad type for grouping physical and conceptual entities related to the domain of molecular imaging.
Imageable Probe	A broad type for any highly specific agent (such as a radiolabelled drug or conjugated antibody) used in imaging to report on an event.
Imageable Target	A broad type for any target (usually a protein or gene product) that interacts with an imageable molecular probe.
Amplification Technique	A method for increasing imageable signal.
Imaging Instrument	A device for determining the presence, measure, and time/spatial distribution of a quantity under observation

In addition, the GO's HTML browser, AmiGO!,[10] was used to search for terms that existed in our terminology. GO terms that mapped to our terminology were noted in the 'GO Code' slot for that particular concept.

4   Results

The Glossary of Molecular Imaging Terminology[2] contained a total of 197 terms and definitions. Forty-eight terms were related specifically to molecular imaging. Sixteen additional terms were obtained from the literature. In addition to terms, the molecular imaging glossary included six abbreviations that were specific to molecular imaging.

A frame-based knowledge model, based on the Medical Entities Dictionary, was created as the representation for the terminology. A directed acyclic graph (DAG) formed the hierarchical structure. Figure 1 represents a subset of the information we found relevant to the domain. The top-level parent is the concept *Molecular Imaging Entity.*

Direct descendants correspond to the four key elements of molecular imaging: *Imageable Targets, Imageable Probes, Amplification Techniques* and *Imaging Instruments.*   Concepts were assigned named attributes that may contain corresponding values. Eight named attributes were assigned: 'Name', 'Synonyms', 'Abbreviations', 'Definition', 'GO Code', 'NonSemanticID', 'DescendantOf'. Hierachical relationships between concept nodes are 'is a' type. Figure 2 depicts a concept node frame with associated slot.

A search of all molecular imaging specific terms revealed 11 terms that were GO related. All GO codes were mapped through the named attribute of 'GO Code' in the concept frame. All of the GO terms found were mapped to our molecular imaging terminology through the general class of 'Imageable Targets'.   The mapping of the gene ontology with the concepts of molecular imaging is listed in Table 2.

Table 2. Linking of GO concepts to our molecular imaging terminology

Concept Name	Molecular Imaging NonSemantic ID	GO Code
Enzyme	10	0003824
Cell Surface Receptor	37	0007166
Thymidine Kinase	32	0004797
Creatine Kinase	33	0004111
Tyrosinase	34	0009309
Somatostatin Receptor	35	0004994
Cytosine Deaminase	36	0004131
Beta-Galactosidase	38	0004565
Dopamine Receptor	39	0004952
NADPH-ferrihemoprotein reducatase	40	0003985
Gastrin Receptor	41	0015054

618

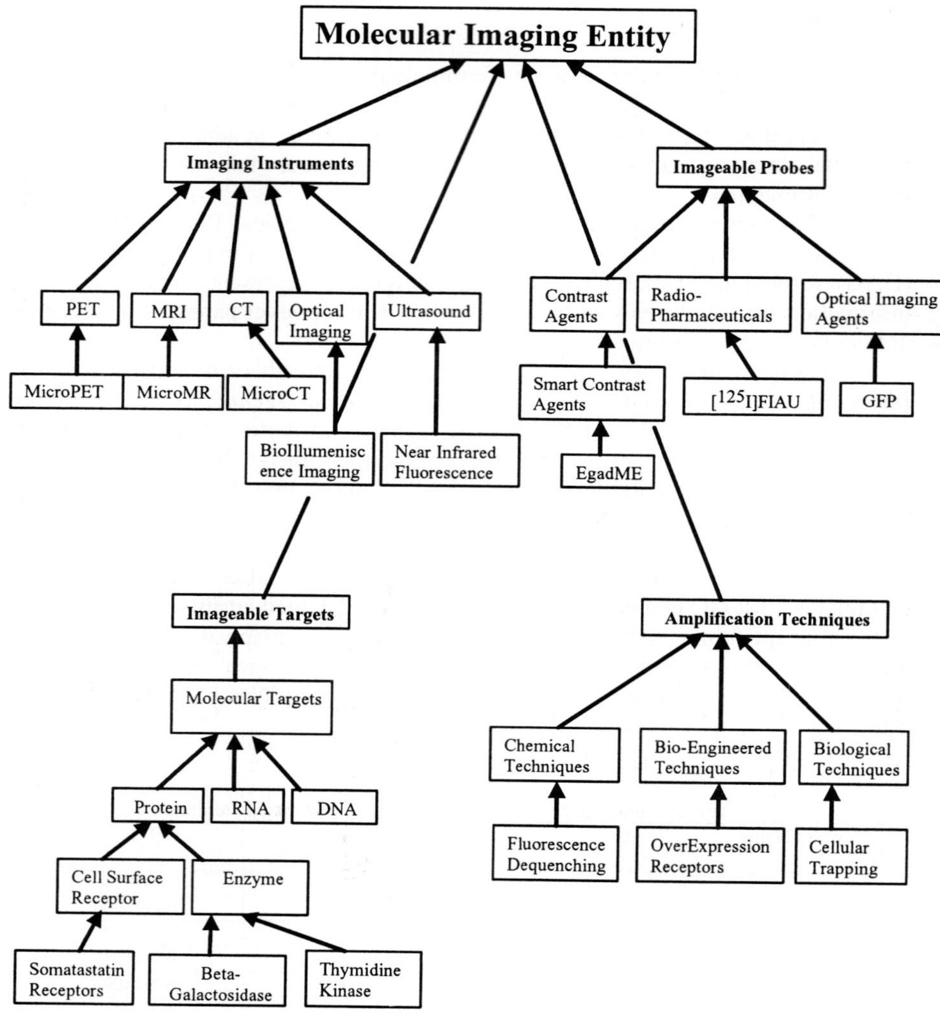

Figure 1. Molecular Imaging Representation. The direct descendants of the four key elements of Molecular Imaging (Imageable Probes, Imageable Targets, Imaging Instruments, and Amplification Techniques) and a subset of children nodes are depicted.

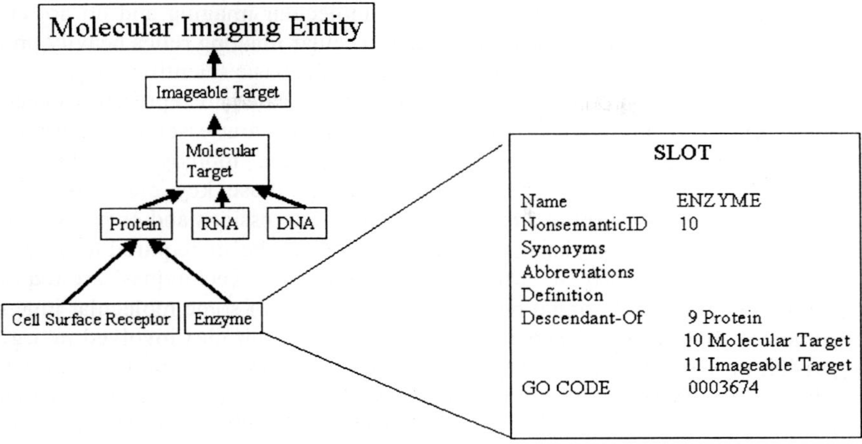

Figure 2. Concept Node Frame with Associated Slots and Slot Values

5   Discussion

The Glossary of Molecular Imaging Terminology developed by Wagenaar is a first attempt in defining terms for the molecular imaging domain.[2] Although not a comprehensive glossary, it does provide a foundation for the development of a molecular imaging specific terminology. The glossary is composed of terms relevant to the fields of molecular imaging and also the molecular sciences. One source of difficulty in extracting and classifying terms that exist in the realm of molecular imaging is that the 'language' used by basic biological scientists is also used in molecular imaging. The same terms may have several different meanings. For instance, the term, *amplification*, is defined in biology as 'an increase in the number of copies of a specific DNA fragment'. In imaging, however, amplification refers to an increase, not in the copies of DNA, but in the imageable signal.[3,24] Another interesting overlap of terms is the notion of a *reporter gene*. The *E. coli* beta-galactosidase gene has been extensively used in basic science research as a reporter gene. A reporter gene is defined as a gene that encodes an easily assayed and detectable protein. In molecular imaging, the beta-galactosidase gene is now used as an imaging reporter gene (also referred to as a *marker gene* in biology and a *imaging marker gene* in molecular imaging).[25] Its product is referred to as an *imaging reporter product* (also referred to as an *imaging molecular target*). These issues of concept ambiguity are resolved in our terminology through the use of nonsemantic identifiers as described in the Desiderata guidelines.[19] We uniquely identify concepts that correspond to a single, coherent meaning. These examples demonstrate that classifying terms specific to molecular imaging and do not cross into other related fields, is challenging since molecular imaging relies heavily on the information and resources previously developed in the basic scientific areas.

Despite the overlapping terms, concepts were created to represent molecular imaging terms as unambiguously as possible. There are four essential properties of molecular imaging. The direct descendants of the topmost concept node, 'Molecular Imaging Entity', are defined to reflect those key elements.[3,4] An 'Imageable Target' is essential since it is the target expressions and or pathway that *in vivo* imaging attempt to visualize. An 'Imageable probe' is required to probe for or locate the target of interest. Currently, 'Amplification Techniques' are required to increase the imageable signal because the resolutions of some molecular imaging instruments cannot detect the relatively small size of molecules involved in a gene expression or a molecular pathway. 'Imaging Instruments' are required to detect the 'Imageable Probe'. Other concept nodes can be added as the field matures; 'Smart Contrast Agents' and 'MicroPET' have been added to reflect the recent developments in molecular imaging.

Our controlled terminology uses a model similar to that found in the MED. In selecting this model, we expect that our terminology would follow the guidelines enumerated in the Desiderata. In terms of domain coverage, we have added concepts that we considered reflective of the current state of the domain at the time of its development. Molecular imaging is a dynamic and continuously evolving field. As such, the content of the molecular imaging terminology will also be expanding and evolving. The current structure supports the addition of new terms and the rearrangement of old terms. In addition, we provided each concept node with a slot 'NonsemanticID' that contains a meaningless, unique identifier. The directed acyclic graph structure supports multiple parents, which satisfies the Desiderata's notion of a polyhierarchy.

For example, in our terminology, 'green fluorescent protein' is considered both a child of 'optical imaging agent' and 'molecular target' because green fluorescent protein is an auto-fluorescent molecule;[26] that is, it requires no additional agent for its presence to be detected by imaging techniques. Therefore, concepts in our terminology can have multiple classifications. Unfortunately, the one desideratum not satisfied in our terminology is the notion of multiple levels of granularities. As the field progresses, new concepts and relationships will emerge that will contribute to the increase in finer granularity of the terminology.

The mappings in Table 2 demonstrate several salient points. Figure 3 is an example from Table 2 of the linking of the imageable target 'thymidine kinase' to the GO concept 'thymidine kinase'. The gene ontology mapped this item into two concepts. One is the concept of thymidine kinase as a subclass of kinases. In addition, 'thymidine kinase' mapped to the higher-level concept of 'transferase' in GO. The gene ontology links kinases and transferases to external gene databases. The gene databases contain the relevant gene information including the organism from which the genes were derived and the relevant literature associated with the gene. In one imaging strategy, the gene expression of thymidine kinase from the Herpes Simplex Virus 1 is imaged.[27] A possible inference that can be made from this mapping is that all transferases and kinases are imageable targets. In addition, it may be possible to image thymidine kinase expression from other organisms as well. Therefore, the possibility of such hypotheses being utilized and tested by imaging experts may prove useful. In addition to the generation of testable hypotheses, the linking of the two terminologies provides a connection between information found throughout research articles in both domains. The development of specialized computer applications that utilize this connection can facilitate automatic information retrieval from the enormous number of publications in molecular imaging and genomics. Automatic information retrieval applications are being developed for genomics. Our terminology may provide the foundation for such applications in molecular imaging.

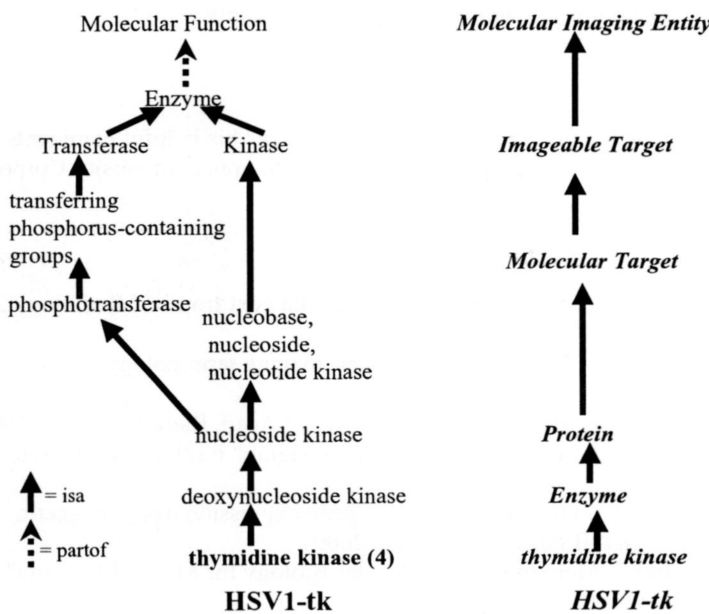

Figure 3. The association of the Herpes Virus Simplex 1 thymidine kinase gene product from GO to our molecular imaging terminology

Two commonly used imaging targets were not originally described in GO. Green fluorescent protein (GFP), isolated from coelenterates such as the Pacific jellyfish, is an imageable target used in optical imaging.[28] There is no reference to this protein and it's function in the GO.[12] Another protein, luciferase (from the firefly Photinus pyralis), is a bioluminescence imageable target.[26] A query of the current version of the GO's gene products for luciferase did not retrieve it.[12]

6 Conclusion

We have developed a controlled terminology of molecular imaging to represent the concepts appropriately within the domain. We have linked our molecular imaging terminology to the concepts in the Gene Ontology. We proposed that this linking of concepts facilitates knowledge sharing among molecular imaging and molecular scientists. Our terminology may provide the foundation for use of automatic information retrieval applications in the emerging field of molecular imaging. Future work will include the expansion of the terminology and the determination of the complex relationships between concepts.

Acknowledgments

We wish to acknowledge our colleague, Neil Sarkar, for his helpful comments and insightful discussions. Travel support provided by Columbia University Corporate Affiliates Program.

References

1. R. Weissleder, "Molecular imaging: exploring the next frontier" Radiology 212, 609 (1999)
2. D.J. Wagenaar et al., "Glossary of molecular imaging terminology" Acad. Radiol. 8, 409 (2001)
3. M.G. Pomper, "Molecular imaging: an overview" Acad. Radiol. 8, 1141 (2001)
4. R. Weissleder and U. Mahmood, "Molecular imaging" Radiology 219, 316 (2001)
5. A.Y. Louie et al., "In vivo visualization of gene expression using magnetic resonance imaging" Nat. Biotechnol. 18, 321 (2000)
6. A. Hengerer and T. Mertelmeier, "Molecular Biology for Medical Imaging" electromedica 69, 44 (2001)
7. A. Rzhetsky et al., "A knowledge model for analysis and simulation of regulatory networks" Bioinformatics 16, 1120 (2000)
8. S. Schulze-Kremer, "Ontologies for molecular biology" Pac. Symp. Biocomput., 695 (1998)
9. R. Stevens et al., "Building a bioinformatics ontology using OIL" IEEE Trans. Inf. Technol. Biomed. 6, 135 (2002)
10. The Gene Ontology Consortium, "Creating the gene ontology resource: design and implementation" Genome Res. 11, 1425 (2001)

11.   M. Ashburner et al., "Gene ontology: tool for the unification of biology. The Gene Ontology Consortium" Nat. Genet. 25, 25 (2000)
12.   http://www.geneontology.org, 2002
13.   C.G. Chute et al., "A framework for comprehensive health terminology systems in the United States: development guidelines, criteria for selection, and public policy implications. ANSI Healthcare Informatics Standards Board Vocabulary Working Group and the Computer-Based Patient Records Institute Working Group on Codes and Structures" J. Am. Med. Inform. Assoc. 5, 503 (1998)
14.   D.C. Berrios, "Automated indexing for full text information retrieval" Proc. AMIA Symp., 71 (2000)
15.   J.J. Cimino et al., "Supporting infobuttons with terminological knowledge" Proc. AMIA Annu. Fall Symp., 528 (1997)
16.   W.D. Bidgood, Jr. et al., "Image acquisition context: procedure description attributes for clinically relevant indexing and selective retrieval of biomedical images" J. Am. Med. Inform. Assoc. 6, 61 (1999)
17.   D.A. Lindberg et al., "The Unified Medical Language System" Methods. Inf. Med. 32, 281 (1993)
18.   P.L. Elkin et al., "Guideline for health informatics: controlled health vocabularies--vocabulary structure and high-level indicators" Medinfo. 10, 191 (2001)
19.   J.J. Cimino, "Desiderata for controlled medical vocabularies in the twenty-first century" Methods Inf. Med. 37, 394 (1998)
20.   J.J. Cimino et al., "Knowledge-based approaches to the maintenance of a large controlled medical terminology" J. Am. Med. Inform. Assoc. 1, 35 (1994)
21.   J.J. Cimino et al., in Secondary "Designing an introspective, multipurpose, controlled medical vocabulary" (Washington, DC, 1989)
22.   C.H. Tung et al., "Preparation of a cathepsin D sensitive near-infrared fluorescence probe for imaging" Bioconjug. Chem. 10, 892 (1999)
23.   Y. Yu et al., "Quantification of target gene expression by imaging reporter gene expression in living animals" Nat. Med. 6, 933 (2000)
24.   D. Hogemann and J.P. Basilion, ""Seeing inside the body": MR imaging of gene expression" Eur. J. Nucl. Med. Mol. Imaging 29, 400 (2002)
25.   C. Bremer and R. Weissleder, "In vivo imaging of gene expression" Acad. Radiol. 8, 15 (2001)
26.   M.J. Hickey et al., "Luciferase in vivo expression technology: use of recombinant mycobacterial reporter strains to evaluate antimycobacterial activity in mice" Antimicrob. Agents Chemother. 40, 400 (1996)
27.   R.G Blasberg and Tjuvajev J.G , "Herpes simplex virus thymidine kinase as a marker/reporter gene for PET imaging of gene therapy", Q J Nucl Med 43(2):163-9 (1999)
28.   M. Chalfie et al., "Green fluorescent protein as a marker for gene expression" Science 263, 802 (1994)

A METHODOLOGY TO MIGRATE THE GENE ONTOLOGY TO A DESCRIPTION LOGIC ENVIRONMENT USING DAML+OIL

C.J. WROE, R. STEVENS, C. A. GOBLE

Department of Computer Science, University of Manchester,
Oxford Rd, Manchester, M13 9PL, UK
{cwroe robert.stevens}@cs.man.ac.uk

M. ASHBURNER

EMBL – European Bioinformatics Institute,
Wellcome Trust Genome Campus,
Hinxton, Cambridge CB10 1SD, UK

The Gene Ontology Next Generation Project (GONG) is developing a staged methodology to evolve the current representation of the Gene Ontology into DAML+OIL in order to take advantage of the richer formal expressiveness and the reasoning capabilities of the underlying description logic. Each stage provides a step level increase in formal explicit semantic content with a view to supporting validation, extension and multiple classification of the Gene Ontology. The paper introduces DAML+OIL and demonstrates the activity within each stage of the methodology and the functionality gained.

1   Introduction

The Gene Ontology Consortium set out to provide 'a structured precisely defined common controlled vocabulary for describing the roles of genes and gene products in any organism'.[1] The resulting, publicly available, Gene Ontology (GO) has become the *defacto* standard used to provide ~250,000 annotations for entries in at least 14 major bioinformatics databases. GO has been successful in supporting the needs of molecular biologists due to its comprehensive coverage in a relatively simple but consistent structure acceptable to the biological communities. However, its growing success and size now leads to several challenges for ongoing manual curation.

The Gene Ontology Next Generation project (GONG) aims to demonstrate that, in principle, migrating to a finer grained formal conceptualization will allow computation techniques such as description logics to aid in the curation and delivery of the ontology. This migration must be practical. Providing a fine-grained conceptualization in a formal language is a significant knowledge acquisition process and it is unrealistic to approach it as a one-off effort. We aim to prove the exercise can be undertaken in a staged manner, both in terms of number and granularity of formal concept definitions, with useful benefits received at each increment. The paper is organized as follows: This section continues with an introduction to the existing structure and use of GO, and the challenges it faces. Section 2 provides an overview of the ontology language DAML+OIL. Section 3 provides an overview of the methodology we propose and then a detailed look at each stage examining its aim, procedure and results. We conclude with a discussion of the current status of the project and plans for the future.

1.1 Current structure and use of the Gene Ontology

The Gene Ontology (GO) is split into three orthogonal sub-ontologies containing a total of about 11,000 concepts. The 'cellular component' ontology is used to annotate the location at which a gene product acts. 'Molecular function' terms are used to annotate the specific capabilities of a gene product, while 'biological process' terms capture the higher order processes in which the gene product is involved. GO is more than a controlled vocabulary. The aim is to associate a textural definition to each term to promote an explicit shared understanding and currently 60% of terms have such a definition. Each term is also placed in a directed acyclic graph (DAG) allowing multiple parents both along 'is-a' relationships and 'part-of' relationships. The hierarchical arrangement of terms is primarily used by humans rather than software to accomplish three main tasks:

1. **Query/ browse bioinformatics databases.** GO can act as an index into databases. GO Browsers, e.g. AmiGO (http://www.godatabase.org) allow users to link directly from the hierarchical view of the ontology to database entries annotated with those terms.
2. **Interpret results.** GO annotations provide biologists with more meaningful yet concise alternatives to the cryptic abbreviations used to label experimental results and so help in interpretation of the large data sets, e.g. microarray data.[1]
3. **Aggregate information.** A GO Slim is a non-overlapping subset of high-level GO terms. Aggregating all entries annotated with hierarchical descendants of each GO Slim term can produce useful summary statistics. Several GO Slims have been created to aggregate different sets of annotations for different purposes.[a] The 'GO summary' feature of the AmiGO browser demonstrates how this information is used to provide a high level view of GO annotation statistics.

The range of applications of GO is constantly growing,[b] which places increasingly exacting requirements on GO's internal structure as detailed below:

1. **Multiple classification and consistency.** There are multiple ways to organize terms in a classification. The exact choice depends on the task at hand. Multiple classification within GO is currently maintained by hand but experience from the medical domain has shown that numerous parent-child links are omitted in such hand crafted, phrase based controlled vocabularies.[2] While of less importance to manual interpretation, machine interpretation will falter in the face of such inconsistencies.
2. **Extension.** There is a growing desire to extend the content of processes such as embryonic development. The effort to manually pre-enumerate and maintain the

---

[a] A list of GO Slims can be found at ftp://www.geneontology.org/pub/go/GO_slims

[b] A list of published uses of GO can be found at http://www.geneontology.org/doc/GO.biblio.html

626

cross product of developmental processes against all anatomical structures in every organism would be immense.

3. **Machine interpretation.** Biologists are able to interpret information both within term names and the lexical definitions. However this implicit information is inaccessible to computer applications. The hierarchical structure of GO has been used for automated processing.[3] However, the definition of a concept is only implicitly and incompletely encoded by its hierarchical position, e.g. for 'protein kinase C' (GO:0004697), its parentage implies it is an 'enzyme function' that can 'transfer a phosphorous containing group to an alcohol group', is a 'serine/threonine kinase', and is a 'phorbol ester receptor'. However, we cannot synthesize a complete formal definition from this information.

Many formal ontology representation languages have been developed in the AI community to capture formal concept descriptions including frame-based systems, conceptual graphs and description logics (DLs). DLs offer a new paradigm in modeling vocabulary. Rather than annotate manually classified concepts with additional properties, explicit concept definitions actually form the basis for calculating a classification or checking the logical consistency of an existing classification.

2 DAML+OIL

DAML+OIL arose from EU and US DARPA research programs and is currently undergoing standardization through the W3C WebOnt activity,[a] to become the Ontology Web Language (OWL). Irrespective of its reasoning capabilities it is becoming a standard language for ontology interchange. As an interchange language it has been designed to encode a wide range of ontologies from taxonomies, frame based ontologies, to ontologies that include logic based concept definitions. This flexibility allows the staged evolution of an ontology within a single representation, greatly simplifying the process.

Within a DAML+OIL ontology each concept is represented as a class. At its simplest, DAML+OIL allows each class to be placed in a taxonomy with the use of the subclass relationship e.g.

> **class** isocitrate dehydrogenase (NAD+) (GO:0004449) [b]
> **subClassOf** 'oxidoreductase, acting on the CH-OH group of donors, NAD or NADP as acceptor' (GO:0016616)

Classes can be further described (or restricted in DAML+OIL terms) by their attributes, specified as property/value pairs, e.g.

> **class** isocitrate dehydrogenase (NAD+) (GO:0004449)
> **restriction onProperty** has_substrate **hasClass** isocitrate

---

[a] http://www.w3c.org/2001/sw/WebOnt/

[b] Space precludes the reproduction of DAML+OIL in its standard XML format. An abridged non-standard format is used the purposes of illustration.

Both universal and existential quantification can be used to represent such definitions, as 'carbohydrate metabolism is the metabolism of some carbohydrate and only carbohydrate'.

```
class carbohydrate metabolism (GO:0005975) defined
    subClassOf metabolism
    restriction onProperty acts_on hasClass carbohydrate
    restriction onProperty acts_on toClass carbohydrate
```

Each restriction can also be associated with numerical cardinality constraints:

```
class tricarboxylic acid defined
    subClassOf organic acid
    restriction onProperty has_part 3 (carboxyl group or carboxylate group)
```

The above class 'tricarboxylic acid' can be specified as *defined* because the description completely captures its definition and as such its place in the classification can be inferred by merit of its definition using description logic reasoners such as FaCT.[4] Note also the use of anonymous embedded expressions and logical operators '(carboxyl group or carboxylate group)', which provides greatly increased expressive power with respect to standard frame-based languages.

Horrocks[4] gives a more detailed description of the capabilities of DAML+OIL and Stevens[5] describes the use of DAML+OIL in capturing molecular biology domain knowledge with a high degree of fidelity.

3   Methodology

The methodology is designed to embrace evolution not revolution. We have therefore partitioned development into well-defined stages. At each stage we increase both the quantity and complexity of the explicit semantic content by incremental extension of class descriptions. Figure 1 illustrates the five steps involved and the resources involved at each stage. Step 0 is a foundation stage in which GO is translated into a DAML+OIL ontology. Step 1 uses DL reasoning to group related components based on part-of relationships specified in the current GO. Step 2 programmatically creates partial class descriptions from existing structured information in bioinformatics databases, enabling the grouping of existing terms under abstractions which could form a novel GO Slim. Step 3 manually completes the partial descriptions of step 2 to enable the reasoner to check the consistency of the existing hierarchy and detect missing is-a relationships. Step 4 allows annotation applications to dynamically extend GO as required. At each stage it should be possible to re-express a subset of the information within the DAML+OIL ontology in the original GO XML format enabling existing applications to take advantage of reorganized hierarchies and additional concept information. The feedback of results as a static snap-shot is similar to the creation of thesauri from description logic ontologies described by Bechhofer *et al.*[6]

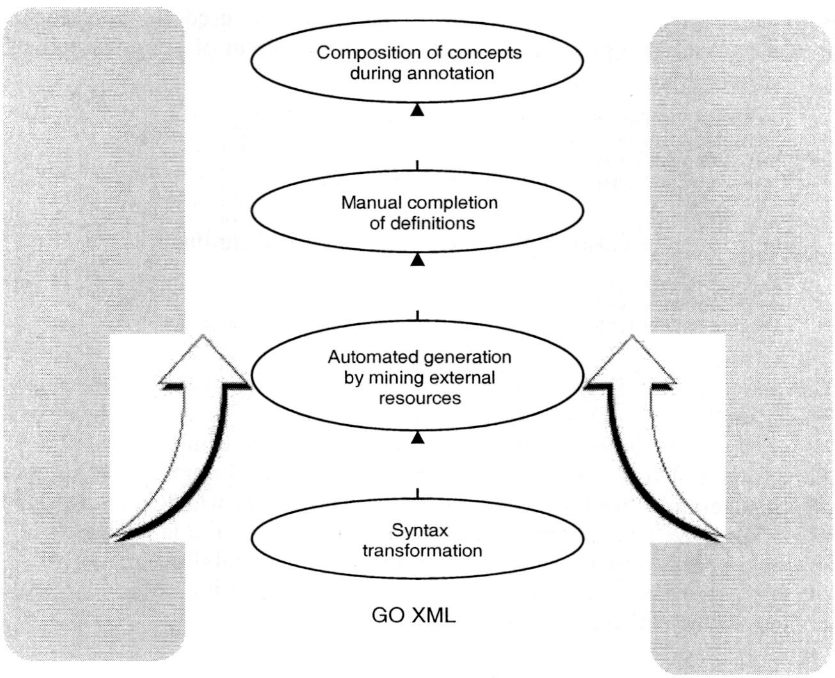

Figure 1. Overview of the staged migration described in this paper.

3.1   Materials

The XML version of GO released January 2002 (http://www.godatabase.org/dev/
database/archive/2002-01-01/) was used throughout the work described in the paper
and all references are to that version. OilEd version 3.4 (http://oiled.man.ac.uk) was
used to edit DAML+OIL ontologies, and provided the DAML+OIL data structures
manipulated by scripts. OilEd also provided the GO XML to DAML+OIL
conversion capability. The COHSE ontology server (http://cohse.semanticweb.org)
provided an API to link to the FaCT reasoner, and provided a server to demonstrate
client side composition of ontology concepts. DAGEdit version 1.302
(http://www.geneontology.org/#tools) was used to browse the current Gene
Ontology in its native format.

The KEGG enzyme database (downloaded 17/05/02 from http://www.genome
.ad.jp/kegg) was used to extract enzyme substrate, product and cofactor information.
BioPython (http://www.biopython.org) was used to parse KEGG enzyme flat file
format and load into a MySQL database (http://www.mysql.com). UMLS
knowledge sources 2002AA (http://www.nlm.nih.gov/research/umls) loaded into a

MySQL database were used as a source of the MeSH chemical taxonomy. Lexical tools bundled with UMLS version 2002AA were used to lexically normalise chemical terms in the KEGG enzyme database.

Jython version 2.1 (Java Python, http://www.jython.org) was used as the scripting environment, with which to integrate large-scale programmatic manipulation of DAML+OIL ontologies, database queries and lexical tools.

Step 0. Transforming GO XML into DAML+OIL

GO is not currently published in DAML+OIL, so the first stage of any migration must be a syntactical transformation from an available format, e.g. GO XML into DAML+OIL. The transformation involves the simple mapping of XML elements to equivalent constructs in DAML+OIL as shown in Table 1.

Table 1. Mapping between GO XML and DAML+OIL

GO XML	DAML+OIL
`<go:term>`	`<daml:Class>`
`<go:isa>`	`<daml:subClassOf><daml:Class>`

`<go:part-of>`	`<daml:subClassOf><daml:Restriction>`
	`<daml:onProperty><daml:ObjectProperty`
	`rdf:resource="go:part-of"/>`
	`<daml:hasClass><daml:Class>`

Not all GO terms have a subsumption relationship (orphan terms), but instead are related to another term by a part-of relationship. Formal ontologies require the majority of concepts to be at least a kind of one other concept, as the is-a (subsumption) network forms a key substrate on which reasoning occurs. At this stage our solution is simply to add three additional abstractions during the transformation: 'part_of_cellular component', 'part_of_molecular function' and 'part_of_biological process'. Orphan terms become a kind of one of these respective abstractions.

This purely syntactic step paves the way for future work, but there is no additional functionality gained at this stage.

Step 1. Reasoning over existing semantic information

In the previous stage we placed all orphan terms under at least one parent, e.g. 'part-of_cellular component'. When viewed as a hierarchy these orphan terms form a long unorganized list in which it is difficult to associate related terms. Native Gene Ontology browsers, such as AmiGO, circumvent this problem by presenting both the 'is-a' and 'part-of' relationships as parent-child links within the same tree structure. Terms subsumed by nothing (orphan terms) can still be visually related to the structures or processes that contain them (Fig. 2a). To replicate this organization within a pure is-a hierarchy requires the addition of abstractions that group together terms that are 'part-of' of a common structure. This would be a laborious task to undertake by hand, but it can be straightforward to achieve using a DL reasoner.

To demonstrate this step, we manually identified 20 biologically significant cellular structures that have numerous components specified in the cellular component ontology. We then added 20 corresponding DAML+OIL classes to the ontology in order to group those components, e.g.

> 'component of mitochondrion' **defined**
> **subClassOf** cellular component
> **restriction onProperty** part-of **hasClass** mitochondrion

Only the definitions were manually created. The grouping of terms was achieved by submitting the ontology with newly defined classes to the FaCT reasoner, which automatically inferred the required is-a links. This creates a similar organization to that displayed in native GO browsers, but is based purely on is-a links. Figure 2 shows the evolution of the hierarchy focused on the GO term 'TCA cycle enzyme complex' (GO:0030062). These novel abstractions are for organization only and should not be used for annotation. Therefore metadata should be applied to these abstractions, to prevent their use in annotation tools.

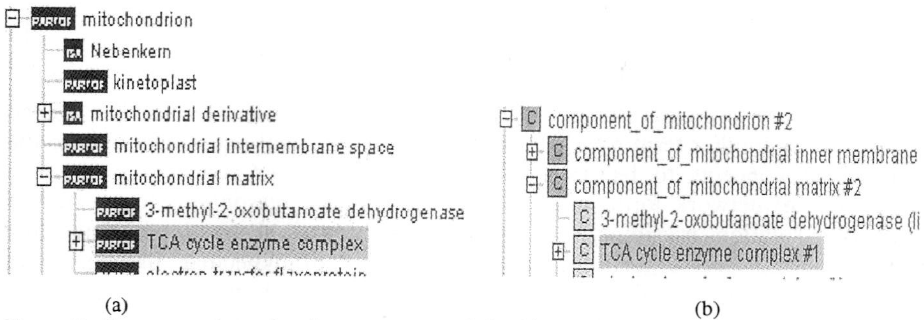

(a)   (b)

Figure 2. Two screenshots showing an extract of the hierarchical position of 'TCA cycle enzyme complex' (GO:0030062) as shown in (a) DAG edit and (b) OilEd after addition of 'component of' abstractions and inference of new subsumption relationships using the FaCT reasoner.

Step 2. Programmatically adding partial descriptions from other sources

Step 1 allowed the classification of GO terms using existing 'part-of' information. The creation of further novel abstractions grouping current GO terms in alternative ways requires the addition of the relevant explicit concept information on which the reasoner can operate. For example, the descendants of 'enzyme' (GO:0003824) are manually organized from a biochemical point of view derived from the Enzyme

Classification (EC).[7] Biologists from other disciplines may prefer to group enzyme functions by the type of chemical substances they function on rather than detailed chemical substructures involved in the reactions. Given there will inevitably be effort required to manually create the information required to support this alternative classification, it is advisable to first investigate the reuse of existing structured information from other sources.

There are numerous bioinformatics resources available that contain structured information characterizing various aspects of enzymes. To support the reclassification described above, we need to capture the substrates and products of that reaction and any cofactors involved. To do this we used the enzyme database published as part of the Kyoto Encyclopedia of Genes and Genomes (KEGG).[8] Each substrate, product and cofactor entry in the relevant KEGG enzyme record (cross referenced by EC identifier) was converted into an existential restriction on the relevant DAML+OIL class as shown in section 2.

Of the 2960 enzyme functions in GO, 2513 were annotated with an EC identifier and so could be linked to external databases. Of these 1596 had a corresponding entry in the KEGG enzyme database. The reasoner is unable to classify enzyme functions based on chemical class specified in these restrictions unless we provide a classification of said chemicals. Chemical thesauri do exist, the most widely known is that embedded within the Medical Subject Headings MESH *et seq.*[9] We therefore represented the relevant subset of MeSH as a DAML+OIL ontology and linked the chemicals specified in enzyme description with this MESH chemical ontology. No direct cross-reference exists between the KEGG enzyme database and MeSH. Linking based on an exact term name match, yields links for only 4% (106/2443). Use of lexical tools and synonym information available with the Unified Medical Language System,[10] led to the resolution of three sources of mismatch, resulting in an increase in matches to 35% (856/2443):

1. Syntactic differences   e.g. Divalent cation --> Cations, Divalent
2. Abbreviations   e.g. dUMP --> 2'-deoxyuridylic acid
3. Synonyms   e.g. 20-Hydroxyecdysone --> Ecdysterone

Of the remaining 1587 unmatched chemicals, most covered specializations of concepts within MESH, e.g. 'manganese^{2+} ion' as opposed to the term 'manganese' present in MESH. This points to the need for ontology integration tools that interleave related concepts rather than provide just an exact mapping between equivalent terms. Ontology integration tools do exist such as Chimera and PROMPT,[11,12] and the next phase of the project will evaluate their utility for this task.

The reasoner can now group enzyme functions by the class of chemicals they involve, (as shown in fig. 3) providing those chemicals are linked to the MeSH ontology.

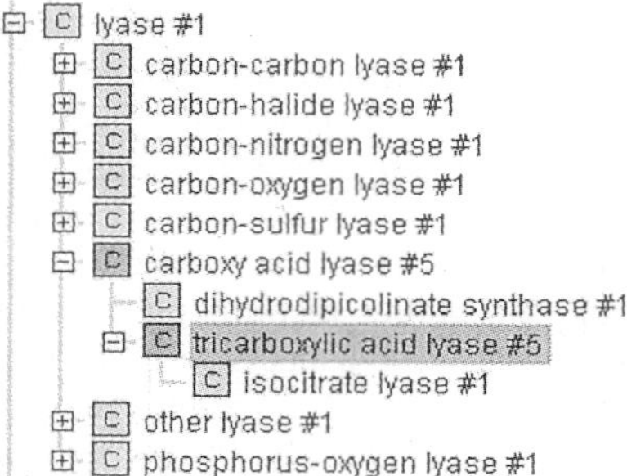

Figure 3. Automated grouping of 'isocitrate lyase' under novel abstractions 'tricarboxylic lyase' and 'carboxy acid lyase' using the FaCT reasoner.

Step 3. Manually adding semantic information to support validation of existing classification.

The previous step added partial semantic information in a shallow and broad manner that can be used to index specific leaf node terms along additional axes of classification. In most cases the partial definition mined from existing resources must then be completed and checked by hand. Only then can they be used to verify the existing classification. The resulting definitions can be simple, such as GO metabolism concepts, or complex such as GO enzyme function concepts.

The majority of 'metabolism' (GO:008152) descendants convey only two aspects of information: the subtype of metabolism ('catabolism' (GO:009056), 'biosynthesis' (GO:009058)), and the chemical being metabolized. This results in many metabolism concepts having multiple is-a parents. As mentioned in the introduction, manual maintenance of multi-axial hierarchies is known to be prone to error.[2] To detect and resolve possible omissions we explicitly represented the chemical involved and metabolism subtype as DAML+OIL restrictions, for 250 descendants of 'carbohydrate metabolism' (GO:0005975). Chemicals referenced by these restrictions were defined in a separate ontology derived from MeSH (see stage 2). The resulting class definitions were submitted to the FaCT reasoner and any additional inferred is-a relationships recorded. The reasoner inferred 22 new is-a relationships, e.g. 'fructosamine catabolism' (GO:0030392) is now inferred to be a kind of 'carbohydrate catabolism' (GO:0016052). The set of inferred 'is-a' links have been reviewed by the editorial team resulting in 17 additional is-a links in the published Gene Ontology.

A complete definition of a metabolism term is usually simple. However, complete descriptions of, for example, enzyme functions are far more complex. Reaction process terms such as 'oxidising' and 'reducing' need to be specified for each enzyme function term, and these in turn require properties detailing which chemical the reaction acts on and at what site. Reaction sites need specifying in terms of chemical substructure such as 'carboxyl group', and each chemical must include a description of which and how many substructures in contains. It is at this stage that the full expressive power of DAML+OIL is brought to bear. For example, exact cardinality constraints in combination with logic operators are used to precisely define 'tricarboxylic acid' as a chemical, which has exactly three 'carboxy' or 'carboxylate anion' groups. The dedicated DAML+OIL ontology editor OilEd[13] was used to support the authoring, but it was not designed to support the complexity of descriptions required for these GO concepts. For example, it took one day to author the nine enzyme function definitions involved in the tricarboxylic acid cycle.[a] This issue was encountered during the construction of the large medical ontologies in the GALEN-IN-USE project,[14] and alternative environments were needed to manage this complexity. A simplified intermediate representation was developed to capture or 'dissect' concept definitions in a single domain (surgical procedures). These were then automatically expanded into the more complex underlying description logic representation for reasoning. The next phase of the project will develop tools to support an intermediate representation for dissecting GO terms.

Even with a simplified representation, adding detailed semantic information by hand is time consuming. Term names often follow a stereotyped pattern. Simple scripts were used to help parse metabolism terms in the work described above, but more sophisticated term mining tools would greatly increase productivity.

Step 4. On-demand composition of new concepts within client annotation software

Realistically, only central maintainers of the ontology can undertake steps 0 to 3. However, the dynamic nature of description logic ontologies enables on demand concept composition to occur in client annotation applications. As discussed in the introduction, cross products are a key future challenge for the Gene Ontology. Many concepts such as metabolism and development are highly compositional in nature, encompassing a process acting on a chemical or anatomical structure. Manually enumerating these cross products is untenable. Automatic enumeration would lead to a huge and largely redundant ontology. Therefore these compositions should only be created if required at the point of use. Experience with medical terminologies, such as SNOMED, shows that manual compositions are not correctly classified and users often find multiple methods for expressing the same concept.[15] Use in an automated environment also opens up the possibility of nonsense compositions. The fine-grained semantic information and constraints added in previous steps together with a DL reasoner can be used to provide a logical basis for such compositions. Such an activity requires an *ontology server* with which client applications can communicate in order to compose new logically consistent concepts and integrate them in the existing classification. A DAML+OIL ontology server has been constructed during the COHSE project[16] and the recently funded GOAT project (Gene Ontology

---

[a] Difficulty authoring complex expressions is compounded by the absence of a human readable syntax for DAML+OIL.

Annotation Tool, http://goat.man.ac.uk) will explore the utility of concept composition within client annotation tools.

4   Discussion

We have shown how DAML+OIL can be used to represent a range of ontologies from taxonomies such as subsets of MESH, to ontologies in which complete formal definitions are given to each term. We have also shown the utility of reasoning in helping to maintain existing hierarchies, and in grouping terms along novel axes of classification.

The aim of the project was to enable the migration to DAML+OIL rather than provide significant DAML+OIL content. However, the stereotypical pattern of many definitions in categories of concept such as metabolism, has enabled us to automatically generate ~1500 metabolism definitions with three weeks programming effort and a further 1600 partial enzyme definitions with a similar effort. We are investigating how successfully this approach can be extended to other categories.

The technical feasibility of a methodology does not necessarily translate into a successful practical solution. We have shown in the later stages of the methodology the complexity of DAML+OIL expressions that may be required to realize the full functionality of such an approach. In the next stage of the GONG project, we plan to develop ontology engineering tools to support these later stages on a large scale with emphasis on ontology integration, automated and assisted content authoring from existing sources, ontology versioning and targeted deployment. Once these tools start to become available we can evaluate the impact of such a methodology in terms of resources and changes to the current maintenance process for GO.

Other Ontology editing environments exist. Protégé 2000, a frame based ontology environment, implements many analogous content creation tasks in the form of tabs.[17] However, there is currently no DAML+OIL reasoning support from within Protégé, although such functionality is in development. Members of the GO consortium are also actively pursuing DAML+OIL as an ontology representation language and are developing a new editing environment to support it called GOET (Gene Ontology Editing Tool) (http://sourceforge.net/projects/gmod/). We are not clear to what extent GOET will use the reasoning capabilities of DAML+OIL, but we hope to inform the design process for the tool and the working practices which evolve around its use.

5   Acknowledgments

This work is supported by the GONG project grant (DARPA DAML subcontract PY-1149 from Stanford University) and the ^{my}Grid eScience pilot project grant (EPSRC GR/R67743). We would like to thank Midori Harris and Jane Lomax of the GO Consortium Editorial Team for their advice and guidance.

References

1. The Gene Ontology Consortium, Gene Ontology: tool for the unification of biology. *Nature Genetics* **25** (2000) pp. 25-29.
2. Rogers J.E., Price C., Rector A.L., Solomon W.D., Smejko N. Validating Clinical Terminology Structures: Integration and Cross-Validation of Read Thesaurus and GALEN. *AMIA Fall Symposium, Orlando, USA* (1998)
3. Chang, J., Raychaudhuri, S., Altman, R., Including Biological Literature Improves Homology Search. *Pacific Symposium on Biocomputing* **6** (2001) pp. 374-383.
4. Horrocks I., DAML+OIL: a reason-able web ontology language. *Proc. of EDBT 2002* (March 2002).
5. Stevens R., Horrocks I., Goble C., Bechhofer S., Building a Reason-able Bioinformatics Ontology Using OIL. *IJCAI'01 Workshop on Ontologies and Information Sharing, Seattle, USA*, (August 2001) pp. 81-90.
6. Bechhofer S., Goble C., Thesaurus construction through knowledge representation. *Data & Knowledge Engineering* **37** (April 2001) pp. 25-45.
7. Enzyme Nomenclature. *Academic Press, San Diego, California, ISBN 0-12-227165-3* (1992).
8. Kanehisa M., Goto, S., KEGG: Kyoto Encyclopedia of Genes and Genomes. *Nucleic Acids Res.* **28** (2000) pp. 27-30.
9. Medical Subject Headings, MESH. http://www.nlm.nih.gov/mesh/
10. McCray A.T., Srinivasan S., Browne A.C., Lexical methods for managing variation in biomedical terminologies. *Proc Annu Symp Comput Appl Med Care* (1994) pp. 235-239.
11. McGuinness, D.L., Fikes R, Rice J., Wilder S., The Chimaera Ontology Environment. *AAAI-2000, Austin, TX* (2000).
12. Noy N.F., Musen. M.A., PROMPT: Algorithm and Tool for Automated Ontology Merging and Alignment. *AAAI-2000, Austin, TX* (2000).
13. Bechhofer S., Horrocks I., Goble C., Stevens R., OilEd: a Reason-able Ontology Editor for the Semantic Web. *Proceedings of KI2001, Joint German/Austrian conference on Artificial Intelligence, Vienna. Springer-Verlag LNAI* **2174** (2001) pp. 396-408.
14. Rogers J.E., Solomon W.D., Rector A.L., Pole P.M., Zanstra P., van der Haring E., Rubrics to Dissections to GRAIL to Classifications. *Medical Informatics Europe, IOS Press* **43** (1997) pp: 241-245.
15. Rector A.L., Clinical terminology: Why is it so hard? *Methods of Information in Medicine, Schattauer, ISSN 0026-1270* **38** (1999) pp. 239-252.
16. Bechhofer S., Carr L., Goble C., Hall W., Conceptual Open Hypermedia = The Semantic Web? *Second International Workshop on the Semantic Web* (2001).
17. Musen M.A., et al., Component-Based Support for Building Knowledge-Acquisition Systems. *IIP 2000, Beijing* (2000).

SPECIAL PAPER

The organizers of this year's Pacific Symposium on Biocomputing received the following request:

```
Subject: Late paper for PSB re: World Trade Center
informatics
Date: Sun, 8 Sep 2002 21:52:45 -0400
From: Howard Cash <howardc@genecodes.com>
To: <psb@SMI.Stanford.EDU>
```

I would not presume to propose a late submission for a paper to a conference if this were not such an extraordinary circumstance. My team has been in charge of developing the informatics tools for doing DNA identification of remains of those killed at the World Trade Center. Using non-standard development techniques under truly extraordinary pressures, we have developed a large integrated software system under contract to the City of New York for combining STR, mtDNA and SNP data to a large number of remains at various levels of compromise from decomposition and heat denaturing. It is not in any way an exaggeration to say that nothing like this system had ever been attempted before.

When we started this project, we agreed to a very strict level of confidentiality. As the anniversary approaches, I have been given permission to present the scientific details on this software development project in professional journals and/or conferences, provided, of course, that no identifying information is presented or published. I believe the PSB is an appropriate venue to first present this information.

If the organizers are willing to consider this, despite the fact that the deadline has passed, I would be happy to provide an abstract. I regret that I was not able to make this proposal before this time.

```
Sincerely,
Howard Cash
President
----------------------------------------------------------------
Gene Codes Forensics, Inc., a division of Gene Codes
Corporation
775 Technology Drive, Suite 100A, Ann Arbor, MI 48108 USA
phone 734-769-7249 fax 734-769-7074
WWW: http://www.genecodes.com/
```

Following receipt of this message, the organizers quickly decided that it would be in the best interests of our meeting and of society at large to consider the publication of this paper - if the quality of the work met the PSB standards, and if the content of the paper proved to be commensurate with the goals and objectives of the meeting.

A draft of the paper was received on Monday, September 23, 2002. This paper not reviewed in the usual way by anonymous referees, but instead was critically evaluated by the organizers. We concurred that the paper by Cash and co-workers is a fascinating and PSB-appropriate paper of quality, offering significant results in aspects of both program development methodology and algorithm approaches to a meaningful bioinformatics problem. We therefore concluded that the paper should be published.

The following paper by Cash, Hoyle, and Sutton is an extraordinary demonstration that bioinformatics is not only a captivating intellectual pursuit, but can also be of critical importance to the broader society.

Russ Altman
Keith Dunker
Larry Hunter
Teri Klein

DEVELOPMENT UNDER EXTREME CONDITIONS: FORENSIC BIOINFORMATICS IN THE WAKE OF THE WORLD TRADE CENTER DISASTER

HOWARD D. CASH, JONATHAN W. HOYLE, AMY J. SUTTON

Gene Codes Forensics, 775 Technology Drive, Suite 100A,
Ann Arbor, MI 48108, USA

The terrorist attacks of September 11, 2001 resulted in death and devastation in three locations, and extraordinary efforts have been exerted to identify the remains of all victims. As mass fatalities go, this one has been unusual at a policy level because the goal has been not merely to identify remains for every decedent, but to identify every bit of remains found so that even small pieces of tissue can be returned to families for burial. While the human impact at the Pentagon and Shanksville, PA was horrific, the World Trade Center site presented a particularly complex challenge for forensic DNA matching and data handling. A complete and definitive list of all those killed is still elusive, and human remains were crushed and co-mingled by the falling towers. Software tools had never been considered for a problem of this scale and scope. New data handling systems had to be created under extreme software development conditions characterized by incomplete requirements specifications, chaotically changing priorities, truly impossible deadlines and rapidly rolling production releases. Partly because of the company's experience with mtDNA tools built for the Armed Forces DNA Identification Lab starting in 1997, the New York City Office of Chief Medical Examiner [OCME] contacted Gene Codes Corporation in late September as existing data-handling tools began to fail. We began work on the project in mid-October, 2001. Our approach to the problem included:

- *Extreme Programming* [XP] methodology for functional software development,
- On-site time and motion analysis at the OCME for user interface design,
- Evidentiary references between STR, SNP and mtDNA analysis results, and
- Separate data Quality Control [QC] and software Quality Assurance [QA] initiatives.

A substantial software suite was developed called *M-FISys*, an acronym for *Mass-Fatality Identification System*.

1 Background

The New York City Office of Chief Medical Examiner [OCME] faced an unprecedented problem after September 11, 2001. Early estimates suggested that over 10,000 people had been killed in the attacks at the World Trade Center (this number would be revised down to 2,801 by the first anniversary of the disaster). Forensic identification of remains would be complicated by the substantial fragmenting of remains (with individuals recovered in as many as 200 pieces) and the heat and moisture associated with the fighting of jet fuel fires that burned for months. Certainly some fatalities would be identified by classical methods such as viewed remains, fingerprints, dental records and personal property. Because of the crushing and co-mingling of remains, personal belongings found at "Ground Zero" have been considered highly suspect as identifying evidence.[a] The vast majority of individual remains would have to be identified by DNA matching, and the associated data handling is the subject of this paper. Because of the public interest that we all share in this case, attention will be given to the circumstantial pressures that impacted the software development process.

Functionality had to be added to the M-FISys program [pronounced like *emphasis*] on a very fast schedule without sacrificing software quality and testability. In fact, a new release of M-FISys has been delivered to the OCME almost every week since mid-December, 2001 (the 38th iteration is being released at the time of this writing). The need for stringent software quality control is self-evident and the moral magnitude of the task plus the knowledge of the catastrophic impact of any false identification was a constant weight on the shoulders of the developers. Extreme Programming [XP] [14,15,16,17,18] has been an invaluable methodology under the existent pressures. Since almost all studies of software correctness indicate that code reviews are one of the most valuable QA tools, XP programmers work in pairs, with one engineer constantly reviewing the work of the other. Both unit tests and acceptance tests must be written before functionality can be added. In order for new work to be checked back into the source code control system, not only must it pass these tests, but it must pass all other tests that have been written against all functions since the project began. An obvious benefit of this discipline is that it makes it difficult to unknowingly "break" a part of the program while fixing a seemingly unrelated bug. It can also add a cost any time there is a major architectural refactoring because many of the legacy tests will have to be re-written to accommodate the new architecture.

Under the XP methodology, functionality is added each week through observation and direct negotiation with users in the field. One or two staffers have been on-site daily at the OCME since the project began, and the chief designer has traveled back and forth between the user location in Manhattan and the developer site in Ann Arbor each week. Weekly reviews identify opportunities for process improvement. Productivity of the engineering is carefully monitored to the point that one can literally choose any week since the project began, recover the source code as it was during that period, identify what new functionality was scheduled to be added, which engineers were available, initial time estimates for tasks and review the actual programming velocity of the entire team for that iteration.

An important principal of our methodology has been to rigorously resist the urge to propose functionality to the OCME. This is very different from taking the luxury of adding our own "helpful" tools as we would in a commercial design and

---

[a] To illustrate the complexity of the investigation, it is worthwhile to note that a number of remains would be identified by tracing serial number on implants such as cardiac pacemakers and artificial joints, contacting the manufacturer (sometimes overseas), tracing the hospital to which the item had been sold, and then cross-referencing all implantation surgeries against the list of missing persons from the Trade Center.

development project. Our process analysts work through proposals based on the immediate needs and priorities of the forensic biologists, and while we take responsibility for architecture and initial interface design, we do not allow the engineering team to pursue its own vision for the program's functionality. It takes effort for skilled and creative engineers to stick to this discipline, but it is critical to keeping up our development velocity. Outside contractors and advisors have become frustrated and occasionally openly hostile when pet proposals have not made it onto the development list because OCME staff have evaluated those proposals as adding insufficient value. Certainly the emotional content of this development project is a major component of the managerial complexity.

2   Methods of Identification

The most widely used forms of DNA-based human identification involve Short Tandem Repeat [STR] analysis at 13-15 nuclear loci. A second procedure involves sequencing the hypervariable regions of the mitochondrial genome. A relatively new forensic procedure, pioneered by Orchid Biosciences, is based on Single Nucleotide Polymorphisms [SNPs] with well-characterized inheritance and frequency patterns. All of these techniques had to be combined in the M-FISys program.

The theoretical basis of STR analysis in human identification can be traced back to the work of Alec Jeffreys [24]. In modern applications, 13-15 unlinked STR loci are sized, resulting in a "profile."[b]

Table 1: Sample STR data

D3S1358	vWA	FGA	D8S1179	D21S11	D18S51	D5S818
15/16	16/17	26/28	14	30/32.2	15/16	12/13

D13S317	D7S820	D16S539	THO1	TPOX	CSF1PO
12	12	9/12	9.3	6/9	10/11

To orient the reader, the above profile can be read as follows: at the FGA locus, this person inherited 26 repeat-elements from one parent and 28 from the other. The D8S1179 locus is homozygous for 14 repeats. Based on experimentally observed and published frequencies of each repeat value in various ethnic populations, this profile would be expected to be seen no more than one in 10^{17} individuals.

The likelihood value for a given STR profile is the product of the likelihoods of each of the 13 STR loci. As likelihood values are inverses of frequencies, we can

---

[b] In addition to being chosen for genetic distance and independence, the core loci have been selected to be medically uninformative, thereby protecting the medical privacy of individuals in a forensic or criminal investigation.

justify a simple multiplication since the value at each locus is an independent event, unrelated to the other loci.

The benefit of this is that we can have comparative likelihoods across partial profiles. If a policy decision is made to set a threshold likelihood in order to report an identification, for example a likelihood of no less than 10^{10}, an incomplete STR profile's likelihood may still surpass this value if the available alleles are rare enough.

The likelihood value for a given profile will differ depending on ethnic population; however, access to that information is not always available or reliable where badly damaged remains are involved. We therefore assume the "worst case scenario" by examining the likelihood values across four races (Asian, Black, Caucasian & Hispanic) and take the lowest value as the final likelihood. By using this lowest value, we err on the side of preventing false positives.

To determine the likelihood of a particular locus, we take the inverse of its frequency. This frequency can be first approximated by ignoring population structure and using the Hardy-Weinberg proportions:

$$p^2 \quad \text{for homozygous alleles, where p = probability of the allele} \tag{1}$$
$$2pq \quad \text{for heterozygous alleles, where p,q = probabilities of each allele}$$

However, since all humans are related, variance of allele proportions is factored in with the inbreeding coefficient θ (theta), with:

$$p^2 + p(1-p)\theta \quad \text{for homozygous alleles} \tag{2}$$
$$2pq(1-\theta) \quad \text{for heterozygous alleles}^c$$

As stated in *Interpreting DNA Evidence* by West & Weir, "$\theta = 0.03$ is a conservative upper bound on the values appropriate for human populations (National Research Council 1996)."[1] For the WTC project, we use the value of 0.03 for θ.

The probabilities p,q are determined empirically themselves, based upon population. (The values we used are in the STR13.xls spreadsheet provided by Dr. Howard Baum, Deputy Director of Forensic Biology at the New York City OCME. [12])

Let S be any STR profile containing A_k alleles, k<=13. Furthermore let p_k be the probability frequency of the high allele of A_k, and (if heterozygous), let q_k be the probability frequency of the low allele of A_k. Then the Likelihood of S, L(S), is defined as:

$$\tag{3}$$

$$L(S) = \prod_{k \in Alleles} \frac{1}{P(A_k)}$$

---

[c] The Medical Examiner's Office in New York chose to use the equation involving θ only for the homozygous alleles, so that the Likelihood values were always more conservative.

642

where

$$(4)$$

$$P(A) = \begin{cases} p^2 + p(1-p)\theta \mapsto \hom ozygous \\ 2pq \mapsto heterozygous \\ 1 \mapsto non-existent \end{cases}$$

for each k and $\theta = 0.03$.

3 Implementation and Design Decisions

We will first address the issue of matching profiles that are complete (all 13 loci plus a gender locus) or nearly so. Limit this discussion to those cases where profiles have a likelihood of no less than 10^{10}.[d] Even at this level of match stringency, the first task was to address the information glut of having 25, 50 or 100+ individual samples that had identical profiles because they were fragments of the same person.

For the first three weeks after the disaster, attempts to match remains used a software package called CoDIS (Combined DNA Index System). CoDIS was designed as a criminal investigation tool to compare a single DNA profile to a database of profiles from crime scenes and from convicted felons.

CoDIS, the result of a federally funded software development project, is difficult to use under ideal circumstances. Early attempts to apply it to the World Trade Center disaster were described by the forensic scientists as a disaster in itself. The program generated up to 4,000 printed pages of internally consistent matches (matches between samples and between samples and exemplars). Teams of forensic scientists would literally spend days pouring over CoDIS output with highlighter and ballpoint pens in an attempt to sift through the information. The usability issue of reducing the size of this haystack was clearly a first priority.

The first working version of M-FISys, delivered on December 13, had the immediate advantage of collapsing multiple samples with the same profile into an expandable "aggregate" sample. This alone was such an improvement over the CoDIS-based methodology that 55 new ID's were recognized the first day the program was installed. In Figure 1, the line labeled RM #34 (6) indicates that there are six internally consistent profiles in the collapsed set.[e]

---

[d] As suggested earlier, the threshold of 10^{10} was a policy decision. The basis of the selection was that in a population of 5,000 victims (the estimate of fatalities at the time of the discussion), this would reduce the likelihood of even a single misidentification based on a fortuitously shared STR profile to less than 1 in 1000.

[e] RM stands for Reported Missing. Each presumed victim has an RM number. These numbers have been changed and names obscured for purposes of this paper.

Fig. 1. The Master List displays aggregated STR profiles in a searchable and collapsible format.

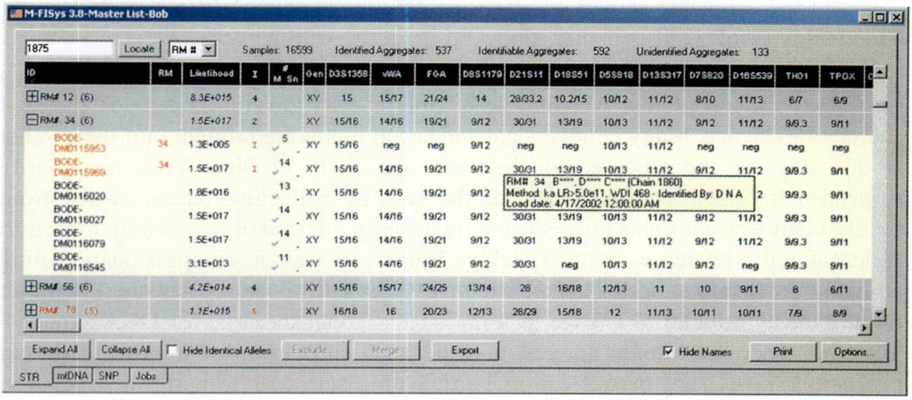

Data reaches the DNA ID group at the OCME in three categories: 1) DNA profiles from victim remains, 2) ante-mortem DNA profiles of personal references from missing persons (e.g., a pre-existing pathology sample, a toothbrush or a lipstick) and 3) profiles from buccal swabs taken from first-order relatives. The first attempt is to match a victim sample to a personal reference exemplar.

Unlike a criminal investigation application where a single sample might be compared to the database, the WTC recovery required repeated all-against-all comparisons as additional samples were recovered and analyzed in the lab. Furthermore, as compromised samples were reanalyzed with hopes of extracting additional information, "Virtual Profiles" needed to be created, showing all accumulated values while keeping track of exactly which laboratory attempt had generated which data points. As data is added and operator decisions are made about assignment of data, a history (necessarily including free-text annotations) is kept for each sample.

The possibility of fortuitous (coincidental) matches increases as match stringencies are lowered. For samples with low partial profiles (likelihoods in the $10^3 - 10^4$ range) one might expect several possible direct matches. This is particularly true if one further allows for the possibility of allelic drop out (loss of one of the two alleles in a locus because of the damaged condition of the sample – allowing an experimentally homozygous 12 to match a 12/16 heterozygous reference). M-FISys helps a forensic scientist to resolve these ambiguous matches by a process we call *iterative pruning*. The operator can confirm or exclude possible matches for any ambiguous sample, annotating reasons along the way. For instance, a right hand with degraded DNA can be excluded from a potential ID on the grounds that a full profile for a right hand has already been reported. As more information is accumulated, more matches can be excluded or confirmed. mtDNA or SNP data may help to confirm or refute a potential match. Finally, experimental errors can also be corrected and annotated as part of an exquisitely meticulous Administrative Review and Quality Control process that is beyond the scope of this paper.

When direct matches to ante-mortem exemplars are not possible, or if the chain of custody for the exemplar cannot be reliably traced[f], kinship analysis is the next source of evidence. Many ethical, legal and social issues are brought into sharp relief at this stage of the process. For instance, the State of New York temporarily suspended Informed Consent rules for the case of collecting buccal swabs from family members. This was not, as some mistakenly inferred, motivated by a need to collect family references more quickly. Rather it was a compassionate ruling allowing for samples to be taken (in many cases from young children) without forcing the contributor to engage in a discussion about the need for DNA samples. There are those who might suggest that the legal suspension of Informed Consent regulations gives tacit permission to participating scientists to perform population genetics research using the database of collected victim and kin samples. This would be a gross and egregious violation of the dignity, privacy and civil rights (under other laws in the State of New York) of the victims and their families. A less extreme issue is the decision on what to do if *false paternity* is ever detected. It seems only reasonable and compassionate to withhold any such findings permanently: What could be more devastating for a father than to learn that not only had a child been lost to a senseless act of terrorism, but that the child had never been his?

As part of the WTC recovery and in collaboration with Dr. Charles Brenner and Dr. Benoit LeClaire, kinship analysis can be tactically approached in two different ways. Using Brenner's approach, family pedigrees are constructed based on the reported relationships to the victim and the set of victim profiles is scanned for promising candidates to fill in the missing (victim's) position in the pedigree. The data can then be revised to allow for incorrectly reported relationships and matches can be further refined. Dr. LeClair takes the conservative approach of assuming that any reported relationship could be wrong (this turns out to be true with unfortunate frequency due to errors in the original collection of data). Each kinship sample is compared to victim samples and tested for likelihood at various relationships such as parent-child, sibling or half-sibling. These are sorted by most-likely match. In an accurate match for a particular victim sample, the correct familial samples will float to the top of the list.

The mathematics of kinship analysis is well established.[1,2,4]   Scores are reported as a hypothesis-based "likelihood ratio," calculated as the likelihood that the

---

[f] Verifying the origin of a reference sample has proven to be an enormous challenge in this process. Because of the chaos and emotional stress of the first weeks following 9-11, many contributed samples were accepted with incomplete or even incorrect paperwork, As a result, all identifications are based on no less than two match modalities.

relationship between the victim sample and the kin sample is as hypothesized (e.g., full sibling in the Hispanic population) divided by the likelihood of seeing the same number of matching alleles in two unrelated persons in the same population.

Similar to the Likelihood in direct STR matches, Kinship Likelihood is the product of Kinship Locus Likelihood across all existing alleles (partial profiles are possible here as well). What is different is that two profiles are being compared instead of just one. In the case of partial profiles, only those loci extant in both profiles can be compared. As in the case of direct STR Likelihood, Kinship Likelihoods were calculated across the four aforementioned races, the lowest value taken as "a worst case scenario".

The Kinship Locus Likelihood takes as input a relationship between the two profiles, which is one of: Parent/Child, Full Sibling, Half Sibling. Formulae exist for First Cousin relationships as well, but it was felt to be less useful for the World Trade Center project, so was never used (swabs from cousins were not taken); however, because it was coded into M-FISys, the formula will be shown here as well. Let p be the frequency probability of the high allele value, q the frequency probability of the low allele value. Then the Kinship Locus Likelihood is defined the following way [13]:

$$(a_2 p_2 + a_1 p_1 + a_0 p_0) \ / \ p_2 \tag{5}$$

where a_i's are proportions based on relationship:

$$\tag{6}$$

Full Sibling:	$a_2 = 1/4$,	$a_1 = 1/2$,	$a_0 = 1/4$
Half Sibling:	$a_2 = 1/2$,	$a_1 = 1/2$,	$a_0 = 0$
Parent-Child:	$a_2 = 0$,	$a_1 = 1$,	$a_0 = 0$
First Cousin:	$a_2 = 3/4$,	$a_1 = 1/4$,	$a_0 = 0$

and the pi's are defined as follows:

$$\tag{7}$$

$p_2 = p^2$ if the victim is homozygous and matches one of the relative's alleles

   $= 2pq$ otherwise

$p_1 = 0$ if relative and victim alleles share no common value

   $= p$ if the relative is homozygous and the victim shares its (homozygous or heterozygous) low value

   $= q$ if the relative is homozygous and the victim shares its (heterozygous) high value

   $= p/2$ if the relative is heterozygous and shares a exactly one value with the victim's (homozygous or heterozygous) low allele

   $= q/2$ if the relative is heterozygous and shares exactly one value with the victim's heterozygous high allele

   $= (p+q)/2$ if relative and victim are identical and heterozygous

$p_0 = 1$ if the relative and victim alleles are identical

   $= 0$ otherwise

The unfortunate truth is that some samples are so severely burnt that full STR profiles will not be available for either direct or kinship analysis. When samples are badly compromised through harsh environmental degradation, many of the loci may be blank and the likelihood of observation can drop to levels where one would expect many instances of shared "partial profiles" in a population of 2,801 victims. At this point, the options for making a positive ID are a) to re-extract and retest DNA with hopes of collecting information at additional loci, or b) to use other techniques that might be more effective on highly degraded samples, even if they are not as discriminating. One of these alternative techniques is mitochondrial DNA analysis.

Mitochondrial typing involves direct sequencing of the highly variable regions of the genome adjacent to the origin of replication. The standard in forensic communities is to report results not in the form of the sequence itself, but in a compact format that only shows the difference between the experimental results and an established and internationally recognized standard known as *The Anderson Sequence*[g].[19] If the sequence being typed is identical to the Anderson reference, the difference report will be null. Point mutations are described as a base position, plus the base that differs from the reference sequence. Deletions are represented as a "D" character (not to be confused with the IUB code for "A, G or T, but not C"). We call this difference list the "delta representation." A typical mitotype might look like this:

```
16093:        C
16224:        D
16311:        C
195:          C
263:          G
315.1:        C
```

Note that the base positions cross the origin of the 16569-base genome. The first lines indicate that this sample has a C at position 16093 where a T is usually found, and a deletion of the base at position 16224. To maintain the integrity of numbering, an insert is indicated as a decimal point position on the base that the insert follows. In this case, "315.1: C" indicated that there is a C insert after position 315 in the reference sequence.

Although standards have been promulgated, there is still the occasional disagreement over nomenclature, particularly as it applies to inserts. For instance, if the reference sequence includes the sequence "TTT" starting at position 16091, and the sample under study has four T's rather than three, it can logically be represented as any of the following:

---

[g] This reference is actually a consensus made up of several sequences and is not necessarily the most common. An alternative reference sequence called The Modified Cambridge Reference Sequence is gaining in popularity. The Anderson Reference is used in the WTC analysis.

```
16090.1:    T
16091.1:    T
16092.1:    T
16093.1:    T
```

The sequence used in mitotyping is not in a coding region and, since the inserted T has no biological significance, it can be referenced as an insert in any of several places. Two of the collaborating labs were not in precise agreement on the standard notation for this insert, even though both were producing accurate mtDNA profiles. The overtly political positioning (almost exclusively from outside advisors not working at either lab or with the software development team) over the "correctness" of one naming representation or another was an example of one of the greatest non-technical challenges of this project. We defined this to be a non-issue for the current project, since our solution was to reverse-translate the difference list to generate a non-ambiguous original sequence and compare those sequences themselves.[h]

mtDNA is very hardy material that can survive intact under conditions where nuclear DNA degrades. However, the variation in the hypervariable d-loop of the mitochondrial genome is not nearly as discriminating as a 13 locus STR profile. In fact, over 5% of the Caucasian population share the same, common mitotype. Mitotyping is most often used to exclude a match but by itself would not normally contain enough information to confirm an identification with a high degree of confidence. Fortunately, we can take advantage of the fact that mitotypes are independent of the STR profiles and use them as additional information to supplement a partial STR match.

Mitochondrial DNA is inherited only through the maternal line. Therefore, many of the samples available to the OCME are useless for mitotype matching. Fathers will not contribute mtDNA to their offspring, and children of male victims (roughly 2/3 of the victims were male) do not have relevant mtDNA for identification. However, as an Operations Research issue, it was found to be more efficient and less expensive to mitotype all kin swabs than to pull out and re-array only those samples that were part of a matrilinear line to a victim.

A proposal was made that mitotype frequency statistics be generated from roughly 5,200 personal effect and kinship samples available. With just a quick review of the database as of September 16, 2002, it was found that at least 24% of the 2,801 victims had no maternally-related kin samples available to the process. This, plus the enormous amount of effort that would be required to confirm the validity of all kin swabs and personal effects, argues that a frequency database from

---

[h] There was a concern that the differences in nomenclature could obscure matches between reported mitotypes. Still, it was sociologically fascinating that a complete and provable alternative that avoided that problem did not allow us to avoid a significant expenditure of resources in the debate over who was more "correct." Several people have observed that this has been an issue primarily for outside advisors who, like all of us, wanted desperately to be of material help in the face of such a tragedy.

the general population might be preferable to one that could be reasonably and reliably created specifically for WTC victim population.

At the time of this writing SNP data has not been applied to the identification process. The ability to collect data from very short sequences makes this an exciting technology which offers great hope for collecting identifiable information from badly degraded samples. SNPs occur within the human genome on average every 100 to 300 bases and are stable from an evolutionary standpoint, making them easier to follow in populations as they do not change much from generation to generation. A panel of 70 SNPs has been characterized by the GeneScreen division of Orchid Biosciences, all from nuclear DNA. Roughly 2 out of every 3 SNPs involves replacing a C with a T, and it is among these biallelic loci that the 70 forensic SNPs are chosen, specifically ones in which C and T are equally likely.

For kinship analysis, it is important to ensure that these loci are genetically independent of each other and not linked to nuclear STR loci used in identifying the same sample. In an ideal world, all loci would be no less than 50MB or 50cM apart. With 70 SNPs plus up to 15 STR loci in a genome that is only 3.5 billion bases long, this is clearly not possible.

In a study by Dr. Ranajit Chakraborty of the Center for Genome Information, the estimated allele frequencies for the current panel ranged between 15.5% to 81.3%, although the great majority were nearly equal (an average heterozygosity of 46% across the three population groups Caucasian, Black and Hispanic). Fortunately, his study showed the allelic dependence being relatively small, 5.71%, comparable to the 5% expected by chance alone. He further concluded that despite the lack of theoretically independent loci, an analysis of 713 genotypes from 3 different populations support the efficacy of using this 70 SNP panel for identification purposes. With a complete profile of 70 SNPs at probability 1/2, the likelihood of match would be 2^{70}, or approximately 10^{21}. Even with a revised probability average of 46%, the likelihood is 10^{18}.

The authors would reject the proposal that this panel of SNPs should be replaced with one that is more evenly spaced. Because of the limitation of the genome size, linkage in a 70 locus panel is a problem intractable in the laboratory but addressable in software. M-FISys can take an arbitrarily large panel of SNP and STR loci and, given a threshold likelihood value, select those loci that create an unlinked set of STRs and SNPs to reach that threshold in a valid match (assuming the data is available). The remaining loci, of course, would still be reviewed and any directly conflicting data highlighted to indicate experimental error or other anomalies that should be resolved by a forensic biologist before a positive ID is reported.

Combining STR, mtDNA and SNP data is as much a human-computer interaction issue as it is a computational one. Keep in mind that the goal is not for the software to make identifications; as a matter of law, neither the developers nor the software have such authority. Rather it is to present data to a qualified and authorized forensic biologist to make it easier for that person to certify an identification. Instead of developing an interface that combined all STR, mtDNA and SNP data in a single view, M-FISys is broken up into STR-*centric*, mito-*centric* and SNP-*centric* views of the data, with indications that other data supports or contradicts a proposed identification.

Fig. 2. STR-centric view of the Master List displays details of STR results as well as indicators of supporting or conflicting experimental evidence using mitotyping or SNP analysis.

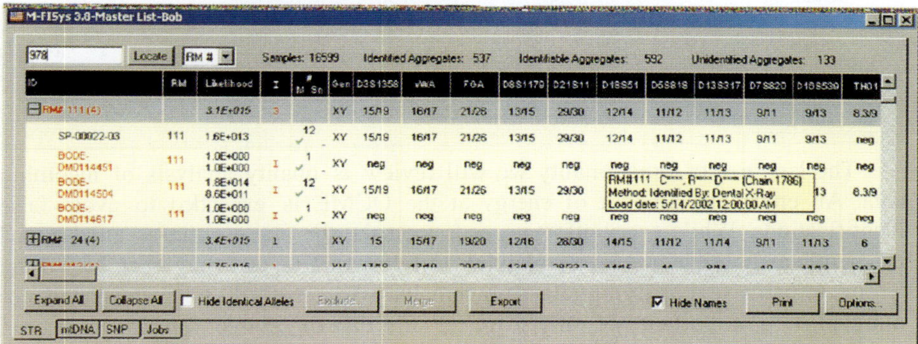

Figure 2 shows the STR-centric view of the master comparison list. The first column shows the names of samples including personal effects (starting with "SP" because they were typed at the State Police lab in Albany) and disaster samples that, following an optional prefix, are coded as DM01nnnnn, where the final 5 digits are the order in which the remains were discovered and checked in at the morgue.[i] The fifth column contains three values. The number at the top is the number of loci (including the gender locus) that yielded results. There is a character on the lower left and the lower right of that column that indicates consistency between the STR match and mito data, and SNP data, respectively. The possible values for mito and SNP data are a green check mark (✓: the data is consistent with the STR results), a red X (X: the data is inconsistent or contradictory), a question mark (?: the data is partially or ambiguously consistent) or a dash (— : this analysis has not yet been performed). Following this is the complete STR profile of the sample.

Similarly, in the mito-centric view of the same data (Figure 3), the operator can see the complete mitotype, plus an indication of whether any STR or SNP data supports or contradicts a mito-based match. This approach allows us to add an arbitrarily large number of different modalities for identification. Individual forensic biologists may be charged with reviewing a particular type of data within their specialty, but are immediately alerted if contradictory data needs to be reviewed.

---

[i] The header "DM01" stands for "Disaster: Manhattan, 2001." There are also "DQ01" samples, referring to the "Disaster: Queens" when flight 587 lost part of its tail and crashed in November of the same year.

Fig. 3. The mito-centric view of the Master List displays mitotypes as well as indicators of supporting or refuting STR and SNP data.

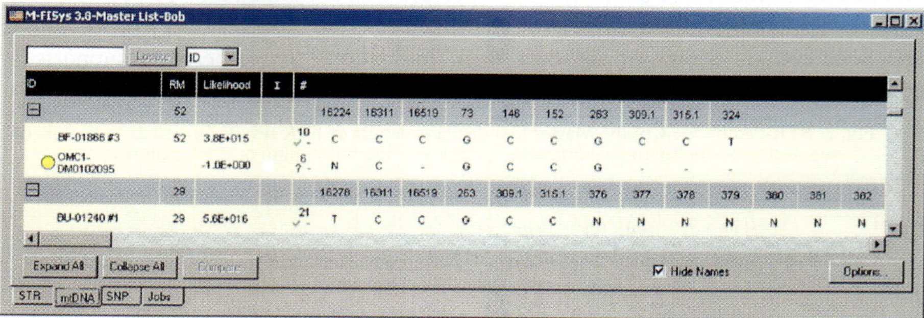

The last area of functionality we will review is quality analysis of incoming data. An enormous amount of energy at the OCME is expended to avoid false identifications. While every effort is made to make the process run as quickly and efficiently as possible, the idea of having to tell a family that they have buried the wrong remains is horrifying. Because remains were crushed, fragmented and co-mingled, it is important not only to make sure that the correct remains have been typed, but that no additional remains are being released when a funeral director accepts custody of human tissue from Memorial Park.

For that reason, the first person to review remains when they arrive at the morgue is not a Medical Examiner, but a Forensic Anthropologist. This person's job is to be as certain as possible that a) the recovered remains are human, and b) that only one person is included in a collection. The policy is that if remains are not physically connected by tissue or sinew, that they should be separated into individual pieces and typed independently.

Fig. 4. Quality Control Report displays experimental results with inconsistent STR profiles.

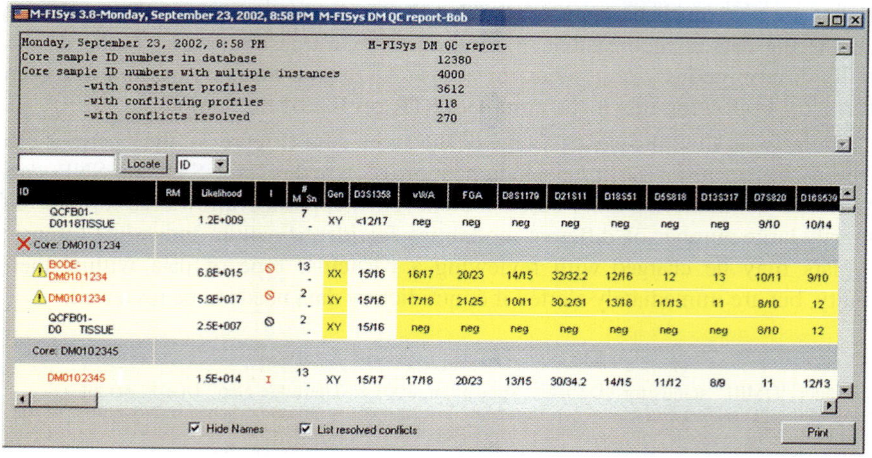

An example of the kind of QC that M-FISys performs is shown in Figure 4. Sample DM0101234 (not the real number) has been typed twice. In one case, the tissue was typed by Myriad Genetics. But bone from the same sample was sent to Bode Technology Group, a well regarded forensics lab in Virginia with a highly developed capacity for extracting DNA from bone.[j] If you look at the values for each locus, several of them (in yellow) are in conflict. To a forensic biologist, these profiles clearly represent two different people. In fact, they are different genders.

There are several possible resolutions to this conflict. It could be that, upon examining the shipping manifests, it is found that a sample has been mislabeled. This is almost never a consideration, but it still must be an option for resolving the problem within the program. Smaller discrepancies can be the result of experimental error or allelic drop out. In this case, a review of the remains showed that bone from one person was embedded in the muscle tissue of another. Certainly a bizarre circumstance in most medical examiner's offices, but a sadly familiar possibility at the World Trade Center. The program will prevent these samples from being identified and released until this conflict of data for a single "victim sample" has been resolved.

Another of the several data-integrity tests in M-FISys compares the profiles of personal effects to the swabs taken from family members. Families of missing persons are able to call the "DNA Hotline" at the OCME and find out if enough DNA has been collected for an identification. For instance, a mother whose husband did not come home after the attacks might bring in the presumed victim's toothbrush as well as the oldest child of the victim to give a buccal swab. However, she might ask that the OCME determine if there is enough DNA available from the personal effect to make an identification without additional kin samples because she does not want to further traumatize her younger children by asking them to provide swabs to help identify their father. A call to the DNA hotline might indicate that the toothbrush gave a full profile and that further samples are not required. But what if the wrong toothbrush was brought in? If the toothbrush matches the DNA for the wife or the oldest child, the full profile is not valid and a swab from additional children might be key to making a final identification.

---

[j] Bone is difficult for several reasons. One issue is that the dust created when a sample is extracted must not contaminate the next sample from the same bench. More significantly, Calcium is a PCR inhibitor. Bode Technology Group has a core competency in overcoming these obstacles that has been invaluable in the WTC project.

4   Summary

When we were brought onto this project several weeks after the September 11 attacks, we developed five goals and two major deliverables. The goals were:

1. Identify individual remains,
2. Reunify partial remains so that they can be returned to families,
3. Collect and warehouse meta-data for administrative review of reference samples,
4. Track samples among collaborating labs, and
5. Create an information management system to report metrics and make problem resolution proposals to supervisors at the OCME.

Not all aspects of the engineering effort have been discussed in this paper, but we hope a broad overview of both the process and the challenges has been conveyed.

The first deliverable has been to create a mass-fatality identification and recovery system for the OCME, creating installed value to address specific data-handling problem on the fastest possible schedule without running the risk of false identifications. This has been a seven-day-a-week job for over eleven months as of this writing, and all of the participants in the project feel that it is the most important thing they will do in their professional lives. The second deliverable will be to create a generalized and portable version of this tool that can be deployed anywhere in the world where there is a massive human tragedy, be it natural or man made.

Acknowledgments

Thanks to the many contributors including the primary process analysis, development and support team: Tracy Beeson, Debra Cash, Peter Fattori, Lucy Hadden, Mike Hennessey, Jeff Ingalls, Anna Khizhnyak, Tom Kubit, Anna Korn, Simon Mercer, Judy Nolan, Gregory Poth, David Relyea, Matt Smith, Francis Sullivan and Bill Wake. Other truly major contributors have included staff at Gene Codes Corporation, Gene Codes Forensics, the New York State Police, AFDIL and the OCME too numerous to name. Special thanks to Chief Medical Examiner Dr. Charles Hirsch who I would follow into war, George Carmody for his always generous technical assistance on short notice, Charles Brenner, Benoit LeClaire, and to the many inspiring families we have met along the way who lost loved ones on September 11, with particular love and admiration for the Cartier, Hannon, Ielpi, Lynch and Weiner families.

References

1. I. Evett and B. Weir, *Interpreting DNA Evidence* (Sinauer Associates, 1998).
2. National Research Council, *The Evaluation of Forensic DNA Evidence* (National Academy Press, 1996).
3. National Research Council, *DNA Technology in Forensic Science* (National Academy Press, 1992).
4. J. Ballantyne et al, *DNA Technology and Forensic Science* (Cold Spring Harbor Laboratory Press, 1989).
5. L. Kirby, *DNA Fingerprinting* (Stockton Press, 1990).
6. R. Saferstein, *Forensic Science Handbook -Volume 1* (2nd Edition, Prentice Hall, 1982, 2002).
7. N. Rudin et al, *Forensic DNA Analysis* (2nd Edition, CRC Press, 2002).
8. J. Butler, *Forensic DNA Typing* (Academic Press, 2001).
9. National Institute of Justice, *The Future of Forensic DNA Testing* (2000).
10. Federal Bureau of Investigations, *State DNA Database Statutes* (1999).
11. Federal Bureau of Investigations, *National DNA Index Systems (NDIS) Procedures Manual* (1999).
12. H. Baum, *PCR Statistics: STR13.xls* (personal communication, Office of Chief Medical Examiner, 6/14/99).
13. G. Carmody, *KinTest - CODIS 13 Core Loci* (personal communication, 2001).
14. W. Wake, *Extreme Programming Explored* (Addison-Wesley, 2002)
15. K. Beck et al, *Planning Extreme Programming* (Addison-Wesley, 2001)
16. J. Newkirk and R. Martin, *Extreme Programming in Practice* (Addison-Wesley, 2001)
17. K. Beck, *Extreme Programming Explained* (Addison-Wesley, 2000)
18. R. Jeffries et al, *Extreme Programming Installed* (Addison-Wesley, 2001)
19. S. Anderson et al, *Sequence and Organization of the Human Mitochondrial Genome* (Nature 290: 457-465, 1981).
20. P. Awadalla et al, *Linkage Disequilibrium and Recombination in Hominid Mitochondrial DNA* (Science 286: 2524-2525, 1999).
21. D. Balding and P. Donnelly, *Inferring Identity from DNA Profile Evidence* (Proc. Natl. Acad. Sci. USA 92: 11741-11745, 1995).
22. B. Budowle et al, *Population Data on the Thirteen CODIS Core Short Tandem Repeat Loci in African-American, U.S. Caucasians, Hispanics, Bahamians, Jamaicans, and Trinidadians* (J. Forensic Sci. 44: 1277-1286, 1999).
23. A. Jeffreys et al, *Hypervariable Minisattellite Regions in Human DNA* (Nature 314: 67-72, 1985).
24. A. Jeffreys et al, *Individual Specific "Fingerprints" of Human DNA* (Nature 316: 75-79, 1985).